Plant Secondary Metabolites

This book explores the pharmacological potentials of various classes of plant secondary metabolites. The initial chapter introduces the chemical diversity, biosynthesis pathways, and ecological functions of plant secondary metabolites. It discusses the pivotal role of transcription factors in regulating their production. Furthermore, this book examines the specific classes of secondary metabolites, such as phenolic compounds, terpenes, and anthocyanins, highlighting their properties and various applications across the food, cosmetics, and pharmaceutical industries. This book delves into emerging areas such as plant metabolites as nutraceuticals, the utilization of horticultural and vegetable crops for secondary metabolite production, and the exploration of staple crops for health improvements. The chapters discuss the applications of fungal endosymbionts and algal sources in shaping plant metabolite compositions. The chapters further review innovative techniques, such as hairy root engineering and recombinant systems, for enhancing metabolite production. This book concludes with a detailed overview of methods for isolating and identifying secondary metabolites. This book is an essential resource for students, researchers, and professionals of ethnopharmacology, pharmaceutical sciences, botany, and biomedical sciences.

Key features

- Covers various classes of plant secondary metabolites and their pharmacological potentials.
- Explores the role of transcription factors in controlling secondary metabolite production in plants.
- Discusses secondary metabolites from horticultural and vegetable crops, highlighting their pharmaceutical applications.
- Presents techniques like hairy root engineering and recombinant systems for enhancing metabolite production.
- Examines lichen metabolites and their applications in human health, as well as metabolites from gymnosperms.

Plant Secondary Metabolites

Sources and Applications

Edited by

Santosh Kumar Upadhyay and
Dev Bukhsh Singh

CRC Press
Taylor & Francis Group
Boca Raton London New York

CRC Press is an imprint of the
Taylor & Francis Group, an **informa** business

Designed cover image: Shutterstock

First edition published 2026
by CRC Press
2385 NW Executive Center Drive, Suite 320, Boca Raton, FL 33431

and by CRC Press
4 Park Square, Milton Park, Abingdon, Oxon, OX14 4RN

CRC Press is an imprint of Taylor & Francis Group, LLC

ISBN: 9781032854762 (hbk)
ISBN: 9781041086154 (pbk)
ISBN: 9781003518358 (ebk)

DOI: 10.1201/9781003518358

Typeset in Times
by codeMantra

Contents

About the Editors

Dr. Santosh Kumar Upadhyay is an Assistant Professor in the Department of Botany, Panjab University, Chandigarh, India. With over 14 years of experience in plant biotechnology, he specializes in functional genomics. His research endeavors have focused on characterizing vital defense-related protein families in bread wheat, including receptor-like kinases, antioxidant enzymes, and cation transporters. Moreover, his group is engaged in elucidating the roles of long non-coding RNAs in abiotic and biotic stress responses. His scholarly contributions are extensive, comprising over 140 publications encompassing research papers in renowned international journals, national and international patents, book chapters, and 12 authored books. His remarkable achievements have garnered recognition, including the NAAS Young Scientist Award (2017–2018) and NAAS Associate (2018) from the National Academy of Agricultural Sciences, India. Additionally, he has been honored with the INSA Medal for Young Scientist (2013) from the Indian National Science Academy, India; the NASI-Young Scientist Platinum Jubilee Award (2012) from the National Academy of Sciences, India; and the Altech Young Scientist Award (2011). In acknowledgment of his expertise and contributions, Dr. Upadhyay has been bestowed with prestigious fellowships such as the DST-INSPIRE Faculty Fellowship (2012) and the SERB-Early Career Research Award (2016) from the Ministry of Science and Technology, Government of India. He has also secured several research grants from prominent funding agencies like SERB, DBT, and CSIR. Furthermore, Dr. Upadhyay actively contributes to the academic community by serving as a member of the editorial board and a reviewer for numerous peer-reviewed international journals.

Dr. Dev Bukhsh Singh is an Assistant Professor at the Department of Biotechnology, Siddharth University, c, Siddharth Nagar, India, a position he has held since 2021. Prior to this, he served as an Assistant Professor in the Department of Biotechnology, Chhatrapati Shahu Ji Maharaj University, Kanpur, India, from 2009 to 2021. He obtained his B.Sc. (Biology) and M.Sc. degrees from the University of Allahabad; an M.Tech from the Indian Institute of Information Technology, Prayagraj, India; and a Ph.D. in Biotechnology from Gautam Buddha University, India. Since 2009, he has been actively involved in teaching the students of B.Sc. and M.Sc. Biotechnology. His research interests lie in bioinformatics, drug design, and Medicinal Biology. He has edited four books and published several research articles in journals of high repute. Additionally, Dr. Singh is a member of various national and international academic bodies and serves as a reviewer for several international journals.

1 Unravelling the Roles of Plant Secondary Metabolites

Biological Functions and Therapeutic Applications

Sounok Sengupta[1,2], *Akash Malhotra*[1,3], *Romita Sharma*[1,3], *Sahil Gautam*[1], *H. Lalhlenmawia*[4], *Nilimanka Das*[5], *and Deepak Kumar*[1]

[1]Department of Pharmaceutical Chemistry, School of Pharmaceutical Sciences, Shoolini University, Solan, India

[2]Department of Pharmacology, School of Pharmaceutical Sciences, Shoolini University, Solan, India

[3]Faculty of Biotechnology and Applied Sciences, Shoolini University, India

[4]Department of Pharmacy, Regional Institute of Paramedical and Nursing Sciences, Aizawl, India

[5]Regional Institute of Pharmaceutical Science and Technology, Agartala, India

1.1 INTRODUCTION

Plant secondary metabolites, a large group of natural products classified under non-essential compounds, i.e. are not directly involved in the basic metabolism but still have been a part of enormous bioactivities enabling the survival of plants in various ecosystems. Secondary metabolites are employed as messengers to help the plant to communicate with its surroundings, as with other plants, animals and microorganisms. The clear-cut division into primary and secondary metabolites is one of the most important aspects concerning plant biochemistry. The primary metabolites, which include carbohydrates, proteins, lipids and nucleic acids, are essential for the existence and normal activities of the plant (Crozier et al., 2010; Crozier et al., 2006) and are useful for the cellular activities such as photosynthesis, respiration and cell division. Meanwhile secondary metabolites have relevant values in an evolutionary sense as they equip the plant to be stronger in defending its own self by competing with other plants for resources as well as for shaping its environment. Although the number of primary metabolites is relatively low, the plants make a great number of secondary metabolites, often unique to certain species or even certain families of plants, and can be classified into three major groups based on their biosynthetic origin, i.e. alkaloids and phenolic together with terpenoids. They are each composed of a very wide range of chemically different compounds, and many of them have been implemented scientifically by virtue of their biological and pharmacological uses (Crozier et al., 2010).

Secondary metabolites in plants have extensively been employed in human nutrition and disease as traditional remedies over centuries, now integrated into current pharmacopoeias. For instance, morphine and quinine alkaloids are used due to their huge pain-relieving and antimalarial qualities, respectively. Terpenoids such as menthol and taxol are valued for use in suffering from respiratory

DOI: 10.1201/9781003518358-1

ailments and cancers, respectively. Flavonoids are among the most studied phenolic compounds, and their antioxidant properties are associated with the defence against chronic diseases, for example, cardiovascular disorders and cancers (Sharma et al., 2022; Upadhyay & Singh 2021, 2023, 2024). Most recently, there has been a growing appreciation of plant secondary metabolites as arising sources of crops and products for agriculture and other industries. They find application in agriculture, for example, as biocontrol agents for use as natural insect repellents, herbicides and for stimulation of plant growth than the synthetic chemical options (Clemensen et al., 2020). These compounds have shown their utility in the cosmetics and food industries due to their antioxidant, antimicrobial and anti-inflammatory properties, and for this reason, find applications in cosmetics, food preservatives and functional foods. Natural product biosynthesis produces secondary metabolites from primary metabolites through a variety of processes including, hydroxylation, methylation, glycosylation and polymerization. Such biosynthetic pathways may be species-specific and are believed to have produced a chemically diverse secondary metabolite. Three key biosynthetic pathways are responsible for the majority of secondary metabolites: For aromatic compounds such as phenolic, shikimate is used, while the terpenoids are synthesised using MVA and methylerythritol phosphate (MEP) pathways that are used to synthesise terpenoids; fatty acid-derived compounds comprising of flavonoids and some alkaloids uses the polyketide pathway (Dhaniaputri et al., 2022).

Plant secondary metabolites are chemically diverse compounds the knowledge of which is of importance to various disciplines, including botany, chemistry, ecology, pharmacology, and biotechnology. In the past decades, facilities from analytical technologies like mass spectra, nuclear magnetic resonance (NMR) spectroscopy and metabolomics have improved capacity to discover and characterize the SMs and thus changed present-day understanding of SM biosynthesis and roles. In addition, many important compounds are able to be produced from plant cell culture and metabolic engineering, which means that these valuable resources are not as completely dependent on wild plant sources, and they are able to cause less distress to the environment (Leiss et al., 2011). However, several difficulties are still present in the research and utilisation of plant secondary metabolites even to this date. Their multiple activities, high molecular weights, and low levels in many plant species account for their difficult extraction, identification, and synthesis due to the chemical intricacy of their structures and biospecificity. Also, while individual secondary metabolites have been known to have pharmacological roles separately, there is increasing evidence that secondary metabolites tend to act collectively; that is, a mixture of compounds is more active than each independent compound. This interaction is especially relevant regarding traditional medicines involving the consumption of mixtures of plant secondary metabolites (Badyal et al., 2020; Wink, 2010). Thus, this chapter provides an understanding of the definition, synthesis, role, and uses of secondary metabolites for the reader and discusses how they can benefit both the environment and man. As more studies attempt to reveal the versatility of these composites, many people can trust that plant secondary metabolites will keep on attracting scientific interest and offer new opportunities in the future.

1.2 MAJOR CLASSES OF PLANT SECONDARY METABOLITES

The categorization of the plant secondary metabolites are into three primary groups: **alkaloids**, **phenolics**, and **terpenoids** (Figure 1.1) and each category includes compounds with diverse chemical structures and functions, contributing to human health in various ways (Table 1.1) (Tiwari & Rana, 2015).

1.2.1 Alkaloids

The nitrogen-containing compounds known as alkaloids have pharmacological activities which include analgesic, antimalarial and antimicrobial effects. Many well-known drugs, such as morphine (*Papaver somniferum*), quinine (*Cinchona* species), and caffeine (*Coffea* species), are alkaloids. As nutraceuticals, alkaloids have potential applications in improving mental alertness, alleviating pain, and enhancing athletic performance (Tiwari & Rana, 2015).

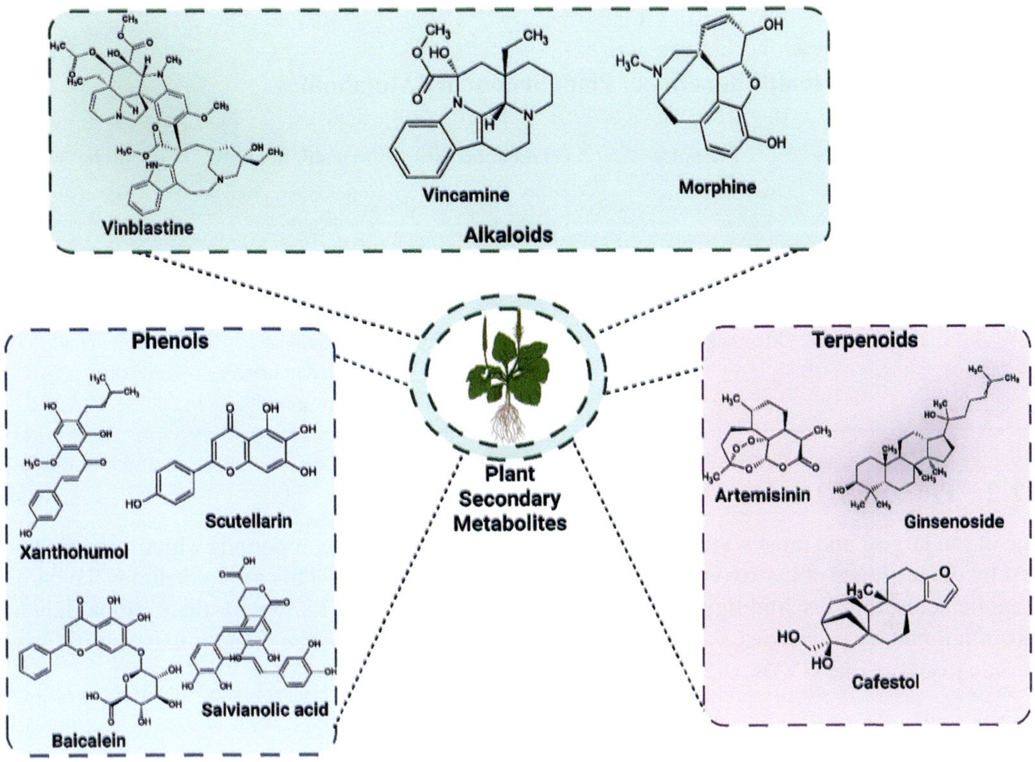

FIGURE 1.1 General classifications of plant secondary metabolites.

TABLE 1.1
Classification and Health Benefits of Plant Secondary Metabolites

Class of PSMs	Sub-Classes	Examples	Plant Sources	Health Benefits	References
Alkaloids	Tropane alkaloids	Atropine, Cocaine	*Atropa belladonna*, *Coca*	Analgesic, anticholinergic, local anaesthetic	Tiwari and Rana (2015)
	Pyridine alkaloids	Nicotine	*Nicotiana tabacum*	Stimulant, improves cognitive function	Elshafie et al. (2023); Tiwari and Rana (2015)
	Indole alkaloids	Reserpine, Vinblastine	*Rauvolfia*, *Catharanthus*	Antihypertensive, anticancer	Kabera et al. (2014); Tiwari and Rana (2015)
	Purine alkaloids	Caffeine, Theobromine	*Coffea*, *Theobroma cacao*	Stimulant, improves mental alertness	Wink (2010)
Phenolics	Flavonoids	Quercetin, Catechins, Anthocyanins	Fruits, Tea, Vegetables	Antioxidant, anti-inflammatory, anticancer	Egamberdieva et al. (2023)
	Phenolic acids	Ferulic acid, Caffeic acid	Whole grains, Coffee, Vegetables	Antioxidant, anti-inflammatory	Egamberdieva et al. (2023)
	Tannins	Ellagitannins, Proanthocyanidins	Tea, Wine, Berries	Antioxidant, antimicrobial, cardioprotective	Tiwari and Rana (2015)

(Continued)

TABLE 1.1 (*Continued*)
Classification and Health Benefits of Plant Secondary Metabolites

Class of PSMs	Sub-Classes	Examples	Plant Sources	Health Benefits	References
Terpenoids	Carotenoids	Beta-carotene, Lutein	Carrots, Spinach, Tomatoes	Antioxidant, promotes eye health, anticancer	Crozier et al. (2010); Sharma et al. (2022)
	Saponins	Ginsenosides, Dioscin	*Panax ginseng*, *Dioscorea*	Immune-boosting, cholesterol-lowering	Sharma et al. (2022)
	Essential oils	Menthol, Limonene	Mint, Citrus fruits	Antimicrobial, anti-inflammatory, digestive health	Crozier et al. (2006)

1.2.2 Phenolics

One of the largest and most widely distributed groups are phenolic compounds which are characterized by the presence of hydroxyl groups attached to aromatic rings. This group includes flavonoids, phenolic acids, tannins, and lignins and these compounds are well known for their antioxidant and anti-inflammatory properties, which are key in preventing oxidative stress and inflammation-driven chronic diseases like CVDs, diabetes, and cancer (Elshafie et al., 2023).

1.2.2.1 Flavonoids

Flavonoids are found in a wide range of fruits, vegetables, and beverages such as tea and wine. They are known for their potent antioxidant activity, which helps neutralize reactive oxygen species (ROS). Flavonoids like quercetin, catechins, and anthocyanins reduce oxidative stress and improve cardiovascular health (Ahmed et al., 2017).

1.2.2.2 Phenolic Acids

Compounds like ferulic acid and caffeic acid are known for their ability to reduce oxidative stress and inflammation. These phenolic acids are commonly found in whole grains, fruits, and vegetables and have been shown to lower the risk of chronic diseases like cancer and cardiovascular disorders (Kabera et al., 2014).

1.2.2.3 Tannins

Tannins are polyphenolic compounds present in tea, coffee, and wine. Their astringency is often associated with their role in plant defence. In humans, tannins exhibit strong anticancer, antimicrobial, and cardioprotective properties (Li et al., 2020; Tiwari & Rana, 2015).

1.2.3 Terpenoids

A class of five-carbon isoprene compounds known as terpenoids, or isoprenoids, contribute to the aromatic properties of plants and have significant medicinal properties. Carotenoids (such as beta-carotene) and saponins are well-known terpenoids used in the nutraceutical industry for their antioxidant, anticancer, and immune-boosting properties (Kabera et al., 2014).

1.3 BIOSYNTHETIC PATHWAYS OF PLANT SECONDARY METABOLITES

They are a diverse and relatively composite class of organic compounds and mainly have no connexion with life's essential activities as well as processes of vital growth and reproduction compared to primary metabolisms, which comprise sugars, lipids, proteins and so on. Nonetheless, they play an important role in regard to the protection of plants against sundry pathogenic attacks and signalization, as well as interactions of plants with the environment. Some of the main identified classifications of secondary metabolites in general emanate from primary metabolic processes such as glycolysis, the tricarboxylic acid (TCA) cycle and the pentose phosphate pathway (J. Li et al., 2014; Reshi et al., 2023; Alami et al., 2022).

1.3.1 Alkaloid Biosynthesis

Alkaloids represent a group of nitrogenous substances which are usually formed from amino acids. This category of compounds is believed to possess pharmacological properties, and most of the alkaloids have medical value. The formation of alkaloids is usually provoked by the decarboxylation of amino acids like tryptophan, tyrosine, or ornithine. For example, nicotine, an acknowledged alkaloid, is biosynthesised through ornithine for producing putrescine after decarboxylation. This compound is then methylated and cyclized to create the final structure of the alkaloid. In some cases the β-carboline is also N-oxide at the terminal nitrogen since many alkaloids are derivatives of this compound (Badri et al., 2019). This is because the particular enzymes participating in these reactions result in change of structure and formation of other kinds of alkaloids, such as quinine, morphine, nicotine, etc. (Kishimoto et al., 2016; Kumar, 2022).

1.3.2 Phenolic Compound Biosynthesis

They include compounds having a hydroxyl group on the aromatic carbon and are, in general, antioxidants in nature. These compounds are mostly synthesised by the shikimate pathway, a pathway that is involved in the synthesis of phenyl alanine, an amino acid that is essential in animals. Its biosynthesis starts with phenylalanine to cinnamic acid conversion through the enzyme phenylalanine ammonia-lyase (PAL). From cinnamic acid the following transformations occur and produce several phenolic products such as flavonoids, lignans, tannins and coumarins. For example, flavonoid biosynthesis begins with the condensation between p-coumaroyl-CoA derived from cinnamic acid of the phenylpropanoid pathway and acetyl-CoA, producing chalcones which are cyclised to flavonoids. These compounds are therefore involved in plant colour, UV cover and also anti-herbivore defence (Hu et al., 2022; Perez de Souza et al., 2020).

1.3.3 Terpenoid Biosynthesis

Terpenes are a large group of plant secondary metabolites of which terpenoids have a backbone typically of isoprene units tagged by C5. Terpenoid biosynthesis occurs via two major pathways: The two other pathways which are also referred to as these pathways include the MVA pathway that takes place in the cytoplasm and the hydrophobic MEP that takes place in plastids. The MVA pathway begins with acetyl CoA and then is converted into me ammonia acid to get isopentenyl diphosphate, or IPP, the terpenoid building block. In both the MEP and the methyl-D-erythritol pathways, pyruvate alongside glyceraldehyde-3-phosphate is used in the synthesis of 1-deoxy-D-xylulose-5-phosphate (DXP) for IPP synthesis. Terpenoids are classified based on the number of isoprene units and contain long-chain isoprenoids or non-oxygen heterocyclic hydrocarbons: monoterpenes, containing ten carbons; sesquiterpenes, containing fifteen carbons; diterpenes, containing twenty carbons; and so on. These compounds fulfil various roles: some are hormones like gibberellins; others are pigments such as carotenoids; a third group consists of volatile compounds that are involved in pollinator attraction or herbivore deterrence (Huang et al., 2022).

1.3.4 Glucosinolate and Polyketide Pathways

Other examples of secondary metabolite pathways are glucosinolate biosynthesis pathway, which is synthesised from amino acids and found in high amounts in species including mustard and broccoli.

These compounds are glucosinolates that are secondary metabolites of glycosidic sulphate esters of some organic compounds derived from sulphur for use in plant defence mechanisms against herbivores and disease-causing organisms. Furthermore, every assembly line utilizes the polyketide pathway that involves the aldol condensation of both acetyl-CoA and malonyl-CoA to produce various products, including anthraquinone and macrolide; despite these few products being used in medicine (Perez de Souza et al., 2020).

1.4 METABOLISM OF PLANT METABOLITES

The digestion, absorption, distribution and excretion of plant secondary metabolites in humans have been identified as follows. When plant-derived active compounds are taken as food or supplements, they go through the absorption process in the gastrointestinal canal first. Whereas the polyphenols and alkaloids are modified by gut microbiota enzymes before serving their purpose in some cases, there are other secondary metabolites that pass through the intestinal wall unaltered. Things like chemical nature and solubility characteristics as well as their compatibility with other foods in the system determine the rate of absorption of these metabolites. After a secondary metabolite has been taken into the body, it circulates in the blood and is biotransformed in the liver. This process is being ranked under two major stages. During phase I, carbohydrates like cytochrome P450 and reactions like oxidation, reduction or hydrolysis takes place, which create new functional groups for example, hydroxyl group, and metabolites of these compounds become more polar. These newly modified metabolites are then taken through Phase II reactions, where the molecules they now have carry conjugation processes with other molecules such as glucuronic acid, sulphate or glutathione to improve solubility still further (Krishnaiah et al., 2007; Wauquier et al., 2023). The liver's biotransformation enhances the solubility of such secondary metabolites, thereby easing their expulsion from the body. After going through the transition, these metabolites are transported in the bloodstream all over the body. In this phase, they get to tissues and organs of the body where they can display biological action, including antioxidant actions, anti-inflammatory impact or alteration of cellular functions. There are those metabolites that circulate in the active form by binding themselves to the tissues which make their duration of activity longer; others circulate in the blood. They are then expelled from the body in the process of excretion, which is through urine or the biliary system and faeces. It could also be during this process that the gut microbiota dephosphorylated some of the compounds, liberating additional bioactive subunits. This microbial metabolism is very central in defining the overall bioactivity of secondary metabolites. For instance, gut bacteria transform ellagitannins into urolithins that act as anti-inflammatory and anti-cancer compounds. Further, some of the compounds may follow enterohepatic circulation, in which the compounds are ejected into the bile and can be reabsorbed in the gastrointestinal tract before returning to the liver. In the end, the ability of a metabolite to enter the body's systemic circulation and be effective is known as bioavailability that also impacts the effectiveness of secondary metabolites (Cosme et al., 2020). The third reason is the plant secondary metabolites that are ingested have low bioavailability because they are not effectively absorbed into the bloodstream and are rapidly metabolized and excreted in the urine. Still, there are some compounds that can pass through the blood barriers and get deposited in tissues so that they can continue to remain biologically active for quite a long time. Therefore, the comprehensive biological activity of plant secondary metabolites requires the assimilation and metabolism of multiple stages of distribution and biotransformation in the liver, tissues, faeces and gut bacteria. They define the pharmacokinetics and pharmacodynamics of the plant-derived active principles in the human body and their beneficial effects, like antioxidant, anti-inflammatory and anticancer effects etc. (Hsiao et al., 2020; Wauquier et al., 2023; Zagoskina et al., 2023).

1.5 MECHANISMS OF ACTION AND HEALTH BENEFITS

1.5.1 Antioxidant Activity

Many plant secondary metabolites, particularly **phenolics**, **flavonoids**, and **terpenoids**, act as antioxidants, which can neutralize free radicals, reduction in oxidative stress, and inhibit lipid peroxidation, and occur by donating hydrogen or electrons to the unstable free radicals. Antioxidants have been essential for the prevention of chronic diseases, including **cancer**, **cardiovascular diseases**, and **neurodegenerative disorders.** PSMs, polyphenols, and flavonoids are presented to act on the cell membrane receptor to stimulate intracellular signalling that involves the protection of cells against OS. Of oxidative stress, the increased generation of ROS results in cellular damage such as that observed in lipid peroxidation. MDA is a breakdown product of lipid peroxidation because of ROS activities, which is an accumulation of ubiquitin and p53 when disrupted with antioxidants. PSMs regulate the production of ROS and therefore limit oxidation injury. These control membrane depolarization UIBs and activate SIRT2 within the intracellular space; the decrease in MDA levels and restriction of ROS synthesis are also facilitated. SIRT2 was identified as deacetylating substrates associated with oxidative stress/apoptosis to reduce cellular damage and enhance stress response and survival. PSMs also incite SIRT1 in the nucleus to control the FOXO transcription factors. FOXO proteins are one of the crucial control elements of antioxidant genes, which include superoxide dismutase (SOD) and catalase (CAT), and thus eliminate ROS formations, neutralizes superoxide radicals (O_2^-) into inactive forms; CAT further decomposes hydrogen peroxide (H_2O_2) into water and oxygen, substantially minimizing oxidative stress and preventing further cell damage. The activation of SOD and CAT through the SIRT1-FOXO by PSMs also develops a strong antioxidant defence mechanism (Figure 1.2). This is, in turn, helpful restoring redox balance within a cell and preventing a range of different damages that can be generated by ROS, including DNA damage, protein oxidation and mitochondrial injuries (Hu et al., 2022; SIREESHA et al., 2019).

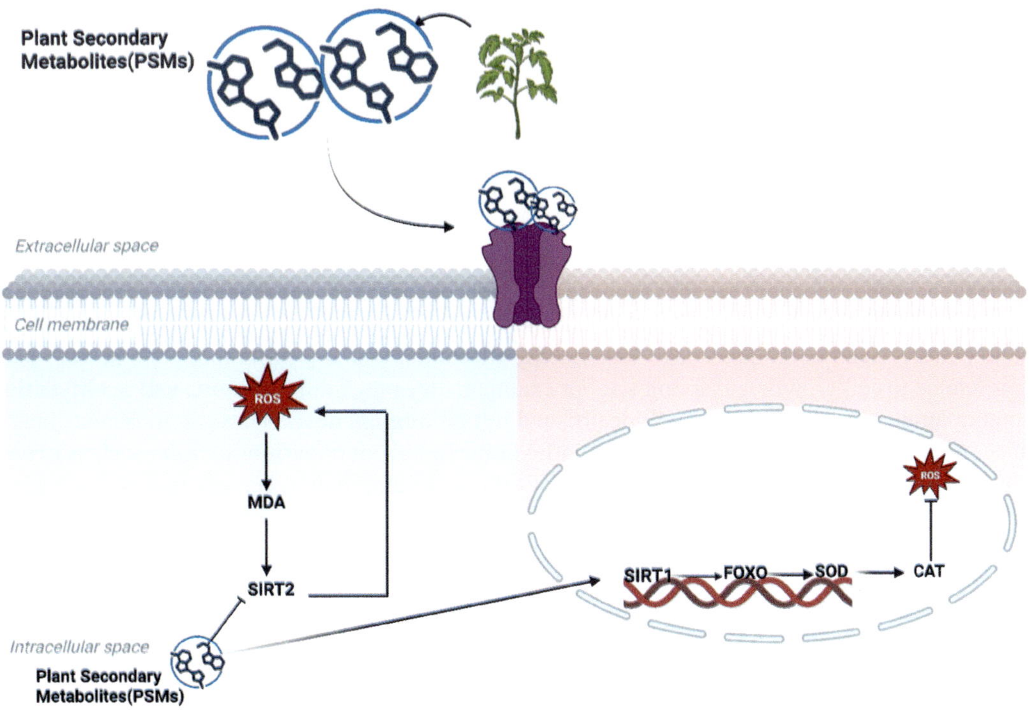

FIGURE 1.2 Antioxidant activities of plant secondary metabolism (Yang et al., 2016).

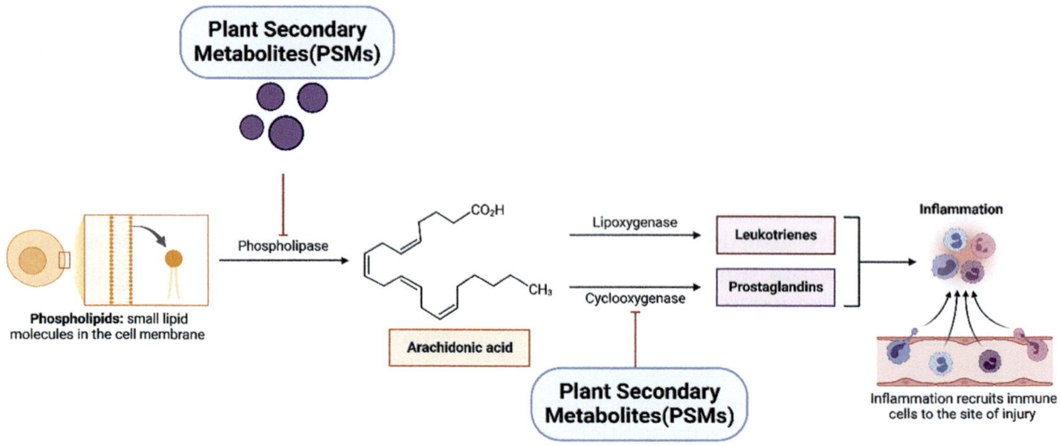

FIGURE 1.3 Anti-inflammatory activities of plant secondary metabolism.

1.5.2 Anti-inflammatory Effects

Flavonoids, **alkaloids**, and **triterpenoids** inhibit the production of pro-inflammatory cytokines (e.g., TNF-α, IL-6) and reduce the activity of enzymes like **cyclooxygenase (COX)** and **lipoxygenase (LOX)** involved in the inflammation pathway (Figure 1.3). The anti-inflammatory properties of these compounds help in the management of conditions like **arthritis**, **asthma**, and **inflammatory bowel diseases** (Indriyani et al., 2023; Roy et al., 2022).

1.5.3 Antimicrobial Action

Plant secondary metabolites like **alkaloids**, **terpenoids**, and **essential oils** can disrupt microbial cell walls, inhibit DNA replication, or interfere with the synthesis of essential proteins in bacteria, fungi, and viruses (Figure 1.4). These antimicrobial properties make them useful in fighting **infections**, reducing the development of **antibiotic resistance**, and promoting wound healing (Compean & Ynalvez, 2014; Huang et al., 2022; Eshboev et al., 2024).

1.5.4 Anticancer Activity

Many plant-derived compounds like **flavonoids**, **alkaloids**, and **saponins** modulate several pathways related to cell growth and apoptosis. For example, they may inhibit **tumour cell proliferation**, promote **apoptosis** (programmed cell death), and inhibit **angiogenesis**. These actions contribute to cancer prevention and therapy, offering potential in the treatment of various cancers such as **breast**, **colon**, and **lung cancer**. Alkaloids interact with the cell membrane receptors and activate signal transduction pathways, primarily the MAPK/ERK1/2/JNK pathway and downstream signalling of important cellular processes including apoptosis, cell cycle regulations and autophagy (Barku et al., 2019). When bound, MAPK (Mitogen-Activated Protein Kinase) gets activated and signalling cascade continues, and other transcriptional factors, including p53 and FOXO1/3a are recruited. They are of great importance in contributing to apoptosis through regulation of other pro-apoptotic genes-BAX and BCL2 proteins. By the MAPK pathway, alkaloids stimulate the p53 tumour suppressor gene. Thus, it eventually causes the cytochrome c to release from the mitochondria, which is central to intrinsic apoptosis (Figure 1.5). Cytochrome c further activates Apaf-1 resulting in the activation of caspase-9, then that of caspase-3 prior to the activation of apoptosis through cleavage of PARP (Damavandi et al., 2023; Ramakrishna et al., 2021).

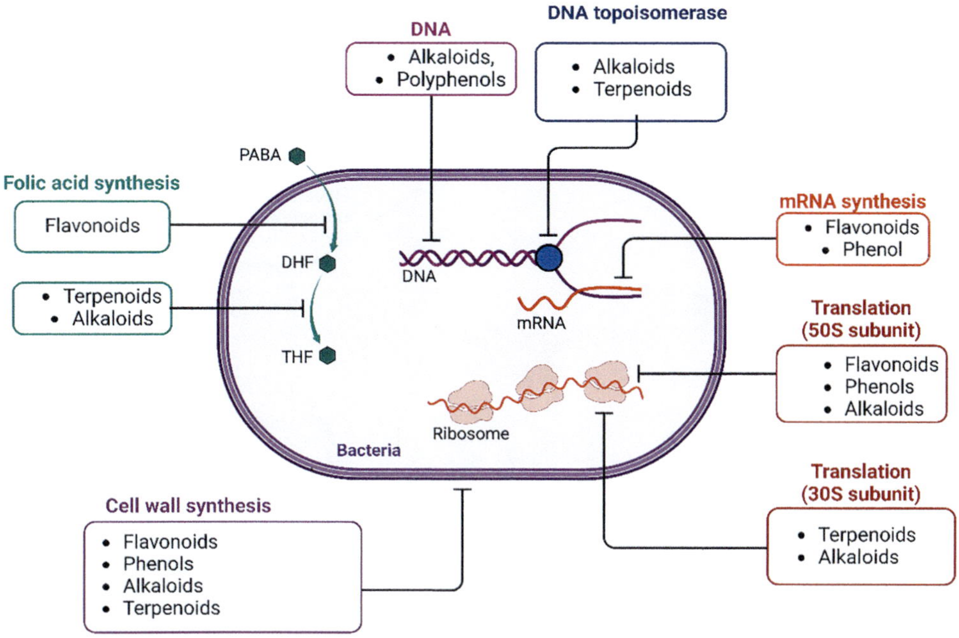

FIGURE 1.4 Antimicrobial activities of plant secondary metabolism.

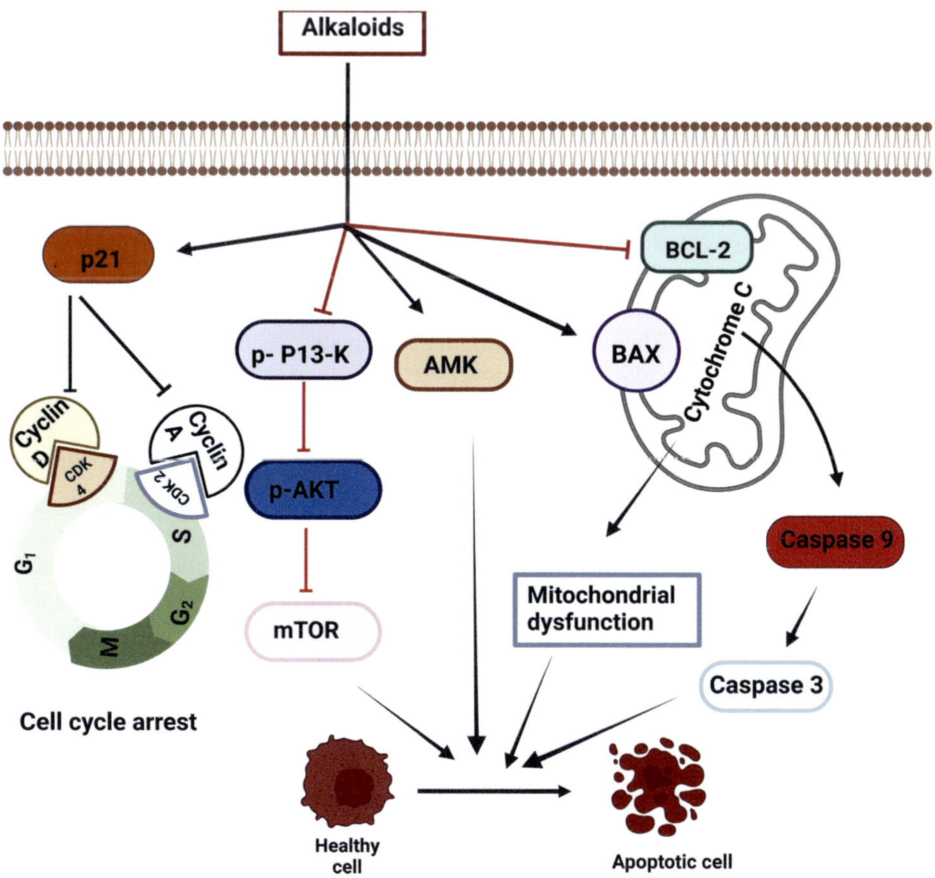

FIGURE 1.5 Anticancer mechanism of plant secondary metabolism.

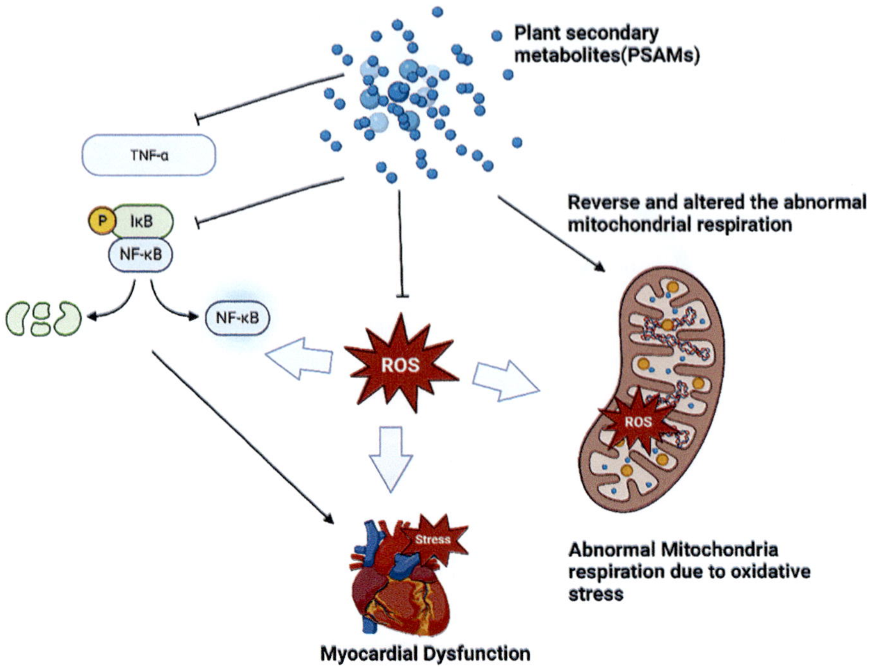

FIGURE 1.6 Cardioprotective activities of plant secondary metabolism.

1.5.5 Cardioprotective Effects

Certain secondary metabolites exert cardioprotective effects by reducing **oxidative stress**, enhancing **nitric oxide production** (leading to vasodilation), and improving **lipid metabolism**. These compounds help in reducing the risk of **atherosclerosis**, **hypertension**, and other cardiovascular disorders. Plant secondary metabolites work to modulate mitochondrial reactions and prevent the development of ROS and oxidative stress. These biological agents help in stabilizing mitochondrial activity, decreasing oxidative injury to cells, and returning a rational account of cell respiration. PSAMs also display anti-inflammatory properties, pointing to inhibition of the NF-κB cascade by blocking IκB phosphorylation. Thus, by repairing the damage to the activity of mitochondria in the myocardium and decreasing oxidative stress levels, PSAMs can prevent myocardial dysfunction (Figure 1.6). The potential for enhanced mitochondrial performance and decreased cellular stress in cardiac tissues exists (Luo et al., 2022; Riaz et al., 2023).

1.5.6 Neuroprotective Action

Secondary metabolites, including **phenolics**, **flavonoids**, and **terpenoids**, modulate neurotransmitter systems (e.g., **dopamine**, **serotonin**), reduce oxidative stress in neuronal cells, and inhibit neuroinflammatory pathways. These properties make them valuable in the prevention or treatment of **neurodegenerative diseases** such as **Alzheimer's** and **Parkinson's disease**. Flavonoids are derived from phenolic compounds and have the ability to chelate with Aβ peptides, thus, inhibiting the formation of toxic plaques. Amyloid-beta plaques are considered to be the main one of the causes of damages to the synaptic connections and neurotoxicity in AD. Since plaque formation is suppressed by phenolics, they also help maintain the health of neurons by preventing the toxicity of amyloids. In Alzheimer's, neurofibrillary tangles made from hyperphosphorylated tau proteins play a role in neurodegeneration. Flavonoids, a subgroup of phenolic compounds, have been demonstrated to modulate enzymes such as glycogen synthase kinase 3-beta and prevent the formation of NFTs due

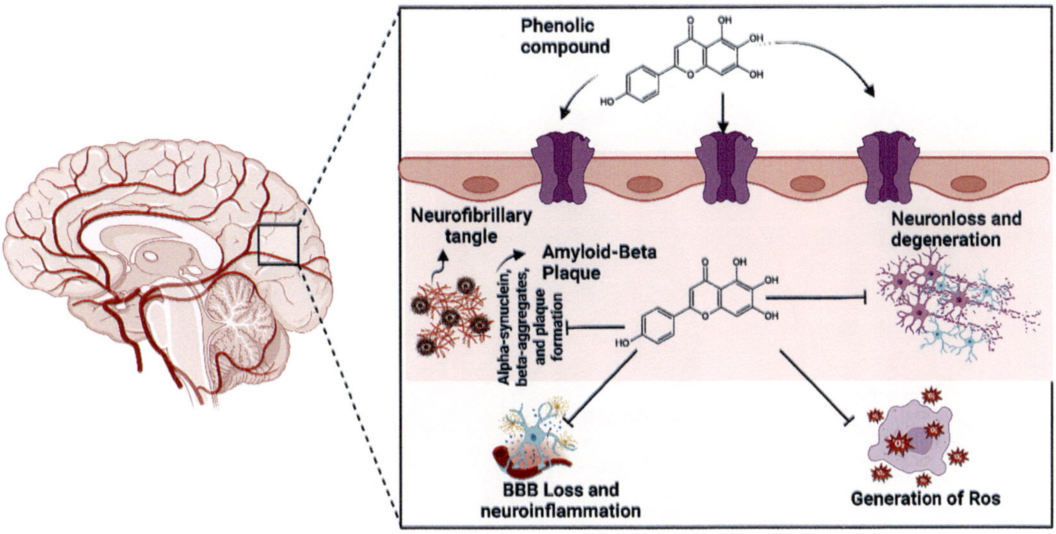

FIGURE 1.7 Neuroprotective activities of plant secondary metabolism.

to tau hyperphosphorylation. It not only ceases the formation of tangles but also stops extra neuronal death. Phenolic compounds have great antioxidant ability that can eliminate the ROS, marking it as one of the crucial activities. Oxidative stress from excessive ROS production is one of the greatest causes of neuronal damage as well as apoptosis in neurodegenerative diseases. Phenolic compounds also influenced the inflammation through the suppression of microglial activation and the inhibition of the synthesis of pro-inflammatory cytokines like IL-1β and TNF-α (Figure 1.7). This helps maintain neuronal structure and decreases inflammation of the neurons, which are other key factors in the progression of the diseases. Phenolic compounds prevent neuronal degeneration by influencing several important survival proteins, including the PI3K/Akt signalling, which has roles in cell survival and differentiation. Not only does the phenol affect the existing neuronal cells by improving their synaptic plasticity, but it also induces the creation of new neurons also providing functional benefits of the product in the form of improved cerebral function (Luo et al., 2022; Abdulwanis Mohamed et al., 2019; Riaz et al., 2023).

1.5.7 Hepatoprotective Activity

Plant metabolites like **saponins**, **flavonoids**, and **terpenoids** can protect liver cells by enhancing antioxidant defences, modulating detoxification enzymes, and reducing lipid peroxidation. They are useful in preventing or treating **liver diseases**, including **hepatitis** and **fatty liver disease**. Plant secondary metabolites can chelate ROS, resulting in low oxidative stress. Prospective plant metabolites may upregulate GSH, CAT, and SOD, all of which are antioxidant enzymes (Figure 1.8) (Gupta & Pandey, 2019). This helps in maintaining the redox balance and preventing further cellular damage. The metabolites produced by the plant may modulate the activity of JNK, thereby suppressing apoptosis and necrosis.

1.6 APPLICATIONS OF PLANT SECONDARY METABOLITES IN HUMAN HEALTH

They are natural products having a number of therapeutic uses in the treatment of human ailments. Some of the identified compounds include alkaloids, flavonoids and terpenoids, phenolics and glycosides and these compounds exert a range of biological activities. Among the functions they have their, main one is that they act as antioxidants, where they donate their electrons to free radicals in a

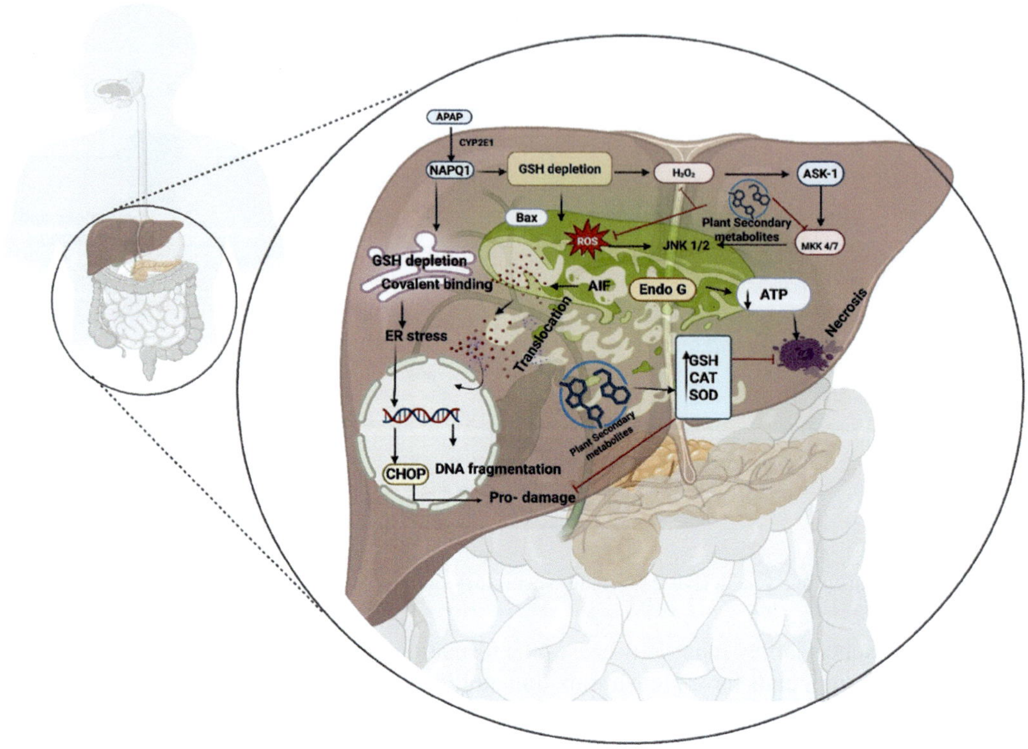

FIGURE 1.8 Hepatoprotective activities of plant secondary metabolism.

step that reduces oxidative stress, which leads to major health conditions, including cardiovascular diseases, neurodegenerative disorders and cancer. For example, flavonoids like quercetin existing in fruits and vegetables and catechins have strong antioxidant capability (Crozier et al., 2010; Luo et al., 2022). They also exhibit favourable anti-inflammatory action by modulating mentioned above inflammatory pathways such as down-regulation of NF-κB, which results in low levels of cytokines. This makes them advantageous in chronic inflammatory diseases, for instance, arthritis, asthma and inflammatory bowel disease. Beverages derived from turmeric and identified as curcumin and from grapes, better known as resveratrol, have been described to have anti inflammatory activity. Furthermore, PSMs are claimed to possess anticancer activity as a result of their effects consisting of blocking cancer cell division, promoting cancer cell death, blood vessel blocking and prevention of spreading of cancer cells. Quadruple helix compounds such as epigallocatechin gallate from green tea and vincristine from periwinkle are shown in cancer treatment and prevention (Sharma et al., 2022; Tiwari & Rana, 2015). Besides, PSMs demonstrate activities against microbial and viral cells through the destruction of microbial cell walls and inhibition of DNA synthesis in pathogens and may be used in the perspective of substitution for antibiotics due to the antimicrobial resistance issue. Berberine alkaloids and phenolic acids have shown inhibitory effects on different bacterial and viral species. In cardiovascular status, PSMs have positive effects on endothelium relaxing factor, blood pressure, the degree of lipid peroxidation, and work effectively in hypertension and atherosclerosis. Cocoa and grapeseed polyphenols, including resveratrol, have been shown to possess cardioprotective properties (Luo et al., 2022).

In diabetes mellitus as well, PSMs are advantageous because they improve insulin sensitivity, control glucose homeostasis, and lessen the enzymes that can cleave glucose to glucose-6-phosphate. There is evidence that flavonoids and saponins at least ginsenosides of ginseng, are useful in modulating blood glucose concentration. Several studies suggest some of the PSMs possess neuroprotective properties such as reducing the oxidation of neurons, balancing inflammation within the brain, and improving brain function, which might provide beneficial in the treatment of neurological disorders including Alzheimer's and Parkinson's Diseases (Abdulwanis Mohamed et al., 2019). For instance, *Ginkgo biloba* extract and curcumin are some of the most popular ones for which neuroprotective effects are investigated. An unhealthy liver benefits from PSMs by eliminating free radicals, increasing the activity of detoxification enzymes, and decreasing liver inflammation in diseases such as hepatitis. Silybin from milk thistle is amongst the best-known categories of plant secondary metabolites utilized for liver protection. PSMs also modulate the immune system and, therefore, could be employed to enhance immune response in immunodeficient patients or treat autoimmune disorders. Echinacea and ginsenosides from ginseng are the examples of immunomodulatory substances (Damavandi et al., 2023; Abdulwanis Mohamed et al., 2019). These compounds are also useful in wound healing, for the increase of collagen synthesis and for tissue regeneration, as has been explained for *Aloe vera* and *Centella asiatica*. Also, PSM way facilitates the hormone receptors or even mimics them in order to control hormonal conditions, including menopause and thyroid problems. Soy isoflavones are used extensively to treat the symptoms of menopause and for bone health. Finally, PSMs address obesity by regulating lipid synthesis and lowering fat storage; capsaicin from chilli peppers and chlorogenic acid from green coffee beans are used in weight loss products. Therefore, PSMs have significant roles in human health with prospective therapeutic values in disease prevention and control. Due to their variable biological properties, they find use in almost all fields of pharmaceutical research as well as in nutraceuticals, and so they continue to be the subject of research on their future prospects for use in medical science (Barku, 2019).

1.7 CHALLENGES AND FUTURE PROSPECTS

Secondary metabolites of plants (PSMs) confront several problems regarding their utilisation for human needs, and, at the same time, they offer extensive opportunities for future developments. One of the greatest difficulties is based in the degrees of isolation and purification of the PSMs as they exist at very low concentrations in the plant matrix, hence making the extraction process very laborious and expensive. This causes an increase in the cost of production of these compounds and hence hinders their utilisation in high quantities. Regarding this, improvement could be made to the production technologies such as supercritical fluid extraction and microwave-assisted extraction. Another challenge is the differences in the content of PSM because of differences in growth conditions and extraction methods, and standardisation is therefore challenging. Such variation is important because therapeutic applications depend on the efficacy, safety and other related characteristics. It will be critically important to standardize and develop robust methods for extraction, purification and testing of plant extracts and pharmaceuticals to have the ability to reproduce the form and results safely across a number of rounds (Feng et al., 2022). Moreover, many of the PSMs have poor bioavailability and poor absorption; besides, they are quickly eliminated from the body and therefore not effective in curing the diseases. The modification of their physicochemical property by the use of nanotechnology and advanced drug delivery systems like nanoparticles as well as liposomes could enhance their bioavailability. The way in which PSMs influence these biological states may also hinder their formulation and incorporation into treatments since the multi-step processes involved within myriad levels of molecular communication can impede biochemical drug production. Other detailed investigations such as network pharmacology and omics approaches might explain these complex functions better. Secondly, the uncontrolled harvesting of medicinal plants threatens the concept of the ability and the diverse environment. Some of the methods that may be used in cultivation include tissue culture and sustainable agriculture (Allwood et al., 2021; Feng et al., 2022). Lastly, harnessing artificial intelligence (AI) in PSM research can accelerate drug discovery by predicting biological activities and optimizing extraction processes. Overall, while challenges remain, the future of plant secondary metabolites is bright, with the potential for significant contributions to healthcare, personalized medicine, and sustainable practices (Sun et al., 2022).

1.8 CONCLUSION

In conclusion, secondary metabolites are important components in plant-plant and plant-herbivore interactions and concluded that they possess therapeutic properties. These diverse compounds, which are emitted as a result of some environmental stressors, have a crucial impact on the defence mechanisms of plants, attraction of pollinators and symbiotic plant mutualism and, therefore plants' survival and reproduction. Furthermore, their pharmacological properties have attracted a lot of attention from researchers in medicine and biotechnology, resulting in the discovery of lots of bioactive compounds for the treatment of various diseases. Studying plant secondary metabolites provides understanding of the impact of plant substances on the environment and is valuable for conservation because of the focus on plant and environment interaction. At the same time, it has become evident that these metabolites have therapeutic effects and, thus, may serve as natural substitutes for synthetic-based pharmaceuticals, opening brand new horizons in drug discovery and design. While there are numerous papers published describing biosynthetic pathways and the actions of these compounds, further investigations should target the effects of environmentally sustainable farming practices and how climate change might affect secondary metabolite production. Lastly, optimisation can enhance ecological conservation and human health, continually proving the potential worth of these compounds in the natural world.

REFERENCES

Abdulwanis Mohamed Z, Mohamed Eliaser E, Mazzon E, Rollin P, Cheng Lian Ee G, Abdull Razis AF. Neuroprotective potential of secondary metabolites from *Melicope lunu-ankenda* (Rutaceae). *Molecules*. **2019**;24(17):3109.

Allwood BW, Byrne A, Meghji J, Rachow A, van der Zalm MM, Schoch OD. Post-tuberculosis lung disease: Clinical review of an under-recognised global challenge. *Respiration*. **2021** Aug 5;100(8):751–63.

Ahmed E, Arshad M, Zakriyya Khan M, Shoaib Amjad M, Mehreen Sadaf H, Riaz I, et al. Secondary metabolites and their multidimensional prospective in plant life. *Journal of Pharmacognosy and Phytochemistry*. **2017**;6(2):205–14.

Alami MM, Ouyang Z, Zhang Y, Shu S, Yang G, Mei Z, Wang X. The current developments in medicinal plant genomics enabled the diversification of secondary metabolites' biosynthesis. *International Journal of Molecular Sciences*. **2022**;23(24):15932.

Badri S, Basu VR, Chandra K, Anasuya D. A review on pharmacological activities of alkaloids. *World Journal of Current Medical and Pharmaceutical Research*. **2019**:230–4.

Badyal S, Singh H, Yadav AK, Sharma S, Bhushan I. Plant secondary metabolites and their uses. *Plant Archives*. **2020**;20(2):3336–40.

Barku VY. Wound healing: Contributions from plant secondary metabolite antioxidants. *Wound Healing-Current Perspectives*. **2019**;13.

Clemensen AK, Provenza FD, Hendrickson JR, Grusak MA. Ecological implications of plant secondary metabolites – Phytochemical diversity can enhance agricultural sustainability. *Frontiers in Sustainable Food Systems*. **2020**;4:547826.

Compean KL, Ynalvez RA. Antimicrobial activity of plant secondary metabolites: A review. *Research Journal of Medicinal Plant*. **2014**;8(5):204–13.

Cosme P, Rodríguez AB, Espino J, Garrido M. Plant phenolics: Bioavailability as a key determinant of their potential health-promoting applications. *Antioxidants*. **2020**;9(12):1263.

Crozier A, Del Rio D, Clifford MN. Bioavailability of dietary flavonoids and phenolic compounds. *Molecular Aspects of Medicine*. **2010**;31(6):446–67.

Crozier A, Yokota T, Jaganath IB, Marks S, Saltmarsh M, Clifford MN. Secondary metabolites in fruits, vegetables, beverages and other plant based dietary components. *Plant Secondary Metabolites: Occurrence, Structure and Role in the Human Diet*. **2006**;7;208–302.

Damavandi MS, Shojaei H, Esfahani BN. The anticancer and antibacterial potential of bioactive secondary metabolites derived from bacterial endophytes associated with Artemisia absinthium. *Scientific Reports*. **2023**;13(1):18473.

Dhaniaputri R, Suwono H, Amin M, Lukiati B. Introduction to plant metabolism, secondary metabolites biosynthetic pathway, and in-silico molecular docking for determination of plant medicinal compounds: An overview. In *7th International Conference on Biological Science (ICBS 2021)*, Yogyakarta, Indonesia, **2022** May 2 (pp. 373–382). Atlantis Press.

Eshboev F, Mamadalieva N, Nazarov PA, Hussain H, Katanaev V, Egamberdieva D, Azimova S. Antimicrobial action mechanisms of natural compounds isolated from endophytic microorganisms. *Antibiotics*. **2024**;13(3):271.

Egamberdieva D, Parray JA, Davranov K, editors. *Plant Endophytes and Secondary Metabolites*. Elsevier; **2023**.

Elshafie HS, Camele I, Mohamed AA. A comprehensive review on the biological, agricultural and pharmaceutical properties of secondary metabolites based-plant origin. *International Journal of Molecular Sciences*. **2023**;24(4):3266.

Feng C, Chen J, Ye W, Liao K, Wang Z, Song X, Qiao M. Synthetic biology-driven microbial production of resveratrol: Advances and perspectives. *Frontiers in Bioengineering and Biotechnology*. **2022**;10:833920.

Gupta A, Pandey AK. Plant secondary metabolites with hepatoprotective efficacy. In Charis M. Galanakis (ed.), *Nutraceuticals and Natural Product Pharmaceuticals* (pp. 71–104). Academic Press; **2019**.

Hsiao YH, Ho CT, Pan MH. Bioavailability and health benefits of major isoflavone aglycones and their metabolites. *Journal of Functional Foods*. **2020**;74:104164.

Hu W, Sarengaowa, Guan Y, Feng K. Biosynthesis of phenolic compounds and antioxidant activity in fresh-cut fruits and vegetables. *Frontiers in Microbiology*. **2022**;13:906069.

Huang W, Wang Y, Tian W, Cui X, Tu P, Li J, Shi S, Liu X. Biosynthesis investigations of terpenoid, alkaloid, and flavonoid antimicrobial agents derived from medicinal plants. *Antibiotics*. **2022**;11(10):1380.

Indriyani NN, Anshori JA, Permadi N, Nurjanah S, Julaeha E. Bioactive components and their activities from different parts of *Citrus aurantifolia* (christm.) Swingle for food development. *Foods*. **2023**;12(10):2036.

Kabera JN, Semana E, Mussa AR, He X. Plant secondary metabolites: Biosynthesis, classification, function and pharmacological properties. *Journal of Pharmacy and Pharmacology*. **2014**;2(7):377–92.

Kishimoto S, Sato M, Tsunematsu Y, Watanabe K. Evaluation of biosynthetic pathway and engineered biosynthesis of alkaloids. *Molecules*. **2016**;21(8):1078.

Krishnaiah D, Sarbatly R, Bono A. Phytochemical antioxidants for health and medicine: A move towards nature. *Biotechnology and Molecular Biology Reviews*. **2007**;1(4):97–104.

Kumar A. Plant alkaloids. In *Phytoconstituents and Antifungals* (pp. 53–64). Academic Press; **2022**.

Leiss KA, Choi YH, Verpoorte R, Klinkhamer PG. An overview of NMR-based metabolomics to identify secondary plant compounds involved in host plant resistance. *Phytochemistry Reviews*. **2011**;10:205–16.

Li Y, Kong D, Fu Y, Sussman MR, Wu H. The effect of developmental and environmental factors on secondary metabolites in medicinal plants. *Plant Physiology and Biochemistry*. **2020**;148:80–9.

Luo Y, Jian Y, Liu Y, Jiang S, Muhammad D, Wang W. Flavanols from nature: A phytochemistry and biological activity review. *Molecules*. **2022**;27(3):719.

Perez de Souza L, Garbowicz K, Brotman Y, Tohge T, Fernie AR. The acetate pathway supports flavonoid and lipid biosynthesis in arabidopsis. *Plant Physiology*. **2020**;182(2):857–69.

Reshi ZA, Ahmad W, Lukatkin AS, Javed SB. From nature to lab: A review of secondary metabolite biosynthetic pathways, environmental influences, and in vitro approaches. *Metabolites*. **2023**. 13: 895 [Internet].

Ramakrishna W, Kumari A, Rahman N, Mandave P. Anticancer activities of plant secondary metabolites: Rice callus suspension culture as a new paradigm. *Rice Science*. **2021**;28(1):13–30.

Riaz M, Khalid R, Afzal M, Anjum F, Fatima H, Zia S, Rasool G, Egbuna C, Mtewa AG, Uche CZ, Aslam MA. Phytobioactive compounds as therapeutic agents for human diseases: A review. *Food Science & Nutrition*. **2023**;11(6):2500–29.

Roy A, Khan A, Ahmad I, Alghamdi S, Rajab BS, Babalghith AO, Alshahrani MY, Islam S, Islam MR. Flavonoids a bioactive compound from medicinal plants and its therapeutic applications. *BioMed Research International*. **2022**;2022(1):5445291.

Sharma A, Sharma S, Kumar A, Kumar V, Sharma AK. Plant secondary metabolites: An introduction of their chemistry and biological significance with physicochemical aspect. In Anil Kumar Sharma and Ajay Sharma (eds.), *Plant Secondary Metabolites: Physico-Chemical Properties and Therapeutic Applications* (pp. 1–45). Springer Nature Singapore; **2022**.

Sun J, Sun W, Zhang G, Lv B, Li C. High efficient production of plant flavonoids by microbial cell factories: Challenges and opportunities. *Metabolic Engineering*. **2022**;70:143–54.

Tiwari R, Rana CS. Plant secondary metabolites: A review. *International Journal of Engineering Research and General Science*. **2015**;3(5):661–70.

Upadhyay SK, Singh SP (2021). *Bioprospecting of Plant Biodiversity for Industrial Molecules*. Wiley. DOI: 10.1002/9781119718017

Upadhyay SK, Singh SP (2023). *Plants as Bioreactors for Industrial Molecules*. Wiley. DOI:10.1002/9781119875116

Upadhyay SK, Singh SP (2024). *Herbal Nutraceuticals: Products and Processes*. Wiley. DOI:10.1002/9781394241576

Vani I, Sireesha R, Mak KK, Rao PM, Prasad KR, Rao MB. Microwave assisted synthesis and antimicrobial and antioxidant activities of dimers of 1, 2, 3-triazole-benzofuran bearing alkyl spacer derivatives. *Chemical Data Collections*. **2021** Feb 1;31:100605.

Wauquier F, Boutin-Wittrant L, Krisa S, Valls J, Langhi C, Otero YF, Sirvent P, Peltier S, Bargetto M, Cazaubiel M, Sapone V. Circulating human metabolites resulting from TOTUM-070 absorption (a plant-based, polyphenol-rich ingredient) improve lipid metabolism in human hepatocytes: Lessons from an original ex vivo clinical trial. *Nutrients*. **2023**;15(8):1903.

Wink M. Introduction: Biochemistry, physiology and ecological functions of secondary metabolites. In M. Wink (ed.), *Annual Plant Reviews*; Wiley-Blackwell: Oxford, UK; **2010**.

Yang D, Tan X, Lv Z, Liu B, Baiyun R, Lu J, Zhang Z. Regulation of Sirt1/Nrf2/TNF-α signaling pathway by luteolin is critical to attenuate acute mercuric chloride exposure induced hepatotoxicity. *Scientific Reports*. **2016**;6(1):37157.

Zagoskina NV, Zubova MY, Nechaeva TL, Kazantseva VV, Goncharuk EA, Katanskaya VM, Baranova EN, Aksenova MA. Polyphenols in plants: Structure, biosynthesis, abiotic stress regulation, and practical applications. *International Journal of Molecular Sciences*. **2023**;24(18):13874.

2 Role of Transcription Factors in the Regulation of Secondary Metabolites Production in Plants

Ruchi Singh[1], *Payal Jain*[2], *Avadhesh Kumar Mishra*[2,3], *Ashu Tiwari*[1,3], *and Dheerendra Kumar*[1,3]

[1]Department of Biotechnology, RR Institute of Modern Technology (Affiliated with AKTU), Lucknow, India

[2]Department of Genetics and Plant Breeding. Chaudhary Charan Singh University, Meerut, India

[3]Integrated Institute of Biotechnology (Approved by AICTE), Lucknow, India

2.1 INTRODUCTION

Plants synthesis an array of diverse compounds throughout their life cycle, which can be classified into two main categories: primary and secondary metabolites. Primary metabolites, including starch, sucrose, amino acids, lipids, and proteins, are essential for the fundamental functions of the plant. In contrast, secondary metabolites serve primarily as a defense mechanism. The synthesis of these secondary metabolites occurs in response to pathogen attacks or as a means of managing stress (Cui et al. 2015). Secondary metabolites can be grouped into three principal categories: terpenes, alkaloids, and phenolic compounds. The production of secondary metabolites is facilitated by various metabolic pathways that enable plants to effectively adapt to stress conditions. These pathways are derived from those of primary metabolites, which yield the necessary precursors for secondary metabolite synthesis. The shikimate pathway initiates the biosynthesis of aromatic amino acids, activated under stress conditions. This pathway leads to the production of tryptophan, tyrosine, and phenylalanine, all of which contribute to the enhanced synthesis of secondary metabolites (Francenia Santos-Sánchez et al. 2019). The shikimate signaling pathways respond to both biotic and abiotic stresses and encompass various downstream defensive signals, including mitogen-activated protein kinase (MAPK) cascades (Pitzschke et al. 2009), reactive oxygen species (ROS) (Camejo et al. 2016), JA, salicylic acid (SA), and ethylene. These pathways, along with their associated crosstalk, subsequently activate TFs that modulate plant defense mechanisms by interacting with the promoter regions of specific target genes (Menke 1999).

Transcription factors play a key role in regulating flux toward the secondary metabolism by controlling either structural pathway gene expression or functional genes. TFs not only affect the synthesis of SMs but also affect the transportation and accumulation of SMs and their coordination with stress management. There are mainly six TF families named AP2/ERF, WRKY, bHLH, bZIP, MYB, and NAC involved in SMs production and stress tolerance (Meraj et al. 2020). TFs assimilate both internal (typically developmental) and external signals, attaching to specific cis-elements frequently located within promoter regions—to either activate or inhibit the expression of genes responsible for encoding enzymes (Yang et al. 2012). Additionally, some of these regulators engage

DOI: 10.1201/9781003518358-2

with other TFs to create a complex. For example, R2R3-MYB-bHLH-WD40 protein complexes have been shown to control multiple enzymatic steps in the flavonoid biosynthetic pathway in *C. sinensis* (Zhao et al. 2013a). Moreover, the MYB36-ERF6/ERF115 module regulates the biosynthesis of tanshinones and phenolic acids in *Salvia miltiorrhiza* hairy roots (Li et al. 2020). This chapter aims to encapsulate recent findings in these domains, focusing particularly on newly discovered TFs and their roles in the regulation of secondary metabolism. We extend our review to summarize the role of TFs in terpene, alkaloid, flavonoid, and phenolic compound biosynthesis. Some important SM biosynthesis and its transcriptional regulation are also discussed in this chapter.

2.2 TRANSCRIPTIONAL REGULATION OF TERPENE BIOSYNTHESIS

Terpenoids represent a vast and varied category of natural compounds that originate from mevalonic acid and are made up of isoprene units. These compounds are essential for various processes in plants, including growth, development, photosynthesis, and protection against environmental stressors. Beyond their biological importance, terpenoids find extensive applications in sectors such as pharmaceuticals, food, and cosmetics, exhibiting properties that include antitumor, anti-inflammatory, antibacterial, and antiviral activities. Furthermore, their intricate structures and diverse mechanisms of action suggest potential uses in areas such as insect resistance, immunoregulation, antioxidation, anti-ageing, and neuroprotection (Yang et al. 2020a). Terpenoids are categorized according to the number of isoprene units they contain, leading to classifications such as monoterpenes, sesquiterpenes, and diterpenes, among others. Notable terpenoids recognized for their medicinal attributes include cannabinoids derived from Cannabis sativa, tanshinone extracted from *S. miltiorrhiza*, and paclitaxel sourced from Taxus media, which is known for its anti-cancer effects. Additionally, artemisinin from *Artemisia annua* is effective in treating malaria, while ginsenoside from *Panax ginseng* offers adaptogenic benefits, and ginkgolide from *Ginkgo biloba* is acknowledged for its neuroprotective qualities (Wang et al. 2023).

2.2.1 Role of Transcription Factors in Biosynthesis of Terpenoids

Numerous prior investigations have highlighted a range of TFs that play significant roles in the regulation of terpenoid biosynthesis, including MYB, WRKY, bHLH, AP2/ERF, and bZIP TFs. In particular, WRKY TFs are essential in plant defense mechanisms, responses to environmental stimuli, and the regulation of secondary metabolism. Specifically, WRKY 8, WRKY 47, and WRKY 33 are known to regulate genes associated with taxol biosynthesis in *Taxus* species (Chen et al. 2022). Additionally, WRKY1 and WRKY4 contribute to the enhancement of artemisinin production, while WRKY1 and WRKY2 are implicated in the regulation of tashinone biosynthesis in *Salvia miltiorrhiza* (Deng et al. 2019). Consequently, WRKY genes present promising opportunities for increasing the yield of medicinal compounds. Furthermore, the biosynthesis of andrographolide is modulated by seven distinct WRKY TFs, including WRKY01, WRKY08, WRKY12, WRKY14, WRKY19, WRKY20, and WRKY50 (Zhang et al. 2021). A sesquiterpene cyclase known as delta-cadinene synthase (CAD1) plays a crucial role in a branch-point reaction during the biosynthesis of phytoalexins, such as gossypol, and is regulated by the transcription factor WRKY1 (Xu et al. 2004).

MYB TFs play a crucial role in regulating terpenoid biosynthesis within medicinal plants. Specifically, MYB3 functions as an activator for the TBT and TS genes involved in paclitaxel production. Additionally, MYB97 and SmMYB98 are known to enhance the biosynthesis of tanshinone in *Salvia miltiorrhiza* (Hao et al. 2020). MYB17 is associated with the promotion of glandular trichome development and the production of artemisinin, while MIXTA may facilitate cannabinoid production through the development of trichomes. Conversely, certain MYB TFs serve as negative regulators; for instance, MYB15 inhibits the expression of ORA and represses artemisinin synthesis, whereas MYB1 represses the promoters of CBCAS and CBDAS, thereby inhibiting cannabinoid synthesis (Wang et al. 2023).

AP2/ERF TFs play a vital role in the biosynthesis of terpenoids across a range of plant species. In Artemisia annua, the TFs ERF1, ERF2, and TAR1 are instrumental in regulating the production of sesquiterpenoids by interacting with the promoters of the amorpha-diene synthase (ADS) and CYP71AV1 genes. In Salvia miltiorrhiza, the TFs ERF6, ERF8, ERF128, and ERF1L1 facilitate the biosynthesis of tanshinones, with ERF1L1 notably increasing tanshinone levels through the upregulation of DXR and direct binding to its promoter. In *Taxus* species, ERF15 is responsible for promoting the biosynthesis of paclitaxel by activating the TASY gene, while AP2L1 enhances cannabinoid production in Cannabis by elevating the expression of CBCAS and THCAS. Additionally, *Andrographis paniculata* produces andrographolide, a natural antibiotic known for its potent anti-inflammatory and antiviral properties. The genes AP2_018, AP2_019, AP2_026, AP2_052, and AP2_083, which are co-expressed with CPS2, may play a role in the biosynthesis of andrographolide in *A. paniculata* (Zhang et al. 2015a). MYC2 and ERF3 play a positive role in the transcriptional regulation of 3-hydroxy-3-methylglutaryl CoA reductase (HMGR2) and AFS. The upregulation of AFS results in an increased production of α-farnesene in apple fruit. MYC2 and ERF3 exert direct control over AFS by interacting with its promoter, which is a key enzyme in the α-farnesene biosynthetic pathway (Wang et al. 2020) (Figure 2.1).

bHLH TFs play a vital role in the regulation of plant growth and the biosynthesis of terpenoids. Specifically, bHLH1 and MYC2 are known to promote the production of artemisinin, while bHLH10 and bHLH148 are involved in activating the enzymes responsible for tashinone biosynthesis. Conversely, certain bHLH factors can inhibit the synthesis of secondary metabolites; for instance, bHLH3 negatively impacts tanshinone biosynthesis, and in the case of Taxus, the factors JAMYC1, JAMYC2, and JAMYC4 serve to suppress paclitaxel production (Hong et al. 2012). In *Medicago truncatula*, the bHLH transcription factor plays a crucial role in regulating the triterpenoid pathway, which is essential for the synthesis of saponins. The overexpression of TSAR2 has been shown to increase

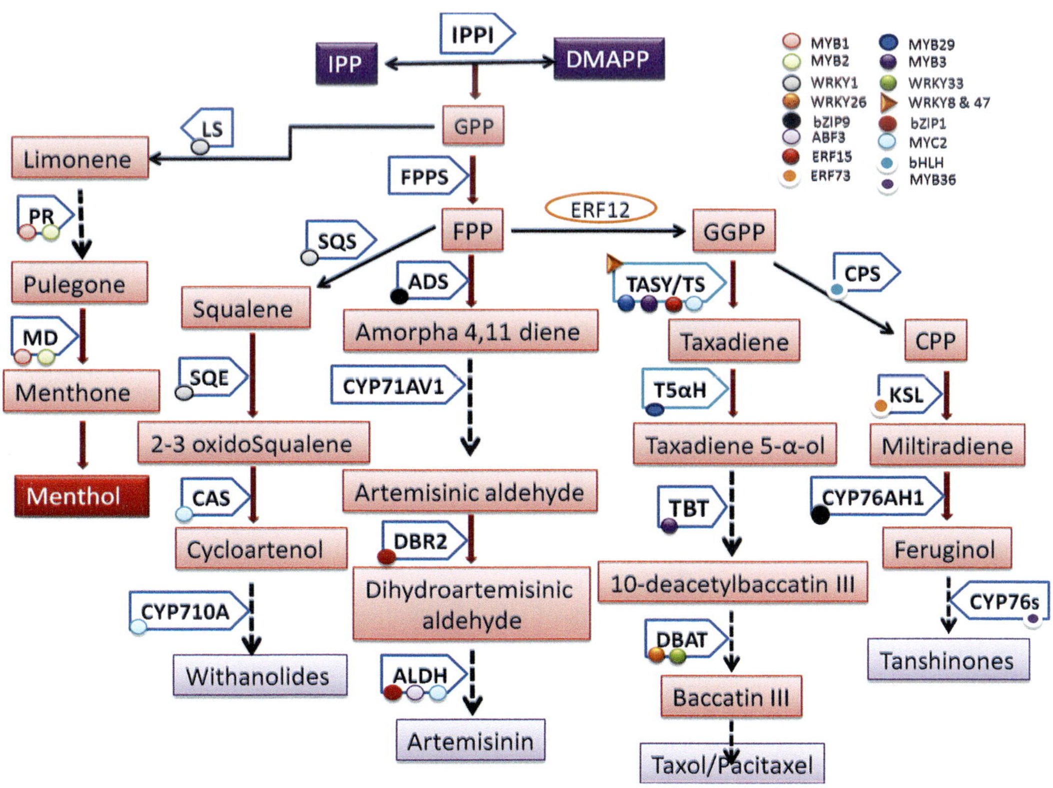

FIGURE 2.1 Role of TFs in terpene biosynthesis.

the production of hemolyticsaponins, while the overexpression of TSAR1 has been found to enhance the biosynthesis of nonhemolyticsaponin (Mertens et al. 2016). Additionally, other TFs such as NAC1, HY5, and bZIP1 play a crucial role in promoting the biosynthesis of artemisinin, while GRAS TFs, including GRAS1 and GRAS2, contribute to the increased levels of tashinone. Although numerous TFs associated with terpenoid biosynthesis have been identified, there are still significant challenges in confirming their specific functions and elucidating their substrate specificity (Wang et al. 2023).

2.2.2 ERF and MYB TF Involved in Capsaicinoid (CAP) Biosynthesis

Capsaicinoids represent the primary constituents of secondary metabolites that confer protection against herbivores and pathogens. The ethylene-inducible ERF2 functions as a positive regulator of capsaicinoid synthesis, playing a crucial role in the biosynthetic pathway by modulating the expression of genes associated with capsaicinoid production. Additionally, other TFs, including MYB31, MYB108, MYB48, and Jerf, also influence this process. Specifically, MYB31 directly targets capsaicinoid biosynthetic genes (CBGs), leading to alterations in their expression, while MYB108 activates the CBG promoter, thereby enhancing the synthesis of capsaicinoids (Wen et al. 2022).

2.2.3 Regulation of Taxol/Pacitaxel Biosynthesis

Taxol, also known as paclitaxel, is a diterpenoid compound predominantly derived from the bark of Taxus species. It is extensively utilized in the treatment of various malignancies, particularly solid tumors, including breast, ovarian, and lung cancers, as well as Kaposi's sarcoma. The mechanism of action of Taxol involves the induction of mitotic arrest and subsequent cell death through the disruption of microtubule dynamics (Gallego-Jara et al. 2020). The concentration of taxol in Taxus species is notably low, and its synthesis is costly, necessitating a comprehensive understanding of the regulatory mechanisms involved in taxol biosynthesis to enhance its yield. Key TFs such as WRKY, MYC, MYB, and ERF are crucial in this biosynthetic pathway. Specifically, MYB3 functions as a positive regulator of taxol biosynthesis by interacting with the promoters of the TS and TBT genes, thereby acting as an activator. Additionally, MYB29 serves as an abscisic acid-dependent activator of taxol biosynthesis, primarily influencing the expression of the TS and T5αH genes. MYB29 is capable of binding to the B1 site within the promoter of the T5αH gene, allowing it to directly regulate its expression (Cao et al. 2022). The ERF TF factors play a significant role in the regulation of genes associated with taxol biosynthesis by modulating the expression of the tasy gene. The tasy gene, which is crucial for the production of taxidiene, is subject to negative regulation by ERF12, while ERF15 functions as an activator of the tasy gene (Zhang et al. 2015a). WRKY TFs constitute a significant family of regulatory proteins that play a crucial role in the modulation of the taxol biosynthesis pathway. Specifically, WRKY8, WRKY20, WRKY47, and WRKY26 function as positive regulators of genes associated with taxol biosynthesis. The expression levels of DBAT, T5H, and ERF15 are enhanced by the activity of these WRKY genes. Notably, WRKY26 serves as an activator for the expression of DBAT. In the absence of JA, WRKY26 and MYC2a are inhibited by JAV3 and JAZ3, respectively. However, when JA is present, both WRKY26 and MYC2a are released from inhibition, allowing them to bind directly to the promoters of DBAT and TASY, thereby activating downstream gene expression (Chen et al. 2022). WRKY8 and WRKY47 function to decrease the expression of ERF12, a known negative regulator of the tasy gene. In addition, WRKY44 and WRKY52 serve as repressors of taxol biosynthesis by suppressing the expression of the DBAT, T5H, and ERF12 genes (Zhang et al. 2018b). WRKY33 plays a crucial role in the regulation of taxol biosynthesis genes, functioning directly by binding to the promoter regions of specific genes, such as the DBAT gene, through interaction with the W box. Additionally, it can exert indirect regulation by influencing other activators, exemplified by its role in the activation of the ERF15 promoter, which in turn affects TASY. Consequently, WRKY33 emerges as a significant target for the modulation of taxol biosynthesis, as it activates a majority of the genes implicated in this metabolic pathway (Chen et al. 2021).

2.2.4 Transcription Factors Involved in Withanolides Synthesis

Withanolides represent a class of pharmacologically significant triterpenoids, exhibiting a range of therapeutic properties, including anti-cancer, cardioprotective, anti-inflammatory, immunomodulatory, anti-coagulant, anti-diabetic, anti-oxidant, and neuroprotective effects. These compounds are synthesized in the *Withania somnifera* plant through the phytosterol pathway. The overexpression of TFs that regulate the biosynthetic pathway of withanolides, particularly under in vivo conditions, can significantly enhance their production. Numerous studies have indicated that various types of TFs, such as WRKY and MYB, are crucial for the biosynthesis of withanolides (Ahmad et al. 2024). WRKY1, which is activated by jasmonate and salicin, directly influences the expression of genes involved in the triterpenoid biosynthetic pathway, specifically squalene synthase and squalene epoxidase, through its binding to the W box within their promoter regions. This transcription factor plays a crucial role in enhancing the biosynthesis of phytosterols and withanolides while also contributing to the plant's defense mechanisms against biotic stress (Singh et al. 2017). MYC2 also positively regulates withanolide production as it upregulates CYP85A, CYP90, CYP710A and CAS. Withanolides and stigmasterol accumulation are elevated as a result of overexpression of WRKY1 as the expression of upregulation of metabolic pathway genes, whereas suppression of WRKY1 leads to a decreased supply of sterol precursor, hence depleted withanolide content (Sharma et al. 2019). The biosynthesis pathway of withanolides is also subject to negative regulation by MYBL1 and MYBL2 (Singh et al. 2024). Two WDR proteins, LWD1 and WUSCHEL, have been identified and may be associated with the biosynthesis of withanolides. The proportion of withaferin A is elevated in lines transformed with LWD1 and WUSCHEL when compared to the control. These proteins play a regulatory role in the processes of organ development as well as in the accumulation of secondary metabolites in *Withania somnifera* (Tripathi et al. 2017).

2.2.5 Transcription Factors Involved in Menthol Biosynthesis

Menthol, a cyclic monoterpene, is widely recognized for its cooling properties and distinctive minty aroma, serving as a primary component of the Mentha genus (Kamatou et al. 2013). Monoterpenoid biosynthesis is primarily facilitated by the enzymes MD and PR. The TFs MYB1 and MYB2 engage with the promoters of MD and PR, thereby enhancing the production of menthol (An et al. 2024). WRKY1 plays a crucial role in the regulation of genes associated with the biosynthesis pathway of monoterpenes, including the production of pulegone. The biosynthesis of pulegone is modulated through the regulation of genes such as IPD, IPR, and LS, which are integral to the monoterpene biosynthetic process (Wang et al. 2024). Monoterpene biosynthesis is inhibited by MYB TFs, which suppress the expression of the large subunit of geranyl diphosphate (GPP) synthase (Reddy et al. 2017). YABBY5 exerts a negative regulatory effect on terpene biosynthesis through the activation of WRKY75, which subsequently downregulates the genes associated with monoterpene biosynthesis in spearmint (Wang et al. 2016).

2.2.6 Transcription Factors Involved in Tanshinones Biosynthesis

Tanshinones represent a category of abietane diterpenoid natural products that are produced in the roots of *S. miltiorrhiza*, commonly referred to as Danshen, which is a prominent traditional Chinese medicinal herb. To date, more than 40 tanshinones and related compounds have been discovered, including cryptotanshinone, tanshinone IIA, dihydrotanshinone, and tanshinone I (Wang and Peters 2022). These compounds are clinically important for cardiovascular and cerebrovascular diseases (Zheng et al. 2021). The biosynthesis and accumulation of tanshinones involve the bHLH TFs bHLH37, bHLH74, and bHLH92 (Zhang et al. 2015b). Additionally, bHLH3 has been bound to the G-box present in the promoters of TAT, HPPR, KSL1, and CYP76AH1, leading to a downregulation of both phenolic acids and tanshinone accumulation (Zhang et al. 2020). Conversely, bHLH10 interacts with G-box motifs in the promoters of DXS2, CPS1, and CPS5, thereby enhancing the accumulation of tanshinones in hairy roots (Xing et al. 2018). ERF73 is known to recognize GCC-box promoter elements of four genes associated with tanshinone production (DXR1, CPS1, KSL1, and CYP76AH3) and activates their expression. Furthermore, ERF73 is upregulated in response to stress elicitors. Moreover, ERF73 is partially mediated by the JA signaling pathway through interaction with JAZ3. This suggests that ERF73 plays a crucial role in coordinating the expression of tanshinone biosynthetic genes, indicating a potential connection between tanshinone production and plant stress responses (Zhang et al. 2018c). Additionally, bZIP7 and bZIP20 may also be involved in the regulation of tanshinone biosynthesis. The overexpression of bZIP1 has been found to positively influence phenolic acid biosynthesis by enhancing the expression of cinnamate-4-hydroxylase (C4H1), with bZIP1 binding to the G-Box-like1 element in the promoter of the C4H1 gene. The MYB36-ERF6/ERF115 module is implicated in the regulation of tanshinone and phenolic acid biosynthesis in hairy roots of *S. miltiorrhiza* (Li et al. 2023). WRKY TFs (specifically WRKY1, 7, 19, 29, 45, 52, 56, 58, and 68) are likely to play a significant role in the biosynthesis of tanshinones and phenolic acids. Notably, WRKY1 has the capacity to enhance the expression of DXR by interacting with its promoter. Furthermore, WRKY1 is involved in the regulation of tanshinone biosynthesis, functioning as a positive regulator by activating DXR within the MEP pathway (Cao et al. 2018).

2.3 TRANSCRIPTIONAL REGULATION OF PHENOLIC COMPOUNDS

The biological importance of bioactive phenolic natural secondary metabolites is substantial, as they serve as antioxidants, antimicrobial agents, and anti-inflammatory substances. These compounds, characterized by the presence of a phenol moiety, consist of a benzene ring that is substituted with a hydroxyl group. Notable phenolic compounds that have been extensively researched include myricetin, gallic acid, quercetin, cyanidin, p-coumaric acid, naringenin, vanillic acid, resveratrol, ferulic acid, and luteolin, all of which contribute distinctive flavors and colors to various fruits (Haminiuk et al. 2012). The phenolic compound group encompasses polyphenols, oligophenols, and monophenols. Simple hydroxyl phenols include catechol, resorcinol, hydroquinone, pyrogallol, hydroxyquinol, and phloroglucinol, while hydroxyl cinnamic acids comprise caffeic acid, ferulic acid, and sinapic acid. Oligophenols are represented by coumarins and their derivatives, such as scopoletin and auraptene. Additionally, flavones, flavonols, chalcones, isoflavones, anthocyanins, lignin, and tannins are classified as polyphenols (H. Al Mamari 2022). There has been a notable enhancement in the biosynthesis of phenolic and polyphenolic compounds, which aids in their adaptation to various abiotic and biotic stresses, including salinity, heavy metals, drought, temperature fluctuations, ultraviolet radiation, and the progression of diseases (Tuladhar et al. 2021). bZIP1 has been shown to positively influence the biosynthesis of phenolic acids by increasing the expression of key biosynthetic genes, including cinnamate-4-hydroxylase (C4H1). Additionally, biochemical analyses demonstrated that bZIP1 interacts with the G-Box-like1 element located in the promoter region of the C4H1 gene. Further investigations identified bZIP1 as a target of ERF1L1 in *Salvia miltiorrhiza* (Deng et al. 2020). bHLH3 is another transcription factor that negatively regulates the biosynthesis of phenolic acids by interacting with the promoters of structural genes involved in the pathway, specifically TAT, HPPR, KSL1, and CYP76AH1. bHLH3 contains a specific binding site for the E/G-box, which may interact with the E/G-box located in the promoter regions of these biosynthetic pathway genes (Zhang et al. 2020). WRKY24 enhances the promoter activity of STS29 and works in conjunction with MYB14 and MYB15 to promote the accumulation of resveratrol in *Vitis vinifera* (Vannozzi et al. 2018). Conversely, WRKY8 inhibits the STS gene by directly interacting with MYB14 (Jiang et al. 2019). Transcriptional regulation of some of the phenolic compounds like anthocyanin, flavonoid, lignin and tannin is extensively studied, which are summarized under the following headings and listed in Table 2.1.

TABLE 2.1
Transcription Factors Associated with Secondary Metabolite Biosynthesis in Different Plant Species

Plant Species	Transcription Factor	Function	References
A. paniculata	WRKY01, 08, 12, 14, 19, 20, 50	Andrographolide biosynthesis	Zhang et al. (2021)
M. domestica	MYB3	Regulation of anthocyanin biosynthesis and flower development.	Vimolmangkang et al. (2013)
	MYC2 and ERF3	Positively co-regulate α-Farnesene biosynthesis	Wang et al. (2020)
A. thaliana	MYC2	Regulating sesquiterpene synthase gene expression	Hong et al. (2012)
	MYB111	Regulating flavonoid biosynthesis	Li et al. (2019a)
	NAC078	Regulates flavonoid biosynthesis under high-light. Plant and cell physiology	Morishita et al. (2009)
	MYB58 and MYB63	Activators of the lignin biosynthetic pathway during secondary cell wall formation	Zhou et al. (2009)
	LACCASE4 and 17	Enhancement of lignin biosynthesis	Berthet et al. (2011)
A. annua	bHLH2 and bHLH3 and MYC2	Negative regulators in artemisinin biosynthesis	Shen et al. (2022)
	bZIP9, 1	Positively regulates the biosynthesis of artemisinin	Shen et al. (2019); Shu et al. (2022)
	ABF3	Positively regulates artemisinin biosynthesis	Zhong et al. (2018)
C. acuminata	bZIP	Regulation of camptothecin biosynthesis	Chang et al. (2019)
	AP2/ERF		Hu et al. (2020)
	AP2/ERF	Involved in the capsaicinoid biosynthesis	Wen et al. (2022)
C. roseus	AP2/ERF transcription factor CR1	Regulates the accumulation of vindoline and serpentine in	Liu et al. (2017)
Gossypium sp.	WRKY1	Regulates the sesquiterpene synthase and gossypol	Xu et al. (2004)
E. californica	WRKY1	Modulation of benzylisoquinoline alkaloid biosynthesis	Yamada et al. (2017)
E. globulus	NAC141	Positively regulates lignin biosynthesis and increases lignin deposition	Sun et al. (2021)
F. tataricum	MYB102 and bHLH4	Accumulation of rutin	Mi et al. (2023)
	MYB	Repress rutin biosynthesis	Zhang et al. (2018a)
F. hybrida,	MYB21, MYC2	Control the expression of terpene synthase genes in flowers	Yang et al. (2020b)
G. biloba	bZIP08	Regulation of flavonoid biosynthesis	Han et al. (2023)
	R2R3-*MYB*	Negative regulator of the flavonoid biosynthesis pathway	Xu et al. (2014)
G. max	MYB100	Negatively regulates plant flavonoid biosynthesis	Yan et al. (2015)
V. vinifera	WRKY8, MYB14	Regulation of resveratrol biosynthesis	Jiang et al. (2019)
	bZIPC22	Regulation of flavonoid biosynthesis	Malacarne et al. (2016)
	WRKY, MYB	Regulation of stilbene synthase	Vannozzi et al. (2018)

(Continued)

TABLE 2.1 (*Continued*)
Transcription Factors Associated with Secondary Metabolite Biosynthesis in Different Plant Species

Plant Species	Transcription Factor	Function	References
M. truncatula	bHLH	Regulate triterpene saponin biosynthesis	Mertens et al. (2016)
M. haplocalyx Briq.	MYB1 & MYB2	Monoterpenoid biosynthesis pathway	An et al. (2024)
M. spicata	YABBY5	Terpene biosynthesis in plants	Wang et al. (2016)
O. pumila	MYB	Regulating specialized metabolisms	Rohani et al. (2016)
	AP2/ERF	Regulation of Specialized metabolism	Udomsom et al. (2016)
	WRKY3	Development and biosynthesis of camptothecin and its precursors	Wang et al. (2019)
	WRKY	Regulating the biosynthesis of the anti-cancer drug camptothecin	Wang et al. (2022)
P. virgatum	NAC and MYB	Involved in regulation of biomass production	Zhong et al. (2015)
P. persica	MYB, R2R3	Regulation of flavonoid biosynthesis	Ravaglia et al. (2013)
O. sativa	WRKY45	Phytoalexin biosynthesis	Akagi et al. (2014)
	WRKY53	Enhanced blast resistance	Chujo et al. (2014)
	bZIP79	Suppresses the production of diterpenoid phytoalexin	Miyamoto et al. (2015)
	bZIP coordinated with TGAP1	Inductive production of diterpenoid phytoalexins	Okada et al. (2009)
	WRKY76	Transcriptional repressor playing opposite roles in blast disease resistance	Yokotani et al. (2013)
	NAC055	Regulates GA-mediated lignin biosynthesis	Liu et al. (2022)
S. miltiorrhiza	WRKY1, WRKY2, WRKY4	Regulation of tashinone biosynthesis	Deng et al. (2019)
	MYB97 and SmMYB98	Enhance the biosynthesis of tanshinone	Hao et al. (2020)
	WRKY1	Positively promotes the biosynthesis of tanshinones	Cao et al. (2018)
	WRKY2	Tanshinone production enhanced	Deng et al. (2019)
	bZIP	Involved in modulating biosynthesis of phenolic acids	Deng et al. (2020)
	SmMYB36-SmERF6/SmERF115	Regulates the biosynthesis of tanshinones and phenolic acids	Li et al. (2023)
	bHLH10	Enhances tanshinones biosynthesis	Xing et al. (2018)
	bHLH3	Repressor of both phenolic acids and tanshinone biosynthesis	Zhang et al. (2020)
	bHLH	Tanshinone biosynthesis	Zhang et al. (2015b)
	bZIP	Tanshinone biosynthesis	Zhang et al. (2018c)
	ERF-VII, ERF73	Coordinately regulates tanshinone biosynthesis in response to stress elicitors	Zheng et al. (2021)
	WRKY1	Monoterpene biosynthesis induced by light intensity	Wang et al. (2024)

(*Continued*)

TABLE 2.1 (*Continued*)
Transcription Factors Associated with Secondary Metabolite Biosynthesis in Different Plant Species

Plant Species	Transcription Factor	Function	References
S. bicolor	*Myb60*	Phenylpropanoid biosynthesis and alters secondary cell wall composition	Scully et al. (2016)
M. spicata	MYB, R2R3	Negatively regulates monoterpene production	Reddy et al. (2017)
F. ananassa	bHLH9-HY5	Positive regulator of anthocyanin biosynthesis	Li et al. (2020)
I. batatas	BBX29	Regulates leaf development and flavonoid biosynthesis	Gao et al. (2023)
F. tataricum	ERF, EAR3	Degradation and rutin biosynthesis	Ding et al. (2022)
	MYB16	Regulate rutin biosynthesis	Li et al. (2019b)
	MYB5/6,GBF1,UFGT163	Promotes rutin biosynthesis	Zhao et al. (2024)
T. chinensis	WRKY26	Taxol biosynthesis	Chen et al. (2022)
	WRKY26, WRKY 33	Promote taxol biosynthesis	Chen et al. (2021, 2022)
	WRKY	Regulation of taxol biosynthesis	Zhang et al. (2018b)
	ERF12 and ERF15	Repressor and activator of tasy gene	Zhang et al. (2015a)
	WRKY 8, WRKY 47, and WRKY 33	Taxol biosynthesis	Chen et al. (2022)
C. sinensis	MYB2 and MYB26	Involved in flavonoid biosynthesis	Wang et al. (2018b)
W. somnifera	MYC2	Regulates the biosynthesis of triterpenoid withanolides and phytosterol	Sharma et al. (2019)
	MYBL1 and MYBL2	Withanolide biosynthesis	Singh et al. (2024)

2.3.1 Transcription Factors Involved in Anthocyanin Biosynthesis

Anthocyanins act as polyphenolic pigments that are ubiquitously found in plants. Anthocyanins are not only essential for plant protection against biotic and abiotic stresses but also have beneficial effects on human health due to antioxidant, anti-mutagenic, and anti-carcinogenic capacities (Khoo et al. 2017). The anthocyanin biosynthesis pathway has been extensively defined, with the relevant genes isolated and characterized in many plants like petunia, snapdragon, and Arabidopsis (Tanaka et al. 2008). The genetic framework for anthocyanin production is comprised of structural and regulatory genes. The TFs involved in the regulation of structural genes determine the pigmentation patterns in plants (Grotewold 2006). A trio of TFs, namely the bHLH, R2R3 MYB, and WD40 proteins, primarily governs the genes involved in the anthocyanin biosynthesis (Allan et al. 2008). MYB TFs have been documented as a key regulator of secondary metabolism, development, signal transduction, and disease resistance in plants (Jin and Martin 1999). R2R3 MYB TFs regulate the anthocyanin biosynthesis pathway at multiple points. Regulation is either suppressive or activating. R2R3 MYB TFs control the all structural gene of anthocyanin biosynthesis in *Perilla* (Saito and Yamazaki 2002) while in *V. vinifera* it regulates only the downstream genes of anthocyanin biosynthesis (Kobayashi et al. 2002). MYB TFs positively influence anthocyanin biosynthesis in Arabidopsis (PAP1, PAP2, MYB113, and MYB114), grape (MYB1a), apple (MYBA, MYB1,10 and 3) and Gerbera (MYB10) (Vimolmangkang et al. 2013) (Table 2.1). However, in strawberry (MYB1), *Antirrhinum* (MYB308), *Arabidopsis* (MYB4, MYBL2) suppress the anthocyanin accumulation (Chen et al. 2019). MYB TFs play a crucial role in the regulation of anthocyanin biosynthesis by influencing various signaling pathways, including the JA signaling pathway, cytokinins (CKs) signaling pathway, temperature-induced responses, light signals, the 26S proteasome pathway, as well as interactions with NAC and bHLH TFs (Li et al. 2022). Additionally, the structural and regulatory genes activated by MYB TFs, along with the target genes that are either activated or repressed by these factors, and the interactions between MYB TFs and bHLH proteins are essential for the effective regulation of anthocyanin biosynthesis (Figure 2.2).

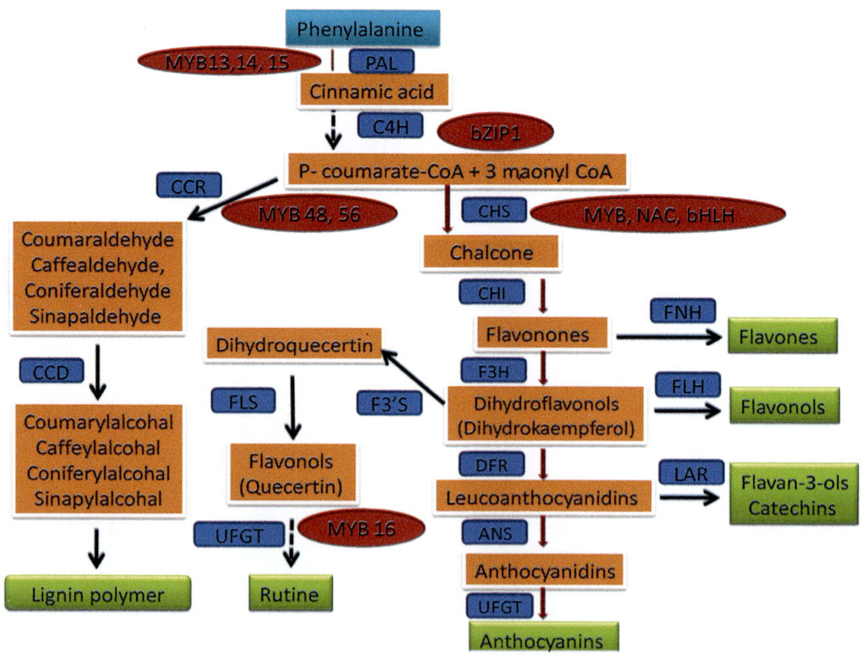

FIGURE 2.2 Transcriptional regulation of phenolic compounds biosynthesis.

ZCT proteins diminish STR activity by obstructing the binding of the transcriptional activator ORCA to the STR promoter. Consequently, ZCT proteins function by either hindering the interaction between transcriptional activators and the TDC and STR promoters or by inhibiting the activity of transcriptional activator proteins already associated with the DNA (Memelink and Gantet 2007).

The expression of RMT1 is stimulated by MeJA and MYC2. RMT1 exerts an indirect negative regulatory effect on MYC2 by binding to the AACGTG sequence within the ORCA3 promoter, which is also a target for MYC2. Consequently, RMT1 inhibits the ORCA3 gene by competing with MYC2 for the identical cis element located in the ORCA3 promoter region. Additionally, GBF1 and GBF2, which are G-box binding factor (GBF) proteins characterized by the presence of the bZIP motif, also negatively influence the biosynthetic genes of MIA, functioning as antagonists to MYC2. When co-expressed with MYC2, both GBF1 and GBF2 establish direct interactions with it (Yamada and Sato 2021).

2.5.4 Regulation of Sesquiterpene Alkaloid

Artemisinin, a sesquiterpene alkaloid obtained from the *Artemisia annua* plant, is used to treat malaria. The enzyme ADS is essential in the first committed step of the artemisinin-specific biosynthetic pathway, where it converts farnesyl pyrophosphate (FPP) into amorpha-4,11-diene. Furthermore, the transcription factor AabZIP9 promotes artemisinin biosynthesis by binding to ACGT cis-elements in the promoter region of the ADS gene. This upregulation of ADS leads to an increased production of artemisinin, dihydroartemisinic acid, and artemisinic acid (Shen et al. 2019).

ALDH1 plays a crucial role in the biosynthesis of artemisinin, facilitating the conversion of dihydroartemisnic aldehyde to dihydroartemisinin acid (DHAA). The TFs ABF3 and bZIP1 modulate the activity of ALDH1 by interacting with its promoter, ALDH1. Specifically, ABF3 binds directly to the promoter of the ALDH1 gene, thereby enhancing its expression, while bZIP1 exerts an indirect regulatory effect on its activity (Zhong et al. 2018). bZIP1 plays an indirect role in the activation of genes involved in artemisinin biosynthesis, such as DBR2 and ALDH17. It interacts with the promoter of MYC2, resulting in the upregulation of its expression. Subsequently, MYC2 is capable of directly binding to the promoter of ALDH1, thereby enhancing its transcription and facilitating an increase in artemisinin production (Shu et al. 2022) (Figure 2.4).

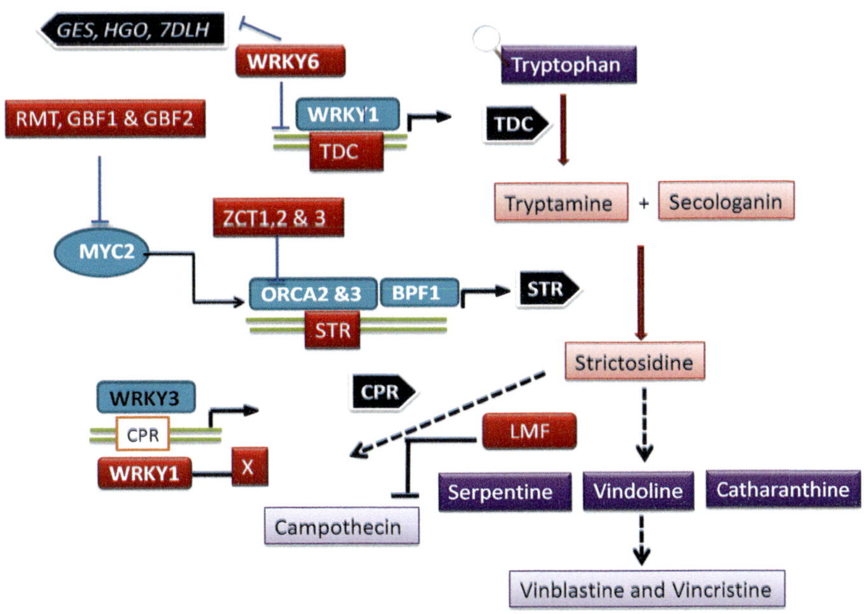

FIGURE 2.4 Transcriptional regulation of tyrosine mediated alkaloid biosynthesis.

bHLH2 and bHLH3 are members of the bHLH family and function as transcriptional repressors. These nuclear proteins possess a DNA-binding domain that is analogous to that of MYC2; however, they do not contain the conserved activation domain. This indicates that the mechanism of repression may involve competition between bHLH2 and bHLH3 and MYC2 for identical cis-regulatory elements (Shen et al. 2022). The process of artemisinin biosynthesis and its storage occurs within the trichomes. A transcription factor belonging to the AP2/ERF family, known as trichome and artemisinin regulator 1 (TAR1), plays a positive role in the development of trichomes and the biosynthesis of artemisinin. The inhibition of TAR1 leads to a reduction in the levels of artemisinin, whereas its overexpression contributes to an increased production of this compound (Tan et al. 2015).

2.5.5 Synthesis and Regulation of Monoterpene Indole Alkaloid (MIA)

Camptothecin (CPT), a monoterpene indole alkaloid (MIA), is derived from the plant *Camptotheca acuminata* and is extensively utilized in the treatment of various cancers, including lung, colorectal, cervical, and ovarian malignancies. CPT is acknowledged for its antitumor properties, primarily due to its distinctive mechanism of action, which involves the inhibition of DNA topoisomerase I. The factor known as LMF (Light-Mediated CPT biosynthesis Factor) is a bZIP transcription factor that responds to light and negatively regulates the genes involved in the biosynthesis of camptothecin (CPT). The overexpression of LMF results in the downregulation of these biosynthetic genes, leading to a reduction in CPT accumulation within the leaves (Chang et al. 2019).

The secologanin component of CPT is produced through the 2-C-methyl-d-erythritol 4-phosphate pathway, while tryptamine is derived from the shikimate pathway. Subsequently, tryptamine and secologanin undergo a transformation into strictosidine, facilitated by the enzyme strictosidine synthase. Further, strictosidine is converted into strictosamide through the action of strictosidine beta-glucosidase. The biosynthetic pathway for CPT is regulated by eight specific genes: AP2/ERF095, AP2/ERF096, AP2/ERF123, AP2/ERF125, AP2/ERF126, AP2/ERF127, AP2/ERF149, and AP2/ERF152. Additionally, key genes involved in the biosynthesis of CPT include iridoid synthase/cyclase (CYC1 and CYC2), TDC1 and TDC2, cytochrome P450 monooxygenase (CYP72A610 and CYP72A565), and 10-hydroxycamptothecin O-methyltransferase (10OMT) (Hu et al. 2020).

2.5.6 TFs Involved in the Synthesis of Isoquinolone Alkaloid

The primary bioactive constituent of *C. chinensis* is identified as isoquinolone alkaloid. (S)-reticuline functions as a pivotal intermediate in the biosynthesis of several isoquinoline alkaloids, including morphine and codeine. Other isoquinolone alkaloids produced through the (S)-reticuline pathway include berberine, coptisine, palmatine, jatrorrhizine, columbamine, and magnoflorine. The TF WRKY2 exerts a positive regulatory effect on CPT. In contrast, jatrorrhizine and coptisine exhibit a negative correlation with the transcription factor WRKY2, while they show a positive correlation with magnoflorine. The WRKY6 functions as a negative regulator of CPT production, as its overexpression results in a reduced accumulation of CPT. Conversely, in lines where WRKY6 has been knocked out, there a showed increase in CPT levels. Additionally, WRKY6 interacts with the promoters of the GES, HGO, 7DLH, and TDC genes, thereby inhibiting their activity (Wang et al. 2022).

WRKY1 also functions as a negative regulator of campothecin biosynthesis through its interaction with the CPR promoter. In contrast, WRKY2 acts as a positive regulator, enhancing the expression of the TDC gene, which distinguishes it from WRKY1 and WRKY6. Additionally, WRKY3 plays a significant role in the biosynthesis of campothecin precursors, including tryptamine, loganin, and secologanin. WRKY3 TF also influences campothecin production by binding to the W box within the CPR gene promoter and by modulating the growth and development of hairy roots. Notably, the overexpression of WRKY3 leads to an increase in the transcripts of both TDC and CPR (Wang et al. 2019).

In addition, the AP2/ERF transcription factor has been regulating the CPT biosynthesis in *Ophiorrhiza pumila*. The ERF2 protein plays a positive role in modulating the expression of genes associated with the 2-C-methyl-D-erythritol 4-phosphate and secologanin-strictosidine metabolic pathways (Udomsom et al. 2016). MYB1 functions as a suppressor of CPT synthesis; its overexpression in the hairy roots of *O. pumila* leads to a reduction in the transcript levels of TDC, which in turn results in a decreased accumulation of CPT (Rohani et al. 2016).

2.5.7 Regulation of BIA Biosynthesis Pathway

WRKY genes play a crucial role in the regulation of biosynthetic genes associated with benzylisoquinoline alkaloids (BIA). In the case of *C. japonica*, the gene WRKY1 has been shown to respond to MeJA and binds to the W-box within the promoters of its target genes, thereby regulating the majority of the genes encoding BIA biosynthetic enzymes. Furthermore, WRKY1 has a positive regulatory effect on berberine biosynthesis; silencing of WRKY leads to a reduction in berberine levels, while its overexpression correlates with an increase in the transcripts of berberine biosynthetic genes (Beaudoin and Facchini 2014).

The biosynthetic enzymes involved in BIA production also possess cis-acting elements that interact with AP2/ERF TFs, indicating that AP2/ERF plays a role in regulating BIA biosynthesis in *C. japonica*. Similar to WRKY TFs, several Group IX AP2/ERF genes exhibit responsiveness to MeJA in *E. californica*. Four Group IX AP2/ERF TFs have been functioning as transcriptional activators for the genes encoding BIA biosynthetic enzymes. *E. californica* is recognized as a source of significant BIA compounds, such as chelirubine. Notably, only the genes CYP719A3, P6H, and two novel OMT genes (G3OMT and G11OMT) demonstrated increased expression in response to the ectopic expression of WRKY1. The gene G11OMT is responsible for the O-methylation of 10-hydroxybenzophenanthridine alkaloids during the biosynthesis of chelirubine, and a correlation has been established between the activity of G11OMT and the accumulation of chelirubine. Additionally, the expression levels of CYP80B1 and BBE were also found to be elevated (Yamada et al. 2017). The opium poppy (*Papaver somniferum L.*) serves as a significant source of BIAs. Among these compounds are morphine, codeine, thebaine, sanguinarine, papaverine, and noscapine. Morphine is utilized in the treatment of sleep disorders, while codeine is effective in alleviating muscle tension. Thebaine functions as a sedative agent, sanguinarine is employed in remedies for inflammation, papaverine is indicated for muscle spasms and vasoconstriction, and noscapine is used for severe cough and certain cancers. Additionally, both papaverine and sanguinarine exhibit inhibitory effects on tumor cells (Aghaali et al. 2024).

The promoter regions of biosynthetic genes associated with BIAs contain numerous w-boxes, which function as binding sites for WRKY TFs. Consequently, WRKY1 TFs play a pivotal role in regulating BIA synthesis. The overexpression of WRKY1 leads to an enhanced accumulation of baine, dihydrosanguinarine and 10-hydroxydihydrosanguinarine. The elevation in thebaine levels is linked to increased transcriptional activity of the TYDC, 6OMT, and 4′OMT2 genes (Apuya et al. 2008). Additionally, the biosynthetic genes for noscapine are influenced by another transcription factor from group II WRKY TFs. Furthermore, a MYB transcription factor has been identified that exhibits high expression levels in the leaves of the Opium poppy (Kakeshpour et al. 2015).

2.5.8 TF Involved in SGA Biosynthesis

Steroidal glycoalkaloids (SGAs) are synthesized from cholesterol and are predominantly found in solanaceous plants. In the case of red ripe fruits, esculeosides are the most abundant SGAs, whereas α-tomatine and dehydrotomatine are the primary SGAs present in green tomatoes. The biosynthesis of SGAs, along with several genes involved in the upstream mevalonate and cholesterol precursor pathways, is regulated by GLYCOALKALOID METABOLISM 9 (GAME9), which belongs to the APETALA2/Ethylene Response Factor family. Additionally, GAME9/JRE4 has been interacting with a homolog of COI1, thereby influencing the biosynthesis of SGAs. The bZIP transcription factor known as HY5 inhibits the activity of HMGR within the mevalonate pathway. Furthermore, HY5, in conjunction with PIF3, can modulate the SGA content by affecting the transcript levels of GAME genes (Wang et al. 2018a).

2.6 CONCLUSION

The plant produces two types of metabolites. The first type is primary metabolite essential for the metabolism of the plant and food for the animals. Another type of metabolite is known as secondary metabolites, produced in response to defense mechanism against abiotic or biotic stress. These secondary metabolites are mainly used as medicine by the humans. These metabolites are synthesized by the specific metabolism process and regulated by the TFs. MYB, WRKY, bZIP, bHLH, NAC, and ERF1 are the TFs involved in the regulation of secondary metabolites like terpene, phenolics and alkaloids. WRKY is the TF involved in the synthesis of an array of secondary metabolites like taxol, tashinone, andographolide, gossypol, phytoalexins, etc. However, MYB is involved with paclitaxel, artemisinin and canabinoids biosynthesis. bHLH, ERF, and MYC promote the artemisinin and capsaicinoid biosynthesis while negatively regulating the tanshinone biosynthesis. Sometimes a single TF positively and negatively regulates the secondary metabolite biosynthesis. For example, WRKY8, WRKY20, WRKY47, and WRKY26 function as positive regulator, while WRKY44 and WRKY52 serve as repressors for taxol biosynthesis. Lignin is a polymer gives strength to cell or plant. Lignin biosynthesis is regulated by MYB and NAC TFs. Flavonoids like anthocyanin biosynthesis are mainly governed by MYB, bZIP and NAC TFs. Many times TF coordinates with other TFs or genes and regulates the secondary metabolite biosynthesis. R2R3-MYB-bHLH-WD40 proteins complexes have been shown to control multiple enzymatic steps in the flavonoid biosynthetic pathway in *C. sinensis* (Zhao et al. 2013a). Moreover, the MYB36-ERF6/ERF115, R2R3-MYB-bHLH-WD40, MYB36-ERF6/ERF115 are the modules involved in the rutin, flavonoid and phenolic acid biosynthesis, respectively. Again, WRKY, MYB and ERF are involved in alkaloid biosynthesis. Thus, TF plays an important role in the secondary metabolic pathway.

AUTHOR'S CONTRIBUTION

RS is responsible for the whole concept and manuscript writing. PJ is involved with figure preparation and writing of the manuscript. AT summerized the data and made the table while DK and AKM analyzed the data and helped in concept preparation. No funding provided and availed for this chapter preparation.

REFERENCES

Aghaali Z, Naghavi MR, Zargar M (2024) Promising approaches for simultaneous enhancement of medicinally significant benzylisoquinoline alkaloids in opium poppy. *Front Plant Sci* 15. https://doi.org/10.3389/fpls.2024.1377318

Ahmad Z, Shareen, Ganie IB, et al (2024) Enhancing withanolide production in the withania species: Advances in in vitro culture and synthetic biology approaches. *Plants* 13:2171. https://doi.org/10.3390/plants13152171

Akagi A, Fukushima S, Okada K, et al (2014) WRKY45-dependent priming of diterpenoid phytoalexin biosynthesis in rice and the role of cytokinin in triggering the reaction. *Plant Mol Biol* 86:171–183. https://doi.org/10.1007/s11103-014-0221-x

Al Mamari H (2022) Phenolic compounds: Classification, chemistry, and updated techniques of analysis and synthesis. In: A. Badria F (ed) *Biochemistry*. IntechOpen, London, UK, pp. 1–22..

Allan AC, Hellens RP, Laing WA (2008) MYB transcription factors that colour our fruit. *Trends Plant Sci* 13:99–102. https://doi.org/10.1016/j.tplants.2007.11.012

Aly SH, Elbadry AMM, El-Shazly M, Hwang T-L (2023) Exploring the potential role of genus *Sophora* in the management of osteoporosis: A phytochemical and biological review. *Front Nat Prod* 2. https://doi.org/10.3389/fntpr.2023.1302371

An X, Liao Y, Yu Y, et al (2024) Effects of *MhMYB1* and *MhMYB2* transcription factors on the monoterpenoid biosynthesis pathway in *l*-menthol chemotype of *Mentha haplocalyx* Briq. *Planta* 260:3. https://doi.org/10.1007/s00425-024-04441-y

Apuya NR, Park J-H, Zhang L, et al (2008) Enhancement of alkaloid production in opium and California poppy by transactivation using heterologous regulatory factors. *Plant Biotechnol J* 6:160–175. https://doi.org/10.1111/j.1467-7652.2007.00302.x

Beaudoin GAW, Facchini PJ (2014) Benzylisoquinoline alkaloid biosynthesis in opium poppy. *Planta* 240: 19–32. https://doi.org/10.1007/s00425-014-2056-8

Berthet S, Demont-Caulet N, Pollet B, et al (2011) Disruption of *LACCASE4* and *17* results in tissue-specific alterations to lignification of *Arabidopsis thaliana* stems. *Plant Cell* 23:1124–1137. https://doi.org/10.1105/tpc.110.082792

Boerjan W, Ralph J, Baucher M (2003) Lignin biosynthesis. *Annu Rev Plant Biol* 54:519–546. https://doi.org/10.1146/annurev.arplant.54.031902.134938

Bondonno NP, Murray K, Cassidy A, et al (2021) Higher habitual flavonoid intakes are associated with a lower risk of peripheral artery disease hospitalizations. *Am J Clin Nutr* 113:187–199. https://doi.org/10.1093/ajcn/nqaa300

Camejo D, Guzmán-Cedeño Á, Moreno A (2016) Reactive oxygen species, essential molecules, during plant–pathogen interactions. *Plant Physiol Biochem* 103:10–23. https://doi.org/10.1016/j.plaphy.2016.02.035

Cao W, Wang Y, Shi M, et al (2018) Transcription factor SmWRKY1 positively promotes the biosynthesis of tanshinones in Salvia miltiorrhiza. *Front Plant Sci* 9:554. https://doi.org/10.3389/fpls.2018.00554

Cao X, Xu L, Li L, et al (2022) TcMYB29a, an ABA-responsive R2R3-MYB transcriptional factor, upregulates taxol biosynthesis in Taxus chinensis. *Front Plant Sci* 13. https://doi.org/10.3389/fpls.2022.804593

Chang C, Liu Z, Wang Y, et al (2019) A bZIP transcription factor, CaLMF, mediated light-regulated camptothecin biosynthesis in *Camptotheca acuminata*. *Tree Physiol* 39. https://doi.org/10.1093/treephys/tpy106

Chen L, Hu B, Qin Y, et al (2019) Advance of the negative regulation of anthocyanin biosynthesis by MYB transcription factors. *Plant Physiol Biochem* 136:178–187. https://doi.org/10.1016/j.plaphy.2019.01.024

Chen L, Wu L, Yang L, et al (2022) TcJAV3-TcWRKY26 cascade is a missing link in the jasmonate-activated expression of taxol biosynthesis gene DBAT in *Taxus chinensis*. *Int J Mol Sci* 23:13194. https://doi.org/10.3390/ijms232113194

Chen Y, Zhang H, Zhang M, et al (2021) Salicylic acid-responsive factor TcWRKY33 positively regulates taxol biosynthesis in *Taxus chinensis* in direct and indirect ways. *Front Plant Sci* 12. https://doi.org/10.3389/fpls.2021.697476

Chujo T, Miyamoto K, Ogawa S, et al (2014) Overexpression of phosphomimic mutated OsWRKY53 leads to enhanced blast resistance in rice. *PLoS One* 9:e98737. https://doi.org/10.1371/journal.pone.0098737

Cui H, Tsuda K, Parker JE (2015) Effector-triggered immunity: From pathogen perception to robust defense. *Annu Rev Plant Biol* 66:487–511. https://doi.org/10.1146/annurev-arplant-050213-040012

Deng C, Hao X, Shi M, et al (2019) Tanshinone production could be increased by the expression of SmWRKY2 in *Salvia miltiorrhiza* hairy roots. *Plant Sci* 284:1–8. https://doi.org/10.1016/j.plantsci.2019.03.007

Deng C, Shi M, Fu R, et al (2020) ABA-responsive transcription factor bZIP1 is involved in modulating biosynthesis of phenolic acids and tanshinones in Salvia miltiorrhiza. *J Exp Bot* 71:5948–5962. https://doi.org/10.1093/jxb/eraa295

Ding M, He Y, Zhang K, et al (2022) JA-induced FtBPM3 accumulation promotes FtERF-EAR3 degradation and rutin biosynthesis in Tartary buckwheat. *Plant J* 111:323–334. https://doi.org/10.1111/tpj.15800

Duan A-Q, Tao J-P, Jia L-L, et al (2020) AgNAC1, a celery transcription factor, related to regulation on lignin biosynthesis and salt tolerance. *Genomics* 112:5254–5264. https://doi.org/10.1016/j.ygeno.2020.09.049

Francenia Santos-Sánchez N, Salas-Coronado R, Hernández-Carlos B, Villanueva-Cañongo C (2019) Shikimic acid pathway in biosynthesis of phenolic compounds. In: Soto-Hernández M, García-Mateos R, Palma-Tenango M (eds) *Plant Physiological Aspects of Phenolic Compounds*. IntechOpen, London, UK, pp. 1–15.

Gallego-Jara J, Lozano-Terol G, Sola-Martínez RA, et al (2020) A compressive review about Taxol®: History and future challenges. *Molecules* 25:5986. https://doi.org/10.3390/molecules25245986

Gao X, Zhang H, Li X, et al (2023) The B-box transcription factor IbBBX29 regulates leaf development and flavonoid biosynthesis in sweet potato. *Plant Physiol* 191:496–514. https://doi.org/10.1093/plphys/kiac516

Godbole RC, Pable AA, Singh S, Barvkar VT (2022) Interplay of transcription factors orchestrating the biosynthesis of plant alkaloids. *3 Biotech* 12:250. https://doi.org/10.1007/s13205-022-03316-x

Grotewold E (2006) The genetics and biochemistry of floral pigments. *Annu Rev Plant Biol* 57:761–780. https://doi.org/10.1146/annurev.arplant.57.032905.105248

Hao, X., Pu, Z., Cao, G., You, D., Zhou, Y., Deng, C., Shi, M., Nile, S.H., Wang, Y., Zhou, W. and Kai, G., (2020). Tanshinone and salvianolic acid biosynthesis are regulated by SmMYB98 in Salvia miltiorrhiza hairy roots. *Journal of Advanced Research, 23*, pp.1–12. https://doi.org/10.1016/j.jare.2020.01.012.

Haminiuk CWI, Maciel GM, Plata-Oviedo MSV, Peralta RM (2012) Phenolic compounds in fruits – An overview. *Int J of Food Sci Tech* 47:2023–2044. https://doi.org/10.1111/j.1365-2621.2012.03067.x

Han H, Dong L, Zhang W, et al (2023) Ginkgo biloba GbbZIP08 transcription factor is involved in the regulation of flavonoid biosynthesis. *J Plant Physiol* 287:154054. https://doi.org/10.1016/j.jplph.2023.154054

Hong G-J, Xue X-Y, Mao Y-B, et al (2012) *Arabidopsis* MYC2 interacts with DELLA proteins in regulating sesquiterpene synthase gene expression. *Plant Cell* 24:2635–2648. https://doi.org/10.1105/tpc.112.098749

Hu Y-T, Xu Z-C, Tian Y, et al (2020) Genome-wide identification and analysis of AP2/ERF transcription factors related to camptothecin biosynthesis in *Camptotheca acuminata. Chin J Nat Med* 18:582–593. https://doi.org/10.1016/S1875-5364(20)30070-4

Jiang J, Xi H, Dai Z, et al (2019) VvWRKY8 represses stilbene synthase genes through direct interaction with VvMYB14 to control resveratrol biosynthesis in grapevine. *J Exp Bot* 70:715–729. https://doi.org/10.1093/jxb/ery401

Jin, H., Martin, C. Multifunctionality and diversity within the plant MYB-gene family. *Plant Mol Biol* 41:577–585 (1999). https://doi.org/10.1023/A:1006319732410

Kakeshpour T, Nayebi S, Rashidi Monfared S, et al (2015) Identification and expression analyses of MYB and WRKY transcription factor genes in *Papaver somniferum* L. *Physiol Mol Biol Plants* 21:465–478. https://doi.org/10.1007/s12298-015-0325-z

Kamatou GPP, Vermaak I, Viljoen AM, Lawrence BM (2013) Menthol: A simple monoterpene with remarkable biological properties. *Phytochemistry* 96:15–25. https://doi.org/10.1016/j.phytochem.2013.08.005

Khoo HE, Azlan A, Tang ST, Lim SM (2017) Anthocyanidins and anthocyanins: Colored pigments as food, pharmaceutical ingredients, and the potential health benefits. *Food Nutr Res* 61:1361779. https://doi.org/10.1080/16546628.2017.1361779

Kobayashi S, Ishimaru M, Hiraoka K, Honda C (2002) *Myb*-related genes of the Kyoho grape (*Vitis labruscana*) regulate anthocyanin biosynthesis. *Planta* 215:924–933. https://doi.org/10.1007/s00425-002-0830-5

Li B, Fan R, Guo S, et al (2019a) The Arabidopsis MYB transcription factor, MYB111 modulates salt responses by regulating flavonoid biosynthesis. *Environ Exp Bot* 166:103807. https://doi.org/10.1016/j.envexpbot.2019.103807

Li C, Yu W, Xu J, et al (2022) Anthocyanin biosynthesis induced by MYB transcription factors in plants. *IJMS* 23:11701. https://doi.org/10.3390/ijms231911701

Li J, Zhang K, Meng Y, et al (2019b) FtMYB16 interacts with Ftimportin-α1 to regulate rutin biosynthesis in tartary buckwheat. *Plant Biotechnol J* 17:1479–1481. https://doi.org/10.1111/pbi.13121

Li Q, Fang X, Zhao Y, et al (2023) The SmMYB36-SmERF6/SmERF115 module regulates the biosynthesis of tanshinones and phenolic acids in *salvia miltiorrhiza* hairy roots. *Hortic Res* 10:uhac238. https://doi.org/10.1093/hr/uhac238

Li Y, Xu P, Chen G, et al (2020) FvbHLH9 functions as a positive regulator of anthocyanin biosynthesis by forming a HY5–bHLH9 transcription complex in strawberry fruits. *Plant Cell Physiol* 61:826–837. https://doi.org/10.1093/pcp/pcaa010

Liu J, Gao F, Ren J, et al (2017) A novel AP2/ERF transcription factor CR1 regulates the accumulation of vindoline and serpentine in *Catharanthus roseus. Front Plant Sci* 8:2082. https://doi.org/10.3389/fpls.2017.02082

Liu Y, Wu Q, Qin Z, Huang J (2022) Transcription factor OsNAC055 regulates GA-mediated lignin biosynthesis in rice straw. *Plant Sci* 325:111455. https://doi.org/10.1016/j.plantsci.2022.111455

Malacarne G, Coller E, Czemmel S, et al (2016) The grapevine VvibZIPC22 transcription factor is involved in the regulation of flavonoid biosynthesis. *EXBOT J* 67:3509–3522. https://doi.org/10.1093/jxb/erw181

Memelink J, Gantet P (2007) Transcription factors involved in terpenoid indole alkaloid biosynthesis in *Catharanthus roseus. Phytochem Rev* 6:353–362. https://doi.org/10.1007/s11101-006-9051-z

Menke FLH (1999) A novel jasmonate- and elicitor-responsive element in the periwinkle secondary metabolite biosynthetic gene *Str* interacts with a jasmonate- and elicitor-inducible AP2-domain transcription factor, ORCA2. *EMBO J* 18:4455–4463. https://doi.org/10.1093/emboj/18.16.4455

Meraj TA, Fu J, Raza MA, et al (2020) Transcriptional factors regulate plant stress responses through mediating secondary metabolism. *Genes* 11:346. https://doi.org/10.3390/genes11040346

Mertens J, Pollier J, Vanden Bossche R, et al (2016) The bHLH transcription factors TSAR1 and TSAR2 regulate triterpene saponin biosynthesis in *Medicago truncatula. Plant Physiol* 170. https://doi.org/10.1104/pp.15.01645

Mi Y, Li Y, Qian G, et al (2023) A transcriptional complex of FtMYB102 and FtbHLH4 coordinately regulates the accumulation of rutin in *Fagopyrum tataricum. Plant Physiol Biochem* 194:696–707. https://doi.org/10.1016/j.plaphy.2022.12.016

Middleton E, Kandaswami C, Theoharides TC (2000) The effects of plant flavonoids on mammalian cells: Implications for inflammation, heart disease, and cancer. *Pharmacol Rev* 52:673–751.

Miyamoto K, Nishizawa Y, Minami E, et al (2015) Overexpression of the bZIP transcription factor OsbZIP79 suppresses the production of diterpenoid phytoalexin in rice cells. *J Plant Physiol* 173:19–27. https://doi.org/10.1016/j.jplph.2014.09.001

Miyamoto K, Shimizu T, Okada K (2014) Transcriptional regulation of the biosynthesis of phytoalexin: A lesson from specialized metabolites in rice. *Plant Biotechnol* 31:377–388. https://doi.org/10.5511/plantbiotechnology.14.0730a

Morishita T, Kojima Y, Maruta T, et al (2009) Arabidopsis NAC transcription factor, ANAC078, regulates flavonoid biosynthesis under high-light. *Plant Cell Physiol* 50:2210–2222. https://doi.org/10.1093/pcp/pcp159

Okada A, Okada K, Miyamoto K, et al (2009) OsTGAP1, a bZIP transcription factor, coordinately regulates the inductive production of diterpenoid phytoalexins in rice. *J Biol Chem* 284:26510–26518. https://doi.org/10.1074/jbc.M109.036871

Pitzschke A, Schikora A, Hirt H (2009) MAPK cascade signalling networks in plant defence. *Curr Opin Plant Biol*12:421–426. https://doi.org/10.1016/j.pbi.2009.06.008

Ravaglia D, Espley RV, Henry-Kirk RA, et al (2013) Transcriptional regulation of flavonoid biosynthesis in nectarine (Prunus persica) by a set of R2R3 MYB transcription factors. *BMC Plant Biol* 13:68. https://doi.org/10.1186/1471-2229-13-68

Reddy VA, Wang Q, Dhar N, et al (2017) Spearmint R2R3-MYB transcription factor MsMYB negatively regulates monoterpene production and suppresses the expression of geranyl diphosphate synthase large subunit (MsGPPS.LSU). *Plant Biotechnol J* 15:1105–1119. https://doi.org/10.1111/pbi.12701

Rohani ER, Chiba M, Kawaharada M, et al (2016) An MYB transcription factor regulating specialized metabolisms in *Ophiorrhiza pumila*. *Plant Biotechnol* 33:1–9. https://doi.org/10.5511/plantbiotechnology.15.1117a

Scully ED, Gries T, Sarath G, et al (2016) Overexpression of *SbMyb60* impacts phenylpropanoid biosynthesis and alters secondary cell wall composition in *Sorghum bicolor*. *Plant J* 85:378–395. https://doi.org/10.1111/tpj.13112

Sharma A, Rather GA, Misra P, et al (2019) Jasmonate responsive transcription factor WsMYC2 regulates the biosynthesis of triterpenoid withanolides and phytosterol via key pathway genes in *Withania somnifera* (L.) Dunal. *Plant Mol Biol* 100:543–560. https://doi.org/10.1007/s11103-019-00880-4

Shen Q, Huang H, Xie L, et al (2022) Basic helix-loop-helix transcription factors AabHLH2 and AabHLH3 function antagonistically with AaMYC2 and are negative regulators in artemisinin biosynthesis. *Front Plant Sci* 13. https://doi.org/10.3389/fpls.2022.885622

Shen Q, Huang H, Zhao Y, et al (2019) The transcription factor Aabzip9 positively regulates the biosynthesis of artemisinin in *Artemisia annua*. *Front Plant Sci* 10:1294. https://doi.org/10.3389/fpls.2019.01294

Shu G, Tang Y, Yuan M, et al (2022) Molecular insights into AabZIP1-mediated regulation on artemisinin biosynthesis and drought tolerance in *Artemisia annua*. *Acta Pharm Sin B* 12:1500–1513. https://doi.org/10.1016/j.apsb.2021.09.026

Singh AK, Kumar SR, Dwivedi V, et al (2017) A WRKY transcription factor from *Withania somnifera* regulates triterpenoid withanolide accumulation and biotic stress tolerance through modulation of phytosterol and defense pathways. *New Phytol* 215:1115–1131. https://doi.org/10.1111/nph.14663

Singh N, Singh D, Dwivedi S, et al (2024) Involvement of MYB family transcription factors, WsMYBL1 and WsMYBL2, in withanolide biosynthesis in *Withania somnifera*, a medicinal plant. *Plant Cell Tiss Organ Cult* 156:60. https://doi.org/10.1007/s11240-024-02681-0

Sun Y, Jiang C, Jiang R, et al (2021) A novel NAC transcription factor from eucalyptus, EgNAC141, positively regulates lignin biosynthesis and increases lignin deposition. *Front Plant Sci* 12:642090. https://doi.org/10.3389/fpls.2021.642090

Takeuchi M, Kegasa T, Watanabe A, et al (2018) Expression analysis of transporter genes for screening candidate monolignol transporters using *Arabidopsis thaliana* cell suspensions during tracheary element differentiation. *J Plant Res* 131:297–305. https://doi.org/10.1007/s10265-017-0979-4

Tan H, Xiao L, Gao S, et al (2015) *Trichome and artemisinin regulator* 1 is required for trichome development and artemisinin biosynthesis in *Artemisia annua*. *Mol Plant* 8:1396–1411. https://doi.org/10.1016/j.molp.2015.04.002

Tanaka Y, Sasaki N, Ohmiya A (2008) Biosynthesis of plant pigments: Anthocyanins, betalains and carotenoids. *Plant J* 54:733–749. https://doi.org/10.1111/j.1365-313x.2008.03447.x

Tripathi S, Sangwan RS, Narnoliya LK, et al (2017) Transcription factor repertoire in Ashwagandha (*Withania somnifera*) through analytics of transcriptomic resources: Insights into regulation of development and withanolide metabolism. *Sci Rep* 7:16649. https://doi.org/10.1038/s41598-017-14657-6

Tsuyama T, Kawai R, Shitan N, et al (2013) Proton-dependent coniferin transport, a common major transport event in differentiating xylem tissue of woody plants. *Plant Physiol* 162:918–926. https://doi.org/10.1104/pp.113.214957

Tuladhar P, Sasidharan S, Saudagar P (2021) Role of phenols and polyphenols in plant defense response to biotic and abiotic stresses. In: *Biocontrol Agents and Secondary Metabolites*. kwang W. Jeon (Ed.), Elsevier, pp. 419–441.

Udomsom N, Rai A, Suzuki H, et al (2016) Function of AP2/ERF transcription factors involved in the regulation of specialized metabolism in *Ophiorrhiza pumila* revealed by transcriptomics and metabolomics. *Front Plant Sci* 7. https://doi.org/10.3389/fpls.2016.01861

Vannozzi A, Wong DCJ, Höll J, et al (2018) Combinatorial regulation of stilbene synthase genes by WRKY and MYB transcription factors in grapevine (*Vitis vinifera* L.). *Plant Cell Physiol* 59:1043–1059. https://doi.org/10.1093/pcp/pcy045

Vimolmangkang S, Han Y, Wei G, Korban SS (2013) An apple MYB transcription factor, MdMYB3, is involved in regulation of anthocyanin biosynthesis and flower development. *BMC Plant Biol* 13:176. https://doi.org/10.1186/1471-2229-13-176

Wang C, Hao X, Wang Y, et al (2022) Identification of WRKY transcription factors involved in regulating the biosynthesis of the anti-cancer drug camptothecin in *Ophiorrhiza pumila*. *Hortic Res* 9:uhac099. https://doi.org/10.1093/hr/uhac099

Wang C, Meng L, Gao Y, et al (2018a) Manipulation of light signal transduction factors as a means of modifying steroidal glycoalkaloids accumulation in tomato leaves. *Front Plant Sci* 9. https://doi.org/10.3389/fpls.2018.00437

Wang C, Wu C, Wang Y, et al (2019) Transcription factor *OpWRKY3* is involved in the development and biosynthesis of camptothecin and its precursors in *Ophiorrhiza pumila* hairy roots. *Int J Mol Sci* 20:3996. https://doi.org/10.3390/ijms20163996

Wang Q, Liu H, Zhang M, et al (2020) MdMYC2 and MdERF3 positively co-regulate α-Farnesene biosynthesis in apple. *Front Plant Sci* 11. https://doi.org/10.3389/fpls.2020.512844

Wang Q, Reddy VA, Panicker D, et al (2016) Metabolic engineering of terpene biosynthesis in plants using a trichome-specific transcription factor *MsYABBY5* from spearmint (*Mentha spicata*). *Plant Biotechnol J* 14:1619–1632. https://doi.org/10.1111/pbi.12525

Wang Q, Zhao X, Jiang Y, et al (2023) Functions of representative terpenoids and their biosynthesis mechanisms in medicinal plants. *Biomolecules* 13:1725. https://doi.org/10.3390/biom13121725

Wang W-L, Wang Y-X, Li H, et al (2018b) Two MYB transcription factors (CsMYB2 and CsMYB26) are involved in flavonoid biosynthesis in tea plant [*Camellia sinensis* (L.) O. Kuntze]. *BMC Plant Biol* 18:288. https://doi.org/10.1186/s12870-018-1502-3

Wang X, Liang Y, Shu J, et al (2024) Transcription factor StWRKY1 is involved in monoterpene biosynthesis induced by light intensity in *Schizonepeta tenuifolia* Briq. *Plant Physiol Biochem* 214:108871. https://doi.org/10.1016/j.plaphy.2024.108871

Wang Z, Peters RJ (2022) Tanshinones: Leading the way into Lamiaceae labdane-related diterpenoid biosynthesis. *Curr Opin Plant Biol* 66:102189. https://doi.org/10.1016/j.pbi.2022.102189

Wen J, Lv J, Zhao K, et al (2022) Ethylene-inducible AP2/ERF transcription factor involved in the capsaicinoid biosynthesis in capsicum. *Front Plant Sci* 13. https://doi.org/10.3389/fpls.2022.832669

Xiao R, Zhang C, Guo X, et al (2021) MYB transcription factors and its regulation in secondary cell wall formation and lignin biosynthesis during xylem development. *IJMS* 22:3560. https://doi.org/10.3390/ijms22073560

Xing B, Yang D, Yu H, et al (2018) Overexpression of SmbHLH10 enhances tanshinones biosynthesis in Salvia miltiorrhiza hairy roots. *Plant Sci* 276:229–238. https://doi.org/10.1016/j.plantsci.2018.07.016

Xu F, Ning Y, Zhang W, et al (2014) An R2R3-MYB transcription factor as a negative regulator of the flavonoid biosynthesis pathway in *Ginkgo biloba*. *Funct Integr Genom* 14:177–189. https://doi.org/10.1007/s10142-013-0352-1

Xu Y-H, Wang J-W, Wang S, et al (2004) Characterization of GaWRKY1, a cotton transcription factor that regulates the sesquiterpene synthase gene (+)-delta-cadinene synthase-A. *Plant Physiol* 135:507–515. https://doi.org/10.1104/pp.104.038612

Yamada Y, Sato F (2013) Transcription factors in alkaloid biosynthesis. In: *International Review of Cell and Molecular Biology*. kwang W. Jeon (Ed.), Elsevier, pp. 339–382.

Yamada Y, Sato F (2021) Transcription factors in alkaloid engineering. *Biomolecules* 11:1719. https://doi.org/10.3390/biom11111719

Yamada Y, Shimada T, Motomura Y, Sato F (2017) Modulation of benzylisoquinoline alkaloid biosynthesis by heterologous expression of CjWRKY1 in *Eschscholzia californica* cells. *PLoS One* 12:e0186953. https://doi.org/10.1371/journal.pone.0186953

Yan J, Wang B, Zhong Y, et al (2015) The soybean R2R3 MYB transcription factor GmMYB100 negatively regulates plant flavonoid biosynthesis. *Plant Mol Biol* 89:35–48. https://doi.org/10.1007/s11103-015-0349-3

Yang C, Fang X, Wu X, et al (2012) Transcriptional regulation of plant secondary metabolism[F]. *JIPB* 54:703–712. https://doi.org/10.1111/j.1744-7909.2012.01161.x

Yang W, Chen X, Li Y, et al (2020a) Advances in pharmacological activities of terpenoids. *Nat Prod Commun* 15:1934578X20903555. https://doi.org/10.1177/1934578X20903555

Yang Z, Li Y, Gao F, et al (2020b) MYB21 interacts with MYC2 to control the expression of terpene synthase genes in flowers of *Freesia hybrida* and *Arabidopsis thaliana*. *J Exp Bot* 71:4140–4158. https://doi.org/10.1093/jxb/eraa184

Yokotani N, Sato Y, Tanabe S, et al (2013) WRKY76 is a rice transcriptional repressor playing opposite roles in blast disease resistance and cold stress tolerance. *J Exp Bot* 64:5085–5097. https://doi.org/10.1093/jxb/ert298

Zhang C, Xing B, Yang D, et al (2020) SmbHLH3 acts as a transcription repressor for both phenolic acids and tanshinone biosynthesis in *Salvia miltiorrhiza* hairy roots. *Phytochemistry* 169:112183. https://doi.org/10.1016/j.phytochem.2019.112183

Zhang K, Logacheva MD, Meng Y, et al (2018a) Jasmonate-responsive MYB factors spatially repress rutin biosynthesis in *Fagopyrum tataricum*. *J Exp Bot* 69:1955–1966. https://doi.org/10.1093/jxb/ery032

Zhang M, Chen Y, Nie L, et al (2018b) Transcriptome-wide identification and screening of WRKY factors involved in the regulation of taxol biosynthesis in *Taxus chinensis*. *Sci Rep* 8:5197. https://doi.org/10.1038/s41598-018-23558-1

Zhang M, Li S, Nie L, et al (2015a) Two jasmonate-responsive factors, TcERF12 and TcERF15, respectively act as repressor and activator of tasy gene of taxol biosynthesis in *Taxus chinensis*. *Plant Mol Biol* 89:463–473. https://doi.org/10.1007/s11103-015-0382-2

Zhang R, Chen Z, Zhang L, et al (2021) Genomic characterization of WRKY transcription factors related to andrographolide biosynthesis in *Andrographis paniculata*. *Front Genet* 11:601689. https://doi.org/10.3389/fgene.2020.601689

Zhang X, Luo H, Xu Z, et al (2015b) Genome-wide characterisation and analysis of bHLH transcription factors related to tanshinone biosynthesis in *Salvia miltiorrhiza*. *Sci Rep* 5:11244. https://doi.org/10.1038/srep11244

Zhang Y, Xu Z, Ji A, et al (2018c) Genomic survey of bZIP transcription factor genes related to tanshinone biosynthesis in *Salvia miltiorrhiza*. *Acta Pharm Sin B* 8:295–305. https://doi.org/10.1016/j.apsb.2017.09.002

Zhao H, Hu M, Fang Y, et al (2024) Regulatory module FtMYB5/6–FtGBF1–*FtUFGT163* promotes rutin biosynthesis in tartary buckwheat. *J Agric Food Chem* 72:12630–12640. https://doi.org/10.1021/acs.jafc.4c01399

Zhao L, Gao L, Wang H, et al (2013a) The R2R3-MYB, bHLH, WD40, and related transcription factors in flavonoid biosynthesis. *Funct Integr Genom* 13:75–98. https://doi.org/10.1007/s10142-012-0301-4

Zhao Q, Nakashima J, Chen F, et al (2013b) *LACCASE* is necessary and nonredundant with *PEROXIDASE* for lignin polymerization during vascular development in *Arabidopsis*. *Plant Cell* 25:3976–3987. https://doi.org/10.1105/tpc.113.117770

Zheng H, Jing L, Jiang X, et al (2021) The ERF-VII transcription factor SmERF73 coordinately regulates tanshinone biosynthesis in response to stress elicitors in *Salvia miltiorrhiza*. *New Phytol* 231:1940–1955. https://doi.org/10.1111/nph.17463

Zhong R, Ye Z-H (2009) Transcriptional regulation of lignin biosynthesis. *Plant Signal Behav* 4:1028–1034. https://doi.org/10.4161/psb.4.11.9875

Zhong R, Yuan Y, Spiekerman JJ, et al (2015) Functional characterization of NAC and MYB transcription factors involved in regulation of biomass production in switchgrass (*Panicum virgatum*). *PLoS One* 10:e0134611. https://doi.org/10.1371/journal.pone.0134611

Zhong Y, Li L, Hao X, et al (2018) AaABF3, an abscisic acid–responsive transcription factor, positively regulates artemisinin biosynthesis in *Artemisia annua*. *Front Plant Sci* 9. https://doi.org/10.3389/fpls.2018.01777

Zhou J, Lee C, Zhong R, Ye Z-H (2009) MYB58 and MYB63 are transcriptional activators of the lignin biosynthetic pathway during secondary cell wall formation in *Arabidopsis*. *Plant Cell* 21:248–266. https://doi.org/10.1105/tpc.108.063321

3 Phytophenolic Compounds and Their Anticancer Properties

Felix Wambua Muema[1,2], *Catherine M. Miller*[2,3], *Lionel Hebbard*[2,4,5], *and Phurpa Wangchuk*[1,2]

[1]College of Science and Engineering, James Cook University, Cairns, Queensland, Australia

[2]Australian Institute of Tropical Health, and Medicine (AITHM), Cairns Campus, James Cook University, Cairns, Queensland, Australia

[3]College of Medicine and Dentistry, James Cook University, Smithfield, Queensland, Australia

[4]Department of Molecular Genetics, Centre for Molecular Therapeutics, James Cook University, Townsville, Queensland, Australia

[5]Storr Liver Centre, Westmead Institute for Medical Research, Westmead Hospital and University of Sydney, Westmead, New South Wales, Australia

3.1 INTRODUCTION

Plant phenolics are among the most crucial secondary metabolites of plants. They occur in all plants, and they are produced by the plant for development and protection, especially against bacterial and fungal infections [1,2]. A characteristic structural component of phenolic compounds is a six-carbon aromatic ring with one or more hydroxyl groups attached. There are more than 8,000 known plant phenolic compounds with structures ranging from simple phenolic to highly polymerized scaffolds [3]. Plants produce these compounds through the pentose phosphate, shikimate, and phenylpropanoid pathways [4]. Due to the nutritional and protective roles of plant phenolics, studies have focussed on their functional roles and biosynthesis pathways and for their use in clinical therapies [1]. Vegetables, fruits and beverages are the basic plant phenolic chemical sources in the human diet [5,6].

Plant polyphenols can inhibit ultraviolet radiation (UVR)-induced alterations of the skin. Consuming these polyphenolic compounds may be a safe and effective way to give ongoing photoprotection [7]. Nobile et al. revealed that polyphenolic extracts of *Rosmarius officinalis* and *Citrus paradisi* decreased UVB and UVA-induced skin alterations through the reduction of skin redness and the production of lipoperoxides [7]. The study further demonstrated that the polyphenolic-rich extracts increased skin elasticity and decreased wrinkledness, with changes observed from week two of treatment, hence illustrating the anti-ageing properties of plant phenolic compounds. Clinical trials of curcumin (trial ID NCT03085680) revealed its ability to improve physical function grip strength, memory and attention when evaluated on ageing adults who are at risk of being disabled [8]. Furthermore, plant phenolic compounds have exhibited their capability in preventing inflammatory diseases that are linked with ageing. Curcumin has been shown to downregulate levels of the inflammatory markers cyclooxygenase-2 (COX-2), nuclear factor-κB (NF-κB), tumour necrosis factor (TNF), and interleukin 6 (IL-6), which are also ageing hallmarks [9–11].

DOI: 10.1201/9781003518358-3

Plant-derived phenolic compounds exhibit *in vitro* and *in vivo* biological anticancer activities via several cellular targets that are involved in carcinogenesis [12,13]. These compounds have shown potential molecular mechanisms such as inhibiting cell proliferation, inducing apoptosis, downregulating matrix metalloproteinases-2 and -9 activities, and cell cycle arrest, among other mechanisms [14–16]. The promising anticancer properties of polyphenols have been demonstrated in cancers including lung, liver, colon, prostate, breast and skin cancer [17,18].

Cancer is defined by uncontrolled cell proliferation, loss of spatial control and the development of metastasis. Oncogene activation and tumour suppressor gene deactivation often result in uncontrolled cell cycle progression and the deactivation of apoptotic mechanisms [19]. Effective apoptotic signalling is critically important to maintaining a healthy balance between cell death and cell survival. Furthermore, appropriate apoptosis supports genome integrity, as pointed to the emergence of apoptosis evasion as a key characteristic of cancer [20]. Chemical compounds that restore apoptosis pathways are considered potential anticancer therapies. Various signalling pathways, for example, PI3K/AKT/mTOR, JAK/STAT, and Ras/Raf/MEK/ERK, associate with growth factor receptors and are involved in cancer development [21]. Phytophenolic compounds that can inhibit these signalling pathways by blocking growth factor communication are potential cancer drugs.

This chapter will discuss several plant phenolic compounds from different chemical classes and their pharmaceutical and medical applications in cancer drug discovery. The information discussed in this chapter is from published studies with search keywords and phrases such as "plant phenolic compounds in cancer application" or the names of different chemical classes followed by application in cancer using search engines such as PubMed, Web of Science, Scopus, and Google Scholar.

3.2 SOURCES AND ISOLATION OF PHENOLIC COMPOUNDS

Phenolic compounds are not universally distributed in plants and have varying stability; hence, the extraction and isolation of these compounds is complex. The diversity of structures is tied to a range of qualities, which are associated with unique tasks in plants, and thus determines their specific distribution. Plants convert approximately 2% of their photosynthetic carbon into flavonoids and related chemicals [22]. Higher plants produce thousands of distinct phenolic chemicals, with more compounds being extracted and isolated. Vascular plant leaves contain various chemical classes such as hydroxycinnamic acid esters, glycosides, and amides. Glycosylated flavonoids such as proanthocyanidins and flavonols have been reported in vascular plant leaves too [23]. Some phenolic chemicals are extremely widespread, such as chlorogenic acid, whereas others are limited to specific plant groups, such as families or genera, plant parts, or even specific developmental stages [24].

Natural structure changes can also affect the existence of phenolic chemicals, causing them to exist as polymers or monomers [25]. Single-step extraction and improper extraction methods may have an impact on the extraction yield of phenolic chemicals from plant samples. Hence, appropriate extraction method selection is crucial for extracting the targeted phenolic chemicals [26]. Pretreatment steps prior to extraction, such as proper drying, grinding, homogenization and filtration hydrolysis, are crucial. Conventional liquid extraction is often employed in crude extraction, among other extraction methods such as ultrasound-assisted, enzyme-assisted, microwave-assisted and supercritical liquids [25,27]. In addition to the extraction methods, the solvents used play a critical role in the class of compounds extracted as well as the yield of the extracts. The commonly used extraction solvents are ethanol, methanol, water, ethyl acetate and acetone [28]. Polyphenols of lower molecular weights are best extracted with methanol, while aqueous acetone is best for higher molecular weight flavanols, though its toxicity limits its use and is hence replaced with ethanol [29]. High temperatures improve extraction efficiency by making cell walls permeable, increasing solubility and diffusion coefficients of substances, and decreasing solvent viscosity, allowing for easier transit through solid substrates. Excessive temperatures can degrade phenolic compounds; hence, traditional solvent extraction rarely exceeds 25°C [30].

Plant phenolic extracts contain a variety of soluble classes of phenolic compounds. To remove undesired phenolics and nonphenolic compounds, further processes may be needed, along with an effective clean-up method. Column chromatography, solid-phase extraction and droplet counter-current chromatography are among the purification and separation techniques applied [31]. To isolate pure phenolic compounds, chromatographic techniques like column chromatography, high-performance liquid chromatography, and stationary and mobile phases are required. Sample stability, extraction efficiency and reducing potential interferences are achieved with proper sampling and sample preparation steps.

3.3 CLASSES OF PHENOLIC COMPOUNDS AND THEIR APPLICATIONS

Phenolic compounds from various medicinal plants have the potential to be employed in pharmaceutical and medical applications, particularly for health promotion. These include antioxidant, anti-bacterial, anticancer, cardioprotective, and anti-inflammatory effects, and skin protection from UV radiation [32]. Chemotherapy is widely used in cancer treatment around the world; however, it has many unwanted effects. Thus, it is interesting to investigate new treatments involving phenolic compounds targeting cancer that have potentially fewer side effects and are less expensive [33,34].

3.3.1 Simple Phenols and Benzoquinones

Simple phenols are single benzene (C_6) rings having at least one hydroxyl group. Some examples of simple phenols isolated from plants include cardonol 17 and cardonol 13 isolated from *Lannea edulis* [35]. Ribeiro et al. isolated five benzoquinones from *Cyperus* species and examined their activity on cell lines A549 and AGS. Four of the isolated compounds, cyperaquinone, dihydrocyperaquinone, scabequinone and hydroxycyperaquinone, exhibited better activity on AGS cells with IC_{50} values ranging from 1.7 to 27 μM compared to A549, whose IC_{50} values ranged from 3.0 to 46.6 μM [36]. The benzoquinones exhibited cytotoxic effects on cancer cells by triggering ER stress and apoptosis through caspase-4 and CHOP activation. Hydroxycyperaquinone, the most effective compound among the five, may cause ER stress by inhibiting proteasomes. This is likely due to an overflow of misfolded proteins in the ER caused by inefficient protein breakdown. Rapanone, a naturally occurring benzoquinone, was determined to be a strong inhibitor of mitochondrial respiration and induced apoptosis [37]. The chemical structures of the bioactive simple phenols and benzoquinones are depicted in Figure 3.1.

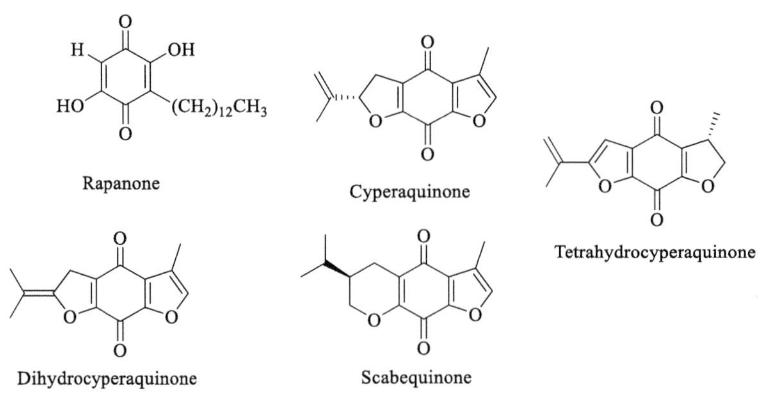

FIGURE 3.1 Chemical structures of bioactive phenols and benzoquinones.

3.3.2 Phenolic Acids and Phenolic Aldehyde

Phenolic acids and phenolic aldehydes are phenolic compounds with a basic skeleton of C_6–C_1. Examples of phenolic acids include gallic acid, syringic acid, salicylic acid among others. Phenolic aldehydes include vanillin, syringaldehyde, salicyladehyde, gallic acid aldehyde and gentisaldehyde [35]. Phenolic acids and phenolic aldehydes have been examined for their potential application in various cancer treatments. A study by Sun et al. investigating the antiproliferative effects of gallic acid on HepG2 and SMMC-7721 hepatocellular carcinoma (HCC) cell lines revealed gallic acid induced apoptosis in SMMC-7721 cells. Furthermore, gallic acid inhibited SMMC-7721 proliferation and HepG2 cells [38]. *In vivo* studies further revealed that gallic and syringic acid decreased tumour size and the levels of marker enzymes in the serum of diethylnitrosamine (DEN)-induced rats [39,40]. Syringic acid presents in olives, dates, spices, red wine, pumpkin, palm oil and grapes, and has been shown to exhibit anticancer activities on colon cancer [15,41–43]. *In vitro* and *in vivo* tests determined its capacity to induce apoptosis through the mitochondrial pathway by elevating caspase 3 and caspase 9 expression, p53, cytochrome c and Bax and decreasing Bcl-2 gene expression [44].

3.3.3 The C_6–C_3 Category of Phenolic Compounds

This category includes polypropene, coumarin, hydroxycinnamic acid, chromones, coumarins and isocoumarins [3]. Hydroxycinnamic acid is widely distributed in plants- and usually esterified with cyclic polyols. Esculetin, a coumarin, dramatically reduced migration, proliferation, and death of HCCLM3 and HUH7 cells. Further, esculetin impacted oxidative stress levels and altered the levels of autophagy and iron metabolism, caused a ferritinophagy-related phenomenon and elevated cellular lipid peroxidation [45]. *In vivo*, esculetin has been shown to reduce tumour volume, enhance NCOA4 and LC3 expression, suppress hydroxyl radical inhibiting capability and GSH, increase Fe^{2+} and MDA levels, and reduce antioxidant protein expression in tumour tissue. In addition, esculetin may enhance iron deposition in tumour tissues, stimulate ferritinophagy, and trigger tumour ferroptosis [45]. Angelicin induced apoptosis and activation of caspases-3 and -9 in cell lines Huh-7 and HepG2, promoted cancer cell apoptosis and inhibited tumour growth *in vivo* [46]. Thus, angelicin demonstrates chemotherapeutic efficacy by triggering cancer cell apoptosis via the caspase-dependent mitochondrial pathways. Caffeic acid inhibited cell growth in HCC cells isolated from marmots. The activity was connected with the mitochondrial integrity loss, culminating in cytochrome c release, caspase 9 activation, and cell death [47]. Caffeic acid phenethyl ester, a caffeic acid derivative, significantly increased apoptosis in Hep3B HCC cells by positively regulating death receptor 5 (DR5), which was mediated by tumour necrosis factor-related apoptosis-inducing ligand (TRAIL) [48]. Ferulic acid has cytotoxic action and induces apoptosis in HepG2 cancer cells via activation of caspases 8 and 9 [49]. Representative chemical structures of bioactive compounds under this category are shown in Figure 3.2.

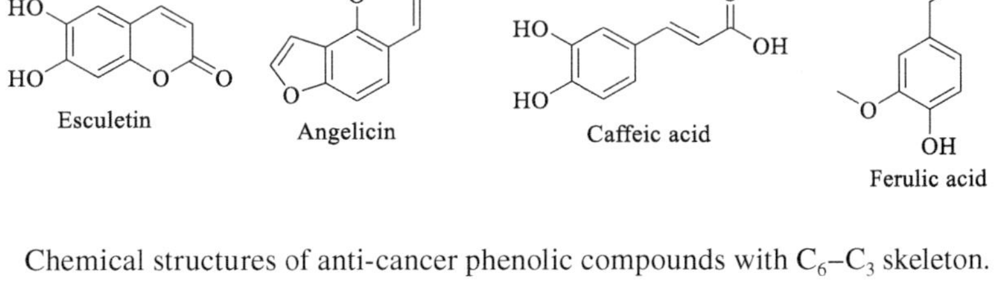

FIGURE 3.2 Chemical structures of anti-cancer phenolic compounds with C_6–C_3 skeleton.

3.3.4 Naphthoquinones

Naphthoquinones (C_6–C_4) bear a naphthalene ring and depending on the positions of the carbonyls, they can either be sub-grouped as 1,2 naphthoquinones (the carbonyls at ortho positions) or 1,4 naphthoquinones (carbonyls at para positions) [50]. The physicochemical differences between these isomers result in different pharmacological actions [51]. Naphthoquinones have been reported and isolated in higher plants in families such as Bignoniaceae, Boraginaceae, Verbenaceae, Proteaceae, and are widely present in the genus *Tabebuia* [50]. Shikonin, a naphthoquinone isolated from Chinese herbal medicine zicao prepared from *Lithospermum erythrorhizon, Arnebia euchroma,* or *A. guttata,* exhibits strong anticancer activities through cell cycle arrest induction, autophagy and cell death [52]. Inhibitory studies on A549, MDA-MB-231, PANC-1, and U2OS cell lines revealed that shikonin inhibited growth with IC_{50} values ranging 1–2 µM. Furthermore, shikonin induced a G2/M phase cell cycle arrest, inhibited the ERK growth signalling pathway and activated the protective autophagy pathway [52,53].

Juglone, isolated from *Juglans mandshurica,* was shown to induce apoptosis by downregulating androgen receptor (AR) expression in LNCaP cells [54]. Inhibition of AR has been demonstrated to suppress cell proliferation; hence, juglone exhibits its potential as an androgen-sensitive prostate cancer drug. Furthermore, juglone suppresses PIK3, Akt, and mTOR phosphorylation leading to apoptosis [55]. Plumbagin is mainly isolated from the plant species of the Plumbaginaceae family, such as *Plumbago capensis* and *Plumbago zeylanica.* Plumbagin demonstrates significant *vitro* and *in vivo* activities, in single use and in conjunction with existing chemotherapy and radiotherapy [56]. Plumbagin was revealed to induce apoptosis in A549, H460 and H292 cell lines and increased the activity of caspases-3 and 9, upregulated Bax and Bak expression and downregulated expression of Bcl-2 [57]. A study by Wu et al. established that plumbagin exhibits anticancer activity through P13K/Akt-mTOR pathway inhibition [58]. The naphthoquinones chemical structures are depicted in Figure 3.3.

3.3.5 Xanthones (C_6–C_1–C_6)

Xanthones are secondary metabolites with three rings commonly present in higher plants and microbes. They have a very broad pharmacological profile, including anticancer, anti-hypertensive, anti-thrombotic and anti-oxidative activities [59]. The broad range of biological activity is dependent on their diverse chemical structures due to substitutions on the ring system. α-Mangostin, isolated from *Garcinia mangostana,* was demonstrated to significantly inhibit the growth rate of human leukaemia via activation of caspase-3-dependent apoptosis [60]. Psorospermin, isolated from *Psorospermum febrifugum,* showed promising anticancer activities both *in vitro* and *in vivo* [59,61]. Zhao et al. established that Xipshantone H, isolated from *Garcinia xipshuanbannaensis,* inhibited A549 cell proliferation by targeting the STAT3 signalling pathway and reduced metastasis through the FAK/MMP-2 signalling pathway [62]. A review by Kurniawan et al. showed that several xanthone derivatives, such as dihydroxanthone and tetrahydroxanthone, isolated from *Garcinia oligantha* leaves, ananixanthone and caloxanthone B isolated from *Calophyllum* species, and

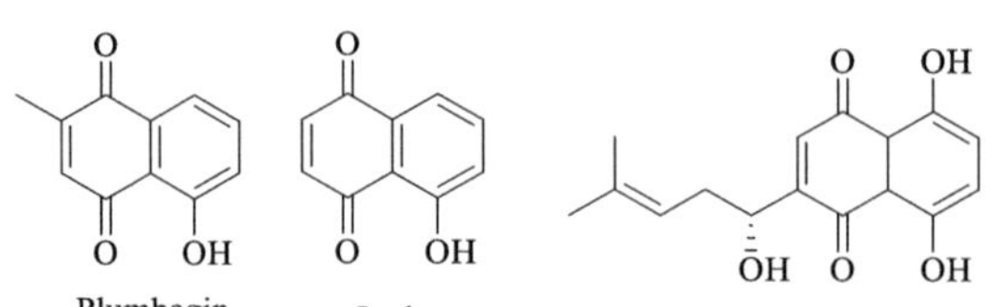

FIGURE 3.3 Anti-cancer naphthoquinones.

schomburgone A isolated from *Garcinia schomburgkiana,* demonstrated strong anticancer activities on different cell lines, including HepG2, HeLa, S-3, K562 and PC-3 [63]. Figure 3.4 shows the chemical structures of xanthones (Table 3.1).

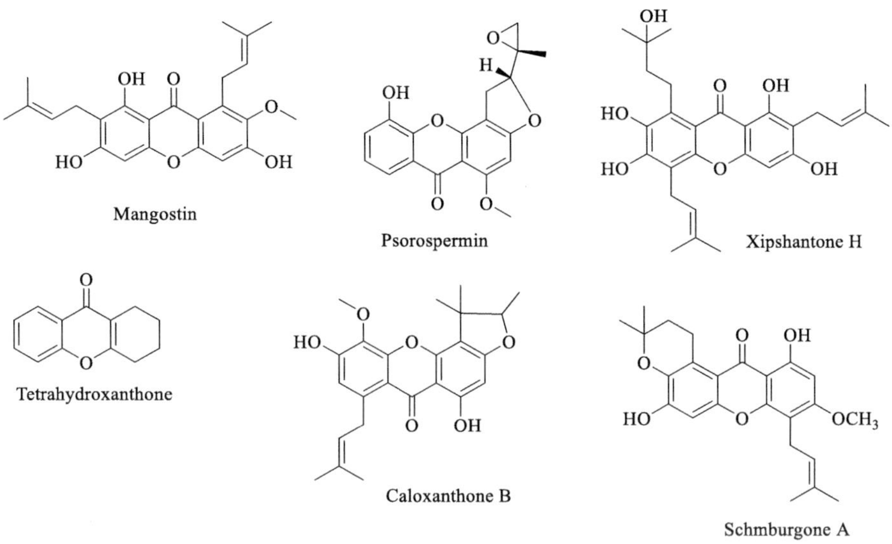

FIGURE 3.4 Chemicals structures of bioactive anti-cancer xanthones.

TABLE 3.1
Bioactive Plant Phenolic Compounds and Their Mechanism of Action

Phenolic Class	Compound Name	Mechanisms	References
C_6	Cyperaquinone, tetrahydrocyperaquinone, dihydrocyperaquinone, scabequinone, hydroxycyperaquinone	Induction of apoptosis, triggering ER stress	[36]
	Rapanone	Induces apoptosis and inhibits mitochondrial respiration	[37]
C_6 C_3	Caffeic acid	Caffeic acid inhibits HCC by reducing ROS generation and induces apoptosis through DNA oxidation. It also reduces VEGF-induced vascularization and suppresses MMP-2 and MMP-9 expression.	[48,64]
	Ferulic acid	Apoptosis induction in HepG2, activation of caspases 8 and 9.	[49]
	Syringic acid	Induces apoptosis via the mitochondrial pathway, upregulation of p53, caspases-3 and 9, downregulation of cell cycle proteins cyclins B, C and CDK4,6	[44,65,66]
C_6–C_4	Shikonin	Cell cycle arrest at G2/M phase, P21 upregulation, induces apoptosis, and upregulates autophagy genes	[52,53]

(Continued)

TABLE 3.1 (*Continued*)
Bioactive Plant Phenolic Compounds and Their Mechanism of Action

Phenolic Class	Compound Name	Mechanisms	References
	Juglone	Induces cell apoptosis, TNF activation, inhibits PI3K/Akt/mTOR signalling	[55,67]
	Plumbagin	Induces apoptosis, suppressed proliferation, and cell cycle arrest. Increased activity of ROS, caspases 3 and 9. Downregulation of P13K/Akt-mTOR, IL-6, p53.	[57,58]
C_6–C_1–C_6	α-Mangostin	Induce caspase-3-dependent apoptosis	[60]
	Xipshantone H	Regulates STAT3 pathway	[62]
C_6–C_3–C_6	Apigenin	Induced apoptosis and protective autophagy, target mTOR/PI3K/AKT signalling pathway, inhibit cell proliferation, suppress the AMPK pathway and autophagy flux.	[68]
	Isovitexin & vitexin	Induce apoptosis and autophagy via endoplasmic reticulum (ER) stress	[69,70]
	Chrysin	Induction of apoptosis through ROS and ER stress	[71,72]
	Quercetin	Suppresses tumour, metastasis, and glycolysis via AKT/mTOR-mediated autophagy,	[73]
$(C_6$–$C_3)_2$	Methyl (E)-6-hydroxy-2-(4-hydroxyphenyl)-5-(prop-1-en-1-yl) benzofuran-3-carboxylate	upregulates bax, caspases-3 and 9, downregulates bcl-2 expression	[74]
$(C_6$–C_3–$C_6)_2$	Hinokiflavone	Induction of apoptosis through activation of mitochondrial ROS/JNK/caspase pathway and inhibition of NF-κB	[75]
	Robustaflavone A	Induction of ferroptosis via the mitochondrial pathway	[76]
	7-O-methyldelicaflavone	Induce apoptosis, suppression of anti-apoptotic XIAP and surviving proteins expression,	[77]
	Amentoflavone	Induction of apoptosis, tumour growth inhibition	[78]

3.3.6 Anthraquinones and Stilbene

Plant-isolated chemical compounds from the classes stilbene and anthraquinones, with a basic skeleton chemical structure C_6–C_2–C_6, have demonstrated pharmacological activities for example, anticancer, anti-bacterial, anti-malarial, anti-diabetic, and anti-viral properties [79]. Rubiadin, isolated from *Rubia cardifolia,* exhibited antiproliferative activity in HeLa cell lines [80], while chrysophanol suppressed cancer cell proliferation, and blocked the cell cycle at the G1/S phase in MCF-7 and MDA-MB-231 cell lines [81]. The study further established that chrysophanol promotes apoptosis through the NF-κB/cyclin D1 and NF-κB/Bcl-2 signalling cascades. Wijesekara et al. established that physcion induces apoptosis by downregulating Bcl-2 expression, upregulating Bax expression, and caspase-3 pathway activation. Furthermore, physcion induced ROS formation in HeLa cells [82]. Aloe-emodin has been established to exhibit anticancer activity on various cancer types. Guo et al. established that aloe-emodin inhibits HeLa cells in dose- and time- dependent manner [83]. Piceatannol, isolated from many plants such as *Passiflora edulis, Aiphanes asuleata* and *Cassia garretiana,* exhibited chemopreventive activity against several cancers [84,85]. Piceatannol displays anticancer activities by regulating pro-apoptotic and anti-apoptotic proteins, as well as

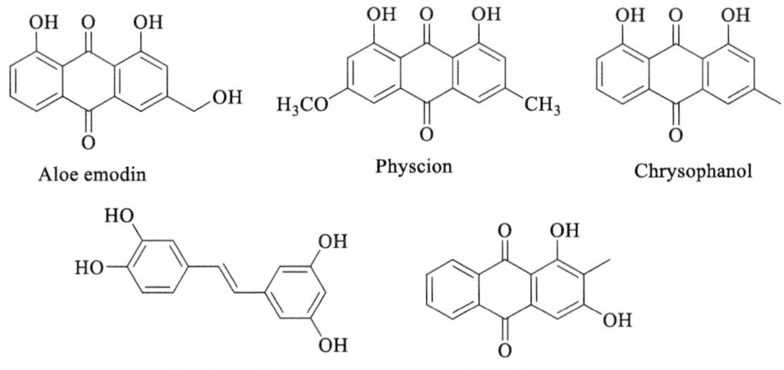

FIGURE 3.5 Chemical structures of anthraquinones and stilbene.

regulating the JAK-STAT3, NF-κB and Akt/mTOR pathways. It also has potent anti-metastatic actions by inhibiting MMPs, Wnt signalling, PI3K/Akt/mTOR, and epithelial to mesenchymal transition [86,87]. The bioactive anthraquinones discussed in this study are shown in Figure 3.5.

3.3.7 Flavonoids and Isoflavonoids

Flavonoids are found in plants as glycosylated derivatives, and the basic structure contains a flavan nucleus [1]. The different classes of flavonoids are flavones, flavan-3-ols, flavanones, flavonols, isoflavones, flavanonol, and anthocyanidins, which are widely found in both food and medicine [88]. They were once thought to be of tremendous medicinal value and had a key therapeutic influence. Flavonoids have significant biological activities on numerous chronic diseases, including cancer, and have enormous development potential [89,90]. Flavonoids' anticarcinogenic mechanisms include proliferation inhibition, the induction of apoptosis, necrosis, autophagy, cell cycle arrest, cell migration and invasion impairment and the inhibition of tumour angiogenesis [91,92]. Apigenin, one of the most ubiquitous flavonoids isolated from plants due to its strong anti-oxidant and anti-inflammatory activity, has been of great interest in cancer prevention studies [93]. It has been applied in studies of various cancers, including the breast, cervix, colon, liver, pancreatic, prostate, brain, skin, lung, and ovary. Apigenin exhibited anticancer activity on colorectal cancer by inhibiting cell proliferation and migration. Furthermore, by inhibiting Wnt/β-catenin signalling pathway, apigenin significantly inhibited colon cancer cell proliferation, migration, invasion and organoid growth [94]. Apigenin caused apoptosis by activating the mitogen-activated protein kinase (MAPK) signalling pathway and down-regulating Srx expression in cutaneous squamous cell carcinoma cell lines [95].

Isovitexin and vitexin have been used for treating a variety of diseases in traditional Chinese medicine [96,97]. Vitexin and isovitexin have anti-neoplastic actions in several *in vitro* and *in vivo* models through promoting apoptosis and autophagy while inhibiting proliferation and migration [98]. The two flavonoids displayed anticancer activity by inducing apoptosis in Hep G2 liver cells, HCT116 colorectal, MCF-7 human breast cancer cells, HeLa cervical cells, HEp-2 larynx cells, and BHK-21 baby hamster kidney cells [99–101]. Vitexin and isovitexin treatment induced apoptosis in U937 cells, as well as reduced mitochondrial membrane potential levels. In addition, treatment with vitexin and isovitexin decreased Bcl-2 protein levels while increasing caspase-3 and -9 protein production in U937 cells [102,103].

Chrysin, a dihydroflavone widely distributed in plant species, demonstrates a variety of pharmacological activities, including anticancer. Ryu et al. established that chrysin inhibited the P13K/AKT and activated the ERK1/2 MAPK and p38 MAPK pathways which mediate tumour development,

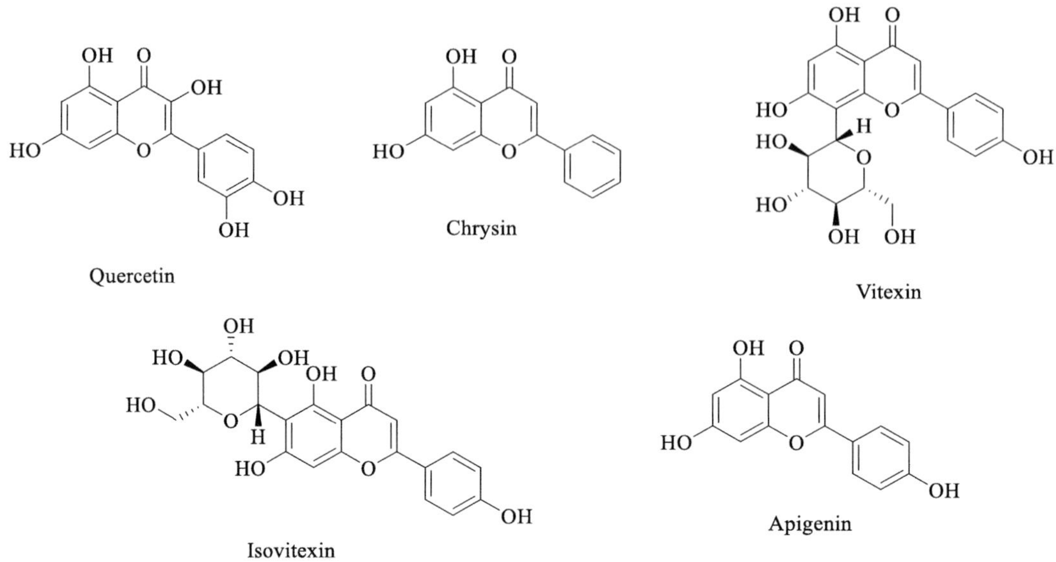

FIGURE 3.6 Bioactive flavonoids and isoflavonoids.

metastases and cellular metabolism, hence inducing apoptosis [71]. Furthermore, chrysin increases ER stress regulatory protein expression in a dose-wise manner in prostate cancer DU145 and PC-3 cell lines and stimulates ROS production [72]. Quercetin is a flavonol that has been examined for its pharmacological application in treating a variety of diseases and for anticancer applications. Quercetin exerts its anticancer effects by modulation of dysregulated signalling pathways such as apoptosis and autophagy [104]. Khorsandi et al. established that quercetin may efficiently reduce the survival and proliferation of MCF-7 cells by activating both apoptosis and necroptosis signalling pathways [105]. The study further determined apoptosis was produced by raising the expression of Bax and caspase-3 while lowering the expression of Bcl-2 genes. Necroptosis was produced by increasing the expression of RIPK1 and RIPK3. Quercetin may decrease glycolysis and inhibit tumour cell migration by lowering acidity in the tumour microenvironment. Furthermore, Jia et al. discovered that quercetin inhibited the Akt-mTOR pathway, leading to autophagy. When combined with the autophagy inhibitor 3-MA and the Akt-mTOR pathway inducer Insulin Growth Factor-1 (IGF-1), quercetin inhibited cell mobility and glycolysis via the Akt/mTOR pathway mediation of autophagy induction [73]. The chemical structures of phytophenolic compounds under this category are shown in Figure 3.6.

3.3.8 $(C_6–C_3)_2$ Lignans, Neolignans

Lignans are a broad and rather diverse category of derivatives of phenylpropane that are prevalent in vascular plants [106]. Their pharmacological actions have garnered a great deal of attention. While certain lignans have been developed into approved treatments, others are lead structures for new pharmaceuticals [107]. Rao et al. isolated three neolignans and lignans from *Cinnnamomum bejolghota* which exhibited significant growth inhibition activity in A549 (lung), HCT-116 (colon) and MDA-MB-231(breast) cell lines [108]. Neolignans bejolghotins A-C exhibited moderate anticancer activity when tested on HCT-116, A549 and MDA-MB-231. The study further revealed isolated lignans bejolghotin G, (7′S,8′R,8R)-lyoniresinol-9-O-(E)-feruloyl ester and (7′S,8′R,8R)-lyoniresinol-9,9′-di-O-(E)-feruloyl ester demonstrated strong inhibition on the three cell lines with IC_{50} values ranging from 0.78 ± 0.09 to 2.79 ± 0.11 μM. Jiang et al. isolated bioactive neolignans from *Mappianthus iodoies,*

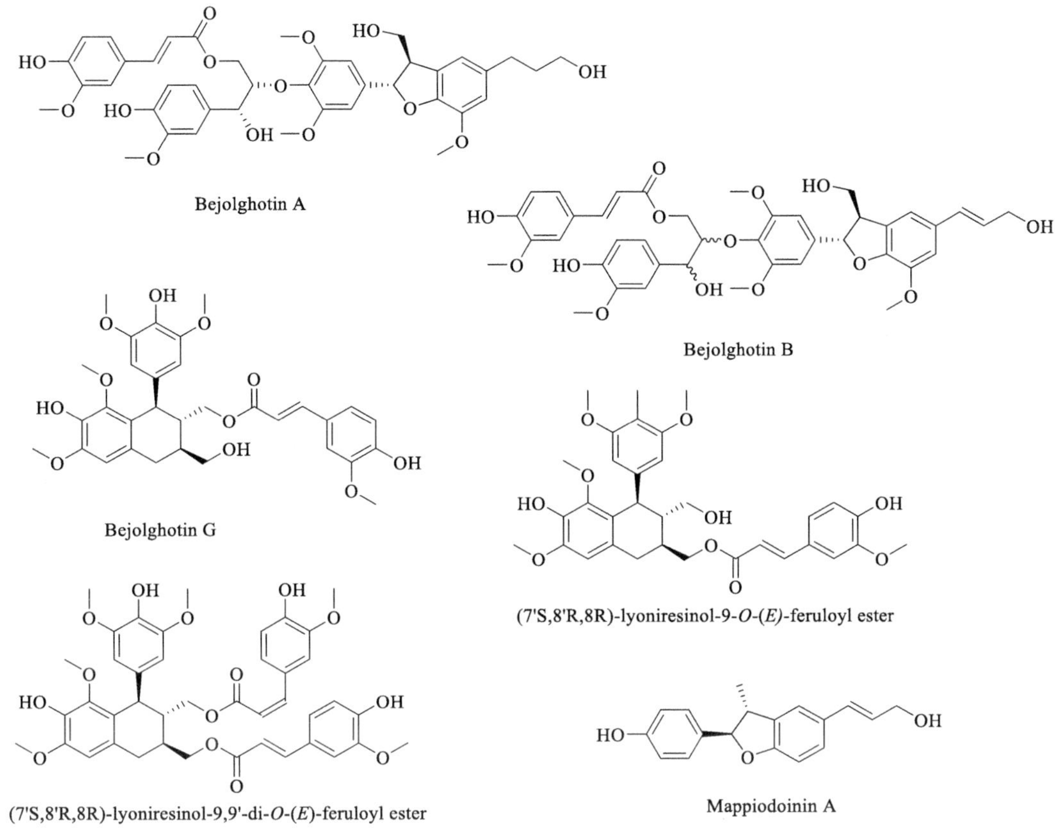

FIGURE 3.7 Chemical structures of lignans and neolignans with anti-cancer activities.

and subsequent cytotoxic evaluation on the cancer cell lines HL-60, A-549, SW-480, SMMC-7721 and MCF-7 revealed that they had strong inhibitory activity [109]. Mappiodoinin A exhibited stronger inhibitory activity than the positive control drug, cisplatin, with IC_{50} values ranging from 0.16±0.06 to 2.19±0.10 μM [109]. Methyl (E)-6-hydroxy-2-(4-hydroxyphenyl)-5-(prop-1-en-1-yl) benzofuran-3-carboxylate isolated from *Daphniphyllum macropodum* was established to inhibit cell proliferation in A549 and H460 cell lines and induce apoptosis [74]. Tran et al. isolated lignans from *Austrobaileya acandens* and evaluated their anticancer activity on the PC3 prostate cancer cell line. The study revealed that ten isolated compounds induced apoptosis via G2/M phase arrest, topoisomerase II enzyme inhibition and disruption of microtubules [110]. Altogether, both lignans and neolignans (chemical structures shown in Figure 3.7) have been found to exhibit significant anticancer activity on different cancer cell lines and are of interest in cancer drug discovery.

3.3.9 $(C_6–C_3–C_6)_2$ Biflavonoids

Bioflavonoids are a naturally occurring class of polyphenol compounds that are formed as a result of oxidative coupling of two chalcone units and modification of the C3 central units [1]. The chemical structures are shown in Figure 3.8. Bioflavonoids are limited to several plant species and have been revealed to exhibit anti-inflammatory, anti-protozoal, anticancer, and anti-oxidant [111]. Hinokiflavone, a bis-apigenyl ether first isolated from the leaves of *Chamaecyparis obtuse,* has been isolated in many other plant species and found to exhibit anticancer properties on different cancer

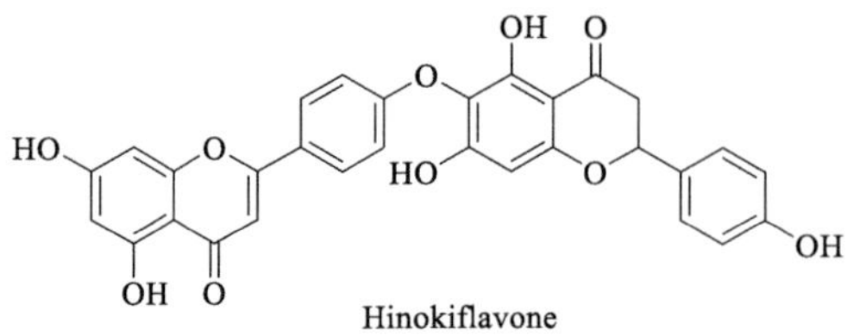

Hinokiflavone

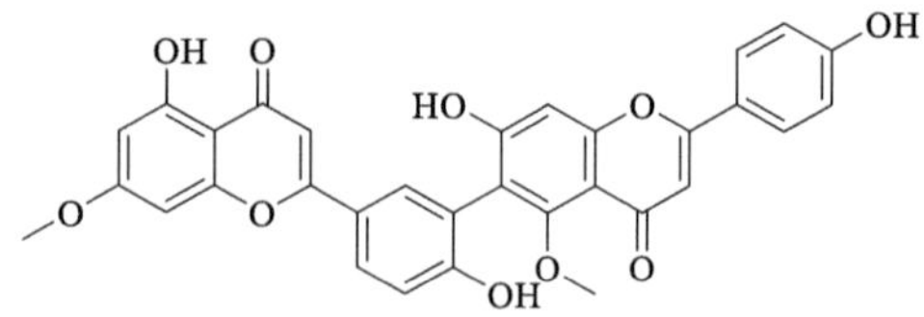

Robustaflavone A

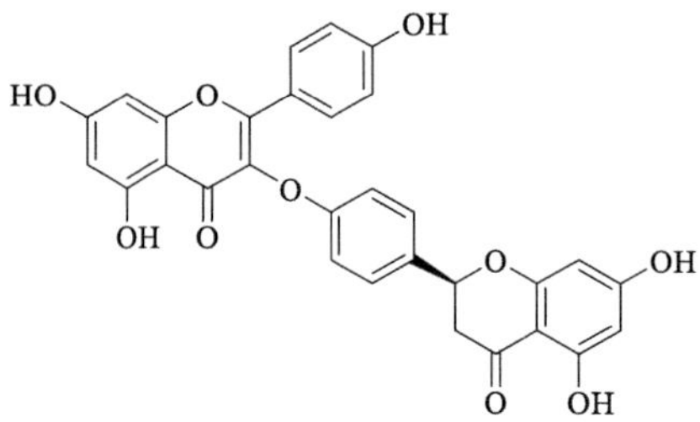

(2"S)-2",3"-dihydrodelicaflavone

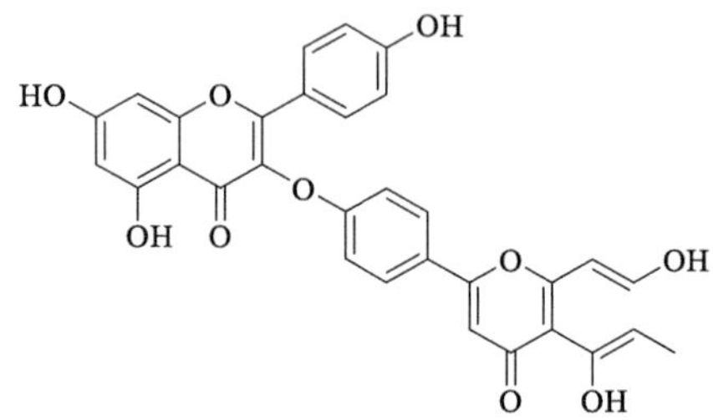

6"-methyldelicaflavone

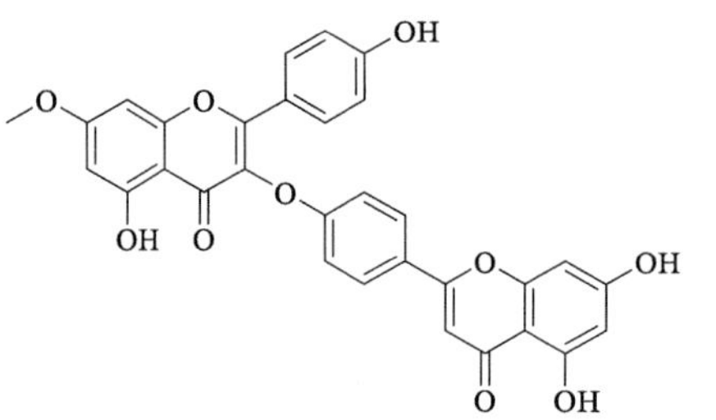

7-O-methyldelicaflavone

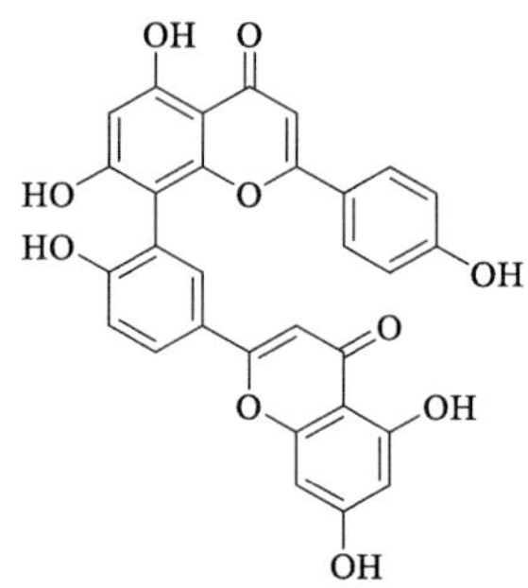

Amentoflavone

FIGURE 3.8 Bioactive biflavonoids with anti-cancer activities.

cell lines [112]. Hinokiflavone induces apoptosis in colon cancer cells by up-regulating the protein Bax, down-regulating Bcl-2, and causing a drug-induced loss of mitochondrial potential [113]. In hepatocellular carcinoma, hinokiflavone inhibits NFκB signalling to suppress the expression of anti-apoptotic genes. This reinforces the compound's direct pro-apoptotic effect by up-regulating phospho-JNK [75]. Hinokiflavone, isolated from *Selaginella doederleinii,* demonstrated strong anticancer activity *in vitro* and *in vivo* via the inhibition of the PI3K/AKT/mTOR and MAPK signalling cascades [114,115]. Robustaflavone A, isolated from *Selaginella trichoclada,* exhibited strong activity by reducing cell viability, with an IC_{50} of 11.89 μM. Furthermore, robustaflavone A caused non-apoptotic cell death in MCF-7 cells by ferroptosis, by increasing VDAC2 channel expression and decreasing Nedd4 E3 ubiquitin ligase expression, resulting in lipid peroxidation and ROS generation [76]. Bioflavonoids (2″S)-2″,3″-dihydrodelicaflavone, 6″-methyldelicaflavone and 7-O-methyldelicaflavone isolated from *Selaginella doederleinii* demonstrated strong activity on cell line A549 [77]. The three compounds induced apoptosis via caspase-3/cleaved-caspase-3-dependent mechanisms. 7-O-methyldelicaflavone further arrested lung cancer cells at the S phase. Tsai et al. established that amentoflavone improves sorafenib's ability to decrease tumour growth, ERK/AKT phosphorylation, and anti-apoptotic protein production [78]. Furthermore, amentoflavone triggered sorafenib-induced apoptosis via both extrinsic and intrinsic mechanisms, hence enhancing the therapeutic efficacy of sorafenib and inducing apoptosis in HCC *in vivo.*

3.4 CONCLUSION AND FUTURE PERSPECTIVES

Plants contain bioactive phenolic chemicals, which are categorized into various classes based on their structure. These bioactive chemicals exhibit immunomodulatory and antioxidant characteristics, rendering them potential anticancer drugs. The various structural classes of plant phenolic compounds have seen their wide application in anticancer drug discovery. This chapter noted significantly more research reported on *in vitro* studies compared to *in vivo* studies. This is attributed to some isolated phenolic compounds not demonstrating significant anticancer activities *in vitro,* hence not proceeding to *in vivo* screening. To identify more bioactive phytophenolics with anticancer properties, more isolation studies are required to be conducted. This will in turn allow more *in vivo* studies for those compounds that exhibit strong anti-proliferative activity as well as testing on different cancer cell lines.

REFERENCES

1. Pratyusha, S., Phenolic compounds in the plant development and defense: An overview. *Plant Stress Physiology: Perspectives in Agriculture*, 2022. **11**, pp. 125–140.
2. Lin, D., et al., An overview of plant phenolic compounds and their importance in human nutrition and management of type 2 diabetes. *Molecules*, 2016. **21**, DOI: 10.3390/molecules21101374
3. Tsimogiannis, D. and V. Oreopoulou, Classification of phenolic compounds in plants, in *Polyphenols in Plants* (Second Edition), R. R. Watson, Editor. 2019, Massachusetts, United States: Academic Press. pp. 263–284.
4. Vuolo, M.M., V.S. Lima, and M.R. Maróstica Junior, Phenolic compounds: Structure, classification, and antioxidant power, in *Bioactive Compounds*, M. R. S. Campos, Editor. 2019, Cambridge, England: Woodhead Publishing. pp. 33–50.
5. Pu, F., X.-L. Ren, and X.-P. Zhang, Phenolic compounds and antioxidant activity in fruits of six *Diospyros kaki* genotypes. *European Food Research and Technology*, 2013. **237**: pp. 923–932.
6. Fiorentino, A., et al., Isolation and structure elucidation of antioxidant polyphenols from quince (*Cydonia vulgaris*) peels. *Journal of Agricultural and Food Chemistry*, 2008. **56**(8): pp. 2660–2667.
7. Nobile, V., et al., Skin photoprotective and antiageing effects of a combination of rosemary (*Rosmarinus officinalis*) and grapefruit (*Citrus paradisi*) polyphenols. *Food & Nutrition Research*, 2016. **60**(1): p. 31871.
8. Pereira, Q.C., et al., The molecular mechanism of polyphenols in the regulation of ageing hallmarks. *International Journal of Molecular Sciences*, 2023. **24**(6): p. 5508.
9. Zahedipour, F., et al., Phytochemicals as inhibitors of tumor necrosis factor alpha and neuroinflammatory responses in neurodegenerative diseases. *Neural Regeneration Research*, 2022. **17**(8): pp. 1675–1684.
10. Jiang, A., et al., Curcumin reactivates silenced tumor suppressor gene RARβ by reducing DNA methylation. *Phytotherapy Research*, 2015. **29**(8): pp. 1237–1245.
11. Du, L., et al., Reactivation of RASSF1A in breast cancer cells by curcumin. *Nutrition and Cancer*, 2012. **64**(8): pp. 1228–1235.
12. Khan, H., et al., Anti-cancer effects of polyphenols via targeting p53 signaling pathway: Updates and future directions. *Biotechnology Advances*, 2020. **38**: p. 107385.
13. Khan, H., et al., Targeting NF-κB signaling pathway in cancer by dietary polyphenols. *Critical Reviews in Food Science and Nutrition*, 2020. **60**(16): pp. 2790–2800.
14. Soltani, A., et al., Apoptosis-inducing plant-based phenolic compounds are effective on leukemia cell lines. *Current Pharmaceutical Design*, 2023. **29**(14): pp. 1092–1104.
15. Huang, W.-Y., Y.-Z. Cai, and Y. Zhang, Natural phenolic compounds from medicinal herbs and dietary plants: Potential use for cancer prevention. *Nutrition and Cancer*, 2009. **62**(1): pp. 1–20.
16. Hanif, R., et al., Curcumin, a natural plant phenolic food additive, inhibits cell proliferation and induces cell cycle changes in colon adenocarcinoma cell lines by a prostaglandin-independent pathway. *Journal of Laboratory and Clinical Medicine*, 1997. **130**(6): pp. 576–584.
17. Carocho, M. and I. CFR Ferreira, The role of phenolic compounds in the fight against cancer–A review. *Anti-Cancer Agents in Medicinal Chemistry (Formerly Current Medicinal Chemistry-Anti-Cancer Agents)*, 2013. **13**(8): pp. 1236–1258.
18. Bonta, R.K., Dietary phenolic acids and flavonoids as potential anti-cancer agents: Current state of the art and future perspectives. *Anti-Cancer Agents in Medicinal Chemistry*, 2020. **20**(1): pp. 29–48.

19. Sarkar, S., et al., Cancer development, progression, and therapy: An epigenetic overview. *International Journal of Molecular Sciences*, 2013. **14**(10): pp. 21087–21113.
20. Goldar, S., et al., Molecular mechanisms of apoptosis and roles in cancer development and treatment. *Asian Pacific Journal of Cancer Prevention*, 2015. **16**(6): pp. 2129–2144.
21. Höpfner, M., D. Schuppan, and H. Scherübl, Growth factor receptors and related signalling pathways as targets for novel treatment strategies of hepatocellular cancer. *World Journal of Gastroenterology: WJG*, 2008. **14**(1): p. 1.
22. Robards, K. and M. Antolovich, Analytical chemistry of fruit bioflavonoidsA review. *Analyst*, 1997. **122**(2): pp. 11R–34R.
23. Lattanzio, V., Phenolic compounds: Introduction 50. *Natural Products*, 2013: pp. 1543–1580.
24. Cheynier, V., Phenolic compounds: From plants to foods. *Phytochemistry Reviews*, 2012. **11**(2): pp. 153–177.
25. Alara, O.R., N.H. Abdurahman, and C.I. Ukaegbu, Extraction of phenolic compounds: A review. *Current Research in Food Science*, 2021. **4**: pp. 200–214.
26. Shi, L., et al., Extraction and character'tion of phenolic compounds and their potential antioxidant activities. *Environmental Science and Pollution Research*, 2022. **29**(54): pp. 81112–81129.
27. M'hiri, N., et al., Extraction methods of citrus peel phenolic compounds. *Food Reviews International*, 2014. **30**(4): pp. 265–290.
28. Meneses, N.G.T., et al., Influence of extraction solvents on the recovery of antioxidant phenolic compounds from brewer's spent grains. *Separation and Purification Technology*, 2013. **108**: pp. 152–158.
29. Liu, Q. and H. Yao, Antioxidant activities of barley seeds extracts. *Food Chemistry*, 2007. **102**(3): pp. 732–737.
30. Escribano-Bailon, M., Polyphenol extraction from foods, in *Methods in Polyphenol Analysis,* Gary Williamson, (Editor) Gateshead, England: Athenaeum Press Ltd. 2003.
31. Santos-Buelga, C., et al., Extraction and isolation of phenolic compounds. *Natural Products Isolation*, 2012. 864: pp. 427–464.
32. Tungmunnithum, D., et al., Flavonoids and other phenolic compounds from medicinal plants for pharmaceutical and medical aspects: An overview. *Medicines*, 2018. **5**, DOI: 10.3390/medicines5030093
33. Ahmed, S.I., et al., Pharmacologically active flavonoids from the anticancer, antioxidant and antimicrobial extracts of *Cassia angustifolia* Vahl. *BMC Complementary and Alternative Medicine*, 2016. **16**: pp. 1–9.
34. Mishra, A., et al., Bauhinia variegata leaf extracts exhibit considerable antibacterial, antioxidant, and anticancer activities. *BioMed Research International*, 2013. **2013**(1): p. 915436.
35. Kougan, G.B., et al., Simple phenols, phenolic acids, and related esters from the medicinal plants of Africa, in *Medicinal Plant Research in Africa,* Victor Kuete (Editor). 2013, Elsevier. pp. 225–249.
36. Ribeiro, V., et al., Benzoquinones from *Cyperus* spp. trigger IRE1α-independent and PERK-dependent ER stress in human stomach cancer cells and are novel proteasome inhibitors. *Phytomedicine*, 2019. **63**: p. 153017.
37. Pardo Andreu, G.L., et al., Rapanone, a naturally occurring benzoquinone, inhibits mitochondrial respiration and induces HepG2 cell death. *Toxicology in Vitro*, 2020. **63**: p. 104737.
38. Sun, G., et al., Gallic acid as a selective anticancer agent that induces apoptosis in SMMC-7721 human hepatocellular carcinoma cells. *Oncology Letters*, 2016. **11**(1): pp. 150–158.
39. Jagan, S., et al., Antiproliferative potential of gallic acid against diethylnitrosamine-induced rat hepatocellular carcinoma. *Molecular and Cellular Biochemistry*, 2008. **319**(1): pp. 51–59.
40. Lavanya, M., P. Srinivasan, and R. Padmini, Unveiling the anticancer effect of syringic acid and its derivatives in hepatocellular carcinoma. *International Journal of Applied Pharmaceutics*, 2023. **15**(4): pp. 114–124.
41. Srinivasulu, C., et al., Syringic acid (SA)–A review of its occurrence, biosynthesis, pharmacological and industrial importance. *Biomedicine & Pharmacotherapy*, 2018. **108**: pp. 547–557.
42. Chander, M., Anticancer efficacy of some plant Phenolics-a recent scenario. *IJCMAS*, 2018. **7**: pp. 1746–1768.
43. Ha, S.J., et al., Syringic acid prevents skin carcinogenesis via regulation of NoX and EGFR signaling. *Biochemical Pharmacology*, 2018. **154**: pp. 435–445.
44. Gheena, S. and D. Ezhilarasan, Syringic acid triggers reactive oxygen species–mediated cytotoxicity in HepG2 cells. *Human & Experimental Toxicology*, 2019. **38**(6): pp. 694–702.
45. Xiu, Z., et al., Inhibitory effects of esculetin on liver cancer through triggering NCOA4 pathway-mediation ferritinophagy in vivo and in vitro. *Journal of Hepatocellular Carcinoma*, 2023. **10**: pp. 611–629.

46. Wang, F., et al., Angelicin inhibits liver cancer growth in vitro and in vivo. *Molecular Medicine Reports*, 2017. **16**(4): pp. 5441–5449.
47. Wilkins, L.R., et al., Cinnamic acid derivatives enhance the efficacy of transarterial embolization in a rat model of hepatocellular carcinoma. *Cardiovascular and Interventional Radiology*, 2017. **40**: pp. 430–437.
48. Dilshara, M.G., et al., Caffeic acid phenethyl ester enhances TRAIL-mediated apoptosis via CHOP-induced death receptor 5 upregulation in hepatocarcinoma Hep3B cells. *Molecular and Cellular Biochemistry*, 2016. **418**: pp. 13–20.
49. ElKhazendar, M., et al., Antiproliferative and proapoptotic activities of ferulic acid in breast and liver cancer cell lines. *Tropical Journal of Pharmaceutical Research*, 2019. **18**(12): pp. 2571–2576.
50. Pereyra, C.E., et al., The diverse mechanisms and anticancer potential of naphthoquinones. *Cancer Cell International*, 2019. **19**: pp. 1–20.
51. Silva, M.N.d., V.F. Ferreira, and M.C.B. de Souza, An overview of the chemistry and pharmacology of naphthoquinones with emphasis on beta-lapachone and derivatives. *Quimica Nova*, 2003. **26**: pp. 407–416.
52. Wang, F., et al., Shikonin inhibits cancer through P21 upregulation and apoptosis induction. *Frontiers in Pharmacology*, 2020. **11**: p. 861.
53. Wang, F., et al., Synthesis, biological function and evaluation of Shikonin in cancer therapy. *Fitoterapia*, 2019. **134**: pp. 329–339.
54. Xu, H., et al., Juglone, isolated from Juglans mandshurica Maxim, induces apoptosis via down-regulation of AR expression in human prostate cancer LNCaP cells. *Bioorganic & Medicinal Chemistry Letters*, 2013. **23**(12): pp. 3631–3634.
55. Xu, H.L., et al., Juglone, from Juglans mandshruica Maxim, inhibits growth and induces apoptosis in human leukemia cell HL-60 through a reactive oxygen species-dependent mechanism. *Food and Chemical Toxicology*, 2012. **50**(3–4): pp. 590–596.
56. Tripathi, S.K., M. Panda, and B.K. Biswal, Emerging role of plumbagin: Cytotoxic potential and pharmaceutical relevance towards cancer therapy. *Food and Chemical Toxicology*, 2019. **125**: pp. 566–582.
57. Xu, T.P., et al., Plumbagin from plumbago zeylanica l induces apoptosis in human non-small cell lung cancer cell lines through NF Kb inactivation. *Asian Pacific Journal of Cancer Prevention*, 2013. **14**(4): pp. 2325–2331.
58. Wu, H., X. Dai, and E. Wang, Plumbagin inhibits cell proliferation and promotes apoptosis in multiple myeloma cells through inhibition of the PI3K/Akt-mTOR pathway. *Oncology Letters*, 2016. **12**(5): pp. 3614–3618.
59. Na, Y., Recent cancer drug development with xanthone structures. *Journal of Pharmacy and Pharmacology*, 2010. **61**(6): pp. 707–712.
60. Nakagawa, Y., et al., Characterized mechanism of α-mangostin-induced cell death: Caspase-independent apoptosis with release of endonuclease-G from mitochondria and increased miR-143 expression in human colorectal cancer DLD-1 cells. *Bioorganic & Medicinal Chemistry*, 2007. **15**(16): pp. 5620–5628.
61. Fellows, I.M., et al., Determination of the importance of the stereochemistry of psorospermin in topoisomerase II–induced alkylation of DNA and in vitro and in vivo biological activity. *Molecular Cancer Therapeutics*, 2005. **4**(11): pp. 1729–1739.
62. Zhao, Y., et al., A natural xanthone suppresses lung cancer growth and metastasis by targeting STAT3 and FAK signaling pathways. *Phytomedicine*, 2022. **102**: p. 154118.
63. Kurniawan, Y.S., et al., An update on the anticancer activity of xanthone derivatives: A review. *Pharmaceuticals*, 2021. **14**, DOI: 10.3390/ph14111144
64. Espíndola, K.M.M., et al., Chemical and pharmacological aspects of caffeic acid and its activity in hepatocarcinoma. *Frontiers in Oncology*, 2019. **9**: p. 541.
65. Abaza, M.-S., et al., Syringic acid from *Tamarix aucheriana* possesses antimitogenic and chemo-sensitizing activities in human colorectal cancer cells. *Pharmaceutical Biology*, 2013. **51**(9): pp. 1110–1124.
66. Periyannan, V. and V. Veerasamy, Syringic acid may attenuate the oral mucosal carcinogenesis via improving cell surface glycoconjugation and modifying cytokeratin expression. *Toxicology Reports*, 2018. **5**: pp. 1098–1106.
67. Tang, Y.T., et al., Molecular biological mechanism of action in cancer therapies: Juglone and its derivatives, the future of development. *Biomedicine & Pharmacotherapy*, 2022. **148**: p. 112785.
68. Chen, Z., et al., Apigenin combined with gefitinib blocks autophagy flux and induces apoptotic cell death through inhibition of HIF-1α, c-Myc, p-EGFR, and glucose metabolism in EGFR L858R+ T790M-mutated H1975 cells. *Frontiers in Pharmacology*, 2019. **10**: p. 260.

69. Bhardwaj, M., et al., Vitexin induces apoptosis by suppressing autophagy in multi-drug resistant colorectal cancer cells. *Oncotarget*, 2017. **9**(3): pp. 3278–3291.
70. Lv, S.-X. and X. Qiao, Isovitexin (IV) induces apoptosis and autophagy in liver cancer cells through endoplasmic reticulum stress. *Biochemical and Biophysical Research Communications*, 2018. **496**(4): pp. 1047–1054.
71. Ryu, S., et al., Chrysin induces death of prostate cancer cells by inducing ROS and ER stress. *Journal of Cellular Physiology*, 2017. **232**(12): pp. 3786–3797.
72. Liao, C.-L., et al., Pine (*Pinus morrisonicola* Hayata) needle extracts sensitize GBM8901 human glioblastoma cells to temozolomide by downregulating autophagy and O^6-methylguanine-DNA methyltransferase expression. *Journal of Agricultural and Food Chemistry*, 2014. **62**(43): pp. 10458–10467.
73. Jia, L., et al., Quercetin suppresses the mobility of breast cancer by suppressing glycolysis through Akt-mTOR pathway mediated autophagy induction. *Life Sciences*, 2018. **208**: pp. 123–130.
74. Ma, X.H., et al., Cytotoxic neolignans from the traditional Chinese medicine *Daphniphyllum macropodum* Miq. *RSC Advances*, 2017. **7**(83): pp. 52970–52976.
75. Mu, W., et al., Hinokiflavone induces apoptosis via activating mitochondrial ROS/JNK/caspase pathway and inhibiting NF-κB activity in hepatocellular carcinoma. *Journal of Cellular and Molecular Medicine*, 2020. **24**(14): pp. 8151–8165.
76. Xie, Y., et al., Identification of a new natural biflavonoids against breast cancer cells induced ferroptosis via the mitochondrial pathway. *Bioorganic Chemistry*, 2021. **109**: p. 104744.
77. Kang, F., et al., Biflavonoids from *Selaginella doederleinii* as potential antitumor agents for intervention of non-small cell lung cancer. *Molecules*, 2021. **26**, DOI: 10.3390/molecules26175401
78. Tsai, J.-J., et al., Amentoflavone enhances the therapeutic efficacy of sorafenib by inhibiting antiapoptotic potential and potentiating apoptosis in hepatocellular carcinoma in vivo. *Anticancer Research*, 2018. **38**(4): pp. 2119–2125.
79. Siddamurthi, S., et al., Anthraquinone: A promising scaffold for the discovery and development of therapeutic agents in cancer therapy. *Future Medicinal Chemistry*, 2020. **12**(11): pp. 1037–1069.
80. Paarakh, P.M., et al., In vitro antiproliferative and in silico acytivity of Rubiadin isolated from roots of Rubia cordifolia. *Mintage Journal of Pharmaceutical and Medical Sciences*, 2016. **5**(1): pp. 20–23.
81. Ren, L., et al., Chrysophanol inhibits proliferation and induces apoptosis through NF-κB/cyclin D1 and NF-κB/Bcl-2 signaling cascade in breast cancer cell lines. *Molecular Medicine Reports*, 2018. **17**(3): pp. 4376–4382.
82. Wijesekara, I., et al., Physcion from marine-derived fungus *Microsporum* sp. induces apoptosis in human cervical carcinoma HeLa cells. *Microbiological Research*, 2014. **169**(4): pp. 255–261.
83. Guo, J.-M., et al., Anticancer effect of aloe-emodin on cervical cancer cells involves G2/M arrest and induction of differentiation. *Acta Pharmacologica Sinica*, 2007. **28**(12): pp. 1991–1995.
84. Makena, P.S. and K.-T. Chung, Effects of various plant polyphenols on bladder carcinogen benzidine-induced mutagenicity. *Food and Chemical Toxicology*, 2007. **45**(10): pp. 1899–1909.
85. Sáez, V., et al., Oligostilbenoids in *Vitis vinifera* L. Pinot Noir grape cane extract: Isolation, characterization, *in vitro* antioxidant capacity and anti-proliferative effect on cancer cells. *Food Chemistry*, 2018. **265**: pp. 101–110.
86. Banik, K., et al., Piceatannol: A natural stilbene for the prevention and treatment of cancer. *Pharmacological Research*, 2020. **153**: p. 104635.
87. Su, C., W. Wang, and C. Wang, IGF-1-induced MMP-11 expression promotes the proliferation and invasion of gastric cancer cells through the JAK1/STAT3 signaling pathway. *Oncology Letters*, 2018. **15**(5): pp. 7000–7006.
88. Kozlowska, A. and D. Szostak-Wegierek, Flavonoids-food sources and health benefits. *Roczniki Państwowego Zakładu Higieny*, 2014. **65**(2): pp: 79–85.
89. Wang, S.-F., et al., Autophagy modulators from traditional Chinese medicine: Mechanisms and therapeutic potentials for cancer and neurodegenerative diseases. *Journal of Ethnopharmacology*, 2016. **194**: pp. 861–876.
90. Chahar, M.K., et al., Flavonoids: A versatile source of anticancer drugs. *Pharmacognosy Reviews*, 2011. **5**(9): p. 1.
91. Fang, D., et al., Chemopreventive mechanisms of galangin against hepatocellular carcinoma: A review. *Biomedicine & Pharmacotherapy*, 2019. **109**: pp. 2054–2061.
92. Liao, C.-Y., et al., Novel investigations of flavonoids as chemopreventive agents for hepatocellular carcinoma. *BioMed Research International*, 2015. **2015**(1): p. 840542.
93. Singh, P., et al., Acute exposure of apigenin induces hepatotoxicity in Swiss mice. *PLoS One*, 2012. **7**(2): p. e31964.

94. Xu, M., et al., Apigenin suppresses colorectal cancer cell proliferation, migration and invasion via inhibition of the Wnt/β-catenin signaling pathway. *Oncology Letters*, 2016. **11**(5): pp. 3075–3080.
95. Wang, W., et al., Apigenin induced apoptosis by downregulating Sulfiredoxin expression in cutaneous squamous cell carcinoma. *Oxidative Medicine and Cellular Longevity*, 2022. **2022**(1): p. 8172866.
96. Fu, Y., et al., Preparative separation of vitexin and isovitexin from pigeonpea extracts with macroporous resins. *Journal of Chromatography A*, 2007. **1139**(2): pp. 206–213.
97. Zhang, Y., et al., Determination of flavone C-glucosides in antioxidant of bamboo leaves (AOB) fortified foods by reversed-phase high-performance liquid chromatography with ultraviolet diode array detection. *Journal of Chromatography A*, 2005. **1065**(2): pp. 177–185.
98. Ganesan, K. and B. Xu, Molecular targets of vitexin and isovitexin in cancer therapy: A critical review. *Annals of the New York Academy of Sciences*, 2017. **1401**(1): pp. 102–113.
99. Ninfali, P., et al., Characterization and biological activity of the main flavonoids from Swiss Chard (*Beta vulgaris* subspecies *cycla*). *Phytomedicine*, 2007. **14**(2–3): pp. 216–221.
100. Zu, Y.-g., et al., Chemical composition of the SFE-CO_2 extracts from *Cajanus cajan* (L.) Huth and their antimicrobial activity *in vitro* and *in vivo*. *Phytomedicine*, 2010. **17**(14): pp. 1095–1101.
101. Abdelhady, M.I. and A.A. Motaal, A cytotoxic C-glycosylated derivative of apigenin from the leaves of *Ocimum basilicum* var. *thyrsiflorum*. *Revista Brasileira de Farmacognosia*, 2016. **26**(6): pp. 763–766.
102. Lee, C.-Y., et al., Apoptosis triggered by vitexin in U937 human leukemia cells via a mitochondrial signaling pathway. *Oncology Reports*, 2012. **28**(5): pp. 1883–1888.
103. da Silva, D.A., et al., Antiproliferative activity of *Luehea candicans* Mart. et Zucc. (Tiliaceae). Natural Product Research, 2012. **26**(4): pp. 364–369.
104. Asgharian, P., et al., Potential mechanisms of quercetin in cancer prevention: Focus on cellular and molecular targets. *Cancer Cell International*, 2022. **22**(1): p. 257.
105. Khorsandi, L., et al., Quercetin induces apoptosis and necroptosis in MCF-7 breast cancer cells. *Bratisl Lek Listy*, 2017. **118**(2): pp. 123–128.
106. Barker, D., *Lignans*. 2019, MDPI. Basel, Switzerland. p. 1424.
107. Zhang, J., et al., New lignans and their biological activities. *Chemistry & Biodiversity*, 2014. **11**(1): pp. 1–54.
108. Rao, L., et al., Lignans and neolignans with antioxidant and human cancer cell proliferation inhibitory activities from *Cinnamomum bejolghota* confirm its functional food property. *Journal of Agricultural and Food Chemistry*, 2020. **68**(33): pp. 8825–8835.
109. Jiang, Z.H., et al., Cytotoxic dihydrobenzofuran neolignans from *Mappianthus iodoies*. *Bioorganic Chemistry*, 2017. **75**: pp. 260–264.
110. Tran, T.D., et al., Lignans from the Australian endemic plant *Austrobaileya scandens*. *Journal of Natural Products*, 2016. **79**(6): pp. 1514–1523.
111. Gontijo, V.S., M.H. Dos Santos, and C. Viegas Jr, Biological and chemical aspects of natural biflavonoids from plants: A brief review. *Mini Reviews in Medicinal Chemistry*, 2017. **17**(10): pp. 834–862.
112. Goossens, J.-F., L. Goossens, and C. Bailly, Hinokiflavone and related C–O–C-type biflavonoids as anti-cancer compounds: Properties and mechanism of action. *Natural Products and Bioprospecting*, 2021. **11**(4): pp. 365–377.
113. Zhou, J., et al., Antitumor activity in colorectal cancer induced by hinokiflavone. *Journal of Gastroenterology and Hepatology*, 2019. **34**(9): pp. 1571–1580.
114. Yao, W., et al., Delicaflavone induces ROS-mediated apoptosis and inhibits PI3K/AKT/mTOR and Ras/MEK/Erk signaling pathways in colorectal cancer cells. *Biochemical Pharmacology*, 2020. **171**: p. 113680.
115. Yao, W., et al., Delicaflavone induces apoptosis via mitochondrial pathway accompanying G2/M cycle arrest and inhibition of MAPK signaling cascades in cervical cancer HeLa cells. *Phytomedicine*, 2019. **62**: p. 152973.

4 Terpenes and Their Role in Pharmaceuticals

Gonçalo P. Rosa[1], *Wilson R. Tavares*[1], *Maria Carmo Barreto*[1], *and Ana M. L. Seca*[1,2]

[1]University of Azores, Faculty of Sciences and Technology, Centre for Ecology, Evolution and Environmental Changes (cE3c), Azorean Biodiversity Group & Global Change and Sustainability Institute (CHANGE), Ponta Delgada, Portugal

[2]Associated Laboratory for Green Chemistry (LAQV) of the Network of Chemistry and Technology (REQUIMTE), University of Aveiro, Aveiro, Portugal

4.1 INTRODUCTION

4.1.1 Overview

The term "terpene" has its origin in the expression "*Balsamum terebinthinae,*" which refers to the resin extracted from species of the Pinaceae family, commonly known as turpentine. This resin is a complex mixture of structurally similar compounds collectively termed terpenes. Terpenes, also referred to as isoprenoids and terpenoids, represent a diverse family of specialized metabolites derived from secondary metabolism. These compounds primarily consist of carbon and hydrogen (*e.g.* pinene, abietane and squalene, Figure 4.1), but also include oxygen (*e.g.* carvacrol, thymol, abietic acid, phytol, triptolide, betulinic acid and glycyrrhizin, Figure 4.1) and, more rarely nitrogen, such as in terpenoid-alkaloids commipholactam A and buxrugulines H (Figure 4.1) (Xie et al. 2024; Luo et al. 2024; Wang et al. 2024b). Some authors (Boncan et al. 2020; Masyita et al. 2022) consider that the term terpenoids should apply only to functionalized terpene structures as opposed to the term terpenes, which would include only the basic structures made up solely of carbon and hydrogen. However, in general and in this chapter, these two terms are interchangeable.

They are mainly found in terrestrial and marine plant species (Pichersky & Raguso 2018; Martignago et al. 2023; Xie et al. 2024), as well as in microorganisms (Ma et al. 2024) and animals (Tholl et al. 2023). For example, the *Inula* and *Artemisia* genera are very rich in sesquiterpene lactones (Tavares & Seca, 2019; Trendafilova et al. 2021) while red algae produce halogenated linear terpenoids (Steele et al. 2024), and the less uncommon sesterterpene ophiobolin A (Figure 4.1) is produce by the fungal genus *Bipolaris* (Evidente 2024).

4.1.2 Biosynthesis of Terpenes

This family of organic compounds is renowned for its structural diversity, with estimates of 50,000–90,000 identified terpenes, depending on the bibliographic source (Christianson, 2017; Sorokina et al. 2021; Hosseini & Pereira, 2023, Ma et al. 2024). However, this variability does not stem from the structural diversity of their precursors. In fact, the biosynthesis of all terpenes is initiated from two activated isomeric molecules with a C_5H_8 skeleton (C_5 isoprene unit), the IPP (isopentenyl diphosphate) and the DMAPP (dimethylallyl diphosphate) molecules (Chen et al. 2024). These two molecules are originated by two separate paths: (i) the mevalonic acid pathway (MVA), which takes

 DOI: 10.1201/9781003518358-4

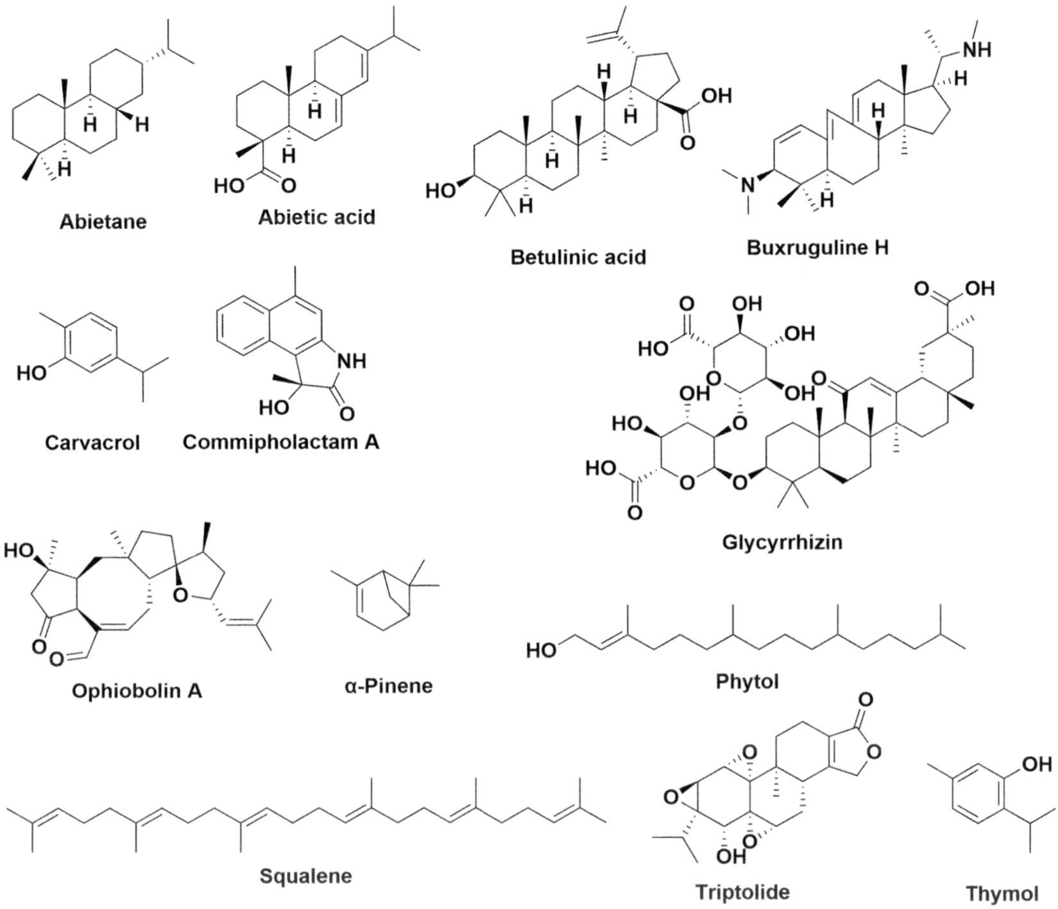

FIGURE 4.1 Chemical structure of some terpenes' examples. (The figure was prepared by the authors using the Licenced Chemdraw 22.0.0.)

place in plant cytosols, fungi, archaea, some bacteria, and animals, and has as its initial step the merging of two acetyl-CoA molecules; and (ii) the methyl-erythritol-phosphate (MEP or DOXP) pathway, which occurs in green algae, plant plastids, and most bacteria, which begins with the merging of G3P (glyceraldehyde-3-phosphate) and the pyruvate anion (Ninkuu et al. 2021; Tian et al. 2022; Xie et al. 2024). The precursors DMAPP and IPP are thus synthesized by these distinct pathways, after a different number of steps and by the catalytic action of different enzymes. Furthermore, IPP and DMAPP (Figure 4.2) are enzymatically interconvertible.

The next step in terpene biosynthesis consists of a *head-to-tail* condensation, *i.e.*, C-1 of the DMAPP unit to C-4 of the IPP unit, a condensation that is catalyzed by *trans*-isoprenyl diphosphate synthase (a terpene synthase enzyme (PTs)), producing linear and non-chiral geranyl diphosphate (GPP, C_{10}), an intermediate to all monoterpenes (Figure 4.2) (Zhou & Pichersky 2020). GPP is the subtract of all monoterpene synthases and to other PTs enzymes, such as FPP-synthases, which, by the condensation of more IPP units, will originate the carbon chain extension intermediary FPP (farnesyl diphosphate), the base molecule of all sesquiterpenes (C_{15}) (Figure 4.2), and the geranylgeranyl diphosphate (GGPP), the intermediary of all diterpenes (C_{20}) (Figure 4.2) (Nagel et al. 2019; Zhou & Pichersky 2020; Ninkuu et al. 2021; Masyita et al. 2022; Ma et al. 2024; Luo et al. 2024). The geranylfarnesyl diphosphate (GFPP), which also results from a *head-to-tail* linkage between GPP and FPP molecules, is the activated C_{25} unity intermediate of sesterterpenes (Figure 4.2),

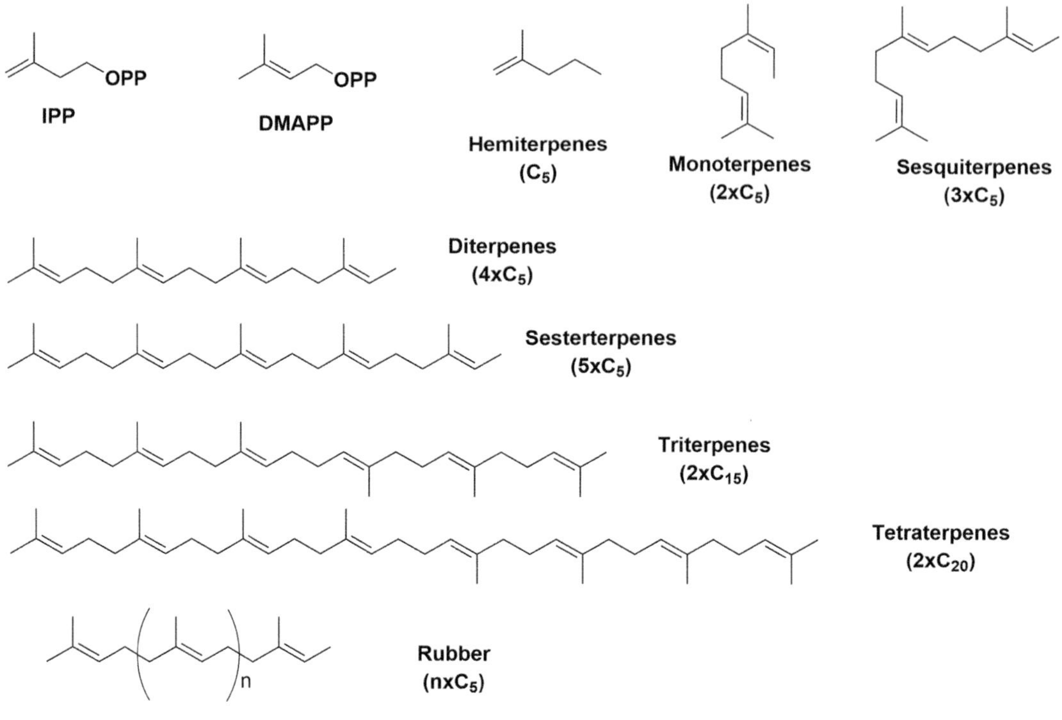

FIGURE 4.2 Chemical structures of precursors and intermediates of terpenes biosynthesis. (The figure was prepared by the authors using the Licenced Chemdraw 22.0.0.)

a lesser-known and less numerous class of terpenes, and whose GFPP synthase was more recently identified in plants such as *Leucosceptrum canum* and *Arabidopsis* sp, in fungi (*e.g. Aspergillus* sp.), and marine sponges (*e.g. Luffariella variabilis*) (Zhou and Pichersky 2020; Li and Gustafson 2021).

Additionally, the presence of squalene synthase (SQS) enzyme catalyzes a C-1 to C-1 linkage between two FPP units (a *head-to-head* condensation), giving rise to squalene (C_{30}), the base molecule of all triterpenes (Figure 4.2), while two GGPP units, through the action of another terpene synthase enzyme, phytoene synthase (PSY), can be condensed, also by a 1→1 linkage, to form tetraterpenes (C_{40}) (Figure 4.2), the intermediate of all carotenoids, and even more complex terpenes such as polyterpenes ($(C_5)_n$ with $n \geq 9$, such as natural rubber) (Figure 4.2) (Abdallah & Quax 2017; Boncan et al. 2020; Zhou et al. 2022; Yu et al. 2023).

Terpenoid alkaloids (*e.g.* commipholactam A, Figure 4.1) also originate from one of the terpene intermediates, IPP, DMAPP, GPP, FPP or GGPP, where a nitrogen atom incorporation, in the form of methylamine, ethylamine, or aminoethanol can take place at several stages of the process, *i.e.* before, after and during the cyclization phase (Cherney & Baran 2011; Wang et al. 2024a; Wang et al. 2024b).

The biosynthetic pathways mentioned above originate from terpenes in which the carbon chain has five carbons or multiples thereof, which are called canonical terpenes. However, several terpenes described in the literature do not follow this "rule of 5." They are called noncanonical terpenoids, which are biosynthesized mainly by microbial sources by several mechanisms, such as chain degradation of canonical terpenes (*e.g.* TMTT (4,8,12-trimethyltrideca-1,3,7,11-tetraene), (Figure 4.3), biosynthetic pathways using alternative substrates (*e.g.* propionyl-CoA in lepidopteran mevalonate pathway), the existence of a large number of specific enzymes (*e.g.* geranyllinalool synthase or 2-methyl-GPP transferase) (Zhou & Pichersky 2020; Peng et al. 2023; Ma et al. 2024).

The terpenoid alkaloids and noncanonical terpenes contribute substantially to the structural diversity of the terpene family.

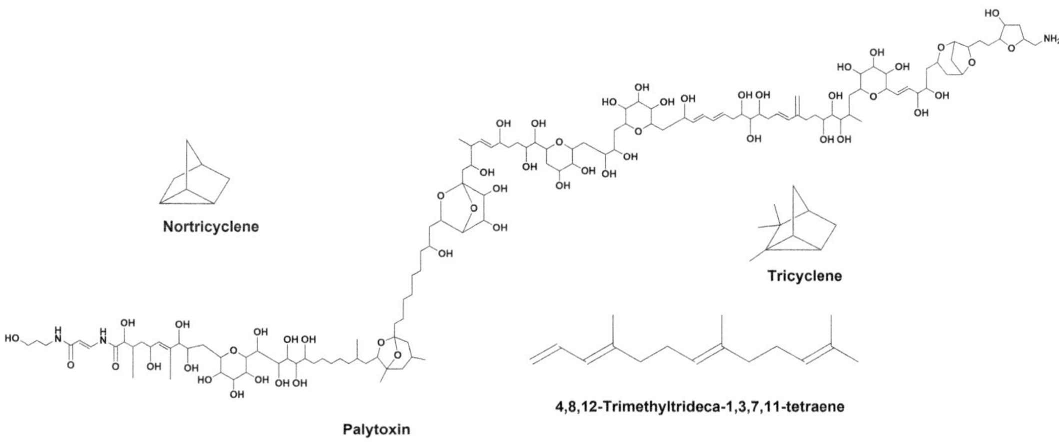

FIGURE 4.3 Chemical structures of canonical and noncanonical terpenes. (The figure was prepared by the authors using the Licenced Chemdraw 22.0.0.)

The third most significant step in terpene synthesis involves structural modifications introduced into the basic linear chain of each terpene class: hemiterpenes, monoterpenes, sesquiterpenes, diterpenes, sesterterpenes, triterpenes, and tetraterpenes. These structural changes mainly include multiple and stereoselective cyclizations of acyclic terpenes, catalyzed through the action of numerous terpenoid cyclase enzymes. After the formation of an atypical carbocationic intermediary, there is an almost stochastic reorganization of chemical bonds, originating a multitude of products with variable numbers of fused rings and chiral centers from the one substrate (Christianson 2017; Luo et al. 2024).

The initial acyclic structures (Figure 4.2) and the generated cyclic ones can also be modified by the action of a broad spectrum of other types of enzymes, which are effective on multiple substrates, such as terpene synthase and cytochrome P450 monooxygenases. They are able to cause a cascade of structural rearrangements and functionalization of carbon chains, such as migration of methyl groups, intra-reactions, methylation, acetylation, hydrogenation, oxygenation, and dehydrogenation (Rudolf & Chang 2020; Zhou & Pichersky 2020; Luo et al. 2024; Xie et al. 2024).

It is also worth mentioning that each species has an immense pool of enzymes, some of which are common to other species, but also has the ability to express specific enzymes depending on the phase of the life cycle and its evolutionary and physiological states, which further contributes to the production of terpenes with remarkably novel and unique structures in nature. Additionally, the peculiarities of terpene biosynthesis have given rise to terpenes that are specific to a species or genus; therefore, several groups of terpenes function as taxonomic markers (Mitić et al. 2017; Malhocká & Švábová 2023; Lin et al. 2024).

As a consequence of biosynthesis, the molecular mass of the natural terpenes identified to date varies from 60 Da, as is the case of hemiterpene tricyclene and the noncanonical terpene nortricyclene (Figure 4.3), reaching up to values of 2,680 Da, as in the case of meroterpene glycosides palytoxin (Figure 4.3), a potent non-peptide toxin identified in dinoflagellates and *Palythoa* soft corals, formed through processes involving terpenoid and polyketide pathways (Hosseini & Pereira 2023). However, 2,680 may be a value surpassed in the future in view of the continuous optimization of experimental techniques.

Terpene volatility also varies within the terpene family. The volatility of terpenes decreases as the size of the carbon chain and molecular mass of the terpenes increase. Thus, terpenes with a molecular mass of approximately up to 300 Da, which mainly include hemiterpenes and monoterpenes, but also some sesquiterpenes and diterpenes with lower molecular mass, constitute the most volatile fraction of terpenes, also known as essential oils, which means they have higher vapor pressure (Rosenkranz et al. 2021). In contrast, non-volatile terpenes such as tetraterpenes, triterpenes, highly functionalized diterpenes, and sesquiterpenes are solids or oils at room temperature. Terpenes are lipophilic compounds that arise from their biosynthesis through the isoprenoid pathway, and their lipophilicity is highest for triterpenes and tetraterpenes. Therefore, they are poorly soluble in water and are mainly transported in the glycosylated form.

4.1.3 Ecological and Physiological Roles of Terpenes

Terpenes are vital specialized metabolites with significant ecological roles that enable species to react dynamically to the environment and adapt to biotic and abiotic stresses. These compounds are relevant in primary metabolism (*e.g.* intervene in the biosynthesis of chlorophyll) and serve as chemical defense against a variety of herbivores and pathogens. For example, the elevated cannabinoid content in *Cannabis sativa* enhances resistance to mites, while *Artemisia annua* produces artemisinin (Figure 4.4) to increase resistance to pathogenic bacteria and fungi and aphid attacks (Ninkuu et al. 2021; Wang et al. 2023). They also control species development through the production of phytohormones (*e.g.* gibberellic acid, Figure 4.4) and reproduction by attracting pollinators with their distinct aromatic profiles. Moreover, they participate in allelopathic interactions, inhibiting the growth of seeds, competing species, and soil bacteria (Yazaki et al. 2017; Boncan et al. 2020; Ninkuu et al. 2021; Castro-Camba et al. 2022; Wang et al. 2023). Some species use an intracellular signal transduction mediated by terpenes, changing their synthesis and/or accumulation to trigger responses that increase abiotic stress tolerance (Boncan et al. 2020; Rosenkranz et al. 2021; Zhan et al. 2022; Lo et al. 2024). For example, the terpene content gets bigger in *Ginko biloba* after UV and cold stresses, while prolonged salt stress causes a decrease in the artemisinin (Figure 4.4) content in *Artemisia annua* (Wang et al. 2023). This diverse functionality highlights the evolutionary importance of terpenes in preserving species vitality and fostering an ecological equilibrium.

4.1.4 Terpenes Applications

In addition to the crucial physiologic and ecological roles of terpenes, they have captured human interest since ancient times due to their extensive use as perfume (*e.g.* limonene, geraniol and pinene), flavoring (*e.g.* menthol and limonene), coloring (*e.g.* astaxanthin, carotene and lycopene), preservative (*e.g.* eugenol), natural pesticide (*e.g.* azadirachtin and pyrethrin) (Figure 4.5) (Tetali 2019; Kim et al. 2020; Masyita et al. 2022), with particular emphasis on their use as medicines (Chen et al. 2021; Weston-Green et al. 2021; Wang et al. 2023; Ekasari et al. 2023; Liu et al. 2024).

Several terpenes have already played a widely recognized role in the development and introduction of new pharmaceuticals and are well known for their ability to interact with specific cellular targets, including enzymes, receptors, and membranes. These interactions facilitate terpenes to modulate biochemical pathways, exert cytotoxic effects on cancer cells, inhibit microbial growth, or reduce inflammation (Câmara et al. 2024). Some of these compounds act through previously unknown, innovative, and efficient mechanisms of action. For instance, artemisinin (Figure 4.4)

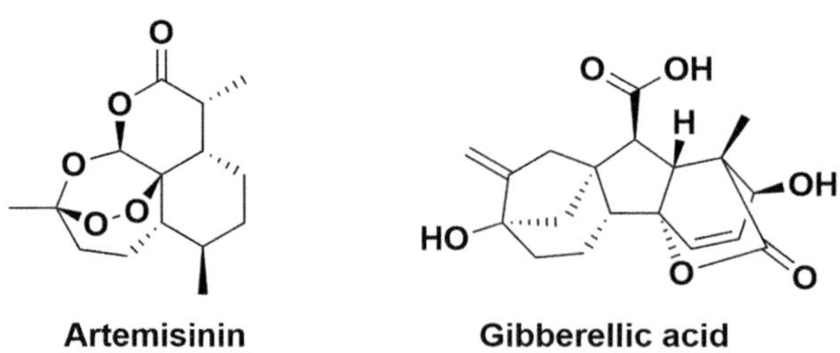

FIGURE 4.4 Two terpenes with ecological and physiological roles. (The figure was prepared by the authors using the Licenced Chemdraw 22.0.0.)

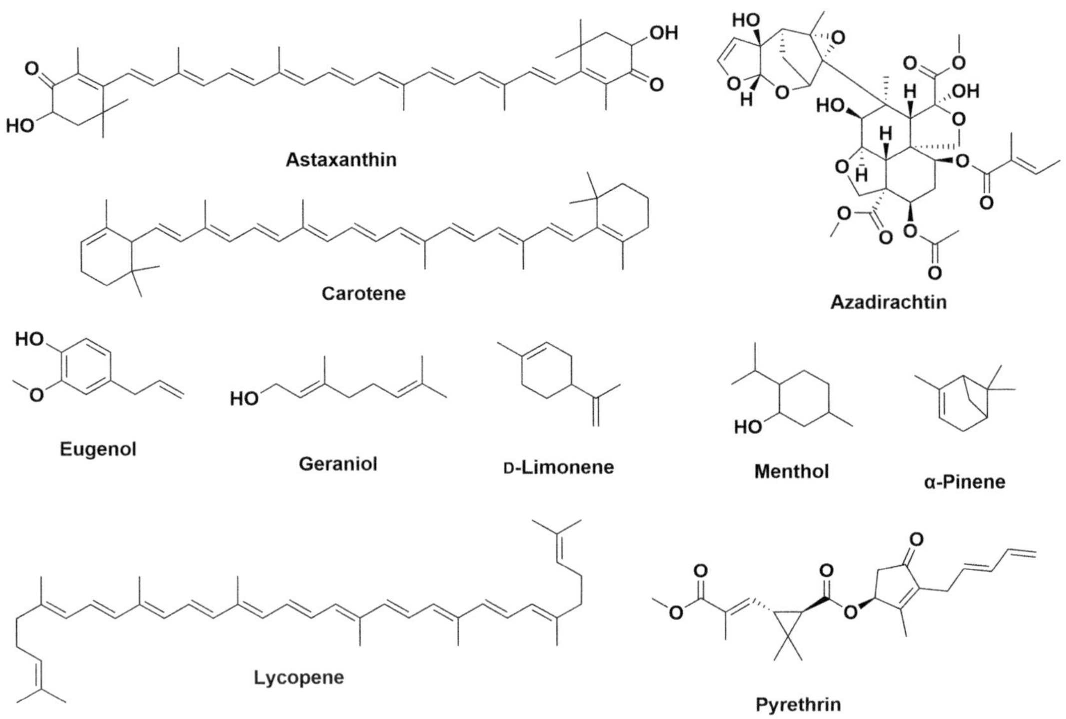

FIGURE 4.5 Chemical structures of terpenes with commercial applications. (The figure was prepared by the authors using the Licenced Chemdraw 22.0.0.)

exerts its antimalarial effects through interactions with iron to produce reactive oxygen species (ROS), damaging the parasite's cellular integrity (Siqueira-Neto et al. 2023), while the anticancer efficacy of paclitaxel (Figure 4.6) stems from its ability to stabilize microtubules, preventing cell division (Zhang et al. 2023). These two terpenes are among the most promising and effective agents for therapeutic applications and represent an important pillar in the development of innovative treatments. Other examples of terpenes with relevant roles in pharmaceuticals include cannabidiol (Figure 4.6) with analgesic effects (Secondulfo et al. 2024), and ginkgolide and tanshinone II (Figure 4.6) to improve the cardiovascular system (Sarkar et al. 2020; Ansari et al. 2021).

Concerning recent data on the role of terpenes as drugs, Yamaguchi et al. (2021), after evaluating 3,999 compounds that reached clinical trials in the United States, Europe, and Japan, estimated an approval rate of 12.8%, which is within the yearly average. As an example of drugs that achieved success, in 2023 the United States FDA approved 55 new molecules, 10 of which were natural products, comprising Filsufez™ a mixture of pentacyclic birch terpenes (Filsufez ™), namely 72%–88% of betulin, 2.4%–5.7% of lupeol, 2.6%–4.2% of betulinic acid (2.6%–4.2%), 0.5%–1.2% of erythrodiol, and 0.3%–0.8% of oleanolic acid (OA) (Figure 4.6), included in a topical gel designed for treating partial-thickness wounds in individuals with junctional and dystrophic epidermolysis bullosa (JEB and DEB), marking the first approved treatment for wounds related to JEB. Omaveloxolone (Skyclarys™) (Figure 4.6), also a terpene, has been approved for the treatment of Friedrich's ataxia (Torre & Albericio 2024). These authors consider triterpenes, together with steroids, to be among the most important small-molecule drugs that regularly achieve approval for pharmaceutical use. Considering that terpenes are the major group of specialized metabolites, with an extensive array of biological effects, they have a therapeutic potential that still has a lot to be discovered in the search for new medicines (Dash et al. 2022).

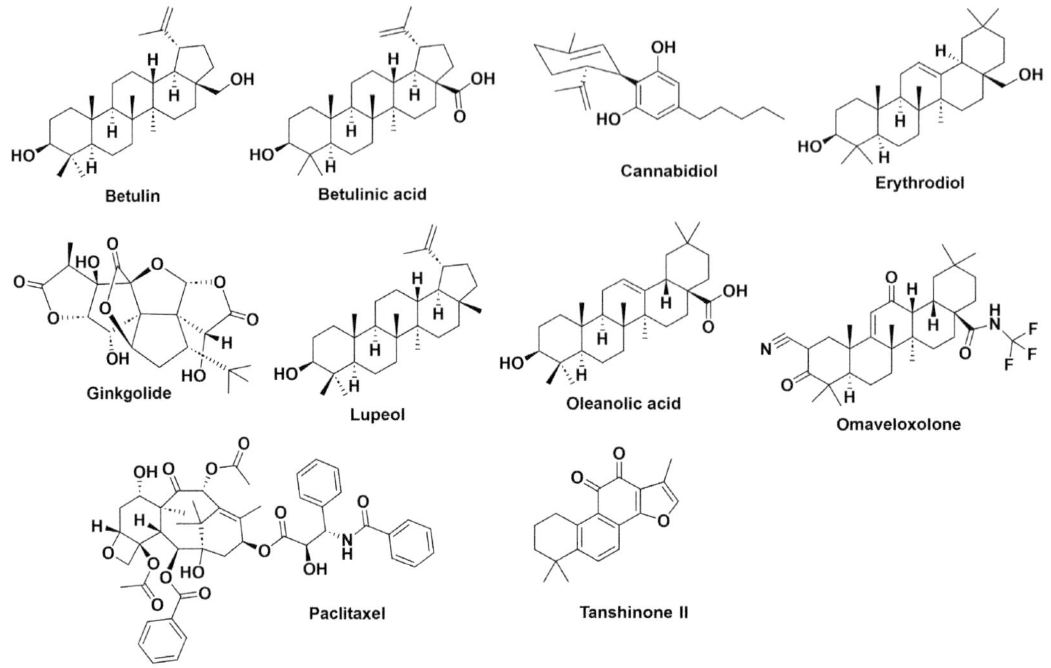

FIGURE 4.6 Chemical structures of terpenes with medicinal uses. (The figure was prepared by the authors using the Licenced Chemdraw 22.0.0.)

4.1.5 Need for New Pharmaceuticals

Although significant advances have been made in the field of modern therapeutic drugs, the need to develop new drugs is as relevant as ever due to the appearance of new diseases or the resurgence of diseases thought to be extinct, the increase in resistant pathogenic agents, the increase in pathologies related to sedentary lifestyles, environmental degradation, and population aging, and the persistence of diseases such as cancers and neurodegenerative diseases for which no satisfactory therapeutic answers have been found until the present moment.

Infectious diseases caused either by transmission from other animal hosts or by the re-emergence and/or geographical expansion of established diseases may have dramatic consequences, even leading to widespread morbidity and mortality, such as the recent COVID-19 pandemic (Morens & Fauci 2020). Infections caused by multidrug-resistant microbial pathogens, due to excessive and/or improper use of antibiotics, may reach an annual death rate of 10 million by 2050, more than the deaths caused by cancer (Samreen et al. 2021). In 2019, 4.95 million deaths associated with antimicrobial resistance (AMR) in 2019, including 1.27 million deaths attributable to bacterial AMR, were estimated by Murray et al. (2022), which is corroborated by the World Health Organization, estimates of annual deaths due to AMR as 1.3 million (WHO 2024). These infectious diseases pose a challenge in the search for therapeutic options that can fight pathogenic agents, particularly molecules with mechanisms of action that allow them to bypass resistance mechanisms.

According to the World Health Organization (WHO), from 2015 to 2050, the percentage of the world's population over 60 years of age will nearly double from 12% to 22% (WHO 2024). There is a clear relation between aging and metabolic dysfunction (Fountain et al. 2024), leading to the appearance of diseases such as those that are clustered into what is known as metabolic syndrome, such as glucose intolerance impairment, dyslipidemia, obesity, and high blood pressure (Nishikawa et al. 2021), and inflammatory processes, which are highly related to the onset of cardiovascular diseases (Liberale et al. 2022). Aging is also correlated with the appearance of neurodegenerative diseases such as Alzheimer's and related dementias (Breijyeh & Karaman 2020; Gonzales et al. 2022), and even with cancer (Montégut et al. 2024). Although there are several breakthroughs in the development of drugs to treat or at least delay the appearance of these pathologies associated with aging, their efficacy is insufficient, and many of these diseases are still responsible for high morbidity and mortality worldwide and often have negative side effects. Taking this into consideration, discovering new drugs with specificity, efficacy and low negative effects is mandatory.

4.2 PHARMACOLOGICAL PROPERTIES OF EMERGING TERPENES

Terpenes, as explored in the preceding section, represent one of the most thoroughly investigated groups of natural products. They are well-known for their wide-ranging structural variations and extensive biological effects. Each year, researchers uncover new terpenes, expanding the chemical repertoire and revealing potential applications across diverse biological fields. Their diverse molecular frameworks not only define their ecological roles but also enable them to interact with biological systems in highly specific ways. These interactions are often supported by their ability to selectively bind to proteins, modulate cellular signaling pathways, and alter membrane dynamics, making terpenes highly versatile pharmacological agents. Moreover, their generally favorable pharmacokinetic profiles, including good membrane permeability and low cytotoxicity, further increase their potential as therapeutic candidates.

Among their many applications, terpenes have demonstrated extraordinary promise in antimicrobial and anticancer therapies. Their facility to target bacterial membranes, disrupt quorum sensing, and inhibit critical enzymes has established them as potent antimicrobials (Kolesnikova et al. 2024; Wiart et al. 2023). Concurrently, many terpenes have been identified as modulators of apoptosis, angiogenesis, and oxidative stress, pivotal in anticancer strategies (Kamran et al. 2022). These dual roles highlight their therapeutic versatility and underscore the need for further research. This section focuses on terpenes that have shown great promise in *in vivo* studies published over the past 5 years, emerging as potential candidates for future therapeutic development for anticancer and antimicrobial activities.

4.2.1 Anticancer Activity

Terpenes have demonstrated significant promise as anticancer agents due to their diverse structures and mechanisms of action, offering a broad spectrum of therapeutic possibilities. These compounds have shown the ability to target multiple hallmarks of cancer, including promoting apoptosis, inhibiting cell proliferation, suppressing metastasis, and interfering with angiogenesis (Kamran et al. 2022). Moreover, terpenes display synergistic effects when used alongside conventional chemotherapeutic agents, enhancing efficacy and reducing side effects. Many terpenes have been shown to exhibit low toxicity *in vivo*, making them suitable candidates for integration into cancer treatment regimens (Kamran et al. 2022). This is particularly important in the context of developing new therapies that can be used in conjunction with existing treatments to enhance their efficacy while minimizing adverse effects. Consequently, this group of compounds continues to attract significant attention as potential candidates for innovative and effective anticancer therapies.

Recent research has shown the anticancer properties of different terpenes, especially concerning certain types of cancer. The findings from key studies published in the past 5 years are outlined in Table 4.1.

The results summarized in Table 4.1 demonstrate that terpenes have significant potential as anticancer agents via multiple mechanisms. Extensive *in vivo* studies have highlighted their ability to target cancer cell survival, induce apoptosis, modulate key signaling pathways, and enhance the efficacy of conventional chemotherapy.

TABLE 4.1
Summary of *in vivo* Mechanistic Results for Anticancer Activity of Selected Terpenes

Compound	Model	Mechanisms of Action	References
Abietic acid	Skin cancer Lung cancer	Modulates IKKβ/NF-κB signaling, crucial for tumor cell survival and proliferation.	Liu et al. (2019)
Betulinic acid	Hepatocellular carcinoma Lung cancer Ovarian cancer	Induces mitochondrial apoptosis and inhibits angiogenesis. Induces autophagy	Jiang et al. (2021c)
Acetyl-11-keto-β-boswellic acid	Glioblastoma Colorectal cancer Pancreatic cancer Hepatocellular carcinoma	Promotes cell-cycle arrest by inhibiting cyclin-dependent kinases, enhances chemotherapy effects.	Schmiech et al. (2021); Elgendy et al. (2019); Lv et al. (2020)
Carvacrol	Melanoma metastasis Colorectal cancer Breast cancer Sarcoma Gastric cancer	Induces apoptosis via caspase activation and ROS generation; inhibits angiogenesis and cell invasion.	Chiu et al. (2023); Fatima et al. (2022); Rojas-Armas et al. (2020); Gunes-Bayır et al.(2022); Sampaio et al. (2021); Ghorbanzadeh et al. (2022)
Carnosol	Prostate cancer Skin cancer	Modulates ROS levels and regulates STAT3 signaling pathway	Jiang et al. (2021b); Alsamri et al. (2021); Chan et al. (2022); O'Neill et al. (2020)
Cedrol	Glioblastoma	Inhibits angiogenesis and induces apoptosis via VEGFR2 and MAPK signaling pathways.	Yun et al. (2022); Chien et al. (2022)
D-Limonene	Breast cancer Lung cancer	Modulates Akt signaling, inhibits angiogenesis and metastasis, regulates lipid metabolism.	Zhou et al. (2021); Mandal et al. (2022); Zhu et al. (2021, 2022); Chebet et al. (2021)
Farnesol	Colorectal cancer Lung cancer	Induces apoptosis and inhibits inflammatory pathways.	Ali et al. (2024)
Oleanolic acid	Colon cancer	Induces apoptosis, arrests cell cycle via PI3K/Akt/mTOR pathway, inhibits EMT.	Wang et al. (2019); Kaushik et al. (2021)
Perillyl alcohol	Breast cancer Pancreatic cancer Liver cancer	Inhibits farnesyl transferase, sensitizes tumors to radiation and chemotherapy.	Chebet et al. (2021)
Phytol	Breast cancer	Reduces cell migration and induces apoptosis.	Alencar et al. (2019)
Rosmanol	Colorectal carcinoma Breast cancer	Enhances chemotherapeutic efficacy by modulating cellular stress responses. Induction of apoptosis by PI3K/AKT and STAT3/JAK2 signaling	Chan et al. (2022); Jiang et al. (2021a)
Thymoquinone	Pancreatic cancer xenograft	Inhibits NF-κB signaling, promotes apoptosis, and modulates oxidative stress pathways.	Almajali et al. (2021)

(*Continued*)

TABLE 4.1 (*Continued*)
Summary of *in vivo* Mechanistic Results for Anticancer Activity of Selected Terpenes

Compound	Model	Mechanisms of Action	References
Ursolic acid	Colorectal cancer Breast cancer	Targets glycolysis, NF-κB, and JNK pathways; enhances chemotherapy outcomes.	Wang et al. (2021b); Zhao (2023)
β-Elemene	Lung cancer Gastric cancer Ovarian cancer	Induces cell cycle arrest and apoptosis by downregulating CDK2, CDK4, and Cyclin D3.	Zhang et al. (2020); Feng et al. (2022); Wang et al. (2021a)

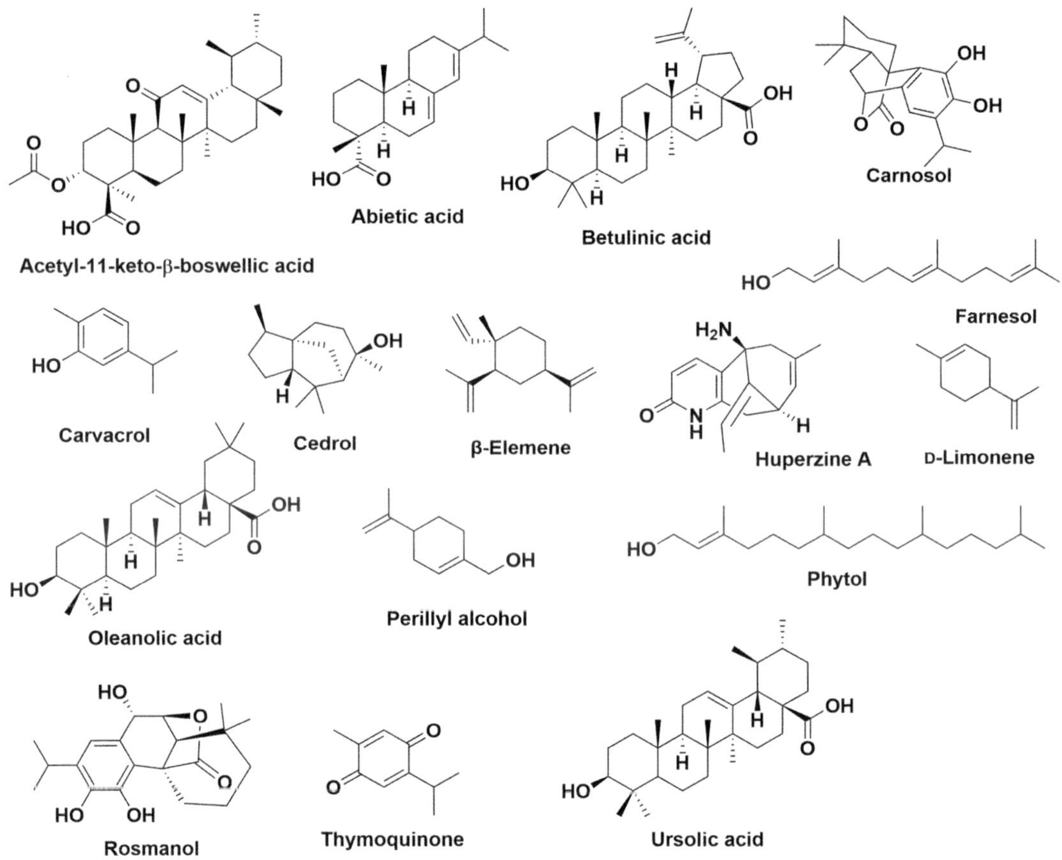

FIGURE 4.7 Chemical structure of terpenes with *in vivo* anticancer activity. (The figure was prepared by the authors using the Licenced Chemdraw 22.0.0.)

Diterpenoids such as abietic acid and carnosol (Figure 4.7) have shown notable anticancer activities. Abietic acid, mainly extracted from pine resin, shows considerable effectiveness against skin cancer and non-small-cell lung cancer (NSCLC) by modulation of the IKKβ/NF-κB signaling pathway, essential for the survival and growth of tumor cells (Liu et al. 2019).

Similarly, carnosol, a diterpenoid abundant in rosemary, has been observed to reduce tumor multiplicity and restrict the invasion of cancer cells in skin and breast cancer models by regulating ROS levels and regulating the STAT3 pathway (O'Neill et al. 2020; Jiang et al. 2021b; Alsamri et al. 2021; Chan et al. 2022).

Among the monoterpenoids, carvacrol and D-limonene (Figure 4.7) are notable for their chemopreventive effects. Carvacrol, a phenolic compound classified as a monoterpenoid, is predominantly found in oregano and thyme and has gained considerable interest due to its anticancer effects observed in various animal studies. A multitude of research has shown its ability to inhibit tumor growth and trigger apoptosis across several cancer types. For example, it was noted that the administration of carvacrol significantly decreased tumor incidence and the count of colonic aberrant crypt foci in male rats that were exposed to 1,2-dimethylhydrazine (DMH)-induced colon carcinogenesis, highlighting its protective role against the onset of colon cancer (Chiu et al. 2023).

In a study by Rojas-Armas et al. (2020), it was noted that carvacrol demonstrated antitumor properties in a breast cancer model induced by 7,12-dimethylbenzaanthracene (DMBA) in female rats, further indicating its promise as a chemopreventive agent (Rojas-Armas et al. 2020). Additionally, carvacrol effectively slowed the growth of Sarcoma-180 cells *in vivo*, achieving over 70% reduction in tumor growth at elevated doses, which underscores its strong antitumor effects (Fatima et al. 2022). Carvacrol also reduced gastric carcinogenesis caused by N-methyl-N′-nitro-N-nitrosoguanidine (MNNG) in Wistar rats, implying that its antioxidant properties play a role in its cancer-protective effects (Gunes-Bayır et al. 2022). Furthermore, research has shown that carvacrol can trigger apoptosis by activating caspases and producing ROS, which are crucial for initiating cell death in cancerous cells (Sampaio et al. 2021).

D-limonene can arrest the cell cycle in the G2/M phase by inhibiting cyclin B1/CDK1, a critical mechanism for tumor regression, by modulating key signaling pathways such as the Akt pathway in colon cancer cells (Mandal et al. 2022). Moreover, research involving animals has demonstrated that D-limonene can significantly slow down the development of mammary tumors, indicating its promise as a potential chemopreventive substance for breast cancer (Zhou et al. 2021). Aside from its ability to induce apoptosis, D-limonene has been found to hinder both angiogenesis and metastasis in various human cancer cell lines, further reinforcing its function as an antitumor agent (Zhou et al. 2021). Its ability to modulate lipid metabolism has also been highlighted, with some works indicating that D-limonene can regulate lipid droplet accumulation in lung cancer cells, thereby inhibiting tumor aggressiveness (Zhou et al. 2021; Zhu et al. 2022). The safety profile of D-limonene is also noteworthy, as it has been reported to be well tolerated in animal studies, with minimal adverse effects observed during treatment (Chebet et al. 2021).

Triterpenoids, including OA and ursolic acid (UA) (Figure 4.7), have received significant attention due to their diverse mechanisms for combating cancer. OA functions by triggering apoptosis and halting the cell cycle via the PI3K/Akt/mTOR pathway. Additionally, it prevents epithelial-mesenchymal transition (EMT), a crucial process for cancer spread, and works synergistically with traditional chemotherapy agents (Wang et al. 2019). This inhibition is thought to be facilitated by the enhancement of nitric oxide synthase dimerization, which is involved in cellular signaling pathways that control cell movement and invasion (Wang et al. 2019). Furthermore, OA's anti-inflammatory effects add to its anticancer capabilities, since inflammation is a recognized promoter of tumor development (Kaushik et al. 2021).

UA enhances chemotherapy outcomes by targeting glycolytic metabolism and pathways such as NF-κB and JNK, positioning it as a valuable compound for combination therapies against cancers like breast and colorectal malignancies (Wang et al. 2021b; Zhao, 2023).

In vivo research has shown that acetyl-11-keto-β-boswellic acid (AKBA) (Figure 4.7) successfully inhibits tumor progression in colorectal and pancreatic cancer models, emphasizing its potential as a therapeutic option (Schmiech et al. 2021).

Furthermore, this compound was shown to enhance the efficacy of conventional chemotherapeutics, such as cisplatin, by promoting synergistic effects against hepatocellular carcinoma in animal models (Elgendy et al. 2019). Additionally, AKBA also inhibits cyclin-dependent kinases, which induces cell-cycle arrest in tumor cells, showing its multifaceted role in cancer therapy (Elgendy et al. 2019; Lv et al. 2020).

Cedrol (Figure 4.7), a sesquiterpenoid extracted from *Juniperus* sp., inhibits angiogenesis and induces apoptosis by targeting VEGFR2 and MAPK signaling pathways. Its therapeutic potential has been demonstrated in glioblastoma and NSCLC models, where it enhances the efficacy of standard therapies (Yun et al. 2022; Chien et al. 2022).

Other notable terpenoids include perillyl alcohol (POH) and β-elemene (Figure 4.7). POH, a metabolite of D-limonene, shows efficacy against breast, pancreatic, and liver cancers by inhibiting farnesyl transferase and sensitizing tumors to radiation and chemotherapy (Chebet et al. 2021).

β-Elemene, originating from *Curcuma zedoaria*, has been shown to effectively inhibit tumor progression by triggering cell cycle arrest and apoptosis across various cancer types, such as lung, gastric, and ovarian cancers. For example, research indicates that it reduces the levels of cell cycle proteins like CDK2, CDK4, and Cyclin D3, thus preventing cell division (Zhang et al. 2020). Additionally, it possesses the capability to hinder tumor metastasis and enhance apoptotic pathways, emphasizing its potential as a treatment option for cancer (Feng et al. 2022), particularly due to its low toxicity profile, which further endorses its use in clinical applications (Wang et al. 2021a).

The diverse mechanisms and significant preclinical evidence supporting the anticancer activities of these compounds show their therapeutic potential. The encouraging findings suggest that terpenes could be important options for cancer treatment; however, additional clinical trials are needed to confirm their safety, effectiveness, and relevance for human populations.

4.2.2 Antibacterial Activity

The increasing prevalence of antibiotic resistance urges the exploration of alternative antimicrobial agents, with particular attention directed toward natural products. Terpenes have demonstrated substantial antibacterial potential, making them promising candidates for combating multidrug-resistant pathogens. Recent studies have showcased the ability of terpenes to disrupt microbial membranes, inhibit biofilm formation, and synergize with conventional antibiotics. The findings from the most relevant studies published in the last 5 years are summarized in Table 4.2, highlighting the potential of these compounds as alternative or adjunctive therapies in combating microbial infections.

TABLE 4.2
Summary of *in vivo* Mechanistic Results for Antibacterial Activity of Selected Terpenes

Compound	*In vivo* Model	Results (Comparison with Controls)	Mechanisms of Action	References
α-Pinene	Pneumonia (*Escherichia. coli*)	Bacterial count ↓60% (Enhanced ciprofloxacin by 25%).	Disruption of bacterial membranes, inhibition of biofilm formation.	Leite-Sampaio et al. (2022)
Bisabolol	*Clostridium difficile* colitis	Toxin production ↓70% (Comparable to fidaxomicin).	Inhibition of toxin production and biofilm formation, immunomodulation	Venkataraman et al. (2022)
Bornyl acetate	UTI (*E. coli*)	Bacterial load ↓65% (Improved nitrofurantoin efficacy).	Disruption of bacterial membranes, synergistic effects with citral.	Wang et al. (2021c)
Camphor	Skin infections (*Staphylococcus aureus*) *Salmonella cholerae*	Wound healing accelerated by 50%.	Disruption of bacterial membranes, inhibition of bacterial growth, modulation of gut microbiota.	Zou et al. (2022); Seidel et al. (2022); Fazmiya et al. (2022), Poudel et al. (2021)
Carvacrol	COPD and *Klebsiella pneumoniae* infections *Staphylococcus* sp.	Reduced inflammation, bacterial load ↓ (Enhanced survival rates).	Membrane disruption, reduction in ATP synthesis, increased antibiotic permeability, immune modulation.	Abed et al. (2021); Magi et al. (2015); Hosseini et al. (2023); Al-Tawalbeh et al. (2024)

(*Continued*)

TABLE 4.2 (*Continued*)
Summary of *in vivo* Mechanistic Results for Antibacterial Activity of Selected Terpenes

Compound	*In vivo* Model	Results (Comparison with Controls)	Mechanisms of Action	References
Citral	Chronic infections (MRSA) *E. coli* *Salmonela typhimurium* *Vibrio parahaemolyticus*	Reduced bacterial counts, ↓ inflammation (TNF-α), enhanced drug efficacy.	Membrane penetration, disruption of metabolic processes, inhibition of virulence factor expression.	Oliveira et al. (2021); Yoplac et al. (2021); Kim et al. (2023); Long et al. (2019)
Geraniol	Wound infection (*S. aureus*) MRSA	Bacterial load ↓3-log CFU (Comparable to mupirocin with better biofilm disruption). Inhibition of biofilm formation, enhanced drug efficacy	Membrane disruption, inhibition of biofilm formation.	Babaei et al. (2024) Gu et al. (2022)
Myrcene	Dental infections (*Streptococcus mutans*)	Growth ↓75% (Comparable to amoxicillin).	Membrane disruption, inhibition of biofilm formation.	Henley-Smith et al. (2024)
Nerol	*S. aureus*	Bacterial growth ↓75% (Comparable to chlorhexidine).	Disruption of bacterial membranes, increased permeability and cell death, broad-spectrum activity.	Shahdadi (2023)
Pulegone	Biofilm-related infections	Biofilm ↓85% (Enhanced effect of gentamicin).	Inhibition of biofilm formation and disruption of bacterial membranes.	Gong et al. (2021)
Sabinene	UTI models (*K. pneumoniae*)	CFU ↓60% in urine (Comparable to cephalexin).	Membrane disruption, inhibition of bacterial adherence, nhibition of biofilm formation	Ghazal et al. (2022)
Spathulenol	Bloodstream infections (MRSA)	CFU ↓80% (Enhanced daptomycin effects).	Disruption of bacterial membranes, synergistic effects with antibiotics.	Cazella et al. (2019)
Thymol	Murine model (*S. aureus*) Broiler chickens (*S. typhimurium*)	Bacterial count ↓	Disruption of bacterial cell membranes, increased permeability, modulation of gut microbiota.	Shanaida et al. (2021); Ibrahim et al. (2021); Yin et al. (2022); Valerio et al. (2021)
Thujone	*E. coli* *Salmonella spp.*	Bacterial count ↓	Disruption of bacterial cell membranes, increased permeability.	Zhang et al. (2021); Bekka-Hadji et al. (2022)
β-Caryophyllene	MRSA	Colonization ↓65% (Comparable to metronidazole).	Membrane disruption, modulation of inflammatory responses, enhanced antibiotic efficacy.	Zhang et al. (2021)
β-Selinene	Lung infections (*Pseudomonas aeruginosa*)	CFU ↓65% (Comparable to levofloxacin).	Disruption of bacterial membranes, inhibition of biofilm formation.	Chebbac et al. (2023)

↓ - Decrease

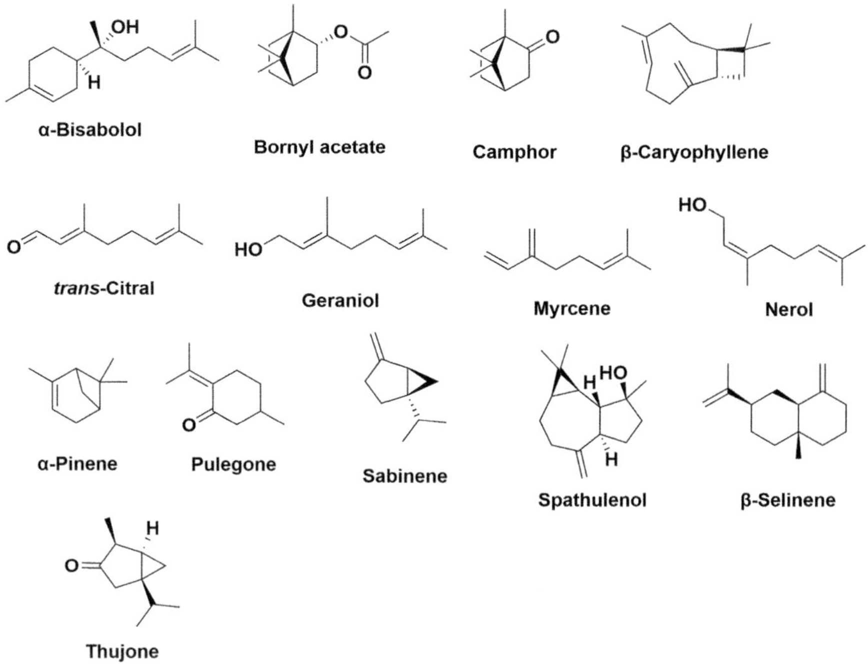

FIGURE 4.8 Chemical structure of terpenes with *in vivo* antimicrobial activity. (The figure was prepared by the authors using the Licenced Chemdraw 22.0.0.)

Thymol (Figure 4.1) has been shown to protect broiler chickens from *Salmonella typhimurium*, enhancing chickens' growth and modulating the gut microbiota (Ibrahim et al. 2021). In another study involving channel catfish, thymol was potent against drug-resistant *Streptococcus iniae*, with minimum inhibitory concentrations (MIC) as low as 128 μg/mL (Yin et al. 2022).

Valerio et al. (2021) reported significant antibacterial effects of thymol in a murine model infected with *Staphylococcus aureus*, with a notable decrease in bacterial counts. Additionally, research conducted by Shanaida et al. (2021) indicated that thymol exhibits enhanced antibacterial effectiveness when used in conjunction with other components of essential oils. The mechanism of action involves the disruption of bacterial cell membranes, which results in heightened permeability and eventual cell death (Shanaida et al. 2021).

In a study by Abed et al. (2021), carvacrol (Figure 4.7) showed notable antibacterial effectiveness against zoonotic, multidrug-resistant *Staphylococcus* species sourced from both livestock and humans, highlighting its potential as a natural antibiotic (Abed et al. 2021). Furthermore, carvacrol exhibited substantial *in vivo* antimicrobial effects against carbapenemase-producing *Klebsiella pneumoniae* in Swiss mice, leading to increased survival rates and reduced bacterial loads (Abed et al. 2021). This suggests that carvacrol not only inhibits bacterial growth but may also enhance the immune response in animal models.

Carvacrol disrupts bacterial cell membranes, damaging the cell walls of methicillin-resistant *Staphylococcus aureus* (MRSA) in a concentration-dependent manner, which causes a reduction in ATP synthesis and increased membrane permeability to other antibiotics (Al-Tawalbeh et al. 2024). This membrane-disrupting capability is crucial for its antibacterial efficacy, as it compromises the integrity of bacterial cells, resulting in cell death (Magi et al. 2015). In studies involving guinea pigs with chronic obstructive pulmonary disease (COPD), carvacrol administration resulted in reduced inflammation and oxidative stress, indicating its role in modulating immune responses (Hosseini et al. 2023). This anti-inflammatory effect may enhance the overall health and recovery of animals suffering from bacterial infections.

Trans-Citral (Figure 4.8), a natural monoterpenoid compound predominantly found in lemongrass and various citrus oils, has garnered attention for its antibacterial properties, particularly in animal models. The antibacterial efficacy of citral has been demonstrated against a range of pathogenic bacteria, including MRSA, *Escherichia coli*, and *Salmonella typhimurium*, among others (Long et al. 2019; Yoplac et al. 2021). Furthermore, citral inhibits the production of virulence factors in *Vibrio parahaemolyticus*, a foodborne pathogen, thereby attenuating its pathogenicity (Sun et al. 2019; Oliveira et al. 2021). This dual action of citral – both as an antibacterial agent and a modulator of virulence – highlights its potential utility in clinical settings. Research involving MRSA-infected mice demonstrated that citral not only reduced bacterial counts but also decreased inflammatory markers such as TNF-α, indicating its anti-inflammatory properties alongside its antibacterial effects (Oliveira et al. 2021). Additionally, citral's incorporation into nanostructured lipid carriers has been explored to enhance its antimicrobial activity, namely against *Pseudomonas aeruginosa* and *Staphylococcus aureus* in chronic infection models (Oliveira et al. 2021; Oliveira et al. 2022; Yoplac et al. 2021). The antibacterial action of citral is attributed to its ability to penetrate bacterial membranes and disrupt metabolic processes (Kim et al. 2023). Bornyl acetate has been less extensively studied *in vivo*, but recent findings by Wang et al. (2021c) demonstrated that bornyl acetate, when combined with citral, exhibited enhanced antibacterial activity against various pathogens, including *E. coli* in a urinary tract infection (UTI) model.

Thujone (Figure 4.8), a compound found in several essential oils, has shown promising antibacterial activity *in vivo*. It is reported that thujone significantly reduced bacterial counts in animal models infected with *E. coli* and *Salmonella* spp. (Bekka-Hadji et al. 2022; Zhang et al. 2021). The antibacterial action is thought to involve the disruption of bacterial cell membranes, resulting in heightened permeability and the death of the cells. Additionally, thujone's anti-inflammatory properties may enhance its therapeutic potential in treating bacterial infections (Zhang et al. 2021).

Camphor (Figure 4.8), a type of monoterpene ketone sourced from the wood of the camphor tree (*Cinnamomum camphora*), has shown notable antimicrobial effects, which are linked to its capability to disrupt the membranes of bacterial cells and hinder their growth (Poudel et al. 2021; Fazmiya et al. 2022). Research conducted by Zou et al. (2022) indicated that camphor can suppress the proliferation of *Salmonella cholerae* in animal studies. Additionally, the impact of camphor on gut microbiota has been explored, indicating its potential to modulate microbial populations favorably (Seidel et al. 2022). Moreover, the synthesis of antimicrobial compounds derived from camphoric acid further illustrates the ongoing exploration of camphor's antibacterial properties and its potential applications in developing new antimicrobial agents (Mikláš et al. 2021).

Nerol (Figure 4.8), a monoterpene alcohol, can significantly reduce bacterial counts in animal models infected with pathogens such as *S. aureus* (Shahdadi, 2023). The mode of action entails the disruption of bacterial cell membranes, which results in heightened permeability and ultimately leads to cell death. Its efficacy against both Gram-positive and Gram-negative bacteria indicates its potential as a broad-spectrum antimicrobial agent (Shahdadi, 2023). *In vivo* research has demonstrated that β-caryophyllene can effectively inhibit the growth of a range of bacterial pathogens, including MRSA (Zhang et al. 2021). Its mechanism of action is believed to involve disruption of bacterial membranes and modulation of inflammatory responses, highlighting its potential therapeutic applications. Furthermore, β-caryophyllene (Figure 4.8) has been shown to enhance the efficacy of certain antibiotics, indicating its potential as an adjunct therapy in treating multidrug-resistant infections (Zhang et al. 2021).

These results emphasize the extraordinary potential of terpenes as versatile antimicrobial agents, able to target various bacterial pathogens, including those resistant to multiple drugs, through mechanisms like disrupting membranes, modulating virulence factors, boosting immune responses, and serving as adjuvants for antibiotics. This review also suggests that terpenes are a promising resource for creating innovative and sustainable approaches to tackle the increasing global issue of antimicrobial resistance.

4.3 CURRENT RESEARCH IN CLINICAL TRIALS

The information presented above regarding the properties of terpene compounds and their derivatives in *in vivo* assays demonstrates their significant pharmaceutical potential, with some of these compounds already in clinical trial phases, being used, or their derivatives in drugs, alone or in mixtures with other compounds. A search on ClinicalTrials.gov (ClinicalTrials, 2024), revealed a total of 40 clinical trials that were terpene-related. The current relevance of this topic is accentuated by the fact that, of the 40 clinical trials, five are approved but not yet recruiting participants, nine are recruiting participants and are set to start shortly, and one study is already active and ongoing. Unfortunately, eight clinical studies were terminated, withdrawn, or their progress status is unknown. In addition, from the 17 completed trials, only two have indications of having results, but they are not published in any peer-reviewed journals. Fortunately, with the purpose of having a better overview of this topic, more research was done across the internet, and it was possible to find other clinical trials that were not listed on this database. The next section (Table 4.3) provides a summary of clinical trials conducted in the past 15 years that evaluated a terpene (the structures are shown in Figure 4.9), either on its own or in combination with other drugs or compounds for treating various diseases, with their results published in peer-reviewed journals. Studies that involved extracts and/or plants recognized for being abundant in one or more terpenes or that exhibited questionable experimental design were excluded.

4.3.1 α-Bisabolol

In order to evaluate if a topical combination product presented better effects over its single constituents, Arenberger and colleagues (Arenberger et al. 2011) carried out a two-centre, double-blind, randomized, reference-controlled, prospective, and four-armed parallel group clinical trial involving 278 patients with atopic dermatitis. Patients were divided into four treatment groups, each receiving a specific topical cream: α-Bisabolol and heparin blend (group A; $n=79$), α-Bisabolol only (group B; $n=80$), heparin only (group C; $n=78$), and the base cream without any active ingredients (group D; $n=41$). The topical combination product was formulated to correspond with the commercially authorized cream Sensicutan®, containing 0.3 g of α-Bisabolol and/or 20,000 IU of heparin per 100 g of cream. The assigned treatment was applied to the affected areas of eczematous skin twice daily for a duration of 8 weeks, and its effectiveness was evaluated using the SCORAD (SCORing Atopic Dermatitis) and VAS (Visual Analogue Scale) indices. The findings indicated a significantly greater improvement in SCORAD scores for group A compared to groups B, C, and D ($p<7\times10^{-8}$), with patients in this group reporting considerable relief from itching. In fact, VAS values for group A decreased by 41.3 mm, while group B showed a decrease of only 13.3 mm and group C a decrease of 21.3 mm, with patients in group D experiencing a slight worsening of their condition reflected in a VAS value increase of 0.6 mm. Considering these results, the authors (Arenberger et al. 2011) conclude that the combination of α-Bisabolol and heparin offers a superior treatment for atopic dermatitis compared to each ingredient used separately. Despite the positive outcomes, the study's strength is somewhat diminished as the control group (group D) had a lower number of participants compared to the other groups, which affects any conclusions drawn from it.

Continuing with adjunctive treatments, a cream formulation (0.05% retinaldehyde, 1% α-Bisabolol, 3% arbutin and 4% nicotinamide) was used in a prospective, single-arm, single-center, open-label study (Crocco et al. 2015), aiming to evaluate its efficacy and safety in the treatment of epidermal melasma. During 60 days, 33 participants applied the study cream dose (once daily) to the entire cleaned face. When comparing with baseline values, the obtained results reported a statistical significance in the reduction of Melasma Area and Severity Index Score ($p<0.0001$). Through medical imaging software, clinical photograph analysis revealed that the mean total melasma surface area was significantly reduced ($p<0.0001$) from 1398.5 mm^2 at day zero to 923.4 mm^2 at day sixty. In addition, even though not statistically significant, reduction in pigmentation of the skin

TABLE 4.3
Summary of Results from Clinical Studies Involving Terpenes and Published in Peer-Reviewed Journals

Compound/ Drug	Disease	Number Participants	Intervention	Comparison	Period of Intervention	Main Outcomes	Reference

α-Bisabolol	Atopic Dermatitis	278	Four Topical 100 g Cream Treatments (2x/day to the Affected Sites):
			• Group A = 0.3 g α-Bisabolol & 20,000 IU Heparin
			• Group B = 0.3 g α-Bisabolol
			• Group C = 20,000 IU Heparin
			• Group D = Cream Base with no Active Ingredients (Control)
				Baseline Values & between Groups	8 Weeks	• Greater ↓ Itching & ↓ SCORAD scores in Group A than in Other Groups ($p < 7 \times 10^{-8}$)
						• VAS values:
						  • Group A ↓ 41.3 mm
						  • Group B ↓ 13.3 mm
						  • Group C ↓ 21.3 mm
						  • Group D ↑ 0.6 mm
							Arenberger et al. (2011)

	Epidermal Melasma	33	Cream Formulation 1x/day to the Entire Face (0.05% Retinaldehyde, 1% α-Bisabolol, 3% Arbutin and 4% Nicotinamide)	Baseline Values	60 Days	• Significantly ($p < 0.0001$) MASI Score ↓ • Significantly ($p < 0.0001$) Mean Total Melasma Surface Area ↓, Day 0 = 1398.5 mm^2 to Day 60 = 923.4 mm^2 • ↓ Skin Pigmentation • ↑ Brightness, Facial Texture, Hydrated & Oiled Skin = ↑ Overall Appearance & Satisfaction	Crocco et al. (2015)

	Atopic Dermatitis	30	Cream Formulation 2x/day to the Affected Sites (α-Bisabolol, Glycyrrhetinic Acid, Starch and Zinc Oxide)	Baseline Values	6 Weeks	• Significantly ($p = 0.0001$) 51% TESS ↓, after 3 weeks and 74% ↓ after 6 weeks
						• Similar ↓ for single area ESS values:
						  • Faces (72% ↓ at 3 & 92% ↓ at 6 Weeks)
						  • Upper Limbs (48% ↓ at 3 & 81% ↓ at 6 Weeks)
						  • Lower Limbs (45% ↓ at 3 & 73% ↓ at 6 Weeks)
							Licari et al. (2017)

(Continued)

TABLE 4.3 (*Continued*)
Summary of Results from Clinical Studies Involving Terpenes and Published in Peer-Reviewed Journals

Compound/Drug	Disease	Number Participants	Intervention	Comparison	Period of Intervention	Main Outcomes	Reference

	Chronic Venous Leg Ulcers	29	Two Spray Treatments (1x/day Over the Wounds):
			• Study Group = Spray Formulation (1% α-Bisabolol, 15% Ozonated Sunflower Oil & 84% Caprylic/Capric Triglyceride)
			• Control Group = Epithelialization Cream (300,000 IU Vitamin A, 30 mg Vitamin E, 5 g Talc, 5 g Zinc Oxide & 100 g Vaseline®)
				Baseline Values & between Groups	30 Days	• Significant ($p < 0.05$) ↓ Wound Area Ratio in the Study Group:
						  • Study Group = 34% ↓ at 7, 59% ↓ at 14 & 73% ↓ at 30 Days
						  • Control Group = 5% ↓ at 7, 8% ↓ at 14 & 13% ↓ at 30 Days
						• Complete Wound Healing:
						  • Study Group = 25%
						  • Control Group = 0%
							Solovăstru et al. (2015)

Borneol	Post-Operative Pain	122	Two Solution Treatments (Only 1 Application on Skin Around Suture):
			• Borneol Group (25% Borneol Solution)
			• Placebo Group (Ethanol)
				Baseline Values & between Groups	30 to 60 Minutes	• Significant ($p = 0.0004$) ↓ Average Pain Score in Borneol Group:
						  • Borneol Group VAS Value ↓ by 32 mm
						  • Placebo Group VAS Value ↓ by 19.6 mm
						• Experienced ↓ in Pain Intensity of ≥ 50%:
						  • Borneol Group = 66% of Patients
						  • Placebo Group = 35% of Patients
							Wang et al. (2017)

(*Continued*)

TABLE 4.3 (*Continued*)
Summary of Results from Clinical Studies Involving Terpenes and Published in Peer-Reviewed Journals

Compound/Drug	Disease	Number Participants	Intervention	Comparison	Period of Intervention	Main Outcomes	Reference
Cineole	Acute Bronchitis	242	Two Capsule Treatments (3x/day): • Cineole Group (200 mg Capsule of Cineole) • Placebo Group (200 mg Capsule of Placebo, not specified)	Baseline Values & between Groups	10 Days	• After Only 4 Days, Significantly (p=0.0383) Improvements of Bronchitis-Sum-Score on Cineole Group	Fischer and Dethlefsen (2013)

Glycyrrhizin	Chronic Hepatitis C	379	Phase I – 3 groups, 5 Ampoules/ week:
			• Group 1=5 Ampoules of 20 mL Glycyrrhizin (2 mg/mL)
			• Group 2=3 Ampoules of 20 mL Glycyrrhizin (2 mg/ mL)+2 Ampoules of 20 mL Placebo (Not Specified)
			• Group 3=5 Ampoules of 20 mL Placebo (Not Specified)
			Phase II – 2 groups, 5 or 3 Ampoules/week:
			• Group 1=5 Ampoules of 20 mL Glycyrrhizin (2 mg/mL)
			• Group 2=3 Ampoules of 20 mL Glycyrrhizin (2 mg/mL)
				Baseline Values & between Groups	Phase I=12 Weeks
					Phase II=40 Weeks
						Phase I
						• ALT Levels ↓:
						  • Group 1=↓ by 32.8±41.2 U/L
						  • Group 2=↓ by 26.9±31.5 U/L
						  • Group 3=↑ by 0.6±51.0 U/L
						• Significantly (p<0.0001) ALT Levels ↓≥50% on Group 1&2:
						  • Group 1=28.7% of Patients
						  • Group 2=29.0% of Patients
						  • Group 3=7.0% of Patients
						Phase II
						• Further ALT Levels ↓:
						  • Group 1=↓ by 14.2±33.2 U/L
						  • Group 2=↓ by 8.2±37.5 U/L
						• Necro-inflammation Score Improvement with no Further Deterioration:
						  • Group 1=63.3% of Patients
						  • Group 2=61.0% of Patients
							Manns et al. (2012)

(*Continued*)

TABLE 4.3 (*Continued*)
Summary of Results from Clinical Studies Involving Terpenes and Published in Peer-Reviewed Journals

Compound/Drug	Disease	Number Participants	Intervention	Comparison	Period of Intervention	Main Outcomes	Reference
Limonene	Breast Cancer	43	Two Capsules (1 g Limonene each) per Day	Baseline Values	2–6 weeks (Till Surgery)	• High Limonene Concentrations (41.3 μg/g of breast tissue) • Statistically Significant (p=0.002) ↓ Cyclin D1 Expression in Tumour Tissue by 22%	Miller et al. (2013)
Silexan (36.8% Linalool and 34.2% Linalyl Acetate)	Anxiety	1297	One Capsule per Day: • Silexan 80 mg Capsules • Silexan 160 mg Capsules • Placebo Capsules (1/1,000 the Amount of Lavender Oil in Silexan Capsules) • Paroxetine 20 mg Capsules • Lorazepam 0.5 mg Capsules	Baseline Values & between Groups	9, 11 or 12 Weeks	• 80 mg Silexan Treatment = Paroxetine Treatment • Same Effect in ↓ Anxiety • 160 mg Silexan Treatment Higher ↓ HAS Score *vs.* Placebo and Other Treatments	Yap et al. (2019)

Rowatinex® (31% α-Pinene, 15% Camphene, 10% Borneol, 4% Anethole, 4% Fenchone, and 3% Cineole)	Urolithiasis	222	3 x 2 Capsules/day:
			• Rowatinex® Group
			• Placebo Group (Not Specified)
				Baseline Values & between Groups	12 Weeks	• Significantly (p=0.0004) More Stone-Free Patients in Rowatinex® Group
						  • Rowatinex® Group=78.4% of Stone-Free Patients
						  • Placebo Group=52.2 % of Stone-Free Patients
						• ↑ Effectiveness in Treatment of Bigger Stones (> 8 mm) in Rowatinex® Group
						  • Rowatinex® Group=65.2% of Stone-Free Patients
						  • Placebo Group=41.9 % of Stone-Free Patients
						• Significantly (p=0.0028) ↓ Median Time to achieve stone-free status in Rowatinex® Group
						  • Rowatinex® Group=56 Days
						  • Placebo Group=85 Days
							Romics et al. (2011)

(*Continued*)

TABLE 4.3 (*Continued*)
Summary of Results from Clinical Studies Involving Terpenes and Published in Peer-Reviewed Journals

Compound/Drug	Disease	Number Participants	Intervention	Comparison	Period of Intervention	Main Outcomes	Reference
Ursolic acid	Metabolic Syndrome	24	One 150 mg Caspule/day • Ursolic Acid Group • Placebo Group (Calcined Magnesia)	Baseline Values & between Groups	12 Weeks	• Temporary Remission of Metabolic Syndrome (p=0.005) in Ursolic Acid Group • Ursolic Acid Group=50% of Patients Achieved Remission of Disease • Placebo Group=0% of Patients Achieved Remission of Disease • Ursolic Acid Treatment Improved General Health Parameters • Significant (p=0.002) ↓ Fasting Plasma Glucose Levels (Week 0=6.0±0.5 mmol/L to Week 12=4.7±0.4 mmol/L) • Significant (p=0.002) ↓ Body Weight (Week 0=75.7±11.5 kg to Week 12=71.0±11.0 kg) • Significant (p=0.008) ↓ Waist Circumference (Week 0=93.0±8.9 cm to Week 12=83.0±8.6 cm) • Significant (p=0.049) ↓ BMI (Week 0=29.9±3.6 kg/m^2 to Week 12=24.9±1.2 kg/m^2)	Ramírez-Rodríguez et al. (2017)

↑Higher or Increase; ↓Lower or Decrease; ALT, Alanine Aminotransferase; BMI, Body Mass Index; ESS, Eczema Severity Score; HAS, Hamilton Anxiety Scale; MASI, Melasma Area and Severity Index; SCORAD, SCORing Atopic Dermatitis; TESS, Total Eczema Severity Score; VAS, Visual Analogue Scale

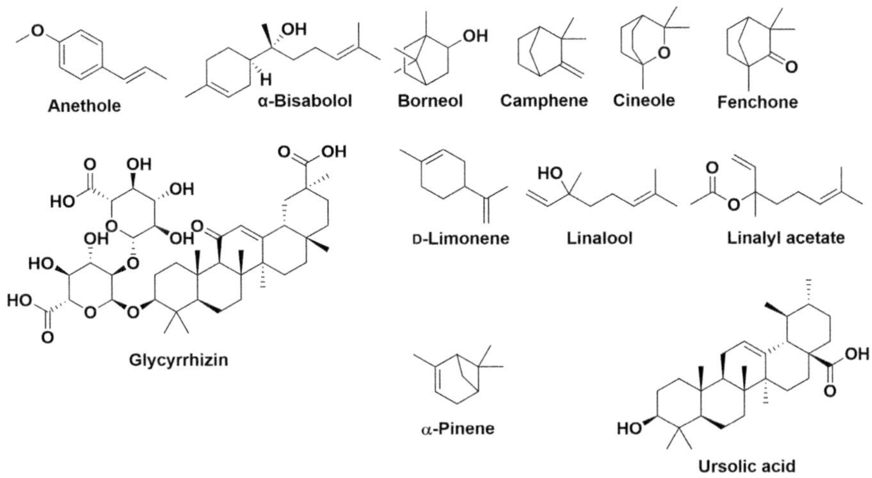

FIGURE 4.9 Chemical structures of terpenes in clinical trials. (The figure was prepared by the authors using the Licenced Chemdraw 22.0.0.)

was confirmed with *in vivo* reflectance confocal microscopy analysis. Additionally, the authors emphasized that the participants' satisfaction was a key element, with overall improvements noted in brightness, skin texture, hydration, and oiliness following the treatment. Indeed, of the 75.8% of subjects who were unhappy with their skin's appearance initially, just 15.2% continued to feel displeased by day sixty.

A different cream formulation combining α-Bisabolol, glycyrrhetinic acid, starch, and zinc oxide was applied twice daily during a 6-week, assessor-blinded, prospective clinical trial across three centers to assess its effectiveness in treating atopic dermatitis in children ($n=30$; 12 females and 18 males; mean age 5 years) (Licari et al. 2017). The findings indicate that applying the cream to the affected areas resulted in a significant reduction ($p=0.0001$) in the Total Eczema Severity Score compared to baseline values, with a decrease of 51% after 3 weeks and 74% after 6 weeks. Comparable reductions were noted in a single area Eczema Severity Score values across participants' faces (a reduction of 72% at 3 weeks and 92% at 6 weeks), upper limbs (a reduction of 48% at 3 weeks and 81% at 6 weeks), and lower limbs (a reduction of 45% at 3 weeks and 73% at 6 weeks). The authors noted a significant improvement in lesions on the trunk but did not provide specific values. Overall, the product was well tolerated, with only one participant experiencing a mild burning sensation at the application site. Regardless of this, all subjects completed the clinical trial. Considering the study outcomes (Licari et al. 2017), the authors endorse the use of similar formulations for managing chronic atopic dermatitis in children.

A novel spray formulation consisting of ozonated sunflower oil (15%), α-Bisabolol (1%), and caprylic/capric triglyceride (84%) was utilized in a randomized, controlled clinical study (Solovăstru et al. 2015) as a topical remedy for chronic venous leg ulcers. Participants were randomly allocated into two groups, the study group ($n=15$; totaling 20 ulcers; treated once daily with the spray formulation) and the control group ($n=14$; totaling 17 ulcers; treated once daily with an epithelialization cream that included vitamin A (300,000 IU), vitamin E (30 mg), talc (5 g), zinc oxide (5 g), and Vaseline® (100 g)), and were given the respective treatments for 30 days. The wound area ratio significantly improved throughout the treatment duration in the study group, showing a continuous decrease after 7 (34%), 14 (59%), and 30 (73%) days of initiating the therapy, whereas the control group exhibited a wound area ratio reduction of 5%, 8%, and 13%, respectively. The differences in the results obtained for the two groups were confirmed as statistically significant ($p<0.05$). Ultimately, at the conclusion of 30 days of topical treatment, 5 patients (25%) in the study group achieved full ulcer healing, whereas none of the patients in the control group reached complete healing.

4.3.2 Borneol

To verify the analgesic effect of a topical borneol application on postoperative pain, a placebo-controlled, double-blind, randomized clinical study (Wang et al. 2017) with 122 patients was carried out. Two randomized groups were formed; one would receive the treatment (borneol 25% solution; $n=62$) and the other group would receive the placebo (ethanol; $n=60$). After an initial pain intensity assessment on a 100 mm VAS between 0 (no pain) and 100 (worst possible pain), each solution would be locally applied once to the skin around the suture of each patient (accordingly to what group the patient belonged) for approximately 30–60 minutes, followed for another pain intensity assessment. Initially, the borneol group had an average pain score of 62.8 mm on the VAS, while the placebo group indicated a score of 62.3 mm. Following treatment, there was a notable difference ($p=0.0004$) in pain scores between the two groups, with patients in the borneol group demonstrating significantly more pain relief than those in the placebo group. In fact, the average decrease in VAS score was 32.0 mm for the borneol group, compared to only 19.6 mm for the placebo group ($p=0.0001$). Furthermore, when looking at the percentage of patients who experienced a reduction in pain intensity of at least 50%, only 35% from the placebo group reported this, while 66% from the borneol group did ($p=0.0006$).

4.3.3 Cineole

Over a period of 10 days, 242 patients with confirmed acute bronchitis enrolled in a multicenter, double-blind, placebo-controlled clinical study (Fischer & Dethlefsen, 2013) in order to assess the mucolytic, bronchodilator, and anti-inflammatory properties of cineole. All patients were randomly assigned to a group to receive 3 times a day a 200 mg capsule of cineole ($n=121$) or a placebo ($n=121$). The main outcome from the trial was that, after only 4 days of treatment, the patient group treated with cineole showed significantly more improvements in the bronchitis-sum-score (which summarises the relevant symptoms of acute bronchitis) than those of the placebo group ($p=0.0383$).

4.3.4 Glycyrrhizin

Manns and coleagues (2012) conducted a randomized, double-blind, placebo-controlled, phase III trial to evaluate the therapeutic potential of glycyrrhizin in patients with chronic hepatitis C who did not respond to standard treatments. In this study, 379 participants were randomly assigned to three groups, each receiving either five, three, or none 20 mL ampoules of glycyrrhizin (2 mg/mL) weekly, along with zero, two, or five 20 mL ampoules of an unspecified placebo for a duration of 12 weeks. This was followed by a randomized, open-label comparison of glycyrrhizin administered five times versus three times a week for a period of 40 weeks. Alanine aminotransferase is an indicator of liver biochemical necro-inflammatory activity, with consistently elevated levels correlating to disease progression, which can result in complications such as cirrhosis and hepatocellular carcinoma (Chen & Huang, 2021). After the initial 12 weeks, the alanine aminotransferase mean levels decreased from the baseline values by 32.8±41.2 U/L in the glycyrrhizin 5x/week group and by 26.9±31.5 U/L in the glycyrrhizin 3x/week group, while increasing by 0.6±51.0 U/L in the placebo 5x/week group. In addition, when compared with the placebo 5x/week group (7.0%), the percentage of patients showing more than a 50% reduction in alanine aminotransferase levels was notably greater ($p<0.0001$) in the treatment groups (glycyrrhizin 3x/week=29.0%; glycyrrhizin 5x/week=28.7%). Following an additional 40 weeks of treatment, a further decrease in alanine aminotransferase levels was noted, with the glycyrrhizin 5x/week group exhibiting a reduction of 14.2±33.2 U/L and the glycyrrhizin 3x/week group showing a reduction of 8.2±37.5 U/L. Moreover, the glycyrrhizin groups at both frequencies demonstrated an improvement in the necro-inflammation score, with no further worsening of 63.3% for the 5x/week group and 61.0% for the 3x/week group. Based on these findings, the authors (Manns et al. 2012) recommend that future treatments for chronic hepatitis patients using glycyrrhizin should involve less frequent doses of the medication.

4.3.5 LIMONENE

In a single-arm, open-label study (Miller et al. 2013), 43 women diagnosed with operable breast cancer consumed 2 g of Limonene daily for a duration of 2–6 weeks (until surgery) to evaluate its anti-proliferative effect. Elevated limonene levels (averaging 41.3 μg/g of breast tissue) were detected in breast tissue from all participants, which correlated with a statistically significant decrease ($p=0.002$) in cyclin D1 expression within tumor tissue by 22%. The overexpression of cyclin D1 accelerates the transition through the G1 phase of the cell cycle; therefore, a reduction in cyclin D1 expression could result in cell-cycle arrest at the G1 phase, consequently hindering tumor cell proliferation (Montalto & De Amicis, 2020).

4.3.6 SILEXAN

Several clinical trials have explored the effect of essential oils rich in linalool on human participants. The effectiveness and safety of silexan (lavender oil capsules contain 36.8% linalool and 34.2% linalyl acetate), were evaluated in a recent network meta-analysis (Yap et al. 2019) comprising 1,297 individuals with anxiety disorders across five separate clinical studies, with 645 subjects taking silexan capsules (80 mg, $n=524$; 160 mg, $n=121$) and 652 subjects taking placebo (1/1,000 the amount of lavender oil in silexan capsules to match the smell of silexan capsules), paroxetine (20 mg) or lorazepam (0.5 mg) capsules. The meta-analysis found that treatment with the 80 mg silexan capsules resulted in the same effect in decreasing the anxiety of patients as the paroxetine treatment. On the other hand, the 160 mg silexan capsules resulted in a higher decrease of the Hamilton Anxiety Scale score in comparison to the placebo and other treatments (*i.e.*, 80 mg silexan capsules, paroxetine and lorazepam).

4.3.7 ROWATINEX®

Rowatinex® is a medical product that consists of an essential oil preparation combining seven naturally occurring terpenes (31% α-pinene, 15% camphene, 10% borneol, 4% anethole, 4% fenchone, and 3% cineole). In a randomized, double-blind, multicenter, therapeutic, parallel group clinical trial (Romics et al. 2011), this drug was analyzed to test its capacity as a pharmaceutical treatment of urolithiasis. After extracorporeal shockwave lithotripsy, 222 patients were randomized to receive Rowatinex® (3×2 capsules/day) or placebo (not specified), during 12 weeks. At weeks 1, 4, 8, and 12, as well as after 2 weeks of follow-up, all participants underwent a physical examination, and the presence of kidney stones was verified through X-ray, intravenous pyelogram, or ultrasound of the kidney, ureter, and bladder, with stone-free status defined as the total expulsion of stones or their fragments. The findings indicate that a significantly greater proportion ($p=0.0004$) of patients receiving the terpene combination treatment were stone-free by the end of the trial (78.4%) compared to the placebo group (52.2%). Additionally, Rowatinex® proved to be more effective at treating larger stones (over 8 mm) than the placebo, with a stone-free rate of 65.2% compared to 41.9% in those receiving the placebo. Moreover, while the placebo group had a median duration of 85 days to reach stone-free status, the group treated with the terpene combination achieved a significantly shorter median time of only 56 days ($p=0.0028$).

4.3.8 URSOLIC ACID

A double-blind, placebo-controlled, randomized clinical trial conducted by Ramírez-Rodríguez et al. in 2017 examined the impact of UA on inflammation, insulin sensitivity, and overall health measures related to metabolic syndrome. Twenty-four individuals diagnosed with metabolic syndrome were randomly divided into two groups of 12, with each group receiving either 150 mg of UA or 150 mg of calcined magnesia (placebo) orally once daily for a duration of 12 weeks. Following the intervention, the placebo group showed no significant changes when compared to their baseline measurements. In contrast, those receiving UA exhibited a notable reduction in fasting plasma glucose levels (from 6.0±0.5 to 4.7±0.4 mmol/L; $p=0.002$), along with decreases in body weight (from 75.7±11.5 to 71.0±11.0kg; $p=0.002$), waist circumference (from 93.0±8.9 to 83.0±8.6cm; $p=0.008$), and body mass index (from 29.9±3.6 to 24.9±1.2 kg/m^2; $p=0.049$). Furthermore, a temporary remission of metabolic syndrome was observed in 50% of the patients receiving ursolic acid, while none in the placebo group achieved remission ($p=0.005$).

4.4 KEY POINTS AND FUTURE PERSPECTIVES

Terpenes are a structurally very diverse family of specialized metabolites (more than 80,000 known terpenes), due to the multiplicity and specificity of the enzymes involved in their biosynthesis and the almost stochastic result of some structural transformations they can catalyze. They have a wide distribution and variable abundance, having already been identified in animals, plants, fungi, and microorganisms. Terpenes are involved in complex physiological and ecological mechanisms, such as controlling the species' development stages, protection against predators, facilitating fertilization, resistance, and adaptation to abiotic stress. Additionally, they exhibit an enormous diversity of biological activities of pharmaceutical interest that can greatly contribute to human well-being. Various terpenes are already available on the market as highly successful medicines for the treatment of malaria, cancer, cardiovascular diseases, and microbial infections.

The *in vivo* recent studies shown that exploration of terpenes in cancer and antibacterial therapy represents a promising frontier in oncological and microbial multidrug-resistance research. Among the most promising terpenes, abietic acid, carvacrol, and D-limonene stand out for their potent anticancer activities, while thymol, carvacrol, and citral emerged as leading antibacterial terpenes. The diverse mechanisms through which these compounds exert their effects, coupled with their favorable safety profiles, show their potential as effective adjuncts or alternatives to conventional cancer treatments.

The outcomes of the different clinical trials outlined above provide evidence to support the application of terpenes either independently or in combination with other substances or medications for addressing various diseases. Among the compounds studied, α-bisabolol has emerged as the most extensively examined terpene in recent years, establishing itself as a leading candidate, particularly for treating skin conditions such as atopic dermatitis. Additionally, it is important to highlight the impact of UA on metabolic syndrome, as its ability to significantly enhance insulin sensitivity and overall health metrics in patients has been demonstrated.

This chapter shows how terpenes are extraordinary in their various facets: in structural terms, functions performed, mechanisms in which they are involved, and above all, their ability to serve as medicines. Knowledge about this family of compounds and their role in the development of pharmaceuticals is far from being a story with an end in sight.

Future studies should focus on elucidating the precise molecular targets of terpenes, optimizing their formulations for clinical use, and conducting rigorous clinical trials to validate their efficacy in several diseases.

REFERENCES

Abdallah II, Quax WJ (2017) A glimpse into the biosynthesis of terpenoids. In *International Conference on Natural Resources and Life Sciences (NRLS-2016)* (pp. 81–98). KnE Life Sciences. DOI: 10.18502/kls.v3i5.981

Abed AH, Hegazy EF, Omar SA, El-Baky RMA, El-Beih AA, Al-Emam A, Menshawy AMS, Khalifa E (2021) Carvacrol essential oil: A natural antibiotic against zoonotic multidrug-resistant *Staphylococcus* species isolated from diseased livestock and humans. *Antibiotics*. 10, 1328. DOI: 10.3390/antibiotics10111328

Alencar MVOB, Islam MT, De Lima RMT, Paz MFCJ, Dos Reis AC, Da Mata AMOF, Filho JWGO, Cerqueira GS, Ferreira PMP, Sousa JMC, Mubarak MS, Melo-Cavalcante AAC (2019) Phytol as an anticarcinogenic and antitumoral agent: An *in vivo* study in swiss mice with DMBA-induced breast cancer. *IUBMB Life*. 71, 200–212. DOI: 10.1002/iub.1952

Ali A, Obireddy SR, Lai W-F (2024) Farnesol as a multifunctional candidate for treatment development. *BIOCELL*. 48, 163–171. DOI: 10.32604/biocell.2023.043839

Almajali B, Al-Jamal HAN, Taib WRW, Ismail I, Johan MF, Doolaanea AA, Ibrahim WN (2021) Thymoquinone, as a novel therapeutic candidate of cancers. *Pharmaceuticals*. 14, 369. DOI: 10.3390/ph14040369

Alsamri H, El Hasasna H, Baby B, Alneyadi A, Al Dhaheri Y, Ayoub MA, Eid AH, Vijayan R, Iratni R (2021) Carnosol is a novel inhibitor of p300 acetyltransferase in breast cancer. *Frontiers in Oncology*. 11, 664403. DOI: 10.3389/fonc.2021.664403

Al-Tawalbeh D, Alkhawaldeh Y, Sawan HM, Al-Mamoori F, Al-Samydai A, Mayyas A (2024) Assessment of carvacrol-antibiotic combinations' antimicrobial activity against methicillin-resistant *Staphylococcus aureus*. *Frontiers in Microbiology*. 14, 1349550. DOI: 10.3389/fmicb.2023.1349550

Ansari MA, Khan FB, Safdari HA, Almatroudi A, Alzohairy MA, Safdari M, Amirizadeh M, Rehman S, Equbal MJ, Hoque M (2021) Prospective therapeutic potential of Tanshinone IIA: An updated overview. *Pharmacological Research*. 164, 105364. DOI: 10.1016/j.phrs.2020.105364

Arenberger P, Arenbergerová M, Drozenová H, Hladíková M, Holcová S (2011) Effect of topical heparin and levomenol on atopic dermatitis: A randomized four-arm, placebo-controlled, double-blind clinical study. *Journal of the European Academy of Dermatology and Venereology*. 25, 688–694. DOI: 10.1111/j.1468-3083.2010.03950.x

Babaei P, Farahpour MR, Tabatabaei ZG (2024) Fabrication of geraniol nanophytosomes loaded into polyvinyl alcohol: A new product for the treatment of wounds infected with methicillin-resistant *Staphylococcus aureus*. *Journal of Tissue Viability*. 33(1), 116–125. DOI: 10.1016/j.jtv.2023.11.002

Bekka-Hadji F, Bombarda I, Djoudi F, Bakour S, Touati A (2022) Chemical composition and synergistic potential of *Mentha pulegium* L. and *Artemisia herba alba* Asso. essential oils and antibiotic against multi-drug resistant bacteria. *Molecules*. 27, 1095. DOI: 10.3390/molecules27031095

Boncan DAT, Tsang SSK, Li C, Lee IHT, Lam H-M, Chan T-F, Hui JHL (2020) Terpenes and terpenoids in plants: Interactions with environment and insects. *International Journal of Molecular Sciences*. 21, 7382. DOI: 10.3390/ijms21197382

Breijyeh Z, Karaman R (2020) Comprehensive review on Alzheimer's disease: Causes and treatment. *Molecules*. 28, 5789. DOI: 10.3390/molecules25245789

Câmara JS, Perestrelo R, Ferreira R, Berenguer CV, Pereira JAM, Castilho PC (2024) Plant-derived terpenoids: A plethora of bioactive compounds with several health functions and industrial applications – A comprehensive overview. *Molecules*. 29, 3861. DOI: 10.3390/molecules29163861

Castro-Camba R, Sánchez C, Vidal N, Vielba JM (2022) Plant development and crop yield: The role of gibberellins. *Plants*. 11, 2650. DOI: 10.3390/plants11192650

Chan EWC, Wong SK, Chan HT (2022) An overview of the chemistry and anticancer properties of rosemary extract and its diterpenes. *Journal of Herbmed Pharmacology*. 11, 10–19. DOI: 10.34172/jhp.2022.02

Cazella LN, Glamoclija J, Soković M, Gonçalves JE, Linde GA, Colauto NB, Gazim ZC (2019) Antimicrobial activity of essential oil of *Baccharis dracunculifolia* DC (Asteraceae) aerial parts at flowering period. *Frontiers in Plant Science*. 10. DOI: 10.3389/fpls.2019.00027

Chebbac K, Benziane OZ, El Moussaoui A, Chalkha M, Lafraxo S, Bin Jardan YA, Nafidi H-A, Bourhia M, Guemmouh, R (2023) Antimicrobial and antioxidant properties of chemically analyzed essential oil of *Artemisia annua* L. (Asteraceae) native to mediterranean area. *Life*. 13(3), 807. DOI: 10.3390/life13030807

Chebet JJ, Ehiri JE, McClelland DJ, Taren D, Hakim IA (2021) Effect of D-limonene and its derivatives on breast cancer in human trials: A scoping review and narrative synthesis. *BMC Cancer*. 21, 902. DOI: 10.1186/s12885-021-08639-1

Chen R, Wang M, Keasling JD, Hu T, Yin X (2024). Expanding the structural diversity of terpenes by synthetic biology approaches. *Trends in Biotechnology*. 42, 699–713. DOI: 10.1016/j.tibtech.2023.12.006

Chen S, Huang H (2021) Clinical non-invasive model to predict liver inflammation in chronic hepatitis B with alanine aminotransferase≤2 upper limit of normal. *Frontiers in Medicine*. 8, 661725. DOI: 10.3389/fmed.2021.661725

Chen Y, Zhu Z, Chen J, Zheng Y, Limsila B, Lu M, Gao T, Yang Q, Fu C, Liao W (2021) Terpenoids from curcumae rhizoma: Their anticancer effects and clinical uses on combination and versus drug therapies. *Biomedicine & Pharmacotherapy*. 138, 111350. DOI: 10.1016/j.biopha.2021.111350

Cherney EC, Baran PS (2011) Terpenoid-alkaloids: Their biosynthetic twist of fate and total synthesis. *Israel Journal of Chemistry*. 51, 391–405. DOI: 10.1002/ijch.201100005

Chien J-H, Chang K-F, Lee S-C, Lee C-J, Chen Y-T, Lai H-C, Lu Y-C, Tsai N-M (2022) Cedrol restricts the growth of colorectal cancer *in vitro* and *in vivo* by inducing cell cycle arrest and caspase-dependent apoptotic cell death. *International Journal of Medical Sciences*. 19, 1953–1964. DOI: 10.7150/ijms.77719

Chiu K-W, Chen H-Y, Chen C-L, Hsieh C-P, Huang Y-F (2023) Attenuation of endoplasmic reticulum stress enhances carvacrol-induced apoptosis in osteosarcoma cell lines. *Life*. 13, 744. DOI: 10.3390/life13030744

Christianson DW (2017) Structural and chemical biology of terpenoid cyclases. *Chemical Reviews*. 117, 11570–11648. DOI: 10.1021/acs.chemrev.7b00287

ClinicalTrials.gov (2024) ClinicalTrials.gov database of privately and publicly funded clinical studies conducted around the world. Available online: https://clinicaltrials.gov/ (accessed on 11/09/2024).

Crocco EI, Veasey JV, Boin MFFC, Lellis RF, Alves RO (2015) A novel cream formulation containing nicotinamide 4%, arbutin 3%, bisabolol 1%, and retinaldehyde 0.05% for treatment of epidermal melasma. *Cutis*. 96, 337–342. https://www.mdedge.com/dermatology/article/104449/pigmentation-disorders/novel-cream-formulation-containing-nicotinamide-4

Dash DK, Tyagi CK, Sahu AK, Tripathi V (2022) Revisiting the medicinal value of terpenes and terpenoids. In *Revisiting Plant Biostimulants*, V Meena, HP Parewa, S K Meena Eds., 130. DOI: 10.5772/intechopen.102612

Ekasari W, Mahardiani A, Putri NT (2023) Terpenoids from Euphorbiaceae as a source of antimalarial medicines: A literature review. *Pharmacy Education*. 23, 294–299. DOI: 10.46542/pe.2023.234.294299

Elgendy AE, Abo Zaid OR, And Elhamid OM, Mansour SZ (2019) Synergistic curative effect of Boswellic acid and Cisplatin against Diethyl nitrosamine-induced hepatocellular carcinoma. *Benha Veterinary Medical Journal*. 36, 256–263. DOI: 10.21608/bvmj.2019.14254.1042

Evidente A (2024) The incredible story of ophiobolin A and sphaeropsidin A: Two fungal terpenes from wilt-inducing phytotoxins to promising anticancer compounds. *Natural Product Reports*. 41, 434–468. DOI: 10.1039/d3np00035d.

Fatima K, Luqman S, Meena A (2022) Carvacrol arrests the proliferation of hypopharyngeal carcinoma cells by suppressing ornithine decarboxylase and hyaluronidase activities. *Frontiers in Nutrition*. 9, 857256. DOI: 10.3389/fnut.2022.857256

Fazmiya MJA, Sultana A, Rahman K, Heyat MBB, Akhtar F, Khan S, Appiah SSCY (2022) Current insights on bioactive molecules, antioxidant, anti-inflammatory, and other pharmacological activities of *Cinnamomum camphora* Linn. *Oxidative Medicine and Cellular Longevity*. 2022, 9354555. DOI: 10.1155/2022/9354555

Feng Y, Li C, Liu S, Yan F, Teng Y, Li X, Sun, Y (2022) β-Elemene Restrains PTEN mRNA degradation to restrain the growth of lung cancer cells via METTL3-mediated N6 methyladenosine modification. *Journal of Oncology*. 2022, 3472745. DOI: 10.1155/2022/3472745

Fischer J, Dethlefsen U (2013) Efficacy of cineole in patients suffering from acute bronchitis: A placebo-controlled double-blind trial. *Cough*. 9, 25. DOI: 10.1186/1745-9974-9-25

Fountain WA, Bopp TS, Bene M, Walston JD (2024) Metabolic dysfunction and the development of physical frailty: An aging war of attrition. *GeroScience*. 46, 3711–3721. DOI: 10.1007/s11357-024-01101-7

Ghazal TSA, Schelz Z, Vidács L, Szemerédi N, Veres K, Spengler G, Hohmann, J (2022) Antimicrobial, multidrug resistance reversal and biofilm formation inhibitory effect of *Origanum majorana* extracts, essential oil and monoterpenes. *Plants*. 11, 1432. DOI: 10.3390/plants11111432

Ghorbanzadeh V, Aljaf KAH, Wasman HM, Pirzeh L, Azimi S, Dariushnejad H (2022) Carvacrol enhance apoptotic effect of 5-FU on MCF-7 cell line via inhibiting p-glycoprotein: An in-silico and in-vitro study. *Drug Research*. 72, 203–208. DOI: 10.1055/a-1766-5491

Gong H, He L, Zhao Z, Mao X, Zhang C (2021) The specific effect of (R)-(+)-pulegone on growth and biofilm formation in multi-drug resistant *Escherichia coli* and molecular mechanisms underlying the expression of pgaabcd genes. *Biomedicine and Pharmacotherapy*. 134, 111149. DOI: 10.1016/j.biopha.2020.111149

Gonzales MM, Garbarino VR, Pollet E, Palavicini JP, Kellogg Jr DL, Kraig E, Orr ME (2022) Biological aging processes underlying cognitive decline and neurodegenerative disease. *Journal of Clinical Investigation*. 132, e158453. DOI: 10.1172/JCI158453

Gu K, Ouyang P, Hong Y, Dai Y, Tang T, He C, Shu G, Liang X, Tang H, Zhu L, Xu Z, Yin L (2022) Geraniol inhibits biofilm formation of methicillin-resistant *Staphylococcus aureus* and increase the therapeutic effect of vancomycin *in vivo*. *Frontiers in Microbiology*. 13. DOI: 10.3389/fmicb.2022.960728

Gunes-Bayır A, Güler E, Bilgin M, Ergün I, Koçyiğit A, Dadak, A (2022) Anti-inflammatory and antioxidant effects of carvacrol on N-methyl-N′-nitro-N-nitrosoguanidine (MNG) induced gastric carcinogenesis in wistar rats. *Nutrients*. 14, 2848. DOI: 10.3390/nu14142848

Henley-Smith CJ, Kok A-M, Botha FS, Baker C, Lall, N (2024) The effect of a poly-herbal plant extract on the adhesion of *Streptococcus mutans* to tooth enamel. *BMC Complementary Medicine and Therapies*. 24(1). DOI: 10.1186/s12906-024-04707-8

Hosseini M, Arab Z, Beheshti F, Anaeigoudari A, Shakeri F, Rajabian A (2023) *Zataria multiflora* and its constituent, carvacrol, counteract sepsis-induced aortic and cardiac toxicity in rat: Involvement of nitric oxide and oxidative stress. *Animal Models and Experimental Medicine*. 6, 221–229. DOI: 10.1002/ame2.12323

Hosseini M, Pereira DM (2023) The chemical space of terpenes: Insights from data science and AI. *Pharmaceuticals*. 16, 202. DOI: 10.3390/ph16020202

Ibrahim D, Abdelfattah-Hassan A, Badawi M, Ismail TA, Bendary MM, Abdelaziz AM, Mosbah RA, Mohamed DI, Arisha AH, El-Hamid MIA (2021) Thymol nanoemulsion promoted broiler chicken's growth, gastrointestinal barrier and bacterial community and conferred protection against *Salmonella typhimurium*. *Scientific Reports*. 11, 7742. DOI: 10.1038/s41598-021-86990-w

Jiang D, Xu J, Liu S, Nasser MI, Wei W, Mao T, Liu X, Zou X, Li J, Li X (2021a) Rosmanol induces breast cancer cells apoptosis by regulating PI3K/AKT and STAT3/JAK2 signaling pathways. *Oncology Letters*. 22, 631 DOI: 10.3892/ol.2021.12892

Jiang S, Liu H, Zhang J, Zhang F, Fan J (2021b) Carnosol induces cell apoptosis and inhibits invasion and proliferation in skin epidermoid cancer in vitro through MMP11-EGFR pathway. *Research Square*. Preprint. DOI: 10.21203/rs.3.rs-148241/v1

Jiang W, Li X, Dong S, Zhou W (2021c) Betulinic acid in the treatment of tumour diseases: Application and research progress. *Biomedicine & Pharmacotherapy*. 142, 111990. DOI: 10.1016/j.biopha.2021.111990

Kamran S, Sinniah A, Abdulghani MAM, Alshawsh MA (2022) Therapeutic potential of certain terpenoids as anticancer agents: A scoping review. *Cancers*. 14, 1100. DOI: 10.3390/cancers14051100

Kaushik S, Dar L, Kaushik S, Yadav JP (2021) Anti-dengue activity of super critical extract and isolated oleanolic acid of *Leucas cephalotes* using in vitro and in silico approach. *BMC Complementary Medicine and Therapies*. 21, 227. DOI: 10.1186/s12906-021-03402-2

Kim S, Yu S, Kim J, Khaliq NU, Choi WI, Kim H, Sung D (2023) Facile fabrication of α-bisabolol nanoparticles with improved antioxidant and antibacterial effects. *Antioxidants*. 12, 207. DOI: 10.3390/antiox12010207

Kim T, Song B, Cho KS, Lee I-S (2020) Therapeutic potential of volatile terpenes and terpenoids from forests for inflammatory diseases. *International Journal of Molecular Sciences*. 21, 2187. DOI: 10.3390/ijms21062187

Kolesnikova AI, Kayumov AR, Gilfanov IR, Frolova LL, Nikitina LE, Trizna EY (2024) Antimicrobial activity of terpens and oxygen-containing terpenoids against *Staphylococcus aureus*. *Microbiology*. 93, 385–388. DOI: 10.1134/s002626172360427x

Leite-Sampaio NF, Gondim CNFL, Martins RAA, Siyadatpanah A, Norouzi R, Kim B, Sobral-Souza CE, Gondim GEC, Ribeiro-Filho J, Coutinho HDM (2022) Potentiation of the activity of antibiotics against ATCC and MDR bacterial strains with (+)-α-pinene and (-)-Borneol. *Biomed Research International*. 2022, 1–10. DOI: 10.1155/2022/8217380

Li K, Gustafson KR (2021) Sesterterpenoids: Chemistry, biology, and biosynthesis. *Natural Product Reports*. 38, 1251–1281. DOI: 10.1039/d0np00070a

Liberale L, Badimon L, Montecucco F, Lüscher TF, Libby P, Camici GG (2022) Inflammation, aging, and cardiovascular disease: JACC review topic of the week. *Journal of the American College of Cardiology*. 79, 837–847. DOI: 10.1016/j.jacc.2021.12.017

Licari A, Ruffinazzi G, DE Filippo M, Castagnoli R, Marseglia A, Agostinis F, Puviani M, Milani M, Marseglia GL (2017) A starch, glycyrretinic, zinc oxide and bisabolol based cream in the treatment of chronic mild-to-moderate atopic dermatitis in children: A three-center, assessor blinded trial. *Minerva Pediatrica*. 69, 470–475. DOI: 10.23736/S0026-4946.17.05015-0

Lin X-Y, Chen G-F, Jiang Z-X, Xing Y-T, Liu Y-Y (2024) New terpenoids from *Callicarpa nudiflora* and their chemotaxonomic significance. *Biochemical Systematics and Ecology*. 116, 104872. DOI: 10.1016/j.bse.2024.104872

Liu X, Chen W, Liu Q, Dai J (2019) Abietic acid suppresses non-small-cell lung cancer cell growth via blocking IKKβ/NF-κB signaling. *OncoTargets and Therapy*. 12, 4825–4837. DOI: 10.2147/OTT.S199161

Liu X, Xin J, Sun Y, Zhao F, Niu C, Liu S (2024) Terpenoids from marine sources: A promising avenue for new antimicrobial drugs. *Marine Drugs*. 22, 347. DOI: 10.3390/md22080347

Lo M-M, Benfodda Z, Molinié R, Meffre P (2024) Volatile organic compounds emitted by flowers: Ecological roles, production by plants, extraction, and identification. *Plants*. 13, 417. DOI: 10.3390/plants13030417

Long N, Tang H, Sun F, Lin L, Dai M (2019) Effect and mechanism of citral against methicillin-resistant *Staphylococcus aureus in vivo*. *Journal of the Science of Food and Agriculture*. 99, 4423–4429. DOI: 10.1002/jsfa.9677

Luo P, Huang J-H, Lv J-M, Wang G-Q, Hu D, Gao H (2024) Biosynthesis of fungal terpenoids. *Natural Products Reports*. 41, 748–783. DOI: 10.1039/d3np00052d

Lv M, Shao S, Zhang Q, Zhuang X, Qiao T (2020) Acetyl-11-keto-β-boswellic acid exerts the anti-cancer effects via cell cycle arrest, apoptosis induction and autophagy suppression in non-small cell lung cancer cells. *OncoTargets and Therapy*. 13, 733–744. DOI: 10.2147/ott.s236346

Ma M, Li M, Wu Z, Liang X, Zheng Q, Li D, Wang G, An T (2024) The microbial biosynthesis of noncanonical terpenoids. *Applied Microbiology and Biotechnology*. 108, 226. DOI: 10.1007/s00253-024-13048-y

Magi G, Marini E, Facinelli B (2015) Antimicrobial activity of essential oils and carvacrol, and synergy of carvacrol and erythromycin, against clinical, erythromycin-resistant group a Streptococci. *Frontiers in Microbiology*. 6, 165. DOI: 10.3389/fmicb.2015.00165

Malhocká A, Švábová M (2023) Diversity of the terpene synthesis in the *Thuja* species – A comparative chemotaxonomic study. *Biochemical Systematics and Ecology*. 110, 104703. DOI: 10.1016/j.bse.2023.104703

Mandal D, Patel P, Verma SK, Sahu BR, Parija T (2022) Proximal discrepancy in intrinsic atomic interaction arrests G2/M phase by inhibiting cyclin B1/CDK1 to infer molecular and cellular biocompatibility of D-limonene. *Scientific Reports*. 12, 18184. DOI: 10.1038/s41598-022-21364-4

Manns MP, Wedemeyer H, Singer A, Khomutjanskaja N, Dienes HP, Roskams T, Goldin R, Hehnke U, Inoue H, European SNMC Study Group (2012) Glycyrrhizin in patients who failed previous interferon alpha-based therapies: Biochemical and histological effects after 52 weeks. *Journal of Viral Hepatitis*. 19, 537–546. DOI: 10.1111/j.1365-2893.2011.01579.x

Martignago CCS, Soares-Silva B, Parisi JR, Silva LCS, Granito RN, Ribeiro AM, Renno ACM, de Sousa LRF, Aguiar ACC (2023) Terpenes extracted from marine sponges with antioxidant activity: A systematic review. *Natural Products and Bioprospecting*. 13, 23. DOI: 10.1007/s13659-023-00387-y

Masyita A, Sari RM, Astuti AD, Yasir B, Rumata NR, Emran TB, Nainu F, Simal-Gandara J (2022) Terpenes and terpenoids as main bioactive compounds of essential oils, their roles in human health and potential application as natural food preservatives. *Food Chemistry:* X. 13, 100217. DOI: 10.1016/j.fochx.2022.100217

Mikláš R, Miklášova N, Bukovský, M (2021) Synthesis and correlation of aggregation and antimicrobial properties of homochiral quaternary ammonium bromides derived from camphoric acid. *European Pharmaceutical Journal*. 68, 10–16. DOI: 10.2478/afpuc-2020-0017

Miller JA, Lang JE, Ley M, Nagle R, Hsu C-H, Thompson PA, Cordova C, Waer A, Chow H-HS (2013) Human breast tissue disposition and bioactivity of limonene in women with early-stage breast cancer. *Cancer Prevention Research*. 6, 577–584. DOI: 10.1158/1940-6207.CAPR-12-0452

Mitić ZS, Jovanović SČ, Zlatković BK, Nikolić BM, Stojanović GS, Marin PD (2017) Needle terpenes as chemotaxonomic markers in *Pinus*: Subsections *Pinus* and *Pinaster*. *Chemistry & Biodiversity*. 14, e1600453. DOI: 10.1002/cbdv.201600453

Montalto FI, De Amicis F (2020) Cyclin D1 in cancer: A molecular connection for cell cycle control, adhesion and invasion in tumor and stroma. *Cells*. 9, 2648. DOI: 10.3390/cells9122648

Montégut L, López-Otín C, Kroemer G (2024) Aging and cancer. *Molecular Cancer*. 23, 106. DOI: 10.1186/s12943-024-02020-z

Morens DM, Fauci AS (2020) Emerging pandemic diseases: How we got to COVID-19. *Cell*. 183, 837. DOI: 10.1016/j.cell.2020.10.022

Murray CJL, Ikuta KS, Sharara F, Swetschinski L, Aguilar GR, Gray A., Han C, Bisignano C, Rao P, Wool E, Johnson SC, Browne AJ, Chipeta MG, Fell F, Hackett S, Haines-Woodhouse G, Hamadani BHK, Kumaran EAP, McManigal, Achalapong S, … Naghavi M (2022) Global burden of bacterial antimicrobial resistance in 2019: A systematic analysis. *Lancet*. 399, 629–55. DOI: 10.1016/S0140-6736(21)02724-0

Nagel, R; Schmidt, A; Peters, RJ (2019) Isoprenyl diphosphate synthases: The chain length determining step in terpene biosynthesis. *Planta*. 249, 9–20. DOI: 10.1007/s00425-018-3052-1

Ninkuu V, Zhang L, Yan J, Fu Z, Yang T, Zeng H (2021) Biochemistry of terpenes and recent advances in plant protection. *International Journal of Molecular Sciences*. 22, 5710. DOI: 10.3390/ijms22115710

Nishikawa H, Asai A, Fukunishi S, Nishiguchi S, Higuchi K (2021) Metabolic syndrome and Sarcopenia. *Nutrients*. 13, 3519. DOI:10.3390/nu13103519

O'Neill E, Hartogh D, Azizi K, Tsiani E (2020) Anticancer properties of carnosol: A summary of *in vitro* and *in vivo* evidence. *Antioxidants*. 9, 961. DOI: 10.3390/antiox9100961

Oliveira D, Borges A, Saavedra M, Borges F, Simões M (2022) Screening of natural molecules as adjuvants to topical antibiotics to treat *Staphylococcus aureus* from diabetic foot ulcer infections. *Antibiotics*. 11, 620. DOI: 10.3390/antibiotics11050620

Oliveira HBM, Selis NDN, Sampaio BA, Júnior MNS, De Carvalho SP, De Almeida JB, Almeida PP, Da Silva IBS, Oliveira CNT, Brito TLS, Da Silva LDO, Teixeira MM, Coelho HILN, Barbosa CD, Andrade YMFDS, Bittencourt RDS, Viana JCS, Campos GB, Timenetsky J, Marques LM (2021) Citral modulates virulence factors in methicillin-resistant *Staphylococcus aureus*. *Scientific Reports*. 11, 16482. DOI: 10.1038/s41598-021-95971-y

Peng XR, Unsicker SB, Gershenzon J, Qiu MH (2023) Structural diversity, hypothetical biosynthesis, chemical synthesis, and biological activity of *Ganoderma* meroterpenoids. *Natural Product Reports* 40, 1354–1392. DOI: 10.1039/d3np00006k

Pichersky E, Raguso RA (2018) Why do plants produce so many terpenoid compounds? *The New Phytologist.* 220, 692–702. DOI: 10.1111/nph.14178

Poudel D, Rokaya A, Ojha P, Timsina S, Satyal R, Dosoky N, Setzer, W (2021) The chemical profiling of essential oils from different tissues of *Cinnamomum camphora* L. and their antimicrobial activities. *Molecules.* 26, 5132. DOI: 10.3390/molecules26175132

Ramírez-Rodríguez AM, González-Ortiz, M, Martínez-Abundis E, Ortega NA (2017) Effect of ursolic acid on metabolic syndrome, insulin sensitivity, and inflammation. *Journal of Medicinal Food.* 20, 882–886. DOI: 10.1089/jmf.2017.0003

Rojas-Armas J, Arroyo-Acevedo J, Palomino-Pacheco M, Herrera-Calderón O, Ortiz-Sánchez J, Rojas-Armas A, Calva J, Castro-Luna A, Hilario-Vargas J (2020) The essential oil of *Cymbopogon citratus* Stapt and carvacrol: An approach of the antitumor effect on 7,12-dimethylbenz-[α]-anthracene (DMBA)-induced breast cancer in female rats. *Molecules.* 25, 3284. DOI: 10.3390/molecules25143284

Romics I, Siller G, Kohnen R, Mavrogenis S, Varga J, Holman E (2011) A special terpene combination (Rowatinex®) improves stone clearance after extracorporeal shockwave lithotripsy in urolithiasis patients: Results of a placebo-controlled randomised controlled trial. *Urologia Internationalis.* 86, 102–109. DOI: 10.1159/000320999

Rosenkranz M, Chen Y, Zhu P, Vlot AC (2021) Volatile terpenes – Mediators of plant-to-plant communication. *Plant Journal.* 108, 617–631. DOI: 10.1111/tpj.15453

Rudolf JD, Chang CY (2020) Terpene synthases in disguise: Enzymology, structure, and opportunities of non-canonical terpene synthases. *Natural Product Reports.* 37, 425–463. DOI: 10.1039/C9NP00051H

Sampaio L, Pina L, Serafini M, Tavares D, Guimarães A (2021) Antitumor effects of carvacrol and thymol: A systematic review. *Frontiers in Pharmacology.* 12, 702487. DOI: 10.3389/fphar.2021.702487

Samreen IA, Malak HA, Abulreesh HH (2021) Environmental antimicrobial resistance and its drivers: A potential threat to public health. *Journal of Global Antimicrobial Resistance.* 27, 101–111. DOI: 10.1016/j.jgar.2021.08.001

Sarkar C, Quispe C, Jamaddar S, Hossain R, Ray P, Mondal M, Mohamed ZA, Jaafaru MS, Salehi B, Islam MT, Razis AFA, Martorell M, Pastene-Navarrete E, Sharifi-Rad J (2020) Therapeutic promises of ginkgolide A: A literature-based review. *Biomed Pharmacotherapy.* 132, 110908. DOI: 10.1016/j.biopha.2020.110908

Schmiech M, Ulrich J, Lang SJ, Büchele B, Paetz C, St-Gelais A, Syrovets T, Simmet T (2021) 11-Keto-α-Boswellic Acid, a novel triterpenoid from *Boswellia* spp. with chemotaxonomic potential and antitumor activity against triple-negative breast cancer cells. *Molecules.* 26, 366. DOI: 10.3390/molecules26020366

Secondulfo C, Mazzeo F, Pastorino GMG, Vicidomini A, Meccariello R, Operto FF (2024) Opioid and cannabinoid systems in pain: Emerging molecular mechanisms and use in clinical practice, health, and fitness. *International Journal of Molecular Sciences.* 25, 9407. DOI: 10.3390/ijms25179407

Seidel D, Walker J, Musser J, Lourenço J, Welch C, Whitney T, Callaway, T (2022) Impact of concentrations of camphor on the in vitro mixed ruminal microorganism fermentation from goats selected for consumption of low and high levels of *Juniperus* spp.-2. *Translational Animal Science.* 6, txac098. DOI: 10.1093/tas/txac098

Shahdadi F (2023) *Mentha longifolia* essential oil and pulegone in edible coatings of alginate and chitosan: Effects on pathogenic bacteria in lactic cheese. *Molecules.* 28, 4554. DOI:10.3390/molecules28114554

Shanaida M, Hudz N, Białoń M, Kryvtsowa M, Svydenko L, Filipska A, Wieczorek, P (2021) Chromatographic profiles and antimicrobial activity of the essential oils obtained from some species and cultivars of the *Mentheae tribe* (Lamiaceae). *Saudi Journal of Biological Sciences.* 28, 6145–6152. DOI: 10.1016/j.sjbs.2021.06.068

Siqueira-Neto JL, Wicht KJ, Chibale K, Fidock DA, Winzeler EA (2023) Antimalarial drug discovery: Progress and approaches. *Nature Reviews Drug Discovery.* 22, 807–826. DOI: 10.1038/s41573-023-00772-9

Solovăstru LG, Stîncanu A, De Ascentii A, Capparé G, Mattana P, Vâţă D (2015) Randomized, controlled study of innovative spray formulation containing ozonated oil and α-bisabolol in the topical treatment of chronic venous leg ulcers. *Advances in Skin & Wound Care.* 28, 406–409. DOI: 10.1097/01.ASW.0000470155.29821.ed

Sorokina M, Merseburger P, Rajan K, Yirik MA, Steinbeck C (2021) COCONUT online: Collection of open natural products database. *Journal of Cheminformatics.* 13, 2. DOI: 10.1186/s13321-020-00478-9

Steele TS, Burkhardt I, Moore ML, de Rond T, Bone HK, Barry K, Bunting VM, Grimwood J, Handley LH, Rajasekar S, Talag J, Michael TP, Moore BS (2024) Biosynthesis of haloterpenoids in red algae via microbial-like type i terpene synthases. *ACS Chemical Biology*. 19, 185–192. DOI: 10.1021/acschembio.3c00627

Sun Y, Guo D, Hua Z, Sun H, Zheng Z, Xia X, Shi, C (2019) Attenuation of multiple *Vibrio parahaemolyticus* virulence factors by citral. *Frontiers in Microbiology*. 10, 894. DOI: 10.3389/fmicb.2019.00894

Tavares WR, Seca AML (2019) *Inula* L. secondary metabolites against oxidative stress-related human diseases. *Antioxidants*. 8, 122. DOI: 10.3390/antiox8050122

Tetali SD (2019) Terpenes and isoprenoids: A wealth of compounds for global use. *Planta*. 249, 1–8. DOI: 10.1007/s00425-018-3056-x

Tholl D, Rebholz Z, Morozov AV, O'Maille PE (2023) Terpene synthases and pathways in animals: Enzymology and structural evolution in the biosynthesis of volatile infochemicals. *Natural Product Report*. 40, 766–793. DOI: 10.1039/d2np00076h

Tian S, Wang D, Yang L, Zhang Z, Liu Y (2022) A systematic review of 1-Deoxy-D-xylulose-5-phosphate synthase in terpenoid biosynthesis in plants. *Plant Growth Regulation*. 96, 221–235. DOI: 10.1007/s10725-021-00784-8

Torre BG, Albericio F (2024) The pharmaceutical industry in 2023: An analysis of FDA drug approvals from the perspective of molecules. *Molecules*. 29, 585. DOI:10.3390/molecules29030585

Trendafilova A, Moujir LM, Sousa PMC, Seca AML (2021) Research advances on health effects of edible *Artemisia* species and some sesquiterpene lactones constituents. *Foods*. 10, 65. DOI: 10.3390/foods10010065

Valerio F, Mezzapesa G, Ghannouchi A, Mondelli D, Logrieco A, Perrino E (2021) Characterization and antimicrobial properties of essential oils from four wild taxa of *Lamiaceae* family growing in Apulia. *Agronomy*. 11, 1431. DOI: 10.3390/agronomy11071431

Venkataraman B, Almarzooqi S, Raj V, Dudeja PK., Bhongade BA, Patil RB, Ojha SK, Attoub S, Adrian TE, Subramanya, SB (2022) α-Bisabolol mitigates colon inflammation by stimulating colon PPAR-γ transcription factor: *in vivo* and in vitro study. *PPAR Research*. 2022, 1–22. DOI: 10.1155/2022/5498115

Wang H, Zhong W, Zhao J, Zhang H, Zhang Q, Yuan L, Liang Y, Chen S, Liu H, Zong S, Tian Y, Zhou H, Sun T, Liu Y, Yang C (2019) Oleanolic acid inhibits epithelial–mesenchymal transition of hepatocellular carcinoma by promoting inos dimerization. *Molecular Cancer Therapeutics*. 18, 62–74. DOI: 10.1158/1535-7163.mct-18-0448

Wang J, Xu C, Chen Y, Shao L, Li T, Fan X, Yu L, Zhang R, Chen B, Chen H, Sui X, Leung ELH, Wu Q (2021a) β-elemene enhances the antitumor activity of erlotinib by inducing apoptosis through AMPK and MAPK pathways in TKI-resistant H1975 lung cancer cells. *Journal of Cancer*. 12, 2285–2294. DOI: 10.7150/jca.53382

Wang Q, Zhao X, Jiang Y, Jin B, Wang L (2023) Functions of representative terpenoids and their biosynthesis mechanisms in medicinal plants. *Biomolecules*. 13, 1725. DOI: 10.3390/biom13121725

Wang S, Chang X, Zhang J, Li J, Wang N, Yang B, Pan B, Zheng Y, Wang X, Ou H, Wang Z (2021b) Ursolic acid inhibits breast cancer metastasis by suppressing glycolytic metabolism via activating SP1/Caveolin-1 signaling. *Frontiers in Oncology*. 11, 745584. DOI: 10.3389/fonc.2021.745584.

Wang S, Zhang D, Hu J, Jia Q, Xu W, Su D, Song H, Xu Z, Cui J, Zhou M, Yang J, Xiao J (2017) A clinical and mechanistic study of topical borneol-induced analgesia. *EMBO Molecular Medicine*. 9, 802–815. DOI: 10.15252/emmm.201607300

Wang W, Ren Z, Wang L, Cai Y, Ma H, Fang L, Su J (2021c) Nanoparticle-stabilized encapsulation of borneol and citral: physicochemical characteristics, storage stability, and enhanced antibacterial activities. *Journal of Food Science*. 86, 4554–4565. DOI:10.1111/1750-3841.15910

Wang X, Xin J, Sun L, Sun Y, Xu Y, Zhao F, Niu C, Liu S (2024b) Exploring the biomedical potential of terpenoid alkaloids: Sources, structures, and activities. *Molecules*. 29, 1968. DOI: 10.3390/molecules29091968

Wang YH, Cui LT, Sun YJ, Li YY, Luo J, Kong LY (2024a) Structurally diverse triterpenoid alkaloids from *Buxus rugulosa*. *Journal of Asian Natural Products Research*. 26, 18–25. DOI: 10.1080/10286020.2023.2297071

Weston-Green K, Clunas H, Naranjo CJ (2021) A review of the potential use of pinene and linalool as terpene-based medicines for brain health: Discovering novel therapeutics in the flavours and fragrances of cannabis. *Frontiers in Psychiatry*. 12, 583211. DOI: 10.3389/fpsyt.2021.583211

WHO (2024) WHO at 75 years: combatting 21st century health challenges. Available online: https://www.who.int/about/accountability/results/who-results-report-2022-mtr/who-at-75-years-combatting-21st-century-health-challenges (accessed 25/11/2024).

Wiart C, Kathirvalu G, Raju CS, Nissapatorn V, Rahmatullah M, Paul AK, Rajagopal M, Sathiya SJS, Rusdi NA, Lanting S, Sulaiman M (2023) Antibacterial and antifungal terpenes from the medicinal angiosperms of Asia and the Pacific: Haystacks and gold needles. *Molecules*. 28, 3873. DOI: 10.3390/molecules28093873

Xie C, Gu J, Zhu S (2024) Progress in research on terpenoid biosynthesis and terpene synthases of Lauraceae species. *Forests*. 15, 1731. DOI: 10.3390/f15101731

Yamaguchi S, Kaneko M, Narukawa M (2021) Approval success rates of drug candidates based on target, action, modality, application, and their combinations. *CTS Clinical and Translational Science*. 14, 1113–1122. DOI: 10.1111/cts.12980

Yap WS, Dolzhenko AV, Jalal Z, Hadi MA, Khan TM (2019) Efficacy and safety of lavender essential oil (Silexan) capsules among patients suffering from anxiety disorders: A network meta-analysis. *Scientific Reports*. 9, 18042. DOI: 10.1038/s41598-019-54529-9

Yazaki K, Arimura G, Ohnishi T (2017) 'Hidden' terpenoids in plants: Their biosynthesis, localization and ecological roles. *Plant Cell Physioogy*. 58, 1615–1621. DOI: 10.1093/pcp/pcx123

Yin L, Liang C, Wei W, Huang S, Ren Y, Geng Y, Ouyang, P (2022) The antibacterial activity of thymol against drug-resistant *Streptococcus iniae* and its protective effect on channel catfish (*Ictalurus punctatus*). *Frontiers in Microbiology*. 13, 914868. DOI: 10.3389/fmicb.2022.914868

Yoplac I, Vargas L, Robert P, Hidalgo A (2021) Characterization and antimicrobial activity of microencapsulated citral with dextrin by spray drying. *Heliyon*. 7, 06737. DOI: 10.1016/j.heliyon.2021.e06737

Yu H, Chen B, Li J, Dong N, Chang X, Wang J, Peng H, Zha L, Gui S (2023) Identification and functional characterization of two *trans*-isopentenyl diphosphate synthases and one squalene synthase involved in triterpenoid biosynthesis in *Platycodon grandiflorus*. *Planta*. 258, 115. DOI: 10.1007/s00425-023-04273-2

Yun H, Jeoung D, Jin S, Park J, Lee E, Lee H, Choi Y, Kim B, Kwon H (2022) Induction of cell cycle arrest, apoptosis, and reducing the expression of mcm proteins in human lung carcinoma A549 cells by cedrol, isolated from *Juniperus chinensis*. *Journal of Microbiology and Biotechnology*. 32, 918–926. DOI: 10.4014/jmb.2205.05012

Zhan X, Qian Y, Mao B (2022) Metabolic profiling of terpene diversity and the response of prenylsynthase-terpene synthase genes during biotic and abiotic stresses in *Dendrobium catenatum*. *International Journal of Molecular Sciences*. 23, 6398. DOI: 10.3390/ijms23126398

Zhang R, Pan T, Xiang Y, Zhang M, Feng J, Liu S, Duan T, Chen P, Zhai B, Chen X, Wang W, Chen B, Han X, Chen L, Yan L, Jin T, Liu Y, Li G, Huang X, … Sui X (2020) β-Elemene reverses the resistance of p53-deficient colorectal cancer cells to 5-fluorouracil by inducing pro-death autophagy and cyclin D3-dependent cycle arrest. *Frontiers in Bioengineering and Biotechnology*. 8. DOI: 10.3389/fbioe.2020.00378

Zhang S, Ye T, Liu Y, Hou G, Wang Q, Zhao F, Li F, Meng Q (2023) Research advances in clinical applications, anticancer mechanism, total chemical synthesis, semi-synthesis and biosynthesis of paclitaxel. *Molecules*. 28, 7517. DOI: 10.3390/molecules28227517

Zhang X, Zhang G, Chai M, Yao X, Chen W, Chu P (2021) Synergistic antibacterial activity of physical-chemical multi-mechanism by TiO_2 nanorod arrays for safe biofilm eradication on implant. *Bioactive Materials*. 6, 12–25. DOI: 10.1016/j.bioactmat.2020.07.017

Zhao M (2023) Anti-inflammatory and antioxidant activity of ursolic acid: A systematic review and meta-analysis. *Frontiers in Pharmacology*. 14, 1256946. DOI: 10.3389/fphar.2023.1256946

Zhou F, Pichersky E (2020) More is better: The diversity of terpene metabolism in plants. *Current Opinion in Plant Biology*. 55, 1–10. DOI: 10.1016/j.pbi.2020.01.005

Zhou J, Azrad M, Kong L (2021) Effect of limonene on cancer development in rodent models: A systematic review. *Frontiers in Sustainable Food Systems*. 5. DOI: 10.3389/fsufs.2021.725077.

Zhou X, Rao S, Wrightstone E, Sun T, Lui ACW, Welsch R, Li L (2022) Phytoene synthase: The key rate-limiting enzyme of carotenoid biosynthesis in plants. *Frontiers in Plant Science*. 13, 884720. DOI: 10.3389/fpls.2022.884720

Zhu T, Li Y, Feng T, Yang Y, Zhang K, Gao J, Quan X, Qian Y, Yu H, Qian B (2022) D-limonene inhibits the occurrence and progression of LUAD through suppressing lipid droplet accumulation induced by $PM_{2.5}$ exposure *in vivo* and *in vitro*. *Respiratory Research*. 23, 1380. DOI: 10.1186/s12931-022-02270-9

Zou W, Huang H, Wu H, Cao Y, Lu W (2022) Preparation, antibacterial potential, and antibacterial components of fermented compound Chinese medicine feed additives. *Frontiers in Veterinary Science*. 9, 808846. DOI: 10.3389/fvets.2022.808846

5 Anthocyanins Sources and Applications

Sanchita Gupta[1], *Priyamvada Mishra*[1], *Prabhatam Mishra*[1], *and Mehar H. Asif*[1]
[1]Biotechnology Division, CSIR-National Botanical Research Institute, Lucknow, India

5.1 INTRODUCTION

Anthocyanins are natural organic compounds and a subclass of flavonoids. These are the group of water-soluble phenolic compounds. The anthocyanins are present in different fruits, vegetables and grains, imparting their variability in pigmentation [1]. These are a type of pigment found in the cell vacuoles of almost every part (fruits, flowers, stems, leaves and roots) of the plants. The color of these pigments ranges from red, purple to greenish-yellow, depending on their acidic, neutral, and basic pH, respectively. During the autumn season, the leaves of some trees turn red due to the presence of anthocyanins. Different legumes like red kidney beans, peas, and black soybeans also have anthocyanins responsible for their varied colors. It is reported that more than 700 anthocyanins are identified in nature [2]. The plants of different families, for example Vitaceae, Rosaceae, Eriacaceae, Saxifragaceae, Caprifoliaceae, Cruciferae, and Fabaceae contain anthocyanins [1]. Based on the changes in functional groups in the basic phenolic structure, anthocyanins have different derivatives, such as cyanidin, delphinidin, malvidin, pelargonidin, peonidin, and petunidin. Recently, China has recommended that the daily dietary intake of anthocyanins be 50 mg/day [3]. Nowadays, anthocyanins have gained immense popularity for their usefulness as natural colorants. Thus, the research on enhancing anthocyanin production should be substantially focused. After consumption, the anthocyanins are metabolized majorly in the distal lower bowel of the gastrointestinal tract and get absorbed in the blood vessels very quickly [4]. The degradation of anthocyanins starts in the mouth, and the level of degradation depends on the structure and microbiome of the oral cavity. After degradation, it gets absorbed throughout the gastrointestinal tract. Anthocyanins also have their contribution in fruits to be attracted by animals when they ripen as well as flowers to be pollinated when it is bloomed. The biological activity of anthocyanins is governed by both chemical structure and cellular location. The anthocyanins contain three carbon rings and a sugar at the C3 position. The sugar can be glucose, arabinose, rhamnose or galactose. Due to various pharmaceutical and pharmacological usefulness, anthocyanins are essential compounds in nutrition and medicine. This chapter aims to describe every aspects of anthocyanins.

5.2 CHEMICAL STRUCTURE OF ANTHOCYANIN PIGMENT

The basic chemical structure of anthocyanidins is a diphenylpropanoid (flavylium cation) consisting of two 6-carbon aromatic rings (A and B) linked by a 3-carbon chain forming an oxygenated heterocyclic (Pyryllum) ring (C) (Figure 5.1).

The anthocyanins are classified based on the number and position of hydroxyl (–OH) and methoxyl (–OCH3) groups. Different types of anthocyanins, Pelargonidin, Cyanidin, and Delphinidin, have one, two, and three hydroxyl (–OH) groups on the B ring of flavylium cation, respectively [5]. Petunidin has two –OH and one –OCH3 groups. Peonidin has one –OH and one –OCH3 group.

DOI: 10.1201/9781003518358-5

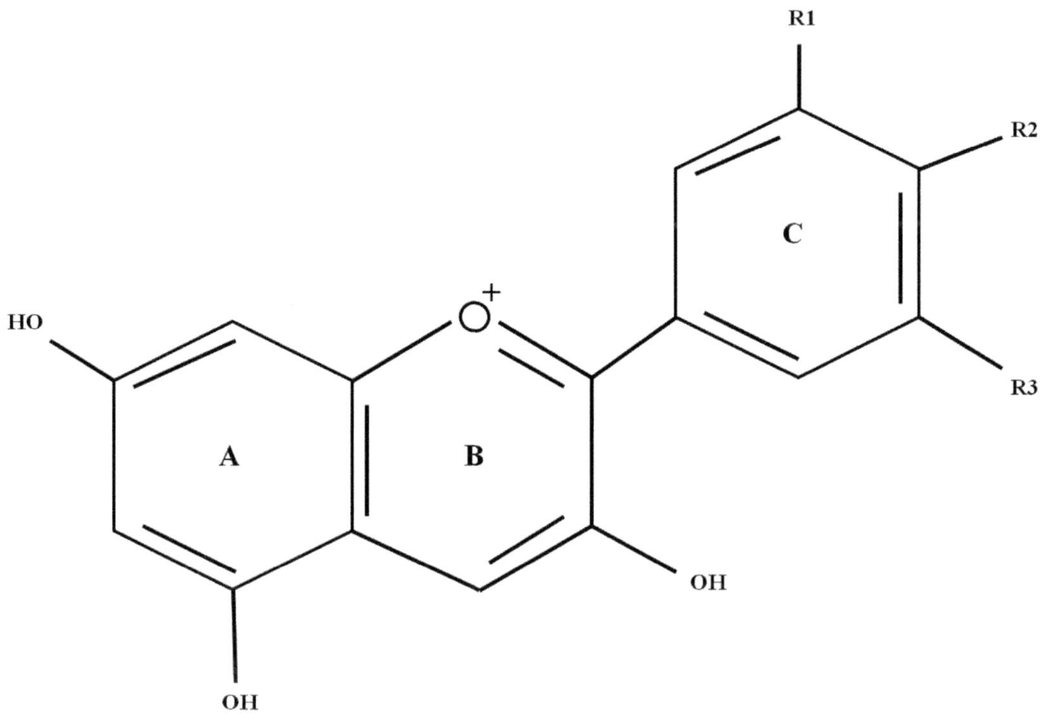

FIGURE 5.1 Basic chemical structure of anthocyanidin.

Malvidin has one –OH and two –OCH3 groups. More is the methoxyl group more will be red color and more is the hydroxyl group more will be blue color. Six major types of anthocyanins are present in the plant kingdom, namely Cyanidin, Peonidin, Pelargonidin, Delphinidin, Malvidin, and Petunidin [6]. The anthocyanins without sugar are called anthocyanidins. The anthocyanins are formed after glycosylation of anthocyanidins (aglycones) at the hydroxyl moiety ring C. The glycosylation of anthocyanidins produces their corresponding 3-*O*-*β*-glucosides (Table 5.1).

The change in the color of anthocyanins depends on their pH levels due to different chemical constituents. At acidic pH, they appear red and are in the form of flavylium cations (oxonium-charged oxygen); at neutral pH, they are purple and are in the form of uncharged quinones; and at basic pH, they are blue and are in the form of stable anionic quinones. The color intensity of anthocyanins also depends on the number of hydroxyl (–OH) and methoxyl (–OCH3) groups in their structure, as the hydroxyl group leads to bluish color while methoxyl groups lead to red color [7–10].

5.3 ANTHOCYANIN BIOSYNTHESIS

Anthocyanins accumulate in the cells' vacuoles after being synthesized in the endoplasmic reticulum. The biosynthesis of anthocyanins starts with L-phenylalanine through the phenylpropanoid biosynthetic pathway. L-phenylalanine converts to cinnamate, followed by 4-coumarate and 4-coumaryl-CoA through the enzymatic reactions. The enzymes that are involved in the reaction are phenylalanine ammonia-lyase (PAL), cinnamate 4-hydroxylase (C4H), and 4-coumarate-CoA ligase (4CL). Then 4-coumaryl-CoA enters into flavonoid biosynthesis and gets converted to Naringenin chalcone followed by Dihydrokaempferol (DHK) with the help of chalcone synthase (CHS), chalcone isomerase (CHI) and flavanone 3-hydroxylase (F3H). DHK is further converted to either dihydroquercetin (DHQ) by flavonoid 3′ hydroxylase (F3′H) or dihydromericetin by flavonoid 3′5′-hydroxylase (F3′5′H). In the next step, all three DHK, dihydroquercetin (DHQ), and dihydromericetin are converted to different leucoanthocyanidins such as leucopelargonidin, leucocyanidin,

TABLE 5.1
List of Chemical Structure and Physical Characteristics of Different Anthocyanidins

Name of Anthocyanidins	Chemical Structure	Name of Anthocyanins	Colour	Names of Plants
Cyanidin (Cy)	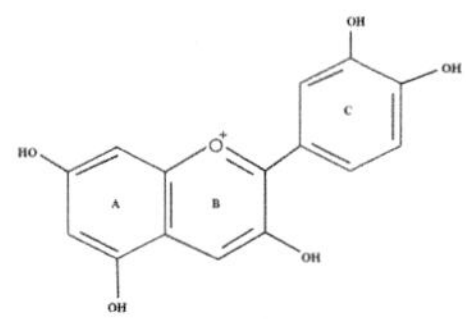	Chrysanthemin	Orange-red Magenta and crimson (Usenik, 2009) [91]	Camellia sinensis, *Viburnum rafinesquianum*, *Malus domestica, Pyrus communis*, *Prunus domestica*
Peonidin (Pn)	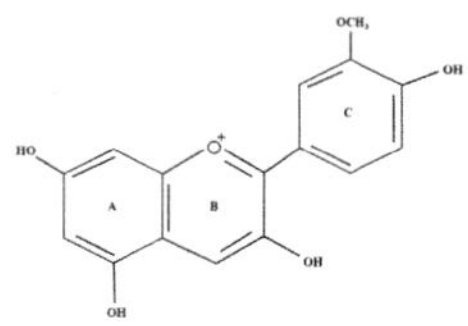	Peonidin-3-*O*-glucoside	Orange-red Magenta	*Vitis vinifera,* red onion, purple corn, *Mangifera indica*
Pelargonidin (Pg)	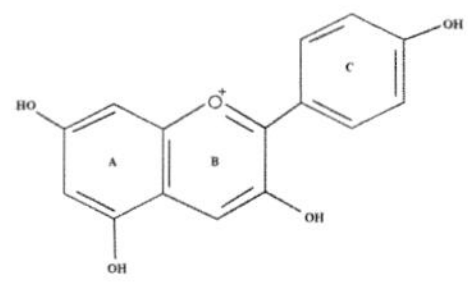	Callistephin	Orange salmon colour (Deroles, 2009 [92]; Fang [5]; Bueno et al. [7])	*Ipomoea purpurea, Vaccinium, Fragaria, Musa, Raphanus sativus, Solanum tuberosum*
Malvidin (Mv)	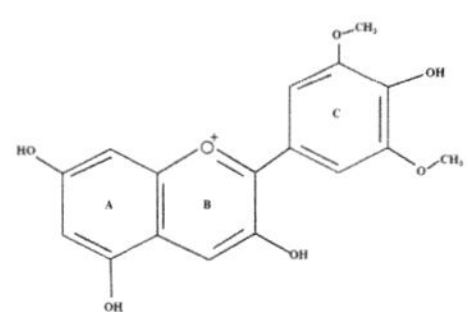	Malvin	Blue-red Purple	*Vitis vinifera, Vaccinium corymbosum, Amelanchier alnifolia* *Vaccinium myrtillus*
Petunidin (Pt)	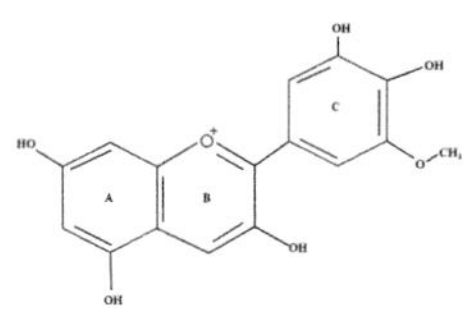	Petunidin-3-*O*-glucoside	Blue-red Purple	*Vitis vinifera, red wine* *Vaccinium myrtillus, Vaccinum* spp.
Delphinidin (Dp)	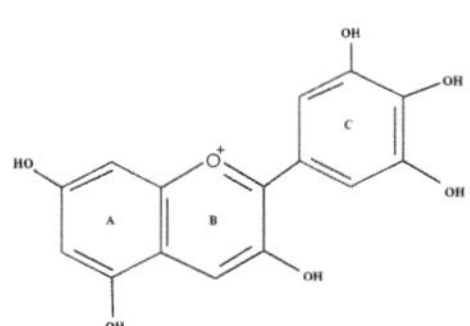	Myrtillin	Blue-red Purple, mauve	Pomegranate *Solanum melongena* *Phaseolus vulgaris*

and leucodelphinidin, respectively, by dihydroflavonol 4-reductase (DFR). Finally, these leucoanthocyanidins form different anthocyanidins (cyanidin, pelargonidin, delphidin) through catalysis by anthocyanidin synthase/leucoanthocyanidin dioxygenase (ANS/LDOX). After that, the anthocyanidins enter into the anthocyanin biosynthetic pathway and are modified into various water-soluble pigments known as anthocyanins through UDP-glucose flavonoid glucosyltransferase (UFGT), acetylase, O-methyltransferase (OMT), and anthocyanin transferase (AT). The types of anthocyanins vary in different plants due to changes in anthocyanin metabolism. Enhancement in the biosynthesis of anthocyanins occurs due to adverse environmental conditions, suggesting their involvement in biotic and abiotic stress conditions (Figure 5.2).

Any phytomolecule biosynthetic pathway consists of two types of genes: structural and regulatory. The structural genes are directly involved in the pathway leading to the formation of the corresponding phytomolecules. Whereas regulatory genes are responsible for the modulation of specific

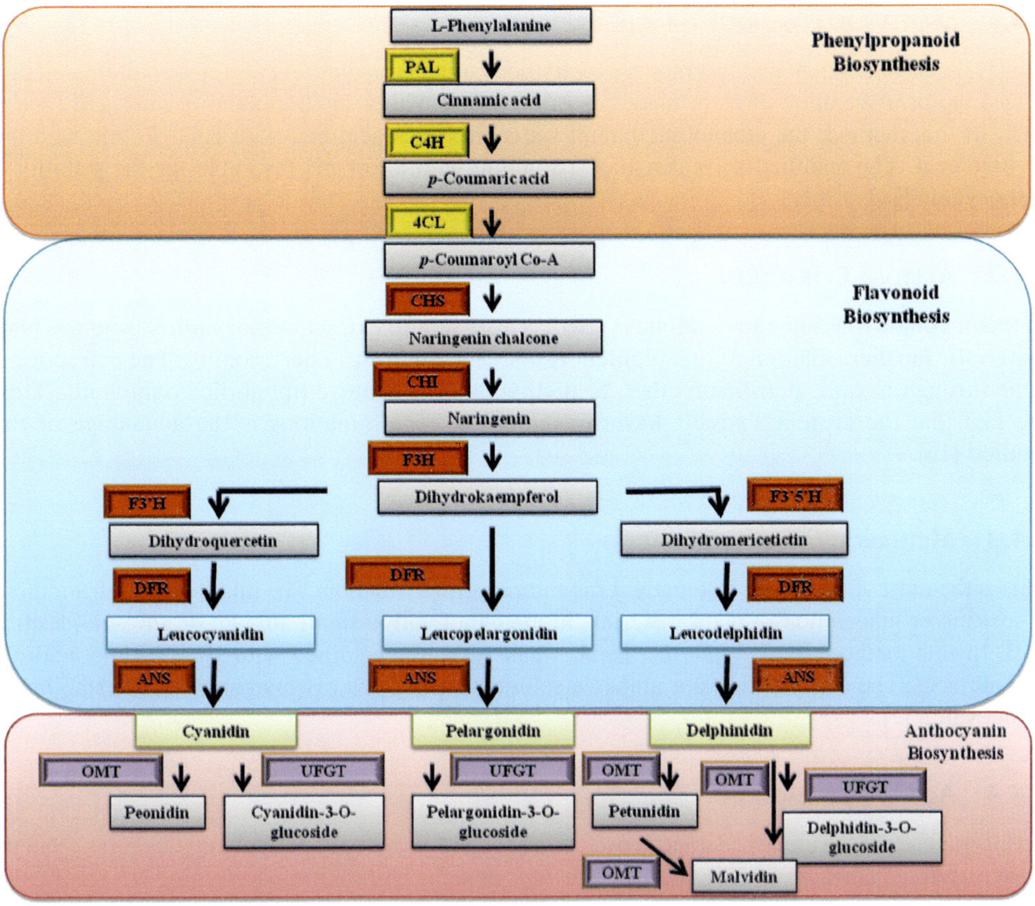

FIGURE 5.2 Schematic representation of biosynthetic pathway of anthocyanins. The enzymes involved are shown in boxes. PAL, phenylalanine ammonialyase;C4H, cinnamate 4-hydroxylase; 4CL, 4-coumarate CoA ligase; CHS, chalcone synthase; CHI, chalcone isomerase; F3H, flavonol 3-hydroxylase; DFR, dihydroflavonol-4-reductase; ANS, anthocyanidin synthase; UFGT, UDP-glucose:flavonoid-3-O-glycosyltransferase.

steps of the pathway, such as genes encoding transcription factors, MYB, bHLH, and WD40 are the important TFs known for regulating the genes involved in anthocyanin biosynthesis [11]. Many MYB TF genes have been reported to modulate anthocyanin biosynthesis and accumulation. In potatoes, the StMYBA1 gene is reported to regulate anthocyanin biosynthesis by activating the expression of two structural genes, CHS and CHI [12]. Similar results were reported in apple [13], Arabidopsis [14], and eggplant [15].

5.4 ANTHOCYANIN ESTIMATION: ANALYTICAL METHODS

Based on their complexity and specificity different strategies have been applied to extract the anthocyanins based on their complexity and specificity. Some methods are solvent extraction, distillation, pressing, and sublimation. An up-to-date report on various techniques used to extract Anthocyanins and anthocyanidins from biological matrices has been reported [16,17]. The selection of the extraction method depends on the nature of anthocyanin and the source material. Anthocyanins from organic sources can be extracted using different solvents.

5.4.1 Methanol/Ethanol Extraction

This is the most efficient and fast method for anthocyanin extraction. The initial step of this method is soaking or maceration of plant materials in methanol with a small amount of any, acid such as HCl. In this method, the ethanol/methanol: water mixtures acidified with HCl, formic acid and organic acid. The acidified solvent makes a solution with a low pH favorable for the stability of anthocyanin [18].

5.4.2 Acetone Extraction

In this method, 70% aqueous acetone is used as a solvent to extract anthocyanins from the plant materials. Further isolation of this pigment is done by chloroform partitioning. The extraction is done through hexane, petroleum ether, or diethyl ether to remove lipophilic compounds. Thus, the fact that the acetone extracts have no lipophilic contaminations is the advantage of this method [16].

5.4.3 Methanol/Ethanol Extraction

This is the most efficient and fast method for anthocyanin extraction. The initial step of this method is soaking or maceration of plant materials in methanol with a small amount of any acid, such as HCl. In this method, the ethanol/methanol: water mixtures acidified with HCl, formic acid and organic acid. The acidified solvent makes a solution having low pH favorable for the stability of anthocyanin [2].

5.4.4 Acetone Extraction

In this method 70% aqueous acetone is used as a solvent for the extraction of anthocyanins from the plant materials. Further isolation of this pigment is done by chloroform partitioning. The extraction is done through hexane, petroleum ether or diethyl ether to remove lipophilic compounds. Thus, the acetone extracts having no lipophilic contaminations is the advantage of this method [3].

5.4.5 Ultrasonic Extraction

This is a non-thermal insulation method in which the thermal decomposition of heat-sensitive compounds is prevented. In this method, a lipophilic organic solvent with ultrasonication is used. The ultrasonic extraction method leads to fast processing, resulting in higher yield. This method is used extensively in the production of food anthocyanins at the industrial level [19]. The main advantages of using the ultrasonic extraction method are that the operation is very easy and safe as well, and the cost of investment and maintenance is meager. This method is used in the case of anthocyanin extraction from seeds.

5.4.6 Natural Deep Eutectic Solvents (NADES) Method

This is a green alternative method for the extraction of anthocyanins. The composition of NADES solvent is a 1:1 ratio of choline chloride and proline. This extraction method is reported to be safe for the food industry.

After extraction, purification is the next step used for anthocyanin characterization. The purification procedure mainly differentiates green (chlorophyll component) and red (anthocyanin component) pigment extracts. After the estimation of chlorophyll, the extracts are used further for different anthocyanin purification methods. Purification is performed by removing unwanted compounds through various analytical techniques such as NMR spectra and HPLC separation.

- **Spectrophotometric measurements:** The determination of anthocyanins in diverse biological samples has been reported. The spectrophotometric quantification of anthocyanins is based on the light absorbance characteristics, which depend on the ionization state of samples.
- **Chromatographic analysis:** The high-performance and ultra-performance liquid chromatography (HPLC and UHPLC) methods are used for anthocyanin estimation. This method utilizes a reverse-phase column employing aqueous/organic gradients using acidified water and acetonitrile or methanol as an organic solvent.

5.5 HEALTH BENEFITS OF ANTHOCYANINS

Accumulation of free radicals in our body leads to oxidative stress, which results in increased risk of various diseases such as cancer and heart disease. Anthocyanins have antioxidant activities and fight the harmful free radicals (ROS and NOS) resulting in decreased stress. The role of anthocyanins in reducing the chronic effects of many diseases has already been reported. The therapeutic roles of anthocyanins at the molecular level have been reviewed [2]. In *vitro*, animal, and human studies have evaluated the biological and pharmacological activities of anthocyanins. No side effects of anthocyanins have been reported; hence, their use in prevention or treatment can be possible even at higher doses.

5.5.1 Antioxidants

Anthocyanins have greater antioxidant properties due to their unique chemical structure. The antioxidant capacity of anthocyanin compounds can be attributed to their ability to chelate metal ions involved in the production of free radicals, thereby reducing metal-induced peroxidation [8,20]. This property of anthocyanins protects vital biomolecules, such as proteins, lipids, and DNA, from oxidative damage, contributing to aging and various diseases [8,21]. Many commonly occurring anthocyanidins and anthocyanins exhibit higher scavenging activity compared to well-known potent antioxidants. For example, cyanidin has an antioxidant capacity of up to 4.4 times greater than ascorbic acid and the vitamin E analog [22].

5.5.2 Inflammation

Anthocyanins are reported to reduce inflammation. The anti-inflammatory effect of anthocyanins occurs due to different pathways. Nuclear Factor kB Pathway (NF-kB): Anthocyanins reduce the external stimulation by inhibiting the activation of the NF-kB signaling pathway. Anthocyanins exhibit anti-inflammatory effects by blocking the release of pro-inflammatory cytokines, lowering the expression of Toll-like receptor 4 (TLR4), and inhibiting the activation of the nuclear factor kappa B (NF-κB) and mitogen-activated protein kinases (MAPKs) signaling pathways. At the same time, they reduce the production of nitric oxide (NO), ROS, and prostaglandin E2 (PGE2), helping to prevent excessive stimulation [23]. An in vivo study by Zhang et al. showed that anthocyanins extracted from goji berries effectively lowered uric acid levels in mice, helping to prevent gouty arthritis [24]. The application of blueberry anthocyanins in mice demonstrated a notable reduction in the expression of cyclooxygenase-2 (COX-2) and NF-κB, providing additional evidence of the anti-inflammatory properties of these anthocyanins via the NF-κB pathway [25].

5.5.3 Diabetes

Diabetes (hyperglycemia) occurs due to the loss of insulin sensitivity to trigger the signal transduction pathways in some body organs such as the liver, muscles, and adipose tissues [26]. The use of dietary anthocyanins against diabetes to enhance insulin sensitivity is widely accepted in alleviating the disorders of glucose metabolism. In traditional Chinese medicinal prescriptions, the Cornus fruits rich in anthocyanins are reported to be used to treat diabetes. The main anthocyanins present in Cornus fruit are cyanidin, delphinidin, and pelargonidin. The glucosides of cyanidin and delphinidin are reported to help insulin secretion from pancreatic β cells in rodents [27]. Pelargonidin has the highest capability of inducing insulin secretion among all anthocyanins. The ability to secrete insulin increases with several hydroxyl (–OH) groups present on the B-ring of the flavylium cation of anthocyanins. Pelargonidin has also been reported to increase insulin secretion at average glucose levels in humans [28]. Few reports indicate that bilberry anthocyanins can lower hyperglycemia and enhance insulin sensitivity by activating adenosine monophosphate-activated protein kinase (AMPK) in the skeletal muscle, liver, and white adipose tissue of mice with type 2 diabetes [29]. The findings indicated that administering a high dose of blueberry anthocyanins substantially lowered blood glucose levels in rats and enhanced insulin sensitivity in obese rats [30].

5.5.4 Cancer

It is reported that anthocyanins prevent damage to DNA due to oxidative stress at the onset of tumor formation. It also inhibits the proliferation and migration of cancerous cells. The signaling pathways AMP-activated protein kinase, PI3K/AKT/mTOR, and JAK-STAT have been reported to be inhibited due to anthocyanins. The anticancer activities of anthocyanins have been studied extensively [31]. Anthocyanins may also initiate the apoptosis of cancer cells, which are not present in the tumor cells, via two different pathways: the internal mitochondrial pathway and the external death receptor pathway [32]. Anthocyanins can inhibit cancerous cells by reducing the activity of matrix metalloprotease 1 (MMP1), a transcriptional target of JNK signaling (c-Jun N-terminal kinase), and reducing cancer metastasis [33]. Another study by Liao et al. [34] revealed that anthocyanins extracted from Thai purple rice could inhibit the proliferation of the human colorectal adenocarcinoma cell line (Caco-2). An earlier study by Stepien et al.; 2023 [34] revealed that an extract (Anthocyanin rich) extracted from *Sambucus nigra* was found to hinder the proliferation and metastasis of cells, and the anthocyanins obtained from grapes have shown apoptotic properties in human breast cancer cells. This is because they can block the formation of carcinogen-DNA adducts, regulate the activity of carcinogen-metabolizing enzymes, and inhibit ROS in non-cancerous human breast cells. A study by Kuntz and his fellows has shown that if the plasma of healthy individuals who consumed anthocyanin-rich fruits were put into pancreatic cancer cells, it may inhibit the mRNA expression of matrix metalloproteinases-9 (MMP-9), thus reducing the decay of extracellular matrix proteins and causing trouble to the migration of cancer cells [35].

5.5.5 Neurodegenerative Diseases

Alzheimer's disease (AD), Huntington's disease (HD), and Parkinson's disease are the most critical neurodegenerative diseases in which misfolding/unfolding of intraneuronal or extraneuronal proteins occurs. The onset of these diseases happens with aging in humans. Various pathways such as apoptosis, autophagy, mitochondrial dysfunction, or oxidative DNA damage and repair are involved in these diseases with corresponding mechanisms. These diseases can be cured by protecting the neurons from oxidative stress, suppressing neuroinflammation, and attenuating cell signaling pathways. Thus, anthocyanins can protect neurons through natural preventive or therapeutic measures; from the reports, it is found that anthocyanins are effective for memory improvement.

Few reports claimed that animals that were given purple sweet potato extract, along with blackcurrant and bilberry extract (rich in Anthocyanin), showed improvements in spatial learning and memory in those animals [36]. Few clinical trial findings indicate that blueberry supplementation enhances nerve response and alleviates neurological disorders [37]. One other clinical trial report by Kent et al. [38] showed that the individuals with AD who consumed daily supplementation of anthocyanin-rich cherry juice showed significant improvements in memory and cognitive function after 12 weeks of treatment. These collective reports show the considerable role of anthocyanin in neurodegenerative diseases.

5.5.6 Visual Health

Traditionally, it is known that anthocyanins are suitable for healthy eyes. The anthocyanins in purple corn are reported to prevent cataracts in rats, possibly due to decreased oxidative stress [39]. The therapeutic application of anthocyanins has shown the potential for improvement in eye health. The in vivo study represented the effect of anthocyanins on rhodopsin regeneration and smooth muscle relaxation [40]. A recent report by Oliinyk et al. [41] claimed that Blueberry anthocyanins could play a role in preventing the advancement of diabetic retinopathy and may be viable candidates for clinical trials aimed at drug development.

5.5.7 Antimicrobial

Earlier studies have focused on the antibacterial activity of anthocyanins against a broad range of microorganisms. This antimicrobial property is attributed to several mechanisms, including the destabilization and permeabilization of the plasma membrane and the induction of cell damage and deformation by compromising the integrity of the cell wall, plasma membrane, and intercellular matrix. Few reports show that anthocyanin-rich extracts may inhibit the activity of many pathogenic bacteria (both in gram-positive and gram-negative but more potent against gram-negative bacteria due to structural dissimilarity).

5.5.8 Cardiovascular Disease

Anthocyanins have a pivotal role in reducing cardiovascular diseases in animals and Humans. A study by Cassidy et al. [42] involving animals has shown that rats fed a diet with 20% maize seeds high in anthocyanins for 8 weeks exhibited a decreased vulnerability to ischemia-reperfusion injury. The anthocyanin-rich extract from black rice has been shown to lower serum total cholesterol and non-high-density lipoprotein (non-HDL) cholesterol levels, thereby improving the stability of atherosclerotic plaques [43]. A clinical study by Xu et al. revealed that daily supplementation of anthocyanins at doses ranging from 80 to 320 mg over 12 weeks improved cholesterol efflux in individuals with dyslipidemia. Research spanning the last 16 years on the health of postmenopausal women shows that regularly consuming foods rich in anthocyanins is linked to a decrease in both the incidence and mortality rates of coronary heart disease [44].

5.5.9 Additional Benefits

Some other beneficial effects of anthocyanins have also been reported. Anthocyanins help in combating aging-associated changes and chronic diseases like neurodegenerative diseases, cancer, and osteoporosis, as well as vision-related problems like glaucoma. One of the studies associated with aging shows that anthocyanins present in pomegranate have a remarkable anti-aging effect and enhance skin permeation in older people [45]. Moreover, anthocyanins are potentially active in combating oxidative stress, inflammation, fat oxidation, and fatigue. Several reports suggest that consuming fruit-derived anthocyanins at doses up to 3,600 mg/day for at least 8 weeks decreases the stress factors and enhances the NO production for muscle oxygenation [45]. The antimicrobial effects of anthocyanin pigment were reported against the infections from *the Staphylococcus aureus* bacterium. The anthocyanin pigments also protect our skin from UV rays. A diet rich in anthocyanins has significant effects on losing body fat.

5.6 ROLE OF ANTHOCYANIN IN PLANTS

Anthocyanins are known to perform various key roles in plants' stress response. A study by Ahmed et al. [46] on barley (*Hordeum vulgare*) shows that increased drought stress tolerance has been linked to a higher accumulation of anthocyanin biosynthetic genes (PAL, CAD, and CHI). This upregulation in anthocyanin content reduces DNA damage and aids in stress mitigation. The HSFA8a protein from apples has been shown to enhance anthocyanin biosynthesis by binding to the promoter regions of the MYB12, FLS, and ANS genes [47]. Additionally, the CIPK protein from pigeon peas has been reported to confer drought stress tolerance by increasing the expression of anthocyanin biosynthesis genes, specifically CHS and CHI [48]. Salinity stress tolerance has been linked to the activation of several anthocyanin biosynthesis genes, including CHS, PAL, C4H, 4CL, DFR, F3H, and FLS, which help mitigate the adverse effects of stress [49,50]. Few studies have shown that colored wheat genotypes maintain more excellent dry matter under salt stress due to increased anthocyanin accumulation [51]. Anthocyanins also have a role in cold stress, like anthocyanin accumulation in epidermal cells, lowering the cell's osmotic potential, delaying freezing, and protecting the plants from cold stress [52]. Elevated levels of anthocyanin accumulation have been noted in grapevines and Arabidopsis when subjected to phosphorus-deficient conditions [53]. A recent study by Kaur et al. [54] showed that anthocyanins offer protective benefits in various colored wheat seedlings under phosphorus-deficient conditions, contributing to greater stress tolerance than white wheat. Anthocyanins are known to play a role in UV damage response in plants. When mature leaves of red Coleus cultivars containing epidermal anthocyanins were exposed to UV light, they experienced less damage than variants that did not accumulate anthocyanin. In a recent study by Zhang et al. [24] in *Populus yunnanensis*, under UV stress, BBX18 and BBX12 regulate the anthocyanin biosynthesis to resist damage. Apart from several vital roles in abiotic stress, little is known about the role of anthocyanins in plant biotic stress. Few studies have shown that anthocyanins possess antifungal and antibacterial properties. Several studies have highlighted the activation of anthocyanin biosynthesis genes during infections, which help restrict the spread of pathogens within host cells [55].

5.6.1 Role of Anthocyanin in Drought Stress

Anthocyanin is known for its essential roles in living organisms. A study by Nakabayashi et al. [56] has shown that the level of anthocyanins increased upon drought stress in *Arabidopsis thaliana*. As a result, the ROS level decreased. Additionally, UGT79B2 and UGT79B3 have been identified as anthocyanin rhamnose-transferases regulated by CBF1 in *A. thaliana*. These enzymes may enhance drought stress tolerance by influencing anthocyanin accumulation [57]. Abiotic stress phytohormone, namely Abscisic acid, is also critical in anthocyanin synthesis under drought stress [58]. Many earlier studies in *Arabidopsis* [56], *Nicotiana* [59], and *Solanum* [60] have shown the positive role of anthocyanin accumulation on drought stress. In tobacco, overexpression of the snapdragon bHLH transcription factor *Delila* (*Del*) results in the upregulation of anthocyanin, consequently providing a positive correlation with anthocyanin accumulation and drought stress [59].

5.6.2 Role of Anthocyanin in Salt Stress

Plant salt stress plays a vital role in their growth and development, as salt stress may disturb the Na^+–K^+ balance [61] and induce ROS production [50]. Salt stress also plays a role in anthocyanin accumulation. For example, in colored cuticles, *Eremochloa ophiuroides* are more resistant to salt stress than green cuticles due to their anthocyanin content difference [57]. Mbarki et al. [51] show that salt stress-induced high dry matter accumulation in wheat also caused upregulation of anthocyanin content. A transcription factor from *Vitis vinifera*, namely *VvMYBA6*, was overexpressed in *A. thaliana*, showing the positive regulation of anthocyanin-synthesizing genes and improved tolerance against salt stress [62].

5.6.3 Role of Anthocyanins in High Light and UV Stress

Zhang et al., 2019 showed the regulation of lettuce's anthocyanin synthesis genes under UV control. Under high UV stress, the expression of anthocyanin synthesis genes like *C4H, 4CL, CHS, FLS, DFR*, and H_2O_2 and NO accumulation were also seen [63]. The UV-B-responsive B-box transcription factor, namely *MdBBX20*, is involved in anthocyanin synthesis in apples [5]. Anthocyanins were also known to save plants damage from high light stress by scavenging ROS and inhibiting the entry of high sunlight into leaves, causing damage [64]. ANS also regulates the photoprotective role of anthocyanins, and the shading from high light via anthocyanins is also essential in plants, as plants don't have any movement to save themselves from high light and other stresses [65]. Recently, a few studies have also shed light on the role of anthocyanins in nitrogen deficiency [66], heavy metals [62,67,68], and under low temperatures [69,70].

5.6.4 Role of Anthocyanin in Biotic Stress

Although few recent reports exist, anthocyanins' involvement in biotic stress is not well known. Sivankalyani et al. [71] have shown that mangoes with red cuticles have high resistance against pathogens. The maize gene, namely the Lc gene, was overexpressed in apples and showed excellent resistance against apple Blackstar and fire blight diseases [72]. Anthocyanins were also known to have a role in reducing tomato susceptibility against gray mold [73]. The colored potato with high anthocyanins gave better resistance against soft rot bacterial disease than other light-colored potato varieties [74]. Recently, it has been shown that higher anthocyanins may inhibit the growth of *S. aureus* [75].

5.7 SOURCES OF ANTHOCYANINS

5.7.1 Fruits: Reservoir of Anthocyanins

Red and blue fruits and vegetables are commonly found to contain anthocyanins, and the amount of these compounds in plants varies significantly between species based on several factors, including cultivar or variety, growing region, climate, farming practices, harvest time, ripening, seasonal variability, processing and storage, temperature, and light exposure. Anthocyanins are abundant in berries, including strawberries, blueberries, blackberries, redcurrants, and raspberries, ranging from approximately 100 to 700 mg/100g of fresh product. However, elderberries and chokeberries have the highest anthocyanin content, with some berries containing as much as 1.4–1.8g of anthocyanins per 100g of product. Purple maize, cherries, plums, pomegranates, eggplant, wine, grapes, and red/purple veggies, including black carrots, red cabbage, and purple cauliflower, are additional excellent sources of anthocyanins. Purple corn, cherries, plums, pomegranates, eggplant, wine, grapes, and red/purple vegetables, including black carrots, red cabbage, and purple cauliflower, are other excellent sources of anthocyanins, which can range from a few milligrams to 200–300 mg/100g of product [63,76]. More recently, anthocyanins have been found in many berries, including açai [77], myrtle [78], and Chilean superfruit maqui berries (*Aristotelia chilensis*) [79], whose production and consumption are constantly rising. Plants have two hydroxyl groups on their B-ring, which makes cyanidin the most abundant pigment. Cyanidin-3-O-β-glucoside is the most abundant anthocyanin found in edible plants, with pelargonidin, peonidin, and delphinidin glucosides following [6]. Generally speaking, methylation results in a red hue and increases stability, while hydroxylation produces a blue hue and decreases stability. The process of glycosylation can alter the hydroxyl groups. It is possible to modify the hydroxyl groups through acylation or glycosylation. Both changes alter anthocyanins' chemical and physical characteristics and the molecules' polarity and chemical reactivity [80].

Fruits are the most common dietary source of anthocyanins, supplying up to 70% of daily intake, mainly from apples, pears, berries, stone fruits, and grapes [81]. However, the evaluation of daily intake of anthocyanin could be more precise, primarily due to the need for more data on the anthocyanin levels in food. Given the health-promoting properties of anthocyanins, encouraging fresh fruit and vegetable consumption may be desired to provide a sufficient level of antioxidant and protective chemicals at the plasma level. Eating a diet rich in fruits and vegetables daily contributes to a healthy lifestyle and helps guard against degenerative and chronic illnesses. For example, following a Mediterranean diet high in fruits, berries, vegetables, beans, and cereals high in anthocyanins has decreased inflammation markers and reduced the risk of obesity, diabetes, cancer, and cardiovascular disease [82]. On the other hand, poor consumption of fruits and vegetables is thought to be responsible for around 1.7 million fatalities worldwide, with gastrointestinal cancer contributing to 14% of these deaths, ischemic heart disease contributing to 11%, and stroke contributing 9% [4].

5.7.2 Anthocyanins as Organic Food and Drink Coloring Agents

Numerous chemicals are used as food colorants in the food business. However, this use has some issues, primarily related to health hazards. Synthetic dyes may have negative neurological and behavioral impacts [83]. Because they are nontoxic and may even benefit health, anthocyanins are a desirable substitute for synthetic materials. Indeed, the United States, Japan, Europe (E163), and many other nations [76] allow the widespread use of anthocyanins as food colorings in foods and beverages. Anthocyanin inclusion could benefit soft drinks, syrups, jams, jellies, candies, bakery or dairy items, and powders. In addition to giving food color, anthocyanins have the potential to offer two additional benefits. Anthocyanin inclusion could benefit soft drinks, syrups, jams, jellies, candies, bakery or dairy items, and powders. In addition to giving food color, anthocyanins have the potential to offer two additional benefits. In addition to providing food with a unique quality by boosting its nutritional value and having positive health impacts on consumers, they can also function as antioxidants, safeguarding the food to which they are added.

Nevertheless, there are several issues with using anthocyanins as a natural food coloring. First, their stability is not ideal because of their rapid degradation caused mainly by light, oxygen, metals, enzymes, other oxidants, pH, and temperature. In addition, their cost is higher than that of synthetic compounds. Thirdly, they may result in off-flavors in food items, as red radish anthocyanins have been known to do [63,83]. Several strategies and tactics, such as microencapsulation, oxygen exclusion, co-pigment addition with colorless molecules in solution, and chemical derivatization like acylation, have been used to increase the stability of anthocyanins. When maltodextrins and beta-glucans are utilized, microencapsulation or complexation with biopolymers offers an effective defense [84–87]. One can use freeze- or spray-drying to create microcarriers. Nanoformulations, like nanoliposomes or nanoemulsions, are an alternative to microencapsulation [88]. Grape skin, radishes, red potatoes, red cabbage, purple sweet potatoes, black carrots, black beans, chokeberries, Thymus modern, prunes, *Hibiscus sabdariffa*, and many more are sources of anthocyanins that may be used as food colorants [4,89]. Although some fruits, like elderberries and chokeberries, can be used to extract large amounts of nonacylated anthocyanins at a low cost, making them potentially useful in the food industry, acylated anthocyanins are generally preferred as food colorants because they are more stable than nonacylated anthocyanins [63].

5.8 CONCLUSION AND FUTURE PERSPECTIVES

Anthocyanins are considered antioxidative phytonutrients in most fruits and vegetables, imparting vivid colors. The anthocyanins have the industrial use of synthetic colorants in the food industry. Defense against microorganisms, fungicidal activity, absorbance of excess visible and UV light, scavenging ROS due to stress, and attraction of pollinators are the main physiological functions of anthocyanins [90]. The diseases associated with oxidative stress, such as cardiovascular, neurodegenerative, metabolic diseases and cancer, are prevented through anthocyanins. Many studies have been performed to enhance anthocyanin production at the gene expression level in various plant species. Thus, plant-based foods consist of many bioactive compounds, such as anthocyanins, capable of modulating multiple metabolic pathways and processes in human beings. Anthocyanins help in combating aging-associated changes and chronic diseases like neurodegenerative diseases, cancer, and osteoporosis, as well as vision-related problems like glaucoma. One of the studies associated with aging shows that anthocyanins present in pomegranate have a remarkable anti-aging effect and enhance skin permeation in older people [45]. Moreover, anthocyanins are potentially active in combating oxidative stress, inflammation, fat oxidation, and fatigue. Several reports suggest that consuming fruit-derived anthocyanins at doses up to 3,600 mg/day for at least 8 weeks decreases the stress factors and enhances the NO production for muscle oxygenation [45].

REFERENCES

1. G. Mazza, *Anthocyanins in Fruits, Vegetables, and Grains*, CRC Press, Boca Raton, 2018.
2. B. Salehi, J. Sharifi-Rad, F. Cappellini, Ä. Reiner, D. Zorzan, M. Imran, B. Sener, M. Kilic, M. El-Shazly, N.M. Fahmy, E. Al-Sayed, M. Martorell, C. Tonelli, K. Petroni, A.O. Docea, D. Calina, A. Maroyi, The therapeutic potential of anthocyanins: Current approaches based on their molecular mechanism of action, *Front Pharmacol* 11 (2020) 1–20.
3. C.N. Association, *Chinese Dietary Reference Intakes, Beijing, China*, China Light Industry Publishing House, Beijing, 2001.
4. R. Mattioli, A. Francioso, L. Mosca, P. Silva, Anthocyanins: A comprehensive review of their chemical properties and health effects on cardiovascular and neurodegenerative diseases, *Molecules* 25(17) (2020) 1–42.
5. J. Fang, Classification of fruits based on anthocyanin types and relevance to their health effects, *Nutrition* 31(11–12) (2015) 1301–1306.
6. H.E. Khoo, A. Azlan, S.T. Tang, S.M. Lim, Anthocyanidins and anthocyanins: Colored pigments as food, pharmaceutical ingredients, and the potential health benefits, *Food Nutr Res* 61 (2017) 1–21.
7. J.M. Bueno, P. Sáez-Plaza, F. Ramos-Escudero, A.M. Jiménez, R. Fett, A.G. Asuero, Analysis and antioxidant capacity of anthocyanin pigments. Part II: Chemical structure, color, and intake of anthocyanins, *Crit Rev Anal Chem* 42(2) (2012) 126–151.
8. J. Martín, M.J. Navas, A.M. Jiménez-Moreno, A.G. Asuero, Anthocyanin pigments: Importance, sample preparation and extraction, Phenolic compounds-natural sources, importance and applications. In *Phenolic Compounds: Natural Sources, Importance and Applications*, Marcos Soto-Hernandez, Mariana Palma-Tenango and Maria del Rosario Garcia-Mateos, Intech Open, London, UK, 2017, pp. 117–152.
9. A. Mortensen, Carotenoids and other pigments as natural colorants, *Pure Appl Chem* 78(8) (2006) 1477–1491.
10. Y. Tanaka, N. Sasaki, A. Ohmiya, Biosynthesis of plant pigments: Anthocyanins, betalains and carotenoids, *Plant J* 54(4) (2008) 733–749.
11. A. Lloyd, A. Brockman, L. Aguirre, A. Campbell, A. Bean, A. Cantero, A. Gonzalez, Advances in the MYB-bHLH-WD Repeat (MBW) pigment regulatory model: Addition of a WRKY factor and co-option of an anthocyanin MYB for betalain regulation, *Plant Cell Physiol* 58(9) (2017) 1431–1441.
12. X.J. Zhao, H.L. Zhang, T.F. Liu, Y.N. Zhao, X.X. Hu, S.X. Liu, Y. Lin, B.T. Song, C.Z. He, Transcriptome analysis provides *StMYBA*1 gene that regulates potato anthocyanin biosynthesis by activating structural genes, *Front Plant Sci* 14 (2023) 1–11.
13. R.V. Espley, R.P. Hellens, J. Putterill, D.E. Stevenson, S. Kutty-Amma, A.C. Allan, Red colouration in apple fruit is due to the activity of the MYB transcription factor, MdMYB10, *Plant J* 49(3) (2007) 414–427.
14. S.K. Mondal, S. Roy, Genome-wide sequential, evolutionary, organizational and expression analyses of phenylpropanoid biosynthesis associated MYB domain transcription factors in Arabidopsis, *J Biomol Struct Dyn* 36(6) (2018) 1577–1601.
15. L. Zhou, Y.J. He, J. Li, Y. Liu, H.Y. Chen, CBFs function in anthocyanin biosynthesis by interacting with MYB113 in eggplant (*Solanum melongena* L.), *Plant Cell Physiol* 61(2) (2020) 416–426.
16. L.E. Rodriguez-Saona, R.E. Wrolstad, Extraction, isolation, and purification of anthocyanins, *Curr Protocols Food Anal Chem* 1 (2001) F1. 1.1–F1. 1.11.
17. N. Tena, A.G. Asuero, Up-to-date analysis of the extraction methods for anthocyanins: Principles of the techniques, optimization, technical progress, and industrial application, *Antioxidants (Basel)* 11(2) (2022) 1–40.

18. E.M. Azman, D. Charalampopoulos, A. Chatzifragkou, Acetic acid buffer as extraction medium for free and bound phenolics from dried blackcurrant (*Ribes nigrum* L.) skins, *J Food Sci* 85(11) (2020) 3745–3755.
19. T. Belwal, H. Huang, L. Li, Z. Duan, X. Zhang, H. Aalim, Z. Luo, Optimization model for ultrasonic-assisted and scale-up extraction of anthocyanins from *Pyrus communis* 'Starkrimson'fruit peel, *Food Chem* 297 (2019) 124993.
20. L.P. Dai, X.J. Dong, H.H. Ma, Antioxidative and chelating properties of anthocyanins in Azolla imbricata induced by cadmium, Polish J Environ Stud 21(4) (2012) 837–844.
21. E. Pojer, F. Mattivi, D. Johnson, C.S. Stockley, The case for anthocyanin consumption to promote human health: A review, *Compr Rev Food Sci Food Saf* 12(5) (2013) 483–508.
22. K.S. Gould, S.O. Neill, T.C. Vogelmann, *A Unified Explanation for Anthocyanins in Leaves?*, Advance in Botanical Research, Academic Press, Cambridge, 2002.
23. Y.Z. Zheng, G. Deng, Y.C. Zhang, Multiple free radical scavenging reactions of flavonoids, *Dyes Pigm* 198 (2022) 1–7.
24. Q. Zhang, D.X. Ma, Z.X. Hu, D. Zong, C.Z. He, *PyunBBX18* is involved in the regulation of anthocyanins biosynthesis under UV-B stress, *Genes* 13(10) (2022) 1–21.
25. Z.L. Ma, B. Du, J. Li, Y.D. Yang, F.M. Zhu, An insight into anti-inflammatory activities and inflammation related diseases of anthocyanins: A review of both in vivo and in vitro investigations, *Int J Mol Sci* 22(20) (2021) 1–17.
26. T. Belwal, S.F. Nabavi, S.M. Nabavi, S. Habtemariam, Dietary anthocyanins and insulin resistance: When food becomes a medicine, *Nutrients* 9(10) (2017) 1–22.
27. B. Jayaprakasam, S.K. Vareed, L.K. Olson, M.G. Nair, Insulin secretion by bioactive anthocyanins and anthocyanidins present in fruits, *J Agr Food Chem* 53(1) (2005) 28–31.
28. G.B. Christison, H.A. Mackenzie, Laser photoacoustic determination of physiological glucose-concentrations in human whole-blood, *Med Biol Eng Comput* 31(3) (1993) 284–290.
29. D. Marko, N. Puppel, Z. Tjaden, S. Jakobs, G. Pahlke, The substitution pattern of anthocyanidins affects different cellular signaling cascades regulating cell proliferation, *Mol Nutr Food Res* 48(4) (2004) 318–325.
30. Y.-H. Lou, T. Huang, X. Zhang, L.-M. Tian, W.-B. Bai, Progress in preparation and application of anthocyanin-starch complexes: A review, *Food Science*, 44(13) (2023) 317–327.
31. K. Asadi, L.R. Ferguson, M. Philpott, N. Karunasinghe, Cancer-preventive properties of an anthocyanin-enriched sweet potato in the APC(MIN) mouse model, *J Cancer Prev* 22(3) (2017) 135–146.
32. L.S. Wang, G.D. Stoner, Anthocyanins and their role in cancer prevention, *Cancer Lett* 269(2) (2008) 281–290.
33. H. Zhang, J. Guo, L.P. Mao, Q.Q. Li, M.N. Guo, T. Mu, Q.H. Zhang, X.L. Bi, Up-regulation of miR-24-1-5p is involved in the chemoprevention of colorectal cancer by black raspberry anthocyanins, *Br J Nutr* 122(5) (2019) 518–526.
34. A.E. Stępień, J. Trojniak, J. Tabarkiewicz, Health-Promoting properties: Anti-inflammatory and anticancer properties of *Sambucus nigra* L. flowers and fruits. Molecules. 28(17) (2023) 6235.
35. **S.** Kuntz, C. Kunz, S. Rudloff, Inhibition of pancreatic cancer cell migration by plasma anthocyanins isolated from healthy volunteers receiving an anthocyanin-rich berry juice, Eur J Nutr 56(1) (2017) 203-214.
36. Y. Ma, Y.H. Feng, T.W. Diao, W.L. Zeng, Y. Zuo, Experimental and theoretical study on antioxidant activity of the four anthocyanins, *J Mol Struct* 1204 (2020) 1–7.
37. P.H.L. Tran, T.T.D. Tran, Blueberry supplementation in neuronal health and protective technologies for efficient delivery of blueberry anthocyanins, *Biomolecules* 11(1) (2021) 1–20.
38. K. Kent, K. Charlton, S. Roodenrys, M. Batterham, J. Potter, V. Traynor, H. Gilbert, O. Morgan, R. Richards, Consumption of anthocyanin-rich cherry juice for 12 weeks improves memory and cognition in older adults with mild-to-moderate dementia, *Eur J Nutr* 56(1) (2017) 333–341.
39. P. Thiraphatthanavong, J. Wattanathorn, S. Muchimapura, W. Thukham-Mee, P. Wannanon, T. Tong-Un, B. Suriharn, K. Lertrat, Preventive effect of *Zea mays* L.(purple waxy corn) on experimental diabetic cataract, *Biomed Res Int* 2014 (2014) 1–8.
40. Y. Nomi, K. Iwasaki-Kurashige, H. Matsumoto, Therapeutic effects of anthocyanins for vision and eye health, *Molecules* 24(18) (2019) 1–18.
41. S. Oliinyk, V. Humeniuk, P. Oliinyk, A. Rybachuk, Nutraceutical components of blueberries for the prevention and treatment of diabetic retinopathy, *An Mechnikov's Inst* 1 (2024) 3–12.
42. A. Cassidy, K.J. Mukamal, L. Liu, M. Franz, A.H. Eliassen, E.B. Rimm, High anthocyanin intake is associated with a reduced risk of myocardial infarction in young and middle-aged women, *Circulation* 127(2) (2013) 188–196.

43. K.S. Shaughnessy, I.A. Boswall, A.P. Scanlan, K.T. Gottschall-Pass, M.I. Sweeney, Diets containing blueberry extract lower blood pressure in spontaneously hypertensive stroke-prone rats, *Nutr Res* 29(2) (2009) 130–138.
44. W. Xu, Q. Zhou, Y. Yao, X. Li, J.L. Zhang, G.H. Su, A.P. Deng, Inhibitory effect of Gardenblue blueberry (*Vaccinium ashei* Reade) anthocyanin extracts on lipopolysaccharide-stimulated inflammatory response in RAW 264.7 cells, *J Zhejiang Univ Sci B* 17(6) (2016) 425–436.
45. S.K. Panchal, O.D. John, M.L. Mathai, L. Brown, Anthocyanins in chronic diseases: The power of purple, *Nutrients* 14(10) (2022) 1–30.
46. I.M. Ahmed, U.A. Nadira, N. Bibi, F.B. Cao, X.Y. He, G.P. Zhang, F.B. Wu, Secondary metabolism and antioxidants are involved in the tolerance to drought and salinity, separately and combined, in Tibetan wild barley, *Environ Exp Bot* 111 (2015) 1–12.
47. N. Wang, W. Liu, L. Yu, Z. Guo, Z. Chen, S. Jiang, H. Xu, H. Fang, Y. Wang, Z. Zhang, X. Chen, HEAT SHOCK FACTOR A8a modulates flavonoid synthesis and drought tolerance, *Plant Physiol* 184(3) (2020) 1273–1290.
48. D. Meng, B. Dong, L. Niu, Z. Song, L. Wang, R. Amin, H. Cao, H. Li, Q. Yang, Y. Fu, The pigeon pea CcCIPK14-CcCBL1 pair positively modulates drought tolerance by enhancing flavonoid biosynthesis, *Plant J* 106(5) (2021) 1278–1297.
49. S. Ben-Abdallah, W. Zorrig, L. Amyot, J. Renaud, A. Hannoufa, M. Lachâal, N. Karray-Bouraoui, Potential production of polyphenols, carotenoids and glycoalkaloids in *Solanum villosum* Mill. under salt stress, *Biologia* 74(3) (2019) 309–324.
50. D. Kim, S.J. Jeon, S. Yanders, S.C. Park, H.S. Kim, S. Kim, MYB3 plays an important role in lignin and anthocyanin biosynthesis under salt stress condition in Arabidopsis, *Plant Cell Rep* 41(7) (2022) 1549–1560.
51. S. Mbarki, O. Sytar, M. Zivcak, C. Abdelly, A. Cerda, M. Brestic, Anthocyanins of coloured wheat genotypes in specific response to salstress, *Molecules* 23(7) (2018) 1–15.
52. Q.Q. Pang, W.B. Yu, E. Sadeghnezhad, X.Q. Chen, P.J. Hong, T. Pervaiz, Y.H. Ren, Y.P. Zhang, T.Y. Dong, H.F. Jia, J.G. Fang, Omic analysis of anthocyanin synthesis in wine grape leaves under low-temperature, *Sci Hortic* 307 (2023) 1–12.
53. H.Z. Zheng, H. Wei, S.H. Guo, X. Yang, M.X. Feng, X.Q. Jin, Y.L. Fang, Z.W. Zhang, T.F. Xu, J.F. Meng, Nitrogen and phosphorus co-starvation inhibits anthocyanin synthesis in the callus of grape berry skin, *Plant Cell Tiss Org* 142(2) (2020) 313–325.
54. S. Kaur, A. Kumari, N. Sharma, A.K. Pandey, M. Garg, Physiological and molecular response of colored wheat seedlings against phosphate deficiency is linked to accumulation of distinct anthocyanins, *Plant Physiol Bioch* 170 (2022) 338–349.
55. R. Jan, M.A. Khan, S. Asaf, Lubna, J.R. Park, I.J. Lee, K.M. Kim, Flavonone 3-hydroxylase relieves bacterial leaf blight stress in rice via overaccumulation of antioxidant flavonoids and induction of defense genes and hormones, *Int J Mol Sci* 22(11) (2021) 1–19.
56. R. Nakabayashi, K. Yonekura-Sakakibara, K. Urano, M. Suzuki, Y. Yamada, T. Nishizawa, F. Matsuda, M. Kojima, H. Sakakibara, K. Shinozaki, A.J. Michael, T. Tohge, M. Yamazaki, K. Saito, Enhancement of oxidative and drought tolerance in Arabidopsis by overaccumulation of antioxidant flavonoids, *Plant J* 77(3) (2014) 367–379.
57. P. Li, Y.J. Li, F.J. Zhang, G.Z. Zhang, X.Y. Jiang, H.M. Yu, B.K. Hou, The Arabidopsis UDP-glycosyltransferases UGT79B2 and UGT79B3, contribute to cold, salt and drought stress tolerance via modulating anthocyanin accumulation, *Plant J* 89(1) (2017) 85–103.
58. J. González-Villagra, J.D. Cohen, M.M. Reyes-Díaz, Abscisic acid is involved in phenolic compounds biosynthesis, mainly anthocyanins, in leaves of plants (Mol.) subjected to drought stress, *Physiol Plantarum* 165(4) (2019) 855–866.
59. A.H. Naing, K.I. Park, T.N. Ai, M.Y. Chung, J.S. Han, Y.W. Kang, K.B. Lim, C.K. Kim, Overexpression of snapdragon *Delila* (*Del*) gene in tobacco enhances anthocyanin accumulation and abiotic stress tolerance, *BMC Plant Biol* 17(1) (2017) 65.
60. M. Waseem, X. Rong, Z. Li, Dissecting the role of a basic helix-loop-helix transcription factor, *SlbHLH22*, under salt and drought stresses in transgenic *Solanum lycopersicum* L, *Front Plant Sci* 10 (2019) 734.
61. A.M. Ismail, T. Horie, genomics, physiology, and molecular breeding approaches for improving salt tolerance, *Annu Rev Plant Biol* 68 (2017) 405–434.
62. S.S. Sun, A.R. Liu, Z. Li, T.M. Guo, S.C. Chen, G.J. Ahammed, Anthocyanin synthesis is critical for melatonin-induced chromium stress tolerance in tomato, *J Hazard Mater* 453 (2023) 1–12.

63. L. Zhang, F. Gong, Y. Song, K. Liu, Y. Wan, De novo transcriptome analysis of lettuce (Lactuca sativa L.) and the identification of structural genes involved in anthocyanin biosynthesis in response to UV-B radiation. *Acta Physiologiae Plantarum*, 41 (2019) 1–16.
64. Y.X. Luo, S.L. Teng, H.X. Yin, S.P. Zhang, X.Y. Tuo, L.S.P. Tran, Transcriptome analysis reveals roles of anthocyanin- and jasmonic acid-biosynthetic pathways in rapeseed in response to high light stress, *Int J Mol Sci* 22(23) (2021) 1–21.
65. X.T. Zheng, Y.L. Chen, X.H. Zhang, M.L. Cai, Z.C. Yu, C.L. Peng, ANS-deficient is sensitive to high light due to impaired anthocyanin photoprotection, *Funct Plant Biol* 46(8) (2019) 756–765.
66. J. Liang, J. He, Protective role of anthocyanins in plants under low nitrogen stress, *Biochem Biophys Res Commun* 498(4) (2018) 946–953.
67. X. Li, G.J. Ahammed, X.N. Zhang, L. Zhang, P. Yan, L.P. Zhang, J.Y. Fu, W.Y. Han, Melatonin-mediated regulation of anthocyanin biosynthesis and antioxidant defense confer tolerance to arsenic stress in *Camellia sinensis* L, *J Hazard Mater* 403 (2021) 123922.
68. O.Y. Shoeva, E.K. Khlestkina, Anthocyanins participate in the protection of wheat seedlings against cadmium stress, *Cereal Res Commun* 46(2) (2018) 242–252.
69. H. Fang, Y. Dong, X. Yue, J. Hu, S. Jiang, H. Xu, Y. Wang, M. Su, J. Zhang, Z. Zhang, N. Wang, X. Chen, The B-box zinc finger protein MdBBX20 integrates anthocyanin accumulation in response to ultraviolet radiation and low temperature, *Plant Cell Environ* 42(7) (2019) 2090–2104.
70. M. Xiang, W. Ding, C. Wu, W. Wang, S. Ye, C. Cai, X. Hu, N. Wang, W. Bai, X. Tang, C. Zhu, X. Yu, Q. Xu, Y. Zheng, Z. Ding, C. Lin, Q. Zhu, Production of purple Ma bamboo (*Dendrocalamus latiflorus* Munro) with enhanced drought and cold stress tolerance by engineering anthocyanin biosynthesis, *Planta* 254(3) (2021) 50.
71. V. Sivankalyani, O. Feygenberg, S. Diskin, B. Wright, N. Alkan, Increased anthocyanin and flavonoids in mango fruit peel are associated with cold and pathogen resistance, *Postharvest Biol Technol* 111 (2016) 132–139.
72. H. Flachowsky, I. Szankowski, T.C. Fischer, K. Richter, A. Peil, M. Höfer, C. Dörschel, S. Schmoock, A.E. Gau, H. Halbwirth, M.V. Hanke, Transgenic apple plants overexpressing the *Lc* gene of maize show an altered growth habit and increased resistance to apple scab and fire blight, *Planta* 231(3) (2010) 623–635.
73. Y. Zhang, E. Butelli, R. De Stefano, H.J. Schoonbeek, A. Magusin, C. Pagliarani, N. Wellner, L. Hill, D. Orzaez, A. Granell, J.D. Jones, C. Martin, Anthocyanins double the shelf life of tomatoes by delaying overripening and reducing susceptibility to gray mold, *Curr Biol* 23(12) (2013) 1094–1100.
74. C.B. Wegener, G. Jansen, Soft-rot resistance of coloured potato cultivars (*Solanum tuberosum* L.): The role of anthocyanins, *Potato Res* 50(1) (2007) 31–44.
75. X.H. Sun, T.T. Zhou, C.H. Wei, W.Q. Lan, Y. Zhao, Y.J. Pan, V.C.H. Wu, Antibacterial effect and mechanism of anthocyanin rich Chinese wild blueberry extract on various foodborne pathogens, *Food Control* 94 (2018) 155–161.
76. V. Neveu, J. Perez-Jimenez, F. Vos, V. Crespy, L. du Chaffaut, L. Mennen, C. Knox, R. Eisner, J. Cruz, D. Wishart, A. Scalbert, Phenol-Explorer: An online comprehensive database on polyphenol contents in foods, *Database* 2010 (2010) bap024.
77. M.J. Aliaño-González, M. Ferreiro-González, E. Espada-Bellido, C. Carrera, M. Palma, J.A. Álvarez, J. Ayuso, G. F. Barbero, Extraction of anthocyanins and total phenolic compounds from açai (*Euterpe oleracea* mart.) using an experimental design methodology. Part 1: Pressurized liquid extraction, *Agronomy* 10(2) (2020) 183.
78. A.V.G. de Peredo, M. Vázquez-Espinosa, E. Espada-Bellido, A. Jiménez-Cantizano, M. Ferreiro-González, A. Amores-Arrocha, M. Palma, C.G. Barroso, G.F. Barbero, Development of new analytical microwave-assisted extraction methods for bioactive compounds from myrtle (*Myrtus communis* L.)., *Molecules* 23(11) (2018) 1–16.
79. M. Vázquez-Espinosa, E. Espada-Bellido, A.V.G. de Peredo, M. Ferreiro-González, C. Carrera, M. Palma, C.G. Barroso, G.F. Barbero, Optimization of microwave-assisted extraction for the recovery of bioactive compounds from the Chilean superfruit (*Aristotelia chilensis* (Mol.) Stuntz), *Agronomy* 8(11) (2018) 1–15.
80. J.A. He, M.M. Giusti, Anthocyanins: Natural colorants with health-promoting properties, *Annu Rev Food Sci Technol* 1 (2010) 163–187.
81. R. Zamora-Ros, V. Knaze, L. Luján-Barroso, N. Slimani, I. Romieu, M. Touillaud, R. Kaaks, B. Teucher, A. Mattiello, S. Grioni, F. Crowe, H. Boeing, J. Förster, J.R. Quirós, E. Molina, J.M. Huerta, D. Engeset, G. Skeie, A. Trichopoulou, V. Dilis, K. Tsiotas, P.H.M. Peeters, K.T. Khaw, N. Wareham, B. Bueno-de-Mesquita, M.C. Ocké, A. Olsen, A. Tjonneland, R. Tumino, G. Johansson, I. Johansson,

E. Ardanaz, C. Sacerdote, E. Sonestedt, U. Ericson, F. Clavel-Chapelon, M.C. Boutron-Ruault, G. Fagherazzi, S. Salvini, P. Amiano, E. Riboli, C.A. González, Estimation of the intake of anthocyanidins and their food sources in the European Prospective Investigation into Cancer and Nutrition (EPIC) study, *Br J Nutr* 106(7) (2011) 1090–1099.

82. V. Tosti, B. Bertozzi, L. Fontana, Health benefits of the mediterranean diet: Metabolic and molecular mechanisms, *J Gerontol A-Biol* 73(3) (2018) 318–326.
83. D. McCann, A. Barrett, A. Cooper, D. Crumpler, L. Dalen, K. Grimshaw, E. Kitchin, K. Lok, L. Porteous, E. Prince, E.J.S. Sonuga-Barke, J.O. Warner, J. Stevenson, Food additives and hyperactive behaviour in 3-year-old and 8/9-year-old children in the community: A randomised, double-blinded, placebo-controlled trial, *Lancet* 370(9598) (2007) 1560–1567.
84. N. Lourith, M. Kanlayavattanakul, Improved stability of butterfly pea anthocyanins with biopolymeric walls, *J Cosmet Sci* 71(1) (2020) 1–10.
85. S.A. Mahdavi, S.M. Jafari, E. Assadpoor, D. Dehnad, Microencapsulation optimization of natural anthocyanins with maltodextrin, gum Arabic and gelatin, *Int J Biol Macromol* 85 (2016) 379–385.
86. E. Pieczykolan, M.A. Kurek, Use of guar gum, gum arabic, pectin, beta-glucan and inulin for microencapsulation of anthocyanins from chokeberry, *Int J Biol Macromol* 129 (2019) 665–671.
87. B. Yousuf, K. Gul, A.A. Wani, P. Singh, Health benefits of anthocyanins and their encapsulation for potential use in food systems: A review, *Crit Rev Food Sci* 56(13) (2016) 2223–2230.
88. B.H. Chen, B. Stephen Inbaraj, Nanoemulsion and nanoliposome based strategies for improving anthocyanin stability and bioavailability, *Nutrients* 11(5) (2019) 1–21.
89. D.B. Rodriguez-Amaya, Update on natural food pigments – A mini-review on carotenoids, anthocyanins, and betalains, *Food Res Int* 124 (2019) 200–205.
90. Y. Ma, X. Ma, X. Gao, W. Wu, B. Zhou, Light induced regulation pathway of anthocyanin biosynthesis in plants, *Int J Mol Sci* 22(20) (2021) 1–16.
91. V. Usenik, F. Štampar, R. Veberič, Anthocyanins and fruit colour in plums (Prunus domestica L.) during ripening, Food Chem 114 (2009) 529–534.
92. S. Deroles, Anthocyanin biosynthesis in plant cell cultures: A potential source of natural colourants. *Anthocyanins: Biosynthesis, functions, and applications*, (2009) 108–167.

6 Exploration of Various Conventional and Contemporary Approaches to Enhance Secondary Metabolites Production in Plants

Devanshi Singh[1], *Yashasvi Saxena*[1], *Gurumayum Suraj Sharma*[1], *Divya Mohanty*[1], *and Nupur Mondal*[2]
[1]Department of Botany, Hindu College, University of Delhi, Delhi, India
[2]Department of Botany, ARSD College, University of Delhi, Delhi, India

ABBREVIATIONS

BAD	Betaine aldehyde dehydrogenase
CE	Capillary electrophoresis
CRISPR	Clustered regularly interspaced short palindromic repeats
FDS	Farnesyl diphosphate synthase
FPS	Farnesyl pyrophosphate synthase
GC	Gas chromatography
GE	Genetic engineering
GUS	Beta-glucuronidase
Gy	Gray
HPLC	High Performance Liquid Chromatography
KEGG	Kyoto Encyclopedia of Genes and Genomes
METLIN	Metabolite Mass Spectral Library of Standards
MS	Mass spectrometry
NMR	Nuclear magnetic resonance
nptII	Neomycin phosphotransferase gene
PDS	Phytoene desaturase
SMs	Secondary metabolites
UHPLC-ESI-MS/MS	Ultrahigh-performance liquid chromatography–electrospray ionization tandem mass spectrometry
UPLC-Q-TOF-MS/MS	Ultra Performance Liquid Chromatography coupled with a Hybrid Quadrupole Orthogonal Time-of-Flight Mass Spectrometer
XCMS	eXtensible Computational Mass Spectrometry

DOI: 10.1201/9781003518358-6

6.1 INTRODUCTION

Humans have relied on medicinal plants as the earliest form of health care. Traditional herbal medicine has been a cornerstone of health care for centuries, with various cultural groups developing their own methods to enhance well-being. The World Health Organization (WHO) reports that more than 80% of the world's population relies on plant-based treatments, which are the main source of healthcare for many communities worldwide. In the early times, drugs were derived from plants such as aspirin from *Salix alba*, digoxin from *Digitalis purpurea*, morphine from*Papaver somniferum*, quinine from *Cinchona officinalis*, and pilocarpine from *Pilocarpus microphyllus*. Currently, it is estimated that over 50% of available drugs are derived from medicinal plants (Jamshidi-Kia et al., 2017; Upadhyay and Singh, 2024). Thus, plants are an important source of biologically active molecules that help create novel semi-synthetic medications that can be used to treat and prevent a variety of illnesses(Petrisor et al., 2022).

The biologically active compounds of plants are known as phytochemicals and are categorized into primary and SMs based on their metabolic roles. In plants, the sources of mostSMs are three main pathways: shikimate, isoprenoid and the polyketide pathway (Bocso and Butnariu, 2022; Hussein and El-Anssary, 2019; Pagare et al., 2015; Upadhyay and Singh, 2021).

SMs isolated from medicinal plants can be classified according to their chemical structures (Figure 6.1) into different classes:

- **Terpenoids**:

  Terpenes are chemically derived from 5-carbon isoprene units (C5H8)n assembled in various ways, forming hemiterpenoids, monoterpenoids, sesquiterpenoids, diterpenoids, sesterterpenoids, triterpenoids, tetraterpenoids, and polyterpenoids(Boncan et al., 2020). Medicinally, terpenes exhibit analgesic, antibacterial, antifungal, antiprotozoan, diaphoretic, antirheumatic, and anti-inflammatory properties. Compounds such as camphor and menthol are monoterpenes, whereas vitamin K1, an antihemorrhagic compound discovered in plants in 1929, is a diterpenoid. Vitamin A is also a diterpenoid, and its related

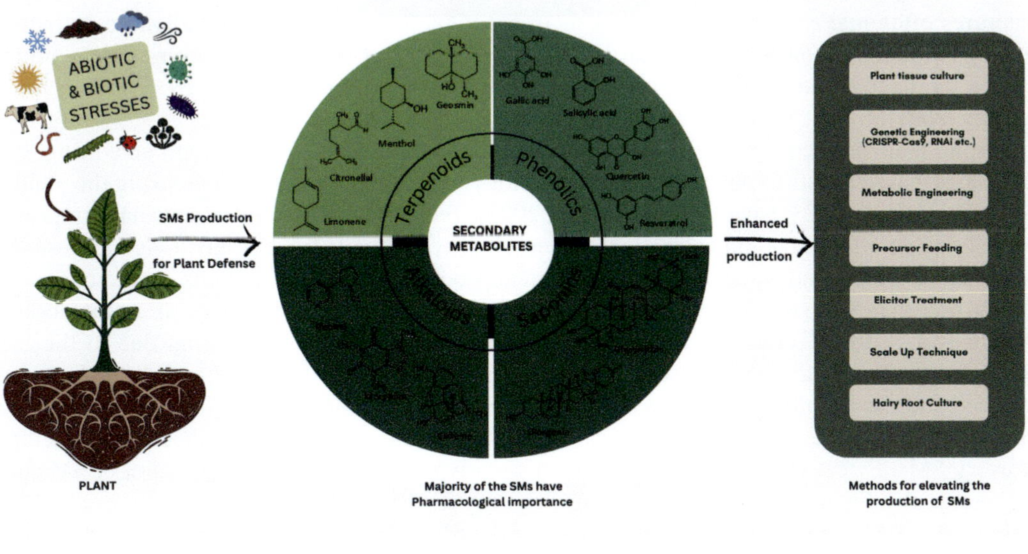

FIGURE 6.1 Diagrammatic representation of different types of secondary metabolites and the varied biotechnological tools employed to enhance their production.

compounds are known as carotenes (Pagare et al., 2015). A sesquiterpene, Artemisinin is used as an effective therapy against malaria-causing *Plasmodium falciparum* strains (Boncan et al., 2020).

- **Phenolics**:
  Phenolics are likely the largest group of plant SMs, derived from the shikimic acid pathway, which is common in bacteria, fungi, and plants. These are aromatic compounds, composed of more than one hydroxyl group that is methylated or glycosylated (Mahajan et al., 2020). They can be categorized based on their structures into several subtypes: phenolics, tannins, coumarins, flavonoids, chromones, stilbenes, and lignans. Some phenolics, like quercetin, are valued for their anti-inflammatory effects, while others, like silybin, are known for their antihepatotoxic properties (Hussein and El-Anssary, 2019).
- **Alkaloids**:
  Alkaloids are organic compounds that contain at least one nitrogen atom within a heterocyclic ring. Alkaloids can be categorized into true alkaloids, protoalkaloids, and pseudoalkaloids. They exhibit a wide range of pharmacological effects, including pain relief, local anesthesia, cardiac and respiratory stimulation, muscle relaxation, vasoconstriction, and toxicity. Additionally, alkaloids have antineoplastic, hypertensive, and hypotensive properties (Hussein and El-Anssary, 2019).
- **Saponins**:
  Saponins are compounds characterized by a polycyclic aglycone moiety, which can be either steroidal (steroidal saponins) or triterpenoidal (triterpenoidal saponins), linked to a carbohydrate unit (monosaccharide or oligosaccharide chain). They are often concentrated in the roots of species like *Dioscorea villosa*, *D. purpurea,Eleutherococcus senticosus*, *Gentiana lutea*, *Glycyrrhiza*spp., and *Panax ginseng*. Some saponins exhibit antitumor, anti-inflammatory, sedative, expectorant, and analgesic properties (Hussein and El-Anssary, 2019).

The production of SMs from medicinal plants faces several challenges. The concentration of these compounds in wild plants can vary significantly due to both genetic and environmental factors. Most of the SMs occur in low quantities in their native plants, and their complex structures often make biochemical synthesis impractical and costly. Additionally, the extinction rates of medicinal plant species jeopardize the availability of these natural compounds, underscoring the need for sustainable production alternatives (Han and Miao, 2024; Wawrosch and Zotchev, 2021).

Despite their high cost, modern methods such as metabolic engineering, genome editing, biotechnology-based substitutions, and synthetic biology offer crucial tools for increasing the yield of SMs (Selwal et al., 2023; Upadhyay and Singh, 2023). Technological advancements have improved the sustainable utilization and extraction of medicinal compounds from plants. Ongoing research is expected to further advance the biotechnological production of specific plant-derived chemicals by boosting the expression of genes that are involved with the biosynthesis of these metabolites (Pagare et al., 2015; Guerriero et al., 2018). However, this is often challenging due to the limited understanding of these processes. Optimizing SMs production through classical methods like plant cell and tissue culture as well as advanced ones like metabolic engineering and heterologous expression in microbes remains a significant research hurdle, as they haven't been optimized yet for every SMs. Effective use of biotic and abiotic elicitors to trigger the upregulation of SMs biosynthesis pathways in plant *in vitro* systems requires further investigation (Cardoso et al., 2019; Reshi et al., 2023). In this study, a comprehensive detailing of the classical and advanced biotechnological methods used for SMs production has been attempted, along with their drawbacks and the future perspectives.

6.2 BIOTECHNOLOGICAL APPROACHES FOR ELEVATING SECONDARY METABOLITE LEVELS

Stress induction by applying physical stress, such as light intensity or temperature changes, has been well known to enhancethe yield of different plant-derived SMs. Thus, a lot of work has been reported that utilizes various biotechnological approaches as effective strategies. These techniques include the following:

6.2.1 Plant Tissue Culture Techniques

Tissue culture techniques allow the controlled production of SMs in controlled conditions by manipulating the growth conditions, media composition, etc. Techniques such as callus culture, suspension culture, organ culture and hairy root culture facilitate the yields of SMs from plants.

6.2.1.1 Callus Culture

Callus culture is the growth of undifferentiated plant cells (callus) from explants (plant tissue samples) on a nutrient medium containing comparatively high auxin or cytokinin and auxin combinations under *in vitro* conditions.There are two main types of callus: embryogenic and non-embryogenic. Non-embryogenic callus cultures do not have the potential to form somatic embryos. Instead, it may differentiate into other plant tissues or remain as an undifferentiated mass. They are used for the enhanced production of plant SMs. This technique provides a large amount of biomass that can be further manipulated for SMs production (Vanisree et al., 2004). A study was reported in *Catharanthus roseus,* whose callus cultures were used to enhance the production of anti-cancerous compounds, vincristine and vinblastine, wherein the culture conditions were optimized to significantly increase the yield of these alkaloids (Patil and Shahnawaz, 2021). This group also reported increased latex production from *Papaver somniferum*, which is a commercial source of analgesics, morphine and codeine. The callus production of *P. somniferum*was optimized as an alternative means for the production of these SMs. Additionally, (Das et al., 2020) reported the increased production of hypericin, an antidepressant compound, from callus cultures of *Hypericum perforatum*, with optimized light and nutrient conditions leading to higher yields of the SMs.

6.2.1.2 Suspension Culture

Suspension culture is a technique in plant tissue culture where plant cells or tissues are grown in a liquid medium, allowing for the suspension of cells rather than anchoring them to a solid surface. Its advantages in scalability, uniformity, and control make it particularly suitable for the enhancement of SMs. The low yield from the natural source and the slow growth of the tree necessitate the exploration of alternative production methods for SMs (Table 6.1).

Cell suspension cultures of *Taxus* species have been investigated as a potential source of Taxol. It boosted taxol production through nutrient optimization, including fructose and phenylalanine, and by adding elicitors. Yields up to 14.78 mg/L were achieved, with high accumulation in specific growth phases. Similarly, diosgenin has been produced from cell suspension cultures of *Dioscorea doryophora*. The diosgenin content in the tuber-derived cell suspensions was comparable to that in the intact tuber, demonstrating the potential of this approach (Vanisree et al., 2004).

Other examples of SMs obtained using suspension cultures include stigmasterol from *Abutilon indicum*; gymnemic acids from *Gymnema sylvestre*; catechin from *Camellia sinensis*; terpenoid indole alkaloids from *Catharanthus roseus*; triterpenoids from *Ocimum basilicum*; artemisinin from *Artemisia annua*; plumbagin from *Plumbago europaea* and *P. zeylanica*; bacoside from *Bacopa monnieri*; hydrolyzable tannin from *Phyllanthus debilis*; atropine from *Hyoscyamus muticus*; triterpenic acids from *Thymus persicus*; saponin from *P. notoginseng*; resveratrol from *Vitis vinifera;* and withanolides (withaferin A and withanolide A) from *Withania coagulans*(Ozyigit et al., 2023; Yue et al., 2016).

TABLE 6.1
Secondary Metabolites Produced at Higher Levels by *in vitro*Plant Callus or Suspension Culture Compared to Whole Plant (Alamgir, 2017)

Secondary Metabolite	Plant Species	Family	Yield (% Dry Weight)		Culture Type
			Tissue Culture	Whole Plant	
Shikonin	*Lithospermum erythrorhizon* Siebold and Zucc.	Boraginaceae	20	1.5	Suspension
Ginsenoside	*Panax ginseng* C.A. Mey.	Araliaceae	27	4.5	Callus
Anthraquinones	*Morinda citrifolia* L.	Rubiaceae	18	0.3	Suspension
Ajmalicine	*Catharanthus roseus* (L.) G. Don	Apocynaceae	1.0	0.3	Suspension
Terpentine	*Catharanthus roseus* (L.) G. Don	Apocynaceae	1.8	0.5	Callus
Rosmarinic acid	*Coleus blumei* Benth.	Lamiaceae	15	3	Suspension
Ubiquinone-10	*Nicotiana tabacum* L.	Solanaceae	0.036	0.003	Suspension
Diosgenin	*Dioscorea deltoidea* Wall. ex Griseb.	Dioscoreaceae	2	2	Suspension
Benzylisoquinoline alkaloids	*Coptis japonica* (Thunb.) Makino	Ranunculaceae	11	5-10	Suspension
Berberine	*Thalictrum minus* L.	Ranunculaceae	10	0.01	Suspension
Berberine	*Coptis japonica* (Thunb.) Makino	Ranunculaceae	10	2-4	Suspension
Anthraquinones	*Galium verum* L.	Rubiaceae	5.4	1.2	Suspension
Anthraquinones	*Galium aparine* L.	Rubiaceae	3.8	0.2	Suspension
Nicotine	*Nicotiana tabacum* L.	Solanaceae	3.4	2.0	Callus
Glutathione	*Nicotiana tabacum* L.	Solanaceae	5.0	2.1	Callus
Bisoclaurine	*Stephania cephalantha* Hayata	Menispermaceae	2.3	0.8	Suspension
Tripdiolide	*Tripterygium wilfordii* Hook. f.	Celastraceae	0.05	0.001	Suspension

6.2.1.3 Organ Culture

Organ culture is a method in plant tissue culture that involves the cultivation of whole plant organs (such as leaves, stems, roots, or flowers) in a controlled environment, which can be directly induced to produce SMs (Table 6.2). This method preserves the differentiation status of cells, often leading to higher yields of specific metabolites. For instance, shoot cultures of *Salvia miltiorrhiza* were optimized to enhance the production of tanshinones, which have cardiovascular and anticancer activities. By manipulating light, hormones, and nutrients, researchers achieved maximum tanshinone yields in these differentiated cultures (Kai et al., 2011). Root cultures of *Withania somnifera* have similarly been exploited to increase the production of withanolides, which possess anti-inflammatory and anticancer properties (Sivanandhan et al., 2014). Using the root culture technique, several possible pharmacological compounds were obtained, including scopolamine, hyoscyamine, tropane alkaloid, plumbagin, and forskolin(Patil and Shahnawaz, 2021). Alamgir in 2017 reported success in root culture of ginseng (*Panax ginseng*) to produce saponin and other valuable metabolites.

TABLE 6.2
Examples of Secondary Metabolite Production Using Organ Culture in Plants

Plant	Family	Secondary Metabolite	Culture Type	References
Hypericum perforatum L.	Hypericaceae	Hypericin	Shoot	Bhatia and Bera (2015)
Lithospermum L.	Boraginaceae	Shikonin	Hairy root	Bhatia and Bera (2015)
Panax ginseng C.A. Mey.	Araliaceae	Ginsenoside	Adventitious root	Bhatia and Bera (2015)
Bupleurum falcatum L.	Apiaceae	Saikosaponins	Root	Vanisree et al. (2004)
Cephaelis ipecacuanha (Brot.) Willd.	Rubiaceae	Emetic alkaloids	Root	Vanisree et al. (2004)
Rhus javanica L.	Simaroubaceae	Gallotannins	Root	Vanisree et al. (2004)
Silybum marianum (L.) Gaertn.	Asteraceae	Flavonolignan	Root	Vanisree et al. (2004)
Withania somnifera (L.) Dunal	Solanaceae	Withaferin A	Shoot	Karuppusamy (2009)
Primula veris L.	Primulaceae	Saponins	Shoot	Karuppusamy (2009)
Podophyllum hexandrum Royle	Berberidaceae	Podophyllotaxin	Shoot	Karuppusamy, 2009
Ophiorrhiza rugosa var. *decumbens* (Gardner ex Thwaites) Deb & Mondal	Rubiaceae	Camptothecin	Shoot	Karuppusamy (2009)
Corydalis ambigua Cham. and Schltdl.	Papaveraceae	Corydaline	Embryo	Karuppusamy (2009)

6.2.2 Genetic Engineering Approaches

While traditional tissue culture techniques have been valuable for SM production, modern biotechnology tools provide enhanced precision, efficiency, and scalability. The GE approach involves the introduction of foreign genes or the modification of existing genes. *Agrobacterium*-mediated transformation, hairy root culture,CRISPR/Cas-9, RNA interference (RNAi) and transient gene expression are various methods that enable researchers to create a genetically modified plant that ultimately results in elevated production of desired SMs.

6.2.2.1 Transformations Mediated by *Agrobacterium*

Agrobacterium species have a unique ability to conduct horizontal genetic exchange between different kingdoms. Plant transformation mediated by *Agrobacterium tumefaciens*, which causes crown gall disease in various plants, is the most common method for introducing foreign genes into the plant cells. Other pathogenic *Agrobacterium* species include *A. rhizogenes* (hairy root disease), *A. vitis* (tumors on grapevines), and *A. rubi* (cane gall disease). These bacteria transfer their DNA from the bacterium to the host plant,which leads to integration and expression of a portion of a large plasmid [Ti- (tumor inducing) or Ri- (root inducing) plasmid], forming T-DNA. T-DNA traverses the host cell cytoplasm, enters the nucleus and integrates into the host genome. *Agrobacterium* transfers genes to plants with stable integration, low copy number, and consistent expression across generations (Gelvin, 2012; Khan et al., 2009).Mostly, tissue culture forms an integral part of the *Agrobacterium*-mediated transformation protocols.

6.2.2.1.1 By Agrobacterium tumefaciens

The rapid advancement in crop biotechnology is largely due to the creation of effective regeneration and transformation methods mediated by *Agrobacterium* for a range of crop species. This success has also beenapplied to medicinal plants, potentially enhancing their SMs content (Khan et al., 2009). Niazian and coworkers in 2019 reportedintroducing the BADH gene to the medicinal plant ajwain (*Trachyspermum ammi*) in order to increase its resistance to salinity and drought. The BADH gene was transferred using the pBI121 binary vector, which contains both the BADH gene and the kanamycin resistance gene (nptII) in its T-DNA region. This increased the valuable thymol content in the essential oil of the transgenic plants (55.07%) compared to the wildtype (39.20%).

Similarly, using*A. tumefaciens* strain EHA105 with a binary vector containing GUS and npt10 genes, metabolite content was increased in *Bacopa monnieri* (Khan et al., 2009). Transformation of *Artemisia annua L.* with a FDS cDNA under a CaMV 35S promoter using *A. tumefaciens* yielded 45 kanamycin-resistant shoots expressing FDS, with 2–3 times more artemisinin compared to controls (Chen et al., 2000). *Agrobacterium*-mediated transformation has also been successfully applied to other medicinal plants like *Taxus* (yew), *Echinacea*, *Scrophularia*, *Digitalis* and *Thalictrum*, all leading to significant increases in SMs (Park et al., 2003; Sales et al, 2003; Wang and To, 2004).

While many studies have optimized *Agrobacterium*-mediated transformation of reporter genes in medicinal plants, few have focused on genes with valuable agronomic traits. Computational methods like artificial neural networks can help overcome the complex nature of this biological system to develop optimized transformation protocols for metabolic engineering and agronomic trait transfer in medicinal plants (Niazian et al., 2019). While *Agrobacterium*-mediated transformation is a powerful tool for introducing new traits, CRISPR/Cas9 offers greater precision and flexibility for optimizing SM production. The ability to make specific, targeted changes in the plant genome allows researchers to enhance yields and develop plants with improved metabolic profiles more efficiently.

6.2.2.1.2 By Agrobacterium rhizogenes and Hairy Root Cultures

Hairy root culture is a specialized plant tissue culture technique that utilizes roots transformed by the bacterium *Agrobacterium rhizogenes*. It has become a useful technique for producing SMs obtained from plants. A portion of the Ri plasmid DNA, which contains the rol genes (rol A, rol B, rol C, and rol D), is transferred and integrated into the plant genome by *A. rhizogenes* after infection. Fast-growing, genetically stable hairy roots that can accumulate large amounts of beneficial phytochemicals are the result of this change. The capacity to generate high amounts of SMs, genetic and biosynthetic stability, and quick growth on basic media without growth regulators are the main characteristics of hairy root cultures(Abraham and Thomas, 2017).

In transgenic root cultures of *Nicotiana tabacum cv.*xanthii, the mean nicotine level increased drastically from 0.8 mg in non-transformed roots to 86 mg in rolABC transgenic roots. Similar results were found in rol ABC-transformed callus cultures of *Rubia cordifolia* and *Atropa belladonna* where the total anthraquinone content and tropane alkaloid production, respectively, were almost 2 times higher than in non-transformed callus (Sarkar et al., 2018).

6.2.2.1.3 Transient Gene Expression by Agroinfiltrations

Transient gene expression is a method used to introduce and express a gene in a host organism for a short period, typically without permanently integrating the gene into the host's genome. This technique is often employed in research to quickly assess the effects of gene modifications or to produce proteins and metabolites. These systems primarily rely on *Agrobacterium*-mediated delivery and viral expression machinery (Komarova et al., 2010).

Protoplasts, calli, whole plants, isolated plant organs, suspension cultures of plant cells, and specialized plant tissues are among the plant systems utilized for transient expression.Two main types of gene constructs are used (Tyurin et al., 2020):vectors bearing altered plant virus genomic sequences, including those cloned into T-DNA cassettes, and expression vectors built with binary Ti plasmids carrying gene cassettes as T-DNA.

Recent studies have shown that virus-based vectors can be efficiently used for high transient expression of foreign proteins in transfected plants. Additionally, non-*Agrobacterium* bacterial species like *Rhizobium* sp. NGR234, *Sinorhizobium meliloti* and *Mesorhizobium loti*have demonstrated the ability to genetically transform various plant tissues and species, laying the foundation for alternative tools in future plant biotechnology (Khan et al., 2009).

Agroinfiltration, or agroinjection, is a transient expression method that uses *Agrobacterium* to deliver genetic constructs into plant tissues without stable integration. It involves infiltrating a solution containing *Agrobacterium*into the intercellular spaces of leaves using a needleless syringe or vacuum. *Nicotiana benthamiana* and related species are commonly used as host plants for agroinfiltration to produce a variety of plant natural products. Examples include cannabinoids, which stop the SARS-CoV-2 coronavirus from entering human cells, and artemisinin, which is used to treat malaria. The classic transient expression system of tobacco was shown to be feasible in the aforementioned circumstances, but it was not suitable for large-scale production(Tyurin et al., 2020; Yao et al., 2022).

Catharanthus roseus(Madagascar periwinkle) produces over 100 medicinal alkaloids, including the anti-cancer drugs vinblastine and vincristine. However, these valuable bis-indole alkaloids accumulate only in trace amounts in leaves, making extraction costly and supply limited. Researchers showed that in undamaged *C. roseus* leaves, agroinfiltration can effectively overexpress important genes involved in the biosynthesis of terpenoid indole alkaloid (TIA), such as str1 (strictosidine synthase) and tdc (tryptophan decarboxylase). This trustworthy and reasonably priced assay is a useful tool for learning more about the TIA pathway and creating metabolic engineering plans for increased production of therapeutic alkaloids in this significant medicinal plant, such as vinblastine and vincristine(Di Fiore et al., 2004).

The main advantage of agroinfiltration is the avoidance of T-DNA integration, reducing biosafety and regulatory concerns while still allowing efficient transcription and translation of the introduced genes. This transient approach enables rapid, pilot-scale experiments to assess the activity of transformation constructs and validate the production of small amounts of recombinant proteins within 3–5 days after infiltration (Kaur et al., 2021).

Transient expression systems based on *Agrobacterium* infiltration have been successfully used to produce high yields of functional monoclonal antibodies in various plant species like tobacco, *N. benthamiana*.An innovative approach using recombinant viral vectors like TMV (Tobamovirus) and PVX (Potato Virus X) enables efficient co-expression of antibody heavy and light chains in plant cells, demonstrating the potential of plant-based platforms for rapid production of therapeutic antibodies (Komarova et al., 2010).

It has been demonstrated that agroinfiltration, a transient gene expression method, may generate significant amounts of recombinant proteins in a matter of days. During this brief phase, the somatic cells in the infiltrated leaf regions may express the transgenes when regulated by a constitutive promoter such as CaMV-35S(Kaur et al., 2021).

6.2.2.2 CRISPR-Cas9 Gene Editing

Before the CRISPR associated9 (Cas9) system appeared in 2013 as a more straightforward, effective, and adaptable RNA-guided engineered nuclease for examining gene function and creating better crops, zinc finger nucleases (ZFNs) and transcription activator-like effector nucleases (TALENs) were the most common genome editing tools(Bortesi and Fischer, 2015; Liu et al., 2017; Upadhyay 2021, 2024). Guide RNAs are used by CRISPR/Cas9 to target particular DNA sequences for alteration. Numerous plant metabolic pathways, including those that produce alkaloids, terpenoids, flavonoids, phenolics, and saponins, have been extensively engineered using it.

By combining technologies from synthetic biology, systems biology, functional genomics, and next-generation sequencing, this technique makes it possible to create genetically modified medicinal crops with improved features for the manufacturing of pharmaceuticals and nutraceuticals(Dey, 2021). The type II clustered regularly interspaced short palindromic repeat (CRISPR)/Cas9 (CRISPR-associated) system from *Streptococcus pyogenes* is the most utilized system(Bortesi and Fischer, 2015).

CRISPR/Cas9 genome editing has enabled precise manipulation of specialized metabolite biosynthesis in various medicinal plants, allowing researchers to engineer improved pharmaceutical properties. Numerous studies have documented the use of CRISPR to modify metabolic flux via particular pathways. For example, in *Salvia miltiorrhiza*, the rosmarinic acid synthase (SmRAS) gene decreased the production of the essential enzyme, resulting in a decrease in the amounts of the phenolic acids, lithospermic acid and rosmarinic acid. Also, CRISPR was used to edit the MdPGT1 gene involved in phloridzin biosynthesis. Edited lines showed a 76.7%–78.4% reduction in phloridzin without impacting plant growth (Mipeshwaree et al., 2023).

CRISPR has also been used to manipulate alkaloid biosynthesis in *Papaver somniferum*, tanshinone production in *S. miltiorrhiza*, farnesyl pyrophosphate levels in *Dioscorea polystachya*, and hyoscyamine accumulation in *Atropa belladonna* (Dey, 2021; Liu et al., 2017; Zeng et al., 2021).

Despite the potential of the CRISPR/Cas system, it still faces some limitations. One major concern is off-target effects, where unintended genetic modifications may occur, potentially disrupting essential metabolic pathways and leading to unwanted phenotypic changes. Additionally, the complexity of SM biosynthesis pathways can make it challenging to predict the outcomes of genetic edits accurately. Regulatory hurdles and ethical considerations surrounding genetic modification can also pose obstacles to research and commercial applications. Furthermore, the reliance on CRISPR technology may oversimplify the intricate interactions within metabolic networks, risking the loss of beneficial traits or synergies that occur in natural systems. This gap in understanding the genetic underpinnings of specialized metabolite production hinders the full realization of CRISPR's potential for engineering improved medicinal and nutritional traits in crops (Rajyaguru et al., 2021).

6.2.2.3 Gene Silencing by RNA Interference (RNAi)

RNAi is a post-transcriptional gene silencing process that breaks down target mRNA transcripts using short double-stranded RNA (dsRNA) molecules. By introducing dsRNA complementary to pathway genes, researchers can effectively downregulate the expression of competing or undesirable pathways, thereby channeling carbon flux toward the metabolite of interest. Genetic circuits can be designed to respond to different combinations of small interfering RNAs (siRNAs) or microRNAs (miRNAs), enabling complex logic operations to optimize metabolite production. This technology complements other GE strategies, like overexpressing rate-limiting enzymes, to enhance the yields of medicinally and industrially important plant-derived compounds (Borgio, 2009; Pathak et al., 2021).Many works have been reported in recent years like in *Papaversomniferum*, RNAi has been used to target key enzymes in the morphine biosynthetic pathway. Silencing the salutaridinol 7-O-acetyltransferase (SalAT) gene led to the accumulation ofintermediates like salutarydine and salutaridinol, rather than the final product morphine. Similarly, knockdown of the codeinone reductase enzyme activity through DNA-directed RNAi resulted in the accumulation of the upstream precursor (S)-reticuline (an important intermediate for isoquinoline alkaloid biosynthesis) instead of downstream alkaloids. In *Eschscholzia californica*, RNAi-mediated suppression of the Berberine Bridge Enzyme (BBE) also caused the accumulation of (S)-reticuline. These studies demonstrate how RNAi can be leveraged to redirect metabolic flux and accumulate desired medicinal compounds by selectively silencing competing pathways (Borgio, 2009; Ziegler and Facchini, 2008).

Researchers created an RNAi construct that targets the 3-hydroxy-3-methylglutaryl-CoA reductase (HMGR) gene in the medicinal plant *Centella asiatica*. This enzyme is essential for catalyzing the initial step of isoprenoid production in the MVA (mevalonate) pathway, which leads to the formation of triterpenoid saponins. The pHANNIBAL vector was used to build the RNAi-HMGR construct, which was then cloned into the binary vector pART27. Transient expression analysis confirmed the silencing of the endogenous HMGR gene, laying the groundwork for future metabolic engineering efforts (Kalita et al., 2018).

In *Panax notoginseng*, silencing the dammarenediol synthase (DDS) gene using RNAi led to an 84.5% reduction in ginsenoside levels in roots, without causing any obvious morphological changes in the transgenic plants. This demonstrated the critical role of DDS expression in ginsenoside biosynthesis. To further enhance triterpene saponin production in *P. notoginseng*, researchers combined overexpression of FPS with RNAi-mediated silencing of cycloartenol synthase (CAS). This dual manipulation increased triterpene accumulation more effectively than modulating a single gene by redirecting metabolic flux toward saponin biosynthesis (Yang et al., 2017).

Phenylalanine ammonia-lyase (PAL) catalyzes the initial step in the phenylpropanoid pathway in Salvia miltiorrhiza, which is essential for the production of rosmarinic acid.A 20%–70% reduction in overall phenolic content and significantly lower amounts of rosmarinic acid and salvianolic acid B were the outcomes of RNAi silencing of the PAL gene, indicating the crucial function that PAL plays in the manufacture of important water-soluble medicinal chemicals(Song and Wang, 2011).Significant inhibition of nicotine biosynthesis and increased accumulation of scopolamine, the desired tropane alkaloid, were the results of RNAi silencing of the quinolinate phosphoribosyl transferase gene in the nicotine biosynthesis pathway in the species *Duboisia*(Wawrosch and Zotchev, 2021).

A major limitation of using RNAi tools is the potential for off-target effects.RNAi relies on short 21–25 nucleotide sequences to target and suppress gene expression. However, these short homology regions may inadvertently match and silence unintended genes, leading to unpredictable phenotypes. Despite this drawback, RNAi remains a widely used tool for identifying gene functions and manipulating biosynthetic pathways to increase the production of desired metabolites in plants. The advantages it offers, such as the ability to selectively downregulate specific pathway genes, often outweigh the risks of off-target effects, especially when combined with other validation techniques (Pathak et al., 2021).

6.2.3 Metabolic Manipulation

Metabolic manipulation strategies aim to enhance SMs production either by the modulation of the expression of the key enzyme or by providing specific precursors in the particular biosynthetic pathway from which the desired SMs can be obtained. The following methods are employed to optimize the synthesis of the SMs:

6.2.3.1 Metabolic Engineering

Metabolic engineering combines GE and cell physiology to optimize production of desired compounds in organisms like plants. Recent advances in DNA synthesis and high-throughput omics technologies have accelerated progress in this field by providing comprehensive gene, protein, and metabolite data across diverse species. Metabolic engineering of plants enables the purposeful alteration of endogenous biochemical pathways to generate crops with improved commercial, agronomic, or processing traits. Key strategies include increasing precursor flux, blocking competitive pathways, introducing new pathways, overcoming rate-limiting steps, and overexpressing regulatory genes. This allows engineering plants as "cell factories" to produce valuable bioactive compounds (Sharma et al., 2019). Identifying key genes involved in specialized metabolite biosynthesis has been revolutionized by high-throughput sequencing and 'omics' technologies, enabling precise manipulation of specialized metabolism to optimize production of valuable plant-derived compounds. Genome-wide association studies (GWAS) and transcriptomics can pinpoint putative biosynthetic genes, even in species lacking reference genomes. Transcriptomic profiling via RNA-seq enables discovery of uncharacterized pathways and enzyme variants for metabolic engineering applications. Comprehensive datasets like CathaCyc for *C. roseus* and the PhytoMetaSyn project are crucial resources for elucidating specialized metabolite biosynthetic pathways (Wilson and Roberts, 2014).

Plant metabolic engineering is being used to optimize the production of valuable medicinal terpenoids and alkaloids in their native plant hosts, paving the way for improved supply of these important natural products. These studies demonstrate the potential of combining upstream and downstream pathway manipulations to effectively channel carbon flux toward desired plant-derived compounds.

Enhancing the production of important plant-derived chemicals has been achieved through the overexpression of essential enzymes involved in their biosynthesis. Tyrosine aminotransferase (TAT) and PAL are important enzymes in the biosynthesis of rosmarinic acid, and it has been demonstrated that overexpressing these enzymes increases the biosynthesis of rosmarinic acid in a variety of plant species (Hakkim et al., 2019).

Similarly, over expressing the hyoscyamine 6β-hydroxylase (H6H) gene, which encodes a key enzyme in the final steps of scopolamine biosynthesis, has increased the production of the valuable alkaloid scopolamine in *Atropa belladonna*, a plant naturally low in this compound (Wawrosch and Zotchev, 2021).

Over expressing key biosynthetic genes has also been successful in enhancing the production of other valuable plant-derived compounds. In ginseng, overexpressing the PgSQS1 gene led to a 1.6–3-fold increase in total ginsenoside levels. In *Artemisia annua*, over expressing the artemisinin biosynthesis genes ADS, CYP71AV1, and CPR promoted the accumulation of the antimalarial compound artemisinin. In *S. miltiorrhiza*, co-expressing the SmHMGR and SmGGDS genes increased tanshinone production 4.7-fold (Lu et al., 2016).

In *C. roseus*, metabolic engineering efforts have targeted key genes in the MEP pathway, such as DXS and G8O, to enhance the production of specific alkaloids. Over expressing DXS increased levels of ajmalicine, serpentine, and lochnericine, while G8O over expression boosted catharanthine accumulation. Co-overexpressing DXS and G8O had a synergistic effect, significantly increasing ajmalicine by 16%, lochnericine by 31%, and tabersonine by 13% (Tang and Pan, 2017).

Despite significant progress in metabolic engineering approaches to improve Taxol production, several challenges remain in developing sustainable biotechnological platforms for this valuable anti-cancer compound. The Taxol biosynthetic pathway is long and complex, involving 19 steps from the universal precursor GGPP, with several key enzymes, particularly cytochrome P450s, still uncharacterized. Expressing functional Taxol pathway enzymes, especially the CYP450s, in heterologous hosts has proven difficult. While advances in sequencing have enabled identification of some missing genes, their performance in different production hosts is still inconclusive (Mutanda et al., 2021; Sabzehzari et al., 2020).

A key limitation to metabolic engineering of plant specialized metabolism is the incomplete elucidation of many biosynthetic pathways. While 'omics' data has expanded, accurately annotating and functionally characterizing enzymes is critical. The complex interactions and regulatory mechanisms among pathways further complicate engineering efforts to optimize production of valuable plant-derived compounds (Wilson and Roberts, 2014).

6.2.3.2 Precursor Feeding

Feeding biosynthetic precursors to plant cell cultures is a strategy to enhance production of valuable SMs. The premise is that adding exogenous or endogenous compounds that can be converted by the plant cells into the desired SMs via biosynthetic pathways will provide additional substrate and promote accumulation of the target compounds. This approach was pioneered in the 1960s for improving alkaloid production in plant cell cultures. By supplying precursors that feed into the alkaloid biosynthetic pathways, researchers were able to boost yields of these medicinally important compounds in the cultured cells. Precursor feeding strategies can be optimized by first culturing cells in nutrient-rich media to promote growth, followed by transfer to production media supplemented with precursors or elicitors. This two-stage approach allows for efficient biomass accumulation followed by targeted metabolite synthesis (Isah et al., 2018). Many works have been successfully reported for this technique.

In 2013, Patil and co-workers reported that *Digitalis purpurea L.* has been cultivated commercially to obtain cardiotonic glycosides, digitoxin and digoxin. These compounds are widely used to strengthen cardiac function and regulate heart rhythm. Adding progesterone, cholesterol, and squalene to shoot cultures significantly increased digitoxin and digoxin accumulation, with up to 9.1-fold and 11.9-fold enhancements, respectively (Patil et al., 2013). Similarly, exogenous sterols like cholesterol, stigmasterol and mixed sterols added to *Solanum lyratum* cell cultures promoted solasodine, solasonidine and solanine accumulation without affecting biomass (Isah et al., 2018). Feeding squalene to *Centella asiatica* callus cultures enhanced madecassoside and asiaticoside production (Hussain et al., 2012).

Amino acid feeding has also proven beneficial for triterpene production. In *C. asiatica* callus and cell cultures, adding leucine led to a manifold increase in the medicinally important triterpene asiaticoside. Likewise, valine, isoleucine and threonine supplementation boosted adhyperforin levels 3–7-fold and 2-fold respectively in *H. perforatum* shoot cultures. Casein hydrolysate and L-phenylalanine promoted biomass and scopoletin production in *Spilanthes acmella*cultures, with L-phenylalanine being more effective (4.51-fold increase).In *Silybum marianum* hairy root cultures, supplementing the medium with the amino acid L-phenylalanine, a precursor for the flavonolignan silymarin, increased its accumulation. Similarly, L-phenylalanine feeding also boosted production of the anti-cancer compound podophyllotoxin in *Podophyllum hexandrium*cell suspension cultures (Isah et al., 2018; Motyka et al., 2023). Supplementing the MS medium with 500–1,000 mg/L tyrosine increases L-DOPA production 1.6–8.1-fold but has less impact on biomass (Rakesh and Praveen, 2022). Feeding L-phenylalanine to *in vitro* shoot cultures of *Vitex agnus-castus* also increased the accumulation of key flavonoids and phenolic acids like neochlorogenic acid, p-coumaric acid, rutin and casticin (Skrzypczak-Pietraszek et al., 2018).Phenylalanine analogues have been used to select high rosmarinic acid-producing cell lines in *Lavandula vera* cultures. Similarly, p-fluorophenylalanine has enabled selection of high phenolics-yielding *Capsicum annuum* cell lines (Isah et al., 2018).

In plant cell cultures, precursor feeding and gamma radiation can increase the synthesis of important flavonoid and phenolic chemicals. The maximum phenolic content was accumulated by callus cultures produced from seed explants subjected to 25 Gy gamma radiation and treated with 4 mg/L phenylalanine in *Silybum marianum*. Similarly, callus cultures from 25 Gy irradiated seeds treated with 1 mg/L phenylalanine had the highest quantities of flavonoids(Khalifa et al., 2022).

Key limitations include developing cost-effective protocols that efficiently convert cheap byproducts into the desired plant SM using precursor feeding. Optimizing the timing, concentration and combination of precursors is critical for maximizing yields in cell culture systems.

6.2.4 Elicitation

Elicitors are molecules that stimulate the plant's defense responses, leading to increased biosynthesis of specialized metabolites. Elicitors can be derived from pathogens, insects, or other organisms. They are recognized by plant immune receptors, triggering defense signaling cascades (Wen et al., 2023). Elicitors have been successfully used to boost the production of many important SMs, such as berberine in *Thalictrum*, thiophenes in *Tagetes*, taxol in *Taxus*, tanshinones in *Salvia*, and hypericins in *Hypericum*species (Patil et al., 2013).

The process of elicitation includes cell surface receptor recognition and signal transduction pathway activation, which in turn upregulates genes involved in secondary metabolism. The genetic makeup and physiological condition of the plant cells, tissues, or organs being cultivated determine how effective the process is. Elicitation studies and differential gene expression analysis offer important new information about how secondary metabolism pathways are regulated. By comparing the transcriptomes of high- and low-metabolite-producing genotypes, researchers can identify key regulatory genes and elucidate the complex networks controlling specialized metabolism in plants (Isah et al., 2018).

Based on their nature, elicitors can be divided into biotic and abiotic categories. We refer to compounds of non-biological origin as abiotic elicitors. These elements fall into three categories: physical, chemical, and hormonal. Biotic elicitors are substances that have a biological origin. In response to injury and pathogen infection, they function as signaling molecules that trigger the creation of several compounds (Bhaskar et al., 2022).

6.2.4.1 Abiotic Elicitation

Physical elicitors:Various physical elicitors have been shown to enhance the production of valuable SMs in plant cell and tissue cultures, like low-energy ultrasound (US) treatment,which increased shikonin yield 2–3-fold in *Lithospermum erythrorhizon*cell cultures, taxol 1.5–1.8-fold in *Taxus chinensis*, 3-fold in*T. baccata*, and ginsenoside saponins by 75% in *Panax ginseng*. Light exposure increased alkaloid content in *Hyoscyamus albus* roots compared to dark-grown roots. In *C. roseus* suspension cultures, UV-B light boosted the formation of vindoline and catharanthine, whereas in *C. roseus* hairy roots, it increased the total terpenoid indole alkaloids. Capsaicin accumulation in *Capsicum chinensis* cell cultures and steviol glycoside levels in *Stevia rebaudiana* callus and suspension cultures were both increased by osmotic stress brought on by sugar. *C. roseus* hairy roots exhibited higher levels of linolenic acid percentage and total indole alkaloids when the cultivation temperature was lowered from 25°C to 19.5°C. A 5°C temperature increase significantly boosted ginsenoside content in *Panax quinquefolius*roots (Naik et al., 2016a).

Chemical elicitors: These are particular metal ions (Ni, Ag, Fe and Co)that have been shown to effectively stimulate the production of valuable SMs in plant cell and tissue cultures. Cobalt (Co) at concentrations of 5, 25 and 50 μM, as well as low doses of silver (Ag) and cadmium (Cd) at 5 μM, were most effective in enhancing phenolic acid and 3-O-glucosyl-resveratrol levels up to 1.6-fold in *Vitis vinifera* cell suspensions after just 4 hours of treatment. In hairy root cultures of *Brugmansia candida*, it was also shown that calcium (Ca^{2+}), silver (Ag^{+}), and cadmium (Cd^{2+}) ions enhanced the accumulation of tropane alkaloids such as scopolamine and hyoscyamine. *Perovskia abrotanoides* root cultures produced tanshinone in response to silver nitrate ($AgNO_3$). In cultures of *Lepidium sativum*, zinc (Zn^{2+}) at 900 μM increased the amount of lepidine(Wen et al., 2023).

Hormonal elicitors: For elicitation studies, various plant hormones have been extensively used, especially jasmonic acid (JA) and salicylic acid (SA) and their derivatives.Many examples have been reported, like in *Artemisia absinthium* cell suspension cultures, where MeJA and JA combined with gibberellic acid (GA) increased the accumulation of the antimalarial compound artemisinin. Similarly, MeJA enhanced phytoecdysteroid levels in *Ajuga bracteosa*(Nilkanthi) hairy roots and increased phenolic and flavonoid content when used with phenylacetic acid. Other examples include MeJA increasingthe production of the triterpenoid, saponin and bacoside in *Bacopa monnieri*. MeJA yielded the maximum gymnemic acid content, while salicylic acid had a moderate effect in *G. sylvestre*. In *Panax ginseng*, MeJA treatment led to the accumulation of the rare ginsenoside Rg3 (Singh and Dwivedi, 2018).

Salicylic acid (SA) is another important elicitor that has been shown to upregulate specialized metabolism in various medicinal plants. In *C. roseus*, the joint use of MeJA and β-cyclodextrins, along with UV exposure, enhanced extracellular ajmalicine production. SA treatment doubled the accumulation of hypericin and pseudohypericin in *Hypericum perforatum* cell cultures. In *Salvia miltiorrhiza*hairy roots, SA combined with GE greatly increased tanshinone production. Other examples include SA-mediated increase in withanolides in *W. somnifera*, scopolamine in *Brugmansia candida*, and anthraquinones in *Rubia cordifolia* (Naik et al., 2016b).

6.2.4.2 Biotic Elicitation

Fungal elicitors: Fungal elicitors are some of the most important biotic elicitors used to boost the production of valuable plant SMs, and they come from both free-living and endophytic fungi. When fungi and plants interact, defensive mechanisms are activated, resulting in increased synthesis of phytoalexins and other specialized metabolites. Fungal extracts of fungi like *Aspergillus niger,A. flavus, Penicillium notatum*, and *Fusarium oxysporum* have been used as elicitors to boost the accumulation of compounds like anthocyanins in *Daucus carota*and thiophenes in *Tagetes patula. A. niger* extracts in particular were highly effective, causing a 2-fold increase in anthocyanins and an 85% increase in thiophenes compared to controls. Similarly, when *A. niger* was used as an elicitor in *Mentha piperita* cell cultures, menthol output reached 140.8 mg/L, surpassing that of prior non-biological elicitation methods. Typically, hyphal tip culture procedures are used to create pure fungal cultures (Bhaskar et al., 2022).

Yeast extract (YE): is a complex product containing diverse macromolecules and small molecules that provide a rich source of nutrients. Studies have shown that the addition of YE to plant cell cultures can enhance the production of valuable SMs. In case of *Atropa belladonna*, YE increased the levels of the precursor amino acids ornithine and arginine, which are important for alkaloid biosynthesis. YE boosted the levels of the medicinally important tropane alkaloids hyoscyamine and scopolamine, in *B. candida* hairy root cultures. Importantly, the ratio of scopolamine to hyoscyamine was significantly increased (Wen et al., 2023).In *C. roseus* cell cultures, YE supplementation led to increased production of the anticancer terpenoid indole alkaloids vinblastine and vincristine.

Bacterial elicitation: It is the use of bacterial cells, cellular components, or cellular extracts to trigger a response in plants, particularly to stimulate SM synthesis in vitro. In *Panax quinquefolius* cell suspension cultures, culture filtrates of *Pseudomonas monteilii* and *Bacillus circulans* were employed to induce ginsenoside synthesis. *Pseudomonas monteilii* increased 2.5-fold, but *B. circulans* declined slightly after 5 days. Prolonged exposure led to a 20%–30% decrease.*Serratia marcescens*at 1.25%–2.5% concentration increased ginsenoside 1.6–2.5-fold in *Panax sikkimensis* cultures. *Bacillus subtilis* at 2.5% demonstrated a 1.8-fold increase(Bhaskar et al., 2022).

Algal Elicitation: Phycocyanin from *Spirulina*platensis was extracted and added into *Capsicum frutescens L.* and *D. carota L.* cell cultures in five different concentrations (1.5, 3.0, 6.0, 12 and 24 µg/mL). The culture showed increased capsaicin production 2-fold on day 3 and 1.5-fold on day 6 at 3–6 µg/mL (Bhaskar et al., 2022).

Polysaccharides: They are found in all living organisms like plants, animals and microbes. Polysaccharide elicitation is commonly used to elicit antimicrobial metabolites. Pectin, sodium alginate, gum arabic, chitosan and yeast extract were used to elicit anthocyanins and phenolics in *Vitis vinifera*cultures. Pectin gave 1.8-fold higher anthocyanins on day 9. Yeast cell wall extract increased saponins 1.17-fold in *Panax ginseng* hairy roots (Bhaskar et al., 2022). The addition of chitosan in *Azadirachta indica*cultures increased azadirachtin production by approximately 5-fold, while signal compounds like jasmonic acid and salicylic acid further enhanced it by around 6-fold and 9-fold, respectively (Satdive et al., 2007). Chitosan and chitin induce phytoalexins, phenylpropanoids and naphtho dianthrones production in plants.

Although biotic elicitors increase metabolite production, there are a few drawbacks that must be addressed. None of the biotic elicitors have a significant impact on SM production. As a result, the experiment must be thoroughly developed in order to optimize the concentration and type of biotic elicitors utilized. Furthermore, the chemical contents and concentrations in crude extracts are unknown, making it difficult to determine which component of the extract is responsible for the elicitation. Aside from these, biotic elicitors influence physiological parameters by creating a stress environment for plant growth (Bhaskar et al., 2022) (Table 6.3).

TABLE 6.3
Selected Examples of Enhanced Production of Various Secondary Metabolites Produced from Various Modern Techniques

Technique	Plant	Methodology	Result	References
Hairy root	*Catharanthus roseus*	*Peganum harmala* (host) with tryptophane decarboxylase gene using A. rhizogenes	Increased concentration of serotonin	Bhatia and Bera (2015)
	Hyoscymus muticus	*A. belladonna* with 6-hydroxylase gene using A. rhizogenes	Increased enzyme activity and a five-fold higher concentration of scopolamine	Bhatia and Bera (2015)
CRISPR Cas9	*Dioscorea zingiberensis*	Farnesyl pyrophosphate synthase gene (DzFPS) involved in diosgenin synthesis is targeted	Reduced transcription of DzFPS results in reduced concentration of squalene with 60% mutation rate	Rajyaguru et al. (2021)
	Salvia miltiorrhiza	Diterpene synthase gene (SmCPS1) involved in tanshinone biosynthesis is targeted	White roots and reduced tanshinones with 42.3% mutation rate	Rajyaguru et al. (2021)
RNAi	*Eschscholzia californica*	Berberine bridge enzyme (BBE) targeted (either partial or incomplete silence)	Benzophenanthridine alkaloids produced	Borgio (2009)
	Torenia fournieri	Glutatione STransferase 1 and 2, ubiquitin conjugating enzyme, putative cullin, anthocyanidin synthase, putative flowering-time gene, GPT and the glutathione conjugate transporter targeted	Anthocyanin produced	Borgio, 2009
	Papaver somniferum	Salutaridinol 7-O-acetyltransferase (SalAT) gene is targeted	Morphine production enhanced	Borgio (2009)
Precursor feeding	*Camptotheca acuminate*	Addition of tryptophan	Increased concentration of camptothecin	Bhatia and Bera (2015)
	Azadirachta indica	Addition of sodium acetate, Squalaine, IPP, GPP precurors in the culture	Increased concentration of azadirachtin	Bhatia and Bera (2015)
	Taxus cuspidate	Addition of phenylalanine	Increased taxol yields	Alamgir (2017)
Abiotic elicitation	*Glycyrrhiza echinata*	Addition of sodium alginate in the plant culture	Increased production of echinatin	Fazili et al. (2022)
	Datura stramonium	Metal ions: Al^{3+}, Cr^{3+}, Co^{2+}, Ni^{2+}, Cu^{2+}, Zn^{2+} added in plant culture	Sesquiterpenoids production increased	Fazili et al. (2022)
	Glycyrrhiza uralensis	UV stress	Glycyrrhizin production increased	Fazili et al. (2022)
Biotic elicitation	*Cupressus lusitanica*	Fungal elicitor added	Enhanced production of indole alkaloids	Fazili et al. (2022)
	Dioscorea zingiberensis	Fusarium oxysporum (cell wall elicitor) added in suspension culture	Diosgenin accumulation was enhanced and jasmonic acid concentration was increased	Bhatia and Bera (2015)
	Polygonum tinctorium	Carbohydrate elicitor chitosan is added	Enhanced indirubin production	Bhatia and Bera (2015)

6.3 SYNTHETIC BIOLOGY

Synthetic biology is an interdisciplinary field that uses ideas from biology, engineering, genetics, and computer science to create and build new biological parts, devices, and systems. Its major purpose is to construct or redesign organisms that can perform certain functions or generate useful products. It provides an alternate source for producing useful plant-derived specialized metabolites such as alkaloids, terpenoids, and phenolics. Researchers are developing genetically encoded biosensors to visualize these metabolites in plants, enabling mapping of their spatial and temporal distribution. This knowledge is used to build comprehensive models of plant metabolism and its integration with growth and development. Principles of synthetic biology, such as standardization and the Design-Build-Test-Learn cycle, streamline the biosensor development process (Garagounis et al., 2021; Hasanpour et al., 2020). It involves assembling standardized DNA "bio parts" into devices and systems, aiming for a 'plug-and-play' approach. Applications span healthcare, industry and environment, but require considering ethical, safety and security implications (Garner, 2021). The modular structure of SM biosynthetic pathways makes them ideal targets for synthetic biology. Advances in genomics and systems biology have enabled characterizing these pathways and rationally designing new metabolites (Leitão and Enguita, 2021). Pioneering projects have created designer cells producing drugs and smart therapeutic cells for autonomous disease treatment. Examples include optogenetics, engineered probiotics, CART cells, and gut-brain-liver axis engineering. These approaches hold immense potential to revolutionize biomedicine, but significant challenges remain in ensuring efficacy, safety and biocontainment (Ausländer et al., 2017; Yan et al., 2023).It offers exciting strategies to enhance production of valuable plant triterpenoids with therapeutic potential.

Synthetic biology approaches have been used in microbial systems to produce complex terpenoid and alkaloid natural products that are difficult to extract from their native plant sources. For instance, Taxol, an important anti-cancer drug, has seen significant advancements in its biosynthesis through metabolic engineering. Researchers were able to engineer *E. coli* to produce the key precursor compound taxadiene at high titers of 1 g/L, a major improvement over the low-yielding plant cell culture approach. Similarly, the biosynthesis of ginsenosides, triterpene saponins with cancer prevention and anti-aging effects, has been successfully achieved in yeast, reaching titers of 2.2 g/L for ginsenoside Rh2 and 5.0 g/L for ginsenoside compound K. In the case of cannabinoids, the biosynthesis of cannabigerolic acid, a key precursor, was established in yeast at 0.1 g/L (Yan et al., 2023). The semi-synthetic production of artemisinin, an important antimalarial drug, also demonstrates the potential of synthetic biology for pharmaceutical development. This was done by metabolically engineering microbes to produce a late-stage artemisinin precursor, followed by organic synthesis of the final steps.Researchers were able to generate a product indistinguishable from plant-derived artemisinin (Paddon and Keasling, 2014).

Despite these advancements, synthetic biology faces various challenges and limitations. Synthetic biology-based therapeutics face ethical and legal challenges similar to those encountered with transgenic foods and stem cell therapies, despite potentially having better control mechanisms. Extensive *in vivo* testing is needed to ensure efficacy and safety, particularly to prevent adverse immune responses. Biocompatibility challenges, regulatory hurdles, and maintaining microbial viability in the gastrointestinal tract are significant issues. Addressing these through sophisticated engineering and regulatory frameworks is essential for practical applications in medicine and agriculture (Brooks and Alper, 2021).

However, the future of synthetic biology-based therapeutics still remains promising. Future directions in synthetic biology include refining genetic circuit activation for controlled treatment release, particularly for engineered probiotics and mammalian cells used in disease treatment. As new technologies and applications in the biomedical area and highly efficient microbial pharmaceutical manufacture become more viable, synthetic biology is set to make important contributions to tackling human health concerns (Brooks and Alper, 2021; Garner, 2021).

6.4 METABOLOMICS

Metabolomics is the thorough examination of biological sample metabolites to detect changes in metabolite profiles. Metabolomics identifies metabolite biomarkers that indicate targets and pathways linked with the efficacy, side effects, and toxicity of various treatments (Hasanpour et al., 2020). It provides phenotypic and physiological information, and when combined with bioinformatics, it enables interpretation of experimental data for drug development applications (Sinha et al., 2019). The concept of metabolomics was first proposed in 1999 by Professor Jeremy Nicholson, drawing parallels to genomics and proteomics. Metabolomics, an integral aspect of systems biology, focuses on tiny molecules with a relative molecular weight of less than 1000 Daltons. Its goal is to quantitatively examine all metabolites in organisms and determine their link with physiological and pathological changes (Yang et al., 2019).

Metabolomics has emerged as an effective approach for identifying herbal medications and natural compounds. Comprehensive metabolite profiling is enabled by a combination of chromatographic techniques such as HPLC, GC, and Capillary Electrophoresis (CE), as well as spectroscopic approaches such as Mass Spectrometry (MS) and NMR. HPLC is the most extensively used technique for assessing the quality of herbal items. Metabolomics has made it easier to classify medicinal plants based on their geographical origin and to control their quality.

Specific bioactive compounds have been identified using HPLC-MS, Ultra High Performance Liquid Chromatography-Electrospray Ionization-Mass Spectrometry (UHPLC-ESI-MS/MS), and Liquid Chromatography-Linear Trap Quadrupole-Orbitrap (LC-LTQ-Orbitrap). GC-MS has also been applied for quality determination of herbal species (Panda et al., 2018).

Metabolomics has become a valuable tool in understanding the mechanisms of action and potential biomarkers of herbal medicines for diabetes management. Studies have identified disturbed metabolic pathways in diabetic models, such as carbohydrate metabolism, the TCA cycle, lipids, purines, and amino acids, with reproducible biomarkers like glucose, pyruvate, lactate, and amino acids being detected. Techniques like LC-MS and NMR provide detailed profiles of lipids, the TCA cycle, amino acids, and carbohydrates, offering insights into the complex interactions and synergistic effects of multi-component therapies (Hasanpour et al., 2020).

The application of metabolomics in traditional Chinese medicine (TCM) research is particularly promising, despite the inherent differences between TCM and Western medical theories. Metabolomics studies have identified key bioactive compounds and elucidated the anti-inflammatory, anticancer, and neuroprotective mechanisms of various TCM herbs. For example, the anti-inflammatory mechanism of Forsythia extracts and Forsythia oil was studied using Ultra Performance Liquid Chromatography-Quadrupole Time-of-Flight Mass Spectrometry (UPLC-Q-TOF-MS/MS), which revealed that they exert their effects via various metabolic pathways. As TCM and Western medicine continue to converge, metabolomics is emerging as a powerful tool to bridge the gap, enabling a more comprehensive understanding of the efficacy and mechanisms of action of TCM (Yang et al., 2019). This knowledge can guide the modernization and globalization of TCM, providing new avenues for diabetes management.

Metabolomic studies generate vast datasets that require specialized software tools for efficient data processing and analysis. Automated programs like MetAlign, MZmine2, and XCMS facilitate tasks such as peak identification, alignment, and metabolite quantification for MS data. Comprehensive metabolite annotation relies on electronic databases like MassBank, METLIN, and the Human Metabolome Database. Compound and pathway databases, including KNApSAcK, KEGG, and PlantCyc, provide valuable information for peak identification and understanding metabolic processes. Statistical techniques like principal component analysis and partial least squares discriminant analysis are crucial for identifying significant metabolite differences across samples. These informatics resources and analytical methods are essential for extracting meaningful insights from the complex data generated in metabolomic studies (Okazaki and Saito, 2012).By leveraging metabolomics, researchers can develop strategies to enhance the production of valuable SMs in plants, benefiting fields such as agriculture, pharmacology, and biotechnology.

6.5 BIOREACTOR UP-SCALING

In vitro production of plant SMs faces challenges like inconsistent efficiency, sensitivity, slow growth, and low oxygen levels in plant tissues. However, biotechnological innovations like bioreactor cultivation of organ and tissue cultures have improved the synthesis of these valuable compounds. Factors such as optimizing the culture environment and measuring biomass are crucial for scaling up SMs production (Patil and Shahnawaz, 2021).

Bioreactors have emerged as valuable tools for producing plant-derived SMs. These controlled environments allow cultivating plant cells, tissues, or microorganisms under optimized conditions to enhance the synthesis of desired compounds. Bioreactors can operate in various modes to maximize productivity. Key parameters like pH, temperature, nutrient supply, and oxygen levels are carefully monitored and adjusted to support targeted biological reactions and metabolic pathways. Compared to conventional methods, bioreactors offer advantages like independent supply, better environmental control, reproducible product quality, and higher yields of SMs. This technology has become an important platform for the large-scale, sustainable production of valuable plant-derived pharmaceuticals and nutraceuticals. Continuous culture is regarded as the most promising strategy for enhancing bioreactor production. For example, a study using continuous cell culture of Taxus cuspidata and semi-continuous culture of Taxus canadensis enhanced paclitaxel production, respectively, with 100% cell recycling (Jamshidi-Kia et al., 2017).

Bioreactor cultivation of adventitious roots has emerged as an attractive alternative to whole plant, cell, or hairy root cultures for producing valuable SMs. Bioreactor technology enables the large-scale cultivation of these specialized root cultures to overcome the limitations of traditional methods. While just a few plant-derived chemicals, such as paclitaxel, shikonin, ginsenosides, and berberine, are now manufactured commercially using this method, the potential is enormous. Recent research has developed novel root culture methods for a variety of important medicinal plants, including *Morinda citrifolia, Echinacea purpurea, Echinacea angustifolia, Hypericum perforatum,* and *Panax ginseng*, using large-scale bioreactors ranging from 500 to 1,000 L. The ginsenoside production system developed for *Panax ginseng* using 10-ton bioreactors demonstrates the scalability of this technique (Baque et al., 2012).

6.6 CONCLUSIONS

Concisely, the biotechnological strategies for increasing SMs in plants gave positive responses across numerous techniques, such as plant tissue culture, genetic modification, elicitation, synthetic biology, breeding approaches and up-scaling of bioreactors. Plant tissue culture techniqueslike callus, suspension, organ and hairy root cultures have become favorable methodsfor SM production for substances like morphine, taxol and diosgenin. Several genetic engineering techniques, such as Agrobacterium-mediated gene transfer, CRISPR/Cas9 gene editing, RNAi technology haveenabled significant improvements inthe metabolic pathways for increasing the production of thesevaluable compounds. Biotic and abiotic elicitors have been found useful in enhancing the accumulation of these metabolitesby stimulating the plantdefense systems. Modern synthetic biology offers apromisingsolution forvisualizing themetabolic processes, thus providing severalpossibilities forremodeling these metabolic processes for better production and yield. Research also highlights various future challenges in this field, like the further enhancement of these biotechnological techniques with the predisposition to increased yield and effectiveness; the genetic and biosynthetic stability of cells grown in culture; the question of ethics, security, and safety when it comes to using genetic engineering and synthetic biology. Furthermore, it is equally challenging to upscale operating at industrial levels that are capable of providing adequate returns on investment while approaching the situation sustainably. To overcome these challenges and to optimize the usage of biotechnological approaches for the production of SMs, research and development are called for continuously.

REFERENCES

Abraham J, Thomas TD (2017) Hairy root culture for the production of useful secondary metabolites. In: *Biotechnology and Production of Anti-Cancer Compound*, pp. 201–230. https://doi.org/10.1007/978-3-319-53880-8_9

Alamgir ANM (2017) Cultivation of herbal drugs, biotechnology, and in vitro production of secondary metabolites, high-value medicinal plants, herbal wealth, and herbal trade. In: Alamgir ANM (eds) *Therapeutic Use of Medicinal Plants and Their Extracts: Volume 1:Pharmacognosy*,vol. 73, pp. 379–452. https://doi.org/10.1007/978-3-319-63862-1_9

Ausländer S, Ausländer D, Fussenegger M (2017) Synthetic biology –The synthesis of biology. *Angewandte Chemie International Edition56*(23):6396–6419. https://doi.org/10.1002/anie.201609229

Baque MA, Moh SH, Lee EJ, Zhong JJ, Paek KY (2012) Production of biomass and useful compounds from adventitious roots of high-value added medicinal plants using bioreactor. *Biotechnology Advances30*(6):1255–1267.https://doi.org/10.1016/j.biotechadv.2011.11.004

Bhaskar R, Xavier LSE, Udayakumaran G, Kumar DS, Venkatesh R, Nagella P (2022) Biotic elicitors: A boon for the in-vitro production of plant secondary metabolites. *Plant Cell, Tissue and Organ Culture (PCTOC)149*(1):7–24https://doi.org/10.1007/s11240-021-02131-1

Bhatia S, Bera T (2015) Classical and nonclassical techniques for secondary metabolite production in plant cell culture. In: Bhatia S, Sharma K, Dahiya R, Bera T (eds) *Modern Applications of Plant Biotechnology in Pharmaceutical Sciences*, pp.231–291. https://doi.org/10.1016/B978-0-12-802221-4.00007-8

Bocso NS, Butnariu M (2022) The biological role of primary and secondary plants metabolites. *Journal of Nutrition and FoodProcessing5*(3):1–7. https://doi.org/10.31579/2637-8914/094

Boncan DAT, Tsang SS, LiC, Lee IH, Lam HM, Chan TF, Hui JH (2020) Terpenes and terpenoids in plants: Interactions with environment and insects. *International Journal of Molecular Sciences21*(19):7382https://doi.org/10.3390/ijms21197382

Borgio JF (2009) RNA interference (RNAi) technology: A promising tool for medicinal plant research. *Journal of Medicinal Plants Research3*(13):1176–1183.

Bortesi L, Fischer R (2015) The CRISPR/Cas9 system for plant genome editing and beyond. *Biotechnology Advances33*(1):41–52. https://doi.org/10.1016/j.biotechadv.2014.12.006

Brooks SM, AlperHS (2021) Applications, challenges, and needs for employing synthetic biology beyond the lab. *Nature Communications12*(1):1390.https://doi.org/10.1038/s41467-021-21740-0

Cardoso JC, Oliveira MEBD, Cardoso FDC (2019) Advances and challenges on the in vitro production of secondary metabolites from medicinal plants. *HorticulturaBrasileira37*(2):124–132. https://doi.org/10.1590/S0102-053620190201

ChenDH, YeHC, LiGF (2000) Expression of a chimeric farnesyl diphosphate synthase gene in *Artemisia annua* L. transgenic plants via *Agrobacterium tumefaciens*-mediated transformation. *Plant Science155*(2):179–185https://doi.org/10.1016/S0168-9452(00)00217-X

Das A, Sarkar S, Bhattacharyya S, Gantait S (2020) Biotechnological advancements in *Catharanthus roseus* L. G. Don. *Applied Microbiology andBiotechnology104*:4811–4835. https://doi.org/10.1007/s00253-020-10592-1

Dey A (2021) CRISPR/Cas genome editing to optimize pharmacologically active plant natural products. *Pharmacological Research164*:105359.https://doi.org/10.1016/j.phrs.2020.105359

Di Fiore S, Hoppmann V, Fischer R, Schillberg S (2004) Transient gene expression of recombinant terpenoid indole alkaloid enzymes in *Catharanthus roseus* leaves. *Plant Molecular Biology Reporter22*:15–22. https://doi.org/10.1007/BF02773344

Fazili MA, Bashir I, Ahmad M, Yaqoob U, Geelani SN (2022) In vitro strategies for the enhancement of secondary metabolite production in plants: A review. *Bulletin of the National ResearchCentre46*(1):35. https://doi.org/10.1186/s42269-022-00717-z

Garagounis C, Delkis N, Papadopoulou KK (2021) Unraveling the roles of plant specialized metabolites: Using synthetic biology to design molecular biosensors. *NewPhytologist231*(4):1338–1352. https://doi.org/10.1111/nph.17470

Garner KL (2021) Principles of synthetic biology. *Essays inBiochemistry 65*(5):791–811.https://doi.org/10.1042/EBC20200059

Gelvin SB (2012) Traversing the cell: *Agrobacterium* T-DNA's journey to the host genome. *Frontiers in Plant Science 3*:52. https://doi.org/10.3389/fpls.2012.00052

Guerriero G, Berni R, Muñoz-Sanchez JA, Apone F et al. (2018) Production of plant secondary metabolites: Examples, tips and suggestions for biotechnologists. *Genes 9*(6):309. https://doi.org/10.3390/genes9060309

Hakkim FL, Idrees M, Bakshi HA, Mombasawala L, Rashan L (2019) Enhancement of rosmarinic acid content by biotechnological approaches and metabolic engineering. *Natural Bio-active Compounds- Biotechnology, Bioengineering, and Molecular Approaches3*:317–330. https://doi.org/10.1007/978-981-13-7438-8_13

Han T, Miao G (2024) Strategies, achievements, and potential challenges of plant and microbial chassis in the biosynthesis of plant secondary metabolites. *Molecules29*(9):2106. https://doi.org/10.3390/molecules29092106

Hasanpour M, Iranshahy M, Iranshahy M (2020) The application of metabolomics in investigating anti-diabetic activity of medicinal plants. *Biomedicine & Pharmacotherapy128*:110263. https://doi.org/10.1016/j.biopha.2020.110263

Hussain MS, Fareed S, Ansari S, Rahman MA, AhmadI Z, Saeed M (2012) Current approaches toward production of secondary plant metabolites. *Journal of Pharmacy and Bioallied Sciences4*(1):10–20. https://doi.org/10.4103/0975-7406.92725

Hussein RA, El-Anssary AA (2019) Plants secondary metabolites: The key drivers of the pharmacological actions of medicinal plants. *Herbal Medicine*1(3):11–30. https://doi.org/10.5772/intechopen.76139

Isah T, Umar S, Mujib A, Sharma MP, Rajasekharan PE, Zafar N, Frukh A (2018) Secondary metabolism of pharmaceuticals in the plant in vitro cultures: Strategies, approaches, and limitations to achieving higher yield. *Plant Cell, Tissue and Organ Culture132*:239–265. https://doi.org/10.1007/s11240-017-1332-2

Jamshidi-Kia F, Lorigooini Z, Amini-Khoei H (2017) Medicinal plants: Past history and future perspective. *Journal of HerbmedPharmacology 7*(1):1–7. https://doi.org/10.15171/jhp.2018.01

Kai G, Xu H, Zhou C, Liao P, Xiao J, Luo X, You L, Zhang L (2011) Metabolic engineering tanshinone biosynthetic pathway in *Salvia miltiorrhiza* hairy root cultures. *Metabolic Engineering 13*(3):319–327. https://doi.org/10.1016/j.ymben.2011.02.003

Kalita R, Modi MK, Sen P (2018) RNAi mediated silencing of 3-hydroxy-3-methylglutaryl-CoA reductases (HMGR) in *Centella asiatica. Gene Reports11*:52–57. https://doi.org/10.1016/j.genrep.2018.02.004

Karuppusamy S (2009) A review on trends in production of secondary metabolites from higher plants by in vitro tissue, organ and cell cultures. *Journal of Medicinal PlantsResearch3*(13):1222–1239.

Kaur M, Manchanda P, Kalia A, Ahmed FK, Nepovimova E, Kuca K, Abd-Elsalam KA (2021) Agroinfiltration mediated scalable transient gene expression in genome edited crop plants. *International Journal of Molecular Sciences22*(19):10882. https://doi.org/10.3390/ijms221910882

Khalifa AM, Abd-ElShafy E, Abu-Khudir R, GaafarRM (2022) Influence of gamma radiation and phenylalanine on secondary metabolites in callus cultures of milk thistle (*Silybum marianum*L.). *Journal of Genetic Engineering and Biotechnology20*(1):166. https://doi.org/10.1186/s43141-022-00424-2

Khan MY, Aliabbas S, Kumar V, Rajkumar S (2009) Recent advances in medicinal plant biotechnology. *Indian Journal of Biotechnology8*(1):9–22.

Komarova TV, Baschieri S, Donini M, Marusic C, Benvenuto E, Dorokhov YL (2010) Transient expression systems for plant-derived biopharmaceuticals. *Expert Review of Vaccines9*(8):859–876. https://doi.org/10.1586/erv.10.85

Leitão AL, Enguita FJ (2021) Synthetic biology approaches for secondary metabolism engineering. In SinghV (ed) *Microbial Cell Factories Engineering for Production of Biomolecules*, pp.51–64. Academic Press. https://doi.org/10.1016/B978-0-12-821477-0.00022-2

Liu X, Xie C, Si H, Yang J (2017) CRISPR/Cas9-mediated genome editing in plants. *Methods121*:94–102. https://doi.org/10.1016/j.ymeth.2017.03.009

Lu X, Tang K, Li P (2016) Plant metabolic engineering strategies for the production of pharmaceutical terpenoids. *Frontiers in Plant Science7*:1647. https://doi.org/10.3389/fpls.2016.01647

Mahajan M, Kuiry R, Pal PK (2020) Understanding the consequence of environmental stress for accumulation of secondary metabolites in medicinal and aromatic plants. *Journal of Applied Research on Medicinal and AromaticPlants18*:100255. https://doi.org/10.1016/j.jarmap.2020.100255

Mipeshwaree Devi A, Khedashwori Devi K, Premi Devi P, Lakshmipriyari Devi M, Das S (2023) Metabolic engineering of plant secondary metabolites: Prospects and its technological challenges. *Frontiers in Plant Science, 14*:1171154.
https://doi.org/10.3389/fpls.2023.1171154

Motyka S, Jafernik K, Ekiert H, Sharifi-Rad J, Calina D, Al-Omari B, Szopa A, Cho WC (2023) Podophyllotoxin and its derivatives: Potential anticancer agents of natural origin in cancer chemotherapy. *Biomedicine & Pharmacotherapy158*:114145. https://doi.org/10.1016/j.biopha.2022.114145

Mutanda I, Li J, Xu F, Wang Y (2021) Recent advances in metabolic engineering, protein engineering, and transcriptome-guided insights toward synthetic production of taxol. *Frontiers in Bioengineering and Biotechnology 9*:632269. https://doi.org/10.3389/fbioe.2021.632269

Naik PM., Al-Khayri JM (2016a) Impact of abiotic elicitors on in vitro production of plant secondary metabolites: A review. *Journal of Advanced Research in Biotechnology1*(2):7. https://doi.org/10.15226/2475-4714/1/2/00102

Naik PM, Al-Khayri JM (2016b) Abiotic and biotic elicitors-role in secondary metabolites production through in vitro culture of medicinal plants. In: ShankerA, Shanker C (eds) *Abiotic and Biotic Stress in Plants—Recent Advances and Future Perspectives*, vol. 17, pp. 247–77.InTech.https://doi.org/10.5772/61442.

Niazian M, Sadat-Noori SA, Tohidfar M, Galuszka P, Mortazavian SMM (2019) *Agrobacterium*-mediated transformation of ajowan (*Trachyspermum ammi* (L.) Sprague): An important industrial medicinal plant. *Industrial Crops and Products132*:29–40. https://doi.org/10.1016/j.indcrop.2019.02.005

Okazaki Y, Saito K (2012) Recent advances of metabolomics in plant biotechnology. *Plant Biotechnology Reports 6*:1–15. https://doi.org/10.1007/s11816-011-0191-2

Ozyigit II, Dogan I, Hocaoglu-Ozyigit A, Yalcin B, Erdogan A, Yalcin IE, Cabi E, Kaya Y (2023) Production of secondary metabolites using tissue culture-based biotechnological applications. *Frontiers in Plant Science 14*:1132555. https://doi.org/10.3389/fpls.2023.1132555

Paddon CJ, Keasling, JD (2014) Semi-synthetic artemisinin: A model for the use of synthetic biology in pharmaceutical development. *Nature Reviews Microbiology*, *12*(5):355–367. https://doi.org/10.1038/nrmicro3240

Pagare S, Bhatia M, Tripathi N, Pagare S, Bansal YK (2015) Secondary metabolites of plants and their role: Overview. *Current trends in Biotechnology andPharmacy 9*(3):293–304.

Panda A, Parida AK, RanganiJ (2018) Advancement of metabolomics techniques and their applications in plant science: Current scenario and future prospective. In: Ahmad P, Ahanger M, Singh V, Tripathi D, Alam P, Alyemeni M (eds) *Plant Metabolites and Regulation under Environmental Stress*, pp. 1–36. Academic Press.https://doi.org/10.1016/B978-0-12-812689-9.00001-7

Park SU, Chae YA, FacchiniPJ (2003) Genetic transformation of the figwort, *Scrophularia buergeriana Miq.*, an Oriental medicinal plant. *Plant Cell Reports 21*:1194–1198. https://doi.org/10.1007/s00299-003-0639-0

Pathak AR, Patel SR, Joshi AG (2021) RNA interference (RNAi): A genetic tool to manipulate plant secondary metabolite pathways. In: Tang G, Teotia S, Tang X, Singh D (eds) *RNA-Based Technologies for Functional Genomics in Plants*, pp.169–198. Springer, Cham.https://doi.org/10.1007/978-3-030-64994-4_9

Patil JG, Ahire ML, Nitnaware KM, Panda S, Bhatt VP, Kishor PBK, Nikam TD (2013) In vitro propagation and production of cardiotonic glycosides in shoot cultures of *Digitalis purpurea* L. by elicitation and precursor feeding. *Applied Microbiology and Biotechnology 97*:2379–2393. https://doi.org/10.1007/s00253-012-4489-y

Patil VN, Shahnawaz M (2021) Tissue culture approaches to enhance plant secondary metabolites production. In: Shahnawaz M (eds) *Biotechnological Approaches to Enhance Plant Secondary Metabolites*, vol. 1, pp. 89–98. CRC Press. https://doi.org/10.1201/9781003034957-6

Petrisor G, Motelica L, Craciun LN, Oprea OC, Ficai D, Ficai A (2022) *Melissa officinalis*: Composition, pharmacological effects and derived release systems—A review. *International Journal of Molecular Sciences 23*(7):3591. https://doi.org/10.3390/ijms23073591

Rajyaguru R, Maheshala N, Mootapally C, Nathani N, Tomar R, Bhalani H, Sharma P (2021) CRISPR-Cas9 approaches to enhance contents of plant secondary metabolites. In: ShahnawazM (eds) *Biotechnological Approaches to Enhance Plant Secondary Metabolites*, vol. 1, pp. 133–152. CRC Press. https://doi.org/10.1201/9781003034957-9

Rakesh B, Praveen N (2022) Elicitor and precursor-induced approaches to enhance the in vitro production of L-DOPA from cell cultures of Mucuna pruriens. *Industrial Crops and Products* 188:115735. https://doi.org/10.1016/j.indcrop.2022.115735

Reshi ZA, Ahmad W, Lukatkin AS, Javed SB (2023) From Nature to lab: A review of secondary metabolite biosynthetic pathways, environmental influences, and in vitro approaches. *Metabolites13*(8):895. https://doi.org/10.3390/metabo13080895

Sabzehzari M, Zeinali M, Naghavi MR (2020) Alternative sources and metabolic engineering of Taxol: Advances and future perspectives. *Biotechnology Advances43*:107569. https://doi.org/10.1016/j.biotechadv.2020.107569

Sales E, Segura J, Arrillaga I (2003) Agrobacterium tumefaciens-mediated genetic transformation of the cardenolide-producing plant *Digitalis minor* L. *Planta Medica 69*(2):143–147.https://doi.org/10.1055/s-2003-37709

Sarkar S, Ghosh I, Roychowdhury D, Jha S (2018) The effects of rol genes of *Agrobacterium rhizogenes* on morphogenesis and secondary metabolite accumulation in medicinal plants. In: Kumar N (eds) *Biotechnological Approaches for Medicinal and Aromatic Plants: Conservation, Genetic Improvement and Utilization*, pp. 27–51. Springer, Singapore.https://doi.org/10.1007/978-981-13-0535-1_2

Satdive RK, Fulzele DP, Eapen S (2007) Enhanced production of azadirachtin by hairy root cultures of *Azadirachta indica* A. Juss by elicitation and media optimization. *Journal ofBiotechnology 128*(2):281–289https://doi.org/10.1016/j.jbiotec.2006.10.009

Selwal N, Rahayu F, Herwati A, Latifah E, Supriyono, Suhara C, Suastika IBK, Mahayu WM, WaniAK (2023) Enhancing secondary metabolite production in plants: Exploring traditional and modern strategies. *Journal of Agriculture and Food Research14*:100702.https://doi.org/10.1016/j.jafr.2023.100702

Sharma M, Koul A, Sharma D, Kaul S, Swamy MK, Dhar MK (2019) Metabolic engineering strategies for enhancing the production of bio-active compounds from medicinal plants. *Natural Bio-Active Compounds: Biotechnology, Bioengineering, and Molecular Approaches 3*:287–316.https://doi.org/10.1007/978-981-13-7438-8_12

Singh A, Dwivedi P (2018) Methyl-jasmonate and salicylic acid as potent elicitors for secondary metabolite production in medicinal plants: A review. *Journal of Pharmacognosy and Phytochemistry 7*(1):750–757.

Sinha R, Sharma B, Dangi AK, ShuklaP (2019) Recent metabolomics and gene editing approaches for synthesis of microbial secondary metabolites for drug discovery and development. *World Journal of Microbiology and Biotechnology35*:1–14. https://doi.org/10.1007/s11274-019-2746-2

Sivanandhan G, Selvaraj N, Ganapathi A, Manickavasagam M (2014) Enhanced biosynthesis of withanolides by elicitation and precursor feeding in cell suspension culture of *Withania somnifera* (L.) Dunal in shake-flask culture and bioreactor. *PLoS One9*(8):e104005. https://doi.org/10.1371/journal.pone.0104005

Skrzypczak-Pietraszek E, Piska K, Pietrasze kJ (2018) Enhanced production of the pharmaceutically important polyphenolic compounds in *Vitex agnus castus* L. shoot cultures by precursor feeding strategy. *Engineering in Life Sciences 18*(5):287–297.https://doi.org/10.1002/elsc.201800003

Song J, Wang Z (2011) RNAi-mediated suppression of the phenylalanine ammonia-lyase gene in *Salvia miltiorrhiza* causes abnormal phenotypes and a reduction in rosmarinic acid biosynthesis. *Journal of Plant Research 124*:183–192.https://doi.org/10.1007/s10265-010-0350-5

Tang K, Pan Q (2017) Strategies for enhancing alkaloids yield in Catharanthus roseus via metabolic engineering approaches. In: Naeem M, Aftab T, Khan M (eds) *Catharanthus Roseus: Current Research and Future Prospects*, pp.1–16. Springer, Cham. https://doi.org/10.1007/978-3-319-51620-2_1

Tyurin AA, Suhorukova AV, Kabardaeva KV, Goldenkova-PavlovaI V (2020) Transient gene expression is an effective experimental tool for the research into the fine mechanisms of plant gene function: Advantages, limitations, and solutions. *Plants*9(9):1187. https://doi.org/10.3390/plants9091187

Upadhyay SK (2021) *Genome Engineering for Crop Improvement*. Wiley. https://doi.org/10.1002/9781119672425

Upadhyay SK (2024) *Applications of Genome Engineering in Plants*. Wiley.https://doi.org/10.1002/9781394183913

Upadhyay SK, SinghSP (2021)*Bioprospecting of Plant Biodiversity for Industrial Molecules*. Wiley. https://doi.org/10.1002/9781119718017

Upadhyay SK, SinghSP (2023)*Plants as Bioreactors for Industrial Molecules*. Wileyhttps://doi.org/10.1002/9781119875116

Upadhyay SK, SinghSP (2024)*Herbal Nutraceuticals: Products and Processes*. Wiley. https://doi.org/10.1002/9781394241576

Vanisree M, Lee CY, Lo SF, Nalawade SM, Lin CY, Tsay HS (2004) Studies on the production of some important secondary metabolites from medicinal plants by plant tissue cultures. *Botanical Bulletin of AcademiaSinica45*(1):1–22.

Wang HM, To KY (2004) *Agrobacterium*-mediated transformation in the high-value medicinal plant *Echinacea purpurea*. *Plant Science166*(4):1087–1096. https://doi.org/10.1016/j.plantsci.2003.12.035

Wawrosch C, Zotchev SB (2021) Production of bioactive plant secondary metabolites through in vitro technologies—Status and outlook. *Applied Microbiology andBiotechnology105*(18):6649–6668. https://doi.org/10.1007/s00253-021-11539-w

Wen Y, Liao Y, Tang Y, Zhang H, Zhang J, Liao Z (2023) Metabolic effects of elicitors on the biosynthesis of tropane alkaloids in medicinal plants. *Plants12*(17):3050. https://doi.org/10.3390/plants12173050

Wilson SA, Roberts SC (2014) Metabolic engineering approaches for production of biochemicals in food and medicinal plants. *Current Opinion in Biotechnology26*:174–182. https://doi.org/10.1016/j.copbio.2014.01.006

Yan X, Liu X, Zhao C, Chen GQ (2023) Applications of synthetic biology in medical and pharmaceutical fields. *Signal Transduction and Targeted Therapy 8*(1):199. https://doi.org/10.1038/s41392-023-01440-5

Yang Q, Zhang AH, Miao JH, Sun H, Han Y, Yan GL, Wu FF, Wang XJ (2019) Metabolomics biotechnology, applications, and future trends: A systematic review. *RSC Advances 9*(64):37245–37257. https://doi.org/10.1039/C9RA06697G

Yang Y, Ge F, Sun Y, Liu D, Chen C (2017) Strengthening triterpene saponins biosynthesis by over-expression of farnesyl pyrophosphate synthase gene and RNA interference of cycloartenol synthase gene in *Panax notoginseng* cells. *Molecules22*(4):581. https://doi.org/10.3390/molecules22040581

Yao X, Wuzhang K, Peng B, Chen T, Zhang Y, Liu H, Li L, Fu X, Tang, K (2022) Engineering the expression of plant secondary metabolites-genistein and scutellarin through an efficient transient production platform in *Nicotiana benthamiana* L. *Frontiers in Plant Science13*:994792. https://doi.org/10.3389/fpls.2022.994792

Yue W, Ming QL, Lin B, Rahman K, Zheng CJ, Han T, Qin LP (2016) Medicinal plant cell suspension cultures: Pharmaceutical applications and high-yielding strategies for the desired secondary metabolites. *Critical Reviews inBiotechnology 36*(2):215–232. https://doi.org/10.3109/07388551.2014.923986

Zeng L, Zhang Q, Jiang C, Zheng Y, Zuo Y, Qin J, Liao Z, Deng H (2021) Development of *Atropa belladonna* L. plants with high-yield hyoscyamine and without its derivatives using the CRISPR/Cas9 system. *International Journal of Molecular Sciences22*(4):1731. https://doi.org/10.3390/ijms22041731

Ziegler J, Facchini PJ (2008) Alkaloid biosynthesis: Metabolism and trafficking. *Annual Review of Plant Biology 59*(1): 735–769. https://doi.org/10.1146/annurev.arplant.59.032607.092730

7 Metabolites from Horticultural Crops

Nikhil Malhotra[1]
[1]ICAR-National Bureau of Plant Genetic Resources
Regional Station, Shimla, India

7.1 INTRODUCTION

The realm of horticultural crops is vast, encompassing a wide spectrum of plants, each with its unique characteristics and functions. From vegetables, which we consume daily, to the ornamental plants that adorn our gardens, these crops play a vital role in our lives. Many of them serve as essential sources of nutrients, providing us with the vitamins, minerals, and antioxidants necessary for optimal health (Upadhyay & Singh 2021, 2023, 2024). A wide range of chemical substances called metabolites are produced by plants and are necessary for their life, growth, and development. According to Elshafie et al. (2023), these metabolites can be broadly categorized into primary and secondary metabolites (Figure 7.1). The fundamental functions of plant life, including growth, respiration, and photosynthesis, are directly impacted by primary metabolites (Wink, 1988). Examples include carbohydrates, proteins, fats, and alcohols. In contrast, secondary metabolites, commonly referred to as bioactive compounds (BCs),

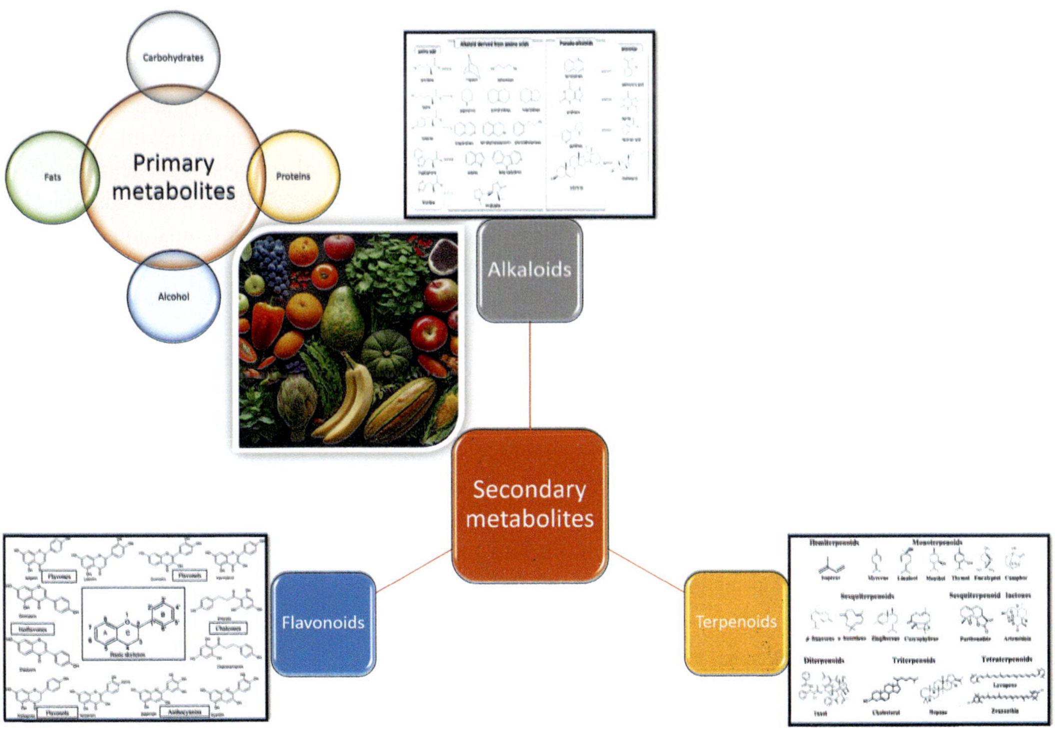

FIGURE 7.1 Horticultural crops and their metabolites. (Metabolite structures based on reviews by Panche et al., 2016; Abdallah and Quax, 2017; Chiocchio et al., 2021.)

DOI: 10.1201/9781003518358-7

are not directly essential for plant survival but can significantly impact their long-term viability. They are essential in mediating interactions between plants and their environments and frequently display species-specific patterns within evolutionary groups (Hill, 1952). For instance, some BCs, like glucosinolates in cruciferous vegetables, act as deterrents to herbivores, protecting the plant from damage. Allelopathy is the term for the occurrence wherein plants produce BCs into the environment that can prevent the growth of other plants. Additionally, BCs can support symbiotic connections between plants and other living things like nitrogen-fixing bacteria or mycorrhizal fungus.

The evolution and domestication of horticultural crops have been shaped by genetic selection for traits that benefit both plant fitness and human preferences. Understanding the biosynthesis and diversification of BCs is essential for developing high-yielding and stress-resistant crop varieties. The evolution and domestication of horticultural crops have been shaped by centuries of genetic selection. Scientists have selected for traits that benefit both the plant and themselves, such as increased yield, improved taste, and resistance to pests and diseases. To fully appreciate the complexity of horticultural crops, it is essential to understand the biosynthesis and diversification of BCs. These compounds often arise from primary metabolic pathways through processes like gene duplication and subsequent sub-functionalization or neo-functionalization. Developments in genomics and analytical methods have significantly benefited plant metabolism research (Naik et al., 2023). However, there are several limitations associated with the wide variety of horticulture crops and the intricacy of their metabolic processes. By integrating genomics, metabolomics, and other approaches, researchers are making strides in unravelling the genetic basis of plant metabolic diversity and developing new crop varieties with desirable traits. Horticultural crops represent a diverse and fascinating group of plants that can play a vital role in ensuring nutritional security (Daniel et al., 2023). Developing robust and sustainable farming methods requires an understanding of the function of BCs in these crops. By exploring the genetic and metabolic underpinnings of plant diversity, we may realize the full potential of horticultural crops and ensure a secure food supply for future generations.

According to Ahmed et al. (2017), BCs—naturally occurring compounds present in plants—are essential to plant biology and function. These substances are divided into three groups according to their chemical makeup and biosynthesis pathways: nitrogenous chemicals like alkaloids, phenolic compounds like flavonoids, and terpenoids like monoterpenes. BCs are particularly abundant in horticultural crops, where they contribute to plant survival and offer benefits to human health. For example, alkaloids, which are widely distributed across various horticultural species, significantly affect a plant's odour, colour, taste, and nutritional properties (Debnath et al., 2018). Flavonoids, another prominent class of bioactive compounds, are abundantly found in horticultural plants and have been extensively studied (Markham and Porter, 1969; Dias et al., 2021). Terpenoids, which represent the largest class of plant metabolites, are especially prevalent in both medicinal and ornamental plants (Bohlmann and Keeling, 2008). These substances are essential for plant defence by providing protection against pathogens and facilitating adaptation to environmental stresses. BCs act as natural pesticides, antioxidants, and signalling molecules, aiding plants in overcoming challenging conditions. This chapter examines the classification, distribution, and functions of imperative BCs, emphasizing their significance in enhancing the stress tolerance of horticultural crops and promoting human health. Future research should further explore the molecular mechanisms through which BCs provide these benefits and investigate their potential applications in plant breeding. A deeper understanding of BCs in plant biology may lead to novel solutions to improve the suppleness and production of horticultural crops while simultaneously contributing to human well-being (Cao et al., 2024).

7.2 CLASSIFICATION AND DISTRIBUTION

Horticultural crops, including fruits, vegetables, and medicinal plants, constitute a fundamental pillar of a balanced diet. These diverse botanicals offer a rich tapestry of flavours, essential nutrients, and potential health-promoting bioactive compounds. As highlighted by Krishnamoorthi et al. (2023), BCs present in these crops, such as alkaloids, flavonoids, and terpenoids, have increasingly captured global attention due to their promising contributions to overall well-being (Table 7.1).

TABLE 7.1
Some Examples of BCs in Horticultural Crops

Class	Compound	Source	Property	References
Alkaloids	Atisine	*Aconitum heterophyllum*	Antidiarrheal	Malhotra et al. (2014, 2016); Malhotra and Sharma (2021)
	Morphine	*Papaver somniferum*	Analgesic	Lahlou (2007)
	Vincamine	*Vinca minor*	Vasodilatory	Al-Rashed et al. (2021)
	Piperine	*Piper officinarum*	Antihyperglycemic	Qiu et al. (2014)
	Caffeine	*Coffea arabica* and *Theobroma cacao*	Antioxidant	Bhambhani et al. (2021)
Flavonoids	Flavonols	*Citrus × paradisi*	Antiviral	Stewart et al. (2000); Cushnie and Lamb (2005)
	Flavones	*Capsicum annuum* and *Solanum lycopersicum*	Antiviral	Cushnie and Lamb (2005)
	Flavonones	*Citrus* sp.	Antibacterial	Miyake et al. (2000)
	Anthocyanidines	*Fragaria × ananassa, Prunus* avium and *Rubus idaeus*	Antioxidant	Stewart et al. (2000); Kong et al. (2003)
	Isoflavones	*Brassica oleracea*	Antimicrobial	Rahmani et al. (2024)
Terpenoids	Azadirachtin	*Azadirachta indica*	Insecticidal	Devrnja et al. (2022)
	Limonene	*Citrus* sp.	Antifungal	Zhou et al. (2004); Young et al. (2013)
	Saponins	*Allium* sp., *Ipomoea batatas* and *Spinacia oleracea*	Hypolipidemic	Ejelonu et al. (2017)
	3-Carene	*Rosmarinus officinalis*, *Cannabis sativa* and *Citrus* sp.	Antimicrobial	Shu et al. (2019)
	Camphene	*Zingiber officinale*	Anticancer	Mahboubi (2019)

7.2.1 Alkaloids

Alkaloids are a large class of naturally occurring chemicals that include nitrogen and are mostly made by plants. These substances start as precursors of amino acids, which go through intricate biosynthesis processes to take on their unique structures. The pyruvate and shikimic acid routes are the two main processes in alkaloid production (Ahmed et al., 2017). According to Gutiérrez-Grijalva et al. (2020), metabolic activities like glycolysis produce amino acids. Alkaloids can be divided into several groups according to their biosynthetic origins according to their varied chemical structures, which are typified by their heterocyclic rings (Singh et al., 2012).

Indole alkaloids, categorized by the existence of serotonin, are a prominent group with over 2,000 identified compounds (Kainsa et al., 2012). Tropane alkaloids, containing an 8-azabicyclo[3.2.1] octane nucleus, are found in plants belonging to the Solanaceae, Brassicaceae, Erythroxylaceae, Convolvulaceae, and Euphorbiaceae families. Important tropane alkaloids include scopolamine, hyoscyamine, cocaine, and atropine, which have significant medicinal applications (Scholl et al., 2001). Quinoline and isoquinoline alkaloids are heterocyclic aromatic compounds formed by the fusion of a benzene ring and a pyridine ring. They are prevalent in horticultural crops such as poppies, cinchona trees, and bitter oranges (Marella et al., 2013). Purine alkaloids, often called xanthines, are formed from adenine and guanine. These include theophylline, aminophylline, theobromine, and caffeine (Ashihara and Crozier, 2001; Ober et al., 2003). Piperidine alkaloids, characterized by a saturated piperidine nucleus, are well-known for their toxicity and are found across the plant kingdom (Greger, 2019). Pyrrolizidine alkaloids, consisting of two fused five-membered rings, are found in plants belonging to the Leguminosae, Compositae, Orchidaceae, Poaceae,

Boraginaceae, and Convolvulaceae families (Majik and Tilve, 2012). Pyrrolidine alkaloids, containing a five-membered ring derived from ornithine or arginine, are another class of alkaloids (Kaur and Arora, 2015). Quinolizidine alkaloids, characterized by two fused six-membered rings, are found in plants such as the cinchona tree and *Remijia pedunculata* (Suzuki et al., 1996).

7.2.2 Flavonoids

The diphenylpropane backbone (C_6–C_3–C_6) of flavonoids, a type of polyphenolic chemical, is made up of two aromatic rings (A and B) joined by a three-carbon chain. The specific arrangement of hydroxyl, methoxy, and glycoside groups on this backbone gives rise to the diverse array of flavonoid subclasses, each with its own unique structural features and biological activities.

Seven primary flavonoid subclasses exist, categorized by the level of oxidation of the core heterocycle: flavonols, flavones, isoflavones, anthocyanins, flavanones, flavanols, and chalcones. Flavonols, characterized by their 3-hydroxy substitution, are prevalent in fruits and vegetables like cauliflower, onions, asparagus, and apples (Panche et al., 2016). Flavones, distinguished by their 4H-amino-4-one structure, are commonly present in oranges, red peppers, tea, and celery (Wong, 1975). Isoflavones, with their B-ring attached to C-3 but not C-2 (Iwashina, 2000), are present in various cereals, nuts, fruits, and veggies (Jain and Ramawat, 2013). Anthocyanins, accountable for the red, blue, and purple colours of plants, are abundant in vegetables and fruits like tomatoes, red cabbage, and blueberries (Harborne and Baxter, 1999). Flavanones, derived from 20-hydroxychalcone (Terahara, 2015), are primarily found in citrus fruits (Agrawal, 2011). Flavanols, characterized by their 3-hydroxylated C ring, are abundant in fruit skins, particularly berries, apricots, plums, cherries, and apples (Ouyang et al., 2021). Chalcones, open-chain flavonoids, are found in the plantain, mulberry, ginger, and legume families (Rodríguez De Luna et al., 2020).

7.2.3 Terpenoids

The five-carbon isopentane unit is the starting point for the varied class of chemical molecules known as terpenoids, which are found across nature. They are made from the condensation products of the main building ingredients, dimethylallyl diphosphate (DMAPP) and isoprene diphosphate (IPP). Two different biosynthetic processes can produce these diphosphates: the methylerythritol phosphate (MEP) system, which is mainly found in plants, bacteria, and some protozoa, and the mevalonic acid (MVA) pathway, which is mainly found in mammals, fungi, and some plants. Acetyl-CoA is converted to mevalonate in the MVA route, which is then followed by a sequence of enzyme events that yield IPP and DMAPP. On the other hand, the MEP pathway uses pyruvate and glyceraldehyde-3-phosphate as precursors to initiate a sequence of redox events that produce IPP and DMAPP (Huang et al., 2021; Mostafa et al., 2022; Chen et al., 2023).

Terpenoids are categorized as follows: monoterpenes (C10), sesquiterpenes (C15), diterpenes (C20), sesterterpenes (C25), triterpenes (C30), and polyterpenes (> C30) according to the number of isoprene units in their structure (Turlings et al., 1995). Monoterpenes, like terpineol, limonene, linalool, and myrcene, are volatile compounds contributing to the fragrance of plants (Magnard et al., 2015). Sesquiterpenes, including farnesol, humulene, and farnesenes, are larger molecules with diverse structures (Kabera et al., 2014). Diterpenes, often non-volatile, have various cyclic forms and are characterized by oxygenated functional groups (Ashour et al., 2010; Abdallah and Quax, 2017). Triterpenes, derived from squalene, encompass a vast array of compounds (Thimappa et al., 2014). Tetraterpenes, such as carotenoids, are composed of eight isoprene units (Sozer et al., 2010; Domonkos et al., 2013). Polyterpenes, with more than 40 carbons, include gutta-percha and natural rubber (Rohmer et al., 1993).

7.2.4 Others

Beyond the commonly recognized alkaloids, flavonoids, and terpenoids, horticultural crops harbour a diverse array of BCs. These include fatty acids, glucosinolates, coumarins, and lignans, each distinguished by its unique structural characteristics and biosynthetic pathways.

The unique flavour and pungency of Brassicaceae plants, such as mustard greens, broccoli, radishes, and cabbage, are attributed to glucosinolates, which are sulfur-containing chemicals. These metabolites, which are derived from amino acids, are created by side chain modification and chain elongation (Mikkelsen et al., 2002). Another type of BC that are common in horticulture plants is fatty acids, which is crucial parts of cell membranes and have roles in signalling and energy storage. They are produced by mechanisms like the fatty acid synthase complex and are found in large quantities in foods including nuts and seeds, avocado, palm oil, and olive oil. Through the phenylpropanoid route, coumarins—aromatic molecules made up of a fused benzene and lactone ring are produced. Coumarins, which are present in a variety of horticulture plants such as grasses, citrus fruits, and cherries, are essential for protecting plants against pests and diseases (Zhu and Jiang, 2018). Lignans are likewise produced from the phenylpropanoid pathway and are complex phenolic compounds with several benzene rings joined by links. Lignans, which are found in plants including flaxseed, sesame seeds, and berries, have been connected to a number of health advantages, such as anti-cancer and antioxidant capabilities.

7.3 RESPONSE AND FUNCTION TO STRESSES

Horticultural crops, like many plants, possess intricate mechanisms to cope with various environmental stresses (Mareri et al., 2022). These responses are essential for survival and productivity in challenging conditions. In order to create sustainable farming methods that might lessen the negative impacts of stress on horticultural crops, it is essential to comprehend these stress reactions and their underlying mechanisms.

7.3.1 Biotic Stresses

Horticultural crops have evolved intricate defence mechanisms to counteract the constant threat posed by biological stressors, such as pathogens and pests (Lal et al., 2023). A crucial component of these defence strategies is the synthesis of BCs. These chemicals are essential for deterring, repelling, or directly killing various organisms that could compromise plant health and productivity. BCs have a wide range of biological functions, including antimicrobial, insecticidal, antifungal, and nematocidal properties. Alkaloids, flavonoids, and terpenoids are notable examples of BCs with these potent effects. By interfering with the physiological processes of pathogens and pests, these compounds can disrupt their life cycles, preventing their establishment and proliferation on the plant. Beyond their direct toxic effects, BCs also contribute to plant defence by strengthening the immune response. They can activate signalling pathways that trigger the production of defence-related proteins and enzymes, enhancing the crop ability to recognize and retort to threats. Additionally, BCs can fortify the structural barriers of a plant, like the cell wall, making it more resistant to penetration by pathogens. The combined effects of these defence mechanisms ensure the optimal growth, development, and productivity of horticultural crops. Researchers and farmers can create plans to take advantage of utilizing BCs to improve crop resilience and lessen dependency on chemical pesticides by comprehending their function in plant defence (Anu et al., 2016).

7.3.2 Abiotic Stresses

Horticultural crops, ranging from fruit and ornamental trees to vegetables and herbaceous plants, are increasingly susceptible to a variety of abiotic stress factors. These environmental challenges, including drought, low temperatures, waterlogging, and elevated salinity, have become more pronounced due to factors such as urbanization, industrialization, and climate change (Kopecká et al., 2023; Yang et al., 2023). The cultivation of such crops in controlled-environment farming systems, while offering certain advantages, can also introduce new challenges, including the proliferation of pests and diseases. The combined effects of abiotic stress and biotic factors can significantly impact both crop yield and quality. To mitigate these adverse effects, researchers have explored the potential of biologically active compounds, such as betacyanins. The stress tolerance and adaptation of horticultural crops have been shown to have improved by BCs, which are naturally occurring pigments present in specific plant species. By activating specific biochemical pathways, BCs can help plants better cope with environmental challenges, promoting growth, development, and ultimately, higher yields (Lal et al., 2023).

7.3.3 Human Health

BCs found in horticultural crops have garnered significant attention for their potential to enhance human health (Zhao et al., 2024). These secondary metabolites, including alkaloids, flavonoids, and terpenoids, exhibit a wide array of biological activities. Extensive research has demonstrated the multifaceted benefits of BCs, encompassing antioxidant, anti-inflammatory, antimicrobial, anticancer, and cardioprotective properties. The incorporation of horticultural crops rich in BCs into the diet is increasingly recognized as a promising strategy for promoting overall health and lowering the chance of developing chronic illnesses. These compounds have proven to neutralize harmful free radicals, alter inflammatory responses, inhibit microbial growth, and interfere with cancer cell proliferation. Furthermore, BCs have been shown to improve lipid profiles and lower blood pressure, among other cardiovascular health benefits. The therapeutic potential of BCs has been recognized for centuries in traditional medicine, where horticultural plants have been used to treat various ailments. This historical evidence, combined with contemporary scientific research, underscores the importance of these compounds in promoting human well-being (Wink, 1988; Lalani and Poh, 2020).

7.4 GENETIC DIVERSITY AND BREEDING STRATEGIES

BCs are secondary metabolites found in horticultural crops that contribute to their nutritional and health-promoting properties. While certain BCs, such as ascorbic acid (vitamin C) and its oxidized derivative dehydroascorbic acid, are ubiquitous in fruits and vegetables, others, like glucosinolates, are more specific to particular plant families or even individual species. This variation in BC composition is largely attributed to the diversity of the plant. The concentration of BCs can vary significantly even within the same species, depending on factors like cultivar, growing conditions, and stage of development. For example, research has shown that different cultivars of broccoli can exhibit substantial differences in their glucosinolate content (Lezoul et al., 2020).

The biosynthetic processes involved in the formation of BCs have been better understood owing to developments in plant genomes, transcriptomics, metabolomics, and bioinformatics (Zhang and Hao, 2020; Manickam et al., 2023). This knowledge has enabled plant breeders to develop molecular markers associated with health-beneficial compounds. These markers can be used to select and breed for cultivars with enhanced nutritional profiles. While traditional breeding methods have historically focused on traits such as yield, disease resistance, and aesthetic appeal, there has been a growing emphasis on improving the nutritional quality of horticultural crops. By employing modern breeding techniques and utilizing molecular markers, it is now possible to create new cultivars that offer increased levels of bioactive compounds, thereby providing consumers with healthier and more nutritious food options.

7.5 PLANT SIGNALING

Terpenoids are one of the most researched classes of volatile BCs, which are used by plants to communicate in response to biotic stress. Terpenoids perform a wide range of roles, including functioning as alarm signals, defence emissions, markers for tracking herbivores, and deterrents by repelling feeding organisms (Parihar et al., 2015). These substances may serve as warning signs, inciting defence mechanisms in nearby plants or in the unaffected tissues of the same plant. This response, activated by volatile terpenoids, can enhance the plant's overall resilience to biotic stressors such as herbivores or pathogens (Xu et al., 2021). Importantly, these signals can function not only within the same species but also across different species. For example, volatile terpenoids are released by wild radish (*Raphanus raphanistrum*) in response to herbivores or invasion by pathogens, particularly glucosinolates. These compounds serve as chemical cues for nearby mustard plants, inducing them to produce defensive substances and bolster their resistance to similar stresses (War et al., 2012). This inter-plant communication facilitates coordinated defence responses, allowing mustard plants to increase their resilience through an early warning system mediated by volatile terpenoids.

For herbivores, several terpenoids also serve as chemo-attractants. One such substance that attracts insect herbivores is 4,8-dimethyl-1,3(E),7-nonatriene, a terpenoid derivative of nerolidol. This attractant function, paradoxically, contributes to reinforcing the plant's defensive capabilities by recruiting herbivores that may inadvertently induce further defence responses in the plant. Moreover, the sagebrush plant *Artemisia tridentata* releases volatile terpenoids into the atmosphere, which include substances like limonene and camphor.

Based on the characteristics of these volatile terpenoids, sagebrush plants are able to differentiate between conspecifics and modify their defensive and growth tactics accordingly. Sagebrush plants have a tendency to decrease their development rates and divert more resources to herbivore defence when they become exposed to their unique volatile signals. This phenomenon suggests a form of kin recognition, where plants modulate their defensive efforts based on signals received from genetically related individuals (Verdeguer et al., 2020). Overall, terpenoids are essential to the intricate web of plant communication, helping plants manage biotic stress through signalling, defence coordination, and kin recognition.

7.6 CONCLUSIONS AND FUTURE PROSPECTS

Agro-pharmaceuticals, plant protection, growth promotion, food preservation, and the creation of new therapies are just a few of the many applications for which plants are an essential source of natural substances. The production of BCs generated from plants has increased recently due to the strong economic interest in these compounds, which has also made it easier to identify new possible biological uses for them. Because of the existence of BCs, horticultural crops among plant species demonstrate an extraordinary capacity to withstand a variety of environmental challenges. These compounds, like alkaloids, flavonoids, and terpenoids, serve as natural defence mechanisms, protecting plants from heat, drought, salinity, ultraviolet radiation, and biotic stressors like pests and diseases. In addition to their protective roles in plants, BCs offer considerable health benefits to humans, demonstrating therapeutic properties in the management of several kinds of diseases. The rising curiosity in unlocking the prospects of BCs in horticultural crops underscores their significance not only in agriculture but also in medicine. Our study aims to shed more light on investigating the multifunctional roles of BCs, particularly in enhancing plant resilience to environmental stresses and their therapeutic applications. Additionally, we aim to determine the industrial prespective of these chemicals, such as their use in biopesticides and pharmaceuticals. We can help create more resilient, productive crop types that are better suited to the needs of an expanding global population and the problems presented by climate change by developing a greater understanding of the processes and advantages of BCs in horticulture crops.

REFERENCES

Abdallah, I. I., & Quax, W. J. (2017). A glimpse into the biosynthesis of terpenoids. In *NRLS Conference Proceedings, International Conference on Natural Resources and Life Sciences (2016)* (pp. 81–98). KnE Life Sciences.

Agrawal, A. D. (2011). Pharmacological activities of flavonoids: A review. *International Journal of Pharmaceutical Sciences and Nanotechnology*, 4, 1394–1398.

Ahmed, E., Arshad, M., Khan, M. Z., Amjad, M. S., Sadaf, H. M., et al. (2017). Secondary metabolites and their multidimensional prospective in plant life. *Journal of Pharmacognosy and Phytochemistry*, 6, 205–214.

Al-Rashed, S., Baker, A., Ahmad, S. S., Syed, A., Bahkali, A. H., Elgorban, A. M., & Khan, M. S. (2021). Vincamine, a safe natural alkaloid, represents a novel anticancer agent. *Bioorganic Chemistry*, 107, 104626.

Anu, A., Sahni, S., Kumar, P., & Prasad, B. D. (2016). Secondary metabolites in horticultural crops. In M. W. Siddiqui & K. Prasad (Eds.), *Plant Secondary Metabolites, Volume One* (pp. 221–234). New York: Apple Academic Press.

Ashihara, H., & Crozier, A. (2001). Caffeine: A well-known but little-mentioned compound in plant science. *Trends in Plant Science*, 6(4), 407–413.

Ashour, M., Wink, M., & Gershenzon, J. (2010). Biochemistry of terpenoids: Monoterpenes, sesquiterpenes and diterpenes. In M. Wink (Ed.), *Annual Plant Reviews Volume 40: Biochemistry of Plant Secondary Metabolism*, Second Edition (pp. 258–303). Oxford, England: Blackwell Publishing Ltd.

Bhambhani, S., Kondhare, K.R., & Giri, A.P. (2021). Diversity in chemical structures and biological properties of plant alkaloids. *Molecules*, 26(11), 3374.

Bohlmann, J., & Keeling, C. I. (2008). Terpenoid biomaterials. *The Plant Journal*, 54, 656–669.

Cao, Y., Li, X., Song, H., Abdullah, M., & Manzoor, M.A. (2024). Editorial: Multi-omics and computational biology in horticultural plants: From genotype to phenotype, volume II. *Frontiers in Plant Science*, 15, 1368909.

Chen, H., Lei, J., Li, S., Guo, L., Lin, J., … & Li, Y. (2023). Progress in biological activities and biosynthesis of edible fungi terpenoids. *Critical Reviews in Food Science and Nutrition*, 63, 7288–7310.

Chiocchio, I., Mandrone, M., Tomasi, P., Marincich, L., & Poli, F. (2021). Plant secondary metabolites: An opportunity for circular economy. *Molecules*, 26(2), 495.

Cushnie, T. P. T., & Lamb, A. J. (2005). Antimicrobial activity of flavonoids. *International Journal of Antimicrobial Agents*, 26, 343–356.

Daniel, M.A., Sebastin, R., Yu, J.-K., Soosaimanickam, M.P., & Chung, J.W. (2023). Enhancing horticultural crops through genome editing: Applications, benefits, and considerations. *Horticulturae*, 9, 884.

Debnath, B., Singh, W. S., Das, M., Goswami, S., Singh, M. K., … & Kumar, S. (2018). Role of plant alkaloids on human health: A review of biological activities. *Materials Today Chemistry*, 9, 56–72.

Devrnja, N., Milutinovic, M., & Savic, J. (2022). When scent becomes a weapon—Plant essential oils as potent bioinsecticides. *Sustainability*, 14, 6847.

Dias, M. C., Pinto, D. C. G. A., & Silva, A. M. S. (2021). Plant flavonoids: Chemical characteristics and biological activity. *Molecules*, 26, 5377.

Domonkos, I., Kis, M., Gombos, Z., & Ughy, B. (2013). Carotenoids, versatile components of oxygenic photosynthesis. *Progress in Lipid Research*, 52, 539–561.

Ejelonu, O. C., Elekofehinti, O. O., & Adanlawo, I. G. (2017). Tithonia diversifolia saponin-blood lipid interaction and its influence on immune system of normal wistar rats. *Biomedicine & Pharmacotherapy*, 87, 589–595.

Elshafie, H. S., Camele, I., & Mohamed, A. A. (2023). A comprehensive review on the biological, agricultural, and pharmaceutical properties of secondary metabolites based-plant origin. *International Journal of Molecular Sciences*, 24, 3266.

Greger, H. (2019). Structural classification and biological activities of Stemona alkaloids. *Phytochemistry Reviews*, 18, 463–493.

Gutiérrez-Grijalva, E. P., López-Martínez, L. X., Contreras-Angulo, L. A., Elizalde-Romero, C. A., & Heredia, J. B. (2020). Plant alkaloids: Structures and bioactive properties. In M. Swamy (Ed.), *Plant-Derived Bioactives* (pp. 85–117). Singapore: Springer.

Harborne, J. B., & Baxter, H. (1999). *The Handbook of Natural Flavonoids* (2 vols.). Chichester: Wiley.

Hill, A. F. (1952). *Economic Botany: A Textbook of Useful Plant and Plant Products*, Second Edition. New York: McGraw-Hill Book Company Inc.

Huang, H., Xing, S., Tang, K., & Jiang, W. (2021). AaWRKY4 upregulates artemisinin content through boosting the expressions of key enzymes in artemisinin biosynthetic pathway. *Plant Cell, Tissue and Organ Culture (PCTOC)*, 146, 97–105.

Iwashina, T. (2000). The structure and distribution of the flavonoids in plants. *Journal of Plant Research*, 113, 287–299.

Jain, N., & Ramawat, K. G. (2013). Nutraceuticals and antioxidants in prevention of diseases. In K. Ramawat & J.- M. Mérillon (Eds.), *Natural Products* (pp. 2559–2580). Berlin, Heidelberg: Springer.

Kabera, J. N., Semana, E., Mussa, A. R., & He, X. (2014). Plant secondary metabolites: Biosynthesis, classification, function, and pharmacological properties. *Journal of Pharmacy and Pharmacology*, 2, 377–392.

Kainsa, S., Kumar, P., & Rani, P. (2012). Medicinal plants of Asian origin having anticancer potential: Short review. *Asian Journal of Biomedical and Pharmaceutical Sciences*, 2, 1–11.

Kaur, R., & Arora, S. (2015). Alkaloids-important therapeutic secondary metabolites of plant origin. *Journal of Critical Reviews*, 2, 1–8.

Kong, J. M., Chia, L. S., Goh, N. K., Chia, T. F., & Brouillard, R. (2003). Analysis and biological activities of anthocyanins. *Phytochemistry*, 64, 923–933.

Kopecká, R., Kameniarová, M., Černý, M., Brzobohatý, B., & Novák, J. (2023). Abiotic stress in crop production. *International Journal of Molecular Sciences*, 24, 6603.

Krishnamoorthi, A., Anushi, G., Varsha Minz, S. D. B., Singh, S. K., Shilpa, Prasad, L., & Jain, S. (2023). Bioactive compounds from horticulture crops and their utilization: A comprehensive review. *International Journal of Plant & Soil Science*, 35(23), 268–277.

Lahlou, M. (2007). Screening of natural products for drug discovery. *Expert Opinion on Drug Discovery*, 2, 697–705.

Lal, M. K., Tiwari, R. K., Altaf, M. A., Kumar, A., & Kumar, R. (2023). Editorial: Abiotic and biotic stress in horticultural crops: Insight into recent advances in the underlying tolerance mechanism. *Frontiers in Plant Science*, 14, 1212982.

Lalani, S., & Poh, C. L. (2020). Flavonoids as antiviral agents for Enterovirus A71 (EV-A71). *Viruses*, 12, 184.

Lezoul, N. E. H., Belkadi, M., Habibi, F., & Guillén, F. (2020). Extraction processes with several solvents on total bioactive compounds in different organs of three medicinal plants. *Molecules*, 25(20), 4672.

Magnard, J. L., Roccia, A., Caissard, J. C., Vergne, P., Sun, P., et al. (2015). Biosynthesis of monoterpene scent compounds in roses. *Science*, 349, 81–83.

Mahboubi, M. (2019). *Zingiber officinale* Rosc. essential oil, a review on its composition and bioactivity. *Clinical Phytoscience*, 5, 6.

Majik, M. S., & Tilve, S. G. (2012). Pyrrolizidine alkaloids pyrrolams A–D: A survey of synthetic efforts, biological activity, and studies on their stability. *Synthesis*, 44, 2673–2681.

Malhotra, N., Chauhan, R. S., & Sood, H. (2016). Transcriptome-wide mining suggests conglomerate of genes associated with tuberous root growth and development in *Aconitum heterophyllum* Wall. *3 Biotech*, 6, 152.

Malhotra, N., Kumar, V., Sood, H., Singh, T. R., & Chauhan, R. S. (2014). Multiple genes of mevalonate and non-mevalonate pathways contribute to high aconites content in an endangered medicinal herb, *Aconitum heterophyllum* Wall. *Phytochemistry*, 108, 26–34.

Malhotra, N., & Sharma, S. (2021). *Aconitum heterophyllum*. In N. Malhotra & M. Singh (Eds.), *Himalayan Medicinal Plants* (pp. 5–25). London, United Kingdom: Academic Press.

Manickam, S., Rajagopalan, V. R., Kambale, R., Rajasekaran, R., Kanagarajan, S., & Muthurajan, R. (2023). Plant metabolomics: Current initiatives and future prospects. *Current Issues in Molecular Biology*, 45(11), 8894–8906.

Marella, A., Tanwar, O. P., Saha, R., Ali, M. R., Srivastava, S., et al. (2013). Quinoline: A versatile heterocyclic. *Saudi Pharmaceutical Journal*, 21, 1–12.

Mareri, L., Parrotta, L., & Cai, G. (2022). Environmental stress and plants. *International Journal of Molecular Sciences*, 23(10), 5416.

Markham, K. R., & Porter, L. J. (1969). Flavonoids in the green algae (chlorophyta). *Phytochemistry*, 8, 1777–1781.

Mikkelsen, M. D., Petersen, B. L., Olsen, C. E., & Halkier, B. A. (2002). Biosynthesis and metabolic engineering of glucosinolates. *Amino Acids*, 22, 279–295.

Miyake, Y., Shimoi, K., Kumazawa, S., Yamamoto, K., Kinae, N., & Osawa, T. (2000). Identification and antioxidant activity of flavonoid metabolites in plasma and urine of eriocitrin-treated rats. *Journal of Agricultural and Food Chemistry*, 48, 3217–3224.

Mostafa, S., Wang, Y., Zeng, W., & Jin, B. (2022). Floral scents and fruit aromas: Functions, compositions, biosynthesis, and regulation. *Frontiers in Plant Science*, 13, 860157.

Naik, B., Kumar, V., Rizwanuddin, S., Chauhan, M., Choudhary, M., Gupta, A.K., Kumar, P., Kumar, V., Saris, P.E.J., Rather, M.A., Bhuyan, S., Neog, P.R., Mishra, S., & Rustagi, S. (2023). Genomics, proteomics, and metabolomics approaches to improve abiotic stress tolerance in tomato plant. *International Journal of Molecular Sciences*, 24(3), 3025.

Ober, D., Harms, R., Witte, L., & Hartmann, T. (2003). Molecular evolution by change of function: Alkaloid-specific homospermidine synthase retained all properties of deoxyhypusine synthase except binding the eIF5A precursor protein. *Journal of Biological Chemistry*, 278, 12805–12812.

Ouyang, Y., Li, J., Chen, X., Fu, X., Sun, S., et al. (2021). Chalcone derivatives: Role in anticancer therapy. *Biomolecules*, 11, 894.

Panche, A. N., Diwan, A. D., & Chandra, S. R. (2016). Flavonoids: An overview. *Journal of Nutritional Science*, 5, e47.

Parihar, P., Singh, S., Singh, R., Singh, V. P., & Prasad, S. M. (2015). Effect of salinity stress on plants and its tolerance strategies: A review. *Environmental Science and Pollution Research*, 22, 4056–4075.

Qiu, S., Sun, H., Zhang, A. H., Xu, H. Y., Yan, G. L., Han, Y., & Wang, X. J. (2014). Natural alkaloids: Basic aspects, biological roles, and future perspectives. *Chinese Journal of Natural Medicine*, 12, 401–406.

Rahmani, M., Esmaeili, A., & Taherkhani, M. (2024). Antimicrobial potential of *Brassica oleracea* extracts (white and broccoli) and their resistance compared to doxycycline against gram-positive and gram-negative bacteria. *Food Science & Nutrition*, 12(10), 7492–7503.

Rodríguez De Luna, S. L., Ramírez-Garza, R. E., & Serna Saldívar, S. O. (2020). Environmentally friendly methods for flavonoid extraction from plant material: Impact of their operating conditions on yield and antioxidant properties. *The Scientific World Journal*, 2020, 6792069.

Rohmer, M., Knani, M., Simonin, P., Sutter, B., & Sahm, H. (1993). Isoprenoid biosynthesis in bacteria: A novel pathway for the early steps leading to isopentenyl diphosphate. *Biochemical Journal*, 295, 517–524.

Scholl, Y., Höke, D., & Dräger, B. (2001). Calystegines in *Calystegia sepium* derive from the tropane alkaloid pathway. *Phytochemistry*, 58, 883–89.

Shu, H., Chen, H., Wang, X., Hu, Y., Yun, Y., Zhong, Q., Chen, W., & Chen, W. (2019). Antimicrobial activity and proposed action mechanism of 3-carene against brochothrix thermosphacta and Pseudomonas fluorescens. *Molecules*, 6, 3246.

Singh, A. K., Chawla, R., Rai, A., & Yadav, L. D. S. (2012). NHC-catalysed diastereoselective synthesis of multifunctionalised piperidines via cascade reaction of enals with azalactones. *Chemical Communications*, 48, 3766–3768.

Sozer, O., Komenda, J., Ughy, B., Domonkos, I., Laczkó-Dobos, H., et al. (2010). Involvement of carotenoids in the synthesis and assembly of protein subunits of photosynthetic reaction centers of Syne chocystis sp. PCC 6803. *Plant and Cell Physiology*, 51, 823–835.

Stewart, A. J., Bozonnet, S., Mullen, W., Jenkins, G. I., Lean, M. E., & Crozier, A. (2000). Occurrence of flavonols in tomatoes and tomato-based products. *Journal of Agricultural and Food Chemistry*, 48, 2663–2669.

Suzuki, H., Koike, Y., Murakoshi, I., & Saito, K. (1996). Subcellular localization of acyltransferases for quinolizidine alkaloid biosynthesis in Lupinus. *Phytochemistry*, 42, 1557–1562.

Terahara, N. (2015). Flavonoids in foods: A review. *Natural Product Communications*, 10, 521–528.

Thimappa, R., Geisler, K., Louveau, T., O'Maille, P., & Osbourn, A. (2014). Triterpene biosynthesis in plants. *Annual Review of Plant Biology*, 65, 225–257.

Turlings, T. C., Loughrin, J. H., McCall, P. J., Röse, U. S., Lewis, W. J., et al. (1995). How caterpillar-damaged plants protect themselves by attracting parasitic wasps. *Proceedings of the National Academy of Sciences of the United States of America*, 92, 4169–4174.

Upadhyay, S. K., & Singh, S. P. (2021). *Bioprospecting of Plant Biodiversity for Industrial Molecules*. Wiley. DOI:10.1002/9781119718017

Upadhyay, S. K., & Singh, S. P. (2023). *Plants as Bioreactors for Industrial Molecules*. Wiley. DOI:10.1002/9781119875116

Upadhyay, S. K., & Singh, S. P. (2024). *Herbal Nutraceuticals: Products and Processes*. Wiley. DOI:10.1002/9781394241576

Verdeguer, M., Sánchez-Moreiras, A. M., & Araniti, F. (2020). Phytotoxic effects and mechanism of action of essential oils and terpenoids. *Plants*, 9, 1571.

War, A. R., Paulraj, M. G., Ahmad, T., Buhroo, A. A., Hussain, B., et al. (2012). Mechanisms of plant defense against insect herbivores. *Plant Signaling & Behavior*, 7, 1306–1320.

Wink, M. (1988). Plant breeding: Importance of plant secondary metabolite for protection against pathogen and herbivores. *Theoretical and Applied Genetics*, 75, 225–233.

Wong, E. (1975). The isoflavonoids. In J. B. Harborne, T. J. Mabry, & H. Mabry (Eds.), *The Flavonoids* (pp. 743–800). Boston, MA: Springer US.

Xu, M., Jiang, Y., Chen, S., Chen, F., & Chen, F. (2021). Herbivory-induced emission of volatile terpenes in Chrysanthemum morifolium functions as an indirect defense against *Spodoptera litura* larvae by attracting natural enemies. *Journal of Agricultural and Food Chemistry*, 69, 9743–9753.

Yang, H., Fang, R., Luo, L., Yang, W., Huang, Q., Yang, C., Hui, W., Gong, W., & Wang, J. (2023). Uncovering the mechanisms of salicylic acid-mediated abiotic stress tolerance in horticultural crops. *Frontiers in Plant Science*, 14, 1226041.

Young, W. K., Ji, K. M., Young, C. B., Yeon, B. D., Seong, K. L., Seul, M. C., Duck, S. L., Myung, C. C., Kyungsil, Y., & Hyung, S. K. (2013). Safety evaluation and risk assessment of d-limonene. *Journal of Toxicology and Environmental Health, Part B*, 6, 17–38.

Zhang, C., & Hao, Y.-J. (2020). Advances in genomic, transcriptomic, and metabolomic analyses of fruit quality in fruit crops. *Horticultural Plant Journal*, 6, 361–371.

Zhao, X., Peng, J., Zhang, L., Yang, X., Qiu, Y., Cai, C., Hu, J., Huang, T., Liang, Y., Li, Z., Tian, M., Liu, F., & Wang, Z. (2024). Optimizing the quality of horticultural crop: Insights into pre-harvest practices in controlled environment agriculture. *Frontiers in Plant Science*, 15, 1427471.

Zhou, J.Y., Tang, F.D., Mao, G.G., & Bian, R.L. (2004). Effect of alpha-pinene on nuclear translocation of NF-kappa B in THP-1 cells. *Acta Pharmacological Sinica*, 25, 480–484.

Zhu, J., & Jiang, J. (2018). Pharmacological and nutritional effects of natural coumarins and their structure–activity relationships. *Molecular Nutrition & Food Research*, 62, 1701073.

8 Secondary Metabolites from Vegetable Crops and Applications

Sibi Kalaivanan[1], *Nitheesh Guptha Surulivel*[1], *Thalapathi Subramani*[1], *Swetha Balaji*[2], *Keerthi Elavarasan*[1], *Janani Sree Sivakumar*[1], *Amit Roy*[3], *Vasanth Krishnan*[1], *and Dhandapani Gurusamy*[2]

[1]Department of Botany, Bharathiar University, Coimbatore, India

[2]Department of Botany, Kongunadu Arts and Science College (Autonomous), Bharathiar University, Coimbatore, India

[3]Faculty of Forestry and Wood Sciences, Department of Forest Genetics and Physiology, Czech University of Life Sciences Prague, Prague, Czechia

ABBREVIATIONS

AMD Age-related Macular Degeneration
RNS Reactive Nitrogen Species
ROS Reactive Oxygen Species
SCFA Short Chain Fatty Acid
UAE Ultrasound-Assisted Extraction

8.1 INTRODUCTION

Secondary metabolites could be described as organic compounds that do not involve direct participation in basic growth, and development or reproduction but also contribute to defense and signaling between the different plants (Agostoni-costa et al., 2012). Adaptation to different abiotic environmental stressors enables considerable interest in these several bioactive compounds and their application in the pharmaceutical, agricultural, and food sectors (Alamgir and Alamgir, 2017). It have several health-benefiting properties of secondary metabolites, alkaloids, flavonoids, terpenoids, and glucosinolates, processes that encompass antioxidant, anti-inflammatory, antimicrobial, and anticancer properties (Isah et al., 2019).

Vegetable crops like tomatoes, carrots, broccoli, and spinach are sources from which countless secondary metabolites have been derived. For instance, lycopene is a powerful antioxidant carotenoid found in tomatoes, able to lower the risks of chronic diseases like cancer and cardiovascular disorders (Giovannucci, 2020; Giovannucci, 1999). Similarly, years of research studies have revealed that glucosinolates present in cruciferous vegetables like broccoli are involved in a mechanism for cancer prevention because these compounds activate phase II detoxifying enzymes and regulate apoptosis and cell cycle arrest in cancerous cells (Murillo et al., 2001). The exploration and use of compounds found in vegetable crops offer significant potential for creating functional foods and nutraceuticals, supporting the overall well-being of individuals (Traka et al., 2011).

 DOI: 10.1201/9781003518358-8

Environmental conditions, agricultural practices, and genetic background all influence the formation of secondary metabolites in vegetable crops. A variety of studies have noted that soil nutrients, water content, temperature, and light all significantly affect the metabolic pathways related to biosynthesis (Treutter, 2021; Ramawat et al., 2023). Additionally, microbial inoculants and UV radiation serve as biotic and abiotic triggers employed to enhance the production of specific secondary metabolites in vegetable crops. This would certainly contribute to the development of optimal farming techniques that maximize yield while preserving the desired quality of secondary metabolites in vegetable crops (Alamgir and Alamgir, 2017).

Our understanding of the biochemical pathways and regulatory networks associated with the accumulated secondary metabolites in vegetable crops has significantly advanced due to recent advancements in omics technologies, including genomics, transcriptomics, proteomics, and metabolomics (Zhaung et al., 2014; Upadhyay and Singh 2023; Dixit et al., 2021).

Through the use of these technologies, important genes, enzymes, and transcription factors that play a role in the biosynthesis of valuable metabolites have been identified. This has opened a new frontier for synthetic biology and metabolic engineering approaches to improve the yield of these metabolites in vegetable crops. These advances represent examples of a sustainable approach to agriculture and food security, besides facilitating the generation of high-value crops with improved nutritional and therapeutic value (Souza et al., 2021).

Apart from their health-related uses, the application of secondary metabolites from vegetable crops reaches into other areas, including sustainable agricultural practices; these may include applications such as biopesticides and herbicides. Secondary metabolites from vegetable crops are valuable for their wide-ranging applications in health, agriculture, and industry. The growing interest in bioactive compounds has driven their potential to contribute to sustainable development and human health. Continuous research in the biosynthesis, regulation, and application of secondary metabolites will pave the way for innovative solutions which may harness the full potential of vegetable crops (Chaudhary et al., 2022).

8.2 NUTRACEUTICALS VALUES OF VEGETABLES

Vegetables contain secondary metabolites that have several health benefits (Kumar et al., 2022; Upadhyay and Singh 2021). These secondary metabolites, including flavonoids, carotenoids, glucosinolates, and phenolics, have attracted much interest because of their nutraceutical benefits. Because they contain antioxidants and anti-inflammatory agents, they help to prevent a wide range of diseases like cancer, cardiovascular conditions, and neurological disorders (Gupta et al., 2015; Singh & Yadav 2022; Upadhyay and Singh, 2024).

Glucosinolates, a group of secondary metabolites, interact with the enzyme myrosinase to generate potent bioactive compounds like sulforaphane, typically found in cruciferous vegetables like broccoli, cabbage, and Brussels sprouts. Once consumed, glucosinolates undergo hydrolysis, producing various bioactive compounds such as sulforaphane and indole-3-carbinol, which have been extensively researched for their chemopreventive properties. Sulforaphane, in particular, activates phase II detoxification enzymes, boosts the body's ability to neutralize carcinogens, and regulates the cell cycle, positioning it as a powerful anticancer compound (Mishra et al., 2023b; Fahey et al., 2001). Research has shown that a diet high in cruciferous vegetables is linked to a lower risk of various cancers, including breast, prostate, and colorectal cancers (Zheng et al., 2025; Wang et al., 2021).

Carotenoids, including β-carotene, lutein, and lycopene, are the pigments responsible for the vibrant orange color in vegetables like carrots and tomatoes. These compounds have the significant characteristic of an antioxidant, which scavenges free radicals, protecting the body from oxidative damage to lipids, proteins, and DNA; lycopene in tomatoes enhances the lipid profile and antioxidant properties. This helps reduce the risk of prostate cancer and cardiovascular diseases while also preventing the oxidation of low-density lipoprotein (LDL) (Rao et al., 2007; Giovannucci et al., 2023). Lutein and zeaxanthin, found in leafy greens, help protect against age-related macular degeneration (AMD) by accumulating in the retina and shielding it from harmful blue light (Johnson et al., 2021; Bernstein et al., 2023).

These secondary metabolites positively influence intestinal health and exhibit antioxidant and anticancer properties (Hunyadi, 2019). Additionally, vegetable dietary fibers and polyphenols act as prebiotics, fostering the growth of beneficial gut bacteria (Tan et al., 2023). SCFAs, such as butyrate, are produced in greater amounts when vegetable-derived chemicals and gut bacteria act symbiotically. These SCFAs are essential for immune response regulation and intestinal health maintenance (Makki et al., 2020; Flint et al., 2022). The variety of the gut microbiota is consequently supported by eating a wide variety of vegetables, which is crucial for maintaining metabolic health and preventing disease (Xu et al., 2023; Sun et al., 2021).

Vegetables are also considered to have high nutraceutical properties because of their high composition of secondary metabolites that have many beneficial effects on health and basic nutrition (Saiwal et al., 2019). These benefits can promote general and overall health enhancement based on a diet rich in diverse vegetables. Research enables us to develop functional foods enhanced by these beneficial metabolites and to gain an understanding of the mechanisms through which these bioactive compounds exert their effects (Patel et al., 2024; Halliwell et al., 2023).

8.3 SECONDARY METABOLITES FROM SOME COMMONLY CONSUMED VEGETABLE CROPS

Common vegetables have significantly contributed to nutritional value and health benefits. Widely consumed vegetables such as tomatoes, drumstick, ladyfinger, beetroot, carrot, bottle gourd, snake gourd, bitter gourd, pumpkin, radish, peas, beans, and green leafy vegetables are abundant in essential nutrients that promote overall health. Secondary metabolites from vegetable crops are enclosed in Tables 8.1 and 8.2.

8.3.1 Tomato (*Solanum lycopersicum*)

An important class of secondary metabolites in tomatoes comprises phenolic compounds, including tannins, phenolic acids, and flavonoids. It is highly recognized that these compounds possess antioxidant properties through their capacity to neutralize free radicals and safeguard cells against oxidative damage. According to various studies, the phenolic makeup of tomatoes is affected by factors such as the cultivar, stage of ripeness, and environmental conditions (Bashir et al., 2020). The primary flavonoids found in tomatoes include quercetin, kaempferol, and naringenin, the last one mostly concentrated in skin healthcare. Flavonoids have been linked to numerous health benefits (Poudel et al., 2021).

Unripe tomatoes contain higher concentrations of tomatine, which gradually decrease as the fruit ripens. Studies have shown that tomatine enhances immune function and can inhibit tumor cell proliferation (Shah et al., 2020). In addition, tomatine acts as a potent natural insecticide, displaying toxicity toward certain insect species and possessing a bitter, unappealing taste (Nguyen et al., 2022).

Carotenoids represent the other major class of secondary metabolites found in tomatoes. Much research has been focused on lycopene because of its potential antioxidant activity and because it imparts the red component in tomatoes. It has been linked to a reduced prevalence of chronic diseases, such as prostate cancer and cardiovascular diseases, and is effective in quenching free radicals, particularly singlet oxygen (Saini et al., 2020; Rao et al., 2003). Food processing methods and dietary lipids significantly influence the bioavailability of lycopene, enhancing its absorption in the body (Rao and Agarwal, 2021).

TABLE 8.1
Secondary Metabolites from Vegetable Crops and Applications

S. No	Secondary Metabolites	Compounds	Uses	Biological Activity	Reference
1	Carotenoid from (Tomato - *Solanum lycopersicum*[a])	Lycopene	Carotenoids are a strong cell reinforcement, liable for the dynamic red tint of ready tomatoes, and people consume much of the time in dietary enhancements food and beauty care products. A forerunner to vitamin A is used in enhancements to assist with forestalling a lack of vitamin A.	Possesses antioxidant properties, which may help lower the likelihood of specific cancers, including prostate cancer, and support heart health.	Rao and Rao (2007)
		β-carotene	Dietary Enhancements: Phytoene is often added to dietary enhancement because of its conceivable help benefits, particularly as a forerunner to carotenoids like lycopene and beta-carotene.	Acts as an antioxidant, aiding in vision and supporting immune function.	Tapiero et al. (2004)
		Phytoene	Like phytoene, phytofluene is added to dietary enhancement for its conceivable medical advantages and as an antecedent to different carotenoids.	Phytoene possesses antioxidant properties that assist in neutralizing free radicals and safeguarding cells from damage caused by oxidative stress. It demonstrates the ability to protect the skin from UV radiation.	Aust et al. (2005); Edge et al. (1997)
		Phytofluene	These mixtures are used in skincare items to safeguard skin from UV radiation and improve skin wellbeing.	Phytofluene displays potent antioxidant properties, protecting cells from oxidative damage. It possesses anti-inflammatory properties that help reduce inflammation and support overall health.	Aust et al. (2005); Hugueney et al. (1995)

(Continued)

TABLE 8.1 (*Continued*)
Secondary Metabolites from Vegetable Crops and Applications

S. No	Secondary Metabolites	Compounds	Uses	Biological Activity	Reference
2.	Phenolic compound	Chlorogenic acid	Phenolic compounds have cell reinforcement properties and are utilized in food and drugs for medical advantages.	Chlorogenic acid: Displays cancer prevention agents, mitigating, and antimicrobial exercises.	Clifford (2000)
		Caffeic acid	Helps to neutralize free radicals from oxidative damage, Reducing inflammation from the various disease	This bioactive is a cancer prevention agent with anticancer properties, anti-inflammatory and antioxidant effects	Kang et al. (2024); Martí et al. (2016)
		Rutin.	Rutin is a Powerful scavenging free radical prevention and is beneficial for Supportive cellular health. Investigations highlight rutin may have the ability to inhibit cell proliferation and damage through epigenetic pathways.	It is a cancer, carcinogen, and oxidative stress prevention agent.	Chen and Chen (2013)
3.	Alkaloid	Tomatine	Tomatine has antifungal properties and is utilized in rural practices to safeguard crops from microbes.	Shows antifungal, antibacterial, and anticancer exercises.	Friedman (2002)
4.	Flavonoids	Quercetin	Flavonoids have health-promoting qualities, they are used in nutraceuticals and dietary supplements.	Cancer prevention agents, antiviral, and anticancer properties.	Kumar and Pandey (2013)
		Kaempferol	Strong antioxidant practices inhibit harmful free radicals, protect against cell damage, and contribute to overall health.	It is used as a cancer prevention agent and makes calming and anticancer impacts.	Montano Calderon et al. (2011)

[a] Tomato (*Solanum lycopersicum*) has several bioactive compounds that has several properties such as antioxidants and anticancer. The compounds are not only utilized in pharmaceutical industries it is also employed in cosmetics industries.

TABLE 8.2
Name of the Vegetables, Type of Secondary Metabolites and Applications

S. No	Plant Name	Present of Secondary Metabolite	Application	Reference
1.	Ladyfinger (*Abelmoschus esculentus*)	Flavonoids (Quercetin, Isoquercitrin), Phenolic compounds (Catechin, Epicatechin)	Neutralizes harmful free radicals. Anti-inflammatory: Reducing inflammation. Promote and regulate blood sugar levels. Prevents the growth of harmful microorganisms.	Roy et al. (2014)
2.	Tomato (*Solanum lycopersicum*)	Carotenoids (Lycopene, Beta-carotene). Phenolic Compounds (Chlorogenic acid). Alkaloids (Tomatine). Flavonoids (Quercetin, Kaempferol).	Control cell damage and oxidative damage. It has been displayed to bring down the prostate disease, cardiovascular. Further inhibits developing heart health by limiting oxidative pressure. Shield the skin from UV radiation and exhibit general skin health.	Rao and Rao (2007)
3.	Drumstick (*Moringa oleifera*)	Flavonoids (Quercetin, Kaempferol), Phenolic compounds (Chlorogenic acid), Alkaloids (Moringinine).	Inhibit oxidative damage. It suppresses the proliferation and dissemination of cancer cells. Reduces inflammation and improves overall health. Acts were against various pathogens. Abundant essential vitamins and minerals provide vital nutrients that promote overall health.	Vergara-Jimenez et al. (2017)
4.	Beetroot (*Beta vulgaris*)	Betalains (Betanin, Vulgaxanthin),	It neutralizes harmful free radicals, preventing oxidative and cell damage.	Milton-Laskibar et al. (2021)
		Phenolic compounds (Caffeic acid).	Exhibit heart health and enhance blood flow. They are used as a natural dye in food protection.	Natella et al (1999)
5.	Carrot (*Daucus carota*)	Carotenoids (Beta-carotene, Alpha-carotene), Phenolic compounds (Chlorogenic acid).	Safeguards cells from oxidative harm. Beta-carotene is processed into vitamin A, which is fundamental for good vision. Further develops skin well-being and safeguards against UV radiation. Those bioactive are used as supplements for their medical advantages. And characteristic color in food items.	Nicolle et al. (2004)
6.	Bottle Gourd (*Lagenaria siceraria*)	Saponins, Flavonoids (Kaempferol), Phenolic compounds (Gallic acid).	It was Shields cells from oxidative harm. Diminishes aggravation. Successful against different microbes. Shields the liver from harm.	Nigam et al. (2015)

(Continued)

TABLE 8.2 (*Continued*)
Name of the Vegetables, Type of Secondary Metabolites and Applications

S. No	Plant Name	Present of Secondary Metabolite	Application	Reference
7.	Snake Gourd (*Trichosanthes cucumerina*)	Flavonoids (Quercetin), Triterpenoids, Phenolic compounds.	Neutralizing harmful free radicals and protects against oxidative damage. Inhibit inflammatory conditions. Against various pathogens, such as bacteria, viruses, etc. Protects the liver from harmful substances including oxidative stress and supports liver health.	Boots et al. (2008)
8.	Bitter Gourd (*Momordica charantia*)	Triterpenoids (Momordicin), Phenolic compounds (Gallic acid), Flavonoids (Quercetin).	Control blood sugar level, and preventing support glucose metabolism. Control oxidative damage. Prevent chronic diseases such as inflammatory disorders, cancer and cardiovascular disease. Inhibiting cancer spread cells, Exhibit apoptosis malignant cells	Grover and Yadav, (2004); Vattem and Shetty (2005)
9.	Pumpkin (*Cucurbita maxima*)	Carotenoids (Beta-carotene, Lutein), Phenolic compounds (Caffeic acid).	Lutein is beneficial for eye health and protects harmful blue light. Preserves skin health and shields against UV radiation damage. Compounds are used for dietary supplements and natural food coloring.	Rezig et al. (2012)
10.	Radish (*Raphanus sativus*)	Glucosinolates, Phenolic compounds (Ferulic acid).	Reducing oxidative damage risk factors, inhibiting chronic disease. Glucosinolates have the potential for anti-cancer properties and promote the flourishing of the gut microbiome, Aiding the preventive treatment of infection.	Beevi et al. (2010)
11.	Peas (*Pisum sativum*)	Flavonoids (Kaempferol), Phenolic compounds (Ferulic acid).	Inhibiting harmful free radicals and promoting chronic disease affection. Exhibit inflammatory supplement pathways to help manage our inflammatory conditions. Used for dietary supplements and numerous health benefits.	Dahl et al. (2012)

Vegetables crops that are in use on daily life has several health benefits and employed to treat several diseases.

Tomatoes contain glycoalkaloids and saponins that are terpenoids, which can also have potential activities in anti-inflammation, antibacterial, and anticancer effects (Fraser et al., 2021). Saponins have an immunomodulatory role, anti-tumor properties, and antiproliferative properties. Meantime, glycoalkaloids such as α-tomatine *have been found to exhibit strong antifungal activity* (Moghe et al., 2020).

8.3.2 Ladyfinger (*Abelmoschus esculentus* L.)

In the Malvaceae family, okra is a widely appreciated vegetable grown chiefly for its health benefits and nutritional value. It is rich in bioactive compounds, vitamins, and minerals, including secondary metabolites like flavonoids, tannins, and saponins. Some of these secondary metabolites contribute to plant defense mechanisms, while others offer general health benefits through their antioxidant, anti-inflammatory, and antimicrobial properties (Kumar et al., 2021; Jain, 2023).

Flavonoids are a group of polyphenolic compounds widely distributed in the pods and leaves of okra. The compound has been proven to show an antioxidant action utilizing scavenging unpaired free radicals formed within the cell to protect them against oxidative damage (Singh et al., 2024). Flavonoids, like quercetin and kaempferol, have been linked to reducing the risk of various chronic diseases, including cardiovascular diseases and certain types of cancers, associated with the consumption of okra (Bhattacharya et al., 2020). More recently, flavonoids from okra have been found to have prospects in diabetes management through carbohydrate-hydrolyzing enzyme inhibition and the improvement of insulin sensitivity (Mishra et al., 2023a ; Chatterjee, 2024).

Another major class of secondary metabolites is tannins, particularly concentrated in the seeds and mucilage. Tannins are known for their astringent properties, which include the precipitation of proteins and their characteristic role in their antimicrobial properties effective against a broad spectrum of pathogens (Gupta et al., 2015). Tannins extracted from okra have demonstrated promising inhibitory effects on pathogenic bacterial growth, like *Escherichia coli* and *Staphylococcus aureus*, suggesting their potential for use in natural antimicrobial agents (Saxena, 2023). Additionally, these tannins have been reported to the properties, which may aid in relieving inflammatory disorders such as joint inflammation (Patel et al., 2022).

Saponins are glycoside compounds that confer the bitter taste of okra and have been associated with some therapeutic effects. Such metabolites from okra have shown pronounced potential for reducing cholesterol levels through the binding of bile acids, hence facilitating their excretion (Tung et al., 2020). Besides, saponins from okra have proven immunomodulatory effects that enhance the immune response of the body from infections and diseases (Kaur et al., 2023). This contains cytotoxic properties of okra saponins and their potential as cancer-preventive agents, particularly in triggering apoptosis in cancerous cells (Sharma et al., 2025; Kaur et al., 2024; Jacob et al.,).

Flavonoids, tannins, and saponins are plant metabolites that make the okra plant a highly nutritious vegetable and a potential source of new natural therapeutic agents. Research on these compounds is ongoing, revealing new health benefits and applications, such as antimicrobial and antioxidant effects and potential anti-diabetic and anti-cancer properties (Bhargava et al., 2022). With the growing interest in natural and plant-based remedies, okra is gaining attention in the new pharmacological agents and functional foods, warranting further studies to explore its full potential.

8.3.3 Drumstick (*Moringa oleifera*)

Moringa oleifera, commonly referred to as the drumstick tree, holds significant value in the fields of nutrition and medicine. The plant contains a variety of secondary metabolites, including alkaloids, flavonoids, tannins, saponins, and phenolic acids, which contribute to its therapeutic properties. These metabolites have been found to demonstrate a range of pharmacological activities, such as anti-inflammatory, antioxidant, antimicrobial, and antidiabetic effects (Gopalakrishnan et al., 2016; Anwar et al., 2007).

Non-essential metabolites *Moringa oleifera* were reported to be rich in leaves, seeds, pods, and roots, and their composition and concentration are highly dependent on the plant part. Most of these compounds are found in large proportions in the leaves, characterized by high polyphenols such as quercetin and kaempferol, including flavonoids, which are strong antioxidants. They also contain alkaloids and saponins known to demonstrate anti-inflammatory and antimicrobial effects (Saini et al., 2016; Verma et al., 2009). The seeds and pods also act as good sources of glucosinolates and isothiocyanates, which have possessed anticancer properties (Leone et al., 2015).

Moringa oleifera secondary metabolites possess various medicinal values. Antioxidant action by flavonoids and phenolic compounds assists in quenching free radicals. Hence, it protects the body from diseases that include cancer and cardiovascular disorders arising out of the degenerative process of oxidative stress (Rahman et al., 2018).

Besides, leaves and roots contain alkaloids and saponins that exhibit antidiabetic activities by regulating blood sugar levels (Mbikay, 2012). These metabolites in Moringa oleifera have also demonstrated antimicrobial activities and effectively combat a range of bacterial and fungal microorganisms (Kasolo et al., 2010; Rahman et al., 2009). *Moringa* oleifera is widely used in both the pharmaceutical and food industries because of its abundant bioactive compounds. Its leaves, in particular, are used as functional food ingredients, offering numerous nutritional and antioxidant benefits. These properties not only enhance the nutritive value of food but also improve its lifestyle (Siddhuraju and Becker, 2003). It has found applications in formulating supplements and medicinal plant properties drugs against diabetes, hypertension, or bacterial infections in the pharmaceutical industry (Fahey, 2005; Kumar et al., 2010).

8.3.4 Beetroot (*Beta vulgaris*)

Beetroot is a root vegetable celebrated for its color and culinary uses but also for its rich content of secondary metabolites, which have garnered significant interest from researchers due to their various biological activities. These secondary metabolites, including betalains, phenolic compounds, and flavonoids, contribute to the health-promoting properties and their potential therapeutic applications.

The primary secondary metabolites unique to the order caryophyllales are pigments called betalains. These pigments are responsible for the distinctive red and yellow hues in beetroot. Betalains are categorized into betacyanins and betaxanthins. Betacyanins produce red-purple tones, and betaxanthins provide a yellow-orange hue. Results from research have shown that betalains are potent antioxidants that neutralize free radicals, thereby safeguarding cells from oxidative stress, which is often the underlying cause of chronic diseases like cardiovascular diseases and cancer (Khan et al., 2023; Nemzer et al., 2022).

It is also an abundant source of phenolic compounds, including phenolic acids and flavonoids, with ferulic and caffeic acids found in beetroot and shown to have antioxidant, antimicrobial, and anti-inflammatory properties (Guldiken et al., 2019). These compounds exert a scavenging action on free radicals, diminish oxidative stress, and hence could be responsible for preventing diseases related to oxidative stress. Other phenolic compounds in beetroot are flavonoids, well known for their cardioprotective effects that reduce atherosclerosis risk by preventing LDL oxidation and improving endothelial function. This indicates that beetroot is a beneficial functional food with properties that enhance health.

8.3.5 Carrot (*Daucus carota* L.)

Carrots are an abundant source of secondary metabolites, representing all the chemicals not participating in the basic metabolic processes of the plant, including growth, development, and reproduction. Such substances often play an ecological role in defending against herbivores, pathogens, and environmental stressors and might contribute to the overall defense of the plant. Carrot plants contain numerous secondary metabolites, including polyacetylenes, flavonoids, phenolic acids, and carotenoids, their specific bioactive function, and potential health benefits.

The most abundant class of secondary metabolites in carrots is polyacetylenes. The intricate nature of their bioactivity has been widely researched, with key compounds such as falcarinol, falcarindiol, and falcarindiol-3-acetate. Polyacetylenes are well-known for their antifungal, antibacterial, and anti-inflammatory properties. Studies suggest that these compounds may also help reduce the risk of certain cancers. For instance, falcarinol has been found to possess anticarcinogenic properties in several *in vitro* studies and animal models, suggesting its potential as an active chemo preventive agent against colon cancer (Metzger et al., 2009). Additionally, carrot polyacetylenes have been implicated in neuroprotection, which may contribute to preventing neurodegenerative diseases (Hansen et al., 2010).

Flavonoids represent another significant class of secondary metabolites in carrots. Flavonoids such as quercetin, luteolin, and kaempferol have been shown to scavenge free radicals and reduce oxidative stress within cells. Oxidative stress is recognized as a major factor in the development of chronic diseases, including cardiovascular diseases and cancers. There is evidence that flavonoids improve endothelial function, modulate inflammation, and are associated with a lower cardiovascular disease risk by consuming carrot-rich food. Furthermore, the flavonoids in carrots have been found to influence various signaling pathways involved in cell proliferation, apoptosis, and angiogenesis, positioning them as potential therapeutic agents for cancer prevention and treatment (Li et al., 2021).

Carotenoids perhaps are among the well-recognized secondary metabolites in carrots, which generally are responsible for their orange color. Carotenoids, including beta-carotene, alpha-carotene, lutein, and zeaxanthin, are crucial to human health due to their role as important precursors in vitamin A synthesis, which is essential for good vision, immunity, and skin health. β-carotene, one of the primary carotenoids found in carrots, has been extensively studied for its antioxidant properties, particularly in protecting cells from oxidative damage. The evidence shows that beta-carotene protects against neoplastic and cardiovascular diseases. Besides carrot carotenoids, compounds like lutein and zeaxanthin protect against AMD and cataracts, highlighting the significance of carrots for eye health.

The great variability of carrot secondary metabolites creates enormous potential for their applications in medicine and the food industry. In medicines, carrot-origin bioactive is studied as a promising candidate for new therapeutic agents against chronic conditions, such as cancer, cardiovascular diseases, and neurodegenerative disorders. The antioxidant and antimicrobial activities of polyacetylenes, flavonoids, and phenolic acids nowadays take part in food preservation and elaboration of functional foods with health-promoting and disease-preventing properties (Dawidowicz et al., 2006).

8.3.6 Bottle Gourd (*Lagenaria siceraria*)

Bottle gourd is a commonly consumed vegetable with a long-standing tradition of use in traditional medicine. The plant is abundant in secondary metabolites such as flavonoids, saponins, and terpenoids, which have significant pharmacological potential. These polyphenolic compounds exhibit strong antioxidant properties. They are crucial in reducing oxidative stress, a factor linked to several chronic diseases, including cardiovascular diseases and cancer (Sharma et al., 2015). Bottle gourd is known for its saponins, which have cholesterol-lowering effects and can improve cardiac health (Kumar et al., 2019). These compounds are bitter-tasting triterpenoids with potential anticancer properties. Cucurbitacins from bottle gourd have been demonstrated to suppress the growth of cancer cells (Raj et al., 2017).

8.3.7 Snake Gourd (*Trichosanthes cucumerina*)

Snake gourd is another gourd vegetable enriched with bioactive compounds. These include triterpenoids, alkaloids, and phenolic acids, which enhance their therapeutic potential. These compounds have been reported to exhibit anti-inflammatory and anticancer properties (Rao et al., 2018). Snake gourd is rich in phenolic acids, known for their antioxidant and antimicrobial properties, which help combat pathogens and oxidative stress (Somegowda et al., 2021). The presence of cucurbitacins further amplifies its anticancer potential.

8.3.8 Bitter Gourd (*Momordica charantia*)

Bitter gourd is renowned for its bitter taste and medicinal value. The fruit is rich in bioactive compounds, including alkaloids, saponins, and glycosides, contributing to its pharmacological importance. A compound with hypoglycemic properties, charantin is crucial in managing diabetes (Kumar et al., 2020). This bitter-tasting triterpene glycoside possesses anti-inflammatory and anticancer effects. Bitter gourd contains vicine, an alkaloid aids to treat antidiabetic properties (Singh and Kaur, 2019).

8.3.9 Pumpkin (*Cucurbita pepo*)

Pumpkin is a nutrient-dense vegetable, with a range of secondary metabolites like carotenoids, flavonoids, and sterols. These pigments possess strong antioxidant properties, supporting eye health and immune function (Hosseini et al., 2017). Pumpkin seeds are abundant in phytosterols, which help to reduce cholesterol levels (Gad et al., 2018). The polyphenolic compounds in pumpkins play a key role in their anti-inflammatory and anticancer effects (Sharma et al., 2020).

8.3.10 Radish (*Raphanus sativus*)

Radish, a root vegetable, is abundant in various bioactive compounds, such as glucosinolates, flavonoids, and phenolic acids. These secondary metabolites contribute to its pungent flavor, nutritional benefits, and medicinal properties. Glucosinolates, such as glucoraphanin and sinigrin, are abundant in radishes. These compounds contribute significantly to the bitter taste and anticancer properties of radish. Known for their ability to break down into isothiocyanates, they display powerful antioxidant and anti-inflammatory effects (Park et al., 2020). Pink and purple radishes are rich in anthocyanins, pigments that offer antioxidant properties and may protect against cardiovascular diseases (Anderson and Major, 2002). Radish contains Various phenolic compounds contribute to its antioxidant capacity, helping prevent oxidative stress-related diseases (Mattila and Hellström, 2007).

8.3.11 Peas (*Pisum sativum*)

Peas are a rich source of proteins and dietary fiber and are packed with secondary metabolites like saponins, flavonoids, and phenolic acids, which contribute to their numerous health benefits. These compounds are recognized for their ability to lower cholesterol levels and enhance immune function. Saponins in peas can also exhibit anti-cancer and anti-inflammatory effects (Elbadrawy and Mostafa, 2024). Peas are a source of isoflavones like genistein and daidzein, which act as phytoestrogens and may help in reducing the risk of hormone-related cancers (Shoff et al., 1998). Phenolic acids, including ferulic acid and p-coumaric acid, exhibit antioxidant properties that help protect cells against oxidative damage (Duodu, 2011). The high saponin content of peas makes them valuable for developing heart-healthy foods that target cholesterol reduction. Isoflavones from peas are also being studied for their role in hormone replacement therapies and cancer prevention (Mazur, 1998). The antioxidant properties of phenolic acids found in peas make them an excellent ingredient in functional foods designed to mitigate oxidative stress (Campos-Vega et al., 2010).

8.3.12 Beans (*Phaseolus vulgaris* L.)

Beans are well known for their high protein content, but they also possess a variety of secondary metabolites, such as flavonoids, tannins, and alkaloids, which play a key role in their health-promoting properties. Beans are rich in flavonoids, particularly quercetin and kaempferol, and are linked to a reduced risk of heart disease and cancer, primarily due to their antioxidant effects (Chung et al., 2002). These compounds are recognized for their astringent properties and their ability to function as antioxidants, promoting cardiovascular health (Sharma et al., 2020).Beans have low concentrations of alkaloids, including trigonelline, which possess anti-diabetic properties (Ali Asgar, 2013).

The potent antioxidant properties of flavonoid-rich extracts from beans make them a growing component in nutraceuticals aimed at combating cardiovascular diseases. Flavonoid-rich extracts from beans are increasingly incorporated into nutraceuticals for treating cardiovascular diseases. Flavonoid-rich extracts from beans are increasingly being used in nutraceuticals targeting cardiovascular diseases owing to their strong antioxidant properties. The presence of tannins also highlights the significance of beans as an important component in foods promoting heart health (Rodríguez et al., 2022). Additionally, beans containing alkaloid content are being explored for their role in diabetes management (Ramdath et al., 2016).

8.3.13 Green Leaves (*Camellia sinensis*)

Green leafy vegetables like spinach, kale, and lettuce are excellent sources of vitamins, minerals, and secondary metabolites, including carotenoids, polyphenols, and flavonoids. They are particularly rich in carotenoids such as lutein and β-carotene, which have antioxidant properties and are beneficial for eye health and immune function (Bungau et al., 2019). Polyphenols like quercetin and rutin are present in green leafy vegetables and are renowned for their anti-inflammatory and antioxidant properties (Thitimuta et al., 2017). Green leafy vegetables like spinach and kale owe much of their anti-inflammatory and cardioprotective effects to their flavonoid content (Alam et al., 2022).

Carotenoids from leafy greens produce dietary supplements for vision power and help reduce the risk of AMD (Wang et al., 2024). The polyphenols in green leaves also offer potential as ingredients in functional foods targeting inflammation and oxidative stress-related diseases (Zeng et al., 2020). Additionally, green leaf extracts containing rich flavonoids are used in heart health supplements due to their ability to improve vascular function and reduce inflammation (Alam et al., 2022). Various secondary metabolites in the vegetables are shown in Figure 8.1.

8.4 SECONDARY METABOLITES, TYPES, AND EXTRACTION METHODS

8.4.1 Secondary Metabolites

Secondary metabolites are natural compounds not directly involved in the growth and development of plants; however, they play crucial roles in plant defense, reproduction, and interaction with the environment. They assume a fundamental part in the important role of resisting herbivores and microbes. The first step in secondary metabolites identifying the compounds responsible for the desired effects of biological activities associated with the biological materials and their extracts. The secondary bioactive is involved in the development and interactions of surrounding living organisms. They are probably essential elements for blocks such as parasites, pathogens, and predators; moreover, they are pollinators and disseminators. Also, higher concentrations of medicinal and aromatic plants contain bioactive compounds that play a significant role in the pharmaceutical sector. These accumulate, which incorporate alkaloids, flavonoids, phenolics, and terpenoids and are of great interest due to their bioactive properties, with notable applications in pharmaceuticals and beauty care products (Andrade, 2010).

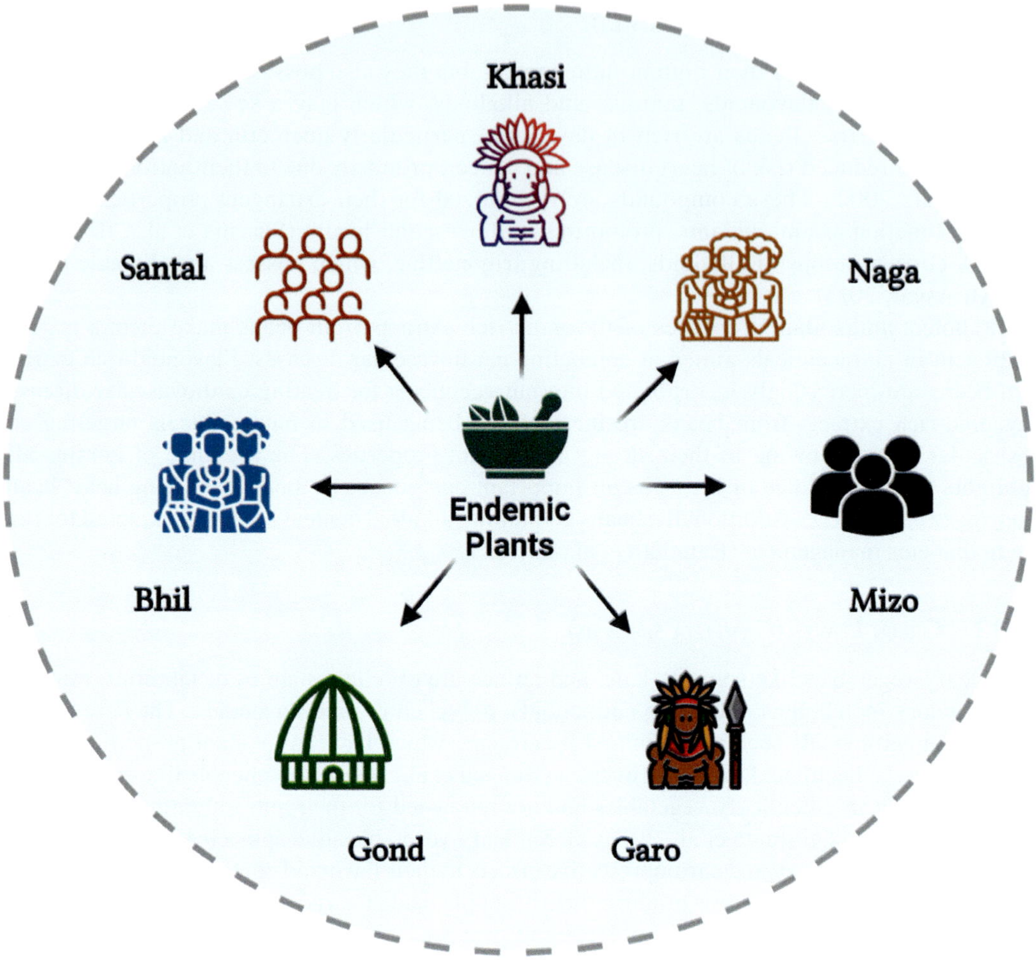

FIGURE 8.1 Distribution of secondary metabolites in various vegetables. The figure shows the secondary metabolites in various vegetables crops that are consumed on daily basis. Generally, carotenoids are rich in yellow and orange coloured vegetables. Flavonoids are abundant in red colour vegetables.

8.4.2 Types of Secondary Metabolites

8.4.2.1 Alkaloids

Nitrogen-containing compounds are known for their therapeutic potential. Examples of alkaloids are bioactive morphine isolated from opium poppy and nicotine isolated from tobacco. Those alkaloids contain compounds used for plant defense mechanisms and physiological effects of interactions with other living organisms (Crozier et al., 2006).

8.4.2.2 Terpenoid

Artemisinin is extracted from the plant *Artemisia annua*, commonly known as sweet wormwood. It has been significantly researched for its potent anti-malarial properties. Highlight its significance in traditional medicine and use for pharmacological research. This provides a credible basis for discussing the pharmacological relevance of natural products in combating diseases like malaria (Sampath Kumar et al., 2020). The important role is to ensure effective communication between pollination and seed dispersers. It helps in the plant defense mechanisms that inhibit herbivores and pathogens (Sharma et al., 2017).

8.4.2.3 Phenolics

Phenolic compounds are naturally occurring chemical compounds found in vegetables and plants like *Camellia sinensis,* adding to their cancer-prevention agent properties. These compounds are classified into two groups: flavonoids and tannins, which are widely used for their antioxidant and antimicrobial properties. Some phenolics, such as lignins, also contribute to the structural rigidity of plants (Crozier et al., 2006).

8.5 EXTRACTION METHODS FOR SECONDARY METABOLITES IN VEGETABLE CROPS

8.5.1 Solvent Extraction

Solvent extraction is a widely utilized method for isolating bioactive compounds from plant materials. This technique involves soaking the plant material in an appropriate solvent to extract the desired metabolites, such as alkaloids, flavonoids, phenolics, and other phytochemicals. Common solvents include ethanol, methanol, and acetone. The plant material is ground using an electric grinder, and the selected solvent dissolves the active metabolites depending on their solubility. After a sufficient extraction period, the solvent containing the dissolved metabolites is separated from the remaining plant material.

The polarity of solvents is crucial in influencing the effectiveness of the extraction process, as different solvents target specific types of compounds. For example: ethanol is suitable for extracting both polar and non-polar compounds. Methanol is highly polar and particularly effective for polar metabolites such as phenolics. Acetone is a versatile solvent capable of extracting polar and non-polar compounds (Sasidharan et al., 2010). Higher temperatures can improve extraction efficiency but may also degrade heat-sensitive compounds, so careful heat control is important. This method is commonly applied to separate essential oils, alkaloids, and phenolics from medicinal plants. It is also used in the industrial-scale production of bioactive compounds for application in nutraceuticals, pharmaceuticals, and cosmetics (Chemat et al., 2012).

8.5.2 Microwave-Assisted Extraction (MAE)

MAE is an innovative technique that uses microwave energy to extract bioactive compounds from plant materials. This energy quickly heats both the solvent and plant material, speeding up the extraction process. Intracellular heating of the microwave heats the water inside plant cells, causing the cells to rupture and release the bioactive compounds into the surrounding extraction solvent (Dias et al., 2021). The plant material is mixed with a polar solvent, like ethanol, methanol, or water, as these solvents efficiently absorb microwave energy. Microwaves cause the solvent molecules to vibrate, rapidly increasing the temperature. The microwave-induced heat causes the plant cells to break open, allowing the bioactive compounds to be released into the solvent. This process is significantly faster than conventional extraction methods (Chemat et al., 2017).

MAE is extensively utilized for extracting polyphenols, flavonoids, essential oils, and other stable bioactive compounds. This technique has applications across various sectors, including food, pharmaceuticals, cosmetics, and nutraceuticals (Lucchesi et al., 2004).

8.5.3 Ultrasound-Assisted Extraction (UAE)

UAE is a method that uses high-frequency sound waves to break down plant cell walls, facilitating the release of metabolites into the solvent. The sound waves create cavitation bubbles in the solvent, which rapidly expand and collapse. This phenomenon creates localized regions of high temperature and pressure, enhancing cell wall degradation and promoting solvent infiltration (Chemat, et al., 2011).

UAE produces high-frequency sound waves that travel through the solvent. Cavitation involves the formation, growth, and sudden collapse of bubbles, creating intense shear forces that break down plant cell walls, promoting the release of bioactive compounds into the solvent (Wen et al., 2018).

UAE is frequently used to extract phenolic compounds, flavonoids, essential oils, and other bioactive metabolites from plant materials. This method is widely applied in the food, cosmetic, nutraceutical, and pharmaceutical industries to obtain valuable plant extracts (Chemat et al., 2017).

8.5.4 Mechanism of Phytochemical Interaction with the Human System

Phytochemicals are bioactive compounds found in plants that enable these compounds to interact with the human body through different molecular mechanisms, aiding in the prevention and treatment of diseases. Flavonoids are among these compounds, alkaloids, polyphenols, terpenoids, and glycosides, which can influence biological processes through multiple pathways (Zhang et al., 2015).

8.5.4.1 Antioxidant Mechanism

One key mechanism through which phytochemicals deliver their beneficial effects is by exhibiting antioxidant activity. They inhibit reactive oxygen species (ROS), which are by-products of cellular metabolism that induce oxidative stress. This stress leads to DNA damage, lipid peroxidation, and protein dysfunction, all of which are linked to chronic diseases such as cancer and cardiovascular disease (Cory et al., 2018). Polyphenolic compounds, including quercetin and resveratrol, neutralize ROS by donating electrons, thereby reducing oxidative stress. This antioxidant capacity is crucial for maintaining cellular homeostasis and protecting the body from oxidative damage (Pisoschi and Pop, 2015).

8.5.4.2 Anti-Inflammatory Mechanism

Phytochemicals also modulate inflammation, a process that underlies many chronic diseases. Flavonoids, such as apigenin and luteolin, inhibit pro-inflammatory enzymes like cyclooxygenase and lipoxygenase, thereby reducing the production of pro-inflammatory cytokines (Caporali et al., 2022). They also suppress nuclear factor kappa B (NF-κB), a crucial transcription factor involved in regulating inflammatory pathways. These interactions help regulate chronic inflammation (Chong et al., 2012).

8.5.4.3 Interaction with Gut Microbiota

The bacteria residing in the gut play a crucial role in metabolizing plant-derived chemicals and converting them into bioactive compounds that the body can utilize. Consequently, these compounds modulate gut microbiota composition, fostering beneficial bacterial growth and inhibiting harmful species like "Reactive Oxygen Species (ROS) and Reactive Nitrogen Species (RNS)" are highly reactive molecules that can cause oxidative and nitrosative stress, etc. (Santhiravel et al., 2022). This interaction enhances overall health by regulating immune function, metabolism, and inflammatory responses (Jayachandran et al., 2017).

8.5.4.4 Cardiovascular Protection

Flavonoids and phenolic acids are present in numerous plants; those plant biochemicals help protect the heart by improving the function of the inner layer of blood vessels (endothelium), lowering blood pressure in circulatory system, and inhibiting blood clot formation and plaque buildup in arteries (Tzounis et al., 2008). Epicatechin, a specific type of polyphenol, enhances the dilation of blood vessels as part of an immune response, thereby improving circulation and reducing the likelihood of heart-related incidents (Hertog et al., 1993).

8.5.4.5 Anti-Cancer Mechanism

Phytochemicals exhibit anti-disease properties by modulating various cellular pathways associated with cell cycle regulation, apoptosis, and metastasis. For instance, green tea extract is rich in epigallocatechin-3-gallate (EGCG), which inhibits the signaling pathway. The epidermal growth factor receptor (EGFR) is a cell surface receptor that plays a key role in regulating cell growth,

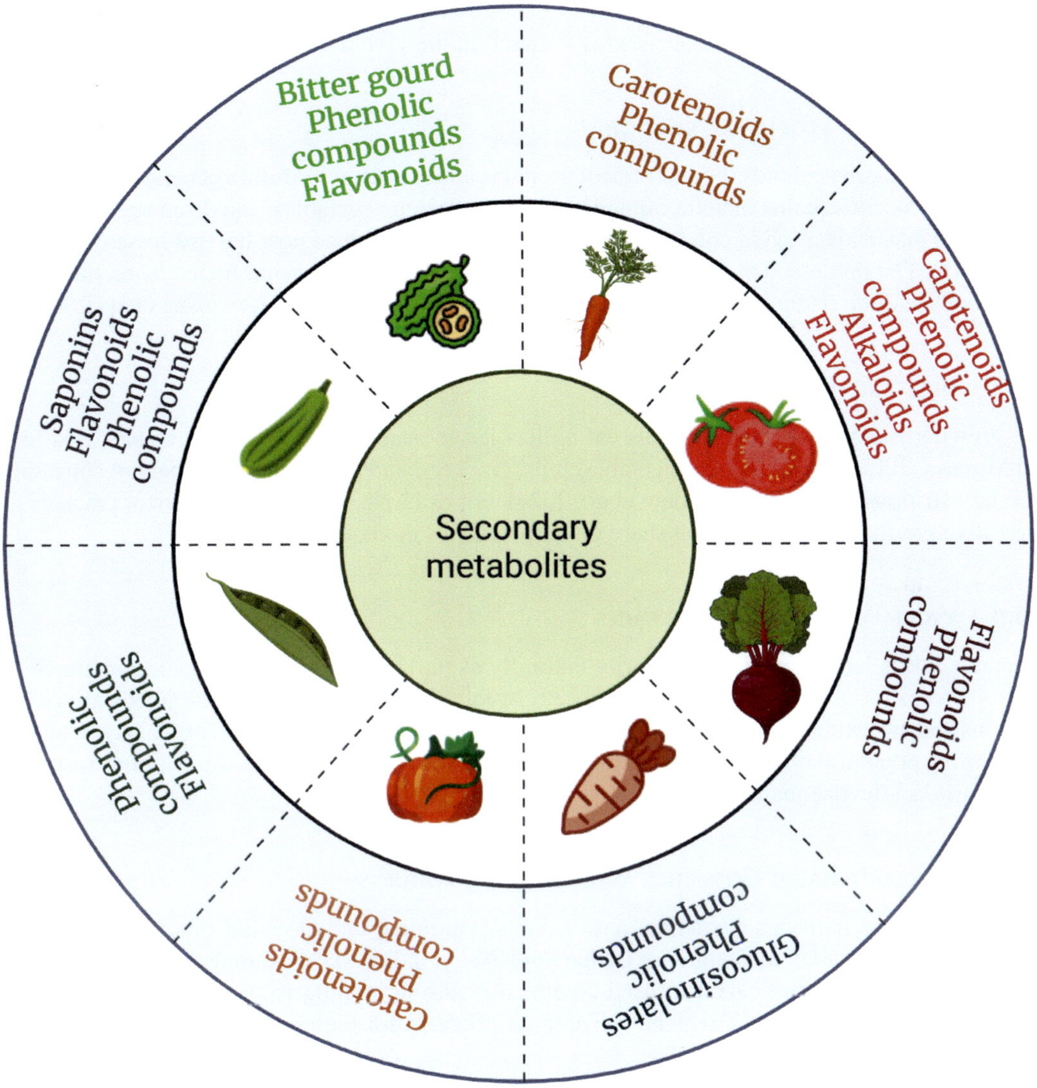

FIGURE 8.2 Plant bioactive molecules treat several chronic diseases. Plant bioactive molecules (Phytochemical constituents) are essential in minimizing chronic diseases by supporting the body's defence systems and boosting the immune response.

survival, and differentiation, thereby reducing cancer cell proliferation. Additionally, sulforaphane, present in cruciferous vegetables, stimulates the Nrf2 pathway, promoting detoxification of carcinogens and aiding in cancer prevention (McWalter et al., 2004). Several phytochemicals that minimize the effect of chronic disease are shown in Figure 8.2.

8.6 MARKETABLE PRODUCTS FROM VEGETABLES

8.6.1 Introduction to Vegetable-Derived Products

Vegetables are a primary source of nutrition and hold immense potential for various value-added products. As consumer awareness of health benefits grows, the industry directs attention to developing innovative vegetable-based products, including dietary supplements and cosmetics

(Trigo et al., 2020). This chapter emphasizes various marketable products derived from vegetables, their economic potential, and their growing demand in the global market.

8.6.2 Processed Vegetable Products

Vegetable processing extends the shelf life of products and is crucial for future generations (Augusto, 2020). The processed items include canned vegetables, frozen vegetables, and dried vegetable products. These materials provide convenience for consumers and reduce post-harvest losses (Majerska et al., 2019). The processed vegetable industry has seen substantial growth, fueled by technological innovations and the rising demand for convenient ready-to-eat products (Kowalska et al., 2017).

8.6.3 Vegetable-Based Snacks

The growing trend of health-conscious eating has led to snacks made from vegetables (Da Silva and Moreira, 2008). Products like vegetable chips, crackers, and puffed snacks appeal to consumers seeking nutritious alternatives (Chobot et al., 2024). For example, kale chips and carrot crackers are innovative products gaining market share globally (Bustos et al., 2023).

8.6.4 Vegetable Juices and Beverages

Vegetable juices have surged in popularity naturally as part of the growing demand for functional foods (Petruzzi et al., 2017). Carrot juice, beetroot juice, and blended vegetable juices are rich in vitamins and antioxidants and have significant health benefits (Zheng et al., 2017). This type of juice is becoming popular with health-conscious people and those looking for healthier options instead of sugary drinks (Bevilacqua et al., 2018).

8.6.5 Vegetable-Based Cosmetics and Skincare Products

Vegetable-derived ingredients such as antioxidants, vitamins, and essential oils are increasingly being used in cosmetic formulations (Athar and Nasir, 2015). For example, cucumber masks, tomato-based serums, and carrot-infused creams are gaining popularity for their natural components (Lacatusu et al., 2020). Farmers utilize these products for their ecological sustainability and minimal use of chemicals (Hill, 2007).

8.6.6 Organic Vegetable Products

Natural vegetables and their products have a unique and growing market (Dorais, 2007). Consumers are ready to spend more on higher prices for organic vegetable products, such as organic juices, canned vegetables, and vegetable chips, because they perceive these products as healthier and more environmentally friendly (Ezov et al., 2020). The organic vegetable market is expanding due to increased awareness of the health dangers related to chemical residues (Dorais et al., 2015).

8.6.7 Functional Foods and Nutraceuticals from Vegetables

Vegetables are abundant in bioactive compounds that can be harnessed to create functional foods and nutraceuticals (Gul et al., 2016). Lycopene from tomatoes, beta-carotene from carrots, and polyphenols from leafy green Vegetables have been widely researched for their health-enhancing benefits (Egbuna et al., 2020). The market for functional foods is projected to reach billions, driven by a shift toward preventive healthcare (Adefegha, 2018).

8.7 APPLICATION OF SECONDARY METABOLITES FROM VEGETABLES

8.7.1 Antioxidant Defence

Secondary metabolites, including carotenoids and polyphenols present in vegetables, play a crucial role in safeguarding the body by neutralizing free radicals and reducing oxidative stress, which can harm cells and lead to chronic diseases. For example, the carotenoids in carrots exhibit powerful antioxidant properties, helping protect cells from damage and supporting overall health (Assous et al., 2014).

8.7.2 Inflammation Reduction

Tomatoes, particularly those rich in lycopene bioactive, can help decrease inflammation by blocking the production of inflammatory cytokines. Research has demonstrated that lycopene effectively reduces inflammation markers. Lycopene, known for its strong antioxidant properties, is associated with a lower risk of certain cancers, highlighting its wide-ranging health benefits (Palozza et al., 2010).

8.7.3 Cancer Prevention

The glucosinolates are found in broccoli and other cruciferous vegetable plants and play an essential role in cancer prevention by modulating oncogenic processes of cancer cell anti-proliferation (Meeran et al., 2010). These compounds are known to influence several biochemical pathways, including enzyme activity, programmed cell death induction, and the suppression of tumor cell proliferation (Soundararajan et al., 2018). Research reveals their potential to reduce the probability of different types of cancers, For instance, breast cancer, prostate cancer, and colorectal cancer, by facilitating the removal of carcinogens and inhibiting tumor growth (Murillo and Mehta, 2001).

8.7.4 Cardiovascular Health

Flavonoid compounds are abundant in vegetables like spinach and play a vital role in promoting cardiovascular health by enhancing endothelial function and reducing blood pressure (Parmenter et al., 2020). These bioactive compounds have antioxidant and anti-inflammatory properties that safeguard blood vessels, improve circulation, and lower cardiovascular disease risk. Numerous research studies has shown that dietary flavonoids can enhance gut endothelial function and promote for the maintenance of vascular health. Additionally, they contribute to the chances of lowering cardiovascular disease by modulating numerous biological pathways, such as improving lipid profiles and decreasing arterial stiffness (Niu et al., 2023).

8.7.5 Antimicrobial Properties

Phenolic compounds introduced in vegetables like garlic demonstrate significant antimicrobial activity against microorganisms. Research has shown their effectiveness in bacterial and parasitic infections, highlighting their crucial importance in various diseases (Goncagul and Ayaz, 2010). These compounds can disrupt microbial cell walls, inhibit enzyme function, and interfere with pathogen replication, thus enhancing the body's defense mechanisms. Phenolic compounds are strong antimicrobial properties found in garlic, emphasizing their potential as natural alternatives to conventional antibiotics and supportive treatments for infectious diseases (Puupponen-Pimia et al., 2001).

8.7.6 Digestive Support

Research has confirmed their beneficial effects on gastrointestinal function, underscoring their potential to enhance digestion and promote overall gut health (Kraft, 1997). Cynarin is especially known for its capacity to assist in fat digestion and support liver function, contributing to more effective digestive processes. Additionally, research indicates that artichokes may help relieve symptoms related to conditions like irritable bowel syndrome (IBS) and bloating, making them a beneficial addition to a diet that promotes digestive health (Ayuso et al., 2024).

8.7.7 Skin Health Improvement

Carotenoids are notable for their valuable antioxidant properties, specifically their role in protecting skin cells from UV radiation and oxidative stress. It shows carotenoids can improve skin elasticity and hydration while reducing photoaging and pigmentation (Darvin et al., 2011). Carotenoid consumption is linked to enhancements in skin color and overall appearance. Studies indicate that individuals with higher carotenoid intake tend to have healthier-looking skin and are frequently seen as more desirable. The findings suggest that carotenoids contribute to improved skin complexion and vibrancy. They also provide protective effects against UV damage and contribute significantly to skin health (Evans et al., 2010).

8.8 COMMERCIAL COMPOUNDS IN THE MARKET

The global vegetable industry is heavily affected by commercial compounds, including fertilizers, pesticides, biostimulants, and post-harvest preservatives. These chemicals are crucial for maintaining crop yield, quality, and shelf life, as shown in numerous studies (Elkins, 1989). However, their application has sparked concerns regarding environmental and health effects, which have been deeply analyzed in the existing literature (Chavarri et al., 2005).

8.8.1 Pesticides in Vegetable Farming

Pesticides function as agents for controlling pests and diseases in vegetable cultivation, resulting in significant yield improvements (Sharma and Singhvi, 2017). However, concerns about pesticide residues on vegetables and their potential health impacts have driven the exploration of organic alternatives (Reeve et al., 2016). Research has demonstrated that biopesticides can effectively reduce pest populations without compromising quality control and yield development (Lengai and Muthomi, 2018). This makes biopesticides a sustainable alternative to conventional chemical pesticides (Sharma and Singhvi, 2017).

8.8.2 Fertilizers

Biofertilizers used for boosting vegetable yield, particularly nitrogen-based fertilizers, have displayed plant growth and development to expand the yield of vegetables over use of nitrogen-based fertilizers can result in soil degradation, reducing its quality and fertility over time. This impacts crop productivity temporarily and causes lasting harm to the ecosystem by disturbing the soil structure and nutrient balance (Fageria et al., 2008). Natural composts have been suggested as a more sustainable alternative, providing the opportunity to promote soil health and enhance the quality of vegetables (Reeve et al., 2016; Elkins, 1989).

8.8.3 Biostimulants

Natural biostimulants promote plant growth, enhance nutrient absorption, and improve resilience to challenging conditions such as drought, salinity, and temperature fluctuations. By fostering root development and stimulating beneficial microbial activity, biostimulants contribute to overall plant health and increase crop productivity, particularly in adverse environmental conditions (Yakhin et al., 2017). They additionally advance development and supplement take-up, with research exhibiting their adequacy in improving vegetable quality (Okon and Baker, 1987).

8.8.4 Post-Harvest Compounds

Post-harvest treatments, including additives and edible coatings, are essential for maintaining the quality of vegetables during preservation (Wang and Teplitski, 2023). Recent studies have focused on using natural preservatives and coatings to extend vegetable shelf life while ensuring safety. These treatments help retain freshness, reduce spoilage, and preserve nutritional value, allowing vegetables to remain consumable for longer periods under diverse storage conditions (Cruz et al., 2015).

8.9 ENDEMIC VEGETABLES UTILIZED BY PARTICULAR TRIBALS

8.9.1 *Momordica cochinchinensis*

Tribal groups: Karen ancestral networks (Thailand and Myanmar)

Momordica cochinchinensis (regularly known as Gac) is an organic product generally involved in the Karen ancestral networks in Thailand and Myanmar. It is esteemed for its rich substance of lycopene and beta-carotene, the two of which are powerful cell reinforcements with potential medical advantages, including hostile-to-malignant-growth properties. The organic product is frequently integrated into rice diets, where its lively red tone and healthful properties (Wimalasiri et al., 2020).

Late examinations have featured Gac's enemy of disease potential, especially its capacity to restrain cancer development and angiogenesis. The natural product's bioactive mixtures have shown adequacy in smothering disease cell multiplication and prompting apoptosis (customized cell demise) in different malignant growth cell lines, including bosom and cellular breakdown in the lung cells (Do et al., 2019). This underutilized natural product is acquiring consideration for its better bio accessibility of carotenoids, looked at than different vegetables like carrots and tomatoes, making it an important expansion to count calories, expecting to support cell reinforcement consumption and lessen disease gambles (Hosseinpour-Jaghdani et al., 2017).

8.9.2 Bamboo (*Bambusa vulgaris*)

Tribal groups: Mizo, Naga, Khasi, Garo (Northeast India)

Different clans in Upper East India broadly consume bamboo shoots. They are plentiful in fiber, nutrients, and minerals. The shoots are commonly gathered during the storm season and are utilized in various dishes, including curries, pickles, and aged food varieties. The clans have conventional strategies for handling bamboo shoots to diminish their poisonousness (because of cyanogenic glycosides) before utilization (Nirmala et al., 2014).

8.9.3 Jungli Karela - *Momordica dioica*

Tribal groups: Gond, Bhil (Central India)

Momordica dioica It tribal people are called "jungli karela," the sharp gourd is a wild relative of the unpleasant gourd. It is plentiful in nutrients, iron, and cancer prevention agents. The ancestral networks in focal India utilize this vegetable for its therapeutic properties, especially in overseeing diabetes and stomach-related messes. It is often cooked as a vegetable dish or used in conventional drugs (Kareti and Rajpoot, 2022).

8.9.4 Amaranthus (*Amaranthus cruentus*)

Tribal groups: Santhal, Oraon, Munda -Jharkhand, Odisha.

Amaranth leaves and seeds are staple food hotspots for genealogical organizations in Eastern India. The plant is highly nutritious, providing protein, vitamins, and minerals. The leaves are used in curries and soups, while the seeds are ground into flour or popped like popcorn. Amaranth is renowned for its resilience in poor soils and ability to thrive in drought conditions (Pradeepkumar et al., 2015).

8.9.5 Taro - *Colocasia esculenta*

Tribal groups: Nicobarese -Andaman and Nicobar Islands, Khasi -Meghalaya

Taro, or "arbi," is a root vegetable with boring corms and palatable leaves. The Nicobarese and Khasi clans use taro in different culinary arrangements, including bubbled, broiled, or cooked in curries. Taro leaves are additionally utilized in making conventional dishes. The plant is a huge wellspring of sugars and is significant in the eating routine of these networks, particularly during the rainstorm season (Blench et al., 2012).

8.9.6 Wild Yam - *Dioscorea villosa*

Tribal groups: Kani (Tamil Nadu, Kerala)

Wild sweet potatoes are a significant food hotspot for the Kani clan in the Western Ghats of South India. These tubers are starch and have been customarily used to ease hunger during food deficiencies. The Kani clan additionally uses wild sweet potatoes for their therapeutic properties. They are used for treating joint inflammation and digestive disorders. The tubers are typically boiled or cooked and consumed to treat consumption (Kumar et al., 2017).

8.9.7 Kandathippili - *Piper longum*

Tribal groups: Irulas (Tamil Nadu)

Piper longum is Known as lengthy pepper, this plant is broadly involved by the Irula clan tribal community in South India. The leafy food materials of the plant are utilized as a flavor and in conventional medicines. This helps for developing absorption, treating respiratory issues, and improving insusceptibility. The Irulas use this plant in conventional recuperating rehearses (Kumar et al., 2011). Tribes use several plants as food and medicine, are enclosed in Figure 8.3.

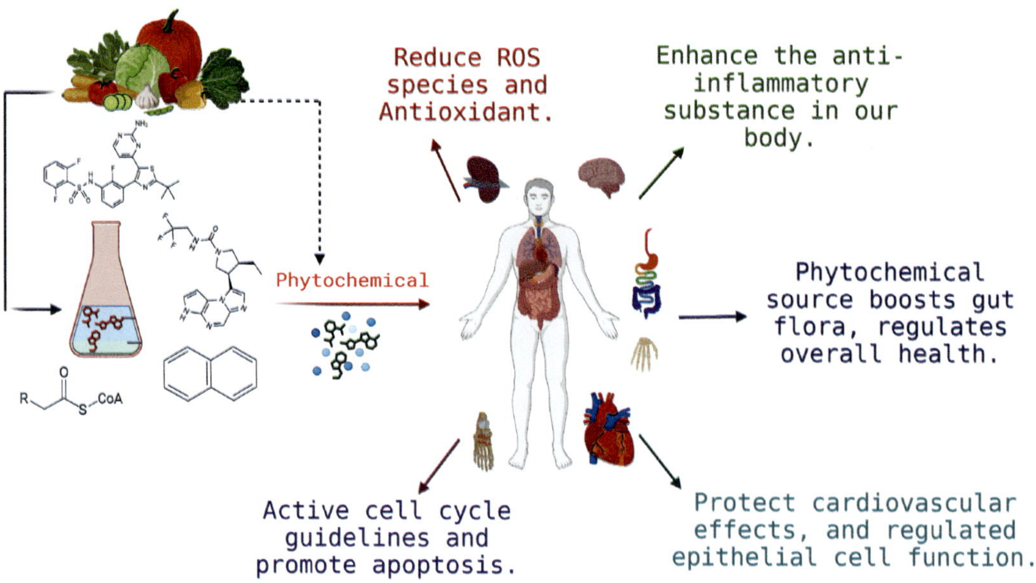

FIGURE 8.3 The various tribal groups utilizing plants as medicine and food. The tribal people in their regions utilize the endemic plant resources for food, medicine, and their traditional practices.

8.10 CONCLUSION

Secondary metabolites derived from vegetable crops are essential for human health and agriculture. Those bioactive metabolites enhance ant-oxidative response, simulative anti-inflammatory, and anticancer, making them valuable components in nutraceuticals and functional foods. Advances in omics technologies have expanded our understanding of the biochemical pathways involved in metabolite synthesis. Their use in sustainable agricultural practices, including bioactive compounds administered as biopesticides and plant growth regulators, offers a promising alternative to synthetic chemicals. The increasing interest in natural, plant-based compounds demonstrates a move toward sustainable development, promoting improved health outcomes and reducing environmental impact. The future of research on secondary metabolites from vegetable crops holds immense potential. With the advancement of omics technologies, there will be deeper insights into the regulatory mechanisms that enhance the biosynthesis of these compounds, paving the way for new possibilities to improve crop yields and the development of metabolite concentrations. Studies on the combined effects of metabolites on human wellness, with a focus on cancer prevention and overall gut health. Additionally, cutting-edge biotechnological tools like CRISPR-Cas9 and synthetic biology could improve the precision of metabolite production in crops. On the agricultural side, optimizing farming practices to boost metabolite yields while reducing environmental impact will be crucial for fostering sustainable food systems. By integrating these advancements, secondary metabolites from vegetable crops will play a critical role in prospective health solutions and supporting sustainable agricultural practices. The application of biotechnological tools enables their elevated production in vegetables by both plant breeding and genetic engineering. Overall, secondary metabolites from vegetable crops are an important bridge linking the health of our plants and their environment to our own well-being and may provide novel applications for the food and pharmaceutical industries.

REFERENCES

Adefegha, S. A. Functional Foods and Nutraceuticals as Dietary Intervention in Chronic Diseases; Novel Perspectives for Health Promotion and Disease Prevention. *J Diet Suppl.* 2018, 15(6): 977–1009.

Agostoni-Costa, T. D. S.; Vieira, R. F.; Bizzo, H. R.; Silveira, D.; Gimenes, M. A. Secondary metabolites. *Chromatography and its Applications.* Brazil In Tech. 2012, 1: 131–164.

Alam, M. J.; Uppulapu, S. K.; Maulik, S. K.; Banerjee, S. K. In Mukherjee, P.-K (Eds). Ethnopharmacological and therapeutic potential of Terminalia arjuna and Camellia sinensis against cardiovascular diseases: Evidence and experimental studies. In Evidence-Based Validation of Herbal Medicine, *Elsevier.* 2022, 2: 651–669.

Alamgir, A. N. M.; Alamgir, A. N. M. Medicinal, Non-medicinal, Biopesticides, Color-and Dye-Yielding Plants; Secondary Metabolites and Drug Principles; Significance of Medicinal Plants; Use of Medicinal Plants in the Systems of Traditional and Complementary and Alternative Medicines (CAMs). *Ther Med Plants Extracts.* 2017, 1: 61–104.

Ali Asgar, M. D. Anti-Diabetic Potential of Phenolic Compounds: A Review. *Int J Food Prop.* 2013, (1): 91–103.

Anderson, J. W.; Major, A. W. Pulses and Lipaemia, Short- and Long-Term Effect: Potential in the Prevention of Cardiovascular Disease. *Br J Nutr.* 2002, 88(S3):263–71.

Andrade, C. Stahl's Essential Psychopharmacology: Neuroscientific Basis and Practical Applications. *Men San Mono.* 2010, 8(1): 146.

Anwar, F.; Latif, S.; Ashraf, M.; Gilani, A. H. *Moringa oleifera*: A Food Plant with Multiple Medicinal Uses. *Phytother Res.* 2007, 21(1): 17–25.

Assous, M. T. M.; Abdel-Hady, M. M.; Medany, G. M. Evaluation of Red Pigment Extracted from Purple Carrots and Its Utilization as Antioxidant and Natural Food Colorants. *Ann Agric Sci.* 2014, 59(1): 1–7.

Athar, M.; and Nasir, S. M. Taxonomic Perspective of Plant Species Yielding Vegetable Oils Used in Cosmetics and Skin Care Products. *Afr J Biotechnol,* 2015, 4(1), 36–44.

Augusto, P. E. Challenges, Trends and Opportunities in Food Processing. *Curr Opin Food Sci.* 2020, 35: 72–78.

Ayuso, P.; Quizhpe, J.; Rosell, M. D. L. Á.; Peñalver, R.; Nieto, G. Bioactive Compounds, Health Benefits and Food Applications of Artichoke (*Cynara scolymus* L.) and Artichoke by-products: A review. *App Sci.* 2024, 14(11): 4940.

Bashir, A. Phenolic Compounds in Tomatoes: Impact on Human Health. *Food Chem*. 2020, 318: 126.

Beevi, S. S.; Narasu, M. L.; Gowda, B. B. Polyphenolics profile, antioxidant and radical scavenging activity of leaves and stem of Raphanus sativus L. *Plant Foods Hum. Nutr.* 2010, 65(1), 8–17.

Bernstein, P. S. Nutritional Modulation of Eye Health: The Role of Lutein and Zeaxanthin. *J Oph*. 2023, 40(1): 45–56.

Bevilacqua, A.; Petruzzi, L.; Perricone, M.; Speranza, B.; Campaniello, D.; Sinigaglia, M.; Corbo, M. R. Nonthermal Technologies for Fruit and Vegetable Juices and Beverages: Overview and Advances. *Compr Rev Food Sci Food Saf.* 2018, 17(1): 2–62.

Bhargava, D. Emerging Roles of Okra in Nutraceuticals. *Phytother Res*. 2022, 36(4): 1023–1035.

Bhattacharya, S. Flavonoids from Okra and Their Potential Health Benefits. *J Nutr Biochem*. 2020, 45: 56–67.

Blench, R. M. Vernacular Names for Taro in the Indo-Pacific Region: Implications for Centres of Diversification and Spread. *Senri Ethnol Stud*. 2012, 78: 21–43.

Boots, A. W.; Haenen, G. R.; Bast, A. Health effects of quercetin: From antioxidant to nutraceutical. *Eur. J. Pharmacol.* 2008, 585(2–3): 325–337.

Bungau, S.; Abdel-Daim, M. M.; Tit, D. M.; Ghanem, E.; Sato, S.; Maruyama-Inoue, M.; Yamane, S.; Kadonosono, K. Health Benefits of Polyphenols and Carotenoids in Age-Related Eye Diseases. *Oxid Med Cell Longev.* 2019, 2019: 1–22.

Bustos, A. Y.; Font, G.; Taranto, M. P. Fruit and Vegetable Snacks as Carriers of Probiotics and Bioactive Compounds: A Review. *Int J Food Sci Tech*. 2023, 58(6): 3211–3223.

Campos-Vega, R.; Loarca-Piña, G.; Dave Oomah, B. Minor Components of Pulses and Their Potential Impact on Human Health. *Food Res Int*. 2010, 43(2):461–82.

Caporali, S.; De Stefano, A.; Calabrese, C.; Giovannelli, A.; Pieri, M.; Savini, I.; Tesauro, M.; Bernardini, S.; Minieri, M.; Terrinoni, A. Anti-Inflammatory and Active Biological Properties of the Plant-Derived Bioactive Compounds Luteolin and Luteolin 7-Glucoside. *Nutrients*. 2022, 14(6): 1155.

Chatterjee, P. Diabetes Management through Plant-based Compounds: Focus on Okra. *J Diabetes Res*. 2024, 18(2): 145–154

Chaudhary, A. Innovations in Plant Secondary Metabolite Research. *Trends Plant Sci*. 2022, 27(5): 402–415

Chavarri, M. J.; Herrera, A.; Ariño, A. The Decrease in Pesticides in Fruit and Vegetables during Commercial Processing. *Int J Food Sci Tech*. 2005, 40(2): 205–211.

Chemat, F.; Rombaut, N.; Meullemiestre, A.; Turk, M.; Perino, S.; Fabiano-Tixier, A.-S.; Abert-Vian, M. Review of Green Food Processing Techniques. Preservation, Transformation, and Extraction. *Innov Food Sci Emerg Technol*. 2017, 41: 357–377.

Chemat, F.; Vian, M. A.; Cravotto, G. Green Extraction of Natural Products: Concept and Principles. *Int J Mol Sci*. 2012, 13(7): 8615–8627.

Chemat, F.; Zill-e-Huma; Khan, M. K. Applications of Ultrasound in Food Technology: Processing, Preservation and Extraction. *Ultrason Sonochem*. 2011, 18(4): 813–835.

Chobot, M.; Kozłowska, M.; Ignaczak, A.; Kowalska, H. Development of Drying and Roasting Processes for the Production of Plant-Based Pro-healthy Snacks in the Light of Nutritional Trends and Sustainable Techniques. *Trends Food Sci Technol*. 2024, 149: 104–553.

Chong, W.; Li, Y.; Liu, B.; Liu, Z.; Zhao, T.; Wonsey, D. R.; Chen, C.; Velmahos, G. C.; deMoya, M. A.; King, D. R.; Kung, A. L.; Alam, H. B. Anti-inflammatory Properties of Histone Deacetylase Inhibitors: A Mechanistic Study. *J Trauma Acute Care Surg*. 2012, 72(2): 347–354.

Chung, Y.-C.; Chang, C. T.; Chao, W. W.; Lin, C. F.; Chou, S. T. Antioxidative Activity and Safety of the 50 Ethanolic Extract from Red Bean Fermented by Bacillus Subtilis IMR-NK1. *J Agric Food Chem*. 2002, 50(8): 2454–2458.

Cory, H.; Passarelli, S.; Szeto, J.; Tamez, M.; Mattei, J. The Role of Polyphenols in Human Health and Food Systems: A Mini-Review. *Front Nutr*. 2018, 5: 87.

Crozier, A.; Clifford, M. N.; Ashihara, H. In Crozier, A.; Clifford, M.-N.; Ashihara, H. (Eds.). Plant Secondary Metabolites. In Occurrence, Structure and Role in the Human Diet. *Blackwell-Publishers*, 2006, 353–372.

Cruz, V.; Rojas, R.; Saucedo-Pompa, S.; Martínez, D. G.; Aguilera-Carbó, A. F.; Alvarez, O. B.; Rodríguez, R.; Ruiz, J.; Aguilar, C. N. Improvement of Shelf Life and Sensory Quality of Pears Using a Specialized Edible Coating. *J Chem*. 2015, 2015: 1–7.

Da Silva, P. F.; Moreira, R. G. Vacuum frying of high-quality fruit and vegetable-based snacks. *LWT-Food Sci. Technol.* 2008, 41(10): 1758–1767.

Dahl, W. J.; Foster, L. M.; Tyler, R. T. Review of the health benefits of peas (Pisum sativum L.). *Br. J. Nutr.* 2012, 108(S1): S3–S10.

Darvin, M. E.; Sterry, W.; Lademann, J.; Vergou, T. The Role of Carotenoids in Human Skin. *Molecules*. 2011, 16(12): 10491–10506.

Dawidowicz, A. L. The Presence of Secondary Metabolites in Carrot Extracts and Their Antioxidant and Antimicrobial Properties. *Plant Foods Human Nutr.* 2006, 61(4): 146–151.

Dias, A. L. B.; de Aguiar, A. C.; Rostagno, M. A. Extraction of natural products using supercritical fluids and pressurized liquids assisted by ultrasound: Current status and trends. *Ultrason. Sonochem.* 2021, 74: 105584.

Dixit, S.; Shukla, A.; Singh, V.; Upadhyay, S. K. *Engineering of Plant Metabolic Pathway for Nutritional Improvement*. Wiley, 2021. DOI:10.1002/9781119672425.ch20

Do, T. V. T.; Fan, L.; Suhartini, W.; Girmatsion, M. Gac (*Momordica cochinchinensis* Spreng) Fruit: A Functional Food and Medicinal Resource. *J Funct Food.* 2019, 62: 103–512.

Dorais, M. Organic Production of Vegetables: State of the Art and Challenges. *Can J Plant Sci.* 2007, 87(5): 1055–1066.

Dorais, M.; Alsanius, B. Advances and Trends in Organic Fruit and Vegetable Farming Research. *Horticul Rev.* 2015, 43(1): 185–268.

Duodu, K. G. Effects of Processing on Antioxidant Phenolics of Cereal and Legume Grains. In *Advances in Cereal Science: Implications to Food Processing and Health Promotion* (pp. 31–54), J. M. Awika; V. Piironen; and S. Bean (Eds.). Washington, DC: American Chemical Society, 2011.

Egbuna, C; Dable Tupas, G. (Eds.). *Functional Foods and Nutraceuticals: Bioactive Components.* Springer International Publishing Springer Nature Switzerland, 2020, 1: 1–632.

Elbadrawy, E.; Mostafa, M. Y. Antioxidant, anti-inflammatory, antimicrobial, and anticancer properties of green broad bean pods (Vicia faba L.). *Foods Raw Mater.* 2024, 12(2): 308.

Elkins, E. R. Effect of Commercial Processing on Pesticide Residues in Selected Fruits and Vegetables. *J AOAC Int.* 1989, 72(3): 533–535.

Evans, J. A.; Johnson, E. J. The Role of Phytonutrients in Skin Health. *Nutrients.* 2010, 2(8): 903–928.

Ezov, A.; Shibzukhov, Z.-G.; Beslaneev, B.; Shibzukhova, Z.; Khantsev, M. Prospects and Technology of Cultivation of Organic Vegetable Production on Open Ground in Southern Russia Conditions. *In E3S Web of Conferences* (vol. 222, p. 02003). EDP Sciences, 2020.

Fageria, N. K.; Baligar, V. C.; Li, Y. C. The Role of Nutrient Efficient Plants in Improving Crop Yields in the Twenty First Century. *J Plant Nutr.* 2008, 31(6): 1121–1157.

Fahey, J. W. Moringa oleifera: A Review of the Medical Evidence for Its Nutritional, Therapeutic, and Prophylactic Properties. *Trees Life J.* 2005, 1(5): 5.

Fahey, J. W.; Zalcmann, A. T.; Talalay, P. The chemical diversity and distribution of glucosinolates and isothiocyanates among plants. *Phytochemistry.* 2001, 56(1): 5–51.

Flint, H. J. The Impact of Vegetables on Gut Microbiota and Metabolic Health. *Nat Rev Gastroenterol Hepatol.* 2022, 19(4): 203–214.

Fraser, P. D. Terpenoids in Tomatoes: Biosynthesis, Regulation, and Role in Plant Development. *Plant Sci.* 2021, 305: 110836.

Gad, H.; Salim, M.; Youssef, S. Phytosterols from Pumpkin Seeds and Their Effect on Cholesterol Levels. *J Lip Sci.* 2018, 13(4): 123–132.

Giovannucci, E. Tomatoes, tomato-based products, lycopene, and cancer: Review of the epidemiologic literature. ***J Natl Cancer Inst.*** 1999, 91(4): 317–331.

Giovannucci, E. Lycopene and Risk of Prostate Cancer: A Review. *Cancer Epidemiol Biomarkers Prev.* 2023, 32(2): 170–179.

Goncagul, G., Ayaz, E. Antimicrobial Effect of Garlic (*Allium sativum*). *Recent Pat Anti-Infect Drug Discov.* 2010, 5(1): 91–93.

Gopalakrishnan, L.; Doriya, K.; Kumar, D. S.; *Moringa oleifera*: A Review on Nutritive Importance and Its Medicinal Application. *Food Sci Hum Well.* 2016, 5(2): 49–56.

Grover, J. K.; Yadav, S. P. Pharmacological actions and potential uses of Momordica charantia: A review. *J. Ethnopharmacol.* 2004, 93(1): 123–132.

Gul, K.; Singh, A. K.; Jabeen, R. Nutraceuticals and Functional Foods: The Foods for the Future World. *Critc Rev Food Sci Nutrt.* 2016, 56(16): 2617–2627.

Guldiken, B.; Toydemir, G.; Nur Memis, K.; Okur, S.; Boyacioglu, D.; Capanoglu, E. Home-Processed Red Beetroot (*Beta vulgaris* L.) Products: Changes in Antioxidant Properties and Bioaccessibility. *Int J Mol Sci.* 2019, 17(6): 858.

Gupta, C.; Prakash, D. Nutraceuticals for geriatrics. ***J Tradit Complement Med.*** 2015, 5(1), 5–14.

Halliwell, B. Antioxidants and Human Health: New Insights from Studies on Vitamin C and Polyphenols. *Nutr Res Rev.* 2023, 36(1): 22–35.

Hansen, S. L.; Purup, S.; Christensen, L. P. Bioactivity of Falcarinol and Related Polyacetylenes in Carrots (*Daucus carota* L.). *J Sci Food Agric.* 2010, 90(1): 55–70.

Hertog, M. G. L.; Feskens, E. J. M.; Kromhout, D.; Hertog, M. G. L.; Hollman, P. C. H.; Hertog, M. G. L.; Katan, M. B. Dietary Antioxidant Flavonoids and Risk of Coronary Heart Disease: The Zutphen Elderly Study. *The Lancet*. 1993, 342(8878): 1007–1011.

Hill, K. Industrial Development and Application of Biobased Oleochemicals. *Pure Appl Chem*. 2007, 79(11): 1999–2011.

Hosseini, S.; Vali, E.; Raza, A.; Carotenoids in Pumpkin: Nutritional and Medicinal Perspectives. *J Nutr Biochem*. 2017, 10(3): 187–195.

Hosseinpour-Jaghdani, F.; Shomali, T.; Gholipour-Shahraki, S.; Rahimi-Madiseh, M.; Rafieian-Kopaei, M. Cornus mas: A Review on Traditional Uses and Pharmacological Properties. *J Complement Integr Med*. 2017, 14(3).

Hunyadi, A. The Mechanism (s) of Action of Antioxidants: From Scavenging Reactive Oxygen/Nitrogen Species to Redox Signaling and the Generation of Bioactive secondary Metabolites. *Med Res Rev*. 2019, 39(6): 2505–2533.

Isah, T. Stress and Defense Responses in Plant Secondary Metabolites Production. *Bio Res*. 2019, 52(1): 39.

Jain, M. Bioactive Compounds in Okra: A Review of Health Benefits. *J Funct Food*. 2023, 85: 104–725.

Jayachandran, M.; Xiao, J.; Xu, B. A Critical Review on Health Promoting Benefits of Edible Mushrooms through Gut Microbiota. *Int J Mol Sci*. 2017, 18(9): 1934.

Johnson, E. J. Lutein and Zeaxanthin and Eye Health. *Nutrients*. 2021, 13(9): 3254.

Jacob, S.; Raj, A.; Gomez, S. Exploring the Nutraceutical Elixirs of Commercial Fruit and Vegetables and Wild Edible Plants with Ethnomedicinal Importance. *Int. J. of Health Sci.* 8(S1): 900–923.

Kareti, S. R.; Rajpoot, V. S. Ethnobotanical Survey of Folk Medicinal Plants Used in Tribal Villages of Amarkantak Region of Central India. *Plant Biosyst*. 2022, 156(4): 1019–1038.

Kasolo, J. N.; Bimenya, G. S.; Ojok, L.; Ochieng, J.; Ogwal-Okeng, J. W. Phytochemicals and Uses of *Moringa oleifera* Leaves in Ugandan Rural Communities. *J Med Plant Res*. 2010, 4(9): 753–757.

Kaur, A.; Sangha, M. K.; Devi, V.; Pathak, M.; Singla, D. Nutritional, antinutritional and antioxidant assessment of immature Abelmoschus pods: Implications for crop improvement from cultivated and wild varieties. ***S Afr J Bot***. 2024, 174: 485–494.

Kaur, K. K.; Allahbadia, G. N.; Singh, M. Inventing the Immunomodulatory Action of Food Plants with Utilization of Metabolomics: A Narrative Review. *ASNH*. 2023, 7 (10): 2582–1423.

Khan, M. I.; Giridhar, P.; Chatterjee, S. Betalains and Their Applications in Food: A Review. *Environ Exp Bot*. 2023, 4(4): 430–444.

Kowalska, H.; Czajkowska, K.; Cichowska, J.; Lenart, A. What's New in Biopotential of Fruit and Vegetable By-Products Applied in the Food Processing Industry. *Trends Food Sci Tech*. 2017, 67: 150–159.

Kraft, K. Artichoke Leaf Extract—Recent Findings Reflecting Effects on Lipid Metabolism, Liver and Gastrointestinal Tracts. *Phytomedicine*. 1997, 4(4): 369–378.

Kumar, A.; Singh, M.; Sharma, R. Saponins. Pharmacological Potential of Bottle Gourd. *Phytochem Rev*. 2019, 12(4): 205–213.

Kumar, R.; Gupta, A.; Sharma, P. Charantin and Its Role in Diabetes Management. *J Diabetes Res*. 2020, 14(2): 235–240.

Kumar, S. Nutritional and Therapeutic Potential of Okra. *J Med Plant Res*. 2021, 15(5): 116–125.

Kumar, S.; Das, G.; Shin, H.-S.; Patra, J. K. Dioscorea spp. (A Wild Edible Tuber): A Study on Its Ethnopharmacological Potential and Traditional Use by the Local People of Similipal Biosphere Reserve, India. *Front Pharmacol*. 2017, 8: 52.

Kumar, S.; Kamboj, J.; Suman; Sharma, S. Overview for Various Aspects of the Health Benefits of Piper Longum Linn. Fruit. *J Acupunct Meridian Stud*. 2011, 4(2): 134–140.

Kumar, S.; Malhotra, R.; Kumar, D. Antidiabetic and Antioxidant Potential of *Moringa oleifera* Leaves in Streptozotocin-Induced Diabetic Rats. *J Pharm Biol Sci*. 2010, 2(4): 271.

Kumar, S.; Saini, R.; Suthar, P.; Kumar, V.; Sharma, R. In Sharma, A.-K., & Sharma, A (Eds.), Plant secondary metabolites: Their food and therapeutic importance. Plant secondary metabolites: Physico-chemical properties and therapeutic applications. Springer Nature Singapore. 2022, 1: 371–413.

Lacatusu, I.; Istrati, D.; Bordei, N.; Popescu, M.; Seciu, A. M.; Panteli, L. M.; Badea, N. Synergism of Plant Extract and Vegetable Oils-Based Lipid Nanocarriers: Emerging Trends in Development of Advanced Cosmetic Prototype Products. *Mater Sci Eng*. 2020, 108: 110412.

Lengai, G. M. W.; Muthomi, J. W. Biopesticides and Their Role in Sustainable Agricultural Production. *J Biol Med*. 2018, 06(06): 7–41.

Leone, A.; Spada, A.; Battezzati, A.; Schiraldi, A.; Aristil, J.; Bertoli, S. Cultivation, Genetic, Ethnopharmacology, Phytochemistry, and Pharmacology of *Moringa oleifera* Leaves: An Overview. *Int J Mol Sci*. 2015, 16(6): 12791–12835.

Li, Y. Flavonoids as Cancer Therapeutic Agents. *Curr Opin Pharmacol.* 2021, 60: 152–161.

Lucchesi, M. E.; Chemat, F.; Smadja, J. Solvent-Free Microwave Extraction of Essential Oil from Aromatic Herbs: Comparison with Conventional Hydro-Distillation. *J Chromatogr A.* 2004, 1043(2): 323–327.

Majerska, J.; Michalska, A.; Figiel, A. A Review of New Directions in Managing Fruit and Vegetable Processing By-Products. *Trends Food Sci Tech.* 2019, 88: 207–219.

Makki, K. Dietary Fibers and Polyphenols as Prebiotics: Impact on the Gut Microbiota. *Front Microbiol.* 2020, 11: 555.

Mattila, P.; Hellström, J. Phenolic Acids in Potatoes, Vegetables, and Some of Their Products. *J Food Compos Anal.* 2007, 20(3–4): 152–60.

Mazur, W. 11 Phytoestrogen Content in Foods. *Baillieres Clin. Endocrinol. Metab.* 1998, 12(4): 729–742.

Mbikay, M. Therapeutic Potential of Moringa oleifera Leaves in Chronic Hyperglycemia and Dyslipidemia: A Review. *Front Pharmacol.* 2012, 3: 24.

McWalter, G. K.; Higgins, L. G.; McLellan, L. I.; Henderson, C. J.; Hayes, J. D.; Song, L.; Thornalley, P. J.; Itoh, K. Yamamoto, M. Transcription Factor, Nrf2 Is Essential for Induction of NAD(P)H:Quinone Oxidoreductase 1, Glutathione S-Transferases, and Glutamate Cysteine Ligase by Broccoli Seeds and Isothiocyanates. *J Nutr.* 2004, 134(12): 3499S–3506S.

Meeran, S. M.; Patel, S. N.; Tollefsbol, T. O. Sulforaphane Causes Epigenetic Repression of hTERT Expression in Human Breast Cancer Cell Lines. *Plos One.* 2010, 5(7): 11457.

Metzger, BT.; Barnes, DM. Polyacetylenes from Carrots as Anti-Carcinogenic Agents. *J Agric Food Chem.* 2009, 57(12): 5459–5464.

Milton-Laskibar, I.; Martínez, J. A.; Portillo, M. P. Current Knowledge on Beetroot Bioactive Compounds: Role of Nitrate and Betalains in Health and Disease. *Foods.* 2021, 10(6): 1314.

Mishra, S.; Sahoo, S.; Sahoo, P. K.; Sahoo, N. K. Recent advancements in the plant and microbial assisted green synthesis of nanomaterials. *Materials Today: Proceedings.* 2023a.

Mishra, S.; Verma, S. S.; Rai, V.; Awasthee, N.; Chava, S.; Hui, K. M.; Gupta, S. C. Long non-coding RNAs are emerging targets of phytochemicals for cancer and other chronic diseases. *Cell. Mol. Life Sci.* 2019b, 76: 1947–1966.

Moghe, G. D. Glycoalkaloids in Tomatoes: Biosynthesis and Potential Applications. *J Agric Food Chem.* 2020, 68(14): 3917–3926.

Murillo, G.; Mehta, R. G. Cruciferous Vegetables and Cancer Prevention. *Nutr Cancer.* 2001, 41(1–2): 17–28.

Natella, F.; Nardini, M.; Felice, M.; Scaccini, C. Benzoic and Cinnamic Acid Derivatives as Antioxidants: Structure-Activity Relation. *J Agric Food Chem.* 1999, 47(4): 1453–1459.

Nemzer, B.; Pietrzkowski, Z.; Spórna, A.; Stalica, P.; Thresher, W.; Michałowski, T.; Wybraniec, S. Betalainic and Nutritional Profiles of Pigment-Enriched Red Beetroot (*Beta vulgaris* L.) Dried Extracts. *Food Chem.* 2022, 127(1): 42–53.

Nguyen, T. T. Tomatine in Tomatoes: A Natural Pesticide for Sustainable Agriculture. *Pestic Biochem Physiol.* 2022, 179: 104–960.

Nicolle, C.; Gueux, E.; Lab, C.; Jaffrelo, L.; Rock, E.; Mazur, A.; Rémésy, C. Lyophilized carrot ingestion lowers lipemia and beneficially affects cholesterol metabolism in cholesterol–fed C57BL/6J mice. *Eur J Nutr* . 2004, 43, 237–245.

Nigam, K.; Suhail, S.; Verma, Y.; Singh, V.; Gupta, S. Molecular characterization of begomovirus associated with leaf curl disease in chilli. *World J Pharm Res.* 2015, 4(3), 1579–1592.

Nirmala, C.; Bisht, M. S.; Laishram, M. Bioactive Compounds in Bamboo Shoots: Health Benefits and Prospects for Developing Functional Foods. *Int J Food Sci Tech.* 2014, 49(6): 1425–1431.

Niu, Y.; Aierken, A.; Feng, L. The Relationship between Dietary Flavonoid Intake and Hypertension: A Cross-Sectional Study from NHANES. 2023, 1: 1–21.

Okon, Y.; Baker, R. Microbial Inoculants as Crop-Yield Enhancers. *Crit Rev Biotech.* 1987, 6(1): 61–85.

Palozza, P.; Parrone, N.; Catalano, A.; Simone, R. Tomato Lycopene and Inflammatory Cascade: Basic Interactions and Clinical Implications. *Curr Med Chem.* 2010, 17(23): 2547–2563.

Park, C. H; Park, S. Y.; Park, Y. J.; Kim, J. K.; Park, S. U. Metabolite Profiling and Comparative Analysis of Secondary Metabolites in Chinese Cabbage, Radish, and Hybrid x Brassicoraphanus. *J Agric Food Chem.* 2020, 68(47): 13711–19.

Parmenter, B. H.; Croft, K. D.; Hodgson, J. M.; Dalgaard, F.; Bondonno, C. P.; Lewis, J. R.; Cassidy, A.; Scalbert, A.; Bondonno, N. P. An Overview and Update on the Epidemiology of Flavonoid Intake and Cardiovascular Disease Risk. *Food Funct.* 2020, 11(8): 6777–6806.

Patel, A. V. Advances in Functional Foods: Bioactive Compounds from Vegetables. *Trends Food Sci Tech.* 2024, 112: 45–60.

Patel, R. Anti-Inflammatory Properties of Okra-Derived Tannins. *J Inflamm Res.* 2022, 15: 2075–2084.

Petruzzi, L.; Campaniello, D.; Speranza, B.; Corbo, M. R.; Sinigaglia, M.; Bevilacqua, A. Thermal Treatments for Fruit and Vegetable Juices and Beverages: A Literature Overview. *Compr Rev Food Sci Food Saf.* 2017, 16(4): 668–691.

Pisoschi, A. M.; Pop, A. The Role of Antioxidants in the Chemistry of Oxidative Stress: A Review. *Eur J Med Chem.* 2015, 97: 55–74.

Poudel, R. P. Antioxidant Flavonoids in Tomatoes and Their Health Benefits. *Nutr Rev.* 2021, 79(3): 201–214.

Pradeepkumar, T.; Indira, V.; Sankar, M. Nutritional Evaluation of Wild Leafy Vegetables Consumed by Tribals in the Wayanad District of Kerala. *Proc. Natl. Acad. Sci. India Sect. B Biol. Sci.* 2015, 85(1): 93–99.

Puupponen-Pimia, R.; Nohynek, L.; Meier, C.; Kahkonen, M.; Heinonen, M.; Hopia, A.; Oksman-Caldentey, K.-M. Antimicrobial Properties of Phenolic Compounds from Berries. *J Appl Microbiol.* 2001, 90(4): 494–507.

Rahman, M. M.; Kabir, M. S.,;Hossain, M. M.; Asaduzzaman, M. Comparative Study of Nutritional Composition and Antibacterial Properties of Fresh and Dry Leaves Crude Extracts of *Moringa oleifera. Int J Pharm Pharm Sci.* 2009, 1(1): 50–53.

Rahman, M. M.; Wasila, H.; Miah, M. N.; Islam, M. M.; Shahriar, S. M. S. Comparative Analysis of Antibacterial Activity of *Moringa oleifera* Leaf Extract with Selected Antibiotics. *South Asian J Exp Biol.* 2018, 8(1): 1–4.

Raj, S.; Patel, A.; Soni, H. Anticancer Potential of Cucurbitacins in Bottle Gourd. *Cancer Res J.* 2017, 9(3): 215–223.

Ramawat, K. G.; Mèrillon, J. M. An Introduction to Specialized Metabolites from Plants, Their Biological and Ecological Roles, and Biotechnological Interventions to Improve the Production. In Plant Specialized Metabolites: Phytochemistry, Ecology and Biotechnology. Cham: Springer Nature Switzerland. 2023, 1–21.

Ramdath, D.; Renwick, S.; Duncan, A. M. The Role of Pulses in the Dietary Management of Diabetes. *Can J Diabetes.* 2016, 40(4): 355–363.

Rao, A. V. Lycopene: Potential Role in Cancer Prevention and Therapeutics. *J Am Coll Nutr.* 2022a, 41(1): 1–11.

Rao, L. G.; Krishnadev, N.; Banasikowska, K.; Rao, A. V. Lycopene I—effect on osteoclasts: Lycopene inhibits basal and parathyroid hormone-stimulated osteoclast formation and mineral resorption mediated by reactive oxygen species in rat bone marrow cultures. *J Med Food.* 2003, 6(2): 69–78.

Rao, L. G.; Mackinnon, E. S.; Josse, R. G.; Murray, T. M.; Strauss, A.; Rao, A. V. Lycopene consumption decreases oxidative stress and bone resorption markers in postmenopausal women. *Osteoporos. Int.* 2007, 18: 109–115.

Rao, A. V.; Agarwal, S. Lycopene and Its Role in Chronic Diseases. *J Nutr.* 2021, 151(2): 282–290.

Rao, S.; Kumar, A.; Mishra, S. Triterpenoids in Snake Gourd and Their Medicinal Applications. *Nat Prod Res.* 2018, 15(2): 303–311.

Rao, A. V.; Rao, L. G. Carotenoids and human health. *Pharmacol. Res.* 2007, 55(3): 207–216.

Reeve, J. R.; Hoagland, L. A.; Villalba, J. J.; Carr, P. M.; Atucha, A.; Cambardella, C.; Davis, D. R.; Delate, K. Organic Farming, Soil Health, and Food Quality: Considering Possible Links. *Adv Agron.* 2016, 137: 319–367.

Rezig, L.; Chouaibi, M.; Msaada, K.; Hamdi, S. Chemical composition and profile characterisation of pumpkin (Cucurbita maxima) seed oil. *Ind. Crops Prod.* 2012, 37(1), 82–87.

Rodríguez, L.; Mendez, D.; Montecino, H.; Carrasco, B.; Arevalo, B.; Palomo, I.; Fuentes, E. Role of *Phaseolus Vulgaris* L. in the Prevention of Cardiovascular Diseases – Cardioprotective Potential of Bioactive Compounds. *Plants.* 2022, 11(2): 186.

Roy, M.; Mukherjee, A.; Mukherjee, S.; Biswas, J. Phytonutrients from Fruits and Vegetables in Breast Cancer Control. *Anticancer Properties of Fruits and Vegetables: A Scientific Review.* 2014, 75–101.

Saini, R. K. Carotenoids in Tomatoes: Their Role in Human Health. *Front Plant Sci.* 2020, 11: 1197.

Saini, R. K.; Shetty, N. P.; Giridhar, P. Carotenoid Content in Vegetative and Reproductive Parts of Commercially Grown *Moringa oleifera* Lam. Cultivars from India by LC-APCI-MS Analysis. *J Food Sci Tech.* 2016, 53(5): 3468–3474.

Saiwal, N.; Dahiya, M.; Dureja, H.; Nutraceutical insight into vegetables and their potential for nutrition mediated healthcare. *Curr Nutr Food Sci.* 2019, 15(5): 441–453.

Sampath Kumar, H. M.; Herrmann, L.; Tsogoeva, S. B. Structural Hybridization as a Facile Approach to New Drug Candidates. *Bioorg Med Chem Lett.* 2020, 30(23): 127–514.

Santhiravel, S.; Bekhit, A. E.-D. A.; Mendis, E.; Jacobs, J. L.; Dunshea, F. R.; Rajapakse, N.; Ponnampalam, E. N. The Impact of Plant Phytochemicals on the Gut Microbiota of Humans for a Balanced Life. *Int J Mol Sci*. 2022, 23(15): 8124.

Sasidharan, S.; Chen, Y.; Saravanan, D.; Sundram, K.; Latha, L. Extraction, Isolation and Characterization of Bioactive Compounds from Plants' Extracts. Complementary and Alternative Medicines. *Afr J Tradit Complement Altern Med*. 2010, 8(1):1–10

Saxena, P. Tannins from Okra: Potential Natural Antimicrobial Agents. *J Ethnopharmacol*. 2023, 306: 114–120.

Shah, R. R. Tomatine and Its Potential Applications in Cancer Therapy. *J Cancer Res Ther*. 2020, 16(3): 567–574.

Sharma, E.; Anand, G.; Kapoor, R. Terpenoids in Plant and Arbuscular Mycorrhiza-Reinforced Defence against Herbivorous Insects." *Ann Bot*. 2017, 119(5): 791–801.

Sharma, K. K.; Sharma, M.; Devi, D.; Gupta, A.; Singh, S. (2025). Phytochemicals Present In Vegetables For Health Promotion. In Plant Metabolites and Vegetables as Nutraceuticals *Apple Academic Press*. 2025, 61–94.

Sharma, N.; Singhvi, R. Effects of Chemical Fertilizers and Pesticides on Human Health and Environment: A Review. *Int J Agric Environ Biotechnol*. 2017, 10(6): 675.

Sharma, P.; Bhandari, A.; Kumar, N. Role of Flavonoids in Health: A Review. *J Clin Nutr*. 2015, 8(1): 102–114.

Sharma, P.; Kaur, G.; Kehinde, B. A.; Chhikara, N.; Panghal, A.; Kaur, H. Pharmacological and Biomedical Uses of Extracts of Pumpkin and Its Relatives and Applications in the Food Industry: A Review. *Int J Veg Sci*. 2020, 26(1): 79–95.

Shoff, S. M.; Newcomb, P. A.; Mares-Perlman, J. A.; Klein, B. E.; Haffner, S. M.; Storer, B. E.; Klein, R. Usual Consumption of Plant Foods Containing Phytoestrogens and Sex Hormone Levels in Postmenopausal Women in Wisconsin. *Nutri can*. 1998, 30(3):207–212.

Siddhuraju, P.; Becker, K. Antioxidant Properties of Various Solvent Extracts of Total Phenolic Constituents from Three Different Agroclimatic Origins of Drumstick Tree (*Moringa oleifera* Lam.) Leaves. *J Agric Food Chem*. 2003, 51(8): 2144–2155.

Singh, P.; Arif, Y.; Mir, A. R.; Alam, P.; Hayat, S. Quercetin-mediated alteration in photosynthetic efficiency, sugar metabolism, elemental status, yield, and redox potential in two varieties of okra. *Protoplasma*. 2024, 261(1): 125–142.

Singh, N.; Kaur, J. Vicine from Bitter Gourd: A Promising Antidiabetic Agent. *Phytomed J*. 2019, 9(1): 67–72.

Singh, N., & Yadav, S. S. A review on health benefits of phenolics derived from dietary spices. *Curr Res Food Sci*. 2022, 5: 1508–1523.

Somegowda, M.; Raghavendra, S.; Sridhara, S.; Rajeshwara, A. N.; N. Pramod, S.; Shivashankar, S.; Lin, F.; El-Abedin, T. K. Z.; Wani, S. H.; Elansary, H. O. Defensive Mechanisms in Cucurbits against Melon Fly (*Bactrocera Cucurbitae*) Infestation through Excessive Production of Defensive Enzymes and Antioxidants. *Molecules*. 2021, 26(21): 6345.

Soundararajan, P.; Kim, J. S. Anti-Carcinogenic Glucosinolates in Cruciferous Vegetables and Their Antagonistic Effects on Prevention of Cancers. *Molecules*. 2018, 23(11): 2983.

Sousa Silva, M.; Cordeiro, C.; Roessner, U.; Figueiredo, A. Metabolomics in crop research—Current and emerging methodologies. *Front Plant Sci*. 2019, 10: 1013.

Sun, M. Vegetables, Gut Microbiota, and Metabolic Health: A Comprehensive Review. *Front Nutr*. 2021, 8: 685–314.

Tan, X.; Cui, F.; Wang, D.; Lv, X.; Li, X.; Li, J. Fermented Vegetables: Health Benefits, Defects, and Current Technological Solutions. *Foods*. 2023, 13(1):38.

Thitimuta, S.; Pithayanukul, P.; Nithitanakool, S.; Bavovada, R.; Leanpolchareanchai, J.; Saparpakorn, P. *Camellia Sinensis* L. Extract and Its Potential Beneficial Effects in Antioxidant, Anti-Inflammatory, Anti-Hepatotoxic, and Anti-Tyrosinase Activities. *Molecules*. 2017, 22(3): 401.

Traka, M. H.; Mithen, R. F. Plant Science and Human Nutrition: Challenges in Assessing Health-Promoting Properties of Phytochemicals. *The Plant Cell*. 2011, 23(7): 248–397.

Treutter, D. Significance of Secondary Metabolites in Plants. *Plant Physiol Biochem*. 2021, 159: 105–110.

Trigo, J. P.; Alexandre, E. M. C.; Saraiva, J. A.; Pintado, M. E. High Value-Added Compounds from Fruit and Vegetable By-Products – Characterization, Bioactivities, and Application in the Development of Novel Food Products. *Crit Rev Food Sci Nutr*. 2020, 60(8): 1388–1416.

Tung, B. T.; Van Thanh, D.; Thanh, N. P. The Effect of Medicinal Plants on the Metabolic Syndrome. ***Metab. Syndr***. 2023, 301–327.

Tzounis, X.; Vulevic, J.; Kuhnle, G. G. C.; George, T.; Leonczak, J.; Gibson, G. R.; Kwik-Uribe, C.; Spencer, J. P. E. Flavanol Monomer-Induced Changes to the Human Faecal Microflora. *Br J Nutr*. 2008, 99(4): 782–792.

Upadhyay, S. K.; Singh, S. P. *Bioprospecting of Plant Biodiversity for Industrial Molecules*. Wiley, 2021. DOI:10.1002/9781119718017

Upadhyay, S. K.; Singh, S. P. *Herbal Nutraceuticals: Products and Processes*. Wiley, 2024. DOI:10.1002/9781394241576

Upadhyay, S. K.; Singh, S. P. *Plants as Bioreactors for Industrial Molecules*. Wiley, 2023 DOI:10.1002/9781119875116

Vattem, D. A.; Shetty, K. Biological Functionality of Ellagic Acid: A Review. *J Food Biochem*. 2005, 29(3): 234–266.

Vats, V.; Yadav, S. P.; Biswas, N. R.; Grover, J. K. Anti-cataract activity of Pterocarpus marsupium bark and Trigonella foenum-graecum seeds extract in alloxan diabetic rats. *J Ethnopharmaco*. 2004, 93(2-3), 289–294.

Vergara-Jimenez, M.; Almatrafi, M.; Fernandez, M. Bioactive Components in Moringa oleifera Leaves Protect against Chronic Disease. *Antioxidants*. 2017, 6(4): 91.

Verma, A. R.; Vijayakumar, M.; Mathela, C. S.; Rao, C. V. *In vitro* and *in vivo* Antioxidant Properties of Different Fractions of *Moringa oleifera* Leaves. *Food Chem Toxicol*. 2009, 47(9): 2196–2201.

Wang, D.; Chen, Y.; Li, J.; Wu, E.; Tang, T.; Singla, R. K.; Shen, B.; Zhang, M. Natural Products for the Treatment of Age-Related Macular Degeneration. *Phytomedicine*. 2024, 130: 155–522.

Wang, L.; Teplitski, M. Microbiological Food Safety Considerations in Shelf-Life Extension of Fresh Fruits and Vegetables. *Curr Opin Biotech*. 2023, 80: 102–895.

Wang, Y. Cruciferous Vegetables and Cancer Prevention: An Update. *Phys Rev*. 2021, 20(3): 567–584.

Wen, C.; Zhang, J.; Zhang, H.; Dzah, C. S.; Zandile, M.; Duan, Y.; Ma, H.; Luo, X. Advances in Ultrasound Assisted Extraction of Bioactive Compounds from Cash Crops – A Review. *Ultrason Sonochem*. 2018, 48: 538–549.

Wimalasiri, D.; Dekiwadia, C.; Fong, S. Y.; Piva, T. J.; Huynh, T. Anticancer Activity of *Momordica cochinchinensis* (red gac) Aril and the Impact of Varietal Diversity. *BMC Complement Med Ther*. 2020, 20(1): 365.

Xu, T.; Wu, X.; Liu, J.; Sun, J.; Wang, X.; Fan, G.; Zhang, Y. The regulatory roles of dietary fibers on host health via gut microbiota-derived short chain fatty acids. Current Opinion in Pharmacology. 2022, 62: 36-42.

Yakhin, O. I.; Lubyanov, A. A.; Yakhin, I. A.; Brown, P. H. Biostimulants in Plant Science: A Global Perspective. *Front Plant Sci*. 2017, 7.

Zeng, M.; Li, M.; Zhang, B.; Li, B.; Kan, Y.; Zheng, X.; Feng, W. Camellia Oil Inhibits Oxidative Stress and Inflammatory Response to Ameliorate LPS-Induced Acute Kidney Injury via Downregulation of TLR4-Mediated Activation of the NF-κB/AP-1/IRF3 and NLRP3 Pathways. *J Funct Food*. 2020, 68: 103–908.

Zhang, Y.-J.; Gan, R.-Y.; Li, S.; Zhou, Y.; Li, A.-N.; Xu, D.-P.; Li, H.-B. Antioxidant Phytochemicals for the Prevention and Treatment of Chronic Diseases. *Molecules*. 2015, 20(12): 21138–21156.

Zheng, S.; Yan, J.; Wang, J.; Wang, X.; Kang, Y. E.; Koo, B. S.; Liu, L. Unveiling the Effects of Cruciferous Vegetable Intake on Different Cancers: A Systematic Review and Dose–Response Meta-analysis. Nutrition Reviews. 2025, 83(5), 842-858.

Zheng, J.; Zhou, Y.; Li, S.; Zhang, P.; Zhou, T.; Xu, D.-P.; Li, H.-B. Effects and Mechanisms of Fruit and Vegetable Juices on Cardiovascular Diseases. *Int J Mol Sci*. 2017, 18(3): 555.

Zhuang, J.; Zhang, J.; Hou, X.-L.; Wang, F.; Xiong, A.-S. Transcriptomic, proteomic, metabolomic and functional genomic approaches for the study of abiotic stress in vegetable crops. Critic reviews in plant sci. 2014, 33(2-3): 225-237.

9 Staple Crops as a Source of Secondary Metabolites for Health Improvements

Keerthi Elavarasan[1], Swetha Balaji[2], Janani Sree Sivakumar[1], Mourian Balachandran[1], Gokul Murugesan[1], Krishna Shajikumar[1], Nivasini Prapakaran[1], Amit Roy[3], Vasanth Krishnan[1], and Dhandapani Gurusamy[2]

[1]Department of Botany, Bharathiar University, Coimbatore, India

[2]Department of Botany, Kongunadu Arts and Science College (Autonomous), Bharathiar University, Coimbatore, India

[3]Faculty of Forestry and Wood Sciences, Department of Forest Genetics and Physiology, Czech University of Life Sciences Prague, Prague, Czechia

ABBREVIATIONS

FAO Food and Agricultural Organization
WHO World Health Organization

9.1 INTRODUCTION

The Food and Agriculture Organization (FAO) states that a staple food is "one that is eaten regularly and in such quantities as to constitute the dominant part of the diet and supply a major proportion of energy and nutrients" to the population (FAO, 2009). Plant-based staples include cereals such as rice, wheat, maize, millet, barley, oats, rye, spelt, emmer, triticale, and sorghum; starchy tubers such as potato, sweet potato, yam, and taro; root vegetables such as cassava, turnip, carrot, and rutabagas; and dried legumes such as lentils and beans. Rice and corn are the primary staple foods in nations around the globe. In several regions worldwide, rice is regarded as a staple food that is eaten frequently by millions of people daily to fulfil their nutritional requirements. Corn is significantly cultivated as the staple food because it is consumed and utilized as animal feed, sweeteners, automobile fuel, and even a type of plastic. Rice, wheat, maize/corn, potatoes and dry soybeans are the most commonly consumed staple crops (Fathima et al., 2017).

There are many benefits associated with starchy staples, especially cereals. The harvested grain crop can be kept for an extended period, but roots are difficult to keep due to their high water content. The amino acid, namely lysine, is the most noticeable nutritional deficiency within cereals and comprises sufficient amounts of legumes. Traditional diets generally consist of starchy staples and legumes, such as wheat and beans in Europe, soybeans and rice in China, or beans and maize in America (Grigg, 1996).

Based on production data 1996, Asia produces more cereals, roots and tubers, dry beans, peanuts, broad beans, chickpeas, lentils, and other staples than any other continent. Cereals, especially wheat, are the main staple of developed nations, whereas rice, corn, millet, and sorghum constitute the staple crop of developing countries. Developing nations also use a lot of plantains,

DOI: 10.1201/9781003518358-9

bananas, and root crops apart from potatoes as staple foods. China has emerged as the world's largest producer of sweet potatoes plus rice. South America, Asia, and Africa are the primary producers of cassava. Africa is the primary producer of taro and yam; however, taro is also cultivated in Asia. Central and Northern America are the most significant producers of soybeans, a key legume. Vegetarians, especially from South Asia, eat plenty of legumes instead of meat. Throughout Asia and South America, more plantains are grown than bananas. Africa is the primary producer of bananas. In a large portion of Tanzania and Uganda, bananas and plantains are the main staple foods (Juliano, 1999).

Rice, wheat, and maize are staple crops because they provide food security and are rich in secondary metabolites, which are beneficial to human health. Though not indispensable for human base metabolic functions, such compounds augment human health. Other metabolites like phenolics, flavonoids, and alkaloids account for antioxidant, anti-inflammatory, and antimicrobial properties in these plants (Tiwari and Cummins, 2013; Isah, 2019; Ti et al., 2014). Staple crops were traditionally valued for their nutritional content and have been demonstrated to be functional foods that avert ailments and promote health. In this regard, staple crops have been instrumental in fighting cardiovascular diseases, diabetes mellitus, and cancer (Xavier et al., 2024). This increased awareness of secondary metabolites in staple crops indicates a new approach towards daily diets for better health outcomes. Bioactive compounds, through extraction from commonly consumed crops, indicate a benefit over basic nutrition, making staple crops an affordable and accessible resource for better public health worldwide (Zaynab et al., 2024).

Recent advancements in agricultural biotechnology have brought new prospects for elevating the concentration and bioavailability of these desirable compounds in staple foods. The level of secondary metabolites, by nature, naturally synthesized in crops is expected to be elevated to combat deficiencies in populations that consume large portions of these foods (Ferguson et al., 2017). Through plant biochemistry innovation in breeding and cultivation techniques, researchers explore how staple crops can serve as caloric sources and vehicles for health-promoting compounds. This chapter comprises nutritional value, secondary metabolites in staple crops, and their medicinal significance, which are beneficial and improve the health of the people.

9.2 NUTRITIONAL VALUES OF STAPLE FOOD CROPS AND THEIR HEALTH BENEFITS

The nutritional composition of these predominant crops plays a role in reducing malnutrition and ensuring food security, especially in regions with limited access to diverse diets.

Rice, wheat, and maize are major food crops that provide people with necessary sustenance around the world. These crops are the foundation of daily diets, providing critical macronutrients like carbs, proteins, and fats and micronutrients like vitamins and minerals, which contribute to overall health and well-being. Nutritional contents of various staple food crops are enclosed in Table 9.1. The health benefits of various staple food crops are enclosed in Figure 9.1.

9.2.1 Rice

Nearly half of the worldwide population is nourished by rice, a vital crop that supplies more than 21% of human calorie requirements while 76% of Southeast Asian caloric consumption. In South Asia, China, Thailand, Japan, and Korea, rice is grown extensively. Rice is primarily divided into four categories based on the colour of its caryopsis: white, black, red, and green. Therefore, rice endosperm, which primarily consists of starch (70%–80%), with proteins of about 7%–10% and lipids in trace amounts less than 1%, determines the nutrient content in polished rice (Fitzgerald et al., 2009; Yang et al., 2019; Zhao et al., 2020) (Table 9.1).

TABLE 9.1
Nutritional Content of Various Staple Food Crops

S. No	Name of the Plant	Genotype or Varieties	Nutritional Values: Calories/100 g	Carbs	Proteins	Lipids	Fiber	Reference
1.	*Oryza sativa*	Red rice	216 kcal	45 g	5.0 g	1.5 g	2.7	https://sprintmedical.in/blog/types-of-rice-and-their-health-benefits
		Black rice	347 kcal	0.50 g	2.80 g	8.90 g	2.80g	Joseph et al. (2024)
		White rice	223 kcal	49.6 g	4.10 g	0.205 g	0.74g	Wasan et al. (2022)
		Basmati rice	160 kcal	33 g	4.0 g	1.5 g	–	
		Jasmine rice	81 kcal	39 g	4.0 g	1.0 g	–	
		Arborio rice	356 kcal	77.8 g	11.1 g	2.22 g	2.22 g	Joseph et al. (2024)
		Wild rice	–	35.0 g	6.0 g	0.5 g	3 g	Wasan et al. (2022)
		Brown rice	232 kcal	49.7 g	4.88 g	1.17 g	3.32 g	
2.	Wheat	*Triticum vulgare*	1927 kcal	75%–80%	9%–18%	1.5-2.5%	–	Ammar et al. (2023); Janssen et al. (2018)
		Triticum durum	339 kcal	70%	12%–16%	1.9%	1.6%	Saini et al. (2023); Mohan and Malleshi (2006)
3.	Maize	*Zea mays* L.	365	74.26	9.42	4.74	7.3	Bathla et al. (2019); Rouf Shah et al. (2016)
4.	Millets	Finger millet	328	81.5%	9.8%	1.3%	1.5%	Gull et al. (2014); Kumar et al. (2021)
		Pearl millet	378	65%–70%	11%–15%	5.6%	15%–20%	Verma and Kumar et al. (2023)
		Proso millet	378	68.2%	11.02%	4.22%	5.4%	Luis et al. (1982); Kaur et al. (2014); McSweeney et al. (2017)
		Little millet	207	67.0%	8.7%	4.7%	7.6%	Gowda et al. (2022); Dhawale et al. (2022)
		Barnyard millet	300	67.5%	11.6%	3.6%	10.1%	Geervani and Eggum (1989)
		Foxtail millet	351	63.2%	12%–15%	4.3%	6.7%	Bano and Singh (2024)
		Kodo millet	353	66.6%	8.3%	1.4%	37%	Abah et al. (2020)
5.	Sorghum	–	329	72.1%	11.3%	3.5%	86.2%	Dicko, et al. (2006)
6.	Tubers	Potato	96–123	9.0%	3.0%	0.1%-0.5%	7.0%	Burgos et al. (2020); Luthra et al. (2009)
		Sweet Potato	351.40	80%	0.89%	0.20%	0.70%	Xi-You et al. (2024); Senthilkumar et al. (2023)
		Yam	80–120	17.5%–28%	1.5%–6%	0.1%–0.2%	0.6%–1.5%	Ferraro et al. (2016)
		Cassava	140	80%—90%	1%—3%	0.1%–-0.5%	0.1–3.7	

TABLE 9.2
Secondary Metabolites and Their Properties in Staple Crops

S. No	Name of the Plant	Parts	Secondary Metabolites	Properties	Reference
1.	Rice *Oryza sativa*	Endosperm and bran	Phenols Caffeic acid Trans-Ferulic acid	Antioxidant activity	Zhang et al. (2014)
		Bran	Flavones Tricin	DPPH radical scavenging activity	Ajitha et al. (2012)
		Coleoptiles	Diterpenoids, Momilacton A	Antifungal activity	Cartwright et al. (1981)
		Bran	Tri terpenoids Gramisterol cis & trans-ferulate	Anti-inflammatory activity	Akihisa et al. (2000)
		Bran	Sterols Stigmasterol cis & trans ferulate	Anti-inflammatory activity	
		Whole rice plant	Alkaloids, Indole 3-acetic acid	Regulation on growth and development	Yamamoto et al. (2017)
		Whole plants	Other metabolites, cis-12-oxo-Phytodienoic acid	Inducible anti-insect activity	Guo et al. (2014)
		Bran	Other metabolites β-Tocotrienol, γ-Tocotrienol, δ-Tocotrienol	Antioxidative, antihypercholesterolemic anticancer, neuroprotective activities	Sookwong et al. (2009)
		Whole rice plant	Sesquiterpenoids, Absicisic acid	Regulation of growth and development	Changan et al. (2018)
2.	Wheat *Triticum aestivum*	Whole plants	Phenolic acids	Antioxidant, anti-inflammatory	Levakova and Lacko-Bartosova et al. (2017)
			Flavonoids	Antioxidant, anticancer	Dalton et al. (2012)
			Phytosterols	Cholesterol lowering agent, Antitumour, Anti-cardiovascular Anti-obesity effect	Rajhi et al. (2020)
			Alkylresorcinols	Anti-fungal, Anti-bacterial	Chrpova et al. (2021)
			Carotenoids	Antioxidant, Photoprotective	

(Continued)

TABLE 9.2 (*Continued*)
Secondary Metabolites and Their Properties in Staple Crops

S. No	Name of the Plant	Parts	Secondary Metabolites	Properties	Reference
3.	Maize *Zea mays* L.	Whole plants	Phenols, carotenoids	Strong anti-oxidant anti-radical	Maraschin et al. (2011)
			Primary carotenoids, xanthophylls, lutein and zeaxanthin	Boasts immunity Prevents breast tumor growth Defence against age related macular degeneration related blindness.	Shahid et al., (2023)
			Mono -, homo - or sesquiterpenes	Facilitate communication between arthropod and plant species.	
			Cefalexins, Kauraalexins and other terpenoid phytoalexins	More root production when the soil is dry.	
			Isoleucine, valine, threonine and 4 amino butanoate.	Employed as prospective metabolic markers in maize breeding to boost the crop's resistance to heat, drought and their combined effects.	
			Alkaloids, Phenolic compounds, terpenes	Protection against Arthropod herbivory, oxidative stress mitigators herbivore deterrents.	
			Benzoxazinoids	Antifeeding and deterring quality	
			Zealexins and kauralexins	Antifungal and Anti-feedant activity.	
			Flavonoids and other polyphenolic secondary metabolites	Protect maize from abiotic stress like Nitrogen deficiency.	
4.	Millets *Eleusine coracana*	Whole plants	Phenolic compounds Flavonoids Tannins	Metal chelating activity, Free radical scavenging activity	Anil et al. (2021)
	Pennisetum glaucam		Polyphenols Carotenoids	Antioxidant activity	Slama et al. (2020)

(*Continued*)

TABLE 9.2 (*Continued*)
Secondary Metabolites and Their Properties in Staple Crops

S. No	Name of the Plant	Parts	Secondary Metabolites	Properties	Reference
	Panicum miliaceum		Phenolic compounds Flavonoids	Antioxidant, Antiproliferative properties	Shen et al. (2018) Saikia et al. (2018)
	Penicum sumatrense		Dicarboxylic acid diterpene alcohol	Antioxidant and anti - inflammatory	Dhawale et al. (2022)
	Echinochloa frumentacea		Polyphenols Tannins ortho-dihydroxy phenol	Antioxidant, Antibacterial	Panwar et al. (2016)
	Setaria italica		Flavonoids and carotenoids, phenolic compounds	Antioxidant, Anti-inflammatory, Anti-microbial	Zhang et al. (2021) Shahidi et al. (2011) Sreerama et al. (2018)
	Paspalum scrobiculatuum		polyphenols, flavonoids	Antioxidant activity, Anti-diabetic, DPPH free radical activity	Dey et al. (2023) Sripriya et al. (1996)
5.	Sorghum *Sorghum bicolor*	Whole plants	Tannins, phenolic acids, anthocyanins, phytosterols, and policosanols.	Anti- diabetic, Anti- Cancer, Antioxidant activity, Detoxifying function, antithrombotic properties	Antarlina et al. (2022) Jiang et al. (2020) Queiroz et al. (2018)
6.	Tubers *Solanum tuberosum*	Whole plants	Solanine, chaconine, chlorogenic acid, rutin	Anticancer, Hypocholesterolemia, Anti-inflammatory, Anti-obesity, antidiabetic	Singh et al. (2009) Mackey et al. (1990) Burgos et al. (2020)
	Ipomoea batatas		Anthocyanins, carotenoids, polyphenols	Free radical scavenging Antioxidative, anti-Mutagenic, anticancer, Antihypertension, Antidiabetes, Antiinflammation and Antibacterial	Liu et al. (2015) Hegde et al. (2024)
	Dioscorea		Diosgenin, steroidal saponins, polyphenols	Anti-mutagenic, Hypoglycemic, Immunomodulatory, Antimicrobial and antioxidant activities	Chen et al. (2011) Kouame et al. (2011) Hegde et al. (2024) Dabonne et al. (2011)

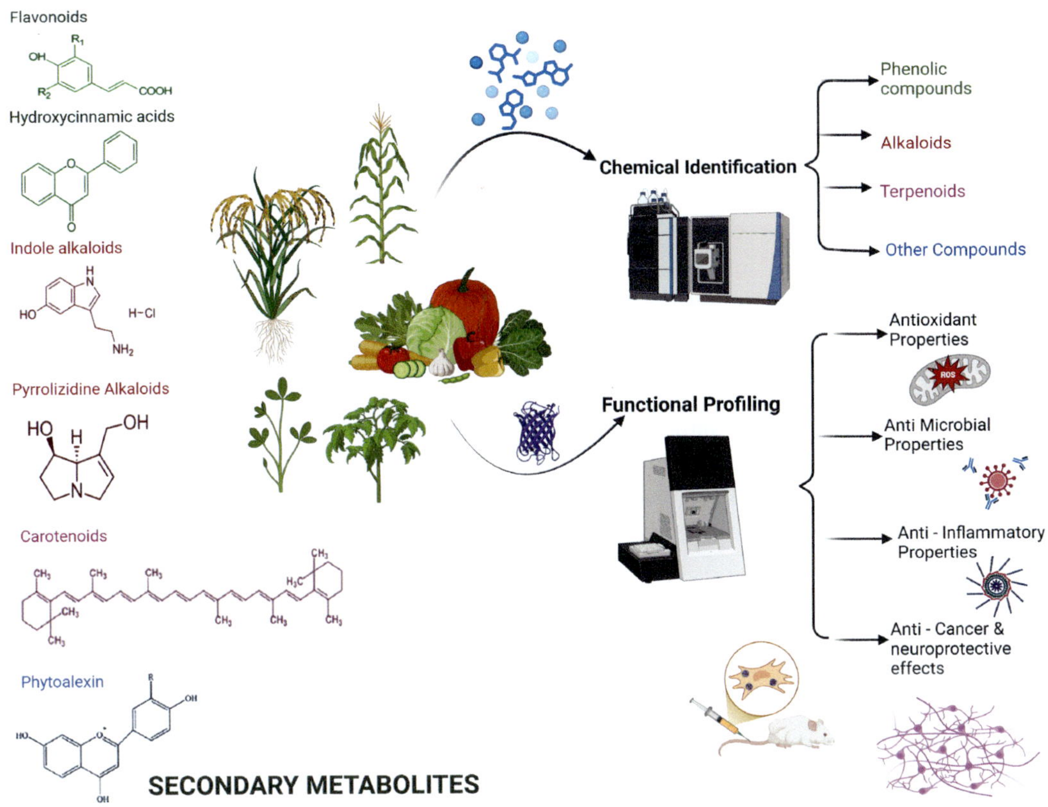

FIGURE 9.1 Health benefits of staple food crops. Staple food crops have several health benefits and also cure various diseases.

One excellent source of complex carbohydrates is rice. A significant portion of glucose produced by the breakdown of carbohydrates serves as energy for physical activity and vital brain fuel. Foods high in carbohydrates play a important role in transporting proteins, vitamins, and micronutrients (Mbabaali, 1998).

It possesses excellent protein digestibility with significant lysine content. The unique advantage of rice protein, which constitutes 8% of the grain, is that it contains eight essential amino acids in a balanced ratio. Naturally, brown rice contains B-complex vitamins, particularly "thiamin, riboflavin, and niacin", promoting healthy skin and nourishment to blood vessels (Lloyd et al., 2000; Chaudhari et al., 2018).

Rice contains several secondary metabolites, including phenolics, flavonoids, and terpenoids (Zhao et al., 2024). The inhabitants of Jammu and Kashmir have long utilized ancient rice types in their Ayurveda and Unani medical systems. According to Ayurveda, rice has aphrodisiac, fattening, oleaginous, diuretic, acrid, tonic, and biliousness-relieving properties. The substance obtained from rice grain bran is used to treat eye conditions, body aches, and neurological problems. Red rice protects against cancer, diabetes, and obesity. It significantly impacts the management of several illnesses, including burns, diarrhoea, fever, bleeding, wounds, and vomiting. It was thought to be an effective remedy for fevers, ulcers, refrigerants, diuretics, improved vision, semen production, spermatophytes, cosmetics, voice quality, tonics, and antitoxic (Bhat and Riar, 2015; Hedge et al., 2013).

During the Chinese dynasty, it was known as "forbidden rice" and was only consumed by royal families. Black rice is called kavuni rice. It prevents diabetes, cancer, atherosclerosis, hypertension, digestive health, osteoporosis, and asthma and reduces stroke, especially in women (Sangma and Parameshwari 2023; Sirisoontaralak et al., 2020).

Red rice improves blood circulation and treats/soothes stomach upsets. It has α-tocopherol and γ-tocopherol that prevent damaging membranes of lipids (Roy et al., 2023).

It is loaded with selenium (Se), manganese (Mn), and magnesium (Mg), which allows for the prevention of numerous ailments, including asthma, colon cancer, and diabetes. (Lee et al., 2019).

9.2.2 Wheat

Two species of wheat, mainly *Triticum aestivum* (bread wheat) and *Triticum durum* (pasta wheat), are the most commonly cultivated. The developed countries rely solely on wheat for making bread, noodles, cakes, pastries, and lactogen (Iqbal et al., 2022).

The most widely grown wheat varieties now are amber, but because of the related health advantages, colourful wheat high in anthocyanins and various phytochemicals is growing in popularity worldwide. The purple wheat originated in the 19th century. Blue wheat emerged in the early 20th century by interspecifically crossing *Triticum aestivm over* multiple wild species, including "*Thinopyrum ponticum, Triticum monococcum* L. spp. *aegilopoides*, and *Th. Bessarabicum*" (Saini et al., 2021).

Over one-third of the global population relies on wheat as an essential, stable food source, and it provides more nutrients and proteins to the global diet than any other cereal crop. Wheat has been recognized as an effective source of dietary fibre, B-group vitamins, minerals, and protein. Wheat accounts for 20% of calories and roughly 55% of carbohydrates. It contains high levels of vitamins (thiamine in addition to vitamin B) and minerals (zinc and iron), as well as carbohydrates (78.10%), protein (14.70%), and fat (2.10%) (Akram et al., 2023).

Additionally, it is an adequate source of trace elements, including magnesium and selenium, vital components for healthy living. Wheat grains contain a high concentration of riboflavin and pantothenic acid, as well as other minerals, carbs, and nutrients. Wheat germ contains rich nutrients and is devoid of cholesterol and sodium. "Vitamin E, magnesium, pantothenic acid, phosphorus, thiamin, niacin, and zinc" are abundant. Additionally, it contains the coenzymes PABA (para-aminobenzoic acid) with Q10 (ubiquinone) (Adams et al., 2002; Shewry 2007; Shewry 2009).

It enables lower death rates resulting from cardiovascular disease and cancer prevention. It also enables treating and curing several diseases, such as tooth disorders, constipation, skin diseases, digestive system disorders, circulatory disorders, boils, scars, pimples and so on (Kumar et al., 2011). Additionally, researchers believe that additional compounds within whole grains can influence hormone levels and reduce the incidence of diseases linked to hormones, such as breast cancer (Shewry and Hey 2015).

9.2.3 Maize

Maize/corn represents a significant annual grain crop in the Poaceae family. The USA is the world's leading maize producer, accounting for around 35% of global production. More than 95% of India's maize production comes from the notable maize-growing states of "Uttar Pradesh, Bihar, Rajasthan, Madhya Pradesh, Punjab, Haryana, Maharashtra, Andhra Pradesh, Himachal Pradesh, West Bengal, Karnataka, and Jammu and Kashmir" (Milind and Isha 2013; Rouf Shah et al., 2016). It provides an adequate source of vitamin B-complex together with carbohydrates. It contains plenty of selenium, beta-carotene, and vitamins C, A, and K, all of which promote healthy thyroid function. In addition, it comprises vitamin C, vitamin E, vitamin K, folic acid, selenium, N-p-coumaryl tryptamine, N-ferrulyl tryptamine, vitamin B1, B2, B, B5 and B6 (Saritha et al., 2020).

Owing to higher B-complex content, it aids joint mobility and treats rheumatism. Potassium is a key component in corn silk, that is a potent diuretic. Stones in the kidney and infections of the urinary tract can be treated using corn silk. However, it remains frequently used in China to treat jaundice and retention of fluids. It treats and cures ulcers, swelling, wounds, jaundice, nausea and vomiting. In addition to enhancing liver function and bile production, corn silk lowers blood pressure (Kumar and Jhariya 2013). It has dietary fibre, carotenoids, flavonoids, phenolic acids (ferulic acid), coumaric acid, along with syringic acid), and minerals such as Mg, P, and K (Mall and Tripathi, 2016). In comparison with both wheat as well as rice, maize contains a lesser amount of protein but has an energy ratio (365 kcal/100 g), carbohydrate (72%), protein (10%), and fat (4%) (Ranum et al., 2014).

9.2.4 Miscellaneous

9.2.4.1 Millets

Throughout the world,

Millets are a diversified group of small-seeded grasses that are widely grown as grain for human use. Millets comprise species such as foxtail millet, finger millet, and pearl millet, each with peculiar nutritional advantages. These grains are often more nutritious than other commonly consumed grains, such as rice and wheat, since they include more dietary fibre, protein, vitamins, and minerals (Pradeep et al., 2018).

Due to its remarkable tolerance to harsh environmental conditions, pearl millet is widely grown in Africa and India. In South India, in addition to East Africa, finger millet is a staple grain known for its high calcium content. Foxtail millet is widely grown throughout East Asia, especially in China. Proso millet exists worldwide, including in China, Russia, and North America (Mall and Tripathi 2016; Huang et al., 2016; Serba et al., 2020; Raut et al., 2023).

Millets are regarded as nutrient-dense, high-energy foods that aid in treating malnutrition. Millet crops possess natural bioactive substances with several health benefits, including high calcium levels, dietary fibre, polyphenols, and protein. Millets are more abundant with fatty acids than rice, corn and wheat, and they are additionally abundant in amino acids, particularly sulfur-containing amino acids, including methionine and cysteine. Foods made from millet are speculated to provide potential health benefits like probiotics in addition to prebiotics (Nithiyanantham et al., 2019).

Treatments for metabolic illnesses like obesity, diabetes, coronary artery disease, and liver problems may benefit from consuming proso millet (Li et al., 2021).

Finger millet reduces the risk of cardiovascular disease, type 2 diabetes, and gastrointestinal cancer. Finger millet consumption prevents mucosal ulceration. A beverage made from finger millet fermented by bacteria that produce lactic acid is utilized for treating diarrhea. Gallic acid, quercetin, and crude polyphenol derived from finger millet work as effective inhibitors of snake venom, hence it has the potential to be employed in treating inflammatory conditions (Chandra et al., 2016).

9.2.4.2 Sorghum

The 5th most significant cereal throughout the global agricultural sector is sorghum. The United States, Nigeria, Mexico, Ethiopia, and India are the primary producers of sorghum, which was grown on 42.3 million hectares of land globally in 2013 and produced 61.5 million metric tonnes. Sorghum is a staple food for over 300 million people across developing nations, serving as their primary source of calories and nutrients. It is also an essential resource of vital nutrients and energy for around a billion people throughout the semi-arid tropical region, similar to millet. Sorghum constitutes a fundamental staple food among many African rural populations, particularly in drought-prone regions (Taylor, 2003; Proietti et al., 2015).

Sorghum contains linolenic acid, protein, amylose, crude fibre, lysine, calcium, niacin, oleic acid, phytic acid, thiamine, etc., (Hossain et al., 2022). The principal component, starch, is mainly found within the endosperm as granules, while some starch additionally exists in the grain's pericarp. Polysaccharide starch usually predominates in sorghum is fructosan, the trisaccharide (raffinose), the disaccharides (sucrose, maltose), and the monosaccharides (glucose, fructose) are the carbohydrates found in sorghum grain (Anglani, 1998; Khan et al., 2013; Khoddami et al., 2023).

Sorghum is a gluten-free substitute for wheat. Antioxidants in sorghum offer several health advantages and may aid in preventing chronic illnesses that now impact people worldwide. Customers are counselled to use whole-grain substitutes in baking to consume the fibre with micronutrients provided by every portion of the grain due to the rising incidence of celiac illnesses and the rising desire for gluten-free products. It enables the treatment of diabetes and regulates blood sugar levels. (McGinnis and Painter 2020).

9.2.4.3 Tubers

Abundant carbohydrate supplies are available in tuber crops and are utilized for various purposes, including cash crops and common food crops. They are becoming increasingly common as animal feed, raw materials in industry, and processed foods to serve humans. Tuber crops make up more than half of the staple foods consumed on a daily basis (Chauhan et al., 2022; Lu et al., 2005). Compared to other tubers, potatoes and yams are rich in protein. It precisely constitutes cysteine and methionine (sulphur-containing amino acids). Vitamin C may be found in sweet potatoes, yams, cassava, and potatoes. β-carotene is found in yellow sweet potato, yam, and cassava variants. Potassium is abundantly found in taro. Although they lack some vitamins and minerals, roots and tubers are good sources of dietary fibre (Chandrasekara and Josheph Kumar 2016; Xue et al., 2014). The juices of sweet potatoes are utilized as a supplement for cancer treatment. Powdered leaves of sweet potatoes are employed as a poison against *S. aureus* bacteria. In tapioca, the contents of iron and copper cure anaemia. Elephant yam is often employed to treat dysentery (Krishnakumar et al., 2022).

9.3 SECONDARY METABOLITES IN STAPLE CROPS

Secondary metabolites are organic compounds produced by plants that play a important roles in ecological interactions, defence mechanisms, and signalling. Secondary metabolites contribute greatly to a plant's adaptability and survival in its environment, unlike primary metabolites, which are necessary for growth and reproduction (Ku et al., 2020). Staple crops such as wheat, rice, maize, and potatoes are fundamental to global food security, providing essential nutrients for billions of people. The secondary metabolites found in these crops contribute to their resilience against pests and diseases (Fan and Shi, 2024). Secondary metabolites from staple food crops are enclosed in Figure 9.2.

9.4 EXTRACTION OF SECONDARY METABOLITES

9.4.1 Rice

Rice (*Oryza sativa*) contains numerous valuable secondary metabolites, including phenolic acids, flavonoids, anthocyanins, and tocopherols, which offer antioxidant, anti-inflammatory, and other health benefits. These metabolites are especially abundant in rice bran and hulls. The extraction methods vary depending on the metabolite and desired yield and include traditional and advanced extraction techniques. Solvent extraction is a fundamental and widely applied method, particularly effective for polar compounds like phenolics and flavonoids. Common solvents include methanol, ethanol, acetone, and water. Rice bran or hulls are ground, mixed with a suitable solvent, and then stirred or sonicated to enhance extraction. The solution is filtered, and the solvent is evaporated under reduced pressure (Wang and Liu, 2009).

The Ultrasound-assisted extraction (UAE) disrupts cell walls to enhance the release of bioactive components including phenolic acids and flavonoids, from rice bran and hulls. This method reduces extraction time and solvent usage. Ground rice bran is mixed with a solvent and subjected to ultrasound waves at an optimal temperature and time. Parameters are adjusted to maximize extraction efficiency (Hiemori et al., 2009; Alam et al., 2016). The enzyme-assisted extraction (EAE) uses enzymes, such as cellulase or pectinase, to break down the cell walls, releasing bioactive

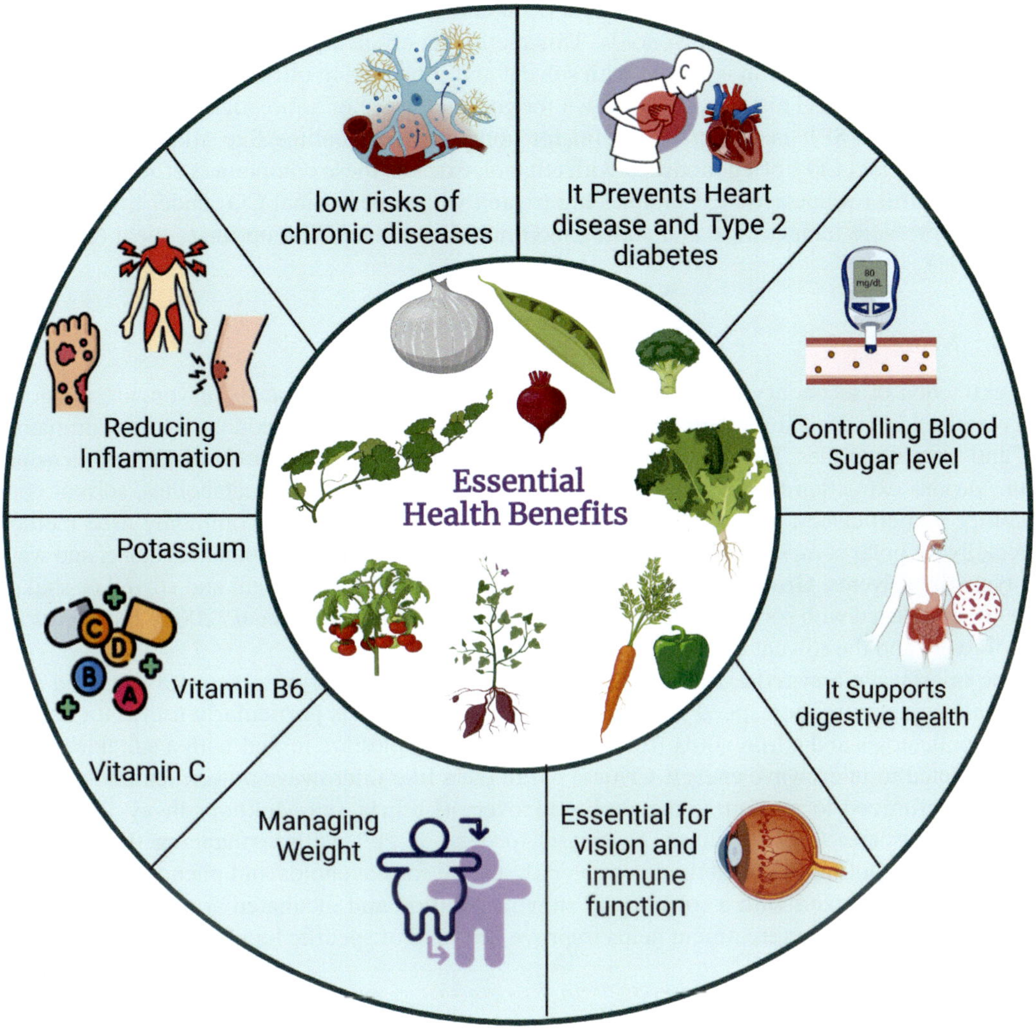

FIGURE 9.2 Secondary metabolites from the staple food crops. Alkaloids, terpenoids, and phenolic compounds are found in staple food crops that treat various diseases and possess several properties such as antioxidant, antimicrobial, anti-inflammatory and anticancer.

compounds like flavonoids and phenolic acids. This technique is particularly effective for rice bran and hulls. Ground rice bran is treated with a solvent-enzyme mixture and incubated under optimal enzyme activity conditions, enhancing the release of certain beneficial components (Maisuthisakul and Suttajit, 2007).

9.4.2 Wheat

In wheat Secondary metabolites, such as phenolics, flavonoids, and alkaloids, provide health benefits and are concentrated in the bran and germ. Extraction methods for these compounds range from traditional solvent extraction to advanced techniques like supercritical fluid extraction (SFE). Solvent extraction is commonly used for isolating polar compounds, such as phenolics and flavonoids, from wheat. The solvents like methanol, ethanol, and acetone are widely applied. Ground wheat bran or germ is mixed with solvent and shaken, sometimes assisted by sonication to improve extraction. The solvent evaporates under reduced pressure to obtain the extract (Zhou and Yu, 2004).

The UAE uses ultrasonic waves to disrupt plant cell walls, enhancing the release of secondary metabolites like phenolics and flavonoids. This method is efficient, requiring less time and lower solvent volumes. Wheat bran is mixed with solvent and subjected to ultrasonic waves at controlled time and temperature. This method is known for good recovery of antioxidant compounds (Wang et al., 2008). The SFE is especially useful for non-polar metabolites like alkaloids and terpenoids. supercritical CO_2, often modified with ethanol, extracts these compounds efficiently without leaving harmful residues. Wheat samples are treated with supercritical CO_2 under high pressure. Adjusting pressure temperature and adding modifiers like ethanol can optimize yield (Galanakis et al., 2021).

9.4.3 Maize

The extraction of secondary metabolites from maize, such as phenolic acids, flavonoids, alkaloids, and carotenoids, is critical for their health advantages, which include antioxidant, anti-inflammatory, and anti-cancer qualities. These metabolites are typically concentrated in the maize kernel, germ, and bran. Various extraction techniques are used based on the nature of the metabolites, solvent compatibility, and efficiency. The solvent extraction is the simplest and most commonly used method, especially for polar compounds like phenolics and flavonoids. Ethanol, methanol, acetone, and water are typically solvents. Ground maize kernels or bran are mixed with solvent and stirred or shaken, sometimes assisted with sonication to enhance extraction (Pascual-Martí et al., 2001). The mixture is then filtered, and the solvent evaporates under reduced pressure.

The microwave-assisted extraction (MAE) uses microwave energy to heat solvents and samples, enhancing extraction speed and efficiency. This technique is particularly useful for extracting phenolic acids and carotenoids from maize. Maize samples are mixed with a suitable solvent and subjected to microwave energy. Critical parameters like microwave power, temperature, and time are optimized to maximize metabolite recovery (Luthria and Mukhopadhyay, 2006). The EAE uses enzymes (e.g., cellulase, pectinase) to degrade cell walls, enhancing the release of secondary metabolites. This method is especially useful for flavonoids and phenolic compounds. Ground maize is mixed with a solvent and enzyme solution and incubated at optimal conditions for enzyme activity. This treatment helps improve the yield of specific bioactive compounds (Cho and Lee, 2018).

9.4.4 Tubers

The extraction of secondary metabolites from tubers, such as phenolic acids, flavonoids, alkaloids, glycoalkaloids, and saponins, is crucial due to their bioactive properties, including antioxidant, anti-inflammatory, and antimicrobial effects. Tubers like potatoes, yams, and sweet potatoes contain various secondary metabolites, predominantly in their skins and outer layers. The choice of extraction method depends on the specific metabolites targeted and their chemical properties. The solvent extraction is the most common and straightforward method for isolating secondary metabolites, especially polar compounds like phenolic acids and flavonoids. Common solvents include ethanol, methanol, acetone, and water. The tuber sample (usually dried and ground) is mixed with a solvent, shaken or stirred, and sometimes subjected to sonication. The mixture is filtered, and the solvent is evaporated under reduced pressure to obtain the extract (Xue et al., 2014).

The UAE uses ultrasonic waves to enhance extraction by disrupting cell walls and facilitating the release of metabolites. This method is efficient and reduces extraction time, especially for phenolic compounds and flavonoids. The tuber sample is mixed with a solvent and subjected to ultrasonic waves for a set period. Parameters such as temperature and time are optimized to maximize extraction yield (Chandrasekara and Shahidi, 2011b). The EAE employs enzymes like cellulase, pectinase, or protease to break down cell walls, improving the release of metabolites. This method is especially effective for extracting phenolic compounds and flavonoids. Tuber samples are mixed with a solvent and enzyme solution and incubated under conditions optimal for enzyme activity, which helps improve the yield of specific bioactive compounds (Ren and Yu, 2015; Richardson et al., 1990).

9.4.5 Millets

Sorghum has a wide range of secondary metabolites, including phenolic acids, flavonoids, tannins, and anthocyanins, all of which have considerable antioxidant, anti-inflammatory, and antibacterial properties. The extraction of these bioactive compounds varies based on the specific metabolite, solvent compatibility, and efficiency. The solvent extraction is a primary but widely used method, particularly for polar compounds like phenolics and flavonoids. Ethanol, methanol, acetone, and water are commonly used solvents. Ground sorghum grain is mixed with the solvent and stirred or sonicated to improve extraction. After filtration, the solvent is evaporated under reduced pressure (Awika and Rooney, 2004).

The UAE employs ultrasonic waves to break down plant cell walls, improving the release of metabolites. This method is particularly effective for extracting phenolic compounds and tannins from sorghum. Sorghum samples are mixed with a solvent and subjected to ultrasound waves for a set period. Optimized temperature and time increase extraction efficiency (Chandrasekara and Shahidi, 2010). The EAE uses enzymes like cellulase or pectinase to break down cell walls, improving the release of secondary metabolites. This method is useful for phenolic compounds and tannins. Sorghum samples are treated with a mixture of solvent and enzymes, which helps release more phenolic compounds and other metabolites (Ashogbon and Akintayo, 2013).

9.5 SECONDARY METABOLITES IN STAPLE FOOD CROPS

Over 80% of people worldwide, according to the "World Health Organization" (WHO), believe in therapeutic plants due to their crucial role in developing human health. Phytoconstituents derived from therapeutic plants are now recognized and widely accepted as key and significant molecules in modern medications (Alves and Rosa, 2005).

9.5.1 Rice

Rice is considered the queen of cereals because of its higher nutritional value, increased digestibility, biological activity, and possible health advantages. Rice has several bioactive compounds such as γ-oryzanol, coumaric acid, phytic acid, tricin etc. Black rice is rich in anthocyanin, whereas red rice contains proanthocyanidin (Verma and Srivastav 2020). Rice comprises caffeic acid, methyl caffeate, cinnamic acid, *p*-hydroxybenzoic acid, *trans*-ferulic acid, vanillic aldehyde, syringic acid, gallic acid, *p*-hydroxy phenylacetic acid, gallic acid and coumaric acid, flavones, flavanones and anthocyanin, that has antioxidant properties (Wang et al., 2018).

9.5.2 Wheat

Wheat has several bioactive compounds, including Linoleic, oleic sitosteryl, α-linolenic acids, β-sitosterol, campesteryl, campesterol stigmasterol, eicosenoic acid and sitostanyl ferulate, to treat several diseases (Akram et al., 2021).

9.5.3 Maize

Corn/maize comprises various phytochemical constituents, namely, particularly carotenoids such as lutein, zeaxanthin, α-carotene, α-cryptoxanthin, β-cryptoxanthin, violaxanthin, and β-carotene

(Chavez-Arias et al., 2022). Phenolic substances, namely quercetin and kaempferol, reduce the risk of numerous diseases, such as heart disease, cataracts and eye infections (Lu et al., 2023).

9.5.4 Millets

Due to the presence of flavonoids such as quercetin, tricin, luteolin, myricetin and catechin, pearl millet, finger millet, and sorghum have cyro-protective, anti-cancerous activity. Flavonol polymers in barnyard millet treat osteo-malacia and rickets (Patel et al., 2023). Proso and foxtail millet containing β-sitosterol and stigma sitosterol prevent the risk of stroke and cardiovascular diseases. Polyphenols such as vanillic acid, coumaric acid and ferulic acid in finger millet reduce inflammation and cure arthritis (Priya et al., 2023; Pradeeo et al., 2018).

9.5.5 Sorghum

Hydrobenzoic acids, hydrocinnamic acids, 3-Deoxyanthocyanidins and proanthocyanidins are the major bioactive compounds found in sorghum. It also contains phenolic acid, flavonoid, procyanidin, and stilbenoids such as caffeic acid, p-coumaric acid, vanillic acidsinapic acid, chlorogenic acid, protocatechuic acid, ferulic acid, salicylic acid, gallic acid, p-hydroxybenzoic acid and syringic acid (Li et al., 2021; Setyorini et al., 2022).

9.5.6 Millets

Millets generally contain several phytochemical constituents, enhancing health benefits and treating numerous diseases. Such bioactive compounds include ferulic acid, coumaric acid, taxifolin, quercetin, vanillic acid, narigin, methyl vanilate, chlorogenic acid and so on (Singh et al., 2022).

9.6 PROPERTIES IN STABLE FOOD CROPS

Antioxidants are vital compounds that help combat oxidative stress in the body, reducing the potential danger of chronic diseases. Many staple food crops are high in antioxidants such as phenolic acids, flavonoids, vitamins, and other bioactive substances. In wheat, antioxidants have high phenolic acid content. The antioxidant activity varies significantly among different wheat varieties, influenced by environmental factors and processing methods (Horvat et al., 2022). Rice has a wide variety of antioxidant chemicals, including phenols and flavonoids. Brown rice has been shown to have higher antioxidant levels than white rice due to the retention of bran layers throughout processing. Maize is another antioxidant-rich grain, especially in its colourful variations. The kernels include a variety of phenolic chemicals and carotenoids, which contribute to their antioxidant properties. Antioxidant activity is also regulated by processing processes, with whole grain maize having higher amounts than processed varieties (Dilworth et al., 2013).

9.7 APPLICATIONS OF STAPLE FOOD CROPS

Rice husk is used as a fuel source and can be converted into various energy products, such as biofuels and biogas. It is used in making bricks, blocks, and panels, among the materials that can be made using it. Rice husk and its derivatives are used in manufacturing activated carbon and other medicinal goods, among other uses in pharmacy and medicine. Because of its unique qualities and potential uses in various disciplines, it is also utilized in nanobiotechnology research (Kordi et al., 2024).

Rice powder is a natural remedy for diaper rash due to its anti-inflammatory and antioxidant properties. Rice powder is used in traditional cosmetics to soften and lighten the skin, protect it

FIGURE 10.1 (a). Foliose lichen *Leptogium corticola* (Taylor) Tuck. grows on *Swietenia mahagoni* bark in Jember District, Indonesia. (b). *Parmelia tinctorum* Nyl grows on *Swietenia mahagoni* bark at Gumitir Mountain, East Java Indonesia.

natural products exploration which comprised of primary and secondary metabolites. This chapter provides insight on applications of primary and secondary metabolites of lichen for medicine and non-medicine uses.

10.2 LICHEN BIOLOGY

The evolutionary history of lichens has been studied extensively, with Linnaeus initially classifying them as part of the algae. However, the application of microscopy in the 1820s revealed the intricate details of the lichen thallus, leading to the recognition of the dual nature of lichens as symbiotic associations between fungi and algae or cyanobacteria (Spribille, Resl, Stanton, & Tagirdzhanova, 2022). Composed of a fungal partner (mycobiont) and a photosynthetic partner (phycobiont), typically an alga or cyanobacterium, lichens have a remarkable ability to inhabit diverse environments, from the harsh Arctic tundra to arid deserts (Oksanen, 2006).

The mycobionts have a network of fungal hyphae that form the structural framework of the lichen, in which they can efficiently absorb water, nutrients, and minerals from the environment to support the lichen, and mycobionts produce secondary metabolites that act as UV-protective pigments, shielding the lichen from environmental stress. Within the lichen thallus, the fungal hyphae as mycobionts provide the structural framework and protection in which they can alter their growth patterns and wall surface properties to suit photobiont needs. On the other hand, the photobiont provides the required nutrients through photosynthesis. The photobiont cells are positioned by the fungal partner to optimize conditions for photosynthesis and gas exchange. This intricate spatial arrangement is a key feature of the lichen symbiosis. A lichen association can be viewed as two separate cellular compartments where mycobionts and photobionts work together to support overall lichen metabolism. The photobiont contributes substantial energy reserves in the form of carbohydrates, which the fungal partner can then use for growth and reproduction (Friedl & Büdel, 2008; Gasulla et al., 2012; Nash, 2008, 2012; Untari, 2024a).

Lichens employ a variety of reproductive strategies, both sexual and asexual, to ensure the continuation of their species. Sexual reproduction in lichens typically involves the production of spores, which are dispersed by wind or water and can establish new lichen colonies upon finding a suitable substrate and algal partner. Asexual reproduction, on the other hand, can occur through the fragmentation of the lichen thallus, the dispersal of specialized reproductive structures called soredia or isidia, or the formation of vegetative propagules (Gauslaa, 2014; Honegger & Scherrer, 2008).

Based on thallus growth form, the lichen thallus can be divided into four growth forms, i.e., crustose, foliose, fruticose, and squamose. Crustose lichens are flat, crust-like lichens that form tight bonds with their substrates e.g., the surface of bark or rock, with the thallus color generally green and greyish white or sometimes pale brown. Foliose lichens are leaf-like lichens with loosely attached lobed thalli, where there is a thallus that later develops into lobes. This lichen is very easy to separate from its substrate because it has rhizines as its thallus attachment system. Fruticose lichens are shrubby, branched lichens that grow outwards from their point of attachment. These lichens can grow upright or can be found hanging from trees and rocks. Squamulose lichens are scaly lichens with small, verlapping, plate-shaped structures with a cup-shaped cup, which can also be known as a fruit (podetia) (Büdel & Scheidegger, 2008; Nash, 2012; Untari, 2024a).

As a unique symbiotic organism, lichens play an important role in the biogeochemical cycling of nutrients in the ecosystem, where lichens contribute to nutrient cycling, soil formation (Büdel & Scheidegger, 2008; Lalley & Viles, 2008; Lázaro et al., 2008; Nash, 2012; Untari, 2024a), air quality monitoring (Delves et al., 2023; Yang, Oh, & Hur, 2023), indicator of climate change (Miller, Villella, Stone, & Hardman, 2020; Nayaka & Rai, 2022; Nelsen & Lumbsch, 2020), and play a vital role in the health of the ecosystem. The presence and diversity of lichens can indicate the air quality and environmental conditions of an area. In addition, certain species of lichens are being investigated for their potential medicinal properties, such as antiviral (Desmarets et al., 2023; Makarevich et al., 2023) and antimicrobial compounds (Aoussar et al., 2020; Maulidiyah et al., 2021), as well as anti-inflammatory (Mendili et al., 2022; Tartouga et al., 2022) and anticancer (Nugraha et al., 2020; Nugraha et al., 2019) compounds. Furthermore, lichens are also commercially harvested for use in the production of dyes (Mendili et al., 2022; Şenol, Gül, & Şimşek, 2022), perfumes (Elkhateeb, El-Ghwas, & Daba, 2022), and other products. Additionally, lichens can act as ecosystem pioneers, food sources, habitats for multiple organisms (Joly, Wasser, & Booth, 2015; Untari, 2024b), and predator camouflage (Fang et al., 2020; Untari, 2024b).

10.3 LICHEN AND ENVIRONMENT INDICATOR

Many of the potential uses of lichens are due to the production of secondary metabolites. Lichen secondary metabolites are diverse and abundant, with over 1,000 unique compounds identified to date. The fungal partner plays a dominant role in the synthesis and accumulation of these metabolites, which are often localized in specific cellular compartments. The production of these compounds is often restricted to a limited number of lichen species, suggesting that they may serve as important taxonomic markers and ecological indicators (Upadhyay, Upadhyaya, Kollanoor-Johny, & Venkitanarayanan, 2014). One of the primary functions of lichen secondary metabolites is their ability to protect the lichen from environmental stressors such as intense sunlight, drought, and temperature extremes (Gasulla et al., 2012; Kranner et al., 2005). For example, the compounds may act as antioxidants, protecting the photosynthetic algae from oxidative damage and maintaining the delicate balance of the symbiotic relationship. In addition, the production of secondary metabolites is thought to play a remarkable role in maintaining the symbiosis.

Lichens' remarkable ability to adapt to a wide range of environmental conditions, including extreme habitats, has made them valuable indicators of environmental health and air quality (Gasulla et al., 2012). One of the key factors contributing to their adaptability is their poikilohydric nature, which means they can tolerate and even thrive in conditions of variable hydration. Numerous studies have investigated the phenotypic plasticity of lichens, revealing how they can modify their morphology, water relations, and photosynthetic capabilities in response to their microenvironment. For example, the lichen *Ramalina capitata* (Ach.) Nyl. has been shown to exhibit variations in the proportion of algal to fungal cells, chlorophyll content, and anatomical structures like thallus shape and thickness in response to different habitat conditions (Pintado, Valladares, & Sancho, 1997).

The ecological significance of lichens cannot be overstated, as they play a crucial role in many terrestrial ecosystems. Lichen communities are often found in areas where vascular plant growth is limited, such as high alpine, arctic, and desert regions (Oksanen, 2006). Their ability to colonize surfaces, including rocks, bark, and soil, makes them important pioneers in the establishment of new habitats. Moreover, lichens serve as food and shelter for a diverse array of organisms, contributing to the overall biodiversity of these ecosystems (Oksanen, 2006).

10.3.1 Lichen Metabolites as Nutrition Source

There are many lichen species being reported with only a few lichens considered edible and prepared as a carbohydrate source. Nevertheless, lichen thallus has been part of more than one billion people's diets throughout the world (Burlingame, 2000). Proper preparation techniques are required to overcome the mild toxicity of lichen secondary metabolites while transforming nutritious polysaccharides into digestible form. Brown horsehair lichen, *Bryoria fremontii,* has been a long source of nutrition for indigenous people of North America (Adenubi et al., 2022). A nutritious porridge or soup was prepared by boiling the thallus whereas marinated thallus was commonly consumed with other edible ingredients (Adenubi et al., 2022). In addition, other genera *Umbilicaria* spp. and *Lasallia* spp. were also prepared as emergency staples in North America (Yang, Devkota, Wang, & Scheidegger, 2021). Fruticose lichen, *Cetraria islandica,* was prepared as porridge, soup or salad by traditional community in northern Europe and Scandinavia (Svanberg & Ægisson, 2012). Lichens have also existed in Indian kitchens as a unique spice due to their distinct chemical constituents. The people of India prepared *Parmotrema perlatum* as an additional ingredient to enrich the aromatic flavor of spice mixes, masala (Svanberg & Ægisson, 2012). In the north of India, indigenous people in the Limbu and Sherpa regions used *Everniastrum cirrhatum, Everniastrum nepalense,* and *Parmotrema cetratum* as important components in preparing traditional sausage (Devkota, Chaudhary, Werth, & Scheidegger, 2017). However, studies on molecule characterization of lichen remain unclear.

10.4 LICHEN METABOLITES AS NATURAL DYES

Color is one of the basic elements in nature in the world in which mankind has always been attracted to discover source of natural dyes, including plants, animals, and minerals. Lichens have served as natural dyes since ancient times throughout the globe due to their unique color (Rather, Jameel, Ali, Bhat, & Yusuf, 2018). Orchil, purple dye in the Roman era, was extracted from lichen (Calà et al., 2019). In the antique era, the dyes production evolved from soaking with ammonia solution from urine for chemical transformation (Calà et al., 2019). Although secondary metabolites of lichen are attributed to their medicinal uses, several constituents are related to their properties as natural dyes in which yellowish, brownish, red, violet, and reddish-brown colors of lichen-based dyes. These natural dyes come from atranorin **1**, atromentin **2**, dermocybin **3**, emodin **4,** evernic acid **5**, gyrophoric acid **6**, grevillin A **7**, grevillin B **8**, grevillin C **9**, lecanoric acid **10**, physcion **11**, pulvinic acid **12**, polyporic acid **13**, salazinic acid **14**, telephoric acid **15,** (+)-usnic acid **16** of lichen (Figure 10.2). A distinct orchil dye from *Evernia furfuracea* was caused by lecanoric acid **10** content (Cardon, 2007; Gill & Steglich, 1987). A yellow-brown color of *Hypogymnia physodes* was due to (+)-usnic acid **16** and other acids in boiling water (Devkota et al., 2017). In addition, *Evernia prunastri* produced an orchil color due to evernic acid **5** and a pale yellow from another acid (Rather et al., 2018).

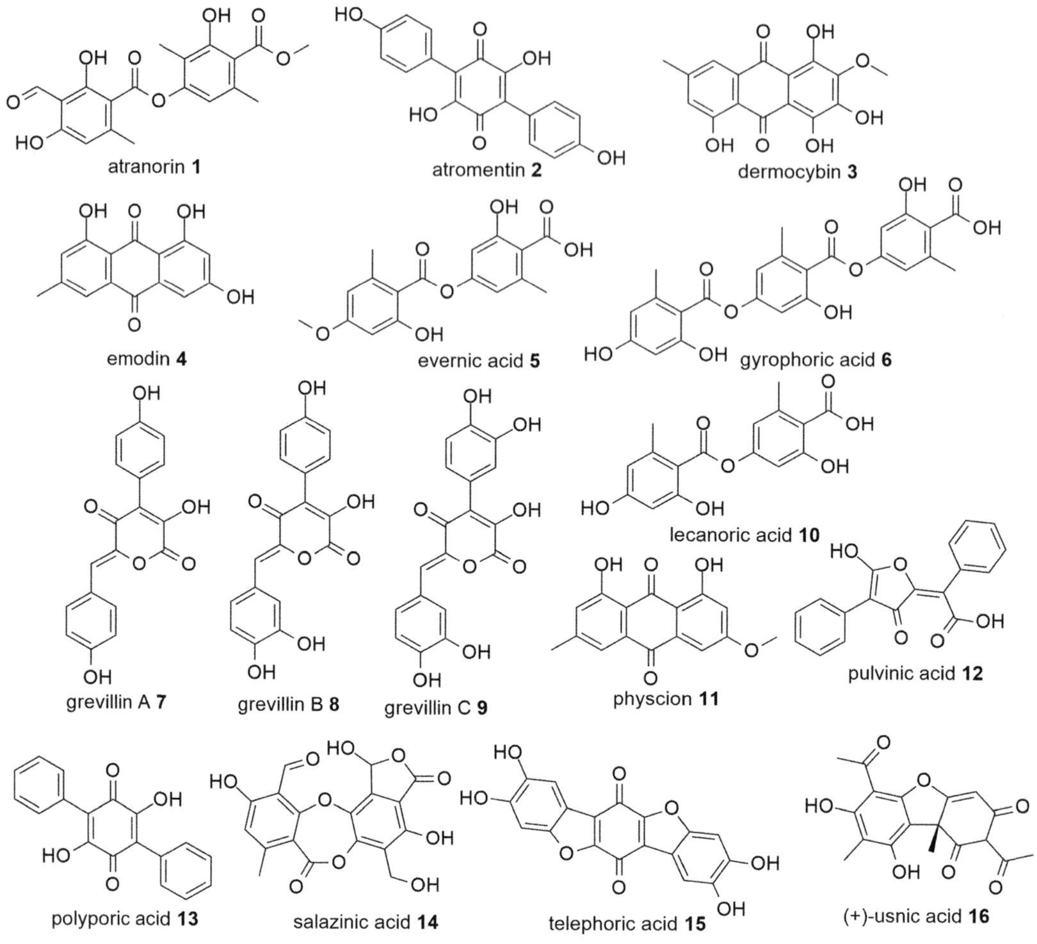

FIGURE 10.2 Chemical components of natural dyes from lichen.

10.5 LICHEN METABOLITES AND THEIR MEDICINAL PROPERTIES

Despite the biomass limitation due to slow-growing nature, several lichens have been prepared as traditional medicine (Nugraha et al., 2020). In Europe, for instance, Indigenous people of Italy use *Lobaria pulmonaria* (L.) Hoffm. (locally called lung lichen), the Peltigeraceae lichen was locally used to treat sciatica and cuts (Guarrera et al., 2008). The same species is also used by Turkish for laxative agents and medicinal tea (Süleyman et al., 2003). Several lichen species were also used traditionally by some communities in the Africa continent including the people of Libya, who commonly prepared fruticose Parmeliaceae lichen (*Evernia furfuracea* (L.) Mann., *Ramalina calicaris* (L.) Fr., and *Ramalina farinacea* (L.) Ach.), for natural dyes, medicine, and food species (De Natale & Pollio, 2012). In the south of the continent, *Usnea barbata* (L.) Weber ex F.H. Wigg was used to treat mammary infection in cattle (Afolayan et al., 2002). On the Asian continent, arrays of lichen biomass were also favored for their medicinal benefits. The Himalayan people of Nepal prepared juice of *Everniastrum cirrhatum* (Fr.) Hale ex Sipman, *Everniastrum nepalense* (Taylor) Hale ex Sipman, *Heterodermia diademata* (Taylor) D.D. Awasthi, *Parmotrema cetratum* (Ach.) Hale, *Thamnolia vermicularis* (Swartz) Ach. ex Schaerer, and *Usnea longissima* Ach. to treat cuts and fresh wound (Devkota et al., 2017). Parmeliacea lichen, *Parmelia perlata* (Huds.) Ach., *Usnea sikkimensis* Biswas were prepared for skin diseases therapy by Indian people (Pradhan & Badola,

2008; Shailajan, Joshi, & Tiwari, 2014). Traditional communities in China are familiar to several medicinally important lichens as tea, which include *Lethariella cashmeriana* Krog, *Lethariella sernanderi* (Motyka) Obermayer, *Lethariella sinensis* Wei & Jiang, *Lobaria isidiophora* Yoshim, *Lobaria kurokawae* Yoshim, *Ramalina conduplicans* Vain, *Ramalina sinensis* Jatta, *Thamnolia subuliformis* (Ehrh.) Culb, *Thamnolia vermicularis* (Swartz) Ach. ex Schaerer (Wang, Narui, Harada, Culberson, & Culberson, 2001). In addition, *Usnea diffracta* Vain was commonly used to alleviate pain and fever by the Japanese (Okuyama, Umeyama, Yamazaki, Kinoshita, & Yamamoto, 1995). Traditionally, lichen was often used as medicine for various diseases or disorders in Indonesia (Kusmoro, Noer, Jatnika, Permatasari, & Partasasmita, 2018). In the archipelago of Indonesia, *Usnea misaminensis* was traditionally prescribed to treat fever, diarrhea, dysentery, aphthous ulcers, abdominal distention inflammation, and hypertension (Nugraha et al., 2019). On the American continent, Parmeliaceae *Bryoria fremontii* (Tuck.) Brodo & D. Hawksw was used to treat warts (Milovanovic, Kostic, Milovanovic, Djordjevic, & Stojanović, 2018). Gastrointestinal problems were alleviated by taking fish broth mixed with lichen flakes, *Lobothallia alphoplaca* (Wahlenb.) Hafellner, *Umbilicaria muhlenbergii* (Ach.) Tuck., and *Xanthoparmelia mexicana* (Gyelnik) Hale (Yeash et al., 2017).

The need for a rational approach on medicine development has led to massive investigation of medicinally important plants used by mankind throughout history by mean medicinal plant bioprospecting. The technology advancement facilitated in investigation chemical constituents of medicinal plants which responsible for the traditional pharmacological claims. Lichens have been explored for their secondary metabolites and have been evaluated for their pharmacological properties against both non-infective and infective diseases.

10.5.1 Secondary Metabolites against Infectious Diseases

10.5.1.1 Lichen as Anti-Bacterial Agents

The growing prevalence of bacterial resistance toward antibiotics has been a serious threat, leading to a heightened urgency in the research on discovering new antibacterial agents. Natural products have been widely researched to find a new antibacterial compound, in which lichen is one of the unique organisms often studied as natural product source. Approximately 50% of lichens that have been researched showed antibacterial properties (Ranković & Kosanić, 2019). Lichen, as a crude extract or isolated metabolite, has been researched for its activity against various bacteria, including resistant bacterial strains. Some of the lichen's extracts and metabolites showed promising antibacterial activity, especially for the isolated lichen compounds (Table 10.1, Figure 10.3).

Lichen metabolites were extensively studied for their biological activity and their mechanism of action. Some of the probable antibacterial mechanisms were cell wall synthesis inhibition, protein synthesis inhibition, cell membrane alteration, nucleic acid synthesis inhibition, and antimetabolite activity (Ranković & Kosanić, 2019).

Usnic acid **16**, as the most widely studied compound, had several proposed antibacterial mechanisms. Pompilio *et al.* studied the effect of usnic acid **16** against Methicillin-Resistant *Staphylococcus aureus* (MRSA) to further understand its antibiofilm effects. The study implied that usnic acid **16** was able to affect the bacterial adhesion and biofilm formation of MRSA, along with decreasing the viability of preformed biofilm (Pompilio et al., 2013). Another study mentioned that usnic acid **16** disrupted the biofilms of *S. pyogenes,* causing pharyngitis, indicating that it could interfere with the thick bacterial exopolysaccharide layer (Nithyanand, Beema Shafreen, Muthamil, & Karutha Pandian, 2015). Usnic acid **16** was also found inhibiting the RNA and DNA synthesis of gram-positive bacteria, in which the mechanism was studied on *B. subtilis* and *S. aureus*. Slight inhibition of RNA was also observed in *V. harveyi*, a gram-negative bacterium. Inhibition of protein synthesis in gram-positive bacteria was delayed, suggesting that it was dependent on RNA inhibition. The report also suggested that DNA synthesis blockage was caused by inhibition on DNA replication elongation (Maciag-Dorszynska, Wegrzyn, & Guzow-Krzeminska, 2014).

TABLE 10.1
Reported Antibacterial Activity of Several Lichen Secondary Metabolites

Bioactive Compound	Source	Model	Result	Reference
1,5-dihydroxy2,4,6-trichloro-7-methylxanthone **17** (–)-usnic acid **18** Didymic acid **19** Condidymic acid **20** Squamatic acid **21** Thamnolic acid **22** Prasinic acid **23**	*Cladonia incrassata*	*S. aureus* (CIP 4.83)	Didymic acid **19**, condidymic acid **20**, and (–)-usnic acid **18** showed antibacterial activity against *S. aureus* (MIC 7.5 µg/mL)	Dieu et al. (2014)
Anziaic acid **24**	*Hypotrachyna sp.*	*B. subtilis* (ATCC 6633) *E. coli* (BAS3023)	Antibacterial activity was observed against *B. subtilis* with MIC 6 µg/mL, and against the membrane permeable-strain of *E. coli* with MIC 12 µg/mL.	Cheng et al. (2013)
(+)-usnic acid **16**	*Usnea steineri*	Clinical isolates: *S. epidermidis* (218 HCRP) *S. aureus* (220 HCRP) *S. haemolyticus* (243 HCRP)	Strong activity against resistant bacterial species *S. epidermidis* (MIC 3.12 µg/mL), *S. aureus* (MIC 12.5 µg/mL), and *S. haemolyticus* (MIC 12.5 µg/mL). Synergistic antibacterial effects were not observed in the combination of (+)-usnic acid **16** with Penicillin or Tetracycline.	Tozatti et al. (2016)
Atranorin **1** Diffractaic acid **25** Sphaerophorin **26** Fumarprotocetraric acid **27** Psoromic acid **28** Tenuiorin **29** Variolaric acid **30** Vicanicin **31**	Chilean lichen species	*S. aureus* ATCC43300 Clinical isolates: *S. aureus* (AQ004SA, AQ013SA, AQ014SA) *S. haemolyticus* (AQ007SH, AQ012SH) *S. warneri* (AQ011SW, AQ012SW)	Antibacterial activities against MRSA were observed from MIC 8 µg/mL for sphaerophorin **26** to MIC 1024 µg/mL for fumarprotocetraric acid **27** and psoromic acid **28**. The lowest MIC (8 µg/mL) was found in sphaerophorin **26** against *S. warneri*.	Celenza et al. (2013)
Divaricatic acid **32**	*Evernia mesomorpha*	*B. subtilis* (KCTC 2189) *M. luteus* (CCARM 0022) *S. epidermidis* (CCARM 3709) *S. mutans* (CCARM 0079) *S. aureus* (CCARM 0027) *E. faecium* (CCARM 5202)	Divaricatic acid **32** showed better antibacterial activity against gram-positive bacteria, with the smallest MIC (7±2 µg/mL) against *B. subtilis*, compared with Vancomycin at 0.78 µg/mL. Vancomycin was effective at MIC 0.78–25.0 µg/mL, while Divariatic acid **32** showed antibacterial activity at 7.0–64/0 µg/mL.	Oh et al. (2018)

(*Continued*)

TABLE 10.1 (*Continued*)
Reported Antibacterial Activity of Several Lichen Secondary Metabolites

Bioactive Compound	Source	Model	Result	Reference
9-(Hydroxymethyl)-1-methyldibenzofuran-3,7-diol **33** isostrepsilic acid **34** 3,7-Dihydroxy-1-methyldibenzofuran-2,9-dicarboxylic acid **35** Methyl β-orsellinate **36** methyl-haematommate **37** sakisacaulon A **38** methyl lobarin **39** esterified lobarin **40** methyl sakisacaulon **41**, esterified sakisacaulon **42** anhydro sakisacaulon A **43** Norlobaric acid **44** lobaric acid **45**	*Stereocaulon paschale*	*P. gingivalis* (ATCC 33277) *S. mutans* (ATCC 25175)	Compounds **33-37** did not show any antibacterial activity against P. *gingivalis* and *S. mutans*, whereas **38-45** had moderate or weak antibacterial effects. The lowest MIC was observed in esterified sakisacaulon A **38** against *S. mutans* (MIC 10 µg/mL)	Carpentier et al. (2017)
3,5-dihydroxybenzoic acid **46** 3,5-dihydroxybenzoate methyl **47** 3,5-dihydroxy-4-methylbenzoic acid **48** 3,5-dihydroxy-4-methoxylbenzoic acid **49** 3-hydroxyorcinol **50** atranol **51** methyl hematommate **52**	*Parmotrema crisuferum*	Antibiotic-resistant *E. faecium* *S. aureus* *A. baumannii*	Metabolites **46–52** at the concentration of 50 µg/mL were inactive against the tested strains.	Do et al. (2022)
Orcinol **53** 4-Chlororcinol **54**	*Ramalina implexa* Nyl. *Roccella phycopsis* Ach.	MRSA ATCC 43300 *E. faecalis* (ATCC 29212) *E. coli* (ATCC 25922) *A. baumannii* (BAA 747) *P. aeruginosa* (ATCC 27853) *K. pneumoniae* (BAA 1705)	Except for *A. baumannii* and *K. pneumoniae*, Orcinol **53** showed antibacterial activities against tested strains with MIC ranged from 9.37 to 300 µg/mL. 4-Chlororcinol **54** exhibited strong antibacterial activity against MRSA with MIC 1.17 µg/mL. 4-Chlororcinol **54** also inhibited the growth of *E. faecalis* at 75 µg/mL, *A. baumannii* and *K. pneumoniae* at 300 µg/mL.	Zorrilla et al. (2022)

(*Continued*)

TABLE 10.1 (*Continued*)
Reported Antibacterial Activity of Several Lichen Secondary Metabolites

Bioactive Compound	Source	Model	Result	Reference
Cryptothecin A **55** Methyl β-orcinol carboxylate **56** Ursolic acid **57**	*Cryptothecia faveomaculata Makhija & Patw*	Gram-negative bacteria *E. coli* *P. aeruginosa* Gram-positive bacteria *B. subtilis* *S. aureus* *A. niger* *F. oxysporum* *S. cerevisiae*	Methyl *β*-orcinol carboxylate **56** was the only compound showing significant antibacterial activity, its MIC against *S. aureus* was observed at 50 μg/mL.	N. T. Nguyen et al. (2021)
(+)-Roccellic acid **58** 4-*O*-Demethylbarbatic acid **59** Psoromic acid **28**	*Roccella phycopsis* *Ramalina siliquosa var. druidarum* *Squamarina cartilaginea*	*S. gordonii* (DL1) *P. gingivalis* (ATCC 33277)	The highest antibacterial activity against *S. gordonii* was found in Psoromic acid **28** with MIC 11.72 μg/mL, followed by 4-*O*-Demethylbarbatic acid **59** at 21.8 μg/mL and (+)-Rocellic acid **59** at 46.9 μg/mL. The lowest MIC against *P. gingivalis* was observed in Psoromic acid **28** with MIC 5.86 μg/mL, followed by 4-O-Demethylbarbatic acid **59** at 10.94 μg/mL and (+)-Rocellic acid **58** at 46.9 μg/mL.	Sweidan et al. (2017)
Parietin **60**	*Xanthoria parietina*	*S. aureus* (ATCC 13709) *E. faecalis* (ATCC 14428)	Gram (+) bacteria were strongly inhibited by Parietin **60**, especially against *S. aureus* and *E. faecalis* with MIC 7.8 μg/mL	Basile et al. (2015)
Protocetraric acid **61**	*Usnea albopunctata*	*M. smegmatis* *K. pneumoniae* *S. thypi*	Protocetraric acid **61** exhibits stronger antibacterial activity than the standard antibacterial agent Ciprofloxacin. The MIC value of Protocetraric acid **61** against *S. typhi* was 0.5 μg/mL, against *K. pneumoniae* was 1 μg/mL, and against *M. smegmatis* was 2 μg/mL.	Nishanth et al. (2015)
Lobaric acid **45** Lobastin **62**	*Stereocaulon alpinum*	*B. subtilis* *S. aureus*	Compounds **45** and **62** showed antibacterial activity against the tested bacteria. MIC of lobaric acid **45** was 88 and 44 μM against *B. subtilis* and 39.6 and 35.2 μM against *S. aureus*.	Bhattarai, Kim, Oh, and Yim (2013)

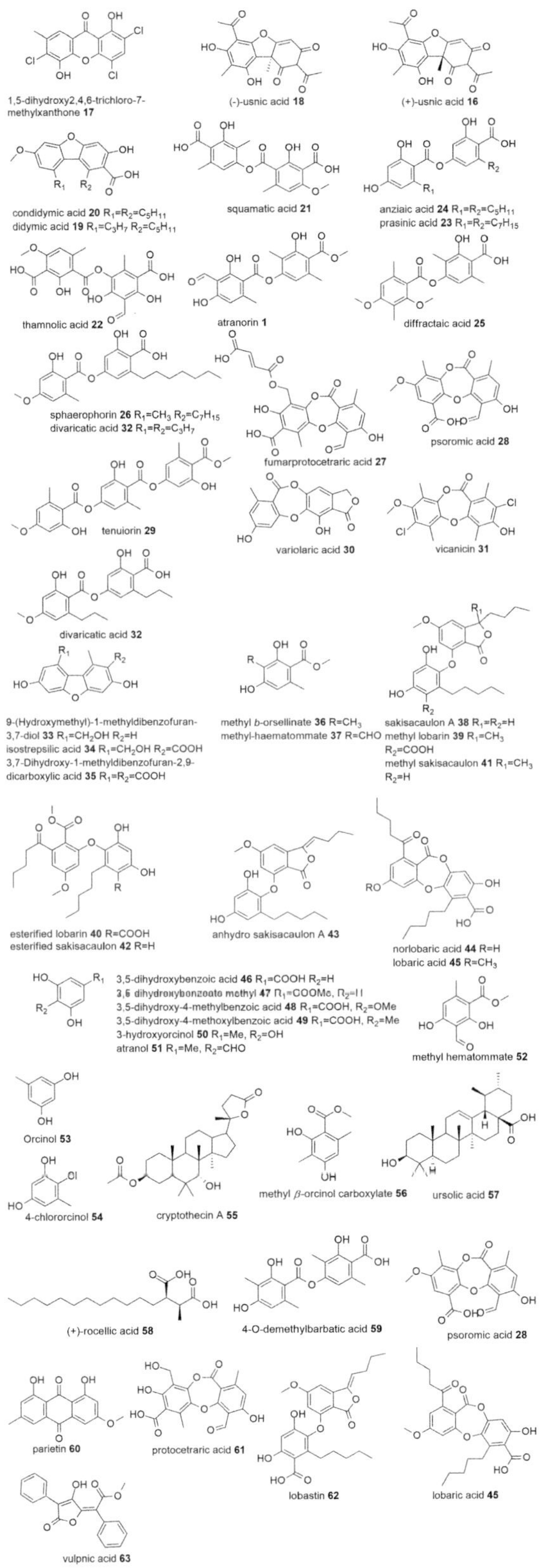

FIGURE 10.3 Lichen metabolites with anti-bacterial activities.

Anziaic acid **24**, a lichen depside, showed potential as a topoisomerase inhibitor for *Y. pestis* and *E. coli* topoisomerase I. Anziaic acid **24** was further discovered to act as a human topoisomerase IIα inhibitor, though it showed weak activity against human topoisomerase I. Activity against human topoisomerase IIα might suggest its potential as an anticancer agent, while activity against human topoisomerase I showed its capability as an antibacterial agent (Cheng et al., 2013). Vulpinic acid **63**, another lichen metabolite, showed antibacterial activity against MRSA with a slow kill rate. Vulpinic acid **63** could damage the cell membrane of MRSA, though it was unable to lyse the cell. It was also suggested that vulpinic acid **63** could interfere with the bacterial cell division process (Shrestha, Thompson, Robison, & St. Clair, 2016).

Considering its potent antibacterial activity, lichen's secondary metabolite was studied in combination with other antibiotics to assess their synergistic effects. According to Segatore (2012), the combination of usnic acid **16** and gentamicin showed synergistic antibacterial activity against clinical isolate strains of MRSA, while its combination with levofloxacin exhibited antagonism activity. Indifference in antibacterial activity was observed in combination with erythromycin (Segatore et al., 2012). Another research on the antibacterial activity of usnic acid **16** in combination with penicillin and tetracycline showed antagonism against *S. haemolyticus* and indifference against *S. aureus, S. epidermidis,* and *E. faecalis* (Tozatti et al., 2016). Antibacterial activity of usnic acid **16** in combination with oxacillin was studied against 28 clinical isolates of MRSA, in which indifference was observed in its activity (Goel, Kalra, Ponnan, Jayaweera, & Kumbukgolla, 2021).

10.5.1.2 Antifungal

Lichen's antifungal activity was often studied, though it was examined less frequently than its antibacterial activity. Research examining both antibacterial and antifungal activity of lichen's metabolites usually showed greater antibacterial activity than its antifungal activity. It was concluded that bacteria were more sensitive to the antimicrobial activity than the fungi, caused by the differences in the cell wall's composition and permeability, in which the cell wall of fungi had less permeability when compared to the cell wall of bacteria (Kosanić, Ranković, & Stanojković, 2012). Published antifungal activities of lichen secondary metabolites are summarized in Table 10.2 (Figure 10.4).

Girardot (2021) researched the effect of polyphenolic compounds from lichen as anti-maturation and anti-biofilm agents against *C. albicans* biofilms. It was found that the depside compounds (squamatic acid **21** and thamnolic acid **22**) exhibited anti-maturation activity at 24 h of contact. The strongest anti-maturation activity was observed on evernic acid **5** (IC_{50} 18 μM), although a longer contact time of 48 h was required. Anti-biofilm activity was evaluated on 24-h-old biofilm, showing evernic acid **5** as a promising anti-biofilm agent with minimal biofilm inhibitory concentration of 12.5 μM that was effective following 48 hours of contact. Evernic acid **5** was proposed as a promising inhibitory agent to counter the biofilm maturation phase and already formed *C. albicans* biofilm with a delayed activity (Girardot et al., 2021).

Vulpinic acid **63**, a lichen-derived secondary metabolite, was mostly found in *L. vulpina*. Vulpinic acid **63** was previously known for its activity on disrupting bacterial membrane and cell division (Shrestha et al., 2016). Kwon (2016) investigated vulpinic acid in cellular level for its antifungal activity. A chemogenomic study was carried out, and it was found that vulpinic acid **63** exerts its antifungal activity by affecting the fungal cell cycle, glycogen metabolism, and transcription and translation phases (Kwon et al., 2016).

Another research by Peralta (2017) proposed that the antifungal activity of usnic acid **16** was associated with oxidative and nitrosative stress on fungal biofilm. The study showed that usnic acid caused alteration in fungal biofilm, in which the biomass and the thickness were reduced. Observations on azole-sensitive and azole-resistant *C. albicans* strains displayed that intracellular and extracellular reactive oxygen species were significantly increased in the presence of usnic acid **16**. These oxidative metabolites might play a role in the inhibition of biofilm by usnic acid **16**. It was then concluded that usnic acid could disrupt the prooxidant-antioxidant balance, causing cell damage and cell death (Peralta, da Silva, Ortega, Cabrera, & Paraje, 2017).

TABLE 10.2
Reported Antifungal Activity of Several Lichen Secondary Metabolites

Bioactive Compound	Source	Model	Result	Reference
Usone **64** Isousone **65**	*Usnea longissima*	*Trichophyton rubrum* spp.	Usone **64** and Isousone **65** showed weak antifungal activity against *Trichophyton rubrum* spp. (MIC 41 µM)	Yu et al. (2016)
Protolichesterinic acid **66**	*Usnea albopunctata*	*A. flavus* *C. albicans* *C. tropicalis* *T. rubrum*	Protolichesterinic acid **66** exhibited good inhibitory effect on tested fungi. Better antifungal activity compared to amphotericin B was observed against *T. rubrum* (0.12 µg/mL).	Sasidharan et al. (2014)
Protocetraric acid **61**	*Usnea albopunctata*	*A. flavus* *C. albicans* *C. tropicalis* *C. glabrata* *C. gastri* *T. rubrum*	Protocetraric acid **61** displayed antifungal activity toward all tested fungi, with better activity than amphotericin B against *T. rubrum* (1 mg/mL)	Nishanth et al. (2015)
Evernic acid **5** Squamatic acid **21** Thamnolic acid **22** Atranorin **1** Gyrophoric acid **6** Norstictic acid **67** Salazinic acid **14** Stictic acid **68** Physodic acid **69** 3-hydroxyphysodic acid **70** (+)-usnic acid **16**	*Evernia prunastri* *Usnea florida* *Cladonia squamosa* *Pseudevernia furfuracea* *Lasallia pustullata* *Pleurosticta acetabulum* *Parmelia saxatilis* *Xanthoparmelia conspersa* *Hypogymnia tubulosa*	*C. albicans*	Compounds **5, 21, 22** exhibited inhibition on biofilm maturation after 48 h of incubation (MMIC <35 µM). Antibiofilm activity was tested on 24-h-old biofilm, with evernic acid **5** showing positive result after 48 h of contact ($MBIC_{50}$ < 40 µM).	Girardot et al. (2021)
Atranorin **1** Fumarprotocetraric acid **27**	*Cladonia furcata* *C. pyxidata* *C. rangiferina*	*A. flavus* *A. fumigatus* *C. albicans* *P. purpurescens* *P. verrucosum*	Atranorin **1** and fumarproto acid **27** exhibited moderate inhibition against all the tested fungi compared to ketoconazole, the standard.	Kosanić, Ranković, Stanojković, Rančić, and Manojlović, (2014)

(*Continued*)

TABLE 10.2 (*Continued*)
Reported Antifungal Activity of Several Lichen Secondary Metabolites

Bioactive Compound	Source	Model	Result	Reference
β-orsellinate **71** Methyl orsellinate **72** Ethyl orsellinate **73** *n*-butyl orsellinate **74**	*Parmotrema tinctorum*	*Colletotrichum* sp. *F. oxysporum* f. sp. *sesami* *P. oryzae*	Compounds **71-74** displayed inhibitions against all the tested fungi at 100 µg/mL, with methyl *β*-orsellinate (**72**) the showing strongest antifungal activity.	T. T. Nguyen et al. (2020)
Physodic acid **69** (+)-Usnic acid **16** Atranorin **1**	*Hypogymnia physodes*	*A. flavus* *A. fumigatus* *C. albicans* *P. purpurescens* *P. verrucosum*	Physodic acid **69**, usnic acid **16**, and atranorin exhibit stronger antifungal activity than ketoconazole. The strongest activity was found in usnic acid with MIC 0.125–0.5 µg/mL against tested fungi.	Ranković, Kosanić, Manojlović, Rančić, and Stanojković (2014)
Atranorin **1** (+)-Usnic acid **16** 2-Hydroxy-4-methoxy 3,6-dimethylbenzoicacid **75**	*Parmelia erumpens Kurok.*	*A. flavus* *C. albicans* *F. oxysporum* *R. solani* *P. expansum* *C. gastricus* *C. tropicalis* *T. rubrum*	All tested compounds showed good inhibition against tested fungi. The strongest activity was displayed by 2-hydroxy-4-methoxy-3,6-dimethylbenzoic acid **75,** in which the compound exhibited better antifungal activity than amphotericin B as the standard fungicide. The best antifungal activity was recorded against *T. rubrum* (MIC 0.12 µg/mL).	Shanmugam, Ponnusamy, Balasubramanian, Nagarajan, and Ganesan (2017)
(+)-Usnic acid **16** Ethyl everninate **76** Scrobiculin **77** Methyl 2,6-dihydroxy-4-methylbenzoate **78** 4-[(2-hydroxy-4-methoxy-6-propylbenzoyl) oxy]-2-methoxy-6-propylbenzoic acid **79**	*Ramalina leiodea* (Nyl.) Nyl.	*C. albicans*	The lichen metabolites showed moderate antifungal activity against *C. albicans*. The strongest activity was displayed by Scrobiculin**77** (MIC 0.01 µg/mL) and 4-[(2-hydroxy-4-methoxy-6-propylbenzoyl)oxy]-2-methoxy-6-propylbenzoic acid **79** (MIC 0.03 µg/mL), in which the antifungal activity was better than Streptomycin as the control.	Bharadwaj and Sastry (2019)
Barbatolic acid **80**	*Bryoria capillaris*	C. krusei F. solani F. moniliforme A. flavus A. parasiticus	Barbatolic acid 80 inhibited the sporulation of A. fumigatus and S. rolfsii. The compound displayed antifungal activity against C. krusei (MIC 50 µg/mL).	Sarıözlü, Yilmaz, Candan, and Tay (2016)

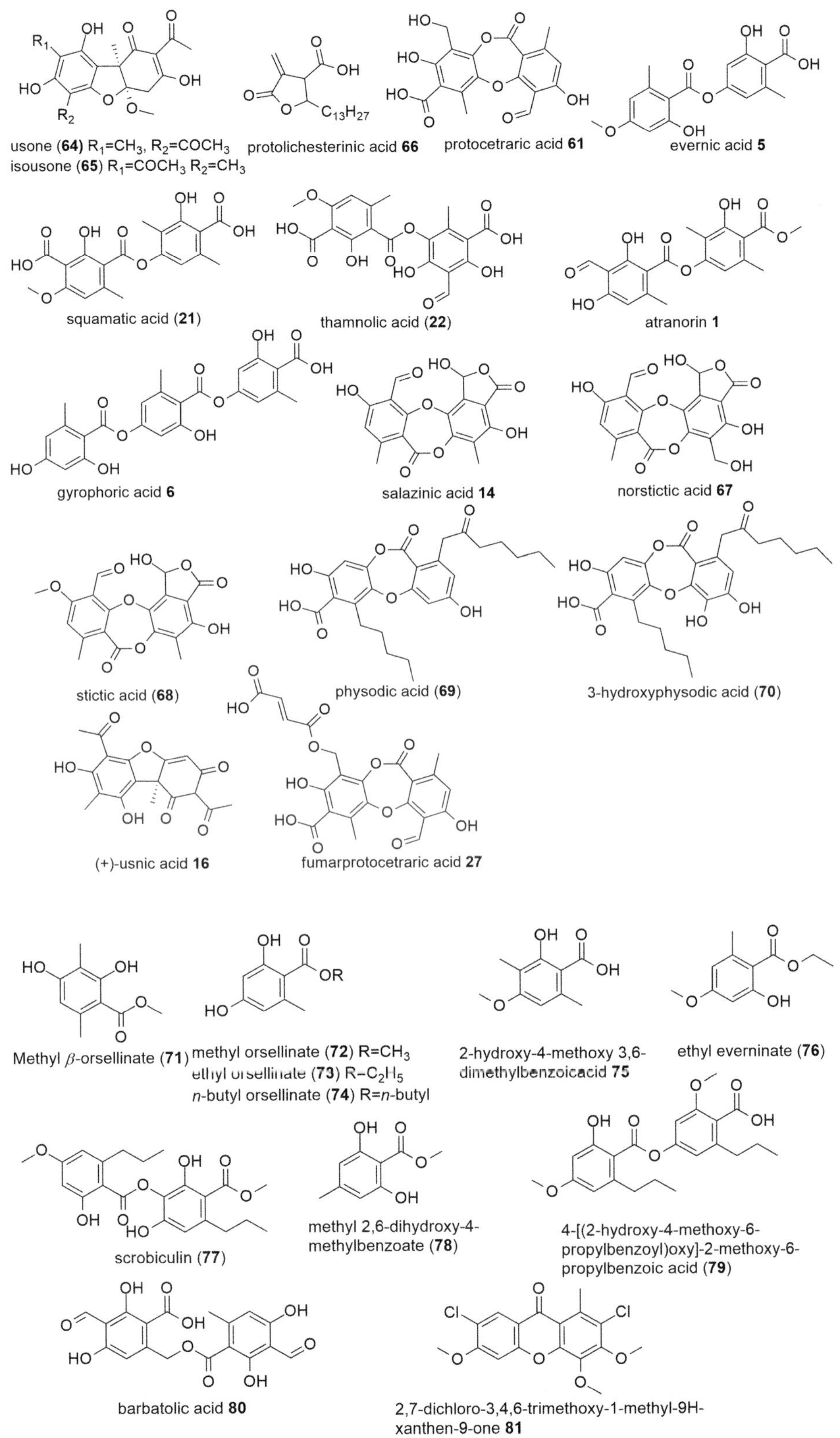

FIGURE 10.4 Lichen metabolites with anti-fungal activities.

Modification of lichen secondary metabolites has become a prominent area of research for finding more effective antifungal agents. The data published by Resende et al. (2018) mentioned that chlorinated xanthones could be a potent antifungal compound. Several chlorination methods were conducted, and the highest antifungal activity was found on 2,7-dichloro-3,4,6-trimethoxy-1-methyl-9H-xanthen-9-one **81**. The compound exhibited potent antifungal activity against clinical strains of *T. rubrum, M. canis, and E. floccosum* with minimum inhibitory concentration (MIC) of 4–8 μg/mL, it also displayed synergy with fluconazole (Resende et al., 2018). Another study carried out by Mallavadhani et al. (2018) was done on stereoselective synthesis of montagnetol, a secondary metabolite of *Rocella montagnei*. The derivative of (+)-montagnetol **28**, especially its congeners (2S, 3S), showed good antifungal activity against *C. albicans* with MIC 0.062 μg/mL (Mallavadhani, Boddu, Rathod, & Reddy Setty, 2018).

10.5.1.3 Lichen as AntiParasitic Agents

Lichens have been screened for their antiparasitic potency against arrays of pathogenic parasites, including plasmodium, nematode, trypanosome, and helminth. Investigation on Japanese origin *Evernia prunastri* L. successfully isolated eight compounds diethyl phthalate **82**, *β*-orcinol carboxylate **56**, ethyl everninate **76**, ethyl hematommate **83**, (+)-usnic acid **16**, everninic acid **84**, ethyl chlorohematommate **85,** friedelin **86** (Figure 10.5) (Ahad et al., 1991). Nematocidal evaluation of compounds revealed compounds with free hydroxyl attached to C4 (*β*-orcinol carboxylate **56**, ethyl hematommate **83**, and ethyl chlorohematommate **84**) exerted significant activity against common zoonosis causal, *Toxocara canis*, with minimum lethal concentration (MLC) values of 410, 100, and 110 μM, respectively. On the other hand, ethyl everninate **76** and everninic acid **84** were inactive. Further investigation suggested a substituent introduction at C2 was favorable for the nematocidal activity (Ahad et al., 1991).

In addition, nematocidal was also explored using the nematode *Meloydogine incognita*, where (+) usnic acid **16**, methyl orsellinate **72**, orcinol **53**, (+) montagnetol **87**, 4–chlororcinol **54** isolated from Italian origins *Ramalina implexa* and *Roccella phycopsis* indicated nematocidal activities with percentage of mortality values of 88.7, 6.2, 89.9, 5.8, and 97.1, respectively (Figure 10.5). An esterified orselinic acid with hydroxylated alkyl indicated an improvement of nematocidal activity, as shown by (+) montagnetol **87** (Figure 10.5) (Zorrilla et al., 2022).

A study on anti-trypanosomal activity against *Trypanosoma cruzi* reported lichens secondary metabolites of Chilean *Pseudocyphellaria coriifolia*, hopan-22-ol **88**, hopan-16b,22-diol **89**, hopan-6a,7b,22-triol **90** to possess a strong activity with percentage of growth inhibition (PGI) values of 80%, 78%, and 84%, respectively while the protozoa were insensitive against methyl evernate **91** and tenuiorin **29** at concentration of of 50 μM (Figure 10.5). The evaluation was conducted after 10 days of incubation with a positive control nifurtimos indicated PGI value of 100% at concertation of 50 μM (Fritis et al., 2013). An investigation on Indian-origin *Cetraria islandica* successfully isolated four lichen acids, protolichesterinic acid **66**, lichesterinic acid **92**, protocetraric acid **61,** and fumarprotocetraric acid **27**. Protolichesterinic acid **65**, lichesterinic acid **92** indicated significant activity against *Trypanosoma brucei* with MIC values of 12.5 and 6.30 μM, respectively, while protocetraric acid **61** and fumarprotocetraric acid **27** were inactive. The lichesterinic acid **92** possessed a stronger trypanosidal activity compared to a standard drug, suramin with an MIC value of 10 μM (Figure 10.5) (Igoli, Gray, Clements, Kantheti, & Singla, 2014).

An experiment on the anti-malarial of lichen compounds showed evernic acid **5**, vulpnic acid **63**, psoromic acid **28**, (+)-usnic acid **16** isolated from *Usnea* species to have varying activities against several *Plasmodium falciparum* strains. The usnic acid **16** indicated no activity against *P. falciparum* FabI, *P. falciparum* FabG, and *P. falciparum* FabZ with an IC_{50} value of >200 μM on all strains. On the other hand, *P. falciparum* FabI and *P. falciparum* FabZ were sensitive against evernic acid **5** with IC_{50} values of 36.1 and 10.7 μM, respectively (Lauinger et al., 2013).

Lichens also constituted molluscicide agents which evernic acid **5** isolated from the Brazil origin *Ramalina aspera,* indicated a strong activity against *Biomphalaria glabrata* with 100%

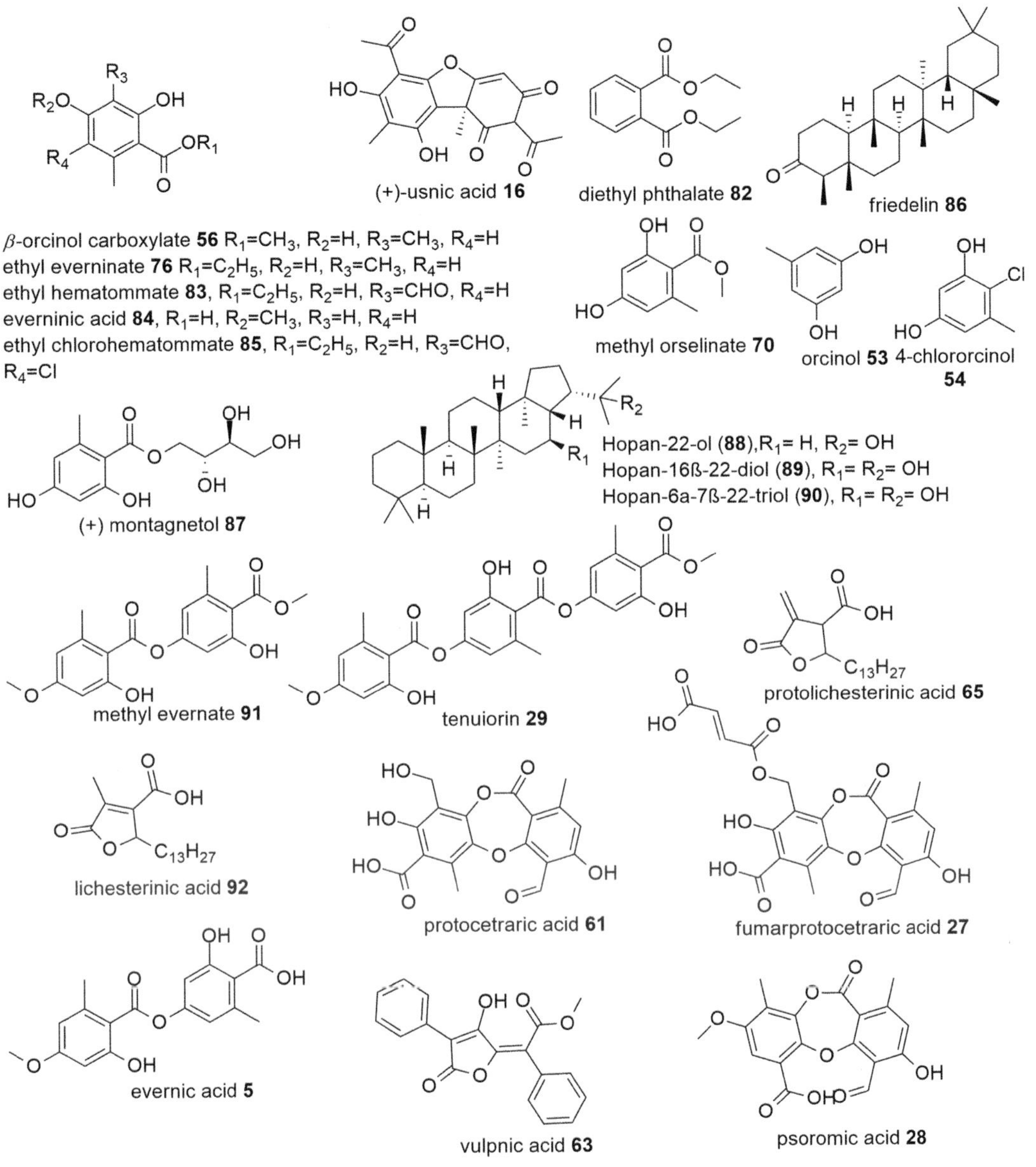

FIGURE 10.5 Lichen secondary metabolites with anti-parasitic activities.

molluscicidal activity at concentration of 5.5 μg/mL (Figure 10.5). Further investigation revealed the acid to have lethal concentration value (LC_{50}) of 3.58 μg/mL. The same acid was also tested against the parasite, *Schistosoma mansoni* and indicated lethal effect at concentration of 0.5 μg/mL causing 100% mortality (Silva et al., 2018).

10.5.1.4 Lichen Metabolites as Antivirus

Anti-viral bioprospecting conducted also on lichens, revealing several notable compounds being evaluated for their anti-viral properties against arrays of pathogenic viruses. Psoromic acid **28**, a common depsidone constituent of lichen detected in several genera, i.e., *Usnea, Psoroma* and *Alectoria,* was investigated for its anti-viral activity against herpes simplex virus type 1 (HSV-1)

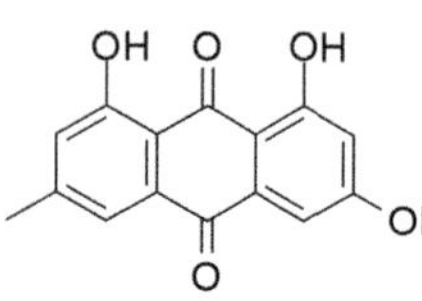

emodin **4**

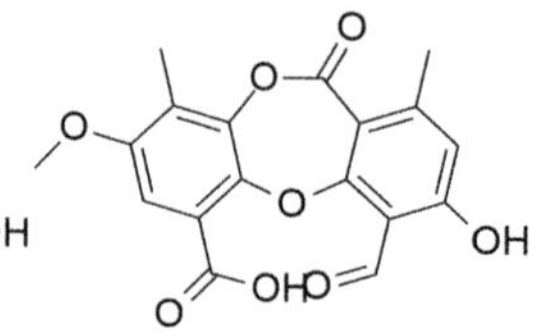

psoromic acid **28**

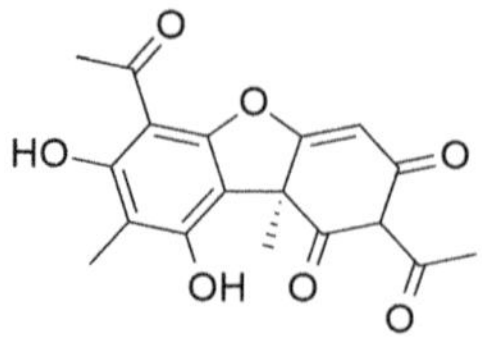

(-)-usnic acid **18**

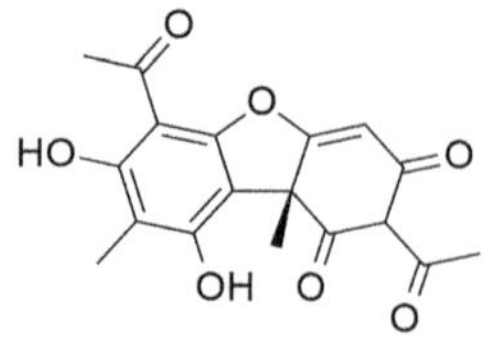

(+)-usnic acid **16**

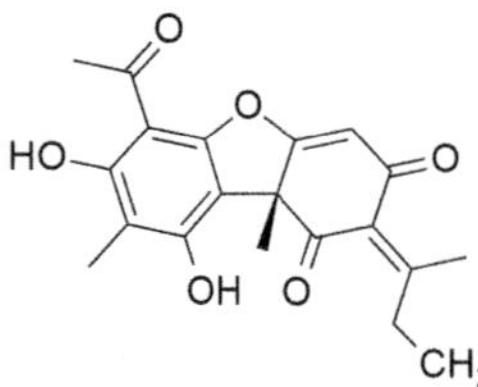

(*R*,*Z*)-6-acetyl-7,9-dihydroxy-8,9b-dimethyl-2-(5,5,6,6-tetrafluorohexan-2-ylidene)dibenzo[*b*,*d*]furan-1,3(2*H*,9b*H*)-dione **93**

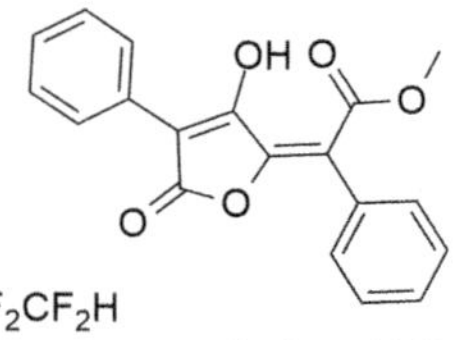

vulpnic acid **63**

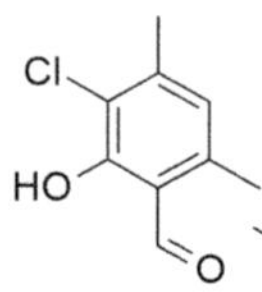

chloroatranol **94**

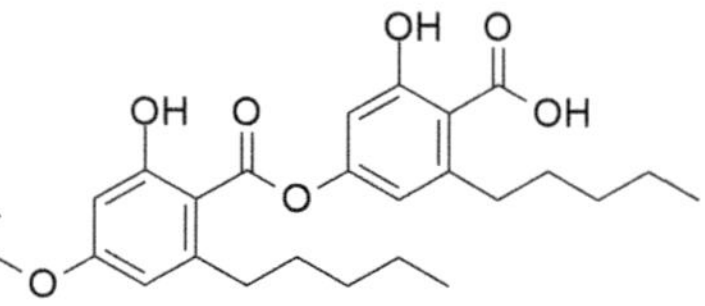

perlatolic acid **95**

FIGURE 10.6 Lichen secondary metabolites with anti-viral activities.

and type 2 (HSV-2) (Figure 10.6). The acid significantly inhibited HSV-1 and HSV-2 replication with respected IC_{50} values of 1.9 and 2.7 μM and respected index values of 163.2, 114.8. The acid **28** possessed comparable activity against a standard drug, acyclovir with IC_{50} values of 2.6 and 2.7 μM, respectively. Further study revealed the acids inhibit the virus replication by inhibiting against HSV-1 DNA polymerase (Hassan, Šudomová, Berchová-Bímová, Šmejkal, & Echeverría, 2019).

The emerging cases of influenza viral infection caused annual epidemics across countries due to traveling. This virus adapts to new environments and threats through mutation, in which anti-viral investigation screened numerous chemical compounds, including usnic acid **16**, a lichen dibenzofuran compound. The usnic acid and derivatives were studied for their anti-viral properties against pandemic influenza virus A(H1N1)pdm09 in Madin-Darby canine kidney (MDCK) cells. The antivirus assay revealed (+) usnic acid **16** was less potent than (–)-usnic acid **18** with ED50 values of 51.7 and 14.5 μM, respectively (positive control ribavirin produced an ED50 value of 24.6 μM). In addition, derivatization was able to produce compounds with higher efficacy with an ED50 value of 1.6 μM, (R,Z)-6-acetyl-7,9-dihydroxy-8,9b-dimethyl-2-(5,5,6,6-tetrafluorohexan-2-ylidene)dibenzo[b,d]furan-1,3(2H,9bH)-dione **93**. Despite the (+) usnic acid **16** was ineffective against the virus, the value was reversed after the chemical modification (Sokolov et al., 2012).

Another interesting study was lichen metabolome screening against human coronavirus HCoV-229E in Huh-7 and Huh-7/TMPRSS2 cells. The investigation indicated chloroatranol **94**, emodin **4**, perlatolic acid **95,** and vulpinic acid **63** to exert potent activity when was added during the replication stage with respective IC_{50} values of 68.86, 59.25, 16.42, and 14.58 μM. Further investigation on SARS-CoV-2 in a Vero-81-derived clone with a GFP reporter probe, perlatolic acid **95** provides the highest activity (Desmarets et al., 2023).

10.5.2 Secondary Metabolites against Non-infectious Diseases

Phytochemical studies revealed several chemicals with pharmacological effects related to non-infectious diseases. Successful exploration on lichen metabolites evaluation for metabolic diseases is summarized below.

10.5.2.1 Lichen as Source for Anti-Oxidant

Oxidative stress has led reactive radical species existent in the cell of living organism, in which prolonged exposure to these species caused functional degradation of living cell system. The situation was correlated to several metabolic diseases depending on organs being affected, including uncontrollable cell mitosis due to genetic mutation or biosynthesis alteration. In the case of endogenous antioxidant was unable to scavenge the radicals, an exogenous antioxidant intake was necessary. Several studies on antioxidant components from plants have been conducted, including antioxidant evaluation of lichen secondary metabolites. Atranorin **1**, didepside constituent of lichen, was measured for its antioxidant activity based on total reactive antioxidant potential (TRAP) and total antioxidant reactivity (TAR) methods (Figure 10.7). The didepside indicated significant antioxidant activity at concentration of 100 µg/mL (Melo et al., 2011).

In vivo antioxidant evaluation on Albino Wistar rats of lichen phenolic, diffractaic acid **25** (Figure 10.7), isolated from *Usnea longissimi*, indicated reduction of antioxidant enzyme activities, catalase (CAT), superoxide dismutase (SOD), glutathione peroxidase (GPx), reduced glutathione (GSH), and lipid peroxidation (LPO) with values of 57.78, 115.29, 5.36, 2.33, and 8.51 nmol/mg tissue, respectively, at 25 mg/kg dose (healthy rats indicated respected values of 73.10, 112.10, 8.62, 3.34, and 12.42 nmol/mg tissue) (Bayir et al., 2006). On the other hand, the acid showed reverse activity of inducible nitric oxide synthases (iNOS) and constitutive nitric oxide synthases (cNOS) of the acid administration at 100 mg/kg dose with values of 0.08, 1.90 nmol/min/mg tissue, respectively (respected values on healthy rat group were 0.16 and 2.52 nmol/min/mg tissue) (Bayir et al., 2006). The study suggested enhancing the antioxidant defense system while reducing neutrophil infiltration.

Major constituents of *Evernia prunastri* and *Pseudoevernia furfuraceae*, evernic acid **5** and physodic acid **69**, were previously isolated, and their antioxidant activities were evaluated (Figure 10.7). Physodic acid **69** showed stronger scavenging capability based on DPPH and superoxide anion

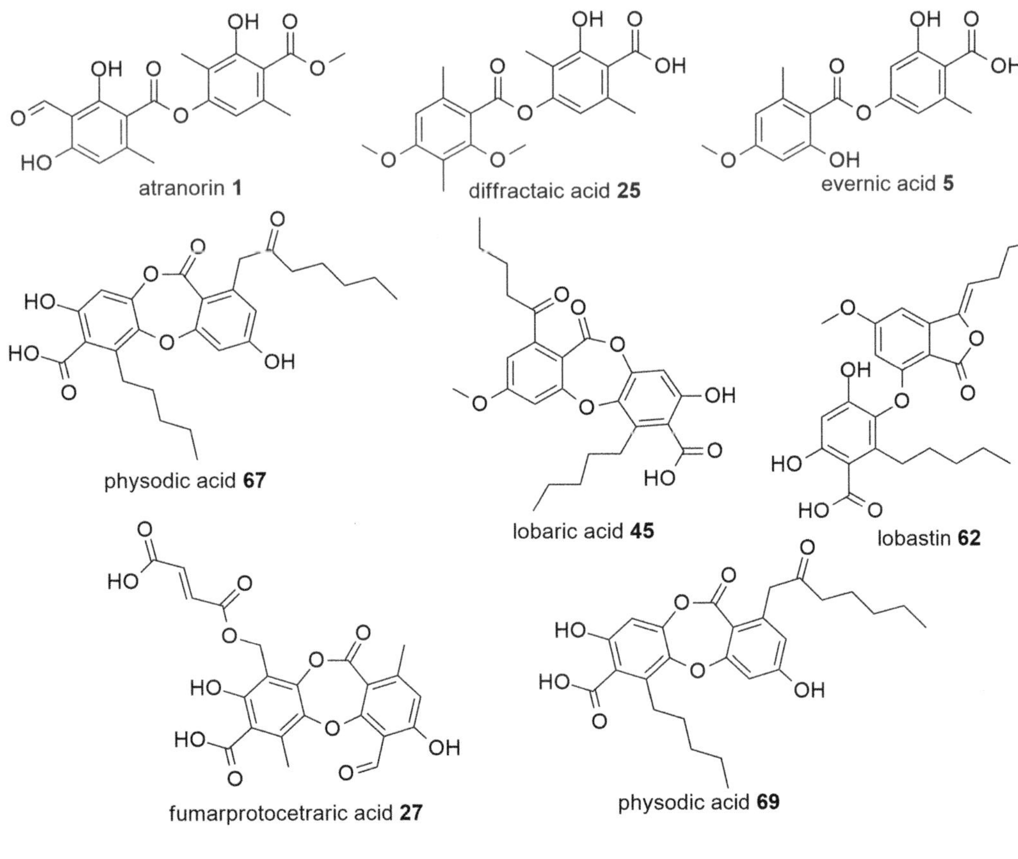

FIGURE 10.7 Lichen metabolites with anti-oxidant activity.

protocol with IC_{50} values of 69.110 and 118.170 μg/mL, respectively. On the other hand, evernic acid **5** indicated respected IC_{50} value scavenging values of 322.44 and 561.76 μg/mL (positive control, vitamin C IC_{50} value of 6.42 and 115.61 μg/mL) (Kosanić, Manojlović, Janković, Stanojković, & Ranković, 2013).

The Antarctic is a home for resilient organisms coping an extreme environment such as cold temperatures, dryness and strong UV light radiation. These conditions require the organism to adapt by altering their biosynthesis and producing unique secondary metabolites, including antioxidant to conform the abiotic stress. A study on the Antarctic lichen, *Stereocaulon alpinum*, revealed antioxidants activity of lichen constituents, lobaric acid **46**, lobastin **62** (Figure 10.7). The depsidone lobaric acid did not show DPPH scavenging activity while depside lobastin **62** indicated medium antioxidant against DPPH with an IC_{50} value of 70.2 μM (positive control BHA, IC_{50} value 26.6 μM) (Bhattarai, Kim, Oh, & Yim, 2013). Nevertheless, another study on depsidone fumarprotocetraric acid **27** of *Cladonia verticillaris* Roddi showed a strong antioxidant activity based on 2,2-diphenyl-1-picrylhydrazyl (DPPH), superoxide anion scavenging (SAS), and reducing power assays (White et al., 2014).

10.5.2.2 Lichen Metabolites as Anti-Inflammatory Agents

Lichen is comprised of unique secondary metabolites with the majority present as depside and depsidone with varying hydroxylated molecular structures. Previous studies on non-fused ring-containing depsides proposed to have significant anti-inflammatory activity by mean cyclooxygenase-1 (COX-1) inhibition (Studzińska-Sroka, 2018). Study on Oak moss (*Evernia prunastri*) revealed evernic acid **5** as major constituent and based on the primary astrocytes experiment, the acid indicated a significant downregulate activity on inflammatory cytokines (IL-1β, IL-6, and TNF-α) and CCL2 (a chemokine) which suggested the anti-inflammatory effect was due to blocking of the NF-κB signaling pathway (Figure 10.8) (Lee et al., 2021). Another lichen metabolite, atraric acid **96** isolated from *Heterodermia hypoleuca* indicated anti-inflammatory based on lipopolysaccharide-stimulated RAW264.7 cells. The experiment indicated atraric acid **96** to downregulates the expression of phosphorylated IκB, extracellular signal-regulated kinases (ERK), and nuclear factor kappa B (NFκB) signaling pathways (Figure 10.8). The anti-inflammatory effect was confirmed through an animal model experiment (Mun et al., 2020).

10.5.2.3 Lichen Metabolites as Analgesic and Antipyretic

Phytochemical investigation on *Usnea diffracta* growing in Japan successfully isolated well-known lichen constituents, (+)usnic acid **16** and diffractaic acid **25** (Figure 10.9). An animal model

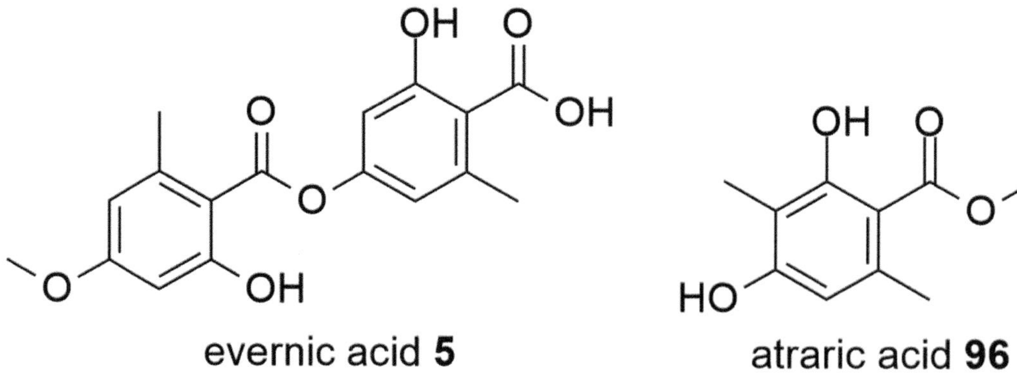

FIGURE 10.8 Lichen metabolites with anti-inflammatory activities.

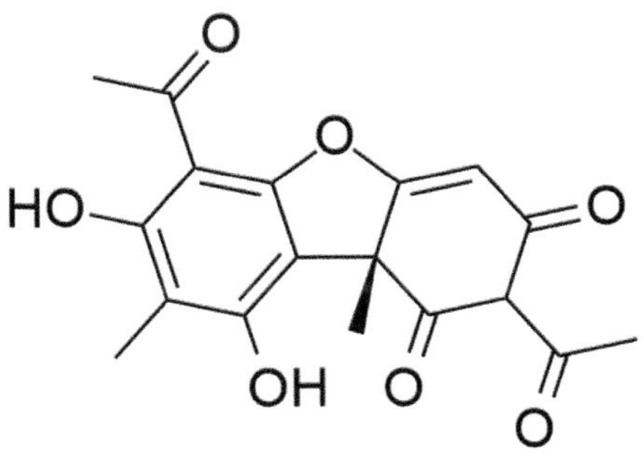

(+)-usnic acid **16**

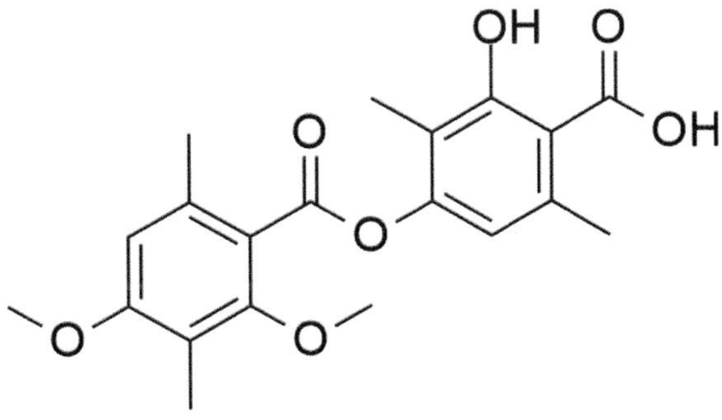

diffractaic acid **25**

FIGURE 10.9 Lichen metabolites with analgesic and antipyretic potency.

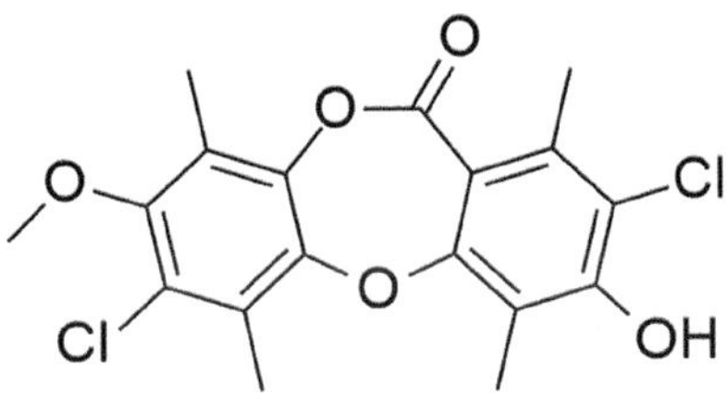

vicanicin **31**

FIGURE 10.10 Lichen metabolites with antidiabetic properties.

experiment revealed usnic acid **16** to possess exerted long-lasting analgesic activity compared to diffractaic acid **25** based on the acetic acid-induced writhing test at the same dose (100 mg/kg bw). On the other hand, diffractaic acid **25** indicated significant hypothermic effect at dose of 200 mg/kg bw with no effect was observed on usnic acid **16** administration (Okuyama et al., 1995).

10.5.2.4 Lichen Metabolites against Diabetes

For example, investigation on Indonesian lichen *Teloschistes flavicansanti* successfully isolated vicanicin **31** with anti-diabetic activities based on α-glucosidase inhibitory activity with an IC_{50} value of 197.04 μg/mL, which is less efficacy than its crude extract (54.05 μg/mL) (Figure 10.10) (Maulidiyah et al., 2020).

10.5.2.5 Lichen as Anti-Cancer Agents

Lichens have been acknowledged for their cytotoxicity activities with many studies have successfully isolated and purified individual cytotoxic compounds. These secondary metabolites are in the form of depsides, depsidones and nitrogen-containing lichen compounds. An intensive study on Indonesian lichens, *Physcia millegrana, Parmelia dilatata*, and *Parmelia aurulenta*, obtained nine depsides and derivatives, evemic acid **97,** methyl 3-hydroxy orsellinate **98,** divaricatic acid **32,** evernic acid **5,** methyl 4-formyl-2,3-dihydroxy-6-methylbenzoate **99,** methyl orsellinate **72,** 2,4-dihydroxy-6-pentylbenzoate **100,** 2-heptyl-6-hydroxy-4-methoxybenzoic acid **101,** stenosporic acid **102** (Figure 10.11) (Nugraha et al., 2020). The 3-(4, 5-dimethylthiazol-2-yl)-2,5-diphenyltetrazolium bromide (MTT) test were conducted on HepG2, A549, HL-60 cell lines, showing A549 as the most

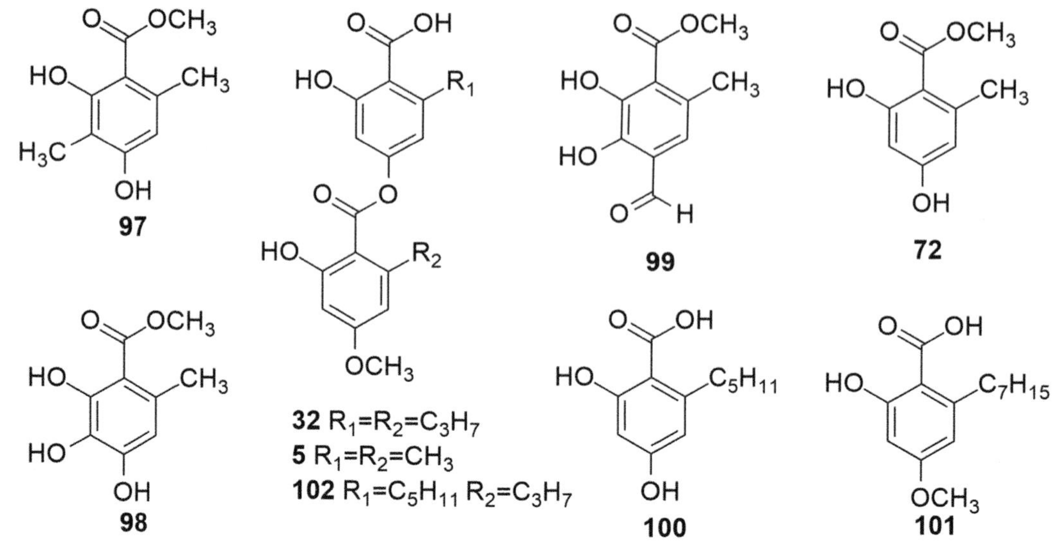

FIGURE 10.11 Lichen metabolites of *Physcia millegrana, Parmelia dilatata*, and *Parmelia aurulenta* with anticancer properties.

insensitive cell lines with compounds **97, 98, 99, 72, 100,** and **101** having IC_{50} values more than 500 µM (Nugraha et al., 2020). Cytotoxicity tests showed di-depside compounds **32, 5** and **102** to possess significant activities against, HepG2, A549, HL-60 cell lines with respected specific IC_{50} values for **31**: 15, 72, and 6 µM; compound **5**, 68, 180, and 197 µM; compound **102**: 15, 28, and 6 µM (Nugraha et al., 2020). Further structure activity relationship analysis suggested an extended alkyl group at C6 contributed to cytotoxicity increase which related to an increase of lipophilicity led to an increase of cellular uptake. In addition, free acid at C1 and methyl group at C3 were unlikely to evoke on depside cytotoxicity (Nugraha et al., 2020).

Prominent antineoplastic lichen metabolomes were obtained from *Xanthoparmelia scabrosa* (Taylor) Hale grown on a coastal cliff face southeast of Batemans Bay, New South Wales, Australia. Five epidithiopiperazinediones were obtained, scabrosin diacetate **103**, scabrosin acetate butyrate **104**, scabrosin dibutyrate **105**, scabrosin acetate hexanoate **106,** and scabrosin butyrate hexanoate **107** in which scabrosin diacetate **103**, and scabrosin acetate butyrate **104** were highly toxic against the P815 mastocytoma cancer cell line with IC_{50} values of 0.56 and 0.50 µM, respectively (Figure 10.12) (standard inhibitor gliotoxin, IC_{50} 2.9 µM) (Ernst-Russell et al., 1999).

Investigation on *Roccella montagnei* collected from tree barks in Tuy Phong district, Binh Thuan province, Vietnam, isolated orsellinylmontagnetols A **108**, orsellinylmontagnetols B **109**, orsellinylmontagnetols C **110**, orsellinylmontagnetols D **111**, and hydrofurylmontagnetol **112**. These compounds were assessed for antineoplastic activity which the results indicated that only orsellinylmontagnetol A **108** showed potential activity against MCF-7 with an IC_{50} value of 68.39 ± 3.46 µM, but it showed no effect on NCI-H460 and HeLa cells. Other compounds were found to be inactive (Figure 10.13) (Duong et al., 2017). Another study on Vietnamese lichen *Parmotrema praesorediosum* (Nyl.) Hale (collected from Nam Cat Tien National Forest Reserve and Intermediate zones in Nam Cat Tien village, Tan Phu district, Dong Nai province, Vietnam) resulted in five compounds being isolated, (+)-vinapraesorediosic acid A **113**, (+)-6-methyl vinapraesorediosate A **114**, (+)-vinapraesorediosic acid B **115**, (+)-6-methyl praesorediosate **116**, (+)-vinapraesorediosic acid C **117** (Figure 10.13). The MTT assay showed (+)-vinapraesorediosic

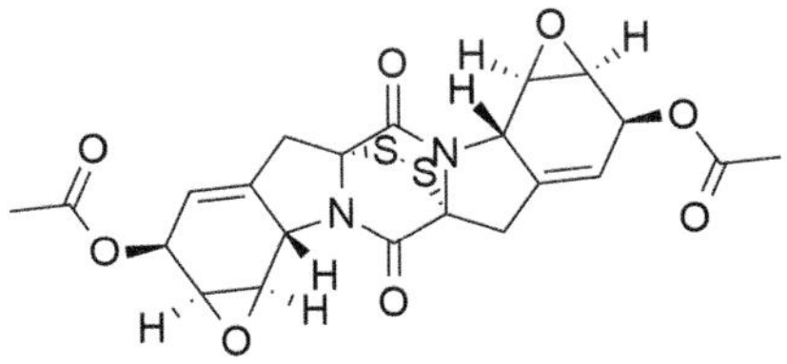

scabrosin diacetate **103**

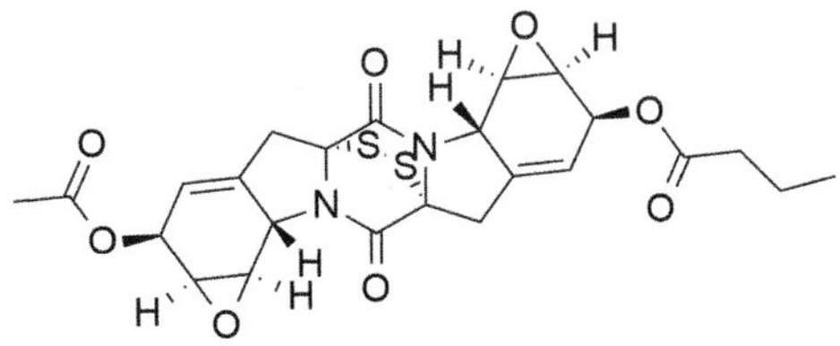

scabrosin acetate butyrate **104**

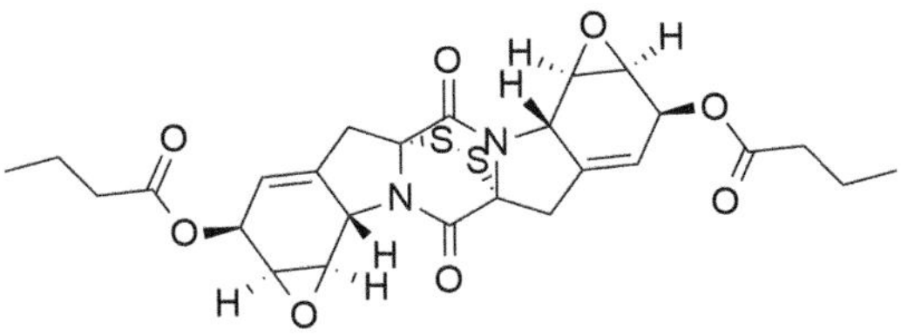

scabrosin dibutyrate **105**

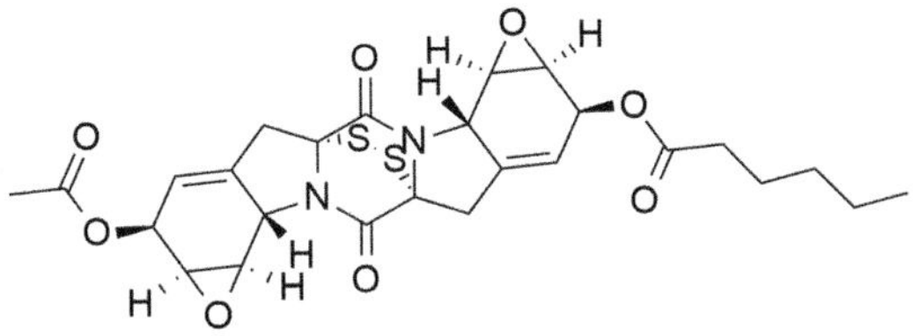

scabrosin acetate hexanoate **106**

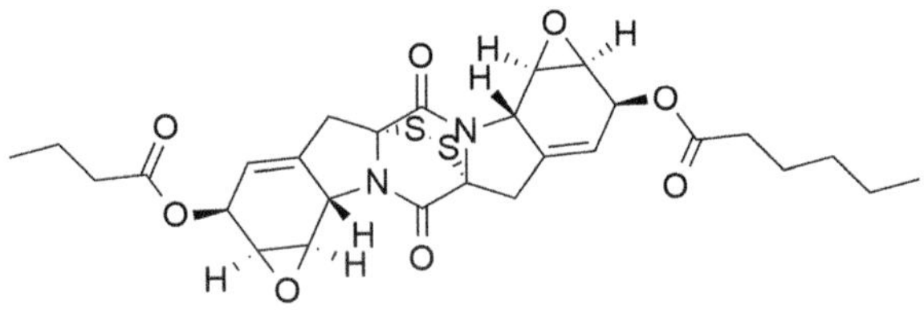

scabrosin butyrate hexanoate **107**

FIGURE 10.12 Lichen metabolites of *Xanthoparmelia scabrosa* with anticancer properties.

acid A **113** exhibited moderate potency against MCF-7 with the inhibition of 52.5%. Nevertheless, other compounds exhibited no cytotoxicity against the cancer cells (Huynh et al., 2016). Seven depsidones were isolated from *Parmotrema tsavoense* (Krog & Swinscow) collected from the southern part of the central of Vietnam, parmosidone A **118**, parmosidone B **119**, parmosidone C **120**, parmosidone D **121**, parmosidone E **122**, parmoether A **123**, parmoether B **124** (Figure 10.13). The compounds showed weak or no activity toward the cancer cells, except Parmoether B **124** that showed moderate activities, with the IC_{50} values (μM) of 44.72±7.33 (HepG2), 36.48±2.34 (NCI-H460), and 22.32±6.51 (MCF-7) (Duong, Chavasiri, Boustie, & Nguyen, 2015).

Anti-cancer bioprospecting on fungi has become scientist's interest, for example, study on spore-derived mycobiont of a crustose *Pyrenula* sp. lichen growing in BiDoup-Nui Ba, Dalat City, Vietnam, producing pyrenulic acid A **125**, pyrenulic acid B **126**, pyrenulic acid C **127**, pyrenulic acid D **128**, pyrenulic acid E **129**, pyrenulic acid F **130**, pyrenulic acid G **131**, pyrenulic acid H **132** (Figure 10.14). Pyrenulic acid A **125** and pyrenulic acid G dose-dependently inhibited the activities of pols α and β. Evaluation of the relative activity of each pol at a set concentration (100 μM) of these compounds showed that the activity ranking was as follows: Pyrenulic acid A **125**>Pyrenulic acid G **131**>Pyrenulic acid D **128**>Pyrenulic acid C **127**> Pyrenulic acid H **132**>Pyrenulic acid F **130**. Pyrenulic acid A **125** showed anticancer activity against the HCT116 human colon carcinoma (IC_{50}=6.4±0.7 μM) (Le, Takenaka, Hamada, Mizushina, & Tanahashi, 2014).

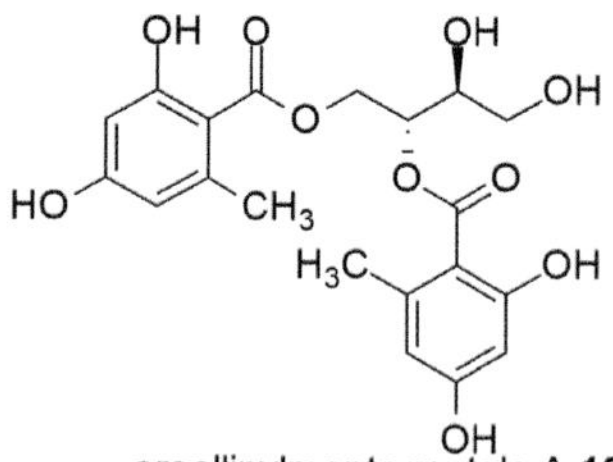
orsellinylmontagnetols A **108**

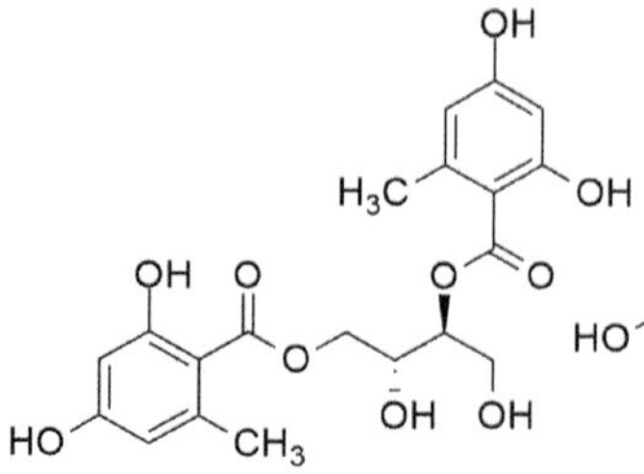
orsellinylmontagnetols B **109**

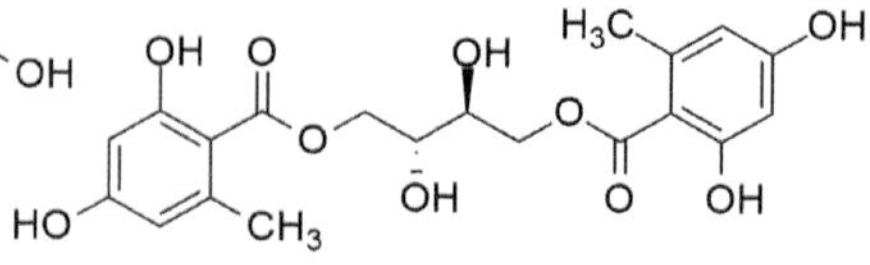
orsellinylmontagnetols C **110**

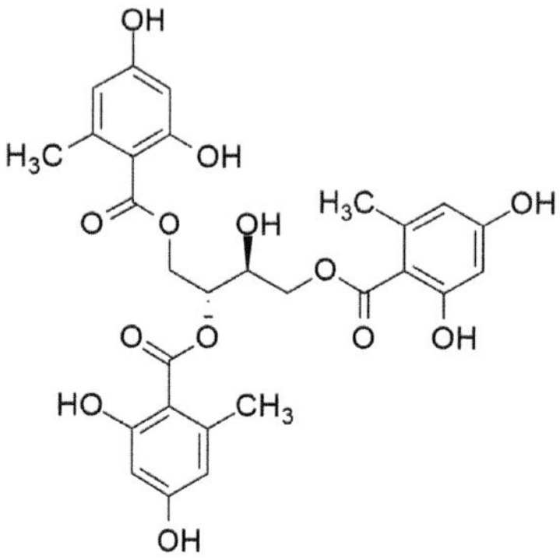
orsellinylmontagnetols D **111**

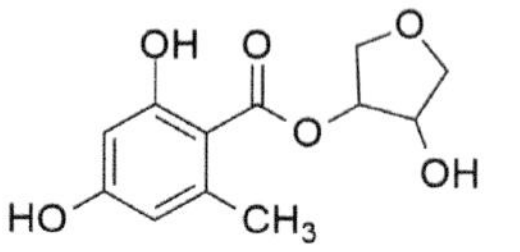
hydrofurylmontagnetol **112**

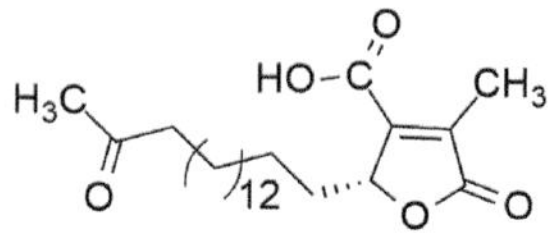
(+)-vinapraesorediosic acid A **113**

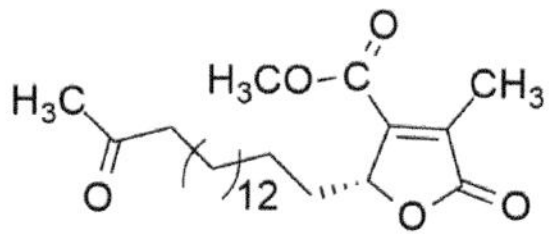
(+)-6-methyl vinapraesorediosate A **114**

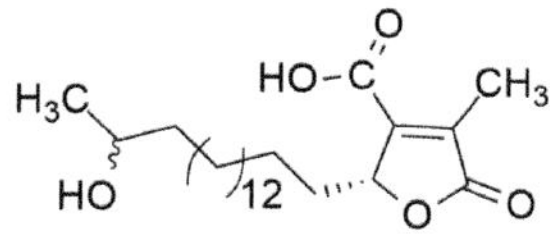
(+)-vinapraesorediosic acid B **115**

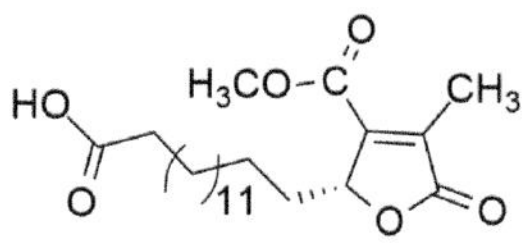
(+)-6-methyl praesorediosate **116**

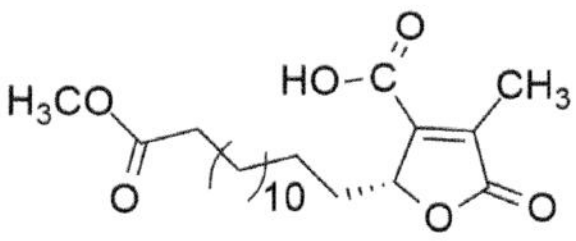
(+)-vinapraesorediosic acid C **117**

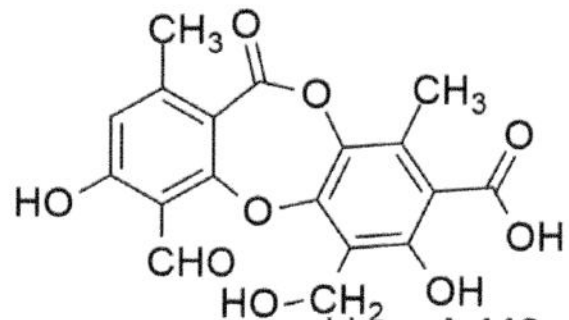
parmosidone A **118**

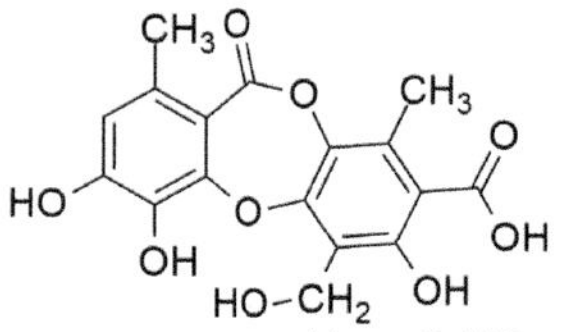
parmosidone B **119**

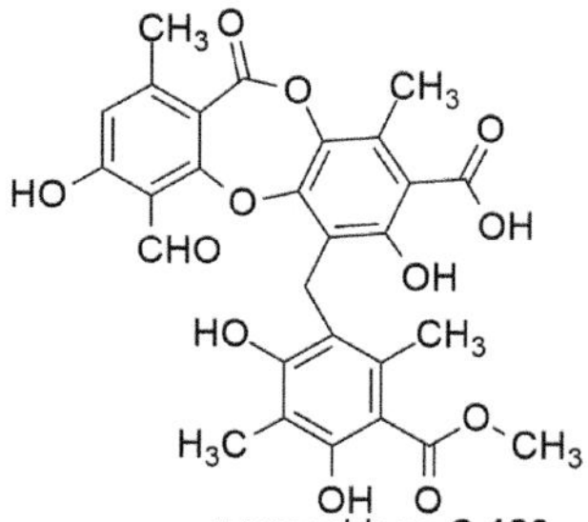
parmosidone C **120**

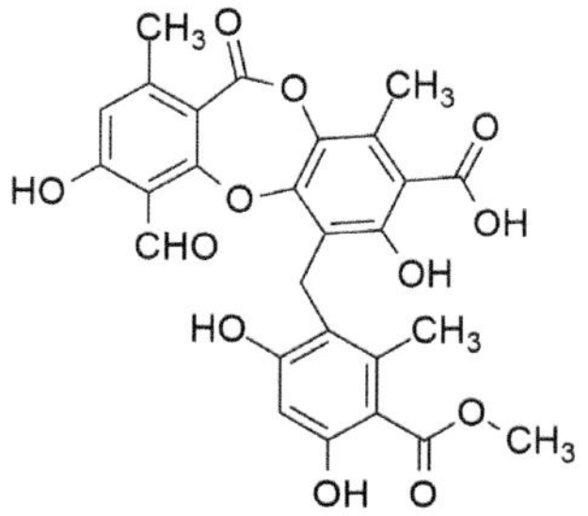
parmosidone D **121**

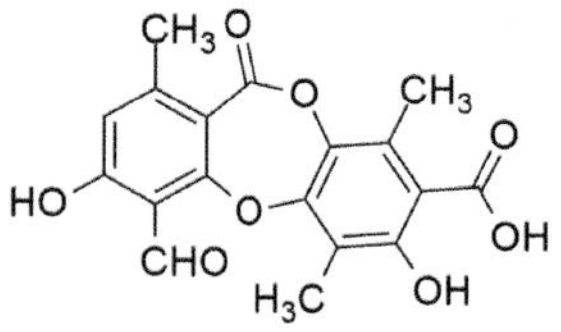
parmosidone E **122**

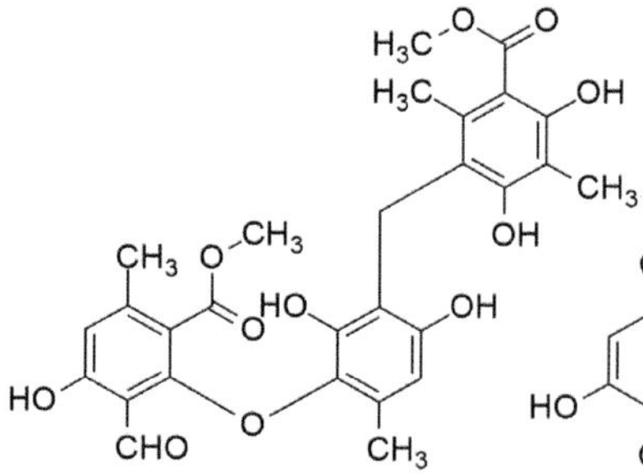
parmoether A **123**

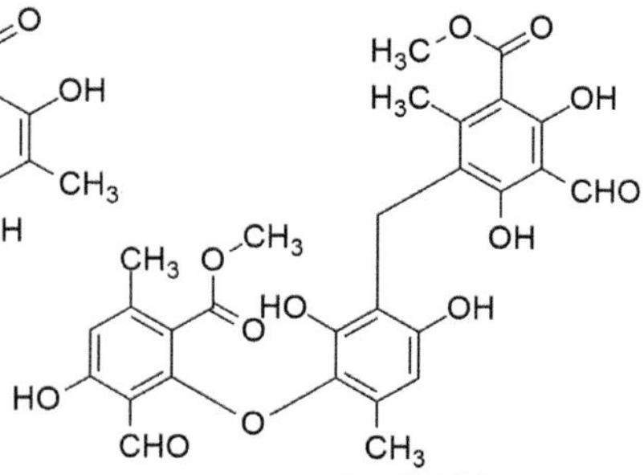
parmoether B **124**

FIGURE 10.13 Lichen metabolites of *Roccella montagnei*, *Parmotrema praesorediosum*, and *Parmotrema tsavoense* with anticancer properties.

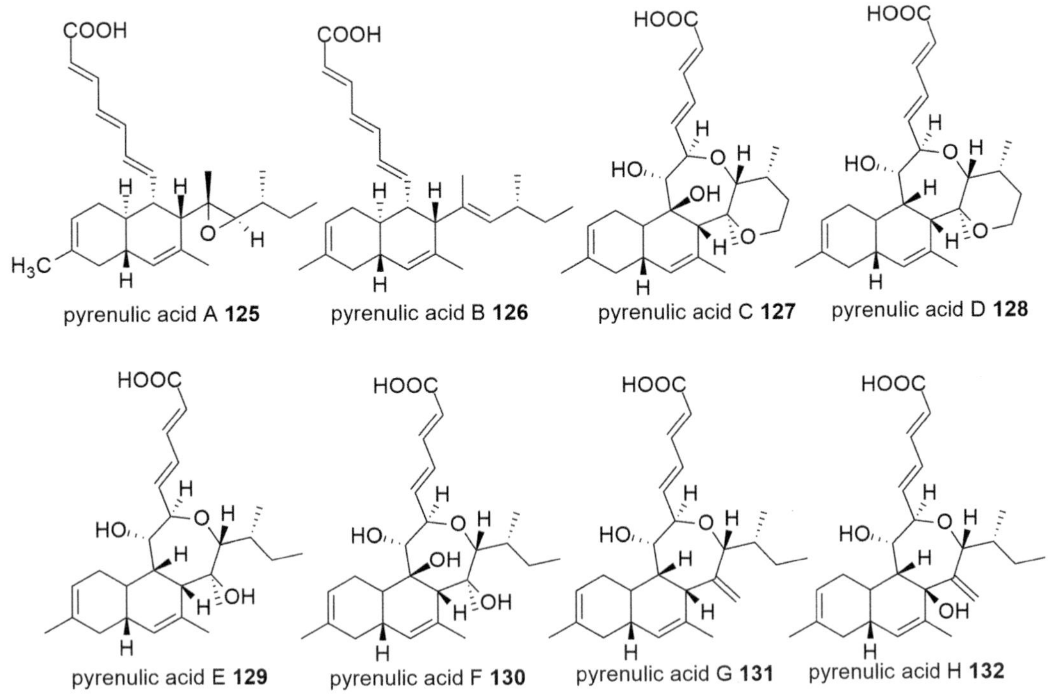

FIGURE 10.14 Lichen metabolites of Vietnamese *Pyrenula* sp. with anticancer properties.

10.6 CONCLUSION

Lichen is a resilient species which is able to cope various biotic and abiotic stresses by different strategies, including producing secondary metabolites. This character was identified and used as environmental quality monitoring. Despite limited biomass, the lichen has been collected and used by mankind throughout the history for different purposes, such as food, dyes, and medicine. Lichen could produce unique color of which chemicals used as natural dyes such as orchil dyes due to evernic acid **5**. Lichen bioprospecting resulted in arrays of pharmacologically active metabolomes. Vulpinic acid **63** was reported for its antimicrobial and antiviral agents. Usnic acid **16** was active against several microbial pathogens. A didepside, diffractaic acid **25,** indicated significant antipyretic properties, while depsidone vicanicin **32** possesses antidiabetic activities. A strong anti-neoplastic was reported on scabrosin diacetate **103,** scabrosin diacetate butyrate **104** from Australian lichen, and didepside stenosporic **102** from Indonesian lichen. Nevertheless, underexplored lichen species remain large and with technology advancement, highly efficient techniques such as the metabolomic approach and artificial intelligence will benefit to lichen bioprospecting by means of dealing with limited biomass.

REFERENCES

Adenubi, O. T., Famuyide, I. M., McGaw, L. J., & Eloff, J. N. (2022). Lichens: An update on their ethnopharmacological uses and potential as sources of drug leads. *Journal of Ethnopharmacology*, *298*, 115657. doi:10.1016/j.jep.2022.115657

Afolayan, A. J., Grierson, D. S., Kambizi, L., Madamombe, I., Masika, P. J., & Jäger, A. K. (2002). In vitro antifungal activity of some South African medicinal plants. *South African Journal of Botany*, *68*(1), 72–76. doi:10.1016/S0254-6299(16)30458-6

Ahad, A. M., Goto, Y., Kiuchi, F., Tsuda, Y., Kondo, K., & Sato, T. (1991). Nematocidal principles in "oakmoss absolute" and nematocidal activity of 2,4-dihydroxybenzoates. *Chemical and Pharmaceutical Bulletin*, *39*(4), 1043–1046. doi:10.1248/cpb.39.1043

Ahmadjian, V. (1995). Lichens are more important than you think. *BioScience*, *45*(3), 124–124. doi:10.1093/bioscience/45.3.124

Aoussar, N., Laasri, F. E., Bourhia, M., Manoljovic, N., Mhand, R. A., Rhallabi, N., . . . Mellouki, F. (2020). Phytochemicalanalysis, cytotoxic, antioxidant, andantibacterial activitie s of lichens. *Evidence-Based Complementary and Alternative Medicine*, *2020*. doi:10.1155/2020/8104538

Basile, A., Rigano, D., Loppi, S., Di Santi, A., Nebbioso, A., Sorbo, S., . . . Bontempo, P. (2015). Antiproliferative, antibacterial and antifungal activity of the lichen *Xanthoria parietina* and its secondary metabolite parietin. *International Journal of Molecular Science*, *16*(4), 7861–7875. doi:10.3390/ijms16047861

Bayir, Y., Odabasoglu, F., Cakir, A., Aslan, A., Suleyman, H., Halici, M., & Kazaz, C. (2006). The inhibition of gastric mucosal lesion, oxidative stress and neutrophil-infiltration in rats by the lichen constituent diffractaic acid. *Phytomedicine*, *13*(8), 584–590. doi:10.1016/j.phymed.2005.07.002

Bharadwaj, T. V., & Sastry, V. G. (2019). Antimicrobial and anti-tubercular activities of isolates and semi-synthetic derivatives of lichen *Ramalina leiodea* (Nyl.) Nyl. *Journal of the Serbian Chemical Society*, *84*(6), 555–562. doi:10.2298/jsc180924003t

Bhattarai, H. D., Kim, T., Oh, H., & Yim, J. H. (2013). A new pseudodepsidone from the Antarctic lichen *Stereocaulon alpinum* and its antioxidant, antibacterial activity. *The Journal of Antibiotics*, *66*(9), 559–561. doi:10.1038/ja.2013.41

Büdel, B., & Scheidegger, C. (2008). Thallus morphology and anatomy. In I. I. I. Nash Thomas H (Ed.), *Lichen Biology* (2nd ed., pp. 40–68). Cambridge: Cambridge University Press.

Burlingame, B. (2000). Wild nutrition. *Journal of Food Composition and Analysis*, *13*(2), 99–100. doi:10.1006/jfca.2000.0897

Calà, E., Benzi, M., Gosetti, F., Zanin, A., Gulmini, M., Idone, A., . . . Aceto, M. (2019). Towards the identification of the lichen species in historical orchil dyes by HPLC-MS/MS. *Microchemical Journal*, *150*, 104140. doi:10.1016/j.microc.2019.104140

Cardon, D. (2007). *Natural Dyes: Sources, Tradition, Technology and Science*: Archetype. London.

Carpentier, C., Queiroz, E. F., Marcourt, L., Wolfender, J.-L., Azelmat, J., Grenier, D., . . . Voyer, N. (2017). Dibenzofurans and pseudodepsidones from the lichen *Stereocaulon paschale* collected in Northern Quebec. *Journal of Natural Products*, *80*(1), 210–214. doi:10.1021/acs.jnatprod.6b00831

Celenza, G., Segatore, B., Setacci, D., Perilli, M., Brisdelli, F., Bellio, P., . . . Nicoletti, M. (2013). Antibacterial activity of selected metabolites from Chilean lichen species against methicillin-resistant staphylococci. *Natural Product Research*, *27*(17), 1528–1531. doi:10.1080/14786419.2012.730043

Cheng, B., Cao, S., Vasquez, V., Annamalai, T., Tamayo-Castillo, G., Clardy, J., & Tse-Dinh, Y.-C. (2013). Identification of anziaic acid, a lichen depside from *Hypotrachyna* sp., as a new topoisomerase poison inhibitor. *PLoS One*, *8*(4), e60770. doi:10.1371/journal.pone.0060770

De Natale, A., & Pollio, A. (2012). A forgotten collection: The Libyan ethnobotanical exhibits (1912–14) by A. Trotter at the Museum O. Comes at the University Federico II in Naples, Italy. *Journal of Ethnobiology and Ethnomedicine*, *8*(1), 4. doi:10.1186/1746-4269-8-4

Delves, J., Lewis, J. E. J., Ali, N., Asad, S. A., Chatterjee, S., Crittenden, P. D., . . . Ellis, C. J. (2023). Lichens as spatially transferable bioindicators for monitoring nitrogen pollution. *Environmental Pollution*, *328*, 121575. doi:10.1016/j.envpol.2023.121575

Desmarets, L., Millot, M., Chollet-Krugler, M., Boustie, J., Camuzet, C., François, N., . . . Séron, K. (2023). Lichen or associated micro-organism compounds are active against human coronaviruses. *Viruses*, *15*(9). doi:10.3390/v15091859

Devkota, S., Chaudhary, R. P., Werth, S., & Scheidegger, C. (2017). Indigenous knowledge and use of lichens by the lichenophilic communities of the Nepal Himalaya. *Journal of Ethnobiology and Ethnomedicine*, *13*(1), 15. doi:10.1186/s13002-017-0142-2

Dieu, A., Millot, M., Champavier, Y., Mambu, L., Chaleix, V., Sol, V., & Gloaguen, V. (2014). Uncommon chlorinated xanthone and other antibacterial compounds from the lichen *Cladonia incrassata*. *Planta Medica*, *80*(11), 931–935. doi:10.1055/s-0034-1382827

Do, T.-H., Duong, T.-H., Nguyen, H. T., Nguyen, T.-H., Sichaem, J., Nguyen, C. H., . . . Long, N. P. (2022). Biological activities of lichen-derived monoaromatic compounds. *Molecules*, *27*(9), 2871. doi:10.3390/molecules27092871

Duong, T.-H., Chavasiri, W., Boustie, J., & Nguyen, K.-P.-P. (2015). New *meta*-depsidones and diphenyl ethers from the lichen *Parmotrema tsavoense* (Krog & Swinscow) Krog & Swinscow, Parmeliaceae. *Tetrahedron*, *71*(52), 9684–9691. doi:10.1016/j.tet.2015.06.107

Duong, T. H., Huynh, B. L. C., Chavasiri, W., Chollet-Krugler, M., Nguyen, V. K., Nguyen, T. H. T., . . . Nguyen, K. P. P. (2017). New erythritol derivatives from the fertile form of *Roccella montagnei*. *Phytochemistry*, *137*, 156–164. doi:10.1016/j.phytochem.2017.02.012

Elkhateeb, W. A., El-Ghwas, D. E., & Daba, W. A. (2022). Lichens uses surprising uses of lichens that improve human life. *Journal of Biomedical Research & Environmental Sciences*, *3*(2), 189–194. doi:10.37871/jbres1420

Ernst-Russell, M. A., Chai, C. L. L., Hurne, A. M., Waring, P., Hockless, D. C. R., & Elix, J. A. (1999). Structure revision and cytotoxic activity of the scabrosin esters, epidithiopiperazinediones from the lichen *Xanthoparmelia scabrosa*. *Australian Journal of Chemistry*, *52*(4), 279–283. doi:10.1071/c99019

Fang, H., Labandeira, C. C., Ma, Y., Zheng, B., Ren, D., Wei, X., . . . Wang, Y. (2020). Lichen mimesis in mid-Mesozoic lacewings. *eLife*, *9*. doi:10.7554/eLife.59007

Friedl, T., & Büdel, B. (2008). Photobionts. In I. I. I. Nash Thomas H (Ed.), *Lichen Biology* (2nd ed., pp. 9–26). Cambridge: Cambridge University Press.

Fritis, M. C., Lagos, C. R., Sobarzo, N. Q., Venegas, I. M., Sánchez, C. S., Altamirano, H. C., . . . Palma, W. Q. (2013). Depsides and triterpenes in *Pseudocyphellaria coriifolia* (lichens) and biological activity against *Trypanosoma cruzi*. *Natural Product Research*, *27*(17), 1607–1610. doi:10.1080/14786419.2012.740033

Gasulla, F., Herrero, J., Esteban-Carrasco, A., Ros-Barceló, A., Barreno, E., Zapata, J. M., & Guéra, A. (2012). Photosynthesis in lichen: Light reactions and protective mechanisms. In M. M. Najafpour (Ed.), *Advances in Photosynthesis* (p. 8). Rijeka: IntechOpen.

Gauslaa, Y. (2014). Rain, dew, and humid air as drivers of morphology, function and spatial distribution in epiphytic lichens. *The Lichenologist*, *46*(1), 1–16. doi:10.1017/S0024282913000753

Gill, M., & Steglich, W. (1987). *Progress in the Chemistry of Organic Natural Products*. Vienna: Springer.

Girardot, M., Millot, M., Hamion, G., Billard, J.-L., Juin, C., Ntoutoume, G. M. A. N., . . . Imbert, C. (2021). Lichen polyphenolic compounds for the eradication of *Candida albicans* biofilms. *Frontiers in Cellular and Infection Microbiology*, *11*, 698883. doi:10.3389/fcimb.2021.698883

Goel, M., Kalra, R., Ponnan, P., Jayaweera, J. A. A. S., & Kumbukgolla, W. W. (2021). Inhibition of penicillin-binding protein 2a (PBP2a) in methicillin resistant *Staphylococcus aureus* (MRSA) by combination of oxacillin and a bioactive compound from *Ramalinaroesleri*. *Microbial Pathogenesis*, *150*, 104676. doi:10.1016/j.micpath.2020.104676

Guarrera, P. M., Lucchese, F., & Medori, S. (2008). Ethnophytotherapeutical research in the high Molise region (Central-Southern Italy). *Journal of Ethnobiology and Ethnomedicine*, *4*(1), 7. doi:10.1186/1746-4269-4-7

Hassan, S. T. S., Šudomová, M., Berchová-Bímová, K., Šmejkal, K., & Echeverría, J. (2019). Psoromic acid, a lichen-derived molecule, inhibits the replication of HSV-1 and HSV-2, and inactivates HSV-1 DNA polymerase: Shedding light on antiherpetic properties. *Molecules*, *24*(16). doi:10.3390/molecules24162912

Hill, M. S. (2014). Production possibility frontiers in phototroph:heterotroph symbioses: Trade-offs in allocating fixed carbon pools and the challenges these alternatives present for understanding the acquisition of intracellular habitats. *Frontiers in Microbiology*, *5*, 357.

Honegger, R., & Scherrer, S. (2008). Sexual reproduction in lichen-forming ascomycetes. In I. I. I. Nash Thomas H (Ed.), *Lichen Biology* (2nd ed., pp. 94–103). Cambridge: Cambridge University Press.

Huynh, B. L. C., Duong, T. H., Do, T. M. L., Pinnock, T. G., Pratt, L. M., Yamamoto, S., . . . Nguyen, K.P.P.. (2016). New γ-Lactone carboxylic acids from the lichen *Parmotrema praesorediosum* (Nyl.) Hale, Parmeliaceae. *Records of Natural Products*, *10*, 332–340.

Igoli, J. O., Gray, A. I., Clements, C. J., Kantheti, P., & Singla, R. K. (2014). Antitrypanosomal activity & docking studies of isolated constituents from the lichen *Cetraria islandica*: Possibly multifunctional scaffolds. *Current Topics in Medicinal Chemistry*, *14*(8), 1014–1021. doi:10.2174/1568026614666140324122323

Joly, K., Wasser, S. K., & Booth, R. (2015). Non-invasive assessment of the interrelationships of diet, pregnancy rate, group composition, and physiological and nutritional stress of barren-ground caribou in late winter. *PLoS One*, *10*(6), e0127586.

Kosanić, M., Manojlović, N., Janković, S., Stanojković, T., & Ranković, B. (2013). *Evernia prunastri* and *Pseudoevernia furfuraceae* lichens and their major metabolites as antioxidant, antimicrobial and anticancer agents. *Food and Chemical Toxicology*, *53*, 112–118. doi:10.1016/j.fct.2012.11.034

Kosanić, M., Ranković, B., & Stanojković, T. (2012). Antioxidant, antimicrobial, and anticancer activity of 3 *Umbilicaria* species. *Journal of Food Science*, *77*(1), T20–T25.

Kosanić, M., Ranković, B., Stanojković, T., Rančić, A., & Manojlović, N. (2014). *Cladonia* lichens and their major metabolites as possible natural antioxidant, antimicrobial and anticancer agents. *LWT-Food Science and Technology*, *59*(1), 518–525.

Kranner, I., Cram, W. J., Zorn, M., Wornik, S., Yoshimura, I., Stabentheiner, E., & Pfeifhofer, H. W. (2005). Antioxidants and photoprotection in a lichen as compared with its isolated symbiotic partners. *Proceedings of the National Academy of Sciences of the United States of America*, *102*(8), 3141–3146. doi:10.1073/pnas.0407716102

Kusmoro, J., Noer, I. S., Jatnika, M. F., Permatasari, R. E., & Partasasmita, R. (2018). Lichen diversity in geothermal area of Kamojang, Bandung, West Java, Indonesia and its potential for medicines and dyes. *Biodiversitas Journal of Biological Diversity, 19*(6), 2335–2343.

Kwon, Y., Cha, J., Chiang, J., Tran, G., Giaever, G., Nislow, C., . . . Kwak, Y. S. (2016). A chemogenomic approach to understand the antifungal action of Lichen-derived vulpinic acid. *Journal of Applied Microbiology, 121*(6), 1580–1591. doi:10.1111/jam.13300

Lalley, J. S., & Viles, H. A. (2008). Recovery of lichen-dominated soil crusts in a hyper-arid desert. *Biodiversity and Conservation, 17*(1), 1–20. doi:10.1007/s10531-007-9153-y

Lauinger, I. L., Vivas, L., Perozzo, R., Stairiker, C., Tarun, A., Zloh, M., . . . Tasdemir, D. (2013). Potential of lichen secondary metabolites against *Plasmodium* liver stage parasites with FAS-II as the potential target. *Journal of Natural Products, 76*(6), 1064–1070. doi:10.1021/np400083k

Lázaro, R., Cantón, Y., Solé-Benet, A., Bevan, J., Alexander, R., Sancho, L. G., & Puigdefábregas, J. (2008). The influence of competition between lichen colonization and erosion on the evolution of soil surfaces in the Tabernas badlands (SE Spain) and its landscape effects. *Geomorphology, 102*(2), 252–266. doi:10.1016/j.geomorph.2008.05.005

Le, D. H., Takenaka, Y., Hamada, N., Mizushina, Y., & Tanahashi, T. (2014). Polyketides from the cultured lichen mycobiont of a Vietnamese *Pyrenula* sp. *Journal of Natural Products, 77*(6), 1404–1412. doi:10.1021/np500143k

Lee, S., Suh, Y. J., Yang, S., Hong, D. G., Ishigami, A., Kim, H., . . . Lee, J. (2021). Neuroprotective and anti-inflammatory effects of evernic acid in an MPTP-induced Parkinson's disease model. *International Journal of Molecular Science, 22*(4). doi:10.3390/ijms22042098

Maciag-Dorszynska, M., Wegrzyn, G., & Guzow-Krzeminska, B. (2014). Antibacterial activity of lichen secondary metabolite usnic acid is primarily caused by inhibition of RNA and DNA synthesis. *FEMS Microbiology Letters, 353*(1), 57–62. doi:10.1111/1574-6968.12409

Makarevich, E., Filippova, E. I., Sedel'nikova, N., Mazurkov, O. Y., Protsenko, M. A., Shishkina, L. N., & Mazurkova, N. A. (2023). Anti-influenza activity of *Cetraria islandica* lichen extracts in *in vitro* experiments. *Bulletin of Experimental Biology and Medicine, 175*(2), 215–218. doi:10.1007/s10517-023-05837-8

Mallavadhani, U. V., Boddu, R., Rathod, B. B., & Reddy Setty, P. (2018). Stereoselective synthesis of the lichen metabolite, (+) montagnetol and its congeners as antimicrobial agents. *Synthetic Communications, 48*(23), 2992–2999. doi:10.1080/00397911.2018.1519076

Maulidiyah, M., Darmawan, A., Asriani, H., Wibowo, D., Salim, L. O. A., Ansharullah, A., . . . Nurdin, M. (2020). Isolation, structure elucidation, and antidiabetic test of vicanicin compound from lichen *Teloschistes flavicans*. *Journal of Applied Pharmaceutical Science*. doi:10.7324/JAPS.2020.10111

Maulidiyah, M., Natsir, M., Nazila, W., Musdalifah, A., Salim, L. O. A., & Nurdin, M. (2021). Isolation and antibacterial activity of diffractic acid compound from lichen *Usnea blepharea* Motyka. *Journal of Applied Pharmaceutical Science, 11*(11), 121–130. doi:10.7324/JAPS.2021.1101116

Melo, M. G. D., dos Santos, J. P. A., Serafini, M. R., Caregnato, F. F., de Bittencourt Pasquali, M. A., Rabelo, T. K., . . . Gelain, D. P. (2011). Redox properties and cytoprotective actions of atranorin, a lichen secondary metabolite. *Toxicology in Vitro, 25*(2), 462–468. doi:10.1016/j.tiv.2010.11.014

Mendili, M., Khadhri, A., Jemâa, J. M.-B., Andolfi, A., Tufano, I., Aschi-Smiti, S., & DellaGreca, M. (2022). Anti-inflammatory potential of compounds isolated from Tunisian lichens species. *Chemistry & Biodiversity, 19*(8), 1–10. doi:10.1002/cbdv.202200134

Miller, J. E. D., Villella, J., Stone, D., & Hardman, A. (2020). Using lichen communities as indicators of forest stand age and conservation value. *Forest Ecology and Management, 475*, 118436. doi:10.1016/j.foreco.2020.118436

Milovanovic, N., Kostic, D., Milovanovic, V., Djordjevic, A., & Stojanović, G. (2018). Antimicrobial activity of the secondary metabolites of the genus Bryoria – A review. *Facta Universitatis – Series: Physics, Chemistry and Technology, 16*, 183–189. doi:10.2298/FUPCT1802183M

Molnár, K., & Farkas, E. (2010). Current results on biological activities of lichen secondary metabolites: A review. *Zeitschrift für Naturforschung C, 65*(3–4), 157–173. doi:10.1515/znc-2010-3-401

Mun, S.-K., Kang, K.-Y., Jang, H.-Y., Hwang, Y.-H., Hong, S.-G., Kim, S.-J., . . . Yee, S.-T. (2020). Atraric acid exhibits anti-inflammatory effect in lipopolysaccharide-stimulated RAW264.7 cells and mouse models. *International Journal of Molecular Sciences, 21*(19). doi:10.3390/ijms21197070

Nash, T. H. (2008). Introduction. In *Lichen Biology* (pp. 1–8): Cambridge: Cambridge University Press,

Nash, T. H. (2012). *Lichen Biology* (T. H. Nash, Editor): Cambridge: Cambridge University Press, Cambridge.

Nayaka, S., & Rai, H. (2022). Antarctic lichen response to climate change: Evidence from natural gradients and temperature enchantment experiments. In N. Khare (Ed.), *Assessing the Antarctic Environment from a Climate Change Perspective: An Integrated Approach* (pp. 235–253). Cham: Springer International Publishing.

Nelsen, M. P., & Lumbsch, H. T. (2020). A data-driven evaluation of lichen climate change indicators in Central Europe. *Biodiversity and Conservation, 29*(14), 3959–3971. doi:10.1007/s10531-020-02057-8

Nguyen, N. T., Kuo, P.-C., Tran Trung, H., Tuong Vi, L. N., Quach, T. H., Dunge, L. T., . . . Thang, T. D. (2021). A new triterpenoid and other compounds from lichens *Cryptothecia faveomaculata* Makhija & Patw. *Natural Product Research, 35*(8), 1349–1356. doi:10.1080/14786419.2019.1648466

Nguyen, T. T., Nguyen, P. D., Mai, V. H., Trang, D. T. X., Danh, L. T., Tran, T. M., . . . Kanaori, K. (2020). Chemical constituents of the lichen *Parmotrema tinctorum* and their antifungal activity. *Chemistry of Natural Compounds, 56*(2), 315–317. doi:10.1007/s10600-020-03017-y

Nishanth, K. S., Sreerag, R. S., Deepa, I., Mohandas, C., & Nambisan, B. (2015). Protocetraric acid: An excellent broad spectrum compound from the lichen *Usnea albopunctata* against medically important microbes. *Natural Product Research, 29*(6), 574–577. doi:10.1080/14786419.2014.953500

Nithyanand, P., Beema Shafreen, R. M., Muthamil, S., & Karutha Pandian, S. (2015). Usnic acid, a lichen secondary metabolite inhibits Group A *Streptococcus* biofilms. *Antonie van Leeuwenhoek, 107*(1), 263–272. doi:10.1007/s10482-014-0324-z

Nugraha, A. S., Laksono, T. A., Firli, L. N., Putri, C. P., Pratoko, D. K., Zulfikar, Z., . . . Keller, P. A. (2020). Anti-cancer evaluation of depsides isolated from Indonesian Folious lichens: *Physcia millegrana, Parmelia dilatata* and *Parmelia aurulenta. Biomolecules, 10*(10). doi:10.3390/biom10101420

Nugraha, A. S., Wangchuk, T., Willis, A. C., Haritakun, R., Sujadmiko, H., & Keller, P. A. (2019). Phytochemical and pharmacological studies on four Indonesian epiphytic medicinal plants: *Drynaria rigidula, Hydnophytum formicarum, Usnea misaminensis*, and *Calymperes schmidtii. Natural Product Communications, 14*(6), 1934578X19856792. doi:10.1177/1934578x19856792

Oh, J. M., Kim, Y. J., Gang, H.-S., Han, J., Ha, H.-H., & Kim, H. (2018). Antimicrobial activity of divaricatic acid isolated from the lichen *Evernia mesomorpha* against methicillin-resistant *Staphylococcus aureus. Molecules, 23*(12), 3068/3061. doi:10.3390/molecules23123068

Oksanen, I. (2006). Ecological and biotechnological aspects of lichens. *Applied Microbiology and Biotechnology, 73*(4), 723–734. doi:10.1007/s00253-006-0611-3

Okuyama, E., Umeyama, K., Yamazaki, M., Kinoshita, Y., & Yamamoto, Y. (1995). Usnic acid and diffractaic acid as analgesic and antipyretic components of *Usnea diffracta. Planta Medica, 61*(2), 113–115. doi:10.1055/s-2006-958027

Pasaribu, N., Atni, O., & Siregar, J. (2023). Diversity and species composition of lichens across altitudinal range in the Batang Toru Forest, North Sumatra, Indonesia. *Biodiversitas Journal of Biological Diversity, 24*. doi:10.13057/biodiv/d240429

Peralta, M. A., da Silva, M. A., Ortega, M. G., Cabrera, J. L., & Paraje, M. G. (2017). Usnic acid activity on oxidative and nitrosative stress of azole-resistant *Candida albicans* biofilm. *Planta Medica, 83*(3/4), 326–333. doi:10.1055/s-0042-116442

Pintado, A., Valladares, F., & Sancho, L. G. (1997). Exploring phenotypic plasticity in the lichen *Ramalina capitata*: Morphology, water relations and chlorophyll content in north- and south-facing populations. *Annals of Botany, 80*(3), 345–353. doi:10.1006/anbo.1997.0453

Pompilio, A., Pomponio, S., Di Vincenzo, V., Crocetta, V., Nicoletti, M., Piovano, M., . . . Di Bonaventura, G. (2013). Antimicrobial and antibiofilm activity of secondary metabolites of lichens against methicillin-resistant *Staphylococcus aureus* strains from cystic fibrosis patients. *Future Microbiology, 8*(2), 281–292. doi:10.2217/fmb.12.142

Pradhan, B. K., & Badola, H. K. (2008). Ethnomedicinal plant use by Lepcha tribe of Dzongu valley, bordering Khangchendzonga Biosphere Reserve, in North Sikkim, India. *Journal of Ethnobiology and Ethnomedicine, 4*(1), 22. doi:10.1186/1746-4269-4-22

Ranković, B., & Kosanić, M. (2019). Lichens as a potential source of bioactive secondary metabolites. In Ranković, B. (Eds.), *Lichen Secondary Metabolites: Bioactive Properties and Pharmaceutical Potential*, Springer, Cham (pp. 1–29).

Ranković, B., Kosanić, M., Manojlović, N., Rančić, A., & Stanojković, T. (2014). Chemical composition of *Hypogymnia physodes* lichen and biological activities of some its major metabolites. *Medicinal Chemistry Research, 23*, 408–416.

Rather, L., Jameel, S., Ali, S., Bhat, K., & Yusuf, M. (2018). Lichen derived natural colorants: History, extraction, and applications. In *Handbook of Renewable Materials for Coloration and Finishing* (pp. 103–114). doi:10.1002/9781119407850.ch6

Resende, D. I. S. P., Pereira-Terra, P., Inacio, A. S., da Costa, P. M., Pinto, E., Sousa, E., & Pinto, M. M. M. (2018). Lichen xanthones as models for new antifungal agents. *Molecules*, *23*(10), 2617/2611. doi:10.3390/molecules23102617

Sarıözlü, N. Y., Yilmaz, M., Candan, M., & Tay, T. (2016). Antimicrobial activity of lichen *Bryoria capillaris* and its compound barbatolic acid. *Biomedical Research-India*, 27, S419–S423.

Sasidharan, N. K., Sreekala, S. R., Lakshmanan, R., Jacob, J., Saraswathy Amma, D. K. B. N., & Nambisan, B. (2014). Protolichesterinic acid: A prominent broad spectrum antimicrobial compound from the lichen *Usnea albopunctata*. *International Journal of Antibiotics*, *2014*(1), 302182.

Segatore, B., Bellio, P., Setacci, D., Brisdelli, F., Piovano, M., Garbarino, J. A., . . . Celenza, G. (2012). *In vitro* interaction of usnic acid in combination with antimicrobial agents against methicillin-resistant *Staphylococcus aureus* clinical isolates determined by FICI and ΔE model methods. *Phytomedicine*, *19*(3–4), 341–347. doi:10.1016/j.phymed.2011.10.012

Şenol, Z. M., Gül, Ü. D., & Şimşek, S. (2022). Bioremoval of Safranin O dye by the identified lichen species called *Evernia prunastri* biomass; biosorption optimization, isotherms, kinetics, and thermodynamics. *Biomass Conversion and Biorefinery*, *12*(9), 4127–4137. doi:10.1007/s13399-020-01216-9

Shailajan, S., Joshi, M., & Tiwari, B. (2014). Hepatoprotective activity of *Parmelia perlata* (Huds.) Ach. against CCl_4 induced liver toxicity in Albino Wistar rats. *Journal of Applied Pharmaceutical Science*, *4*, 70–74. doi:10.7324/JAPS.2014.40212

Shanmugam, P., Ponnusamy, P., Balasubramanian, M. G., Nagarajan, N., & Ganesan, A. (2017). Evaluation of antimicrobial, antioxidant and anticancer activities of few macrolichens collected from Eastern Ghats of Tamil Nadu, India. *International Research Journal of Pharmacy*, *8*(3), 39–43. doi:10.7897/2230-8407.080334

Shrestha, G., Thompson, A., Robison, R., & St. Clair, L. L. (2016). *Letharia vulpina*, a vulpinic acid containing lichen, targets cell membrane and cell division processes in methicillin-resistant *Staphylococcus aureus*. *Pharmaceutical Biology*, *54*(3), 413–418. doi:10.3109/13880209.2015.1038754

Silva, H. A. M. F., Siqueira, W. N., Sá, J. L. F., Silva, L. R. S., Martins, M. C. B., Aires, A. L., . . . Silva, N. H. (2018). Laboratory assessment of divaricatic acid against *Biomphalaria glabrata* and *Schistosoma mansoni* cercariae. *Acta Tropica*, *178*, 97–102. doi:10.1016/j.actatropica.2017.09.019

Sokolov, D. N., Zarubaev, V. V., Shtro, A. A., Polovinka, M. P., Luzina, O. A., Komarova, N. I., . . . Kiselev, O. I. (2012). Anti-viral activity of (-)- and (+)-usnic acids and their derivatives against influenza virus A(H1N1)2009. *Bioorganic & Medicinal Chemistry Letters*, *22*(23), 7060–7064. doi:10.1016/j.bmcl.2012.09.084

Spribille, T., Resl, P., Stanton, D. E., & Tagirdzhanova, G. (2022). Evolutionary biology of lichen symbioses. *New Phytologist*, *234*(5), 1566–1582. doi:10.1111/nph.18048

Studzińska-Sroka, E. (2018). Lichens as a source of chemical compounds with anti-inflammatory activity. *Herba Polonica*, *64*. doi:10.2478/hepo-2018-0005

Süleyman, H., Odabasoglu, F., Aslan, A., Cakir, A., Karagoz, Y., Gocer, F., . . . Bayir, Y. (2003). Anti-inflammatory and antiulcerogenic effects of the aqueous extract of *Lobaria pulmonaria* (L.) Hoffm. *Phytomedicine*, *10*(6–7), 552–557. doi:10.1078/094471103322331539

Svanberg, I., & Ægisson, S. (2012). Edible wild plant use in the Faroe Islands and Iceland. *Acta Societatis Botanicorum Poloniae*, *81*. doi:10.5586/asbp.2012.035

Sweidan, A., Chollet-Krugler, M., Sauvager, A., van de Weghe, P., Chokr, A., Bonnaure-Mallet, M., . . . Bousarghin, L. (2017). Antibacterial activities of natural lichen compounds against *Streptococcus gordonii* and *Porphyromonas gingivalis*. *Fitoterapia*, *121*, 164–169. doi:10.1016/j.fitote.2017.07.011

Tartouga, M. A., Elouar, I., Zeghina, I., Mokhtari, M. B., Gali, L., Rapa, S. F., . . . Bensouici, C. (2022). The evaluation of antioxidant and anti-inflammatory activities of Parm otrema hypotropa lichen extract. *Egyptian Journal of Chemistry*, *65*(9), 109–118. doi:10.21608/EJCHEM.2022.109501.5016

Tozatti, M. G., Ferreira, D. S., Flauzino, L. G. B., da Silva Moraes, T., Martins, C. H. G., Groppo, M., . . . Cunha, W. R. (2016). Activity of the lichen *Usnea steineri* and its major metabolites against gram-positive, multidrug-resistant bacteria. *Natural Product Communications*, *11*(4), 1934578X1601100419. doi:10.1177/1934578x1601100419

Untari, L. F. (2024a). The biology of lichen. In Das, A. K., Sharma, A., Kathuria, D., Ansari, M. J., Bhardwaj, G. (Eds.), *Chemistry, Biology and Pharmacology of Lichen*, John Wiley & Sons Ltd, Hoboken (pp. 13–19).

Untari, L. F. (2024b). Lichen as habitats for other organisms. In Das, A. K., Sharma, A., Kathuria, D., Ansari, M. J., Bhardwaj, G. (Eds.), *Chemistry, Biology and Pharmacology of Lichen*, John Wiley & Sons Ltd, Hoboken (pp. 39–48).

Upadhyay, A., Upadhyaya, I., Kollanoor-Johny, A., & Venkitanarayanan, K. (2014). Combating pathogenic microorganisms using plant-derived antimicrobials: A minireview of the mechanistic basis. *BioMed Research International, 2014*(1), 761741. doi:10.1155/2014/761741

Upadhyay, S. K. & Singh, S. P. (2021). *Bioprospecting of Plant Biodiversity for Industrial Molecules*. Wiley. doi:10.1002/9781119718017

Upadhyay, S. K. & Singh, S. P. (2023). *Plants as Bioreactors for Industrial Molecules*. Wiley. doi:10.1002/9781119875116

Upadhyay, S. K. & Singh, S. P. (2024). *Herbal Nutraceuticals: Products and Processes*. Wiley. doi:10.1002/9781394241576

Wang, L., Narui, T., Harada, H., Culberson, C., & Culberson, W. (2001). Ethnic uses of lichens in Yunnan, China. *The Bryologist, 104*, 345–349. doi:10.1639/0007-2745(2001)104[0345:EUOLIY]2.0.CO;2

White, P. A. S., Oliveira, R. C. M., Oliveira, A. P., Serafini, M. R., Araújo, A. A. S., Gelain, D. P., . . . Santos, M. R. V. (2014). Antioxidant activity and mechanisms of action of natural compounds isolated from lichens: A systematic review. *Molecules, 19*(9), 14496–14527. doi:10.3390/molecules190914496

Yang, J., Oh, S.-O., & Hur, J.-S. (2023). Lichen as bioindicators: Assessing their response to heavy metal pollution in their native ecosystem. *Mycobiology, 51*(5), 343–353. doi:10.1080/12298093.2023.2265144

Yang, M.-X., Devkota, S., Wang, L.-S., & Scheidegger, C. (2021). Ethnolichenology—The use of lichens in the Himalayas and southwestern parts of China. *Diversity, 13*(7). doi:10.3390/d13070330

Yeash, E. A., Letwin, L., Malek, L., Suntres, Z., Knudsen, K., & Christopher, L. P. (2017). Biological activities of undescribed North American lichen species. *The Journal of the Science of Food and Agriculture, 97*(14), 4721–4726. doi:10.1002/jsfa.8340

Yu, X., Guo, Q., Su, G., Yang, A., Hu, Z., Qu, C., . . . Chai, X. (2016). Usnic acid derivatives with cytotoxic and antifungal activities from the lichen *Usnea longissima*. *Journal of Natural Products, 79*(5), 1373–1380. doi:10.1021/acs.jnatprod.6b00109

Zorrilla, J. G., D'Addabbo, T., Roscetto, E., Varriale, C., Catania, M. R., Zonno, M. C., . . . Evidente, A. (2022). Antibiotic and nematocidal metabolites from two lichen species collected on the island of Lampedusa (Sicily). *International Journal of Molecular Sciences, 23*(15), 8471. doi:10.3390/ijms23158471

11 Nutraceutical or Adaptogen or Pharmaceutical

An Overview on the Overlapping Roles of Indian Ginseng (Withania somnifera L.) as an Ayurvedic Rejuvenator

Sumanti Gupta[1] *and Amit Roy*[2]
[1]Department of Botany, Rabindra Mahavidyalaya, Champadanga, Hooghly, India
[2]Faculty of Forestry and Wood Sciences, Forest Molecular Entomology Laboratory, EXTEMIT-K, EVA 4.0, Czech University of Life Sciences Prague, Prague, Czechia

11.1 INTRODUCTION

In India, the history of Traditional Systems of Medicines (TSMs) in treating chronic ailments dates back to approximately 3,000 years, where remission of symptoms and complete cure of any disease were believed to be the outcome of holistic therapy applied over the entire body and mind (Jaiswal and Williams, 2017).TSMs are comprised of Ayurveda, Naturopathy, Homeopathy, Siddha, Unani, and Yoga (Mukherjee,2001). Among these, Ayurvedic doctrine is known to be the most ancient, globally accepted, and widely practiced branch of medicine (Mukherjee et al., 2010). In later years with the gradual development of chemo-informatics and related technological tools and techniques, the medical science became more inclined toward allopathic drugs and their mode of treatment (Varma, 2006). Furthermore, the revolutionary progression of invasive techniques in the form of organ-specific surgery started dominating the medical world (Ouellet et al., 2018). Even then, the TSMs retained their ancestry within a limited group of practitioners and users who adhered to the strong conviction for their respective branches of sciences (Sen and Chakraborty, 2017). Interestingly, since the past few decades which are referred to as the age of new age science and technology, the ancient modes and methods of treatment of TSMs are experiencing a re-emergence of their popularity and acceptance worldwide (Dhandge and Deshmukh, 2023). Beliefs regarding the holistic healing of body and mind are being strengthened (Rastogi et al., 2020).

During ancient times, the TSMs and folk medicines relied mostly on the use of herbal products and byproducts where Indian Ginseng (IG) scientifically named as *Withania somnifera* (L.) was a prominent name. The plant also known as 'Ashwagandha' or Indian winter cherry derived its native nomenclature from 'Ashwa', meaning 'horse' and 'Gandha', meaning 'aroma'. The name symbolizes as being capable of promoting health vitality as that of a 'horse' and is known for its significant 'fragrance' (Singh et al., 2011). The root of the plant is known to be widely used for therapeutic applications. It has been well established as a neuroprotective medicine since ancient traditional Rasayana (Mikulska et al., 2023). In addition, it is known as the Sattvic Kapha Rasayana herb that

 DOI: 10.1201/9781003518358-11

serves as an adaptogen/anti-stress agent (Tripathi, 2022). It not only increases cognitive brain and nerve function but also acts as a memory booster. Besides, it also ensures reproductive health and enhances body resilience against stress by protecting against cell damage which is induced by the generation of free radicals. Improved cell-mediated immunity is also provided by Ashwagandha (Gautam et al., 2020). Traditional uses of IG roots suggest its use as an aphrodisiac, nervine tonic, sedative, diuretic, worm killer, and broad-spectrum stimulant (Połumackanycz et al., 2020). IG is also known to exhibit anti-inflammatory, antibacterial, anti-cancerous, antidiabetic, and cardioprotective properties which all together place this herb as a multifunctional promising candidate for drug discovery (John, 2014). The present study aims to provide a brief narrative of the phenology, chemistry, and traditional uses of IG and the latest understanding of its roles as a pharmaceutical, adaptogen/anti-stress agent and also as a nutraceutical rejuvenator. The probable areas of their molecular overlaps between the diverse roles shall be analyzed in the study. The study shall also discuss about the toxicological effects of IG. Thus, taken together, the study is expected to provide a comprehensive outlook on the potential of IG for future exploration and exploitation in modern herbal-based drug discovery.

11.2 ASHWAGANDHA: THE WONDER SHRUB AT A GLANCE

11.2.1 Morphology, Growth Conditions and Distribution

Withania somnifera(L.), a seed-bearing angiosperm,is taxonomically placed under the division Magnoliophyta, class Magnoliopsida, order Solanales, and family Solanaceae (Saidulu et al., 2014). The plant attains about 30–150cm in height. It is a branched erect, evergreen shrub bearing varied trichomes of glandular and non-glandular nature. The four types of trichomes are glandular capitate, non-glandular branched dendritic, non-glandular two-celled, and non-glandular multicellular. These trichomes are reported to be rich source of alkaloid and phenolic compounds having medicinal values (Munien et al., 2015). Leaves are 8–12cm in length, petiolate, simple with an elliptical to broadly ovate in shape, having alternate phyllotaxy. Besides, leaves are exstipulate, having an acute, cuneate, or connate base which is densely hairy. Venation is of reticulate nature. Leaf arrangement near the inflorescence is opposite and adnate. Pedicellate flowers are 1cm long and 4–6mm in diameter, bisexual, and borne in axillary umbellate cymes. Globose berry-like fruits are 6mm in diameter, orange-red at maturity, and enclosed within an inflated membranous calyx of persistent nature. Seeds (20–25 per fruit) are smooth, discoid, yellow, reniform in shape, and 2.5mm in diameter. The main source of medicinal alkaloids of this plant is its roots, which are stout but fleshy, cylindrical in shape and 1–2cm thick. The unbranched, straight roots bear fiber-like secondary roots. The nodes of the plant show structural variation at the juncture of the leaf petiole. The plant is reported to have a bitter taste and is acrid (Mirjalili et al., 2009; Gaurav et al., 2015). The plant is reported to have chromosome number $2n=48$ (Hawkes et al., 1991).

Ashwagandha is a well-suited plant for semi-arid tropical regions receiving 500–750mm of annual rainfall. Known as a late rainy season kharif crop, it requires dry periods for its growth with 1–2 rainfall spells in winter for proper root development. Well-drained sandy loam or light red soils with pH 7.5–8 are appropriate for its cultivation. A temperature of 20°C–35°C and 600–1,200m altitudinal height is conducive for the crop (Meena et al., 2020).The plant is distributed in the dry and subtropical parts of India, such as in the states of Mumbai, Gujarat, Rajasthan, Madhya Pradesh, Uttar Pradesh, Punjab, and high altitudinal regions (1,800m) of Himachal Pradesh and Jammu and Kashmir. The varieties growing in these above-mentioned regions, however, show diversity due to variable climatic and edaphic factors. The Neemuch district of Madhya Pradesh is the main marketing hub of dried roots of Ashwagandha. Besides India, this plant is also reported to inhabit other parts of the world namely South and East Africa, Madagascar, the Canary Islands, Congo, Spain, Morocco, Egypt, Jordan, Palestine, Afghanistan, and Pakistan. All the varieties growing in different parts of India and worldwide show area-specific variation in terms of root morphometric and chemical properties depending on differential growth conditions (Kumar et al., **2023**).

11.2.2 Chemistry and Traditional Uses

Several high-throughput isolation and analytical techniques have aided the isolation and characterization of active compounds from different parts of Ashwagandha (Tewari et al., 2022). These phytobioactive compounds belong to alkaloids, steroidal lactones, steroids, salts, flavonoids, and nitrogen-containing classes, which were isolated from different parts such as roots, leaves, fruits, and seeds (Dar et al., 2015). Table 11.1 summarizes the list of bioactive compounds found in different parts of Ashwagandha. Among several active compounds, withanolides and alkaloids play predominant pharmaceutical roles. Withanolides are structurally ergostane with a six-membered lactone rind at C-8 and C-9 positions. Withanolides include withapherin A, withanolide A and

TABLE 11.1
Bioactive Compounds of Ashwagandha and Their Uses

Nature of Bioactive Compound	Plant Source	Chemical Nature	Pharmaceutical/ Adaptogen/ Nutracuetical Property	Function
Alkaloids	Root, Leaf, Fruit	Withanine withananine withasomnine Somniferine tropeltigloate somniferinine Somninine Nicotine	Pharmaceutical/ Adaptogen	Anti-inflammatory, Anti-cancerous
Steroidal lactones	Root, Leaf, Seed, Fruit	Withaferin-A withanone, withanolide-E withanolide-F withanolide-A withanolide-G withanolide-H withanolide-I withanolide-J withanolide-K withanolide-L withanolide-M	Pharmaceutical/ Adaptogen/ Nutraceutical	Neuroprotective, Cardioprotective Anti-cancerous Immunomodulatory
Steroids	Root	Cholesterol Beta-sitosterol stigmasterol Diosgenin, Stigmastadien Sitoinosides VII Sitoinosides VIII Sitoinosides IX Sitoinosides X	Pharmaceutical/ Adaptogen/ Nutraceutical	Antioxidant Anticancerous Anti-inflammatory
Salts	Leaf	Cuscohygrine anahygrine Tropine Pseudotropine Anaferine	Pharmaceutical/ Adaptogen	Anti-inflammatory
Flavonoids	Leaf	Kaempferol Quercetin	Adaptogen/Nutraceutical	Neuroprotection, Antimicrobial
Nitrogen-containing compounds	Root	Withanol Somnisol Somnitol	Adaptogen/Nutraceutical	Immunomodulatory and Muscle strength

Source: Adapted from Dar et al. (2015); Bashir et al. (2023); Mikulska et al. (2023).
The table shows the list of bioactive compounds of Ashwagandha, their types, source of origin, chemical nature, and its pharmaceutical/adaptogen/nutracuetical property.

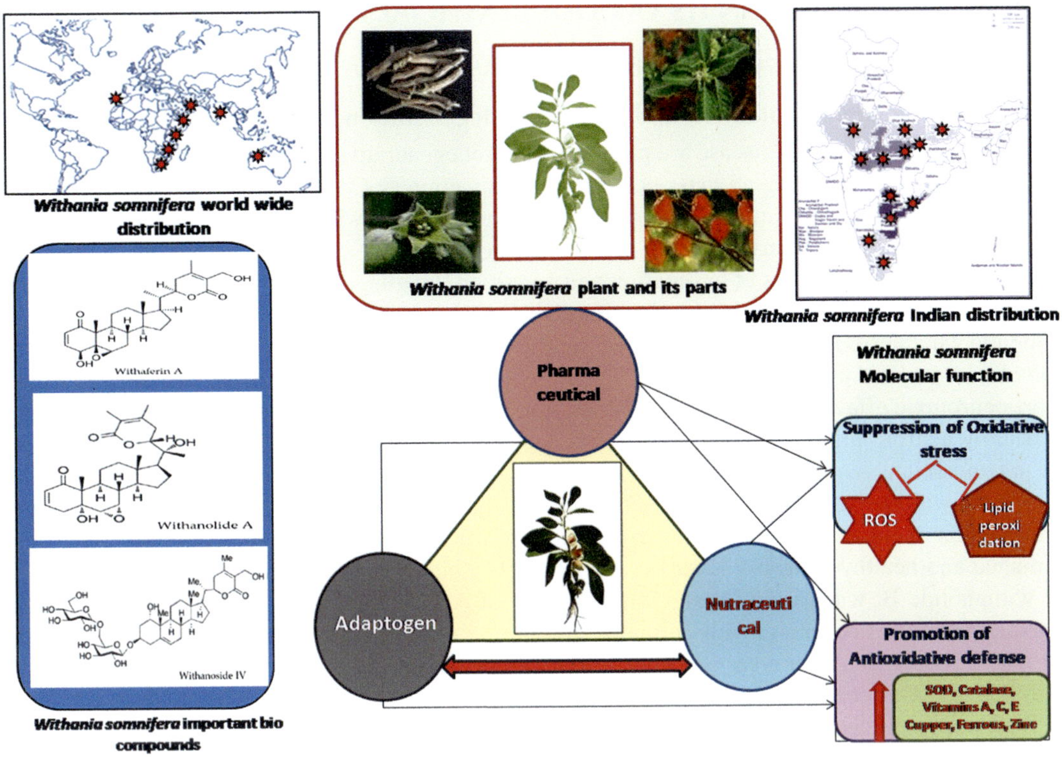

FIGURE 11.1 Schematic figure showing the morphology, global and Indian distribution (marked in red), important chemical constituents, and the central molecular mechanism controlling the diverse roles of *Withania somnifera*.

withanoside IV as the key important withanolides (Mikulska et al., 2023) (Figure 11.1). Withanolides are 30-carbon-containing triterpenoids. Like other terpenoid compounds, they are also synthesized by the classical cytosolic mevalonic acid pathway (MVA) or the alternative plastidial methylerythritol phosphate pathway (MEP) where isoprene units such as isopentenylpyrophosphate (IPP) and dimethyl allyl pyrophosphate (DMAPP) are needed as precursors (Dhar et al., 2015). Recent transcriptomic and proteomic analyses have identified several key biosynthetic genes and proteins of the withanolide biosynthetic pathway that are believed to help in the increased production of the herbal compounds. Besides, several tissue culture-based techniques are also being deployed to enhance the production of the withanolides (Dhar et al., 2015). However, further in-depth research is needed to identify and quantify the different types of bioactive compounds of different varieties of Ashwagandha inhabiting at various locations.

The plant Ashwagandha is an official drug in the Indian Pharmacopoeia. Besides, it was a widely used traditional drug of in Ayurveda and Unani systems of medicines (Uddin et al., 2012). According to Unani medicine, Ashwagandha was a recommended drug for arthritis, lumbago, carbuncle, spermatorrhoea, vitiligo, asthma, general, and sexual debility, depression and anxiety, neurosis, scabies, ulcers and leucorrhoea (Tiwari et al., 2014). *Withania* is also reported as a sleep inducer (Ven Murthy et al., 2010). The dried root of the plant, however, loses its bioactivity after 2 years of treatment. Leaves known for their characteristic bitter taste act as pyrigesic and pain reliever. Flowers are reported to serve as astringent, depurative, aphrodisiac, and diuretic. Seeds possessing anti-helminthic properties, help in removal of the white spots from the cornea and are also known to help in male fertility. Fruits are used in the treatment of tumors, tubercular glands, skin ulcers, and carbuncles (Kaur et al., 2004).

11.3 ASHWAGANDHA IN ITS VARIOUS BENEFICIAL ROLES; AN OVERVIEW

11.3.1 Role of Ashwagandha in the Pharmaceutical Industry

11.3.1.1 IG in Neuroprotection and Anti-neurodegeneration

Dementia and Alzheimer's disease are prominent neurodegenerative diseases associated and affecting almost the maximum aging mass population of any country (Mikulska et al., 2023). The onset of such neurodegeneration starts with mild cognitive imbalances which if ignored lead to dementia syndrome. Dementia is marked by its multicomponent etiology causing memory loss, affecting cortical brain functions, loss of orientation and cognition, language and situational comprehension, and imbalanced emotional control. Alzheimer's is one of the most commonly classified types of dementia which causes the damage of to the central nervous system (CNS) due to abnormal deposition of beta amyloid proteins in the brain. These beta-amyloid proteins when found in fibrillar forms exert neurotoxic effects by inducing the generation of free radicals and obstructing glucose supply to healthy neurons and ultimately causing cell death. The senile plaques formed by beta-amyloids are surrounded by hyperphosphorylated tau proteins. Microglial inflammatory cells are found to be associated with the senile plaques which try to break them. However, in the process, several healthy neuron cells of the brain are also destroyed as these microglial cells fail to differentiate between damaged and healthy neurons (Tiwari et al., 2019). The bioactive compounds, namely withanolide A, withanolide B, witanoside IV and witanoside V of Ashwagandha are known to neutralize the detrimental effects of beta-amyloid proteins, inhibit their production by regulating their aggregate formation, decrease the expression of pro-inflammatory cytokines, nitric oxide, and prevent lipid peroxidation (Das et al., 2021; Dubey et al., 2021).

Parkinson's disease is also a neurodegenerative disease of the CNS which affects both motor and nonmotor functions of the body. Characteristic pathophysiology of the disease states progressive nerve cell death at the substantia nigra region of the midbrain which is known to supply dopamine to the basal ganglia that controls voluntary motor control. Factors related to damage of nigrostriatal cells are reported to be the single or combinatorial effect of genetic prehistory, internal or external toxic factors, neural infections, oxidative stress,and growth malfunctions (Zhou et al., 2023). Ashwagandha root extracts have been reported to improve several biochemical parameters, such as reducing lipid peroxidation-mediated membrane damage, causing increased glutathione concentration, increasing antioxidant defense mechanisms, enhancing dopamine D2 receptor binding,tyrosine hydroxylase expression, etc. (Wongtrakul, et al., 2021).

Huntington's disease is known to be an incurable genetic disease inherited by dominant autosomal inheritance. Mutation of the IT15 gene on chromosome 4 that encodes htt protein causes a conformational change to form the insoluble mutant expanded polyglutamine huntingtin protein. This mutant htt leads to neuronal apoptosis, an imbalance of dopamine, GABA, serotonin, and acetylcholine (Rojas et al., 2022). Ashwagandha, mainly withaferin A, is known to improve the biochemical parameters, reduce the inflammatory responses, and delay neuronal apoptosis (Kumar and Kumar, 2009).

11.3.1.2 IG in the Treatment of Psychological Disorders, Alcohol Addiction and Insomnia

Obsessive-compulsive disorder (OCD) is defined as a psychological disorder where patients show irritative, compulsive behavior which makes them difficult to adjust within society (Leder et al., 2016). It is caused by structural functional defects of the CNS which may have genetic or developmental backgrounds. It is diagnosed as the dysregulation of the serotonergic system which is reported to show prominent improvement during the administration of Ashwagandha biocompounds(Jahanbakhsh et al., 2016). Alcohol-addicted patients are known to exhibit alcohol withdrawal symptoms such as anxiety, irritation, seizures,and depression in the absence of the source of addiction. Ashwagandha shows promising effects during the treatment of alcohol withdrawal syndrome, especially in improving cognition and locomotory functions (Haque et al., 2021). According to latest researchers, insomnia is believed to be a much more complex disorder than previously thought. It is characterized by the difficulty in initiating sleep, maintaining it and also waking up after a short span. Patients suffering from insomnia sleep so less that their sleep-to-work ratio creates a severe imbalance in their daily life and initiates problems in their daytime functioning by creating low concentration, emotional imbalances, cognitive problems, and lack of motivation thus affecting the homeostasis between worklife, personal life and social life (Siemiński et al., 2018).Ashwagandha showed promising results in improving sleep quality and enhancing alertness in patients suffering from with insomnia (Kelgane et al., 2020). Besides, Ashwagandha also showed good results in the treatment of patients with non-restorative sleep (NRS) disorders (Kaushik et al., 2017). Similar positive results were found in the treatment of sleep deprivation when done with root extracts of Ashwagandha (Suganya et al., 2020).

11.3.1.3 IG in the Treatment of Cancer, Diabetes and Arthritis

Cancer is defined as the uncontrolled growth of cells due to mutations in the proto-oncogene that lead to mutated cell cycle proteins. Withanolides isolated from different parts of IG are reported to exhibit anti-cancerous properties and are therefore used singly or in combination with other chemotherapeutic agents for treating cancerous tumors (Mehta et al., 2021). These withanolides induce apoptosis which is effective in arresting cancer of organs such as the breast, colon, lung, and prostrate within specific limits (Singh et al., 2021).Withaferin A prevents the migration of melanoma cells (Nagy et al., 2020). Besides the root extracts of IG are also known to act as a protectant against the adverse effects of radiotherapy-induced organ damage, specifically during liver and spleen cancer therapy (Azab et al., 2022; Mounica et al., 2022).

The anti-diabetic properties of IG roots were demonstrated where withaferin A was reported to lower blood glucose levels and effectively control diabetes Type 1 (Jain et al., 2006). But reports confirming the direct effect of Ashwagandha biocompounds in lowering blood glucose levels are relatively low. On the contrary, reports showed to significant improvement the lipidemic profile of diabetic patients when treated with IG biocompounds (Udayakumar et al., 2009). This result suggests that IG extracts possibly act as an antidiabetic agent in an indirect way.

Arthritis is defined as inflammation of joints. Interestingly, Ashwagandha act as a potent anti-inflammatory agent and protect a wide range of inflammations which may that manifest in any form as arthritis, neuro-inflammation, skin inflammation, or allergy. Reports suggest that IG extracts act as protectants of cartilage and collagen. Besides, IG extracts are also known to suppress the production of lipopolysaccharide-induced pro-inflammatory cytokines such as TNF-a, IL-1b, and IL-12p40, other stress markers like nitric oxide, reactive oxygen species (ROS),which altogether play significant roles in collagen stabilization (Gupta et al., 2018; Kanjilal et al., 2021).

11.3.1.4 IG as a Cardioprotectant

IG serves as a cardioprotectant by decreasing myocardial necrosis which was marked by a decrease in glutathione levels, lipid peroxidation amounts coupled with a decrease in antioxidant enzymes such as catalase, superoxide dismutase (SOD), decrease in amounts of creatinine phosphokinase,lactate dehydrogenase, etc. (Mohanty et al., 2008). Reports also endorsed that withaferin A acted as a cardioprotective drug by inhibiting apoptosis via the mitochondrial anti-apoptotic pathway and restoring oxidative balance in a dose-dependent manner (Guo et al., 2019).

11.3.1.5 IG in Infertility Treatment and Hypothyroidism

Infertility is considered to as a demographic problem that affects both men and women and imposes social and emotional stress. IG was found to show significant good results in men diagnosed with oligospermia over three months of treatment. The semen quality, count, sperm motility, testosterone, and luteinizing hormone levels showed qualitative increment. For men suffering from normozoospermia, Ashwagangha improved quality and semen parameters, thus increasing the chances of zygote formation. An overall improvement in the sexual hormonal profile, an increase in antioxidant enzymes and antioxidant vitamins like Vitamin A, Vitamin C, and Vitamin E in semen plasma were also noticed (Gupta et al., 2013). Lipid peroxidation and protein carbonyl groups (CO) were reduced (Sengupta et al., 2018). However, the relation between lipid peroxidation, accumulation of CO in blood tissues and infertility in men is still not elucidated. Besides, the molecular mechanism of Ashwagandha in regulating the disease also needs concrete conclusions. In females, Ashwagandha root extracts are reported to improve the sexual functioning parameters (Dongre et al., 2015).

Hypothyroidism and diabetes are believed to have an interdependent relationship which is reported to affect a large number of populations. Hypothyroidism causes several symptoms, such as loss of energy, weight gain, decreased appetite, sensitivity to cold, hair loss, skin dryness, blurry vision, memory impairment, lack of concentration, depression, etc. which disturb the functioning of daily life. If untreated, it may lead to worsening and affect vital organs such as the brain, heart, and kidney. Iodine is known to stimulate the thyroid gland and produce hormones Triiodothyronine (T3), Thyroxine (T4), and Thyroid-stimulating hormone (TSH). Ashwagandha root extracts, mainly withaferin A regulates the production of these hormones in patients with subclinical hypothyroidism by regulating the antioxidant system (Abdel-Wahhab et al., 2019). However, in advanced hypothyroidism withaferin A effects are still not conclusive.

11.3.1.6 IG in Skin Diseases

Several skin diseases, such as scleroderma, hyper and hypopigmentation and skin infections caused by viruses are known to show positive results during treatment with withaferin A. Scleroderma, parallelly known as dermal fibrosis is an autoimmune disorder which starts with inflammation and gradually accumulates in the extracellular matrix causing hardening of skin and scars. Withaferin A is reported to function as the antifibrotic agent in therapeutics which controls the inflammation within the tolerable limits and helps in treating dermal fibrosis (Bale et al., 2018).Melanogenesis is a process which regulates the skin color. Importantly, irregularity in the molecular mechanism of melanin formation leads to hypo or hyperpigmentation. The melanin formation is also regulated by redox balances which are controlled by the protein kinase C and MAP kinase pathways (Matsuzawa and Ichijo, 2008). As discussed in previous sections, Ashwagandha root extracts are known for their redox regulatory role which also shows promising results in the treatment of skin disorders (Nakajima et al., 2011). Additionally, withaferin A is reported to act as a binding ligand of Herpes simplex virus and papillomavirus,which plays an important role in arresting the spread of the viruses that are known to cause serious skin infections (Grover et al., 2011; Rotaru et al., 2014).

11.3.1.7 IG in Treatment of COVID Treatment

Research conducted with herbal drugs as prominent antiviral candidates during the COVID-19 pandemic highlighted the importance of active compounds isolated from Ashwagandha in showing antiviral properties. Withanoside and somniferin showed inhibition of SARS-CoV-2M protease. Withanolides showed antiviral activity during Covid treatment (Khanal et al., 2022) while withanone showed antiviral activity by targeting the host's major viral protease and transmembrane TMPRSS2 (Balkrishna et al., 2021).

11.3.1.8 IG as an Antimicrobial and Anti-stress Agent

IG also showed antimicrobial activity against several harmful bacteria such as *Staphylococcus aureus, Enterococcus, Escherichia coli, Proteus mirabilis, Pseudomonas aeruginosa, Salmonella typhi, Citrobactor freundi, Klebsiella pneumonia,*etc., by enhancing immune reactivity, exerting localized cytotoxicity, gene silencing, and arresting bacterial spread (Mwitari et al., 2013). Withanolides induce apoptotic cell death in *Leishmania donovani* (Chandrasekaran et al., 2013). Besides, Ashwagandha is reported to show antifungal effects against a wide range of harmful fungi and malaria parasites as well (Girish et al., 2006; Dikasso et al., 2007). Stress is the marker of any physical as well as psychiatric imbalance. Promisingly, IG root extracts are reported to show consistently positive results in improving stress, anxiety, depression, mood swings, and varied types of distress caused by physical disturbances and/or hormonal-induced psychological imbalances by inhibiting oxidative stress-related pathways (Mikulska et al., 2023).

11.3.2 Adaptogen and/or Nutraceutical Properties of Ashwagandha

Adaptogens are defined as metabolic regulators that build up the body's adaptability to cope with any type of environmental stress and prevent internal damage. However, adaptogens are referred as to be ideal when they are known to pose no side effects on an individual's body and can function as per requirement (Panosian and Wikman, 2009). Studies conducted on rats showed the capacity of Ashwagandha root extract to function as an adaptogen. IG roots were reported to alleviate stress by reducing hyperglycemia, decreasing levels of plasma corticosterone, reducing gastric ulcers, increasing cognitive alertness, and boosting immunity (Bhattacharya and Muruganandam, 2003).

Nutraceutical is referred to as a complex compound which holds the properties of both a nutrient and a pharmaceutical. Furthermore, a nutraceutical is considered to serve as an adaptogen in its full potential which makes the demarcation of nutraceutical and adaptogen blurred. The pharmaceutical importance of IG has been discussed in the above section; however, its classification in the list of nutraceuticals is comparatively new. IG is now a potent health supplement which shows proven effects in building muscle strength, improving gut functioning, and also maintaining a healthy communication between the gut-brain axis. The gut-brain axis is referred to as a bidirectional communication between the CNS and ENS (esophageal nervous system) (Carabotti et al., 2015). IG which enhances anxiolytic properties has been reported to reduce depression, anxiety and promote the gut health by regulating the peristaltic movement and gastric function (Evans and Kalman, 2024).

11.4 CENTRAL MOLECULAR MECHANISM EXPLAINS ABOUT THE PROBABLE OVERLAPS OF DIVERGENT ROLES OF ASHWAGANDHA; A HYPOTHESIS

Several researches conducted with the bioactive root extract of Ashwagandha have proven its importance in the form of an efficient herbal drug for treating various types of diseases. Its role as an adaptogen and nutraceutical has also been realized of late. However, a thorough insight into the molecular mechanism of all the disease etiology pinpoints to the fact that any form of imbalance caused by environmental and/or internal or psychological is coupled with the production of ROS and triggering the oxidative stress pathways. Thus, detoxifying the ill effects of ROS by manipulating the antioxidant defense machinery appears to be the central mechanism of any potential drug component. Interestingly, most of the reports suggest that IG regulates the redox regulatory pathways and prevents any form of inflammation as found in most of the harmful diseases like, neurodegeneration, cardiac complications, cancer, allergy, psychological stress-related disorders, microbial infections, diabetes, arthritis, etc. Similarly, adaptogen and/or nutraceutical compounds are also referred to as efficient redox modulators. Thus the redox regulatory capacity of Ashwagandha has helped the wonder shrub to hold multiple functions of acting as a pharmaceutical, adaptogen, and nutraceutical (Figure 11.1).

11.5 TOXICOLOGICAL ASSESSMENT

The importance and efficacy of Ashwagandha in treating several diseases has been realized since a long time. However, some reports highlighted the potential risk of hepatotoxicity and liver damage due to prolonged use of Ashwagandha specifically in patients with other comorbidity (Philips et al., 2020). Significantly, all the reports of hepatotoxic effects of the herbal drug were reported from countries other than India, indicating that the varietal variation could be an important criterion in deciding the efficacy as well as the toxicity of the plant extracts (Verma et al., 2021). However, like any other drug, the application of Ashwagandha is restricted for pregnant women and/or new mothers providing breastfeeding (Prabu and Panchapakesan, 2015). Besides, reports suggest that Ashwagandha shows contraindications like nervine irregularities such as irritability, anxiety, palpitations, muscle tremors, psychomotor agitations, etc. in patients diagnosed with hyperthyroidism

(Kamal et al., 2022). It is not recommended for the treatment of male sterility in patients already having a prehistory of prostate cancer (Połumackanycz et al., 2020). Administration of this herbal drug is also reported to be hazardous for patients already receiving treatment for autoimmunity and its associated disorders (Savai et al., 2015). Thus, the toxicity assessment of this herbal drug should be performed more meticulously, where plant variety, its location, edaphic, and biospheric conditions are to be taken into account, and its results thoroughly analyzed for making this drug suitable for commercial use.

11.6 PROSPECTS AND THE FUTURE OF ASHWAGANDHA

To summarize, IG is unquestionably a promising agent that could be utilized and applied for the treatment of various diseases and disorders. Additionally, it could be used as an adaptogen and nutritional supplement for boosting immune responses. Even then, some aspects of the wonder plant need to be further investigated. A few such areas of investigation are as follows:

1. How does the efficacy of the different bioactive compounds of IG vary across different geographic locations?
2. How does the efficacy of active compounds of IG vary along with the age of the plant?
3. Does the soil microbiome have any effect on the concentration and effectiveness of the active compounds of IG?
4. How are the active compounds of IG changed with climatic alterations?
5. How do the active compounds of IG behave while treating patients with multiple comorbidities?
6. Since the brain and gut relationship has been well established, how do the IG compounds alter gut microbiota and viceversa?

Extensive experimental answers to the above questions with case-specific substantial toxicity assessment are expected to make IG a stable, consistent, and reliable herbal drug of multidimensional value.

ACKNOWLEDGMENT

The authors duly acknowledge their affiliating institution for providing the professional positions and support. However, no funding was required for conducting the review study.

REFERENCES

Abdel-Wahhab, K.G.; Mourad, H.H.; Mannaa, F.A.; Morsy, F.A.; Hassan, L.K.; Taher, R.F. Role of Ashwagandha methanolic extract in the regulation of thyroid profile in hypothyroidism modeled rats. *Mol. Biol. Rep.* **2019**, 46, 3637–3649.

Azab, K.S.; Maarouf, R.E.; Abdel-Rafei, M.K.; El Bakary, N.M.; Thabet, N.M. *Withania somnifera* (Ashwagandha) root extract counteract acute and chronic impact of radiation on liver and spleen of rats. *Hum. Exp. Toxicol.* **2022**, 41, 9603271221106344.

Bale, S.; Pulivendala, G.; Godugu, C.Withaferin A attenuates bleomycin-induced scleroderma by targeting FoxO3a and NF-κβ signaling: Connecting fibrosis and inflammation. *Biofactors* **2018**, 44, 507–517.

Balkrishna, A.; Pokhrel, S.; Singh, H.; Joshi, M.; Mulay, V.P.; Haldar, S.; Varshney, A. Withanone from *Withania somnifera* attenuates SARS-CoV-2 RBD and Host ACE2 interactions to rescue spike protein induced pathologies in humanized Zebrafish model. *Drug Des. Devel. Ther.* **2021**, 15, 1111–1133.

Bashir, A.; Nabi, M.; Tabassum, N.; Afzal, S.; Ayoub, M.An updated review on phytochemistry and molecular targets of *Withania somnifera* (L.) Dunal (Ashwagandha). *Front. Pharmacol.* **2023**, 29(14),1049334.

Bhattacharya, S.K.; Muruganandam, A.V. Adaptogenic activity of *Withania somnifera*: An experimental study using a rat model of chronic stress. *Pharmacol. Biochem. Behav.* **2003**, 75, 547–555.

Carabotti, M.; Scirocco, A.; Maselli, M.A.; Severi, C. The gut-brain axis: Interactions between enteric microbiota, central and enteric nervous systems. *Ann. Gastroenterol.* **2015**, 28, 203–239.

Chandrasekaran, S.; Dayakar, A.; Veronica, J.; Sundar, S.; Maurya, R. An in vitro study of apoptotic like death in *Leishmania donovani* promastigotes by withanolides. *Parasitol. Int.* **2013**, 62, 253–261.

Dar, N.J.; Hamid, A.; Ahmad, M. Pharmacologic overview of *Withania somnifera*, the Indian Ginseng. *Cell. Mol. Life Sci.* **2015**, 72, 4445–4460.

Das, R.; Rauf, A.; Akhter, S.; Islam, M.N.; Emran, T.B.; Mitra, S.; Khan, I.N.; Mubarak, M.S.Role of withaferin A and its derivatives in the management of Alzheimer's disease: Recent trends and future perspectives. *Molecules* **2021**, 26, 3696.

Dhandge, P.D.; Deshmukh, S. P.A review on role of herbal medicine in daily life. *GSC Bio. Pharma. Sci.* **2023**, 25(3), 179–188.

Dhar, N.; Razdan, S.; Rana, S.; Bhat, W.W.; Vishwakarma, R.; Lattoo, S.K.A decade of molecular understanding of withanolide biosynthesis and in vitro studies in *Withania somnifera* (L.) Dunal: Prospects and perspectives for pathway engineering. *Front. Plant Sci.* **2015**, 27(6), 1031.

Dikasso, D.; Makonnen, E.; Debela, A.; Abebe, D.; Urga, K.; Makonnen, W.; Assefa, A.; Melaku, D.; Makonnen, W.In vivo anti-malarial activity of hydroalcoholic extracts from *Asparagus africanus* Lam In mice infected with *Plasmodium berghei. Ethiop. J. Health Dev.* **2007**, 20, 112–118.

Dongre, S.; Langade, D.; Bhattacharyya, S.Efficacy and safety of Ashwagandha (*Withania somnifera*) root extract in improving sexual function in women: A pilot study. *Biomed Res. Int.* **2015**, 2015, 284154.

Dubey, S.; Kallubai, M.; Subramanyam, R.Improving the inhibition of β-amyloid aggregation by withanolide and withanoside derivatives. *Int. J. Biol. Macromol.* **2021**, 173, 56–65.

Evans, C.; Kalman, D.Addressing GI health through the bidirectional modulation of the gut-brain axis with herbal extracts: A narrative review. *Cureus.* **2024**, 16(8).

Gaurav, N.; Kumar, A.; Tyagi, M.; Kumar, D.; Chauhan, U.K.; Singh, A.P.Morphology of *Withania somnifera* (distribution, morphology, phytosociology of *Withania somnifera* L. Dunal). *Int. J. Current Sci. Res.* **2015**, 1(7), 164–173.

Gautam, S.; Pandey, A.K.Ashwagandha: A rasayana from ayurveda. *Med. Plants-Int. J. Phytomed. Rel. Ind.* **2020**, 12(2), 173–181.

Girish, K.S.; Machiah, K.D.; Ushanandini, S.; Kumar, K.H.; Nagaraju, S.; Govindappa, M.; Vedavathi, M.; Kemparaju, K.Antimicrobial properties of a non-toxic glycoprotein (WSG) from *Withania somnifera* (Ashwagandha). *J. Basic Microbiol.* **2006**, 46, 365–374.

Grover, A.; Agrawal, V.; Shandilya, A.; Bisaria, V.S.; Sundar, D.Non-nucleosidic inhibition of Herpes simplex virus DNA polymerase: Mechanistic insights into the anti-herpetic mode of action of herbal drug withaferin A. *BMC Bioinform.* **2011**, 12, S22.

Guo, R.; Gan, L.; Lau, W.B.; Yan, Z.; Xie, D.; Gao, E.; Christopher, T.A.; Lopez, B.L.; Ma, X.; Wang, Y.Withaferin A prevents myocardial ischemia/reperfusion injury by upregulating AMP-activated protein kinase-dependent B-cell lymphoma2 signaling. *Circ. J.* **2019**, 83, 1726–1736.

Gupta, A.; Mahdi, A.; Shukla, K.K.; Ahmad, M.K.; Bansal, N.; Sankhwar, P.; Sankhwar, S.N.Efficacy of *Withania somnifera* on seminal plasma metabolites of infertile males: A proton NMR study at 800 MHz. *J. Ethnopharmacol.* **2013**, 149, 208–214.

Gupta, M.; Kaur, G.*Withania Somnifera* as a potential anxiolytic and anti-inflammatory candidate against systemic lipopolysaccharide-induced neuroinflammation. *Neuromol. Med.* **2018**, 20, 343–362.

Haque, I.M.; Mishra, A.; Kalra, B.S.; Chawla, S.Role of standardized plant extracts in controlling alcohol withdrawal syndrome—An experimental study. *Brain Sci.* **2021**, 11, 919.

Hawkes, J.G.; Lester, R.N.; Nee, M.; Estrada, N.*Solanaceae III; Taxonomy, Chemistry, Evolution.* The Royal Botanic Gardens, Kew Publishing, London **1991**.

Jahanbakhsh, S.P.; Manteghi, A.A.; Emami, S.A.; Mahyari, S.; Gholampour, B.; Mohammadpour, A.H.; Sahebkar, A.Evaluation of the efficacy of *Withania somnifera* (Ashwagandha) root extract in patients with obsessive-compulsive disorder: A randomized double-blind placebo-controlled trial. *Complement. Ther. Med.* **2016**, 27, 25–29.

Jain, S.; Pandhi, P.; Singh, A.P.; Malhotra, S.Efficacy of standardised herbal extracts in type1 diabetes—An experimental study. *Afr. J. Tradit. Complement. Altern. Med.* **2006**, 3, 23–33.

Jaiswal, Y.S.; Williams, L.L. A glimpse of Ayurveda–The forgotten history and principles of Indian traditional medicine. *J.Trad. Complement.Med.* **2017**, 7(1), 50–53.

John, J.Therapeutic potential of *Withania somnifera*: A report on phyto-pharmacological properties. *Int. J. Pharm. Sci. Res.* **2014**, 5, 2131–2148.

Kamal, H.I.; Patel, K.; Brdak, A.; Heffernan, J.; Ahmad, N.Ashwagandha as a unique cause of thyrotoxicosis presenting with Supraventricular tachycardia. *Cureus* **2022**, 14, e23494.

Kanjilal, S.; Gupta, A.K.; Patnaik, R.S.; Dey, A.Analysis of clinical trial registry of India for evidence of anti-arthritic properties of *Withania somnifera* (Ashwagandha). *Altern. Ther. Health Med.* **2021**, 27, 58–66.

Kaur, K.; Rani, G.; Widodo, N.; Nagpal, A.; Taira, K.; Kaul, S.C.; Wadhwa, R.Evaluation of the anti-proliferative and anti-oxidative activities of leaf extract from in vivo and in vitro raised Ashwagandha. *Food Chem. Toxicol.* **2004**, 42(12), 2015–2020.

Kaushik, M.K.; Kaul, S.C.; Wadhwa, R.; Yanagisawa, M.; Urade, Y.Triethylene glycol, an active component of Ashwagandha (*Withania somnifera*) leaves, is responsible for sleep induction. *PLoS One* **2017**,12, e0172508.

Kelgane, S.B.; Salve, J.; Sampara, P.; Debnath, K.Efficacy and tolerability of ashwagandha root extract in the elderly for improvement of general well-being and sleep: A prospective, randomized, double-blind, placebo-controlled study. *Cureus* **2020**, 12, e7083.

Khanal, P.; Chikhale, R.; Dey, Y.N.; Pasha, I.; Chand, S.; Gurav, N.; Ayyanar, M.; Patil, B.M.; Gurav, S.Withanolides from *Withania somnifera* as an immunity booster and their therapeutic options against COVID-19. *J. Biomol. Struct. Dyn.* **2022**, 40, 5295–5308.

Kumar, A.; Venugopal, S.; Jnanesha, A.C.; Lal, R.K.Agricultural-based challenges, genetic enhancement, and obstacles to an industrially important medicinal plant, ashwagandha (*Withania somnifera* (L.) Dunal): A review. *Eco. Genet. Genom.* **2023**, 28, 100183.

Kumar, R.R.; Reddy, A.P.; Subbaiah, C.J.; Kumar, N.A.; Prasad, N.H.; Bhukya, B.Genetic association among root morphology, root quality and root yield in ashwagandha (*Withania somnifera*). *Genetika.* **2011**, 43(3), 617–624.

Kumar, P.; Kumar, A.Possible neuroprotective effect of *Withania somnifera* root extract against 3-nitropropionic acid-induced behavioral, biochemical, and mitochondrial dysfunction in an animal model of Huntington's disease. *J. Med. Food.* **2009**, 12,591–600.

Leder, S.; Siwiak-Kobayashi, M.; Nerwice, W.B.A.*Psychiatria: podręcznik dla studentów medycyny.* Wydawnictwo Lekarskie PZWL:Warszawa, Poland, **2016**.

Matsuzawa, A.; Ichijo, H.Redox control of cell fate by MAP kinase: Physiological roles of ASK1-MAP kinase pathway in stress signaling. *Biochim. Biophys. Acta* **2008**, **1780**, 1325–1336.

Meena, L.K.; Meena, A.K.; Gupta, J.; Patel, M.; Khan, S.Y.; Kumar, S.; Gupta, A.K.*Ashwagandha (Withania somnifera L.). Medicinal Plants in India: Importance and Cultivation.* Jaya Publishing House: Delhi, India, **2020**.

Mehta, V.; Chander, H.; Munshi, A.Mechanisms of anti-tumor activity of *Withania somnifera* (Ashwagandha). *Nutr. Cancer* **2021**, 73, 914–926.

Mikulska, P.;Malinowska, M.; Ignacyk, M.; Szustowski, P.; Nowak, J.; Pesta, K.; Szeląg, M.; Szklanny, D.; Judasz, E.; Kaczmarek, G.; Ejiohuo, O.P.Ashwagandha (*Withania somnifera*)—Current research on the health-promoting activities: A narrative review. *Pharmaceutics.* **2023**, 15(4), 1057.

Mirjalili, M.H.; Moyano, E.; Bonfill, M.; Cusido, R.M.; Palazón, J.Steroidal lactones from *Withania somnifera*, an ancient plant for novel medicine. *Molecules* **2009**, 14(7), 2373–2393.

Mohanty, I.R.; Arya, D.S.; Gupta, S.K.*Withania somnifera* provides cardioprotection and attenuates ischemia-reperfusion induced apoptosis. *Clin. Nutr.* **2008**, 27, 635–642.

Mounica, P.K.; Devabattula, G.; Godugu, C.Role of Indian ginseng withaferin A in ROS-induced cancer chemoprevention and cancer therapy. In *Handbook of Oxidative Stress in Cancer: Therapeutic Aspects* edited by Sajal Chakraborti, pp. 327–349. Springer Nature Singapore: Singapore, **2022**.

Mukherjee, P.K.; Venkatesh, P.;Ponnusankar, S.Ethnopharmacology and integrative medicine–Let the history tell the future. *J. Ayurved.Int. Med.* **2010**, 1(2), 100.

Mukherjee, P.K.Evaluation of Indian traditional medicine. *Drug Inf. J.* **2001**, 35(2), 623–632.

Munien, P.; Naidoo, Y.; Naidoo, G.Micromorphology, histochemistry and ultrastructure of the foliar trichomes of *Withania somnifera* (L.) Dunal (Solanaceae). *Planta* **2015**, 242, 1107–1122.

Mwitari, P.G.; Ayeka, P.A.; Ondicho, J.; Matu, E.N.; Bii, C.C.Antimicrobial activity and probable mechanisms of action of medicinal plants of Kenya: *Withania somnifera*, *Warbugia ugandensis*, *Prunus africana* and *Plectrunthus barbatus*. *PLoS One* **2013**, 8, e65619.

Nagy, Z.; Cheung, B.B.; Tsang, W.; Tan, O.; Herath, M.; Ciampa, O.C.; Shadma, F.; Carter, D.R.; Marshall, G.M.Withaferin A activates TRIM16 for its anti-cancer activity in melanoma. *Sci. Rep.* **2020**, 10, 1–9.

Nakajima, H.;Wakabayashi, Y.;Wakamatsu, K.; Imokawa, G.An extract of*Withania somnifera* attenuates endothelin-1-stimulated pigmentation in human epidermal equivalents through the interruption of PKC activity within melanocytes. *Phytother. Res.* **2011**, 25, 1398–1411.

Ouellet, C.; Saïas, T.; Sit, V.; Lamothe, L.; Rapinski, M.; Cuerrier, A.; Haddad, P.S.Access to indigenous and allopathic medicines: A systematic review of barriers and facilitators. *Global J. Comm. Psycho. Prac.* **2018**, 9(2).

Panossian, A.; Wikman, G.Evidence based efficacy of adaptogens in fatigue evidence-based efficacy of adaptogens in fatigue, and molecular mechanisms related to their stress-protective activity. *Curr. Clin. Pharmacol.* **2009**, 4, 198–219.

Philips, C.A.; Ahamed, R.; Rajesh, S.; George, T.; Mohanan, M.; Augustine, P.Comprehensive review of hepatotoxicity associated with traditional Indian Ayurvedic herbs. *World J. Hepatol.* **2020**, 12, 574–595.

Połumackanycz, M.; Forencewicz, A.; Wesołowski, M.; Viapiana, A.Ashwagandha (*Withania somnifera* L.)—The plant with proven health-promoting properties. *Farm. Pol.* **2020**, 76, 442–447.

Prabu, P.C.; Panchapakesan, S.Prenatal developmental toxicity evaluation of *Withania somnifera* root extract in Wistar rats. *Drug Chem. Toxicol.* **2015**, 38, 50–56.

Rastogi, S.; Bhattacharya, A.; Singh, R.H.Evidence building in Ayurveda: Generating the new and optimizing the old could be strategic. In *Ayurveda in the New Millennium*, pp. 161–174. Florida, USA: CRC Press, **2020**.

Rojas, N.G.; Cesarini, M.E.; Peker, G.; Da Prat, G.A.; Etcheverry, J.L.; Gatto, E.M.Review of Huntington's disease: From basics to advances in diagnosis and treatment. *J. Neurol. Res.* **2022**, 12, 93–113.

Rotaru, M.; Iancu, G.; Mihalache, M.; Anton, G.; Morariu, S.-H.α-HPV positivity analysis in a group of patients with melanoma and non-melanoma skin cancers. *Rom. Rev. Lab. Med.* **2014**, 22, 471– 478.

Saidulu, C.; Venkateshwar, C.; Rao, S.G.Morphological studies of medicinal plant of *Withania somnifera* (L.) Dunal grown in heavy metal treated (contaminated) soil. *J. Pharma. Phytochem.* **2014**, 3(1), 37–42.

Savai, J.; Varghese, A.; Pandita, N.; Chintamaneni, M.Investigation of CYP3A4 and CYP2D6 interactions of *Withania somnifera* and *Centella asiatica* in human liver microsomes. *Phytother. Res.* **2015**, 29, 785–790.

Sen, S.; Chakraborty, R.Revival, modernization and integration of Indian traditional herbal medicine in clinical practice: Importance, challenges and future. *J.Trad.Complement.Med.* **2017**, 7(2), 234–244.

Sengupta, P.; Agarwal, A.; Pogrebetskaya, M.; Roychoudhury, S.; Durairajanayagam, D.; Henkel, R.Role of *Withania somnifera* (Ashwagandha) in the management of male infertility. *Reprod. Biomed.* **2018**, 36, 311–326.

Siemiński, M.; Skorupa, Ł.; Wi´sniewska-Skorupa, K.Diagnostyka i terapia bezsenności w praktyce ogólnolekarskiej Część I: Epidemiologia, patomechanizm i diagnostyka bezsenności. *Forum Med. Rodz.* **2018**, 12, 242–251.

Singh, N.; Bhalla, M.; de Jager, P.; Gilca, M.An overview on Ashwagandha: A Rasayana (rejuvenator) of ayurveda. *Afr. J. Tradit. Complement. Altern. Med.* **2011**, 8, 208–213.

Singh, N.; Yadav, S.S.; Rao, A.S.; Nandal, A.; Kumar, S.; Ganaie, S.A.; Narasihman, B.Review on anticancerous therapeutic potential of *Withania somnifera* (L.) Dunal. *J. Ethnopharmacol.* **2021**, 270, 113704.

Suganya, K.; Kayalvizhi, E.; Yuvaraj, R.; Chandrasekar, M.; Kavitha, U.; Suresh, K.K.Effect of*Withania somnifera* on the antioxidant and neurotransmitter status in sleep deprivation induced Wistar rats. *Bioinformation* **2020**, 16, 631–637.

Tewari, D.; Chander, V.; Dhyani, A.; Sahu, S.; Gupta, P.; Patni, P., et al.*Withania somnifera* (L.) Dunal: Phytochemistry, structure–activity relationship, and anticancer potential. *Phytomedicine* **2022**, 98, 153949. doi:10.1016/j.phymed.2022.153949

Tiwari, R.; Chakraborty, S.; Saminathan, M.; Dhama, K.; Singh, S.V.Ashwagandha (*Withania somnifera*): Role in safeguarding health, immunomodulatory effects, combating infections and therapeutic applications: A review. *J. Biol. Sci.* **2014**, 14(2), 77–94.

Tiwari, S.; Atluri, V.; Kaushik, A.; Yndart, A.; Nair, M.Alzheimer's disease: Pathogenesis, diagnostics, and therapeutics. *Int. J.Nanomed.* **2019**, 14, 5541–5554.

Tripathi, A.Ashwagandha'an 'adaptogenic herb. *Curr. Trends Med.Chem.* **2022**, 155–169.

Udayakumar, R.; Kasthurirengan, S.; Mariashibu, T.S.; Rajesh, M.; Anbazhagan, V.R.; Kim, S.C.; Ganapathi, A.; Choi, C.W.Hypoglycaemic and hypolipidaemic effects of *Withania somnifera* root and leaf extracts on alloxan-induced diabetic rats. *Int. J. Mol. Sci.* **2009**, 10, 2367–2382.

Uddin, Q.; Samiulla, L.; Singh, V.; Jamil, S.Phytochemical and pharmacological profile of *Withania somnifera*-Dunal: A review. *J. Appl. Pharm. Sci.* **2012**, 2(1), 170–175.

Varma, D.R.From witchcraft to allopathy: Uninterrupted journey of medical science. *Eco. Pol. Weekly* **2006**, 19, 3605–3611.

Ven Murthy, M.R.; K Ranjekar, P.; Ramassamy, C.; Deshpande, M. Scientific basis for the use of Indian Ayurvedic medicinal plants in the treatment of neurodegenerative disorders: 1. Ashwagandha. *Cent. Nerv. Syst. Agents Med. Chem.* **2010**, 10, 238–246.

Verma, N.; Gupta, S.K.; Tiwari, S.; Mishra, A.K. Safety of ashwagandha root extract: A randomized, placebo-controlled, study in healthy volunteers. *Complement. Ther. Med.* **2021**, 57, 102642.

Wongtrakul, J.; Thongtan, T.; Kumrapich, B.; Saisawang, C.; Ketterman, A.J. Neuroprotective effects of *Withania somnifera* in the SH-SY5Y Parkinson cell model. *Heliyon* **2021**, 7, e08172.

Zhou, Z.D.; Yi, L.X.; Wang, D.Q.; Lim, T.M.; Tan, E.K. Role of dopamine in the pathophysiology of Parkinson's disease. *Trans. Neurodegener.* **2023**, 12(1), 44.

12 Hairy Root Engineering for Enhanced Production of Secondary Metabolites

Sompalli Bhavana[1], Wudali Narasimha Sudheer[2], Jameel M. Al-Khayri[3], and Praveen Nagella[2]*

[1]Department of Biotechnology, Surana College (Autonomous), Bangalore, India

[2]Department of Life Sciences, School of Sciences, Christ University, Bengaluru, India

[3]Department of Agricultural Biotechnology, College of Agriculture and Food Sciences, King Faisal University, Al-Ahsa, Saudi Arabia

12.1 INTRODUCTION

Contemporary biotechnology developments have made employing plants sensibly for their therapeutic and commercially valuable applications easier. Many of the plants features are caused by secondary metabolites (SMs), the phytochemicals that are produced by plants (Rabizadeh et al., 2022). The tiny molecules that are the byproducts and intermediates of physiological metabolism are metabolites. SMs are not necessary for living, whereas primary metabolites are (Al-Khayri et al., 2023). The lack of SMs does not immediately restrict an organism's survival but is greatly hampered. The main plant SMs are terpenoids, phenylpropanoids, alkaloids, and their derivatives. These compounds have a wide range of medicinal uses, including hepatoprotective, pain reliever, anti-cancer drugs, anticholinergic, and so on. These plant SMs have been utilized for centuries to promote human health and well-being in several ways, including medicinal, nutraceutical, taste-influencing, antioxidant, fragrant, and pigment-producing properties (Shrihastini et al., 2021; Dixit et al., 2021a, Upadhyay & Singh 2021, 2023, 2024).

Many of the SMs produced by solanaceous plants are frequently extracted from their roots and are of significant medicinal and pharmacological use (Biswas et al., 2023). In plant science, SMs have a variety of functions, such as defense against microbes and herbivores, pollination, attracting compounds for allelopathic agents (chemicals that influence interspecies competition), and seed-dispersing creatures (Divekar et al., 2022). However, because SMs accumulate in small amounts, the yields achieved by using natural plant populations are not high enough to satisfy the commercial need. Furthermore, there is a clear correlation between the age and growth circumstances of the plant and the in-plant production of these significant metabolites. These restrictions have made it possible to investigate other production techniques (Chen et al., 2022). Regardless of seasonal, geographical, or environmental variations, hairy root cultures offer a practical method of producing valuable SMs at an affordable price. Hairy roots are formed with the interaction between *Rhizobium rhizogenes* (*Agrobacterium rhizogenes*) and plants (Pistelli et al., 2010).

With a higher growth rate and metabolite productivity, hairy roots, which resemble the genuine plant's root in terms of phytochemical profile, have been widely explored as a promising substitute for the in vitro production of SMs. As to defend from biotic and abiotic stress, plants have virtually infinite capacity to generate a vast range of metabolic chemicals in various organs of the plant

 DOI: 10.1201/9781003518358-12

body via various biochemical networks. These aromatic compounds find widespread use as agro-chemicals, food additives, insecticides, terpenoids, flavonoids, phenols, alkaloids, or their oxygen-substituted derivatives (Jan et al., 2021; Ramakrishna & Ravishankar, 2011). After the gram-negative soil bacteria *A. rhizogenes* infects explants or in vitro cultures, a hairy root culture develops. It is also referred to as transformed root culture, and it aids in gene analysis as well as the study of plant metabolic processes by yielding useful SMs or recombinant proteins (Singh et al., 2022). Initially, the explants are infected with *A. rhizogenes* and then co-cultivated for two or three days and later moved to a solid medium containing antibiotics, like cefotaxime. They are regarded as an intriguing production system because of their rapid growth and capacity to yield valuable SMs in amounts that are comparable to or occasionally greater than those of adventitious root cultures (Huang et al., 2022). In addition, hairy root cultures (HRCs) are regarded as inexpensive, season-neutral, genetically and biochemically stable systems that may generate valuable SMs even on a large scale. *In vitro* hairy roots demonstrate their capacity to secrete metabolites into the culture medium, increase biomass on a large scale, respond effectively to external stimuli, and create plant SMs consistently (Rency et al., 2019). To further stimulate and/or improve the SM production potential, hairy root cultures can also be coupled with other yield enhancement techniques like precursor feeding, elicitation, and metabolic engineering (Almeida et al., 2022; Alok et al., 2018; Baek et al., 2020; Mahendran et al., 2022; Ru et al., 2016; Shriti et al., 2023; Verma et al., 2009; Wei et al., 2019). In order to boost productivity, researchers are primarily interested in using metabolic engineering and other technologies in hairy root cultivation. Because of biotechnological advancements, hairy root cultures are now using a range of engineering techniques to increase the production of SMs, such as gene co-overexpression, overexpression of transcription factors, overexpression of genes in biosynthetic pathways and silencing, and genome editing (Capriotti et al., 2020). The present chapter emphasizes the details of the hairy root culture's development and addresses the use of modern metabolic engineering techniques to increase their metabolic efficiency. The chapter also looks at future directions for this research.

12.2 SECONDARY METABOLITE PRODUCTION: TRADITIONAL APPROACH

A wide range of chemical substances referred to as SMs can be made from plants. The unique carbon skeleton structures of plant SMs are their fundamental features. However, these phytochemicals play vital roles inside the plant body and significantly influence human health by preventing major sickness (Omokhefe Bruce, 2022). Over the years, several chronic ailments have been successfully treated with natural chemicals and their derivatives derived from plants. However, the isolation and extraction of such phytochemicals may be restricted due to their low usability and environmental constraints (Chaachouay & Zidane, 2024; Vaou et al., 2021). *In vitro* reconstitution of metabolic pathways and genetic modifications of plants and microbes have developed a variety of plant SMs. These metabolites are small molecules that are the intermediates and byproducts of physiological metabolism (Pyne et al., 2019). By serving as protective substances in ecological relationships, symbiosis, competition, etc., they bestow adaptive responsibilities. External forces, growth, cell or body development, and tissue differentiation all affect the production of metabolites (Salam et al., 2023). Many important phytochemicals belonging to various classes, like alkaloids, terpenoids, tannins, phenols, flavonoids, and their derivatives, are made from these aromatic molecules (Alamgir, 2018). The primary role of SMs is to aid in generating organism's survival by serving as predatory agents against other microorganisms, plants, or animals. These substances can also act as sexual hormones, differentiation effectors, scavengers of vital nutrients, and/or initiators and maintainers of symbiotic relationships (Demain & Fang, 2000). Phenolics, terpenoids, alkaloids, and sulfur-containing compounds are the broad categories into which SMs are classified. The most varied category of SMs are terpenes that plants acquire via isopentenyl diphosphate (IPP) and dimethylallyl diphosphate (DMAPP), are five-carbon intermediates (Brahmkshatriya & Brahmkshatriya, 2013; Guerriero et al., 2018). These advantageous bioactive chemicals are only found in medicinal plants, which is why their disappearance is happening so quickly. It has been discovered that to satisfy the need for SMs, medicinal plants covering more than 95% of populations were harvested from the forests. Without taking care of the mother plants, many plant components, including roots, leaves, bark, fruits, and seeds, are being indiscriminately taken from their natural environments (Atanasov et al., 2015). Many of these significant and beneficial species are in danger of going extinct as a result of overexploitation and habitat loss. It is also crucial to consider the concentration of SMs found in plants' natural environment varies greatly depending on a variety of seasonal and environmental circumstances, and it is insufficient to satisfy industrial demand. Many *in vitro* methods, including cell and organ cultures, transformant cultures, and industrial scale cultures, are being increasingly employed for better enhancement of these bioactive compounds in order to permit the best possible production of tailored molecules. Their production was tried and guaranteed in addition to techniques including elicitation, precursor feeding, and biotransformation (Arya et al., 2021; Bapat et al., 2023; Lakshmaiah et al., 2022; Marchev et al., 2020). Table 12.1 provides the various plant phytochemicals or SMs produced by different plants and their medicinal value.

12.3 SECONDARY METABOLITE PRODUCTION: APPLIED BIOTECHNOLOGICAL TECHNIQUES

In vitro plant cultures have the ability to produce a variety of valuable SMs in a controlled, as and when needed manner. Recently, there has been significant progress in identifying cell cultures that can produce specific medicinal compounds at rates equal to or surpassing those of whole plants (Wu et al., 2021). The focus of efforts has been on isolating the biosynthetic activities of cultured cells to obtain high yields appropriate for industrial use. This has been accomplished through the use of transformation methods, precursor supplementation, optimal culture conditions, and the selection of high-yielding strains (Hussain et al., 2012).

The production of plant metabolites from tissue culture has been transformed by transgenic hairy root cultures. They develop quickly, are easier to maintain, and have unique genetic and biosynthetic stability (Giri & Narasu, 2000). New technologies that make it simple to change the source and extract secondary compounds in higher amounts and quality have been made possible by plant tissue culture techniques. Without reducing the impact of the environment, plant tissue culture methods provide a quicker rate of secondary compound production (Ozyigit et al., 2023). Advances in plant cell culture enzymology (Smetanska, 2008), fermentation technology (Willits et al., 2004), and molecular biology (Hayashi, 2009) indicate that these systems are likely to become a reliable source of plant SMs. While plants infected with modified viruses are producing relatively large amounts of desirable chemicals due to genome change, transgenic plants are able to maintain continuous levels of protein synthesis without further support. Large-scale plant tissue culture is seen as an alternative to traditional plantation methods, as it offers a consistent supply of biochemicals regardless of plant availability (Kowalczyk et al., 2020).

Other techniques like suspension culture (Ali et al., 2013; Cetin et al., 2014), elicitation (El-Beltagi et al., 2022; Farjaminezhad & Garoosi, 2021), shoot culture (Kikowska et al., 2020; Haider Abbasi et al., 2020), micropropagation (Amoo et al., 2012; Amoo & Van Staden, 2013), and callus culture (Ahmad et al., 2016; Cabañas-García et al., 2021) are considered as alternatives for enhanced production of metabolites in order to meet the growing demand for medically important natural compounds. These methods include modifying a variety of environmental and physiological factors to establish ideal conditions that encourage the synthesis of certain phyto metabolites. Scientists aim to increase the amount, diversity, and quality of SMs by understanding and using these cultivation methods, expanding their applications in a number of fields (Wilken et al., 2005). On the other hand, if medically important compounds are derived from slow-growing, difficult-to-cultivate, or endangered plants, scientists will need to find other ways to make the compounds. The most effective plan of action is frequently to combine metabolic engineering with plant cell culture (Wawrosch & Zotchev, 2021). It is important to highlight the untransformed root system since it makes it simple to produce valuable pharmacological chemicals without modifying the plants DNA genetically.

TABLE 12.1
Various Secondary Metabolites from Plants and Their Medicinal Properties

Secondary Metabolites	Name of the Metabolite	Plant Source	Pharmacological Activity	References
Terpenes	Eucalyptol	*Eucalyptus globulus* Labill., *Salvia rosmarinus* Spenn., and *Camphora officinarum* Nees	Treatment of several chronic conditions, including chronic obstructive pulmonary disease, allergies and asthma, and bronchial mucus inflation.	Zengin and Baysal (2014)
	Tetrahydrocannabinol	*Cannabis sativa* L.	Treatment for nausea and vomiting brought on by chemotherapy and increased appetite.	Citti et al. (2019); Marks et al. (2009)
	Artemisinin	*Artemisia annua* L.	Antimalarial activity	Bilia et al. (2006); Czechowski et al. (2019)
	β-myrcene, (E)-βocimene, and allo-ocimene	*Machilus velutina* Champ. ex Benth.	Antidiabetic, analgesic, sedative, and antiinflammatory	Ho and Su (2012)
	Rhizathalene A	*Arabidopsis thaliana* (L.) Heynh.	Deterrents or antifeedants, allelochemicals	Vaughan et al. (2013)
	Monoterpenes: linalool, limonene, and β-pinene	*Quercus ilex* L.	Antibacterials	Semwal et al. (2018)
	Terpinolene	*Melaleuca alternifolia* (Maiden & Betche) Cheel	Antimicrobial properties	Brun et al. (2019); Hammer et al. (2012)
Alkaloids	Caffeine	*Coffea arabica* L.	Premature skin aging, dermatoses	de Mello et al. (2023); Saewan et al. (2023)
	Cocaine	*Erythroxylum coca* Lam.	Reduction of fatigue and hunger and antidepressant	Lv et al. (2022)
	Nicotine	*Nicotiana tabacum* L.	Treatment of headache, sinusitis, and conjunctivitis	Ru et al. (2012)
	Quinine	*Cinchona* spp.	Antimalarial, antipyretic, analgesic	Achan et al. (2011); Anugrah et al. (2021)
Phenolic acids	Hydroxycinnamic acids	*Coffea arabica* L., *Zea mays* L.	Antitumor, anti-inflammatory, antimicrobial, and antidiabetic activity	Ali (2024); Ali et al. (2022)
	Caffeic acid	*Coffea arabica* L., *Olea europaea* L.	Anti-inflammatory, anticancer, and antiviral	Gülçin (2006); Tajner-Czopek et al. (2020)
	Chlorogenic acid	*Coffea canephora* Pierre ex A. Froehner *Coffea arabica* L.	Antioxidant and antimutagenic properties	Badmos et al. (2019)
	Ferulic acid	*Triticum aestivum* L., *Hordeum vulgare* L., *Zea mays* L.	Antioxidant activity	Horvat et al. (2020)
	Sinapic acid	*Sinapis alba* L.	Antioxidant activity	Chadni et al. (2021)
	Neochlorogenic acid	*Geranium purpureum* Vill.	Antibacterial, antioxidant, and anti-proliferative activity	Nunes et al. (2017)
Flavonoids	Myricetin	*Comptonia peregrina* (L.) Coult	Anticancer	Akhtar et al. (2021); Berköz et al. (2021)
	Kaempferol	*Aloe vera* (L.) Burm.f., *Euphorbia pekinensis* Rupr., *Ginkgo biloba* L., and *Salvia rosmarinus* Spenn.	Anti-inflammatory and antioxidant activity	Wang et al. (2018)
	Isorhamnetin	*Hippophae rhamnoides* L. *and Ginkgo biloba* L.	Anticancer, anti-inflammatory, and antifungal activity	Sarkar et al. (2023); Unver (2024); Yang et al. (2024)

But there are drawbacks as well, such as the need for a steady supply of auxins, growth in a geotropic environment, and the preference for transformed roots over untransformed cultures due to their reduced branching (Morey & Peebles, 2022). In the field of in vitro metabolite synthesis, the benefits of hairy root cultures—unlimited branching, rapid growth, and genetic stability are highly regarded. Therefore, *Agrobacterium rhizogenes*-induced *in vitro* hairy root cultures are considered a promising alternative for producing valuable SMs (Anuradha et al., 2020).

12.3.1 Hairy Root Transformation Mediated by *Agrobacterium rhizogenes* for Production of Secondary Metabolites

In plant genetic engineering, the bacterial pathogen *A. rhizogenes* plays a key role in transferring target genes into a host plant. The introduced DNA can integrate into the plant's genome within the nucleus, allowing it to be inherited by future generations. Alternatively, the foreign DNA may temporarily reside in the nucleus without integrating into the genome, but still be expressed to produce the desired gene products (Gelvin, 2003). Various SMs, like terpenoids (Gharari et al., 2023), aconites (Mou et al., 2022), flavonoids (Zhao et al., 2014), indole alkaloids (Hanafy et al., 2016), and tropane alkaloids (Abdelkawy et al., 2023), have been shown to be produced by hairy roots. Moreover, antibodies have been expressed using hairy roots (Catellani et al., 2020). Many facets of the relationship between *A. tumefaciens* and plants have been clarified, from the organism's discovery to its widespread use in plant biotechnology. Rapid growth, lateral branching, an abundance of root hairs, and the plagiotropic development of hairy roots are some of their distinguishing traits (Singh et al., 2022). Fine roots are formed when *A. rhizogenes* integrates its T-DNA into the Ri plasmid, resulting in the genetic modification of plant tissue. The bacteria then infiltrate the host plant's tissue, utilizing the chv genes from the bacterial chromosomal DNA and the virulence (vir) genes from the Ri plasmid to transfer T-DNA, a segment of its root-inducing plasmid (Ri-plasmid) (Gelvin, 2003). The primary benefit of the hairy root culture method is that plant growth regulators are not required for media supplementation, as the inserted T-DNA contains the genes necessary for natural auxin production. Hairy roots exhibit extensive branching, and their lack of geotropism may enable them to grow more rapidly than regular roots (Hussain et al., 2022). In addition to the usual metabolites found in roots, hairy roots also produce metabolites typically generated by the plant's aerial parts. Furthermore, like other cell culture techniques, hairy roots maintain stability both physiologically and biochemically (Shi, Liao, et al., 2021). The characteristics of hairy roots make them ideal for research and industry. The entire transformation of plant tissues, establishment and hairy roots growth, and production of SMs mediated by *A. rhizogenes* are depicted in Figure 12.1.

Out of the 18 ORFs in *A. rhizogenes* T-DNA, certain ones are essential for inducing hairy roots. These genes, known as "rol" genes (rolA, rolB, rolC, and rolD), are linked to "rooting locus" genes (Gutierrez-Valdes et al., 2020). These oncogenes affect the metabolism of auxin and cytokinin, phytohormone homeostasis, and signaling in plants. *A. rhizogenes* isolates have been isolated and given some characterization. Studies involving DNA/DNA hybridization and genome mapping reveal that each isolate contains a crucial, highly conserved DNA segment required for the development of hairy roots (Britton et al., 2008). Withstanding the present ambiguity surrounding the categorization of species, *A. rhizogenes* strains may be further divided into several subgroups according to the kind of opioids they generate. The strains of *A. rhizogenes* most commonly identified include agropine-type varieties, which are associated with Ri plasmids such as pRiA4, pRiHRI, pRi1855, pRiLBA9402, and pRi15834; cucumopine-type strains carrying Ri plasmid pRi2659, mannopine-type strains with Ri plasmid pRi8196; and mikimopine-type strains featuring Ri plasmid pRi1724 (Hooykaas & Hooykaas, 2021; Joseph Sahayarayan et al., 2020). The mechanism behind the hairy root development is still mostly unknown. However, the process can generally be divided into four stages: activation, processing, stimulation of root growth and development, and the transfer of bacterial DNA into the plant cell. Because of the established pathogenesis models and related biological processes, the first three stages are the most well-known of the four (Rahmat

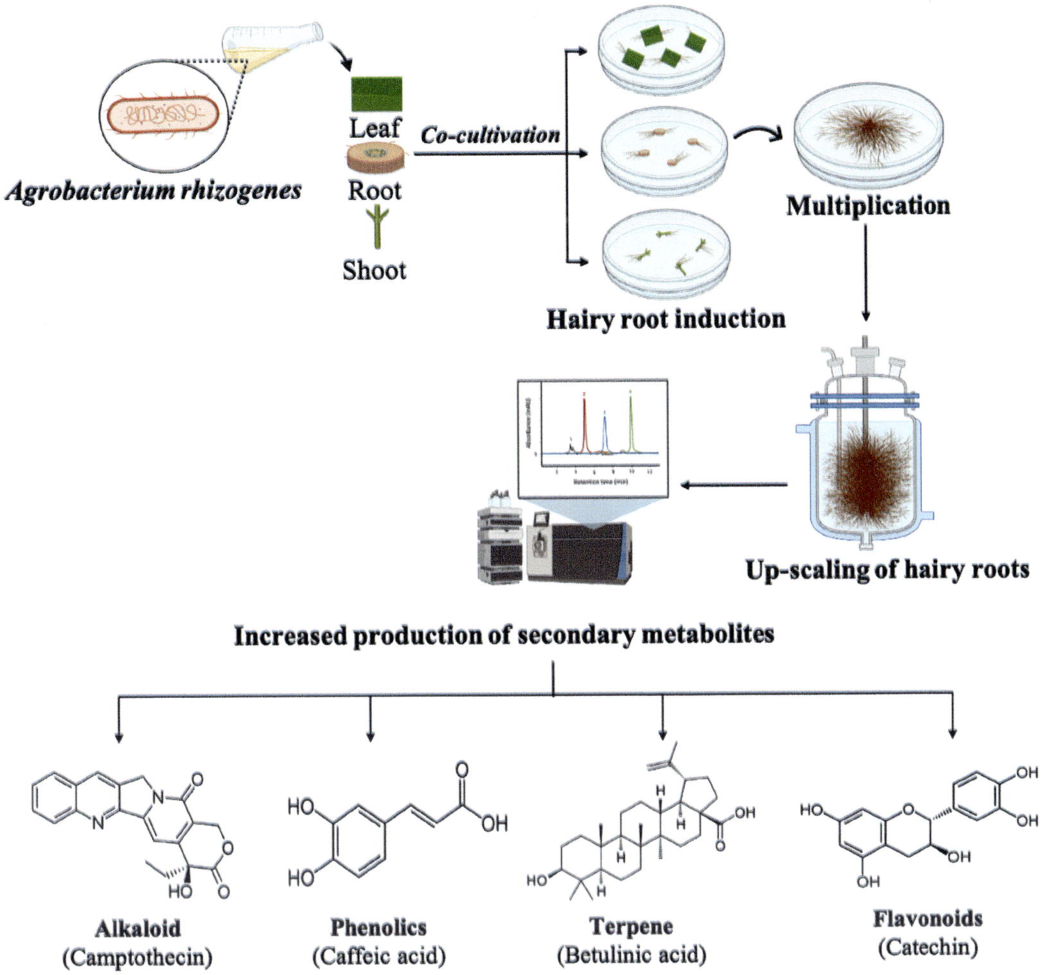

FIGURE 12.1 Hairy root induction in medicinal plants using *Agrobacterium rhizogenes* for production of secondary metabolites.

& Kang, 2019). How the hairy-root phenotype develops and spreads due to *A. rhizogenes* infection and genetic alteration is still unknown. A group of genes linked to oncogenes and opine catabolism is found in wild-type T-DNA. When these genes are expressed in transgenic plant cells, tumorous tissue develops, and opines derivatives of amino acids that bacteria almost exclusively use as a source of carbon and nitrogen are produced (Lankitus et al., 2023; Matthysse, 2006; Rangslang et al., 2018). Genes for digesting T-DNA, attaching and transferring it to plant cells, and catabolizing opiates in transformed plant tissues are included in both the Ri and Ti plasmids (Gordon & Christie, 2014). Over 100 plant species have been cultured using hairy roots to date, including several rare medicinal plants. Furthermore, various techniques were developed to increase the quantities of desired SMs and recombinant proteins, including metabolic engineering, bioreactor design improvements, and T-DNA activation tagging (Dhiman et al., 2018). According to a study, transgenic tobacco hairy root cultures cultivated hydroponically greatly increased the synthesis of the recombinant human-secreted alkaline phosphatase protein, boosting it by five to seven times (Gaume et al., 2003). Similarly, acetylcholinesterase buildup in the hairy roots of *Nicotiana benthamiana* plants was shown to be almost three-fold greater than in plants of the natural type (Woods et al., 2008). The anti-HIV and anticancer protein MAP 30 was produced using hairy root cultures of *N. tabaccum* (Moghadam et al., 2016). *Cichorium intybus* plant SM has a number of pharmacological uses and is high in inulin. In hairy root cultures using *Agrobacterium* strains A4, A13, A7, and ATCC15834, the genetic transformation of chicory to hairy roots by *A. rhizogenes* was studied (Sassi et al., 2005). To induce transgenic hairy roots in *Ocimum basilicum*, *A. rhizogenes* strains were used, and as a result, they showed the greatest amounts of total phenolic content of 62.3±4.95 µg/g DW and rosmarinic acid of 62.05±4.94 µg/g DW (Sathasivam et al., 2022). The highest amount of β-caryophyllene metabolite was produced by hairy root cultures stimulated with *Artemisia vulgaris* utilizing *A. rhizogenes* strains R1000 and A4 on media supplemented with 3 µM farnesyl diphosphate (FDP), resulting in 2.80 and 2.92 mg/ml of β-caryophyllene, respectively (Balasubramani et al., 2021). Research was carried out to investigate the effects of stress-inducing biodegradable ester-based scaffolds, composed of poly lactic acid (PLA) and polyglycerol sebacate, on the growth and deoxyshikonin synthesis in immobilized hairy roots of *Rindera graeca* (Wierzchowski et al., 2022). One such investigation focused on enhancing the production of phenolics and flavonoids in *Lactuca serriola* hairy root cultures transformed with the *A. rhizogenes* strain AR15834 carrying the rolB gene. Transgenic hairy roots exhibited higher levels of total phenolic content (54.8%–96.7%), total flavonoid content (38.1%–76.2%), and total reducing power (56.7%–96.7%) than non-transgenic roots (El-Esawi et al., 2017).

Secondary metabolites with potential medical use from *Semecarpus anacardium* L. have been reported. The plant roots have long been used in ayurvedic medicine for centuries. *In vitro* plant tissues were treated with *A. rhizogenes* strains A4, ATCC15834, and LBA 9402 to induce hairy roots (Panda et al., 2017). Hairy root induction in *F. carica* was carried out by inoculating three-week-old shoots and leaves with various *A. rhizogenes* strains to facilitate efficient SM production. The effects of methyl jasmonate elicitation were also explored (Amani et al., 2020).

A. rhizogenes strain KCCM 11879-infected leaves were used as explants for the induction of hairy root cultures from *Polygonum multiflorum*. The total phenolic content of the HR-01 line increased significantly as well, measuring 26.64 mg DW, compared to 8.36 mg/g DW for non-transgenic roots (Ho et al., 2018). Hairy roots have thus far shown to be an effective endeavor in the production of several SMs from plants. Table 12.2 describes the hairy root induction of plants screened in the past few years for various SM production.

12.3.2 Metabolic Engineering Using Hairy Roots

In plants, metabolic engineering entails altering biosynthetic pathways to boost flux in the direction of certain target metabolites (Dixit et al., 2021b). Its three objectives are to develop a new chemical—one that occurs naturally but is not produced by the host plant to overproduce a desirable molecule and produce less of an undesired component (Sharma et al., 2019). Figure 12.2 depicts genetic engineering techniques such as novel sequence insertion, transcriptional regulation, and gene silencing to enhance metabolite production yield in hairy root systems.

By learning more about biosynthetic pathways and identifying important targets, secondary metabolism may be engineered. For example, it is possible to bioengineer the isoprenoid and shikimate pathways to enable plants to biosynthesize aromatic compounds. Likewise, metabolic engineering may be applied to certain functional groups associated with secondary plant products, including terpenoids, carotenoids (fruit and flower colors), and sesqui-, di-, and triterpenes (plant hormones) (Cui et al., 2014; Wu et al., 2022). Carotenoids are involved in photosynthesis, the manufacture of phytohormones, photoprotection, and pigmentation in photosynthetic organisms. These are isoprenoid pathway-derived SMs. They are lipophilic and often gather in plant organs. It was observed that CRISPR/Cas-mediated editing improved carotenoid accumulation in various plants (Sun et al., 2020; Wurtzel, 2019). While the biosynthetic process for primary metabolism is regulated at the genetic level, extrinsic variables such as environmental pressures can drive specific biosynthesis (Jan et al., 2021). Reactive oxygen species (ROS) cause the creation of apocarotenoids, and these components cause a cascade of events by creating various signaling molecules (Tripathy & Oelmüller, 2012). On the other hand, regulation of secondary metabolism mostly takes place during

TABLE 12.2
Secondary Metabolites Produced by Hairy Root Culture

Explants	Plant	Metabolites	Culture Media	References
Seeds	*Atropa belladonna* L. *and Atropa acuminata* Royle ex Lindl.	Scopolamine	½ MS medium (sucrose 30 g/L) with added elicitors Methyl-β-cyclodextrin and coronatine	Fattahi et al. (2021)
Leaf hypocotyls	*Linum flavum* L.	Podophyllotoxin	Gamborg B5 (B5), Linsmaier and Skoog (LS), Murashige, and Skoog (MS) and Woody Plant Medium.	Renouard et al. (2018)
Seeds	*Hyoscyamus reticulatus* L.	Hyoscyamine, scopolamine	MS Medium	Moharrami et al. (2017)
Seeds	*Nitraria schoberi* L.	Catechin	Dunstan and Short media 500.0 mg/L of cefotaxime	Zheleznichenko et al. (2023)
Seeds	*Scutellaria bornmuelleri* Hausskn. ex Bornm.	Terpenoids	MS medium containing 0.5 mg/L BAP	Gharari et al. (2023)
Roots	*Turbinicarpus lophophoroides* (Werderm.) Buxb. & Backeb.	Feruloyl-glucoside	MS Medium with Feruloyl-glucoside	Solis-Castañeda et al. (2020)
Seeds	Salvia miltiorrhiza Bunge	Diterpenoids and phenolic acids	MS Medium added with fungal elicitors	Xu et al. (2021)
Leaves and petioles	*Centella asiatica* (L.) Urb.	Asiaticoside	MS Medium	Baek et al. (2022)
Leaves	*Rhazya stricta* Decne.	Quebrachamine	Modified Gamborg B5 medium with MeJA elicitation	Akhgari et al. (2019)
Seeds	*Physalis minima* L.	Withaferin A	–	Halder and Ghosh (2022)
Seeds	*Dillenia quercifolia* (C.T.White & Francis ex Lane-Poole) Hoogland	Scopolamine and hyoscyamine	–	Shakeran et al. (2017)
Cotyledon	*Calendula officinalis* L.	Steroids and Triterpenoids	MS Media added with jasmonic acid elicitor	Rogowska et al. (2022)
Leaves, petioles, and stems	*Aristolochia manshuriensis* Kom. Acta Horti Petropol.	Nitrogenous metabolites	Liquid medium	Shkryl et al. (2023)
Leaves	*Atropa komarovii* Blin. & Shalyt	Tropane alkaloids	MS Media added with sucrose	Banihashemi et al. (2020)
Roots	*Tripterygium wilfordii* Hook.fil.	Wilforgine and wilforine	MS media supplemented with 30 g/L Glucose	Zhang et al. (2020)
Shoots	*Salvia viridis* L.	Caffeic acid	MS Media added with 30 g/L sucrose	Grzegorczyk-Karolak et al. (2018)
Leaves	*Cichorium intybus* L.	Esculin and Esculetin	–	Petrova et al. (2023)
Shoots	*Taxus baccata* L.	Taxol	Murashige & Skoog (MS) media supplemented with 15 mg/L acetosyringone	He et al. (2022)
Leaves	*Achyranthes aspera* L.	Hydroxyecdysone	(MS) medium with 4-dichlorophenoxyacetic acid (1 mg/L) and α-naphthaleneacetic acid (1 mg/L).	John et al. (2017)
Seeds	*Brassica juncea* (L.) Czern.	Glucosinolates	(1/2 MS) medium with 2.5 mM MES	Cuong et al. (2018)
Seeds	*Valeriana officinalis* L.	Valepotriates	MS medium containing 500 mg/L cefotaxime	Torkamani et al. (2014)
Seeds	*Ocimum basilicum* L.	Phenolic and flavonoid compounds	MS agar medium	Sathasivam et al. (2022)
Leaves	*Plumbago auriculata* L.	Plumbagin	1/2 MS + Cefotaxime sodium	Zhao, Gao, et al. (2023)
Hypocotyl, leaf, and cotyledons	*Silybum marianum* (L.) Gaertn.	Flavolignans (isosilybin, silychristin, silydianin)	SH medium supplemented with 0.05 mg/L BA and 0.5 mg/L 2,4-D	Rady et al. (2018)

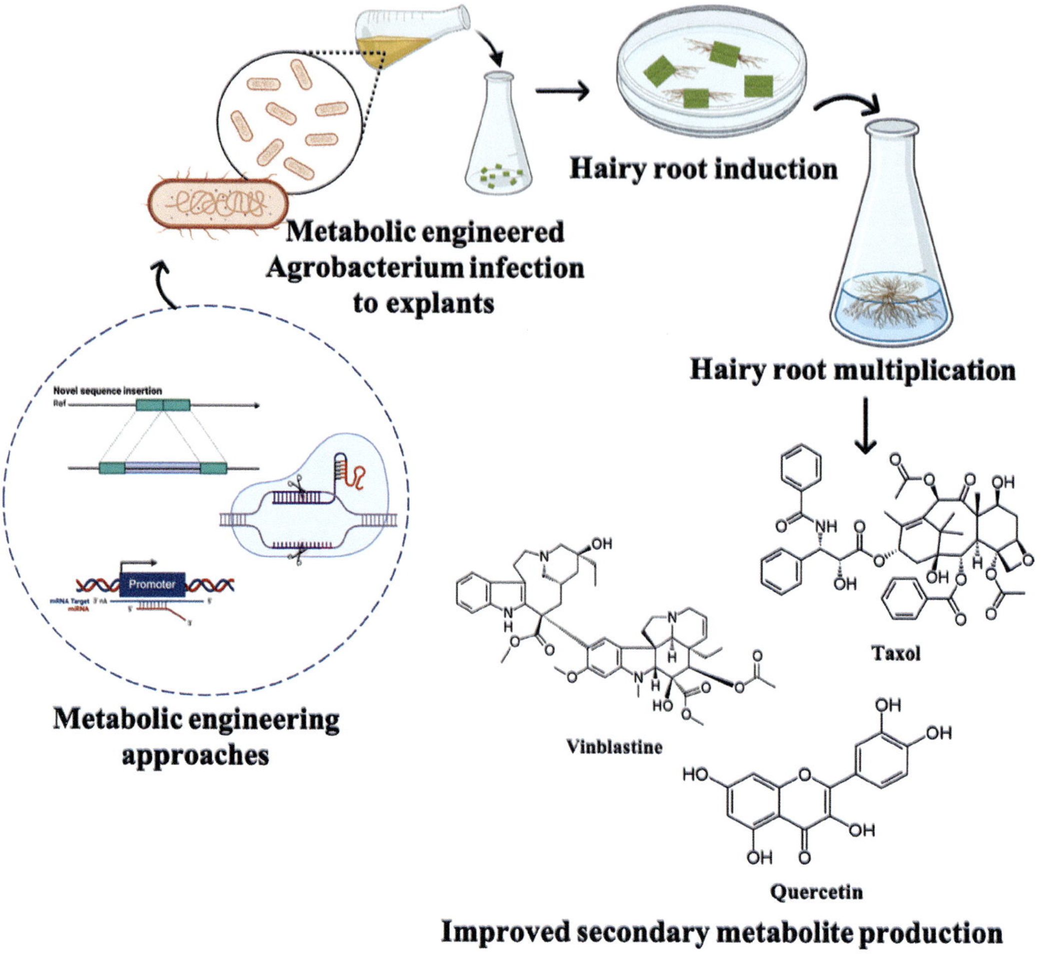

FIGURE 12.2 The establishment of metabolically engineered hairy roots through Agrobacterium infection and secondary metabolite production.

subcellular compartmentation or at the enzyme level (e.g., isoquinoline alkaloid berberine production). Usually, transporter genes or the regulatory genes of the biosynthetic route are the focus of these controls (Zhao, Liu, et al., 2023).

The modification of genes and regulatory components linked to the SM pathway has been a significant use of *A. rhizogenes*-mediated genetic transformation. This has made it easier to understand and adjust important mediators and enzymes in the pathway(s) that produce SMs (Mipeshwaree Devi et al., 2023). The production of recombinant proteins in hairy roots has emerged as another important application of this technique, especially in cases when bacterial, yeast, or other expression systems are unable to produce the protein (Medina-Bolívar & Cramer, 2004). A potent reverse genetic method for examining gene function is RNA interference, or RNAi. RNAi silencing is often used in functional genomic research (Zhai et al., 2009). It can elicit phenotypes in situations when more conventional knockout methods, such as T-DNA insertion, are ineffective. Since these variables affect the stability, structural integrity, and functioning of recombinant proteins synthesized in hairy roots, the procedure necessitates a great deal of parameter tuning. Moreover, maintaining the post-translational integrity of recombinant proteins within a heterologous system is crucial for their proper function and stability (Agrawal et al., 2003). Thus, optimizing recombinant protein production in hairy root cultures requires careful consideration of their tissue and organ specificity, especially with regard to their glycosylation patterns, subcellular localization, and susceptibility to proteolytic degradation (Zhang et al., 2019). Moreover, hairy roots have been utilized in several studies to investigate dsRNA-mediated gene knockdown (Pereira et al., 2023). Using the ex vitro composite plant development technique, dsRNA targeting scorable markers is expressed in *A. rhizogenes*-induced hairy roots, successfully silencing the recombinant β-glucuronidase (GUS) and Green fluorescent protein (GFP) genes in the roots of transgenic plants that stably produce GUS or GFP (Niazian et al., 2023). Transgenics is a valuable tool in the study of molecular features of biological processes and systems, promoter analysis, and the function of genes.

Advancements like the composite plant system and the creation of innovative molecular tools have made hairy root systems a more practical and attractive option for scientists exploring specific genes of interest (Huang et al., 2022). This section describes the use of several metabolites that are produced by the composite plant systems and through hairy roots. It should be noted that hairy root systems can benefit greatly from reporter genes that encode fluorescent proteins (FP) or GUS. In a composite plant model or culture system, reporter gene constructs are primarily used to visually identify roots that express the desired T-DNA among other non-transgenic roots (He et al., 2020; Soboleski et al., 2005). These reporter genes are also useful for investigating the subcellular compartments in which gene products are localized and the sites of gene product expression in promoter analysis investigations (Sassi et al., 2005).

Additionally, due to the advantages of combining suspension cell culture with whole-plant cultivation, hairy roots seem to be among the most popular *in vitro* plant culture methods for generating recombinant proteins. In one study, a calreticulin signal peptide was used to construct a codon-optimized version of the globular adiponectin (gAd) gene. The nptII gene was then added as a selectable marker to a plant expression binary vector containing this sequence. *A. rhizogenes*-mediated transformation produced transgenic hairy roots. By causing ACC phosphorylation in C2C12 muscle cells, the functional yield of recombinant gAd from *Withania somnifera* was 5.1 times more than that of gAd generated from *E. coli* but 45% lower than that of CHO cells (Dehdashti et al., 2020). Hairy root systems are a competitive substitute for the generation of recombinant proteins. It is crucial to maximize the production of hairy root cultures for proteins that are resistant to change. The purpose was to maximize the release of alpha-l-iduronidase (IDUA), a biologically significant substance, from *Brassica rapa* subsp. rapa hairy roots (Gutierrez-Valdes et al., 2022). To enhance product homogeneity and aid in purification, recombinant pharmaceutical proteins produced in hairy root cultures can be released into the medium. However, if the protein is intrinsically unstable or prone to protease degradation, this might lead to substantial degradation (Häkkinen et al., 2014). For the production of SMs, metabolic engineering has paved its way. Terpenes, for instance, can be biosynthesized in any part of the plant; terpene accumulation has been observed in leaves, bark, roots, fruits, and seeds (Masyita et al., 2022). Since terpene biosynthesis is subcellularly compartmentalized in plants, the engineering of the terpenoid biosynthetic pathway is more intricate than in microbes. In plants, monoterpenes and diterpenes are produced by the methylerythritol phosphate (MEP) route in the plastids, whereas sesquiterpenes are produced by the mevalonate (MVA) pathway in the cytoplasm (Li et al., 2023). Another important characteristic is the distribution of terpene synthesis by tissue. The total oil content of mint may increase if the enzymes that control carbon lactate in the monoterpene pathway are overexpressed (Li et al., 2020). The monoterpene profile of mint was modified by altering the expression of an enzyme that redirects metabolites from the menthol biosynthesis pathway, leading to the production of plants with a more commercially valuable essential oil (McConkey et al., 2000). Using the same construct as the one used to convert *Arabidopsis*, the FaNES1 gene in potato plants creates monoterpenes. Wild-type potato plants already release linalool from their leaves, but transgenic strains release far more of the substance (Aharoni et al., 2003). The viability of genetically modifying plants can generate novel terpene chemicals absent from wild-type plants. Usually, heterologous terpene synthase expression is used to accomplish this (Moses et al., 2013). Consequently, it has been possible to effectively design plants to create new diterpenes, sesquiterpenes, and monoterpenes (Bahmani et al., 2022;

Gutensohn et al., 2021; Li et al., 2020). It is interesting to note that *Arabidopsis thaliana* plants transformed with a patchoulol synthase gene derived from patchouli collected a combination of sesquiterpenes (Yu et al., 2015). A buildup of terpenes has been linked to decreased plant development and chlorosis in a number of cases where terpene metabolism in plants has been engineered. This finding suggests potential negative consequences and/or a shift in carbon flow away from vital metabolites (Karunanithi & Zerbe, 2019). Furthermore, manipulating the biosynthesis of terpenes in plants or microbes may offer a practical means of producing novel molecules (Ma et al., 2024; Wang, Cao, et al., 2021). In transgenic hairy root cultures, overexpressing encoding enzymes and several significant transcription-related factors promotes the production of desirable metabolites. The synthesis route of SMs has yielded several intriguing discoveries in the last ten years due to the process of gene cloning for certain enzymes and transcription factors.

Nowadays, metabolic engineering has emerged as a publicity-seeking substitute for increased SM accumulation in therapeutic plants; *Salvia miltiorrhiza* and *Isatis indica* are two such examples (Gai et al., 2015; Xu et al., 2021). By applying metabolic engineering techniques, productivity levels in transformed root culture systems can be reflectively enhanced. This can be achieved through either overexpression of a single gene, which increases the activity of the concerned metabolic enzymes and accumulates the target chemical, or overexpression of multiple biosynthetic genes of the target biosynthetic pathway (Mora-Vásquez et al., 2022). Efforts have been made to increase the production of tropane alkaloids in hairy root cultures of various solanaceous species, such as *Anisodus acutangulus, Atropa baetica, Brugmansia candida, Datura metel, Hyoscyamus muticus,* and some other species. These initiatives have focused on employing metabolic engineering to enhance the expression of genes involved in the biosynthesis of key enzymes, including putrescine N-methyltransferase, hyoscyamine 6β-hydroxylase, and tropinone reductase I. The expression of terpenoid indole alkaloids (TIA) pathway genes like G10h, Tdc, Str, and Sgd, and the respective TIA metabolites like secologanin, strictosidine, and tabersonine, and transcription factors like Orca, Zct, Gbf, Myc2, At-hook, and Wrky1 was investigated in *C. roseus* hairy root cultures (Goklany et al., 2013). *Salvia miltiorrhiza* contains tannins, which are diterpenoids that promote health and have a variety of uses. In a study on *S. miltiorrhiza*, genes like 1-deoxy-D-xylulose-5-phosphate synthase (SmDXSII) and geranylgeranyl diphosphate synthase (SmGGPPS) were added to the hairy roots of *S. miltiorrhiza*. Tanshinone levels are greater in hairy roots where SmGGPPS and SmDXSII overexpression is present compared to control (Shi et al., 2016). Efforts to overexpress the squalene synthase gene in cultivated hairy roots of *W. somnifera* have led to increased production of withanolide A, withanolide B, withaferin A, and withanone (Sivanandhan et al., 2020). A study indicated that overexpressing ORCA3 using a glucocorticoid-inducible promoter in *C. roseus* hairy roots did not affect the overall quantity of TIA measured. Additionally, the research highlighted that RT-qPCR analyses revealed that the overexpression of ORCA3 increased the transcripts of most known TIA pathway genes (Sun & Peebles, 2016). By overexpressing the SalAT gene, which was provided by *P. somniferum*, thebaine was mostly deposited (1.28% DW) in *P. bracteatum* hairy roots. Salutaridinol 7-o-acetyltransferase uses the morphinan pathway to convert salutaridinol to thebaine (Sharafi et al., 2013). In *A. baetica* hairy roots, a high accumulation of scopolamine in the root tissues engineered with the 35S-h6h transcript led to scopolamine leaching into the culture medium, unlike the control roots (Banihashemi et al., 2020).

A. rhizogenes genetically altered the hairy roots of *P. somniferum* to overexpress the 4′OMT2 gene, which improved the amounts of total alkaloid and papaverine (Aghaali et al., 2024). The application of genome editing and CRISPR/Cas9 technologies is beginning to yield promising results in the area of hairy root cultures. Gene silencing using the QPT-RNAi construct was employed to downregulate the quinolinic acid phosphoribosyltransferase (qpt) gene in *Duboisia leichhardtii*, aiming to inhibit resource allocation to the competing pyridine alkaloid pathway. This approach led to an increased synthesis of tropane alkaloids (Singh et al., 2018). Salicylic acid (SA) was introduced as an elicitor to both chalcone synthase (CHS) silencing lines and wild-type lines of *S. miltiorrhiza* hairy roots to investigate the impact of elicitor treatment and genetic modification on phenolic acid production, aiming to suppress CHS transcript levels (Zhang et al., 2015). Additionally, the *N. tabaccum* enzyme ornithine decarboxylase, which generates nicotine, was downregulated, and a 2–3-fold rise in hyoscyamine was produced by the use of RNA interference technology to genetically modify the TRI-PMT gene in the plant *A. acutangulus*. Despite stimulating the wounds, this led to a decrease in nicotine and an increase in anatabine synthesis in the cultured hairy roots of *N. tabaccum* (Hayashi et al., 2020). In order to decrease codeinone reductase (COR) in *P. somniferum* and opium poppy, RNAi-mediated gene silencing was used. Consequently, there was a reported increase in (S)-reticuline concentration and a matching decrease in morphinan alkaloid levels. The accumulation of benzoate, codeine, papaverine, and total alkaloid in *P. somniferum* was negatively impacted by the systematic inhibition of 3′-hydroxyl-N-methylcoclaurine 4′-O-methyltransferase (4′OMT2) (Allen et al., 2004). RNA interference (RNAi) was employed to silence the expression of genes encoding secologanin synthase (SLS) and tryptophan decarboxylase (TDC), two enzymes crucial for the early stages of camptothecin biosynthesis, in the hairy roots of *O. pumila*. Additionally, changes in metabolite profiles were examined (Asano et al., 2013).

The cytochrome P450 gene PlCYP81Q38, which in *P. leptostachya* converts (+)-pinoresinol to (+)-sesamin via (+)-piperitol, was silenced to generate RNAi hairy roots and ascertain its in vivo function in the lignan biosynthesis pathway (Li et al., 2019). There have also been several attempts to boost SM synthesis from non-hairy root cultures, often requiring modifications to their pharmacological properties. Halogenation is a common alteration seen in promising pharmacological prospects, especially in fluorination and chlorination (Matsuura & Fett-Neto, 2015). Mutasynthesis is a technique in which natural biosynthesis is initially halted by genetically silencing the natural precursor and then reinstated by supplying structural analogs of that precursor (Hennrich et al., 2023). In this case, fluorinated Monoterpene indole alkaloids (MIA) analogs were identified when artificial tryptamine 2 analogs were administered to a chemically "silent" background that does not produce alkaloids, along with RNAi-mediated suppression of TDC (Runguphan et al., 2009). In an initial endeavor to modify plants that produce benzoquinoline alkaloids, RNA interference was employed to mute the COR expression in the opium poppy (Larkin et al., 2007; Allen et al., 2004). Other early attempts to increase Benzylisoquinoline alkaloids (BIA) alkaloid output included root cultures of *California poppy* and *Macleaya cordata* that overexpressed the berberine bridge enzyme (BBE). This endeavor produced lower quantities of amino acids and higher levels of downstream alkaloids, but interestingly, the amino acid tyrosine 10, which is used in the manufacture of BIA, remained unchanged (Huang et al., 2018; Park et al., 2002).

In the past, altering a plant's metabolic pathway to achieve a certain objective has proven difficult. Since much of the work on metabolic engineering in plants has been done on single genes or silencing events against the backdrop of natural plant cell metabolism, the field is still in its infancy. Nevertheless, the number of effective attempts to metabolically modify higher plant SM production has increased dramatically in recent years. Table 12.3 provides information on the studies carried out with genetic engineering tools for SM production.

12.3.3 Upscaling of Secondary Metabolites Produced by Hairy Roots in Bioreactors

Hairy roots are a stable and alternate supply of vital substances for pharmaceuticals; nonetheless, they need to be scaled up for use in commercial endeavors. A key prerequisite for developing a scale-up production method for many high-value plant-based goods is bioreactor technology (Mehrotra, Srivastava, et al., 2016). To choose the best industrial-scale bioprocess and the corresponding capital expenditure, the bioreactor must be carefully chosen and designed. Thus, in a range of bioreactor types, such as stirred tanks, conventional airlift, bubble columns, airlift balloons, and nutrient mist bioreactors, scientists from all over the world have boosted the production of SMs from hairy roots of medicinally relevant plants (Mehrotra, Mishra, et al., 2016). The poor

TABLE 12.3
Metabolic Engineering Technique Employed for Secondary Metabolite Production

Plant Metabolite	Engineered Microbe/Plant Species	Technique Used	Increase in Secondary Metabolite Content	Reference
Carotenoid	*Oryza sativa* L.	CRISPR CAS 9	–	Dong et al. (2020)
Paclitaxel	*Saccharomyces cerevisiae*	Recombination	50%	Engels et al. (2008)
p-coumaric acid	*Saccharomyces cerevisiae*	Overexpression	1.9 titer	Rodriguez et al. (2015)
Dopamine	*Solanum lycopersicum* L.	Coexpression	–	Breitel et al. (2021)
Proanthocyanidin	*Camellia sinensis* (L.) Kuntze	Recombination	–	Pang et al. (2013)
Geraniol	*Candida glycerinogenes*	Promoter	1.3-fold	Zhao, Liu, et al. (2023)
Amorphadiene	*Bacillus subtilis*	CRISPR CAS 9	1.4-fold	Song et al. (2021)
b-amyrin	*Saccharomyces cerevisiae*	Genetic modification	12.14-fold	Ahmed et al. (2019)
b-carotene	*Yarrowia lipolytica*	Genetic modification	21 times	Liu et al. (2021)
Lycopene	*Saccharomyces cerevisiae*	Systemic engineering	25%	Chen et al. (2016)
Artemisinin	*Artemisia annua* L.	*Agrobacterium*- mediated transformation	–	Liu et al. (2011)
Isoflavones	*Medicago truncatula* Gaertn.	Genetic modification	–	Deavours and Dixon (2005)
Benzylisoquinoline	*Saccharomyces cerevisiae*	Expression regulation	–	Han et al. (2021)
Polyphenolics	*Lotus japonicus* (Regel) K.Larsen	Recombination	–	Takos and Rook (2014)
Kushenin	*Sophora flavescens* Aiton	Transcriptomics	–	Lei et al. (2021)
Terpenoids	*Salvia miltiorrhiza* Bunge.	Overexpression	3.5-fold	Wei et al. (2019)
Terpenoid indole alkaloids	*Catharanthus roseus* (L.) G.Don	Genome editing	–	Rischer et al. (2006)
Apocarotenoids	*Crocus sativus* L.	CRISPR	4%	Chib et al. (2020)
Alkaloid	*Papaver somniferum* L.	Omics	–	Yucebilgili Kurtoglu and Unver (2021)
Phenolic acids	*Salvia miltiorrhiza* Bunge.	CRISPR	–	Shi, Du, et al. (2021)
Carotenoid	*Dioscorea zingiberensis* C.H. Wright	CRISPR	–	Syombua et al. (2021)
Flavonoids	*Vitis amurensis* Rupr.	OMICS	–	Wang, Lenka, et al. (2021)
Phenolic acids	*Salvia miltiorrhiza* Bunge.	Overexpression	1.43-fold	Sun et al. (2019)
Flavonoids	*Ginkgo biloba* L.	Transcriptome	–	Meng et al. (2019)
Secoiridoids	*Gentiana rigescens* Franch.	Transcriptomic	–	Zhang et al. (2023)
Ginsenoside	*Panax quinquefolius* L.	CRISPR	21.12-fold	Yao et al. (2022)
Flavonoid and sesquiterpene	*Dolomiaea costus* (Falc.) Kasana & A.K.Pandey	Transcriptome	–	Bains et al. (2019)
Tanshinones	*Salvia miltiorrhiza* Bunge.	Overexpression	21-fold	Xing et al. (2018)

yield of secondary chemicals produced commercially utilizing hairy root culture techniques is a major factor in this failure. To boost the synthesis of bioactive chemicals derived from hairy roots, many approaches exist, such as elicitation, metabolic engineering, and upscaling (Kumar et al., 2024). While hairy root cultures hold great potential for the large-scale production of SMs, scaling them up using bioreactors poses several challenges.

The construction of a large-scale culture system and growth analysis are challenging due to the intricate fibrous structure of the roots. The non-uniform development of hairy roots impacts the operation of the reactor (Khan et al., 2019). In an attempt to create a bioreactor that would increase the biomass production of plant organs, the hairy roots of *H. muticus* were cultured in a 5 L bioreactor with an extended baffle assembly installed in the culture tank (Jaremicz et al., 2014). Additionally, the morphology of the hairy roots is very malleable, allowing the roots to adapt to changes in their immediate surroundings. The roots demonstrate resistance to nutrient flow as an interlocking matrix formed by branching (Stepanova et al., 2022). Heterotrophic, respiring creatures with hairy roots depend on oxygen for energy production and other metabolic processes. Understanding the mechanics of oxygen restriction, one of the main obstacles to the large-scale development of hairy root cultures, has advanced significantly (Palavalli et al., 2012). Additionally, the thick layer of root clusters promotes poor liquid mixing, bubble channeling, coalescence, and eventually localized liquid stagnation (Williams & Doran, 2000). This is due to the barrier provided by the periphery of the root mass, which stops the tissues growing toward the core from receiving constant oxygen and nourishment. As a result, the growth matrix consists of distinct phases, with liquid medium and solid (tissue clump) components (Choi et al., 2007).

Hairy root culture bioreactor design should also account for elements like the need for a support matrix and the potential for root mass flow limitation in specific areas of the bioreactor. Several bioreactor designs that allow for the growth of linked tissue that is irregularly distributed across the culture vessel have been described for hairy root culture (Mishra & Ranjan, 2008; Stiles & Liu, 2013). These designs incorporate the criteria mentioned earlier. Hairy root culture reactors are typically classified into three main types: gas-phase, liquid-phase, or hybrid reactors, which combine both phases. Airlift and bubble column bioreactor systems have proven to be the most effective for producing SMs and hairy root cultures (Suresh et al., 2005). In contrast, the mist bioreactor or airlift mesh-draught bioreactor with wire helixes is reported to offer a highly uniform distribution of biological material while minimizing the volume of culture media used (Sivakumar et al., 2010). Additionally, concentrated forms of secreted metabolites are produced. Hairy roots of *Picrorhiza kurroa* (Verma et al., 2015), as well as several shoot cultures of *Glycyrrhiza glabra*, were successfully grown using a similar reactor (Mehrotra et al., 2008). Although there are several challenges, various studies highlight the role of bioreactors for producing SMs, such as nicotine, which is synthesized from *Nicotiana rustica* in an 880 ml air-sparged tank (Röper et al., 1985), tropane alkaloids produced by *Atropa belladonna* are upscaled in a stirred tank with an impeller (Habibi et al., 2015), indole alkaloids produced by *Catharanthus tricophyllus,* on the other hand, are upscaled in a fermenter with mechanical stirring (Thakore & Srivastava, 2021). Airlift bioreactors are employed for the production of diosgenin and saponin metabolites from *Trigonella foenum-graceum* and *Panax ginseng,* respectively (Garagounis et al., 2020; Thanh et al., 2014). Bubble column bioreactors have also been used for upscaling of tropane alkaloids produced from *Scopolia parviflora* (Min et al., 2007), indole alkaloids from *Atropa belladonna* (Kwok & Doran, 1995), liquid two-phase bioreactors are used for thiophene production from *Tagetes patula* (Buitelaar et al., 1991; Suresh et al., 2005), and stirred bioreactors are used for atropine from *Atropa belladonn*a (Lee et al., 1999). Thus, once established, hairy roots can grow rapidly in a medium with a low inoculum. However, the main challenge for the commercial application of hairy root cultures lies in scaling up the process. Developing a suitable bioreactor system, which must account for various physical and chemical factors, is essential for utilizing hairy root cultures as a reliable source of bioactive compounds.

12.3.4 Current Status of Production of Secondary Metabolites by Hairy Root Metabolic Engineering

Metabolic engineering has enabled significant progress in the production of SMs using hairy root cultures. These cultures are particularly valuable for the pharmaceutical, cosmetic, and nutraceutical industries, as they provide a reliable method for producing high-value plant-derived metabolites. Hairy roots grow rapidly and maintain genetic stability. A recent breakthrough in enhancing metabolite production efficiency is the application of CRISPR/Cas9 technology for genome editing. For instance, CRISPR/Cas9 has been successfully utilized to generate α-solanine-free hairy roots in potatoes, and similar approaches have been applied to other plants such as soybeans and Brassica species (Hou et al., 2023; Rasheed et al., 2021). The intended chemical production and purity are improved by these genetic alterations. Compared to previous genetic alteration methods like RNA interference and other targeted nucleases, CRISPR/Cas9 offers several benefits. For instance, CRISPR/Cas9 targets numerous genes and is easier to use and less costly than these methods. Also, when the genomic-editing agents are removed, a gene edit cannot be differentiated from a "naturally occurring" mutation (Khoshandam et al., 2024; Richardson et al., 2023).

Recently, CRISPR-Cas9 nickase (H840A)-reverse transcriptase fusions, guided by prime editing-guide RNAs, have been used to create point insertions, mutations, and deletions in rice and wheat protoplasts. This innovative CRISPR approach holds great potential for identifying the enzymes and pathways that regulate the synthesis of SMs in medicinal plants. It offers the opportunity to enhance yields through precise, large-scale genome editing, further advancing the field of metabolic engineering (Jiang et al., 2022; Lin et al., 2020). Additionally, elicitation methods have been refined to encourage SM buildup. These methods include using bioreactors for large-scale synthesis and adding precursor molecules (Coste et al., 2011; Gadzovska-Simic et al., 2012; Siddiqui et al., 2013). These processes have been further enhanced by the incorporation of computational technologies, such as neural networks and artificial intelligence (AI) (Cembrowska-Lech et al., 2023). This has allowed for greater control over metabolic pathways and better production results. Omics-based approaches have been employed to explore genes, pathways, and bioactive compounds in both medicinal and non-medicinal plants (Kumar et al., 2018; Verma et al., 2014). Similar strategies have been applied to investigate hairy root cultures of medicinal plants such as *Salvia miltiorrhiza*, *Atropa belladonna* (Singh et al., 2021), *A. acutangulus* (Zhou et al., 2024), *Datura stramonium* (Li, Wei, et al., 2021), *and S. baicalensis* (Yao et al., 2024), enabling the discovery of novel genes and metabolic pathways. When analyzing medicinal plants, transcriptomics and genomes have been employed more often than proteomics and metabolomics. Transcriptome analysis of various tissues, elicitor-treated samples, and hairy roots has led to the identification of several new genes associated with tanshinone biosynthesis, along with transcription factors that regulate the production of tanshinone and salvianolic acid (Hao et al., 2020). Additionally, hairy roots may be quickly utilized to characterize gene function in vivo. This method works particularly well for medicinal plants because, in contrast to hairy roots, which may collect useful chemicals in as little as three to four months, many medicinal plants require several years to develop these compounds (Huang et al., 2014). Recently, hairy roots from many plants, including tobacco (Qin et al., 2022), tomato (Jacobs & Martin, 2016), *M. truncatula* (Boisson-Dernier et al., 2001), and *Brassica rapa* (Cardon et al., 2019), as well as medicinal plants like *C. roseus* (Wang et al., 2010), *S. baicalensis* (Fang et al., 2023), and *Monochasma savatieri* (Bai et al., 2023) *Symphytum officinale* (Zakaria et al., 2021) and *Salvia miltiorrhiza* (Zhou et al., 2018) have been effectively employed to rapidly characterize the functions of previously uncharacterized genes. For instance, RNA interference (RNAi) targeting flavone synthase II (FNSII) and cinnamic acid-specific coenzyme A ligase (CLL-7) in *Scutellaria baicalensis* hairy roots demonstrated the critical roles of these enzymes in synthesizing root-specific flavones (Pei et al., 2022). Cas technology has been effectively used in synthetic biology to produce valuable chemicals from bacteria, such as artemisinin, for commercial use (Li, Lei, et al., 2021). The development of genome-editing methods like CRISPR/Cas9, along with advances in "omics-related approaches, should soon identify the primary processes, regulatory mechanisms, and biosynthetic pathways for target drug synthesis (Zhang et al., 2021). Several biosynthetic genes may be inserted into hairy roots" genomes using this method in a single procedure with few indicators. Using CRISPR/Cas9 to generate targeted multiplex mutant hairy roots, which simultaneously disable multiple isoforms within a gene family, enhances the reliability of gene function studies. This method is useful for target molecule metabolic engineering as it will help identify genes and uncover biosynthetic pathways (Das et al., 2024; Upadhyay 2021, 2024).

12.4 CONCLUSION AND FUTURE PROSPECTS

Hairy root cultures offer a highly promising approach for producing SMs, especially pharmaceuticals, due to their numerous advantages. These include rapid growth, ease of cultivation and genetic modification, and, most importantly, a superior ability to synthesize valuable metabolites that surpass those produced by unorganized cells or even conventional plant roots. When using the hairy root culture technique, one should also be aware of potential issues such as co-suppression between endogenous and exogenous genes, chromosomal loss, and variations in synthesis skills across various clones. These have included anything from producing priceless recombinant proteins and scaling them up to the preservation of highly valued medicinal plants under attack. Hairy root cultures have a lot of potential for use in biotechnology because of their quick growth, steady, high-level synthesis of SMs, and adaptability to reactor systems. Furthermore, hairy roots are a desirable substitute for traditional methods, including plant cell suspension cultures, due to their benefits. Drawing on prior analysis and scientific findings, we offer some viewpoints to encourage the biomass engineering of plant hairy roots. First off, the fact that the SM biosynthetic pathway contains very precise spatial and temporal control mechanisms is one of the key explanations for the poor yields of SMs. In the meantime, the control mechanisms of the plant SMs pathway at the DNA, RNA, and protein levels remain unknown. Through the use of a number of cutting-edge methods, including reverse genetics, gene-to-metabolite networks, metabolomics analysis, expressed sequence tags, transcriptomics, proteomics, and cooperation with bioinformatics enhance our comprehension of the connections between biosynthetic pathways and molecular regulation processes. High-throughput mutagenesis methods, such as transposon tagging, T-DNA insertion knockouts, and TILLing, have significantly streamlined the functional analysis of genes. Although they have been shown to be helpful, molecular techniques like promoter analysis, fusion protein constructions, and RNA interference are frequently used only for a small number of genes. All of these methods depend on high-throughput components to enable systematic genome examination, or they considerably benefit from them. Hairy roots and composite plants, derived from *Agrobacterium rhizogenes*, serve as valuable transformation tools that play a pivotal role in advancing plant biology research by enabling high-throughput analysis of gene function. The promise of *A. rhizogenes* as a resource for biotechnological research is further increased by the availability of many molecular tools, such as RNAi and scorable markers, for application in hairy roots.

ACKNOWLEDGMENTS

The authors express appreciation for the support provided by the Deanship of Scientific Research, Vice Presidency for Graduate Studies and Scientific Research, King Faisal University, Saudi Arabia [Grant No. KFU000000].

REFERENCES

Abdelkawy, A. M., Alshammari, S. O., Hussein, H.-A. A., Abou El-Enain, I. M. M., Abdelkhalek, E. S., Radwan, A. M., Kenawy, S. K. M., Maaty, D. A. M., Abed, N. N., Sabry, S., & Mohsen, A. (2023). Effect of silver nanoparticles on tropane alkaloid production of transgenic hairy root cultures of *Hyoscyamus muticus* L. and their antimicrobial activity. *Scientific Reports*, *13*(1), 10397.

Achan, J., Talisuna, A. O., Erhart, A., Yeka, A., Tibenderana, J. K., Baliraine, F. N., Rosenthal, P. J., & D'Alessandro, U. (2011). Quinine, an old anti-malarial drug in a modern world: Role in the treatment of malaria. *Malaria Journal*, *10*, 144.

Aghaali, Z., Naghavi, M. R., & Zargar, M. (2024). Collaboration of hairy root culture and scale-up strategies for enhancing the biosynthesis of medicinal and defensive alkaloids in *Papaver* sp. *Current Plant Biology*, *40*, 100381.

Agrawal, N., Dasaradhi, P. V. N., Mohmmed, A., Malhotra, P., Bhatnagar, R. K., & Mukherjee, S. K. (2003). RNA interference: Biology, mechanism, and applications. *Microbiology and Molecular Biology Reviews: MMBR*, *67*(4), 657–685.

Aharoni, A., Giri, A. P., Deuerlein, S., Griepink, F., de Kogel, W.-J., Verstappen, F. W. A., Verhoeven, H. A., Jongsma, M. A., Schwab, W., & Bouwmeester, H. J. (2003). Terpenoid metabolism in wild-type and transgenic *Arabidopsis* plants. *The Plant Cell*, *15*(12), 2866–2884.

Ahmad, N., Rab, A., & Ahmad, N. (2016). Light-induced biochemical variations in secondary metabolite production and antioxidant activity in callus cultures of *Stevia rebaudiana* (Bert). *Journal of Photochemistry and Photobiology. B, Biology*, *154*, 51–56.

Ahmed, M. S., Ikram, S., Rasool, A., & Li, C. (2019). Design and construction of short synthetic terminators for β-amyrin production in *Saccharomyces cerevisiae*. *Biochemical Engineering Journal*, *146*, 105–116.

Akhgari, A., Laakso, I., Maaheimo, H., Choi, Y. H., Seppänen-Laakso, T., Oksman-Caldentey, K.-M., & Rischer, H. (2019). Methyl jasmonate elicitation increases terpenoid indole alkaloid accumulation in hairy root cultures. *Plants*, *8*(12). https://doi.org/10.3390/plants8120534

Akhtar, S., Najafzadeh, M., Isreb, M., Newton, L., Gopalan, R. C., & Anderson, D. (2021). Anticancer potential of myricetin bulk and nano forms in vitro in lymphocytes from myeloma patients. *Archives of Toxicology*, *95*(1), 337–343.

Alamgir, A. N. M. (2018). Biotechnology, in vitro production of natural bioactive compounds, herbal preparation, and disease management (treatment and prevention). In *Progress in Drug Research* (pp. 585–664). Springer International Publishing.

Ali, A. (2024). Preparation of ultrasound-assisted herbal coffee infusion, bioaccessibility of phenolic compounds and their antioxidant, anti-diabetic, and anti-Alzheimer potential. *Preprints*. https://doi.org/10.20944/preprints202402.1189.v1

Ali, A., Zahid, H. F., Cottrell, J. J., & Dunshea, F. R. (2022). A comparative study for nutritional and phytochemical profiling of *Coffea arabica* (*C. arabica*) from different origins and their antioxidant potential and molecular docking. *Molecules*, *27*(16). https://doi.org/10.3390/molecules27165126

Ali, M., Abbasi, B. H., & Ihsan-ul-haq. (2013). Production of commercially important secondary metabolites and antioxidant activity in cell suspension cultures of *Artemisia absinthium* L. *Industrial Crops and Products*, *49*, 400–406.

Al-Khayri, J. M., Rashmi, R., Toppo, V., Chole, P. B., Banadka, A., Sudheer, W. N., Nagella, P., Shehata, W. F., Al-Mssallem, M. Q., Alessa, F. M., Almaghasla, M. I., & Rezk, A. A.-S. (2023). Plant secondary metabolites: The weapons for biotic stress management. *Metabolites*, *13*(6). https://doi.org/10.3390/metabo13060716

Allen, R. S., Millgate, A. G., Chitty, J. A., Thisleton, J., Miller, J. A. C., Fist, A. J., Gerlach, W. L., & Larkin, P. J. (2004). RNAi-mediated replacement of morphine with the nonnarcotic alkaloid reticuline in Opium poppy. *Nature Biotechnology*, *22*(12), 1559–1566.

Almeida, A., Dong, L., Thorsen, T. H., Raadam, M. H., Khakimov, B., Carreno-Quintero, N., Kampranis, S. C., & Bak, S. (2022). Metabolic engineering of cucurbitacins in *Cucurbita pepo* hairy roots. *Frontiers in Plant Science*, *13*, 1021907.

Alok, A., Kumar, J., & Upadhyay, S. K. (2018). Engineering in hairy roots using CRISPR-/Cas9-mediated editing. In Hairy Roots: *An Effective Tool of Plant Biotechnology*. Springer Nature. https://doi.org/10.1007/978-981-13-2562-5_15

Amani, S., Mohebodini, M., Khademvatan, S., & Jafari, M. (2020). *Agrobacterium rhizogenes* mediated transformation of *Ficus carica* L. for the efficient production of secondary metabolites. *Journal of the Science of Food and Agriculture*, *100*(5), 2185–2197.

Amoo, S. O., Aremu, A. O., & Van Staden, J. (2012). In vitro plant regeneration, secondary metabolite production and antioxidant activity of micropropagated *Aloe arborescens* mill. *Plant Cell, Tissue and Organ Culture*, *111*(3), 345–358.

Amoo, S. O., & Van Staden, J. (2013). Influence of plant growth regulators on shoot proliferation and secondary metabolite production in micropropagated *Huernia hystrix*. *Plant Cell, Tissue and Organ Culture*, *112*(2), 249–256.

Anugrah, F. A., Putra, S. A., Sulisetijono, S., Prabaningtyas, S., & Rusdi, H. O. (2021). Screening of secondary metabolites quinine alkaloid by endophytic bacteria from cinchona plants (*Cinchona ledgeriana* moens.) root. In *International Conference on Life Sciences and Technology (ICoLiST 2020)*, Malang, Indonesia. https://doi.org/10.1063/5.0052924

Anuradha, M., Umalatha, Indu, B. K., Balasubramanya, S., & Pullaiah, T. (2020). In vitro camptothecin production from plants: Prospects and challenges. In *Camptothecin and Camptothecin Producing Plants* (pp. 261–297). Elsevier.

Arya, S. S., Lenka, S. K., Cahill, D. M., & Rookes, J. E. (2021). Designer nanoparticles for plant cell culture systems: Mechanisms of elicitation and harnessing of specialized metabolites. *BioEssays: News and Reviews in Molecular, Cellular and Developmental Biology*, *43*(11), e2100081.

Asano, T., Kobayashi, K., Kashihara, E., Sudo, H., Sasaki, R., Iijima, Y., Aoki, K., Shibata, D., Saito, K., & Yamazaki, M. (2013). Suppression of camptothecin biosynthetic genes results in metabolic modification of secondary products in hairy roots of *Ophiorrhiza pumila*. *Phytochemistry*, *91*, 128–139.

Atanasov, A. G., Waltenberger, B., Pferschy-Wenzig, E.-M., Linder, T., Wawrosch, C., Uhrin, P., Temml, V., Wang, L., Schwaiger, S., Heiss, E. H., Rollinger, J. M., Schuster, D., Breuss, J. M., Bochkov, V., Mihovilovic, M. D., Kopp, B., Bauer, R., Dirsch, V. M., & Stuppner, H. (2015). Discovery and resupply of pharmacologically active plant-derived natural products: A review. *Biotechnology Advances*, *33*(8), 1582–1614.

Badmos, S., Lee, S.-H., & Kuhnert, N. (2019). Comparison and quantification of chlorogenic acids for differentiation of green Robusta and Arabica coffee beans. *Food Research International*, *126*, 108544.

Baek, S., Han, J.-E., Ho, T.-T., & Park, S.-Y. (2022). Development of hairy root cultures for biomass and triterpenoid production in *Centella asiatica*. *Plants*, *11*(2), 148.

Baek, S., Ho, T.-T., Lee, H., Jung, G., Kim, Y. E., Jeong, C.-S., & Park, S.-Y. (2020). Enhanced biosynthesis of triterpenoids in *Centella asiatica* hairy root culture by precursor feeding and elicitation. *Plant Biotechnology Reports*, *14*(1), 45–53.

Bahmani, K., Robinson, A., Majumder, S., LaVardera, A., Dowell, J. A., Goolsby, E. W., & Mason, C. M. (2022). Broad diversity in monoterpene-sesquiterpene balance across wild sunflowers: Implications of leaf and floral volatiles for biotic interactions. *American Journal of Botany*, *109*(12), 2051–2067.

Bai, C., Cao, Y., Zhao, S., Wu, Z., Dai, S., Wang, H., & Fu, C. (2023). Generation of CRISPR/Cas9-mediated mutants in *Monochasma savatieri* using a hairy root system. *Industrial Crops and Products*, *191*, 116008.

Bains, S., Thakur, V., Kaur, J., Singh, K., & Kaur, R. (2019). Elucidating genes involved in sesquiterpenoid and flavonoid biosynthetic pathways in *Saussurea lappa* by de novo leaf transcriptome analysis. *Genomics*, *111*(6), 1474–1482.

Balasubramani, S., Ranjitha Kumari, B. D., Moola, A. K., Sathish, D., Prem Kumar, G., Srimurali, S., & Babu Rajendran, R. (2021). Enhanced production of β-caryophyllene by farnesyl diphosphate precursor-treated callus and hairy root cultures of *Artemisia vulgaris* L. *Frontiers in Plant Science*, *12*, 634178.

Banihashemi, O., Khavari-Nejad, R.-A., Yassa, N., & Najafi, F. (2020). Raise up of scopolamine in hairy roots via *Agrobacterium rhizogenes* ATCC15834 as compared with untransformed roots in *Atropa komarovii*. *Iranian Journal of Pharmaceutical Research: IJPR*, *19*(1), 46–56.

Bapat, V. A., Kavi Kishor, P. B., Jalaja, N., Jain, S. M., & Penna, S. (2023). Plant cell cultures: Biofactories for the production of bioactive compounds. *Agronomy*, *13*(3), 858.

Berköz, M., Yalın, S., Özkan-Yılmaz, F., Özlüer-Hunt, A., Krośniak, M., Francik, R., Yunusoğlu, O., Adıyaman, A., Gezici, H., Yiğit, A., Ünal, S., Volkan, D., & Yıldırım, M. (2021). Protective effect of myricetin, apigenin, and hesperidin pretreatments on cyclophosphamide-induced immunosuppression. *Immunopharmacology and Immunotoxicology*, *43*(3), 353–369.

Bilia, A. R., Melillo de Malgalhaes, P., Bergonzi, M. C., & Vincieri, F. F. (2006). Simultaneous analysis of artemisinin and flavonoids of several extracts of *Artemisia annua* L. obtained from a commercial sample and a selected cultivar. *Phytomedicine: International Journal of Phytotherapy and Phytopharmacology*, *13*(7), 487–493.

Biswas, D., Chakraborty, A., Mukherjee, S., & Ghosh, B. (2023). Hairy root culture: A potent method for improved secondary metabolite production of Solanaceous plants. *Frontiers in Plant Science*, *14*, 1197555.

Boisson-Dernier, A., Chabaud, M., Garcia, F., Bécard, G., Rosenberg, C., & Barker, D. G. (2001). *Agrobacterium* rhizogenes-transformed roots of *Medicago truncatula* for the study of nitrogen-fixing and endomycorrhizal symbiotic associations. *Molecular Plant-Microbe Interactions*, *14*(6), 695–700.

Brahmkshatriya, P. P., & Brahmkshatriya, P. S. (2013). Terpenes: Chemistry, biological role, and therapeutic applications. In *Natural Products* (pp. 2665–2691). Springer, Berlin, Heidelberg.

Breitel, D., Brett, P., Alseekh, S., Fernie, A. R., Butelli, E., & Martin, C. (2021). Metabolic engineering of tomato fruit enriched in L-DOPA. *Metabolic Engineering*, *65*, 185–196.

Britton, M. T., Escobar, M. A., & Dandekar, A. M. (2008). The oncogenes of *Agrobacterium tumefaciens* and *Agrobacterium rhizogenes*. In *Agrobacterium: From Biology to Biotechnology* (pp. 523–563). Springer, New York.

Brun, P., Bernabè, G., Filippini, R., & Piovan, A. (2019). In vitro antimicrobial activities of commercially available Tea tree (*Melaleuca alternifolia*) essential oils. *Current Microbiology*, *76*(1), 108–116.

Buitelaar, R. M., Langenhoff, A. A. M., Heidstra, R., & Tramper, J. (1991). Growth and thiophene production by hairy root cultures of *Tagetes patula* in various two-liquid-phase bioreactors. *Enzyme and Microbial Technology*, *13*(6), 487–494.

Cabañas-García, E., Areche, C., Gómez-Aguirre, Y. A., Borquez, J., Muñoz, R., Cruz-Sosa, F., & Balch, E. P.-M. (2021). Biomass production and secondary metabolite identification in callus cultures of *Coryphantha macromeris* (Engelm.) Britton & Rose (Cactaceae), a traditional medicinal plant. *South African Journal of Botany, 137*, 1–9.

Capriotti, L., Baraldi, E., Mezzetti, B., Limera, C., & Sabbadini, S. (2020). Biotechnological approaches: Gene overexpression, gene silencing, and genome editing to control fungal and oomycete diseases in Grapevine. *International Journal of Molecular Sciences*, *21*(16). https://doi.org/10.3390/ijms21165701

Cardon, F., Pallisse, R., Bardor, M., Caron, A., Vanier, J., Ele Ekouna, J. P., Lerouge, P., Boitel-Conti, M., & Guillet, M. (2019). *Brassica rapa* hairy root based expression system leads to the production of highly homogenous and reproducible profiles of recombinant human alpha-L-iduronidase. *Plant Biotechnology Journal*, *17*(2), 505–516.

Catellani, M., Lico, C., Cerasi, M., Massa, S., Bromuro, C., Torosantucci, A., Benvenuto, E., & Capodicasa, C. (2020). Optimised production of an anti-fungal antibody in *Solanaceae* hairy roots to develop new formulations against *Candida albicans*. *BMC Biotechnology*, *20*(1), 15.

Cembrowska-Lech, D., Krzemińska, A., Miller, T., Nowakowska, A., Adamski, C., Radaczyńska, M., Mikiciuk, G., & Mikiciuk, M. (2023). An integrated multi-omics and artificial intelligence framework for advance plant phenotyping in Horticulture. *Biology*, *12*(10). https://doi.org/10.3390/biology12101298

Cetin, E. S., Babalik, Z., Hallac-Turk, F., & Gokturk-Baydar, N. (2014). The effects of cadmium chloride on secondary metabolite production in *Vitis vinifera* cv. cell suspension cultures. *Biological Research*, *47*(1), 47.

Chaachouay, N., & Zidane, L. (2024). Plant-derived natural products: A source for drug discovery and development. *Drugs and Drug Candidates*, *3*(1), 184–207.

Chadni, M., Flourat, A. L., Reungoat, V., Mouterde, L. M. M., Allais, F., & Ioannou, I. (2021). Selective extraction of sinapic acid derivatives from mustard seed meal by acting on pH: Toward a high antioxidant activity rich extract. *Molecules*, *26*(1). https://doi.org/10.3390/molecules26010212

Chen, D., Mubeen, B., Hasnain, A., Rizwan, M., Adrees, M., Naqvi, S. A. H., Iqbal, S., Kamran, M., El-Sabrout, A. M., Elansary, H. O., Mahmoud, E. A., Alaklabi, A., Sathish, M., & Din, G. M. U. (2022). Role of promising secondary metabolites to confer resistance against environmental stresses in crop plants: Current scenario and future perspectives. *Frontiers in Plant Science*, *13*, 881032.

Chen, Y., Xiao, W., Wang, Y., Liu, H., Li, X., & Yuan, Y. (2016). Lycopene overproduction in *Saccharomyces cerevisiae* through combining pathway engineering with host engineering. *Microbial Cell Factories*, *15*(1), 113.

Chib, S., Thangaraj, A., Kaul, S., Dhar, M. K., & Kaul, T. (2020). Development of a system for efficient callus production, somatic embryogenesis and gene editing using CRISPR/Cas9 in Saffron (*Crocus sativus* L.). *Plant Methods*, *16*, 47.

Choi, Y.-E., Kim, Y.-S., & Paek, K.-Y. (2007). Types and designs of bioreactors for hairy root culture. In *Plant Tissue Culture Engineering* (pp. 161–172). Springer Netherlands.

Citti, C., Linciano, P., Russo, F., Luongo, L., Iannotta, M., Maione, S., Laganà, A., Capriotti, A. L., Forni, F., Vandelli, M. A., Gigli, G., & Cannazza, G. (2019). A novel phytocannabinoid isolated from *Cannabis sativa* L. with an in vivo cannabimimetic activity higher than Δ-tetrahydrocannabinol: Δ-Tetrahydrocannabiphorol. *Scientific Reports*, *9*(1), 20335.

Coste, A., Vlase, L., Halmagyi, A., Deliu, C., & Coldea, G. (2011). Effects of plant growth regulators and elicitors on production of secondary metabolites in shoot cultures of *Hypericum hirsutum* and *Hypericum maculatum*. *Plant Cell, Tissue and Organ Culture*, *106*(2), 279–288.

Cui, Y.-Y., Ling, C., Zhang, Y.-Y., Huang, J., & Liu, J.-Z. (2014). Production of shikimic acid from *Escherichia coli* through chemically inducible chromosomal evolution and cofactor metabolic engineering. *Microbial Cell Factories*, *13*, 21.

Cuong, D. M., Kim, J. K., Bong, S. J., Baek, S. A., Jeon, J., Park, J. S., & Park, S. U. (2018). Comparative analysis of glucosinolates and metabolite profiling of green and red mustard (*Brassica* juncea) hairy roots. *3 Biotech*, *8*(9), 382.

Czechowski, T., Rinaldi, M. A., Famodimu, M. T., Van Veelen, M., Larson, T. R., Winzer, T., Rathbone, D. A., Harvey, D., Horrocks, P., & Graham, I. A. (2019). Flavonoid versus artemisinin anti-malarial activity in whole-leaf extracts. *Frontiers in Plant Science*, *10*, 984.

Das, S., Kwon, M., & Kim, J.-Y. (2024). Enhancement of specialized metabolites using CRISPR/Cas gene editing technology in medicinal plants. *Frontiers in Plant Science*, *15*, 1279738.

de Mello, V., de Mesquita Júnior, G. A., Alvim, J. G. E., Costa, J. de C. da, & Vilela, F. M. P. (2023). Recent patent applications for coffee and coffee by-products as active ingredients in cosmetics. *International Journal of Cosmetic Science*, *45*(3), 267–287.

Deavours, B. E., & Dixon, R. A. (2005). Metabolic engineering of isoflavonoid biosynthesis in Alfalfa. *Plant Physiology*, *138*(4), 2245–2259.

Dehdashti, S. M., Acharjee, S., Nomani, A., & Deka, M. (2020). Production of pharmaceutical active recombinant globular adiponectin as a secretory protein in *Withania somnifera* hairy root culture. *Journal of Biotechnology*, *323*, 302–312.

Demain, A. L., & Fang, A. (2000). The natural functions of secondary metabolites. *Advances in Biochemical Engineering/Biotechnology*, *69*, 1–39.

Dhiman, N., Patial, V., & Bhattacharya, A. (2018). The current status and future applications of hairy root cultures. In *Biotechnological Approaches for Medicinal and Aromatic Plants* (pp. 87–155). Springer Singapore.

Divekar, P. A., Narayana, S., Divekar, B. A., Kumar, R., Gadratagi, B. G., Ray, A., Singh, A. K., Rani, V., Singh, V., Singh, A. K., Kumar, A., Singh, R. P., Meena, R. S., & Behera, T. K. (2022). Plant secondary metabolites as defense tools against herbivores for sustainable crop protection. *International Journal of Molecular Sciences*, *23*(5). https://doi.org/10.3390/ijms23052690

Dixit, S., Shukla, A., Singh, V., & Upadhyay, S. K. (2021a). *Bioprospecting of Natural Compounds for Industrial and Medical Applications*. Wiley. https://doi.org/10.1002/9781119718017.ch3

Dixit, S., Shukla, A., Singh, V., & Upadhyay, S. K. (2021b). *Engineering of Plant Metabolic Pathway for Nutritional Improvement*. Wiley. https://doi.org/10.1002/9781119672425.ch20

Dong, O. X., Yu, S., Jain, R., Zhang, N., Duong, P. Q., Butler, C., Li, Y., Lipzen, A., Martin, J. A., Barry, K. W., Schmutz, J., Tian, L., & Ronald, P. C. (2020). Marker-free carotenoid-enriched rice generated through targeted gene insertion using CRISPR-Cas9. *Nature Communications*, *11*(1), 1178.

El-Beltagi, H. S., Mohamed, H. I., Aldaej, M. I., Al-Khayri, J. M., Rezk, A. A., Al-Mssallem, M. Q., Sattar, M. N., & Ramadan, K. M. A. (2022). Production and antioxidant activity of secondary metabolites in Hassawi rice (*Oryza sativa* L.) cell suspension under salicylic acid, yeast extract, and pectin elicitation. *In Vitro Cellular & Developmental Biology. Plant: Journal of the Tissue Culture Association*. https://doi.org/10.1007/s11627-022-10264-x

El-Esawi, M. A., Elkelish, A., Elansary, H. O., Ali, H. M., Elshikh, M., Witczak, J., & Ahmad, M. (2017). Genetic transformation and hairy root induction enhance the antioxidant potential of *Lactuca serriola* L. *Oxidative Medicine and Cellular Longevity*, *2017*, 5604746.

Engels, B., Dahm, P., & Jennewein, S. (2008). Metabolic engineering of taxadiene biosynthesis in yeast as a first step towards Taxol (Paclitaxel) production. *Metabolic Engineering*, *10*(3–4), 201–206.

Fang, Y., Liu, J., Zheng, M., Zhu, S., Pei, T., Cui, M., Chang, L., Xiao, H., Yang, J., Martin, C., & Zhao, Q. (2023). SbMYB3 transcription factor promotes root-specific flavone biosynthesis in *Scutellaria baicalensis*. *Horticulture Research*, *10*(2), uhac266.

Farjaminezhad, R., & Garoosi, G. (2021). Improvement and prediction of secondary metabolites production under yeast extract elicitation of *Azadirachta indica* cell suspension culture using response surface methodology. *AMB Express*, *11*(1), 43.

Fattahi, F., Shojaeiyan, A., Palazon, J., Moyano, E., & Torras-Claveria, L. (2021). Methyl-β-cyclodextrin and coronatine as new elicitors of tropane alkaloid biosynthesis in *Atropa acuminata* and *Atropa belladonna* hairy root cultures. *Physiologia Plantarum*, *172*(4), 2098–2111.

Gadzovska-Simic, S., Tusevski, O., Antevski, S., Atanasova-Pancevska, N., Petreska, J., Stefova, M., Kungulovski, D., & Spasenoski, M. (2012). Secondary metabolite production in *Hypericum perforatum* L. cell suspensions upon elicitation with fungal mycelia from *Aspergillus flavus*. *Archives of Biological Sciences*, *64*(1), 113–121.

Gai, Q.-Y., Jiao, J., Luo, M., Wang, W., Ma, W., Zu, Y.-G., & Fu, Y.-J. (2015). Establishment of high-productive *Isatis tinctoria* L. hairy root cultures: A promising approach for efficient production of bioactive alkaloids. *Biochemical Engineering Journal, 95*, 37–47.

Garagounis, C., Beritza, K., Georgopoulou, M.-E., Sonawane, P., Haralampidis, K., Goossens, A., Aharoni, A., & Papadopoulou, K. K. (2020). A hairy-root transformation protocol for *Trigonella foenum-graecum* L. as a tool for metabolic engineering and specialised metabolite pathway elucidation. *Plant Physiology and Biochemistry: PPB/Societe Francaise de Physiologie Vegetale, 154*, 451–462.

Gaume, A., Komarnytsky, S., Borisjuk, N., & Raskin, I. (2003). Rhizosecretion of recombinant proteins from plant hairy roots. *Plant Cell Reports, 21*(12), 1188–1193.

Gelvin, S. B. (2003). *Agrobacterium*-mediated plant transformation: The biology behind the "gene-jockeying" tool. *Microbiology and Molecular Biology Reviews: MMBR, 67*(1), 16–37.

Gharari, Z., Bagheri, K., & Sharafi, A. (2023). Enhanced terpenoids production of elicited hairy root cultures of *Scutellaria bornmuelleri. Brazilian Archives of Biology and Technology, 66*. https://doi.org/10.1590/1678-4324-2023210435

Giri, A., & Narasu, M. L. (2000). Transgenic hairy roots: Recent trends and applications. *Biotechnology Advances, 18*(1), 1–22.

Goklany, S., Rizvi, N. F., Loring, R. H., Cram, E. J., & Lee-Parsons, C. W. T. (2013). Jasmonate-dependent alkaloid biosynthesis in *Catharanthus roseus* hairy root cultures is correlated with the relative expression of *Orca* and *Zct* transcription factors. *Biotechnology Progress, 29*(6), 1367–1376.

Gordon, J. E., & Christie, P. J. (2014). The *Agrobacterium* Ti plasmids. *Microbiology Spectrum, 2*(6). https://doi.org/10.1128/microbiolspec.PLAS-0010-2013

Grzegorczyk-Karolak, I., Kuźma, Ł., Skała, E., & Kiss, A. K. (2018). Hairy root cultures of *Salvia viridis* L. for production of polyphenolic compounds. *Industrial Crops and Products, 117*, 235–244.

Guerriero, G., Berni, R., Muñoz-Sanchez, J. A., Apone, F., Abdel-Salam, E. M., Qahtan, A. A., Alatar, A. A., Cantini, C., Cai, G., Hausman, J.-F., Siddiqui, K. S., Hernández-Sotomayor, S. M. T., & Faisal, M. (2018). Production of plant secondary metabolites: Examples, tips and suggestions for biotechnologists. *Genes, 9*(6). https://doi.org/10.3390/genes9060309

Gülçin, I. (2006). Antioxidant activity of caffeic acid (3,4-dihydroxycinnamic acid). *Toxicology, 217*(2–3), 213–220.

Gutensohn, M., Henry, L. K., Gentry, S. A., Lynch, J. H., Nguyen, T. T. H., Pichersky, E., & Dudareva, N. (2021). Overcoming bottlenecks for metabolic engineering of sesquiterpene production in Tomato fruits. *Frontiers in Plant Science, 12*, 691754.

Gutierrez-Valdes, N., Häkkinen, S. T., Lemasson, C., de Groot, J., Ele-Ekouna, J.-P., Guillet, M., Cardon, F., & Ritala, A. (2022). Improving yield of a recombinant biologic in a *Brassica* hairy root manufacturing process. *Biotechnology and Bioengineering, 119*(10), 2831–2841.

Gutierrez-Valdes, N., Häkkinen, S. T., Lemasson, C., Guillet, M., Oksman-Caldentey, K.-M., Ritala, A., & Cardon, F. (2020). Hairy root cultures-A versatile tool with multiple applications. *Frontiers in Plant Science, 11*, 33.

Habibi, P., Piri, K., Deljo, A., Moghadam, Y. A., & Ghiasvand, T. (2015). Increasing scopolamine content in hairy roots of *Atropa belladonna* using bioreactor. *Brazilian Archives of Biology and Technology, 58*(2), 166–174.

Haider Abbasi, B., Ullah, M. A., Nadeem, M., Tungmunnithum, D., & Hano, C. (2020). Exogenous application of salicylic acid and gibberellic acid on biomass accumulation, antioxidant and anti-inflammatory secondary metabolites production in multiple shoot culture of *Ajuga integrifolia* Buch. Ham. ex D. Don. *Industrial Crops and Products, 145*, 112098.

Häkkinen, S. T., Raven, N., Henquet, M., Laukkanen, M.-L., Anderlei, T., Pitkänen, J.-P., Twyman, R. M., Bosch, D., Oksman-Caldentey, K.-M., Schillberg, S., & Ritala, A. (2014). Molecular farming in tobacco hairy roots by triggering the secretion of a pharmaceutical antibody. *Biotechnology and Bioengineering, 111*(2), 336–346.

Halder, T., & Ghosh, B. (2022). Hairy root cultures of *Physalis minima* L. — An alternative source of withaferin A production. *Plant Cell, Tissue and Organ Culture*. https://doi.org/10.1007/s11240-022-02386-2

Hammer, K. A., Carson, C. F., & Riley, T. V. (2012). Effects of *Melaleuca alternifolia* (tea tree) essential oil and the major monoterpene component terpinen-4-ol on the development of single- and multistep antibiotic resistance and antimicrobial susceptibility. *Antimicrobial Agents and Chemotherapy, 56*(2), 909–915.

Han, J., Wu, Y., Zhou, Y., & Li, S. (2021). Engineering to produce plant benzylisoquinoline alkaloids. *aBIOTECH, 2*(3), 264–275.

Hanafy, M. S., Matter, M. A., Asker, M. S., & Rady, M. R. (2016). Production of indole alkaloids in hairy root cultures of *Catharanthus roseus* L. and their antimicrobial activity. *South African Journal of Botany: Official Journal of the South African Association of Botanists, 105*, 9–18.

Hao, X., Pu, Z., Cao, G., You, D., Zhou, Y., Deng, C., Shi, M., Nile, S. H., Wang, Y., Zhou, W., & Kai, G. (2020). Tanshinone and salvianolic acid biosynthesis are regulated by in hairy roots. *Journal of Advertising Research*, *23*, 1–12.

Hayashi, H. (2009). Molecular biology of secondary metabolism: Case study for *Glycyrrhiza* plants. In *Recent Advances in Plant Biotechnology* (pp. 89–103). Springer US.

Hayashi, S., Watanabe, M., Kobayashi, M., Tohge, T., Hashimoto, T., & Shoji, T. (2020). Genetic manipulation of transcriptional regulators alters nicotine biosynthesis in tobacco. *Plant & Cell Physiology*, *61*(6), 1041–1053.

He, J., Plácido, J. P. A., Pateraki, I., Kampranis, S., Favero, B. T., & Lütken, H. (2022). Hairy root induction of *Taxus baccata* L. by natural transformation with *Rhizobium rhizogenes*. *Horticulturae*, *9*(1), 4.

He, Y., Zhang, T., Sun, H., Zhan, H., & Zhao, Y. (2020). A reporter for noninvasively monitoring gene expression and plant transformation. *Horticulture Research*, *7*(1), 152.

Hennrich, O., Weinmann, L., Kulik, A., Harms, K., Klahn, P., Youn, J.-W., Surup, F., & Mast, Y. (2023). Biotransformation-coupled mutasynthesis for the generation of novel pristinamycin derivatives by engineering the phenylglycine residue. *RSC Chemical Biology*, *4*(12), 1050–1063.

Ho, C.-L., & Su, Y.-C. (2012). Composition, antioxidant and antimicrobial activities of the leaf essential oil of *Machilus japonica* from Taiwan. *Natural Product Communications*, *7*(1), 109–112.

Ho, T.-T., Lee, J.-D., Ahn, M.-S., Kim, S.-W., & Park, S.-Y. (2018). Enhanced production of phenolic compounds in hairy root cultures of *Polygonum multiflorum* and its metabolite discrimination using HPLC and FT-IR methods. *Applied Microbiology and Biotechnology*, *102*(22), 9563–9575.

Hooykaas, M. J. G., & Hooykaas, P. J. J. (2021). The genome sequence of hairy root *Rhizobium rhizogenes* strain LBA9402: Bioinformatics analysis suggests the presence of a new opine system in the agropine Ri plasmid. *MicrobiologyOpen*, *10*(2), e1180.

Horvat, D., Šimić, G., Drezner, G., Lalić, A., Ledenčan, T., Tucak, M., Plavšić, H., Andrić, L., & Zdunić, Z. (2020). Phenolic acid profiles and antioxidant activity of major cereal crops. *Antioxidants*, *9*(6). https://doi.org/10.3390/antiox9060527

Hou, X., Guo, X., Zhang, Y., & Zhang, Q. (2023). CRISPR/Cas genome editing system and its application in potato. *Frontiers in Genetics*, *14*, 1017388.

Huang, P., Liu, W., Xu, M., Jiang, R., Xia, L., Wang, P., Li, H., Tang, Z., Zheng, Q., & Zeng, J. (2018). Modulation of benzylisoquinoline alkaloid biosynthesis by overexpression berberine bridge enzyme in *Macleaya cordata*. *Scientific Reports*, *8*(1), 17988.

Huang, P., Lu, M., Li, X., Sun, H., Cheng, Z., Miao, Y., Fu, Y., & Zhang, X. (2022). An efficient -mediated hairy root transformation method in a Soybean root biology study. *International Journal of Molecular Sciences*, *23*(20). https://doi.org/10.3390/ijms232012261

Huang, S.-H., Vishwakarma, R. K., Lee, T.-T., Chan, H.-S., & Tsay, H.-S. (2014). Establishment of hairy root lines and analysis of iridoids and secoiridoids in the medicinal plant *Gentiana scabra*. *Botanical Studies*, *55*(1), 17.

Hussain, M. J., Abbas, Y., Nazli, N., Fatima, S., Drouet, S., Hano, C., & Abbasi, B. H. (2022). Root cultures, a boon for the production of valuable compounds: A comparative review. *Plants*, *11*(3). https://doi.org/10.3390/plants11030439

Hussain, M. S., Fareed, S., Ansari, S., Rahman, M. A., Ahmad, I. Z., & Saeed, M. (2012). Current approaches toward production of secondary plant metabolites. *Journal of Pharmacy & Bioallied Sciences*, *4*(1), 10–20.

Jacobs, T. B., & Martin, G. B. (2016). High-throughput CRISPR vector construction and characterization of DNA modifications by generation of tomato hairy roots. *Journal of Visualized Experiments: JoVE*, *110*. https://doi.org/10.3791/53843

Jan, R., Asaf, S., Numan, M., Lubna, & Kim, K.-M. (2021). Plant secondary metabolite biosynthesis and transcriptional regulation in response to biotic and abiotic stress conditions. *Agronomy*, *11*(5), 968.

Jaremicz, Z., Luczkiewicz, M., Kokotkiewicz, A., Krolicka, A., & Sowinski, P. (2014). Production of tropane alkaloids in *Hyoscyamus niger* (black henbane) hairy roots grown in bubble-column and spray bioreactors. *Biotechnology Letters*, *36*(4), 843–853.

Jiang, Z., Abdullah, Zhang, S., Jiang, Y., Liu, R., & Xiao, Y. (2022). Development and optimization of CRISPR prime editing system in photoautotrophic cells. *Molecules*, *27*(6). https://doi.org/10.3390/molecules27061758

John, R., Shajitha, P. P., Devassy, A., & Mathew, L. (2017). Hairy-root cultures of *Achyranthes aspera* Linn. as a novel route for the production of 20-hydroxyecdysone. *Acta Physiologiae Plantarum/Polish Academy of Sciences, Committee of Plant Physiology Genetics and Breeding*, *39*(12). https://doi.org/10.1007/s11738-017-2555-x

Joseph Sahayarayan, J., Udayakumar, R., Arun, M., Ganapathi, A., Alwahibi, M. S., Aldosari, N. S., & Morgan, A. M. A. (2020). Effect of different strains for hairy root induction, total phenolic, flavonoids contents, antibacterial and antioxidant activity of (*Cucumis anguria* L.). *Saudi Journal of Biological Sciences*, *27*(11), 2972–2979.

Karunanithi, P. S., & Zerbe, P. (2019). Terpene synthases as metabolic gatekeepers in the evolution of plant terpenoid chemical diversity. *Frontiers in Plant Science*, *10*, 1166.

Khan, A. A., Siddiqui, M. H., & Osama, K. (2019). Bioreactors for hairy roots culture: A review. *Current Biotechnology*, *7*(6), 417–427.

Khoshandam, M., Soltaninejad, H., Mousazadeh, M., Hamidieh, A. A., & Hosseinkhani, S. (2024). Clinical applications of the CRISPR/Cas9 genome-editing system: Delivery options and challenges in precision medicine. *Genes & Diseases*, *11*(1), 268–282.

Kikowska, M., Thiem, B., Szopa, A., & Ekiert, H. (2020). Accumulation of valuable secondary metabolites: Phenolic acids and flavonoids in different in vitro systems of shoot cultures of the endangered plant species—*Eryngium alpinum* L. *Plant Cell, Tissue and Organ Culture*, *141*(2), 381–391.

Kowalczyk, T., Wieczfinska, J., Skała, E., Śliwiński, T., & Sitarek, P. (2020). Transgenesis as a tool for the efficient production of selected secondary metabolites from in vitro plant cultures. *Plants*, *9*(2). https://doi.org/10.3390/plants9020132

Kumar, A., Bhamini, K., Singh, R. K., & Singh, D. N. (2024). Production of plant-derived natural compounds through hairy root culture. In *Biosynthesis of Natural Products in Plants* (pp. 213–239). Springer Nature, Singapore.

Kumar, P., Padhan, J. K., Kumar, A., & Chauhan, R. S. (2018). Transcriptomes of *Podophyllum hexandrum* unravel candidate miRNAs and their association with the biosynthesis of secondary metabolites. *Journal of Plant Biochemistry and Biotechnology*, *27*(1), 46–54.

Kwok, K. H., & Doran, P. M. (1995). Kinetic and stoichiometric analysis of hairy roots in a segmented bubble column reactor. *Biotechnology Progress*, *11*(4), 429–435.

Lakshmaiah, V. V., Banadka, A., Thirumoorthy, G., Naik, P. M., Al-Khayri, J. M., & Nagella, P. (2022). In vitro production of bioactive compounds from plant cell culture. In *Nutraceuticals Production from Plant Cell Factory* (pp. 29–67). Springer Nature, Singapore.

Lankitus, D., Zhang, Y., Ariyaratne, M., Barker, D. J., McNulty, S. L., Amstutz, N., Zhao, L., Iaffaldano, B. J., & Cornish, K. (2023). *Agrobacterium rhizogenes*–induced altered morphology and physiology in rubber dandelion after genetic transformation. *Journal of the American Society for Horticultural Science. American Society for Horticultural Science*, *148*(1), 21–28.

Larkin, P. J., Miller, J. A. C., Allen, R. S., Chitty, J. A., Gerlach, W. L., Frick, S., Kutchan, T. M., & Fist, A. J. (2007). Increasing morphinan alkaloid production by over-expressing codeinone reductase in transgenic *Papaver somniferum*. *Plant Biotechnology Journal*, *5*(1), 26–37.

Lee, K.-T., Suzuki, T., Yamakawa, T., Kodama, T., Igarashi, Y., & Shimomura, K. (1999). Production of tropane alkaloids by transformed root cultures of *Atropa belladonna* in stirred bioreactors with a stainless steel net. *Plant Cell Reports*, *18*(7–8), 567–571.

Lei, H., Niu, T., Song, H., Bai, B., Han, P., Wang, Z., & Liu, A. (2021). Comparative transcriptome profiling reveals differentially expressed genes involved in flavonoid biosynthesis between biennial and triennial *Sophora flavescens*. *Industrial Crops and Products*, *161*, 113217.

Li, C., Sarangapani, S., Wang, Q., Nadimuthu, K., & Sarojam, R. (2020). Metabolic engineering of the native monoterpene pathway in spearmint for production of heterologous monoterpenes reveals complex metabolism and pathway interactions. *International Journal of Molecular Sciences*, *21*(17). https://doi.org/10.3390/ijms21176164

Li, C., Zha, W., Li, W., Wang, J., & You, A. (2023). Advances in the biosynthesis of terpenoids and their ecological functions in plant resistance. *International Journal of Molecular Sciences*, *24*(14). https://doi.org/10.3390/ijms241411561

Li, C.-Q., Lei, H.-M., Hu, Q.-Y., Li, G.-H., & Zhao, P.-J. (2021). Recent advances in the synthetic biology of natural drugs. *Frontiers in Bioengineering and Biotechnology*, *9*, 691152.

Li, T., Wei, Z., & Kuang, H. (2021). UPLC-orbitrap-MS-based metabolic profiling of HaCaT cells exposed to withanolides extracted from *Datura metel*.L: Insights from an untargeted metabolomics. *Journal of Pharmaceutical and Biomedical Analysis*, *199*, 113979.

Li, Y., Wang, S., Aioub, A. A. A., Qie, X., Wu, W., & Hu, Z. (2019). Identification and analysis of full-length transcripts involved in the biosynthesis of insecticidal lignan (+)-haedoxan A in *Phryma leptostachya*. *Industrial Crops and Products*, *142*, 111868.

Lin, Q., Zong, Y., Xue, C., Wang, S., Jin, S., Zhu, Z., Wang, Y., Anzalone, A. V., Raguram, A., Doman, J. L., Liu, D. R., & Gao, C. (2020). Prime genome editing in rice and wheat. *Nature Biotechnology*, *38*(5), 582–585.

Liu, B., Wang, H., Du, Z., Li, G., & Ye, H. (2011). Metabolic engineering of artemisinin biosynthesis in *Artemisia annua* L. *Plant Cell Reports*, *30*(5), 689–694.

Liu, L., Qu, Y. L., Dong, G. R., Wang, J., Hu, C. Y., & Meng, Y. H. (2021). Elevated β-carotene production using codon-adapted CarRA&B and metabolic balance in engineered *Yarrowia lipolytica*. *Frontiers in Microbiology*, *12*, 627150.

Lv, Y., Tian, T., Wang, Y.-J., Huang, J.-P., & Huang, S.-X. (2022). Advances in chemistry and bioactivity of the genus *Erythroxylum*. *Natural Products and Bioprospecting*, *12*(1), 15.

Ma, M., Li, M., Wu, Z., Liang, X., Zheng, Q., Li, D., Wang, G., & An, T. (2024). The microbial biosynthesis of noncanonical terpenoids. *Applied Microbiology and Biotechnology*, *108*(1), 226.

Mahendran, G., Verma, N., Singh, M., Shanker, K., Banerjee, S., Kumar, B., & ur Rahman, L. (2022). Elicitation enhances swerchirin and 1,2,5,6–tetrahydroxyxanthone production in hairy root cultures of *Swertia chirayita* (Roxb.) H. Karst. *Industrial Crops and Products*, *177*, 114488.

Marchev, A. S., Yordanova, Z. P., & Georgiev, M. I. (2020). Green (cell) factories for advanced production of plant secondary metabolites. *Critical Reviews in Biotechnology*, *40*(4), 443–458.

Marks, M. D., Tian, L., Wenger, J. P., Omburo, S. N., Soto-Fuentes, W., He, J., Gang, D. R., Weiblen, G. D., & Dixon, R. A. (2009). Identification of candidate genes affecting Delta9-tetrahydrocannabinol biosynthesis in *Cannabis sativa*. *Journal of Experimental Botany*, *60*(13), 3715–3726.

Masyita, A., Mustika Sari, R., Dwi Astuti, A., Yasir, B., Rahma Rumata, N., Emran, T. B., Nainu, F., & Simal-Gandara, J. (2022). Terpenes and terpenoids as main bioactive compounds of essential oils, their roles in human health and potential application as natural food preservatives. *Food Chemistry: X*, *13*, 100217.

Matsuura, H. N., & Fett-Neto, A. G. (2015). Plant alkaloids: Main features, toxicity, and mechanisms of action. In *Plant Toxins* (pp. 1–15). Springer Netherlands.

Matthysse, A. G. (2006). The genus *Agrobacterium*. In *The Prokaryotes* (pp. 91–114). Springer New York.

McConkey, M. E., Gershenzon, J., & Croteau, R. B. (2000). Developmental regulation of monoterpene biosynthesis in the glandular trichomes of peppermint. *Plant Physiology*, *122*(1), 215–224.

Medina-Bolívar, F., & Cramer, C. (2004). Production of recombinant proteins by hairy roots cultured in plastic sleeve bioreactors. *Methods in Molecular Biology*, *267*, 351–363.

Mehrotra, S., Kukreja, A. K., Singh Khanuja, S. P., & Mishra, B. N. (2008). Genetic transformation studies and scale up of hairy root culture of *Glycyrrhiza glabra* in bioreactor. *Electronic Journal of Biotechnology: EJB*, *11*(2), 69–75.

Mehrotra, S., Mishra, S., & Srivastava, V. (2016). Bioreactor technology for hairy roots cultivation. In *Bioprocessing of Plant In Vitro Systems* (pp. 1–25). Springer International Publishing.

Mehrotra, S., Srivastava, V., Goel, M. K., & Kukreja, A. K. (2016). Scale-up of *Agrobacterium rhizogenes*-mediated hairy root cultures of *Rauwolfia serpentina*: A persuasive approach for stable reserpine production. *Methods in Molecular Biology*, 1391, 241–257.

Meng, J., Wang, B., He, G., Wang, Y., Tang, X., Wang, S., Ma, Y., Fu, C., Chai, G., & Zhou, G. (2019). Metabolomics integrated with transcriptomics reveals redirection of the phenylpropanoids metabolic flux in *Ginkgo biloba*. *Journal of Agricultural and Food Chemistry*, *67*(11), 3284–3291.

Min, J. Y., Jung, H. Y., Kang, S. M., Kim, Y. D., Kang, Y. M., Park, D. J., Prasad, D. T., & Choi, M. S. (2007). Production of tropane alkaloids by small-scale bubble column bioreactor cultures of *Scopolia parviflora* adventitious roots. *Bioresource Technology*, *98*(9), 1748–1753.

Mipeshwaree Devi, A., Khedashwori Devi, K., Premi Devi, P., Lakshmipriyari Devi, M., & Das, S. (2023). Metabolic engineering of plant secondary metabolites: Prospects and its technological challenges. *Frontiers in Plant Science*, *14*, 1171154.

Mishra, B. N., & Ranjan, R. (2008). Growth of hairy-root cultures in various bioreactors for the production of secondary metabolites. *Biotechnology and Applied Biochemistry*, *49*(Pt 1), 1–10.

Moghadam, A., Niazi, A., Afsharifar, A., & Taghavi, S. M. (2016). Expression of a recombinant anti-HIV and anti-tumor protein, MAP30, in *Nicotiana tobacum* hairy roots: A pH-stable and thermophilic antimicrobial protein. *PLoS One*, *11*(7), e0159653.

Moharrami, F., Hosseini, B., Sharafi, A., & Farjaminezhad, M. (2017). Enhanced production of hyoscyamine and scopolamine from genetically transformed root culture of *Hyoscyamus reticulatus* L. elicited by iron oxide nanoparticles. *In Vitro Cellular & Developmental Biology. Plant: Journal of the Tissue Culture Association*, *53*(2), 104–111.

Mora-Vásquez, S., Wells-Abascal, G. G., Espinosa-Leal, C., Cardineau, G. A., & García-Lara, S. (2022). Application of metabolic engineering to enhance the content of alkaloids in medicinal plants. *Metabolic Engineering Communications*, *14*, e00194.

Morey, K. J., & Peebles, C. A. M. (2022). Hairy roots: An untapped potential for production of plant products. *Frontiers in Plant Science*, *13*, 937095.

Moses, T., Pollier, J., Thevelein, J. M., & Goossens, A. (2013). Bioengineering of plant (tri)terpenoids: From metabolic engineering of plants to synthetic biology in vivo and in vitro. *The New Phytologist*, *200*(1), 27–43.

Mou, Z., Ye, F., Shen, F., & Zhao, D. (2022). In vitro germination and micropropagation of *Aconitum vilmorinianum*: An important medicinal plant in China. *Phyton*, *91*(10), 2331–2348.

Niazian, M., Belzile, F., Curtin, S. J., de Ronne, M., & Torkamaneh, D. (2023). Optimization of and -mediated hairy root transformation of soybean for visual screening of transformants using RUBY. *Frontiers in Plant Science*, *14*, 1207762.

Nunes, R., Pasko, P., Tyszka-Czochara, M., Szewczyk, A., Szlosarczyk, M., & Carvalho, I. S. (2017). Antibacterial, antioxidant and anti-proliferative properties and zinc content of five south Portugal herbs. *Pharmaceutical Biology*, *55*(1), 114–123.

Omokhefe Bruce, S. (2022). Secondary metabolites from natural products. In *Secondary Metabolites – Trends and Reviews*. IntechOpen.

Ozyigit, I. I., Dogan, I., Hocaoglu-Ozyigit, A., Yalcin, B., Erdogan, A., Yalcin, I. E., Cabi, E., & Kaya, Y. (2023). Production of secondary metabolites using tissue culture-based biotechnological applications. *Frontiers in Plant Science*, *14*, 1132555.

Palavalli, R. R., Srivastava, S., & Srivastava, A. K. (2012). Development of a mathematical model for growth and oxygen transfer in in vitro plant hairy root cultivations. *Applied Biochemistry and Biotechnology*, *167*(6), 1831–1844.

Panda, B. M., Mehta, U. J., & Hazra, S. (2017). Optimizing culture conditions for establishment of hairy root culture of *Semecarpus anacardium* L. *3 Biotech*, *7*(1), 21.

Pang, Y., Abeysinghe, I. S. B., He, J., He, X., Huhman, D., Mewan, K. M., Sumner, L. W., Yun, J., & Dixon, R. A. (2013). Functional characterization of proanthocyanidin pathway enzymes from tea and their application for metabolic engineering. *Plant Physiology*, *161*(3), 1103–1116.

Park, S.-U., Yu, M., & Facchini, P. J. (2002). Antisense RNA-mediated suppression of benzophenanthridine alkaloid biosynthesis in transgenic cell cultures of *California poppy*. *Plant Physiology*, *128*(2), 696–706.

Pei, T., Yan, M., Huang, Y., Wei, Y., Martin, C., & Zhao, Q. (2022). Specific flavonoids and their biosynthetic pathway in *Scutellaria baicalensis*. *Frontiers in Plant Science*, *13*, 866282.

Pereira, B. M., Arraes, F., Martins, A. C. Q., Alves, N. S. F., Melo, B. P., Morgante, C. V., Saraiva, M. A. P., Grossi-de-Sá, M. F., Guimaraes, P. M., & Brasileiro, A. C. M. (2023). A novel soybean hairy root system for gene functional validation. *PLoS One*, *18*(5), e0285504.

Petrova, M., Nikolova, M., Dimitrova, L., Dimitrova, M., & Sergiev, I. (2023). In vitro multiplication and GC/MS-based metabolic profiles of *Cichorium intybus* L. *The Journal of Microbiology, Biotechnology and Food Sciences*, *13*, e9688.

Pistelli, L., Giovannini, A., Ruffoni, B., Bertoli, A., & Pistelli, L. (2010). Hairy root cultures for secondary metabolites production. *Advances in Experimental Medicine and Biology*, *698*, 167–184.

Pyne, M. E., Narcross, L., & Martin, V. J. J. (2019). Engineering plant secondary metabolism in microbial systems. *Plant Physiology*, *179*(3), 844–861.

Qin, S., Liu, Y., Yan, J., Lin, S., Zhang, W., & Wang, B. (2022). An optimized tobacco hairy root induction system for functional analysis of nicotine biosynthesis-related genes. *Agronomy*, *12*(2), 348.

Rabizadeh, F., Mirian, M. S., Doosti, R., Kiani-Anbouhi, R., & Eftekhari, E. (2022). Phytochemical classification of medicinal plants used in the treatment of kidney disease based on traditional Persian medicine. *Evidence-Based Complementary and Alternative Medicine: eCAM*, *2022*, 8022599.

Rady, M. R., Saker, M. M., & Matter, M. A. (2018). In vitro culture, transformation and genetic fidelity of Milk Thistle. *Journal, Genetic Engineering & Biotechnology*, *16*(2), 563–572.

Rahmat, E., & Kang, Y. (2019). Adventitious root culture for secondary metabolite production in medicinal plants: A Review. *Journal of Plant Biotechnology*, *46*(3), 143–157.

Ramakrishna, A., & Ravishankar, G. A. (2011). Influence of abiotic stress signals on secondary metabolites in plants. *Plant Signaling & Behavior*, *6*(11), 1720–1731.

Rangslang, R. K., Liu, Z., Lütken, H., & Favero, B. T. (2018). *Agrobacterium* spp. genes and ORFs: Mechanisms and applications in plant science. *Ciência E Agrotecnologia*, *42*(5), 453–463.

Rasheed, A., Gill, R. A., Hassan, M. U., Mahmood, A., Qari, S., Zaman, Q. U., Ilyas, M., Aamer, M., Batool, M., Li, H., & Wu, Z. (2021). A critical review: Recent advancements in the use of CRISPR/Cas9 technology to enhance crops and alleviate global food crises. *Current Issues in Molecular Biology*, *43*(3), 1950–1976.

Rency, A. S., Pandian, S., Kasinathan, R., Satish, L., Swamy, M. K., & Ramesh, M. (2019). Hairy root cultures as an alternative source for the production of high-value secondary metabolites. In *Natural Bio-Active Compounds* (pp. 237–264). Springer Singapore.

Renouard, S., Corbin, C., Drouet, S., Medvedec, B., Doussot, J., Colas, C., Maunit, B., Bhambra, A. S., Gontier, E., Jullian, N., Mesnard, F., Boitel, M., Abbasi, B. H., Arroo, R. R. J., Lainé, E., & Hano, C. (2018). Investigation of *Linum flavum* (L.) hairy root cultures for the production of anticancer *Aryltetralin Lignans*. *International Journal of Molecular Sciences*, *19*(4). https://doi.org/10.3390/ijms19040990

Richardson, C., Kelsh, R. N., & Richardson, R. J. (2023). New advances in CRISPR/Cas-mediated precise gene-editing techniques. *Disease Models & Mechanisms*, *16*(2). https://doi.org/10.1242/dmm.049874

Rischer, H., Oresic, M., Seppänen-Laakso, T., Katajamaa, M., Lammertyn, F., Ardiles-Diaz, W., Van Montagu, M. C. E., Inzé, D., Oksman-Caldentey, K.-M., & Goossens, A. (2006). Gene-to-metabolite networks for terpenoid indole alkaloid biosynthesis in *Catharanthus roseus* cells. *Proceedings of the National Academy of Sciences of the United States of America*, *103*(14), 5614–5619.

Rodriguez, A., Kildegaard, K. R., Li, M., Borodina, I., & Nielsen, J. (2015). Establishment of a yeast platform strain for production of p-coumaric acid through metabolic engineering of aromatic amino acid biosynthesis. *Metabolic Engineering*, *31*, 181–188.

Rogowska, A., Stpiczyńska, M., Pączkowski, C., & Szakiel, A. (2022). The influence of exogenous jasmonic acid on the biosynthesis of steroids and triterpenoids in plants and hairy root culture. *International Journal of Molecular Sciences*, *23*(20). https://doi.org/10.3390/ijms232012173

Röper, W., Schulz, M., Chaouiche, E., & Meloh, K. A. (1985). Nicotine production by tissue cultures of tobacco as influenced by various culture parameters. *Journal of Plant Physiology*, *118*(5), 463–470.

Ru, M., An, Y., Wang, K., Peng, L., Li, B., Bai, Z., Wang, B., & Liang, Z. (2016). *Prunella vulgaris* L. hairy roots: Culture, growth, and elicitation by ethephon and salicylic acid. *Engineering in Life Sciences*, *16*(5), 494–502.

Ru, Q.-M., Wang, L.-J., Li, W.-M., Wang, J.-L., & Ding, Y.-T. (2012). In vitro antioxidant properties of flavonoids and polysaccharides extract from tobacco (*Nicotiana tabacum* L.) leaves. *Molecules*, *17*(9), 11281–11291.

Runguphan, W., Maresh, J. J., & O'Connor, S. E. (2009). Silencing of tryptamine biosynthesis for production of nonnatural alkaloids in plant culture. *Proceedings of the National Academy of Sciences of the United States of America*, *106*(33), 13673–13678.

Saewan, N., Jimtaisong, A., Panyachariwat, N., & Chaiwut, P. (2023). In vitro and in vivo anti-aging effect of coffee berry nanoliposomes. *Molecules*, *28*(19). https://doi.org/10.3390/molecules28196830

Salam, U., Ullah, S., Tang, Z.-H., Elateeq, A. A., Khan, Y., Khan, J., Khan, A., & Ali, S. (2023). Plant Metabolomics: An overview of the role of primary and secondary metabolites against different environmental stress factors. *Life*, *13*(3). https://doi.org/10.3390/life13030706

Sarkar, S., Das, A. K., Bhattacharya, S., Gachhui, R., & Sil, P. C. (2023). Isorhamnetin exerts anti-tumor activity in DEN+CCl-induced HCC mice. *Medical Oncology*, *40*(7), 188.

Sassi, H. E., Renihan, S., Spence, A. M., & Cooperstock, R. L. (2005). Gene CATCHR--gene cloning and tagging for *Caenorhabditis elegans* using yeast homologous recombination: A novel approach for the analysis of gene expression. *Nucleic Acids Research*, *33*(18), e163.

Sathasivam, R., Choi, M., Radhakrishnan, R., Kwon, H., Yoon, J., Yang, S. H., Kim, J. K., Chung, Y. S., & Park, S. U. (2022). Effects of various *Agrobacterium rhizogenes* strains on hairy root induction and analyses of primary and secondary metabolites in *Ocimum basilicum*. *Frontiers in Plant Science*, *13*, 983776.

Semwal, P., Painuli, S., Badoni, H., & Bacheti, R. K. (2018). Screening of phytoconstituents and antibacterial activity of leaves and bark of *Quercus leucotrichophora* A. Camus from Uttarakhand Himalaya. *Clinical Phytoscience*, *4*(1). https://doi.org/10.1186/s40816-018-0090-y

Shakeran, Z., Keyhanfar, M., & Ghanadian, M. (2017). Biotic elicitation for scopolamine production by hairy root cultures of *Datura metel*. *Molecular Biology Research Communications*, *6*(4), 169–179.

Sharafi, A., Hashemi Sohi, H., Mousavi, A., Azadi, P., Dehsara, B., & Hosseini Khalifani, B. (2013). Enhanced morphinan alkaloid production in hairy root cultures of *Papaver bracteatum* by over-expression of salutaridinol 7-o-acetyltransferase gene via *Agrobacterium rhizogenes* mediated transformation. *World Journal of Microbiology & Biotechnology*, *29*(11), 2125–2131.

Sharma, M., Koul, A., Sharma, D., Kaul, S., Swamy, M. K., & Dhar, M. K. (2019). Metabolic engineering strategies for enhancing the production of bio-active compounds from medicinal plants. In *Natural Bio-Active Compounds* (pp. 287–316). Springer Singapore.

Shi, M., Du, Z., Hua, Q., & Kai, G. (2021). CRISPR/Cas9-mediated targeted mutagenesis of bZIP2 in *Salvia miltiorrhiza* leads to promoted phenolic acid biosynthesis. *Industrial Crops and Products*, *167*, 113560.

Shi, M., Liao, P., Nile, S. H., Georgiev, M. I., & Kai, G. (2021). Biotechnological exploration of transformed root culture for value-added products. *Trends in Biotechnology*, *39*(2), 137–149.

Shi, M., Luo, X., Ju, G., Li, L., Huang, S., Zhang, T., Wang, H., & Kai, G. (2016). Enhanced diterpene tanshinone accumulation and bioactivity of transgenic *Salvia miltiorrhiza* hairy roots by pathway engineering. *Journal of Agricultural and Food Chemistry*, *64*(12), 2523–2530.

Shkryl, Y. N., Tchernoded, G. K., Yugay, Y. A., Grigorchuk, V. P., Sorokina, M. R., Gorpenchenko, T. Y., Kudinova, O. D., Degtyarenko, A. I., Onishchenko, M. S., Shved, N. A., Kumeiko, V. V., & Bulgakov, V. P. (2023). Enhanced production of nitrogenated metabolites with anticancer potential in *Aristolochia manshuriensis* hairy root cultures. *International Journal of Molecular Sciences*, *24*(14). https://doi.org/10.3390/ijms241411240

Shrihastini, V., Muthuramalingam, P., Adarshan, S., Sujitha, M., Chen, J.-T., Shin, H., & Ramesh, M. (2021). Plant derived bioactive compounds, their anti-cancer effects and approaches as an alternative target treatment strategy for breast cancer: An updated overview. *Cancers*, *13*(24). https://doi.org/10.3390/cancers13246222

Shriti, S., Kuldeep Prasad, A., Narasimha Sudheer, W., & Praveen, N. (2023). Hairy roots as a potential source for the production of rosmarinic acid from genus *Salvia*. *Plant Science Today*. https://doi.org/10.14719/pst.2541

Siddiqui, Z. H., Mujib, A., Mahmooduzzafar, Aslam, J., Rehman Hakeem, K., & Parween, T. (2013). *In vitro* production of secondary metabolites using elicitor in *Catharanthus roseus*: A case study. In *Crop Improvement* (pp. 401–419). Springer US.

Singh, M., Suyal, D. C., Dinkar, N., Joshi, S., Srivastava, N., Maurya, V. K., Agnihotri, A., & Agrawal, S. (2022). *Agrobacterium rhizogenes* as molecular tool for the production of hairy roots in *Withania somnifera*. *Journal of Applied Biology & Biotechnology*. https://doi.org/10.7324/jabb.2023.110201

Singh, P., Prasad, R., Tewari, R., Jaidi, M., Kumar, S., Rout, P. K., & Rahman, L. U. (2018). Silencing of quinolinic acid phosphoribosyl transferase (QPT) gene for enhanced production of scopolamine in hairy root culture of *Duboisia leichhardtii*. *Scientific Reports*, *8*(1), 13939.

Singh, S., Pandey, P., Akhtar, M. Q., Negi, A. S., & Banerjee, S. (2021). A new synthetic biology approach for the production of curcumin and its glucoside in *Atropa belladonna* hairy roots. *Journal of Biotechnology*, *328*, 23–33.

Sivakumar, G., Liu, C., Towler, M. J., & Weathers, P. J. (2010). Biomass production of hairy roots of *Artemisia annua* and *Arachis hypogaea* in a scaled-up mist bioreactor. *Biotechnology and Bioengineering*, *107*(5), 802–813.

Sivanandhan, G., Selvaraj, N., Ganapathi, A., & Lim, Y. P. (2020). Up-regulation of Squalene synthase in hairy root culture of *Withania somnifera* (L.) Dunal yields higher quantities of withanolides. *Industrial Crops and Products*, *154*, 112706.

Smetanska, I. (2008). Production of secondary metabolites using plant cell cultures. *Advances in Biochemical Engineering/Biotechnology*, *111*, 187–228.

Soboleski, M. R., Oaks, J., & Halford, W. P. (2005). Green fluorescent protein is a quantitative reporter of gene expression in individual eukaryotic cells. *FASEB Journal: Official Publication of the Federation of American Societies for Experimental Biology*, *19*(3), 440–442.

Solis-Castañeda, G. J., Zamilpa, A., Cabañas-García, E., Bahena, S. M., Pérez-Molphe-Balch, E., & Gómez-Aguirre, Y. A. (2020). Identification and quantitative determination of feruloyl-glucoside from hairy root cultures of *Turbinicarpus lophophoroides* (Werderm.) Buxb. & *Backeb.* (Cactaceae). *In Vitro Cellular & Developmental Biology. Plant: Journal of the Tissue Culture Association*, *56*(1), 8–17.

Song, Y., He, S., Abdallah, I. I., Jopkiewicz, A., Setroikromo, R., van Merkerk, R., Tepper, P. G., & Quax, W. J. (2021). Engineering of multiple modules to improve amorphadiene production in using CRISPR-Cas9. *Journal of Agricultural and Food Chemistry*, *69*(16), 4785–4794.

Stepanova, A. Y., Malunova, M. V., Gladkov, E. A., Evsyukov, S. V., Tereshonok, D. V., & Solov'eva, A. I. (2022). Collection of hairy roots as a basis for fundamental and applied research. *Molecules*, *27*(22). https://doi.org/10.3390/molecules27228040

Stiles, A. R., & Liu, C.-Z. (2013). Hairy root culture: Bioreactor design and process intensification. *Advances in Biochemical Engineering/biotechnology*, *134*, 91–114.

Sun, J., & Peebles, C. A. M. (2016). Engineering overexpression of ORCA3 and strictosidine glucosidase in *Catharanthus roseus* hairy roots increases alkaloid production. *Protoplasma*, *253*(5), 1255–1264.

Sun, M., Shi, M., Wang, Y., Huang, Q., Yuan, T., Wang, Q., Wang, C., Zhou, W., & Kai, G. (2019). The biosynthesis of phenolic acids is positively regulated by the JA-responsive transcription factor ERF115 in *Salvia miltiorrhiza*. *Journal of Experimental Botany*, *70*(1), 243–254.

Sun, T., Tadmor, Y., & Li, L. (2020). Pathways for carotenoid biosynthesis, degradation, and storage. *Methods in Molecular Biology*, 2083, 3–23.

Suresh, B., Bais, H. P., Raghavarao, K. S. M. S., Ravishankar, G. A., & Ghildyal, N. P. (2005). Comparative evaluation of bioreactor design using *Tagetes patula* L. hairy roots as a model system. *Process Biochemistry, 40*(5), 1509–1515.

Syombua, E. D., Zhang, Z., Tripathi, J. N., Ntui, V. O., Kang, M., George, O. O., Edward, N. K., Wang, K., Yang, B., & Tripathi, L. (2021). A CRISPR/Cas9-based genome-editing system for yam (*Dioscorea* spp.). *Plant Biotechnology Journal, 19*(4), 645–647.

Tajner-Czopek, A., Gertchen, M., Rytel, E., Kita, A., Kucharska, A. Z., & Sokół-Łętowska, A. (2020). Study of antioxidant activity of some medicinal plants having high content of caffeic acid derivatives. *Antioxidants, 9*(5). https://doi.org/10.3390/antiox9050412

Takos, A. M., & Rook, F. (2014). Plant-specialized metabolism and its genomic organization in biosynthetic gene clusters in *Lotus japonicus*. In *Compendium of Plant Genomes* (pp. 149–162). Springer, Berlin, Heidelberg.

Thakore, D., & Srivastava, A. K. (2021). Mass scale hairy root cultivation of *Catharanthus roseus* in bioreactor for indole alkaloid production. In *Reference Series in Phytochemistry* (pp. 487–502). Springer International Publishing.

Thanh, N. T., Murthy, H. N., & Paek, K. Y. (2014). Optimization of ginseng cell culture in airlift bioreactors and developing the large-scale production system. *Industrial Crops and Products, 60*, 343–348.

Torkamani, M., Jafari, M., Abbaspour, N., Heidary, R., & Safaie, N. (2014). Enhanced production of valerenic acid in hairy root culture of *Valeriana officinalis* by elicitation. *Open Life Sciences, 9*(9), 853–863.

Tripathy, B. C., & Oelmüller, R. (2012). Reactive oxygen species generation and signaling in plants. *Plant Signaling & Behavior, 7*(12), 1621–1633.

Unver, T. (2024). Isorhamnetin as a promising natural bioactive flavonoid: In vitro assessment of its antifungal property. *International Journal of Agriculture Environment and Food Sciences, 8*(1), 54–61.

Upadhyay, S. K. (2021). *Genome Engineering for Crop Improvement*. Wiley. https://doi.org/10.1002/9781119672425

Upadhyay, S. K. (2024) *Applications of Genome Engineering in Plants*. Wiley. https://doi.org/10.1002/9781394183913

Upadhyay, S. K., & Singh, S. P. (2021). *Bioprospecting of Plant Biodiversity for Industrial Molecules*. Wiley. https://doi.org/10.1002/9781119718017

Upadhyay, S. K., & Singh, S. P. (2023). *Plants as Bioreactors for Industrial Molecules*. Wiley https://doi.org/10.1002/9781119875116

Upadhyay, S. K., & Singh, S. P. (2024). *Herbal Nutraceuticals: Products and Processes*. Wiley. https://doi.org/10.1002/9781394241576

Vaou, N., Stavropoulou, E., Voidarou, C., Tsigalou, C., & Bezirtzoglou, E. (2021). Towards advances in medicinal plant antimicrobial activity: A review study on challenges and future perspectives. *Microorganisms, 9*(10). https://doi.org/10.3390/microorganisms9102041

Vaughan, M. M., Wang, Q., Webster, F. X., Kiemle, D., Hong, Y. J., Tantillo, D. J., Coates, R. M., Wray, A. T., Askew, W., O'Donnell, C., Tokuhisa, J. G., & Tholl, D. (2013). Formation of the unusual semivolatile diterpene rhizathalene by the *Arabidopsis* class I terpene synthase TPS08 in the root stele is involved in defense against belowground herbivory. *The Plant Cell, 25*(3), 1108–1125.

Verma, M., Ghangal, R., Sharma, R., Sinha, A. K., & Jain, M. (2014). Transcriptome analysis of *Catharanthus roseus* for gene discovery and expression profiling. *PLoS One, 9*(7), e103583.

Verma, P. C., Singh, H., Negi, A. S., Saxena, G., Rahman, L.-U., & Banerjee, S. (2015). Yield enhancement strategies for the production of picroliv from hairy root culture of *Picrorhiza kurroa* Royle ex Benth. *Plant Signaling & Behavior, 10*(5), e1023976.

Verma, P. C., Trivedi, I., Singh, H., et al. (2009). Efficient production of gossypol from hairy root culture of cotton (*Gossypium hirsutum*). *Current Pharmaceutical Biotechnology*, 10, 691–700.

Wang, C.-T., Liu, H., Gao, X.-S., & Zhang, H.-X. (2010). Overexpression of *G10H* and *ORCA3* in the hairy roots of *Catharanthus roseus* improves catharanthine production. *Plant Cell Reports, 29*(8), 887–894.

Wang, J., Fang, X., Ge, L., Cao, F., Zhao, L., Wang, Z., & Xiao, W. (2018). Antitumor, antioxidant and anti-inflammatory activities of kaempferol and its corresponding glycosides and the enzymatic preparation of kaempferol. *PLoS One, 13*(5), e0197563.

Wang, Q., Cao, R., Zhang, Y., Qi, P., Wang, L., & Fang, S. (2021). Biosynthesis and regulation of terpenoids from basidiomycetes: Exploration of new research. *AMB Express, 11*(1), 150.

Wang, R., Lenka, S. K., Kumar, V., Sikron-Persi, N., Dynkin, I., Weiss, D., Perl, A., Fait, A., & Oren-Shamir, M. (2021). A synchronized increase of stilbenes and flavonoids in metabolically engineered *Vitis vinifera* cv. Gamay red cell culture. *Journal of Agricultural and Food Chemistry, 69*(28), 7922–7931.

Wawrosch, C., & Zotchev, S. B. (2021). Production of bioactive plant secondary metabolites through in vitro technologies-status and outlook. *Applied Microbiology and Biotechnology*, *105*(18), 6649–6668.

Wei, T., Gao, Y., Deng, K., Zhang, L., Yang, M., Liu, X., Qi, C., Wang, C., Song, W., Zhang, Y., & Chen, C. (2019). Enhancement of tanshinone production in hairy root cultures by metabolic engineering. *Plant Methods*, *15*, 53.

Wierzchowski, K., Kawka, M., Wrzecionek, M., Urbanek, J., Pietrosiuk, A., Sykłowska-Baranek, K., Gadomska-Gajadhur, A., & Pilarek, M. (2022). Stress-induced intensification of deoxy shikonin production in hairy root cultures with ester-based scaffolds. *Plants*, *11*(24). https://doi.org/10.3390/plants11243462

Wilken, D., Jiménez González, E., Hohe, A., Jordan, M., Gomez Kosky, R., Schmeda Hirschmann, G., & Gerth, A. (2005). Comparison of secondary plant metabolite production in cell suspension, callus culture and temporary immersion system. In *Liquid Culture Systems for in vitro Plant Propagation* (pp. 525–537). Springer-Verlag.

Williams, G. R., & Doran, P. M. (2000). Hairy root culture in a liquid-dispersed bioreactor: Characterization of spatial heterogeneity. *Biotechnology Progress*, *16*(3), 391–401.

Willits, M. G., Giovanni, M., Prata, R. T. N., Kramer, C. M., De Luca, V., Steffens, J. C., & Graser, G. (2004). Bio-fermentation of modified flavonoids: An example of in vivo diversification of secondary metabolites. *Phytochemistry*, *65*(1), 31–41.

Woods, R. R., Geyer, B. C., & Mor, T. S. (2008). Hairy-root organ cultures for the production of human acetylcholinesterase. *BMC Biotechnology*, *8*, 95.

Wu, S., Chen, W., Lu, S., Zhang, H., & Yin, L. (2022). Metabolic engineering of shikimic acid biosynthesis pathway for the production of shikimic acid and its branched products in microorganisms: Advances and prospects. *Molecules*, *27*(15). https://doi.org/10.3390/molecules27154779

Wu, T., Kerbler, S. M., Fernie, A. R., & Zhang, Y. (2021). Plant cell cultures as heterologous bio-factories for secondary metabolite production. *Plant Communications*, *2*(5), 100235.

Wurtzel, E. T. (2019). Changing form and function through carotenoids and synthetic biology. *Plant Physiology*, *179*(3), 830–843.

Xing, B., Yang, D., Yu, H., Zhang, B., Yan, K., Zhang, X., Han, R., & Liang, Z. (2018). Overexpression of SmbHLH10 enhances tanshinones biosynthesis in *Salvia miltiorrhiza* hairy roots. *Plant Science: An International Journal of Experimental Plant Biology*, *276*, 229–238.

Xu, W., Jin, X., Yang, M., Xue, S., Luo, L., Cao, X., Zhang, C., Qiao, S., Zhang, C., Li, J., Wu, J., Lv, L., Zhao, F., Wang, N., Tan, S., Lyu-Bu, A. G., Wang, C., & Wang, X. (2021). Primary and secondary metabolites produced in *Salvia miltiorrhiza* hairy roots by an endophytic fungal elicitor from *Mucor fragilis*. *Plant Physiology and Biochemistry: PPB/Societe Francaise de Physiologie Vegetale*, *160*, 404–412.

Yang, X., Wang, H., Shen, C., Dong, X., Li, J., & Liu, J. (2024). Effects of isorhamnetin on liver injury in heat stroke-affected rats under dry-heat environments via oxidative stress and inflammatory response. *Scientific Reports*, *14*(1), 7476.

Yao, D., Xing, J., Tang, Q., Hou, Y., Chen, B., Yao, W., Li, Z., Wang, J., Niu, Y., & Wang, D. (2024). Flavonoid biosynthesis in *Scutellaria baicalensis* Georgi: Metabolomics and transcriptomics analysis. *Agronomy*, *14*(7), 1494.

Yao, L., Zhang, H., Liu, Y., Ji, Q., Xie, J., Zhang, R., Huang, L., Mei, K., Wang, J., & Gao, W. (2022). Engineering of triterpene metabolism and overexpression of the lignin biosynthesis gene PAL promotes ginsenoside Rg accumulation in ginseng plant chassis. *Journal of Integrative Plant Biology*, *64*(9), 1739–1754.

Yu, Z.-X., Wang, L.-J., Zhao, B., Shan, C.-M., Zhang, Y.-H., Chen, D.-F., & Chen, X.-Y. (2015). Progressive regulation of sesquiterpene biosynthesis in *Arabidopsis* and Patchouli (*Pogostemon cablin*) by the miR156-targeted SPL transcription factors. *Molecular Plant*, *8*(1), 98–110.

Yucebilgili Kurtoglu, K., & Unver, T. (2021). Integrated omics analysis of benzylisoquinoline alkaloid (BIA) metabolism in Opium poppy (*Papaver somniferum* L.). In *Oil Crop Genomics* (pp. 291–315). Springer International Publishing.

Zakaria, M. M., Schemmerling, B., & Ober, D. (2021). CRISPR/Cas9-mediated genome editing in comfrey (*Symphytum officinale*) hairy roots results in the complete eradication of pyrrolizidine alkaloids. *Molecules*, *26*(6). https://doi.org/10.3390/molecules26061498

Zengin, H., & Baysal, A. H. (2014). Antibacterial and antioxidant activity of essential oil terpenes against pathogenic and spoilage-forming bacteria and cell structure-activity relationships evaluated by SEM microscopy. *Molecules*, *19*(11), 17773–17798.

Zhai, Z., Sooksa-nguan, T., & Vatamaniuk, O. K. (2009). Establishing RNA interference as a reverse-genetic approach for gene functional analysis in protoplasts. *Plant Physiology*, *149*(2), 642–652.

Zhang, B., Chen, M., Pu, S., Chen, L., Zhang, X., Zhang, J., & Zhu, C. (2020). Identification of secondary metabolites in *Tripterygium wilfordii* hairy roots and culture optimization for enhancing wilforgine and wilforine production. *Industrial Crops and Products*, *148*, 112276.

Zhang, D., Zhang, Z., Unver, T., & Zhang, B. (2021). CRISPR/Cas: A powerful tool for gene function study and crop improvement. *Journal of Advertising Research*, *29*, 207–221.

Zhang, N., Wright, T., Wang, X., Karki, U., Savary, B. J., & Xu, J. (2019). Engineering "designer" glycomodules for boosting recombinant protein secretion in tobacco hairy root culture and studying hydroxyproline-O-glycosylation process in plants. *Plant Biotechnology Journal*, *17*(6), 1130–1141.

Zhang, S., Li, H., Liang, X., Yan, Y., Xia, P., Jia, Y., & Liang, Z. (2015). Enhanced production of phenolic acids in *Salvia miltiorrhiza* hairy root cultures by combing the RNAi-mediated silencing of chalcone synthase gene with salicylic acid treatment. *Biochemical Engineering Journal*, *103*, 185–192.

Zhang, T., Wang, M., Li, Z., Wu, X., & Liu, X. (2023). Transcriptome analysis and exploration of genes involved in the biosynthesis of secoiridoids in. *PeerJ*, *11*, e14968.

Zhao, C., Wang, X.-H., Lu, X.-Y., Zong, H., & Zhuge, B. (2023). Metabolic engineering for sustainable production of geraniol. *ACS Synthetic Biology*, *12*(6), 1836–1844.

Zhao, J.-L., Zou, L., Zhang, C.-Q., Li, Y.-Y., Peng, L.-X., Xiang, D.-B., & Zhao, G. (2014). Efficient production of flavonoids in *Fagopyrum tataricum* hairy root cultures with yeast polysaccharide elicitation and medium renewal process. *Pharmacognosy Magazine*, *10*(39), 234–240.

Zhao, Y., Liu, G., Yang, F., Liang, Y., Gao, Q., Xiang, C., Li, X., Yang, R., Zhang, G., Jiang, H., Yu, L., & Yang, S. (2023). Multilayered regulation of secondary metabolism in medicinal plants. *Molecular Horticulture*, *3*(1), 11.

Zhao, Z., Gao, S., Hu, J., Lei, T., Zhou, Y., Li, Y., & Liu, C. (2023). The hairy root induction efficiency and that plumbagin of *Plumbago auriculata* L. *Plant Cell, Tissue and Organ Culture*, *155*(1), 67–80.

Zheleznichenko, T., Voronkova, M., Asbaganov, S., Kukushkina, T., Filippova, E., Protsenko, M., Mazurkova, N., & Novikova, T. (2023). Impact of different strains on secondary metabolites accumulation in *Nitraria schoberi* L. hairy roots and antiviral activity of their extracts against influenza virus of subtypes A (H5N1) and A (H3N2). *In Vitro Cellular & Developmental Biology. Plant: Journal of the Tissue Culture Association*, *59*(3), 378–392.

Zhou, W., Wang, C., Hao, X., Chen, F., Huang, Q., Liu, T., Xu, J., Guo, S., Liao, B., Liu, Z., Feng, Y., Wang, Y., Liao, P., Xue, J., Shi, M., Maoz, I., & Kai, G. (2024). A chromosome-level genome assembly of anesthetic drug-producing *Anisodus acutangulus* provides insights into its evolution and the biosynthesis of tropane alkaloids. *Plant Communications*, *5*(1), 100680.

Zhou, Z., Tan, H., Li, Q., Chen, J., Gao, S., Wang, Y., Chen, W., & Zhang, L. (2018). CRISPR/Cas9-mediated efficient targeted mutagenesis of RAS in *Salvia miltiorrhiza*. *Phytochemistry*, *148*, 63–70.

13 Biotechnological Advances for Secondary Metabolites Production in Plants

Charu Sharma[1] *and Papiya Mukherjee*[1]
[1]Department of Botany, Panjab University, Chandigarh, India

13.1 INTRODUCTION

Plants have been invaluable sources of medicines since time immemorial, thereby playing a critical role in addressing many health challenges globally. Humans have always counted on medicinal plants for various ailments, as plants often produce valuable compounds that are crucial for their survival and ecological interactions. These compounds, known as secondary metabolites (SMs), do not directly contribute to plant growth; rather, they are essential for its defense, interaction with the environment, and other organisms (Hussain et al. 2012). Although produced in small quantities, SMs are in high demand in several industries, including pharmaceuticals, agrochemicals, and food additives (Balandrin & Klocke 1988). Despite their low abundance, SMs play a vital role in allelopathy, attracting pollinators, and protecting plants against herbivores and pathogens (Bennett & Wallsgrove 1994). The growing demand for these compounds has led to the shifting of many medicinal plants to threatened status, as conventional methods struggle to meet the market demand due to the complex structures of SMs.

Biotechnological methods, like plant tissue culture, have created new opportunities for the synthesis of medicinal and commercially significant specialized metabolites (Hussain et al. 2012). Many strategies, including organ culture, cell culture, elicitation, bioreactor systems, engineering the metabolic pathway, and heterologous synthesis in microorganisms, have been employed to stimulate SM production. The highly valuable bioactive metabolites are produced continuously and sustainably by expanding *in vitro* plant culture on a large scale. Expansion of the plant culture system under *in vitro* conditions is necessary for the sustainable production of industrially significant SM. To meet the growing demand, various plant parts are used for the large-scale synthesis of specialized metabolites in a bioreactor system. Plant suspension culture is a practical method for increasing the production of SM because of its shortened production cycle (Marchev et al. 2020). Furthermore, understanding the molecular basis of their biosynthesis has been made possible by metabolic engineering. This chapter aims to consolidate existing knowledge on the various *in vitro*-based techniques designed for SM synthesis as well as highlight future research directions. By advancing the understanding in this field, other innovative approaches for the sustainable production of these essential compounds can be developed.

13.2 STRATEGIES FOR SECONDARY METABOLITES PRODUCTION

Plants have been employed for their therapeutic properties since ancient times. Their pharmacological potential relies on the phytochemical components as depicted in Figure 13.1, particularly those metabolites that serve as an excellent pool of bioactive chemicals. Specialized metabolites have intricated chemical compositions which are synthesized in reaction to diverse types of stresses. These specialized metabolites have substantial utilization in pharmaceutical sectors, the cosmetic

DOI: 10.1201/9781003518358-13

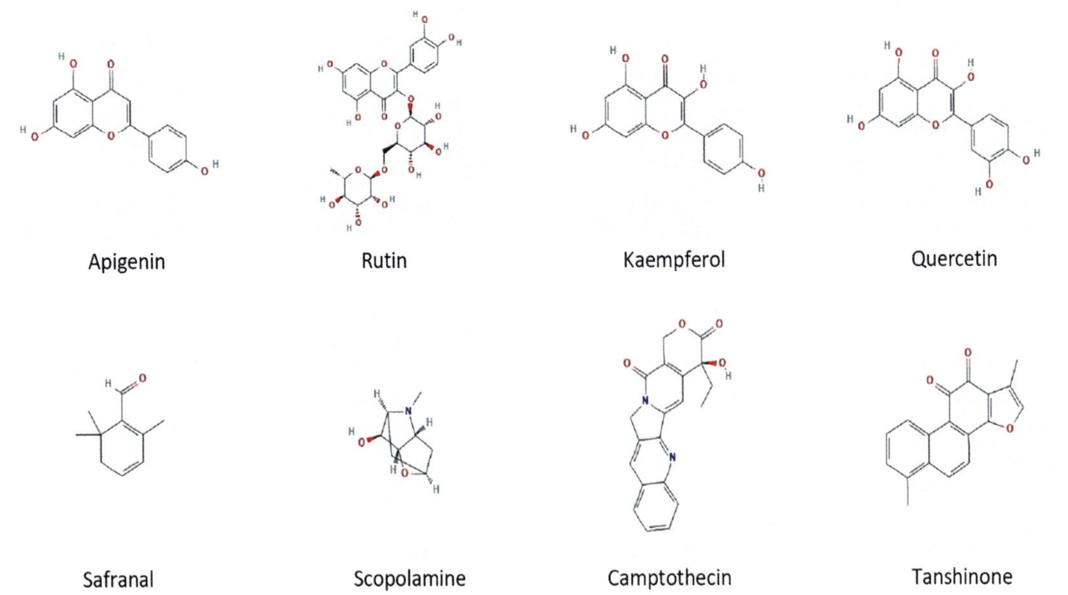

FIGURE 13.1 Chemical structure of major secondary metabolites in plants.

industry, and dietary supplements. This has triggered a necessity to streamline the research for enhancing specialized metabolite production using plant tissue culture techniques. These approaches remain unaffected by climate or geographical conditions and ensure a continuous, sustainable, cost-effective, and feasible production of SM. A plethora of evidence exists in organ culture, cell culture, engineering metabolic pathways, and heterologous production in microorganisms that have proven to be reliable approaches for SM production.

13.2.1 Organ Culture

13.2.1.1 Shoot Culture

The production of specialized metabolites by shoot culture is one of the most significant tissue culture techniques. The process involves either infecting the aboveground plant parts with *Agrobacterium tumefaciens* or inducing shoot proliferation by adding an appropriate concentration of phytoregulators to the culture medium. In *Mentha citrata*, Spencer et al. (1990) created shooty teratomas by inserting T-DNA of *A. tumefaciens* to produce significant quantities of terpenes in mint oil. Contrary to this, Saito et al. (1991) could not synthesize steroidal alkaloids from shooty teratomas of *Solanum tuberosum, Nicotiana tabacum,* and *Atropa belladonna*. SM production, such as solasodine in *Solanum eleagnifolium*, vincristine in *Catharanthus roseus,* and naphthoquinone in *Drosera capensis,* has also been successfully achieved by induction of shooty teratomas (Alvarez et al. 1994; Begum et al. 2009; Begum 2011; Krolicka et al. 2010). In *Scutellaria alpina*, Grzegorczyk-Karolak et al. (2016) cultivated the shoot tips and observed an increase in verbascoside production in young than old plants. As a result, they considered shoots of 3-month-old plants as a potential producer of polyphenolic compounds. According to Sharma et al. (2015), the highest amount of bacoside from *Bacopa monnieri* was achieved when shoot cultures were exposed to elicitors. Elicitation by 50 μM of salicylic acid and 1.0 mg/L of jasmonic acid led to an increase in the bacoside concentration of 8.14 and 8.46 mg/g dry weight (dw), respectively. In *Gynura procumbens,* another method used to achieve the highest yield of flavonoids was temporary immersion bioreactors (Pramita et al. 2018). In this process, the plant material was immersed for 15 min every 12 h in a medium supplemented with indole acetic acid (IAA) and benzyl adenine (BA). Similarly, the shoots of *Verbena officinalis* were cultivated in temporary immersion bioreactors containing 4.92 μM indole butyric acid (IBA). This led to a substantial increase in biomass production accompanied by higher concentrations of essential oils (Kokotkiewicz et al. 2021).

13.2.1.2 Root Culture

13.2.1.2.1 Hairy Root

The cultivation of hairy roots is a versatile biotechnological tool for producing and enhancing the specialized metabolites. The process involves the transformation of desired plant species by using a natural vector, *A. rhizogenes.* In *Arachis hypogaea*, arachidin-1 and arachidin-3 are known for diverse biological activities with potential benefits to human health. The limited availability of the aforementioned compounds under in vivo conditions led to the induction of hairy root cultures followed by their treatment with elicitors. The mutual effect of MJ and methyl-β-cyclodextrin on resveratrol synthase gene expression could explain high arachidin content (Yang et al. 2015). *Linum usitatissimum* is gaining attention in synthesizing lignans, such as secoisolariciresinol diglucoside, secoisolariciresinol, and matairesinol known for their beneficial effects on humans. The hairy root cultures accumulated these lignans when callus was transformed with *A. rhizogenes* on media containing 2,4-Dichlorophenoxyacetic acid (2,4-D) (1.0 mg/L) and Gibberellic acid (GA_3) (0.5 mg/L) or GA_3 (1.0 mg/L) (Gabr et al. 2016). The elicitation of hairy roots is a biotechnological strategy that was successfully implemented in different plants to increase the production of SMs. Iron oxide nanoparticles (FeNPs) have been used as elicitors in hairy roots of *Hyoscyamus reticulatus,* as reported by Moharrami et al. (2017). The highest amount of tropane alkaloids, hyoscyamine, and scopolamine was attained by adding 900 and 450 mg/L of FeNPs for 24 and 48 h duration, respectively. In *Brassica rapa,* the effect of silver nanoparticles (AgNPs) was reported to enhance glucosinolates and phenolic compounds. At the transcript level, enhanced gene expression related to glucosinolates, flavonoids (*MYB34, MYB51, MYB28,* and *MYB29*), and phenolic (*PAL, CHI,* and *FLS*) production was reported in AgNP-elicited HRs. In the future, such approach will help to augment the production of bioactive substances for nutraceutical and pharmaceutical industries (Chun et al. 2018). In *Withania somnifera*, elicitation of HRs with sodium alginate, k-carrageenan, and chitosan was reported by Thilip et al. (2019). Among them, chitosan (100 mg/L) resulted in 4.03-folds increase in Withaferin-A (WA) production than control, which was the maximum among all treatments. This is the first report in *W. somnifera* where natural polysaccharides were proved as potential elicitors for the production of WA on a large scale.

In *Centella asiatica*, leaves and petioles were co-cultured with *A. rhizogenes* for HR induction. As observed, the concentration of triterpenoids in leaves was twice than in petioles, whereas triterpenoid accumulation in petiole-derived cultures was greater than leaf-derived cultures. Being a medicinally important plant, HR-based mass production of triterpenoids fulfills the demand of pharmaceutical and cosmetic industries (Baek et al. 2022). Earlier, Baek et al. (2020) reported the augmentation of triterpenoids in *C. asiatica* by exposing HRs to precursors (squalene, pyruvic acid) and elicitors (methyl jasmonate, salicylic acid). In *Solanum khasianum,* Jacob and Malpathak (2004, 2005) improvised the culture conditions, i.e., light, temperature, CO_2 concentration, and nutrient proportion in the medium to increase the solasodine accumulation. Single-use bioreactors have also received a fair share of attention for regulating the microenvironment to enhance metabolite production. In *Rindera graeca*, the utilization of the WAVE 25 bioreactor system resulted in a significant enhancement in the biomass of Hairy Root Cultures (HRCs), and high production of naphthoquinone (Wierzchowski et al. 2024). The family Solanaceae has demonstrated that undifferentiated cultures do not acquire substantial quantities of tropane alkaloids (Kohnen-Johannsen & Kayser 2019). By using *A. rhizogenes* for metabolic engineering, it was documented that hairy roots do not follow the same pattern. The hairy root line of *Duboisia leichhardtii* when treated with MJ, produced 19.34 mg/g dw of scopolamine, whereas the wild-type line only produced 1.44 mg/g dw (Singh et al. 2018). In transgenic hairy roots of *Ophiorrhiza pumila*, camptothecin production was compared by overexpressing the genes (*OpG10H, OpSLS)* independently and simultaneously. The highest level (3.5 mg/g) was attained by cooperative gene expression as opposed to the single gene (Shi et al. 2020).

13.2.1.2.2 Adventitious Root

It has been shown that metabolite production is usually greater in differentiated plant tissues. Therefore, various protocols have been developed to grow entire plant organs, such as shoots or roots, to obtain significant bioactive compounds under *in vitro* culture conditions. Plant growth, development, and morphogenic adaptation could vary among a wide range of plants due to their distinct nutritional requirements. The medium's strength, concentration, and growth regulators impact the growth and development of adventitious root cultures. Additional significant parameters that affect the culture state are salt concentration, carbon source, pH, explant type, biotic and abiotic elicitors employed along with their inoculum density. The adventitious root formation of various plant species is significantly influenced by the type of media selection. As per the literature, commonly reported media for adventitious root induction is Murashige and Skoog (MS) medium as in *Allamanda cathartica* (Khanam et al. 2018), *Camellia sinensis* (Kim et al. 2013), *Echinacea purpurea* (Wu et al. 2007), ½MS in *Boesenbergia rotunda* (Yusuf et al. 2018), *Vernonia amygdalina* (Khalafalla et al. 2009), and Gamborg's (B5) medium in *Aloe vera* (Lee et al. 2011). The methodology of establishing adventitious root cultures is intricate and involves continuous developmental stages that require various signals. The importance of sucrose in promoting adventitious roots was reported at a concentration of 4 mg/l and 5 mg/l in *Centella asiatica* (Ling et al. 2009). The presence of sucrose in culture media causes osmotic stress which is responsible for SM production. The adventitious roots in optimized culture media produce SM analogous to naturally developing plants. The anthocyanin production was enhanced in *Grevillea ilicifolia* (salt-tolerant) and *Grevillea arenaria* (salt-sensitive) on exposure to saltwater (Kennedy and De Filippis 1999). Similarly, salt stress exhibited elevated levels of vincristine, vinblastine, and phenolics in adventitious root cultures of *Catharanthus roseus* and *Salvia mirzayanii*, respectively (Valifard et al. 2014; Fatima et al. 2015).

In response to stress-causing variables (elicitors), plant begins to synthesize and accumulate the specialized metabolites, when administered in certain amounts. Elicitation is the most efficient among the various biotechnological methods for synthesizing the desired bioactive compound, as mentioned in Table 13.1. In *P. ginseng*, the compound bis-carboxyethyl germanium sesquioxide was employed as a stimulant to augment the production of ginsenosides (Yu et al. 2005). In *Scopolia parviflora* and *Withania somnifera*, the impact of methyl jasmonate and salicylic acid on adventitious root cultures leads to tropane alkaloid and withanolides synthesis, respectively (Kang et al. 2004; Sivanandhan et al. 2012). In *Rubia tinctorum*, the addition of $CuCl_2$ and glutathione leads to the induction of phytochelatin and anthraquinone synthesis (Sato et al. 1997). Biotic elicitors, the compounds derived from living were also reported to modulate the SM synthesis by interacting with receptors and inducing the activation or inhibition of ion channels or enzymes (Thakur et al. 2019). As per the reports, the use of yeast extracts, chitosan, pectin, and bacterial strains in adventitious roots resulted in augmentation of specialized metabolites. The findings indicate increased (1.9 times) quercetin production in *Gynura procumbens* by introducing extract derived from *Saccharomyces cerevisiae* (Faizah et al. 2018). In *Perovskia abrotanoides*, the influence of biotic as well as abiotic elicitors was found to exhibit a multifold increase in cryptotanshinone and tanshinone production (Zaker et al. 2015). The presence of chitosan also resulted in boosting the specialized metabolites in adventitious roots of *Morinda citrifolia* and *Withania somnifera* (Baque et al. 2012; Sivanandhan et al. 2012). In *Hypericum perforatum*, medium containing ½MS supplemented with 0.1 mg/L Kinetin (KIN) in addition to 1.0 mg/L IBA was chosen for growing adventitious roots in a bubble bioreactor. This resulted in the accumulation of 35.01 mg/g dw of total phenolics, 0.97 mg/g dw of flavonoids, and 1.389 mg/g dw of hypericin (Cui et al. 2014).

TABLE 13.1
Some Applications of Elicitors in the Production of Plant Secondary Metabolites

Plant Species	Culture System	Elicitor	Compound	Reference
Fagonia indica	Adventitious root	Methyl jasmonate, Phenylacetic acid	Apigenin, Rutin	Khan et al. (2017)
Momordica charantia	Cell suspension culture	Silver nanoparticles	Hydroxybenzoic, Hydroxycinnamic flavanols	Chung et al. (2018)
Gynura procumbens	Adventitious root	*Saccharomyces cerevisiae*, Copper Sulphate	Kaempferol, Quercetin	Faizah et al. (2018)
Polygonum multiflorum	Adventitious root	Methyl jasmonate, Salicylic Acid, Yeast extract, Chitosan	Phenolics Flavonoids	Ho et al. (2018)
Oldenlandia umbellata	Adventitious root	Yeast extract, Pectin, Xylan	Anthraquinones	Krishnan and Siril (2018)
Panax ginseng	Adventitious root	*Mesorhizobium amorphae*	Ginsenoside	Le et al. (2018)
Stevia rebaudiana	Adventitious root	Methyl jasmonate, Phenyl acetic acid, Melatonin	Steviol glycosides	Kazmi et al. (2019)
Crocus sativus	Cell suspension culture	Ultrasonic waves	Safranal, Crocin	Taherkhani et al. (2019)
Corylus avellana	Cell suspension culture	*Coniothyrium palmarum*, Methyl-β-Cyclodextrin	Paclitaxel	Farhadi et al. (2020)
Panax ginseng	Adventitious root culture	*Alternaria panax*	Ginsenosides	Hao et al. (2020)
Trichosanthes cucumerina	Callus and suspension culture	Chitosan	Bryonolic acid	Lertphadungkit et al. (2020)
Plumbago zeylanica	Root callus	Chitosan, Yeast extract	Plumbagin	Singh et al. (2020)
Bletilla striata	Tissue culture	*Byssochlamys spectabilis*	Phenolic compounds	Chen et al. (2021)
Coelogyne ovalis	Tissue culture	Salicylic acid	Flavonoids, Anthocyanins, Phenolic compounds	Singh and Kumaria (2021)
Gymnema sylvestre	Cell suspension culture	Sodium nitroprusside Salicylic acid	Deacylgymnemic acid, Gymnemagenin, Gymnemic acid IV, Gymnemic acid XVII	Mahendran et al. (2021)
Iberis amara	Cell suspension culture	Chitosan	Total phenol, Flavonoid, Flavanol, Anthocyanin	Taghizadeh et al. (2021)

13.2.2 Cell Culture

Advancements in plant cell culture techniques are expected to offer novel opportunities for the utilization and commercialization of medicinally significant plants as valuable sources. The commercialization of specialized metabolites has generated significant interest in plant cell cultures. The specialized metabolite production under *in vitro* culture conditions necessitates the formation of cell biomass aggregation by exposing them to plant growth regulators. Callus is an amorphous cluster of cells that demonstrates remarkable efficacy in synthesizing SM owing to its rapid proliferation and consistent repeatability. Subsequently, the callus may be utilized to cultivate a suspension culture for synthesizing bioactive compounds. When compared to other strategies, cell culture is a rapid mechanism for multiplication, thus making it an efficient way for the volume production of phytochemical compounds (Kapoor et al. 2018). In *E. platycarpa* cell suspension culture, the addition of Naphthalene acetic acid (NAA) and KIN resulted in a remarkable biomass increase. The suspension also contained phenolic compounds and flavonoids with notable antifungal activity in its dichloromethane extracts (Bernabé-Antonio et al. 2021). Furthermore, the addition of fungal elicitors has been reported to augment the SM synthesis in *Hybanthus enneaspermus* (Velayutham and Karthi 2015). In *Nicotiana tabacum* and *Morinda citrifolia*, the callus formation leads to glutathione and anthraquinones production (Alamgir 2017). Forskolin, a highly effective medicinal chemical, has been discovered in calli derived from the leaves of *Coleus forskohlii* (Mastan et al. 2019). Furthermore, *in vitro* cell cultures of *Gymnema sylvestre* have been employed to enhance the synthesis of gymnemic acid (Veerashree et al. 2012). The use of MJ in cell cultures was found to promote the production of terpinolene and limonene in *Rosa damascena,* alkaloids, and terpenes in *Catharanthus roseus* (Olgunsoy et al. 2017). Callus cultures of *Mentha piperita* have been demonstrated to possess a higher monoterpene content in comparison to plants in their natural habitat (Khan et al. 2019). The cell suspension cultures of *Gingko biloba* also exhibited a higher concentration of bilobalide and gingkolides when treated with biotic elicitors (Kang et al. 2009). In *Hypericum perforatum, in vitro*-derived friable callus was utilized for initiating fresh cell culture suspensions. The highest dry weight of 7–8 g/L of culture medium and flavonoid content of 15–16 mg/g dw was attained (Murch et al. 2003).

In recent times, the use of bioreactors as self-sustaining systems for large-scale production of bioactive compounds has gained significant attention. Bioreactor culture has facilitated the synthesis of several chemicals, including shikonin, a vibrant reddish-purple color employed in lipsticks (Kreis et al. 2001), ginsenosides utilized as additives and bleaching agents (Hibino and Ushiyama 1999; Vanisree et al. 2004), paclitaxel (a potent anti-cancer medication) (Wink et al. 2005), and substances used in the food industry (Ritala et al. 2022). The utilization of bioreactors in suspension culture of *Panax ginseng* resulted in increased biomass and higher levels of ginsenosides (5.4 mg/g) (Jeong et al. 2009). Similarly, the application of salicylic acid to ginseng culture resulted in the augmentation of phenolic compounds (by 62%), flavonoids (by 88%), and ascorbic acid (by 55%) (Ali et al. 2007). In *Ocimum basilicum*, the cambial meristematic cells in bioreactors resulted in 1.75 times the yield of triterpenoids i.e. 3.02 ± 0.76 mg/(l×d) oleanolic acid and 4.79 ± 0.48 mg/(l×d) ursolic acid (Mehring et al. 2020). In the future, bioreactors can be utilized to enhance the production of more valuable plant-derived compounds, which could be a breakthrough in the new era of industrial production, thereby reducing the pressure on medicinal plants in their natural habitat.

13.2.3 Metabolic Engineering

The main objective of plant metabolic engineering is to target specialized metabolites for enhancing the accumulation of a desired product and/or decrease undesirable metabolites. This can be accomplished by regulating the activity of enzymes involved in different metabolic pathways. The enzymes are targeted either in the initial stages of metabolic pathways or in specific pathways for a desired compound. It is also possible to inhibit competing pathways or steps that break down the desired product. Creating storage compartments that accumulate the desired metabolite can also help in synthesizing SM (Farré et al. 2014). In *Atropa belladonna,* the alteration of tropane alkaloid profile served as the initial evidence for the effective manipulation of metabolic processes. The increased gene expression responsible for producing hyoscyamine 6-ß hydroxylase enzyme (*h6h*) facilitated the scopolamine production in *A. belladonna* (Yun et al. 1992). As per the literature, *Duboisia* species are recognized to be grown particularly for scopolamine, nicotine, and pyridine. In the study, Singh et al. (2018) showed the application of RNAi for silencing the quinolinate phosphoribosyl transferase gene which is the part of nicotine biosynthetic pathway. Following this strategy, the production of nicotine was significantly inhibited, and that of scopolamine was increased. Shi et al. (2020) created a transgenic root line of *Ophiorrhiza pumila* that overexpressed the genes, geraniol-10-hydroxylase and secologanin synthetase to improve the production of camptothecin. In *Catharanthus roseus*, the geraniol synthase gene does not induce alkaloid production naturally. Upon stimulation with jasmonic acid, these cell lines generated terpenoid indole alkaloids (TIAs), suggesting the need for the expression of other inducible genes as well (Saiman et al. 2018). In *C. roseus* callus, an extra copy of tryptophan decarboxylase and strictosidine synthase genes was inserted by Sharma et al. (2020). As a result, the total alkaloid content was increased to 1.2 mg/g dw which was twice as obtained in untransformed callus. In another study, an increased expression of the *ORCA3* and strictosidine glucosidase gene in *C. roseus* led to an elevation of 47% in the levels of six specific TIAs that were found in the roots (Sun & Peebles 2016). In *C. roseus*, the *CrTPT2* genes which is responsible for transporting catharanthine, was also successfully overexpressed, for increased catharanthine production (5 mg/g dw) (Wang et al. 2019).

Researchers created suspension cultures using a genetically modified strain of *N. tabacum* that consistently produced the transcription factors *AmRos1* and *AmDel* obtained from *Antirrhinum majus* (Kallam et al. 2017). The cells consistently generated cyanidin 3-O-rutinoside (30 mg/g dw) for 10 years. After introducing an extra gene i.e., 3′,5′-hydroxylase from *Petunia x hybrida*, the suspension culture demonstrated the production of delphinidin 3-O-rutinoside in addition to cyanidin 3-O-rutinoside. Additionally, the gene encoding for anthocyanin 3-O-rutinoside-4‴-hydroxycinn amoyl transferase from *Solanum lycopersicum* was introduced into *AmRos1/AmDel* tobacco line, resulting in the production of anthocyanins with modified acyl groups. The authors showcased the capability of this technique to produce comprehensive and intricately adorned anthocyanins. The aforementioned research emphasizes the significance of elucidating the biosynthetic pathways for SM synthesis to deliberately modify them. Furthermore, the regulation of biosynthetic pathways can be exceedingly intricate and also necessary for the effective manipulation of biosynthetic pathways to increase the production of significant bioactive compounds.

13.2.4 Heterologous Production in Microorganisms

Although it is feasible for various plants to create high-yielding cell suspension and organ lines, they have certain drawbacks that restrict their widespread use in commercial applications. Cell suspension cultures usually display sluggish development and are prone to genetic instability, leading to inconsistent production. Organ cultures exhibit a slower development rate, whereas cultivation on a wide scale is both technically demanding and expensive. There have been many instances where bioactive SM in microbial hosts was successfully produced through heterologous synthesis. Psilocybin, an indole psychotropic alkaloid found in hallucinogenic mushrooms, is generally used for treating depression, post-traumatic stress disorder, and addiction. Fricke et al. (2017) identified four psilocybin biosynthesis enzymes and discovered a way to produce this molecule through enzymatic synthesis by using 4-hydroxy-l-tryptophan. Psilocybin was successfully produced by introducing its biosynthetic genes in *Aspergillus nidulans*, thus resulting in psilocybin titers of 110 mg/L (Hoefgen et al. 2018). A study by Adams et al. (2019) established a platform for the biosynthetic production of psilocybin in *E. coli* through fermentation. Following optimization, psilocybin (1.16 g/L) was produced through the bioconversion of precursors such as 4-hydroxyindole, serine, and methionine. In a separate study, Milne et al. (2020) successfully engineered *Saccharomyces cerevisiae* to synthesize psilocybin and tryptamine derivatives by incorporating genes from *Catharanthus roseus* and *Psilocybe cubensis*. After modifying the genes of the shikimate pathway and improvising the fed-batch fermentation process, the yeast strain produced significant amount of psilocybin (627 mg/L) and psilocin (580 mg/L). Psilocin is the active metabolite of psilocybin that is produced in the human body after oral ingestion of psilocybin. Li et al. (2018) reported a genetically modified strain of *S. cerevisiae*, which included 25 foreign genes from bacteria, plants, mammals, and 6 yeast genes that have been either overexpressed or mutated. By applying strategies such as enzyme engineering pathways, strain engineering, and optimization of fermentation techniques, researchers were able to produce a noscapine titer of 2.21 mg/L. The development of *Corynebacterium glutamicum* strain by Kallscheuer et al. (2016) was specifically engineered to produce plant polyphenols. The genes responsible for breaking down aromatic chemicals were removed, and those encoding for chalcone synthase and chalcone isomerase were introduced. Subsequently, naringenin (35 mg/L) and eriodictyol (37 mg/L) were produced in significant amounts. After elucidating the amino acid metabolism, strains producing naringenin (32 mg/L) or resveratrol (59 mg/L) were successfully acquired. Furthermore, the incorporation of dioxygenase genes into the flavanone-synthesizing strain resulted in the production of kaempferol (23 mg/L) and quercetin (10 mg/L). In a study proposed by Kallscheuer et al. (2017), the production of pterostilbene was increased by introducing O-methyltransferase from grapevine into an earlier engineered resveratrol-producing strain. This phenomenon resulted in a titer of 42 mg/L of the di-O-methylated compound. Li et al. (2019) created co-cultures with various genetically modified strains of *E. coli* to produce rosmarinic acid. By optimizing the ratios of different strains and culture conditions, the rosmarinic acid production was attained up to 172 mg/L. This represents a multifold increase (38 times) compared to the original strain employed for biosynthesis in monoculture. Acetylsalicylic acid, often known as aspirin, is a very notable non-steroidal anti-inflammatory medicine that is manufactured on an industrial scale by acetylating salicylic acid. Ahmadi (2016) added a codon-optimized gluco acyltransferase gene from *Arabidopsis thaliana* to *E. coli* strain that already produced salicylic acid through genetic engineering. The authors were able to produce 2.5 g/L of salicylate 2-O-β-D-glucoside by cultivating two genetically modified strains together. This compound has shown similar characteristics to reduce nitric oxide and reactive oxygen species as aspirin. Several specialized metabolites of cannabis, particularly Δ9-tetrahydrocannabinol, and cannabidiol, have been acknowledged for their therapeutic potential. Drugs utilizing these compounds have been authorized for managing symptoms associated with multiple sclerosis and pediatric epilepsy (Cristiano et al. 2020). During the process of cannabis biosynthesis, geranyl diphosphate and olivetolic acid serve as the primary precursors that are combined to form cannabigerolic acid (CBGA). The intermediate is then transformed into cannabidiolic acid (CBDA), Δ9-tetrahydrocannabinolic acid (Δ9-THCA), and cannabichromenic acid (CBCA) (Flores-Sanchez & Ver poorte 2008). *Pichia pastoris* is a yeast that can synthesize Δ9-THCA, a type of cannabinoid, by expressing Δ9-THCA synthase from *Cannabis sativa* and using CBGA as a substrate in a whole-cell bioconversion test (Zirpel et al. 2015). Later on, this strain of yeast was introduced with prenyltransferase (*NphB)* from *Streptomyces*, thus leading to Δ9-THCA production when provided with olivetolic acid and geranyl diphosphate as supplements (Zirpel et al. 2017). The experiment by Luo et al. (2019) highlighted the significant interest in investigating synthetic cannabinoids due to their enhanced therapeutic potential. *Cannabis* genes responsible for the production of olivetolic acid and cannabinoid synthases were included in their study. The strain with maximum efficiency was achieved while producing 8 mg/L of Δ9-THCA.

The synthesis of the antimalarial drug artemisinin in microbial hosts resulted in the capitalization of "semi-synthetic artemisinin". Artemisinic acid is a precursor compound that was produced by using a modified strain of *S. cerevisiae* and was later transformed into artemisinin (Kung et al. 2018). In another study by Marsafari and Xu (2020), *Yarrowia lipolytica* was selected as a potential microbial host for producing artemisinin. This yeast was genetically modified to store amorphadiene, which is an immediate precursor of artemisinin. By introducing the amorphadiene synthase gene (*Artemisia annua*), an initial low titer of amorphadiene was augmented to 171.5 mg/L. In plants, several investigations have targeted triterpenoid production by heterologous means. Wang et al. (2020) used a synthetic biological platform to elucidate the process of saponin biosynthesis in *Panax notoginseng*. Yeast as a cell factory was engineered to produce ginsenoside compound K which is a specialized metabolite with anticarcinogenic and anti-inflammatory properties. In *S. cerevisiae*, a synthetic pathway was established for the production of cucurbitadienol by Qiao et al. (2019). Initially, the cucurbitadienol synthase gene obtained from *Siraitia grosvenorii* was introduced into a yeast strain known for its elevated levels of squalene. Squalene acts as a precursor in the triterpene biosynthesis pathway. The genes responsible for squalene and ergosterol synthesis were upregulated and downregulated, respectively. As a result, the optimized yeast strain produced 63 mg/L of cucurbitadienol during fed-batch fermentation. In *Saccharomyces cerevisiae*, Zhao et al. (2018) successfully expressed several plant genes, including β-amyrin synthase from *Glycyrrhiza glabra*, oleanolic acid synthase from *Medicago truncatula*, and cytochrome P450 reductase also derived from *M. truncatula*. Additionally, the inactivation of two genes from the galactose regulatory network led to an enhanced oleanolic acid content of 606.9 mg/L.

13.2.5 Conclusion and Future Prospective

Biotechnological interventions offer a sustainable alternative for the production of SMs at a large scale. The regenerated plants under *in vitro* conditions provide uniform, disease-free material that is used for the extraction and purification of bioactive compounds. However, producing significant quantities of SMs through these methods becomes challenging. Therefore, various strategies have been designed to enhance SM production, including organ culture, cell culture, biotransformation, genetic engineering, and expression of desired genes in heterologous organisms. The use of elicitors, optimization of culture conditions, and strategic designing of species-specific techniques are crucial for enhancing the content of bioactive compounds. Genomic research will be useful for studies involving complex gene-level developmental processes, such as organogenesis, somatic embryogenesis, and somaclonal variation. Metabolomic interventions in adventitious root cultures and regenerated tissue-cultured plantlets will facilitate the extraction, enhancement, and characterization of bioactive substances that could benefit the pharmaceutical industries. By utilizing these technologies, the number and quality of significant compounds will strengthen and enhance the medicinal value of plants. These methods should be further investigated to make breakthroughs in technologies involved in plant genetic modification and specialized metabolite synthesis. As the demand for bioactive significant compounds continues to grow, understanding and improving the methods for SM production becomes crucial.

REFERENCES

Adams AM, Kaplan NA, Wei Z, Brinton JD, Monnier CS, Enacopol AL, Jones, JA (2019) In vivo production of psilocybin in *E. coli*. *Metabolic Engineering*. 56, 111–119. https://doi.org/10.1016/j.ymben.2019.09.009

Ahmadi MK (2016) *E. coli* Metabolic Engineering and Green Applications of Natural Products (Doctoral dissertation, State University of New York at Buffalo).

Alamgir ANM (2017) Cultivation of herbal drugs, biotechnology, and in vitro production of secondary metabolites, high-value medicinal plants, herbal wealth, and herbal trade. In *Therapeutic Use of Medicinal Plants and Their Extracts: Volume 1: Pharmacognosy*, pp. 379–452. https://doi.org/10.1007/978-3-319-63862-1_9

Ali MB, Hahn EJ, Paek KY (2007) Methyl jasmonate and salicylic acid induced oxidative stress and accumulation of phenolics in *Panax ginseng* bioreactor root suspension cultures. *Molecules*. 12(3), 607–621. https://doi.org/10.3390/12030607

Alvarez MA, Talou JR, Paniego NB, Giulietti AM (1994) Solasodine production in transformed organ cultures (roots and shoots) of *Solanum eleagnifolium* Cav. *Biotechnology Letters*. 16, 393–396. https://doi.org/10.1007/BF00245058

Baek S, Han JE, Ho TT, Park SY (2022) Development of hairy root cultures for biomass and triterpenoid production in *Centella asiatica*. *Plants*. 11(2), 148. https://doi.org/10.3390/plants11020148

Baek S, Ho TT, Lee H, Jung G, Kim YE, Jeong CS, Park SY (2020) Enhanced biosynthesis of triterpenoids in *Centella asiatica* hairy root culture by precursor feeding and elicitation. *Plant Biotechnology Reports*. 14, 45–53. https://doi.org/10.1007/s11816-019-00573-w

Balandrin MF, Klocke JA (1988) Medicinal, aromatic, and industrial materials from plants. In *Medicinal and Aromatic Plants I*, pp. 3–36. Berlin, Heidelberg: Springer. https://doi.org/10.1007/978-3-642-73026-9_1

Baque MA, Shiragi MHK, Lee EJ, Paek KY (2012) Elicitor effect of chitosan and pectin on the biosynthesis of anthraquinones, phenolics and flavonoids in adventitious root suspension cultures of *'Morinda citrifolia'* (L.). *Australian Journal of Crop Science*. 6(9), 1349–1355.

Begum F (2011) Augmented production of vincristine in induced tetraploids of *Agrobacterium* transformed shooty teratomas of *Catharanthus roseus. Medicinal Plants-International Journal of Phytomedicines and Related Industries.* 3(1), 59–64. https://doi.org/10.5958/j.0975-4261.3.1.007

Begum F, Nageswara Rao SSS, Rao K, Prameela Devi Y, Giri A, Giri CC (2009) Increased vincristine production from *Agrobacterium tumefaciens* C58 induced shooty teratomas of *Catharanthus roseus* G. Don. *Natural Product Research.* 23(11), 973–981. https://doi.org/10.1080/14786410802131153

Bennett, R. N., & Wallsgrove, R. M. (1994). Secondary metabolites in plant defence mechanisms. *New Phytologist*, 127(4), 617–633.

Bernabé-Antonio A, Sánchez-Sánchez A, Romero-Estrada A, Meza-Contreras JC, Silva-Guzmán JA, Fuentes-Talavera FJ, Cruz-Sosa F (2021) Establishment of a cell suspension culture of *Eysenhardtia platycarpa*: Phytochemical screening of extracts and evaluation of antifungal activity. *Plants.* 10(2), 414. https://doi.org/10.3390/plants10020414

Chen J, Li L, Tian P, Xiang W, Lu X, Huang R, Li L (2021) Fungal endophytes from medicinal plant *Bletilla Striata* (Thunb.) Reichb. F. promote the host plant growth and phenolic accumulation. *South African Journal of Botany.* 143, 25–32. https://doi.org/10.1016/j.sajb.2021.07.041

Chun JH, Kim NH, Seo MS, Jin M, Park SU, Arasu MV, Al-Dhabi NA (2018) Molecular characterization of glucosinolates and carotenoid biosynthetic genes in Chinese cabbage (*Brassica rapa* L. ssp. pekinensis). *Saudi Journal of Biological Sciences.* 25(1), 71–82. https://doi.org/10.1016/j.sjbs.2016.04.004

Chung I-M, Rekha K, Rajakumar G, Thiruvengadam M (2018) Elicitation of silver nanoparticles enhanced the secondary metabolites and pharmacological activities in cell suspension cultures of bitter gourd. *3 Biotech.* 8, 1–12. https://doi.org/10.1007/s13205-018-1439-0

Cristiano E, Rojas JI, Alonso R, Pinheiro AA, Bacile EA, Balbuena ME, Correale J (2020) Consensus recommendations on the management of multiple sclerosis patients in Argentina. *Journal of the Neurological Sciences.* 409, 116609. https://doi.org/10.1016/j.jns.2019.116609

Cui XH, Murthy HN, Paek KY (2014) Pilot-scale culture of *Hypericum perforatum* L. adventitious roots in airlift bioreactors for the production of bioactive compounds. *Applied Biochemistry and Biotechnology.* 174, 784–792. https://doi.org/10.1007/s12010-014-1123-8

Faizah H, Tanjung M, Purnobasuki H, Manuhara YS (2018) Biomass and flavonoid production of *Gynura procumbens* (L.). Merr adventitious root culture in Baloon-type Bubble- Bioreactor (BTBB) influenced by elicitation. *Asian Journal of Plant Sciences.* 17, 107–119. https://doi.org/10.3923/ajps.2018.107.119

Farhadi S, Moieni A, Safaie N, Sabet MS, Salehi M (2020) Fungal cell wall and methyl-β–cyclodextrin synergistically enhance paclitaxel biosynthesis and secretion in *Corylus avellana* cell suspension culture. *Scientific Reports.* 10 (1), 5427. https://doi.org/10.1038/s41598-020-62196-4

Farré G, Blancquaert D, Capell T, Van Der Straeten D, Christou P, Zhu C (2014) Engineering complex metabolic pathways in plants. *Annual Review of Plant Biology.* 65(1), 187–223. https://doi.org/10.1146/annurev-arplant-050213-035825

Fatima S, Mujib A, Tonk D (2015) NaCl amendment improves vinblastine and vincristine synthesis in *Catharanthus roseus*: A case of stress signalling as evidenced by antioxidant enzymes activities. *Plant Cell, Tissue and Organ Culture (PCTOC).* 121, 445–458. https://doi.org/10.1007/s11240-015-0715-5

Fricke J, Blei F, Hoffmeister D (2017) Enzymatic synthesis of psilocybin. *Angewandte Chemie International Edition.* 56(40). 12352–12355. https://doi.org/10.1002/anie.201705489

Gabr AM, Mabrok HB, Ghanem KZ, Blaut M, Smetanska I (2016) Lignan accumulation in callus and *Agrobacterium rhizogenes*-mediated hairy root cultures of flax (*Linum usitatissimum*). *Plant Cell, Tissue and Organ Culture (PCTOC).* 126, 255–267. https://doi.org/10.1007/s11240-016-0995-4

Grzegorczyk-Karolak I, Kuźma Ł, Wysokińska H (2016) In vitro cultures of *Scutellaria alpina* as a source of pharmacologically active metabolites. *Acta Physiologiae Plantarum.* 38, 1–9. https://doi.org/10.1007/s11738-015-2024-3

Hao Y-J, An X-L, Sun H-D, Piao X-C, Gao R, Lian M-L (2020) Ginsenoside synthesis of adventitious roots in *Panax ginseng* is promoted by fungal suspension homogenate of *Alternaria panax* and regulated by several signaling molecules. *Industrial Crops and Products.* 150, 112414. https://doi.org/10.1016/j.indcrop.2020.112414

Hibino K, Ushiyama K (1999) Commercial production of ginseng by plant tissue culture technology. In *Plant Cell and Tissue Culture for the Production of Food Ingredients*, pp. 215–224. Boston, MA: Springer US.

Ho TT, Lee JD, Jeong CS, Paek KY, Park SY (2018) Improvement of biosynthesis and accumulation of bioactive compounds by elicitation in adventitious root cultures of *Polygonum multiflorum. Applied Microbiology and Biotechnology.* 102, 199–209. https://doi.org/10.1007/s00253-017-8629-2

Hoefgen S, Lin J, Fricke J, Stroe MC, Mattern DJ, Kufs JE, Valiante V (2018) Facile assembly and fluorescence-based screening method for heterologous expression of biosynthetic pathways in fungi. *Metabolic Engineering.* 48, 44–51. https://doi.org/10.1016/j.ymben.2018.05.014

Hussain MS, Fared S, Ansari S, Rahman MA, Ahmad IZ, Saeed M (2012) Current approaches toward production of secondary plant metabolites. *Journal of Pharmacy and Bioallied Sciences*. 4(1), 10–20. https://doi.org/10.4103/0975-7406.92725

Jacob A and Malpathak N (2004) Green hairy root cultures of *Solanum khasianum* Clarke—A new route to in vitro solasodine production. *Current Science*. 87, 1442–1447.

Jacob A and Malpathak N (2005) Plantlet regeneration enhances solasodine productivity in hairy root cultures of *Solanum khasianum* Clarke. *In Vitro Cellular and Developmental Biology-Plant*. 41, 291–295. https://doi.org/10.1079/IVP2005637

Jeong CS, Murthy HN, Hahn EJ, Lee HL, Paek KY (2009) Inoculum size and auxin concentration influence the growth of adventitious roots and accumulation of ginsenosides in suspension cultures of ginseng (*Panax ginseng* CA Meyer). *Acta Physiologiae Plantarum*. 31, 219–222. https://doi.org/10.1007/s11738-008-0206-y

Kallam K, Appelhagen I, Luo J, Albert N, Zhang H, Deroles S, Martin C (2017) Aromatic decoration determines the formation of anthocyanic vacuolar inclusions. *Current Biology*. 27(7), 945–957.

Kallscheuer N, Vogt M, Bott M, Marienhagen J (2017) Functional expression of plant-derived O-methyltransferase, flavanone 3-hydroxylase, and flavonol synthase in *Corynebacterium glutamicum* for production of pterostilbene, kaempferol, and quercetin. *Journal of Biotechnology*. 258, 190–196. https://doi.org/10.1016/j.jbiotec.2017.01.006

Kallscheuer N, Vogt M, Stenzel A, Gätgens J, Bott M, Marienhagen J (2016) Construction of a *Corynebacterium glutamicum* platform strain for the production of stilbenes and (2S)-flavanones. *Metabolic Engineering*. 38, 47–55. https://doi.org/10.1016/j.ymben.2016.06.003

Kang SM, Jung HY, Kang YM, Yun DJ, Bahk JD, Yang JK, Choi MS (2004) Effects of methyl jasmonate and salicylic acid on the production of tropane alkaloids and the expression of PMT and H6H in adventitious root cultures of *Scopolia parviflora*. *Plant Science*. 166(3), 745–751. https://doi.org/10.1016/j.plantsci.2003.11.022

Kang SM, Min JY, Kim YD, Karigar CS, Kim SW, Goo GH, Choi MS (2009) Effect of biotic elicitors on the accumulation of bilobalide and ginkgolides in *Ginkgo biloba* cell cultures. *Journal of Biotechnology*. 139(1), 84–88. https://doi.org/10.1016/j.jbiotec.2008.09.007

Kapoor S, Raghuvanshi R, Bhardwaj P, Sood H, Saxena S, Chaurasia OP (2018) Influence of light quality on growth, secondary metabolites production and antioxidant activity in callus culture of *Rhodiola imbricata* Edgew. *Journal of Photochemistry and Photobiology B: Biology*. 183, 258–265. https://doi.org/10.1016/j.jphotobiol.2018.04.018

Kazmi A, Khan MA, Mohammad S, Ali A, Kamil A, Arif M, Ali H (2019) Elicitation directed growth and production of steviol glycosides in the adventitious roots of *Stevia rebaudiana* Bertoni. *Industrial Crops and Products*. 139, 111530. https://doi.org/10.1016/j.indcrop.2019.111530

Kennedy B and De Filippis LF (1999) Physiological and oxidative response to NaCl of the salt tolerant *Grevillea ilicifolia* and the salt sensitive *Grevillea arenaria*. *Journal of plant Physiology*. 155(6), 746–754. https://doi.org/10.1016/S0176-1617(99)80092-3

Khalafalla MM, Daffalla HM, El-Shemy HA, Abdellatef E (2009) Establishment of in vitro fast-growing normal root culture of *Vernonia amygdalina*-a potent African medicinal plant. *African Journal of Biotechnology*. 8(21). https://doi.org/10.5897/AJB09.747

Khan MA, Khan T, Ali H (2019) Plant cell culture strategies for the production of terpenes as green solvents. *Industrial Application of Green Solvents*. 50, 1–20. https://doi.org/10.21741/9781644900239-1

Khan T, Abbasi BH, Khan MA, Azeem M (2017) Production of biomass and useful compounds through elicitation in adventitious root cultures of *Fagonia indica*. *Industrial Crops and Products*. 108, 451–457. https://doi.org/10.1016/j.indcrop.2017.07.019

Khanam MN, Anis M, Ahmad S (2018) Establishment of adventitious root cultures of *Allamanda cathartica* L. for the production of iridoid glycosides and its identification using HPTLC MS. *Industrial Crops and Products*. 125, 198–206. https://doi.org/10.1016/j.indcrop.2018.08.044

Kim YD, Kim HG, Kim JC, Sim SJ, Min JY, Hwang JG, Choi MS (2013) Effects of culture media on catechins and caffeine production in adventitious roots of tea tree (*Camellia sinensis* L.). *Journal of Agricultural and Life Sciences Research*. 47(1), 11–20.

Kohnen-Johannsen KL and Kayser O (2019) Tropane alkaloids: Chemistry, pharmacology, biosynthesis and production. *Molecules*. 24(4), 796. https://doi.org/10.3390/molecules24040796

Kokotkiewicz A, Zabiegala B, Kubica P, Szopa A, Bucinski A, Ekiert H, Luczkiewicz M (2021) Accumulation of volatile constituents in agar and bioreactor shoot cultures of *Verbena officinalis* L. *Plant Cell, Tissue and Organ Culture (PCTOC)*. 144, 671–679. https://doi.org/10.1007/s11240-020-01986-0

Kreis W, Baron D, Stoll G (2001) *Biotechnologie der Arzneistoffe: Grundlagen und Anwendungen*. Stuttgart, Germany: Deutscher Apotheker Verlag.

Krishnan SS, Siril EA (2018) Elicitor mediated adventitious root culture for the large-scale production of anthraquinones from *Oldenlandia umbellata* L. *Industrial Crops and Products*. 114, 173–179. https://doi.org/10.1016/j.indcrop.2018.01.069

Krolicka A, Szpitter A, Stawujak K, Baranski R, Gwizdek-Wisniewska A, Skrzypczak A, Kaminski M, Lojkowska E (2010) Teratomas of *Drosera capensis* var. alba as a source of naphthoquinone: Ramentaceone. *Plant Cell, Tissue and Organ Culture (PCTOC)*. 103, 285–292. https://doi.org/10.1007/s11240-010-9778-5

Kung SH, Lund S, Murarka A, McPhee D, Paddon CJ (2018) Approaches and recent developments for the commercial production of semi-synthetic artemisinin. *Frontiers in Plant Science*. 9, 87. https://doi.org/10.3389/fpls.2018.00087

Le KC, Im WT, Paek KY, Park SY (2018) Biotic elicitation of ginsenoside metabolism of mutant adventitious root culture in *Panax ginseng*. *Applied Microbiology and Biotechnology*. 102, 1687–1697. https://doi.org/10.1007/s00253-018-8751-9

Lee YS, Yang TJ, Park SU, Baek JH, Wu S, Lim KB (2011) Induction and proliferation of adventitious roots from *'Aloe vera'* leaf tissues for *'in vitro'* production of aloe-emodin. *Plant Omics*. 4(4), 190–194.

Lertphadungkit P, Suksiriworapong J, Satitpatipan V, Sirikantaramas S, Wongrakpanich A, Bunsupa S (2020) Enhanced production of bryonolic acid in *Trichosanthes cucumerina* L. (Thai Cultivar) cell cultures by elicitors and their biological activities. *Plants*. 9, 709. https://doi.org/10.3390/plants9060709

Li Y, Li S, Thodey K, Trenchard I, Cravens A, Smolke CD (2018) Complete biosynthesis of noscapine and halogenated alkaloids in yeast. *Proceedings of the National Academy of Sciences*. 115(17), E3922–E3931. https://doi.org/10.1073/pnas.1721469115

Li Z, Wang X, Zhang H (2019) Balancing the non-linear rosmarinic acid biosynthetic pathway by modular co-culture engineering. *Metabolic Engineering*. 54, 1–11. https://doi.org/10.1016/j.ymben.2019.03.002

Ling AK, Kok KM, Hussein S, Ong SL (2009) Effects of plant growth regulators on adventitious roots induction from different explants of *Orthosiphon stamineus*. *American-Eurasian Journal of Sustainable Agriculture*. 3(3), 493–501.

Luo X, Reiter MA, d'Espaux L, Wong J, Denby CM, Lechner A, Keasling JD (2019) Complete biosynthesis of cannabinoids and their unnatural analogues in yeast. *Nature*. 567(7746), 123–126. https://doi.org/10.1038/s41586-019-0978-9

Mahendran G, Iqbal Z, Kumar D, Verma SK, Rout PK, ur Rahman L (2021) Enhanced gymnemic acids production in cell suspension cultures of *Gymnema Sylvestre* (Retz.) R. Br. Ex Sm. through Elicitation. *Industrial Crops and Products*. 162, 113234. https://doi.org/10.1016/j.indcrop.2020.113234

Marchev AS, Yordanova ZP, Georgiev MI (2020) Green (cell) factories for advanced production of plant secondary metabolites. *Critical Reviews in Biotechnology*. 40(4), 443–458. https://doi.org/10.1080/07388551.2020.1731414

Marsafari M and Xu P (2020) Debottlenecking mevalonate pathway for antimalarial drug precursor amorphadiene biosynthesis in *Yarrowia lipolytica*. *Metabolic Engineering Communications*. 10, e00121. https://doi.org/10.1016/j.mec.2019.e00121

Mastan A, Bharadwaj RKB, Kushwaha RK, Vivek Babu CS (2019) Functional fungal endophytes in *Coleus forskohlii* regulate labdane diterpene biosynthesis for elevated forskolin accumulation in roots. *Microbial Ecology*. 78(4), 914–926. https://doi.org/10.1007/s00248-019-01376-w

Mehring A, Haffelder J, Chodorski J, Stiefelmaier J, Strieth D, Ulber R (2020) Establishment and triterpenoid production of *Ocimum basilicum* cambial meristematic cells. *Plant Cell, Tissue and Organ Culture (PCTOC)*. 143, 573–581. https://doi.org/10.1007/s11240-020-01942-y

Milne N, Thomsen P, Knudsen N, Rubaszka P, Kristensen M, Borodina I (2020) Metabolic engineering of *Saccharomyces cerevisiae* for the de novo production of psilocybin and related tryptamine derivatives. *Metabolic Engineering*. 60, 25–36. https://doi.org/10.1016/j.ymben.2019.12.007

Moharrami F, Hosseini B, Sharafi A, Farjaminezhad M (2017) Enhanced production of hyoscyamine and scopolamine from genetically transformed root culture of *Hyoscyamus reticulatus* L. elicited by iron oxide nanoparticles. *In Vitro Cellular and Developmental Biology-Plant*. 53, 104–111. https://doi.org/10.1007/s11627-017-9802-0

Murch SJ, Haq K, Rupasinghe HV, Saxena PK (2003) Nickel contamination affects growth and secondary metabolite composition of St. John's wort (*Hypericum perforatum* L.). *Environmental and Experimental Botany*. 49(3), 251–257. https://doi.org/10.1016/S0098-8472(02)00090-4

Olgunsoy P, Ulusoy S, Akçay UÇ (2017) Metabolite production and antibacterial activities of callus cultures from *Rosa damascena* Mill. petals. *Marmara Pharmaceutical Journal*. 21(3), 590–597. https://doi.org/10.12991/marupj.319331

Pramita AD, Kristanti AN, Utami ESW, Manuhara YSW (2018) Production of biomass and flavonoid of *Gynura procumbens* (Lour.) Merr shoots culture in temporary immersion system. *Journal of Genetic Engineering and Biotechnology*. 16(2), 639–643. https://doi.org/10.1016/j.jgeb.2018.05.007

Qiao J, Luo Z, Cui S, Zhao H, Tang Q, Mo C, Ding Z (2019) Modification of isoprene synthesis to enable production of curcurbitadienol synthesis in *Saccharomyces cerevisiae*. *Journal of Industrial Microbiology and Biotechnology*. 46(2), 147–157. https://doi.org/10.1007/s10295-018-2116-3

Ritala A, Heiniö RL, Häkkinen ST, Lille M, Hyytiäinen-Pabst T, Rischer H (2022) Tailoring sensory properties of plant cell cultures for food use. *Food Research International*. 157, 111440. https://doi.org/10.1016/j.foodres.2022.111440

Saiman MZ, Miettinen K, Mustafa NR, Choi YH, Verpoorte R, Schulte AE (2018) Metabolic alteration of *Catharanthus roseus* cell suspension cultures overexpressing geraniol synthase in the plastids or cytosol. *Plant Cell, Tissue and Organ Culture (PCTOC)*. 134, 41–53. https://doi.org/10.1007/s11240-018-1398-5

Saito K, Yamazaki M, Kawaguchi A, Murakoshi I (1991) Metabolism of solanaceous alkaloids in transgenic plant teratomas integrated with genetically engineered genes. *Tetrahedron*. 47(31), 5955–5968. https://doi.org/10.1016/S0040-4020(01)86488-7

Sato K, Kubota H, Goda Y, Yamada T, Maitani T (1997) Glutathione enhanced anthraquinone production in adventitious root cultures of *Rubia tinctorum* L. *Plant Biotechnology*. 14(1), 63–66. https://doi.org/10.5511/plantbiotechnology.14.63

Sharma A, Amin D, Sankaranarayanan A, Arora R, Mathur AK (2020) Present status of *Catharanthus roseus* monoterpenoid indole alkaloids engineering in homo-and hetero-logous systems. *Biotechnology Letters*. 42, 11–23. https://doi.org/10.1007/s10529-019-02757-4

Sharma M, Ahuja A, Gupta R, Mallubhotla S (2015) Enhanced bacoside production in shoot cultures of Bacopa monnieri under the influence of abiotic elicitors. *Natural Product Research*, 29(8), 745–749. https://doi.org/10.1080/14786419.2014.986657

Shi M, Gong H, Cui L, Wang Q, Wang C, Wang Y, Kai G (2020) Targeted metabolic engineering of committed steps improves anti-cancer drug camptothecin production in *Ophiorrhiza pumila* hairy roots. *Industrial Crops and Products*. 148, 112277. https://doi.org/10.1016/j.indcrop.2020.112277

Singh N, Kumaria S (2021) Deciphering the role of stress elicitors on the differential modulation of chalcone synthase gene and subsequent production of secondary metabolites in micropropagated *Coelogyne Ovalis* Lindl., a therapeutically important medicinal orchid. *South African Journal of Botany*. 140, 336–348. https://doi.org/10.1016/j.sajb.2020.06.019

Singh P, Prasad R, Tewari R, Jaidi M, Kumar S, Rout PK, Rahman LU (2018) Silencing of quinolinic acid phosphoribosyl transferase (QPT) gene for enhanced production of scopolamine in hairy root culture of *Duboisia leichhardtii*. *Scientific Reports*. 8(1), 13939. https://doi.org/10.1038/s41598-018-32396-0

Singh T, Sharma U, Agrawal V (2020) Isolation and optimization of plumbagin production in root callus of *Plumbago zeylanica* L. augmented with chitosan and yeast extract. *Industrial Crops and Products*. 151, 112446. https://doi.org/10.1016/j.indcrop.2020.112446

Sivanandhan G, Arun M, Mayavan S, Rajesh M, Mariashibu TS, Manickavasagam M, Selvaraj N, Ganapathi A (2012) Chitosan enhances withanolides production in adventitious root cultures of *Withania somnifera* (L.) Dunal. *Industrial Crops and Products*. 37(1), 124–129. https://doi.org/10.1016/j.indcrop.2011.11.022

Spencer A, Hamill JD, Rhodes MJ (1990) Production of terpenes by differentiated shoot cultures of Mentha citrata transformed with Agrobacterium tumefaciens T37. *Plant Cell Reports*, 8, 601–604. https://doi.org/10.1007/BF00270063

Sun J, Peebles CA (2016) Engineering overexpression of ORCA3 and strictosidine glucosidase in *Catharanthus roseus* hairy roots increases alkaloid production. *Protoplasma*. 253, 1255–1264. https://doi.org/10.1007/s00709-015-0881-7

Taghizadeh M, Nekonan MS, Setorki M (2021) Enhancement production of phenolic compounds in the cell suspension culture of *Iberis amara* L.: The effect of chitosan elicitation. https://doi.org/10.21203/rs.3.rs-785230/v1

Taherkhani T, Asghari Zakaria R, Omidi M, Zare N (2019) Effect of ultrasonic waves on crocin and safranal content and expression of their controlling genes in suspension culture of saffron (*Crocus sativus* L.). *Natural Product Research*. 33, 486–493. https://doi.org/10.1080/14786419.2017.1396598

Thakur M, Bhattacharya S, Khosla PK, Puri S (2019) Improving production of plant secondary metabolites through biotic and abiotic elicitation. *Journal of Applied Research on Medicinal and Aromatic Plants*. 12, 1–12. https://doi.org/10.1016/j.jarmap.2018.11.004

Thilip C, Mehaboob VM, Varutharaju K, Faizal K, Raja P, Aslam A, Shajahan A (2019) Elicitation of withaferin-A in hairy root culture of *Withania somnifera* (L.) Dunal using natural polysaccharides. *Biologia*. 74, 961–968. https://doi.org/10.2478/s11756-019-00236-9

Valifard M, Mohsenzadeh S, Kholdebarin B, Rowshan V (2014) Effects of salt stress on volatile compounds, total phenolic content and antioxidant activities of *Salvia mirzayanii*. *South African Journal of Botany*. 93, 92–97. https://doi.org/10.1016/j.sajb.2014.04.002

Vanisree M, Lee CY, Lo SF, Nalawade SM, Lin CY and Tsay HS (2004) Studies on the production of some important secondary metabolites from medicinal plants by plant tissue cultures. *Botanical Bulletin of Academia Sinica*. 45(1), 1–22.

Veerashree V, Anuradha CM, Kumar V (2012) Elicitor-enhanced production of gymnemic acid in cell suspension cultures of *Gymnema sylvestre* R. Br. *Plant Cell, Tissue and Organ Culture (PCTOC)*. 108, 27–35. https://doi.org/10.1007/s11240-011-0008-6

Velayutham P, Karthi C (2015) GC-MS Profile of in vivo, in vitro and fungal elicited in vitro leaves of *Hybanthus enneaspermus* (L.) F. Muell. *International Journal of Pharmacy and Pharmaceutical Sciences*. 7(1), 260–267.

Wang D, Wang J, Shi Y, Li R, Fan F, Huang Y, Zhang X (2020) Elucidation of the complete biosynthetic pathway of the main triterpene glycosylation products of *Panax notoginseng* using a synthetic biology platform. *Metabolic Engineering*. 61, 131–140. https://doi.org/10.1016/j.ymben.2020.05.007

Wang Y, Yang B, Zhang M, Jia S, Yu F (2019) Application of transport engineering to promote catharanthine production in *Catharanthus roseus* hairy roots. *Plant Cell, Tissue and Organ Culture (PCTOC)*. 139, 523–530. https://doi.org/10.1007/s11240-019-01696-2

Wierzchowski K, Kawka M, Sykłowska-Baranek K, Pilarek M (2024) Intensification of *Rindera graeca* transgenic roots proliferation and deoxyshikonin secretion in wave-agitated disposable bioreactor. *Chemical Engineering and Processing-Process Intensification*. 203, 109905. https://doi.org/10.1016/j.cep.2024.109905

Wink M, Alfermann AW, Franke R, Wetterauer B, Distl M, Windhövel J, Ripplinger P (2005) Sustainable bioproduction of phytochemicals by plant in vitro cultures: Anticancer agents. *Plant Genetic Resources*. 3(2), 90–100. https://doi.org/10.1079/PGR200575

Wu CH, Murthy HN, Hahn EJ, Paek KY (2007) Large-scale cultivation of adventitious roots of *Echinacea purpurea* in airlift bioreactors for the production of chichoric acid, chlorogenic acid and caftaric acid. *Biotechnology Letters*. 29, 1179–1182. https://doi.org/10.1007/s10529-007-9399-1

Yang T, Fang L, Nopo-Olazabal C, Condori J, Nopo-Olazabal L, Balmaceda C, Medina-Bolivar F (2015) Enhanced production of resveratrol, piceatannol, arachidin-1, and arachidin-3 in hairy root cultures of peanut co-treated with methyl jasmonate and cyclodextrin. *Journal of Agricultural and Food Chemistry*. 63(15), 3942–3950.

Yu CT, Chen J, Teng XW, Tong V, Chang TK (2005) Lack of evidence for induction of CYP2B1, CYP3A23, and CYP1A2 gene expression by *Panax ginseng* and *Panax quinquefolius* extracts in adult rats and primary cultures of rat hepatocytes. *Drug Metabolism and Disposition*. 33(1), 19–22. https://doi.org/10.1124/dmd.104.001917

Yun DJ, Hashimoto T, Yamada Y (1992) Metabolic engineering of medicinal plants: Transgenic *Atropa belladonna* with an improved alkaloid composition. *Proceedings of the National Academy of Sciences*. 89(24), 11799–11803. https://doi.org/10.1073/pnas.89.24.11799

Yusuf NA, Rahim NSM, Azhar SZA, Ghani KA, Sommano S, Khalid N (2018) Adventitious root cultures of *Boesenbergia rotunda* as a source of Pinostrobin. *International Journal of Advanced Science Engineering Information Technology*. 8, 377–383.

Zaker A, Sykora C, Gössnitzer F, Abrishamchi P, Asili J, Mousavi SH, Wawrosch C (2015) Effects of some elicitors on tanshinone production in adventitious root cultures of *Perovskia abrotanoides* Karel. *Industrial Crops and Products*. 67, 97–102. https://doi.org/10.1016/j.indcrop.2015.01.015

Zhao Y, Fan J, Wang C, Feng X, Li C (2018) Enhancing oleanolic acid production in engineered *Saccharomyces cerevisiae*. *Bioresource Technology*. 257, 339–343. https://doi.org/10.1016/j.biortech.2018.02.096

Zirpel B, Degenhardt F, Martin C, Kayser O, Stehle F (2017) Engineering yeasts as platform organisms for cannabinoid biosynthesis. *Journal of Biotechnology*. 259, 204–212. https://doi.org/10.1016/j.jbiotec.2017.07.008

Zirpel B, Stehle F, Kayser O (2015) Production of Δ9-tetrahydrocannabinolic acid from cannabigerolic acid by whole cells of *Pichia* (Komagataella) *pastoris* expressing Δ9-tetrahydrocannabinolic acid synthase from *Cannabis sativa* L. *Biotechnology Letters*. 37, 1869–1875. https://doi.org/10.1007/s10529-015-1853-x

14 Potential of Quinoa as a Source of Secondary Metabolites for Human Health

Akriti Thakur[1,2], *Sonia Morya*[1], *Devendra Kumar*[1], *Jasim Ahmed*[3], *Joe Mac Regenstein*[4], *and Domenico Fuoco*[5]

[1]Department of Food Technology and Nutrition, School of Agriculture, Lovely Professional University, Phagwara, India

[2]Department of Nutrition and Dietetics, School of Health Sciences, Amity University, Punjab, India

[3]Research Center, Kuwait Institute for Scientific Research, Safat, Kuwait

4Department of Food Science, Cornell University, Ithaca, New York

[5]Department of Chemical Engineering, Polytechnique Montréal, Montréal, Québec, Canada

14.1 INTRODUCTION

Staple crops form the foundation of global food systems and are vital in feeding billions of people globally. They hold great importance in human nutrition, particularly in low- and middle-income countries, where they often account for more than 70% of daily caloric intake (Fanzo et al., 2020). Intake of crops, including rice, wheat, maize, potatoes, cassava, and legumes, provides daily intake of essential macronutrients (e.g., carbohydrates, proteins, and fats). However, their importance goes beyond basic nutrition. With the growing global population and increasing concerns about food security, staple crops are crucial not only for meeting caloric requirements but also for addressing micronutrient deficiencies, in particular, in developing regions (Oustani et al., 2023; Upadhyay & Singh, 2021, 2023). Although these crops are recognized for their nutritional value, however, these crops are now increasingly appreciated for their bioactive compounds, especially secondary metabolites, which have substantial implications for human health (Upadhyay & Singh, 2024).

Secondary metabolites, often referred to as phytochemicals, are a varied class of organic compounds synthesized by plants that are not directly involved in vital physiological functions such as growth, development, or reproduction. These compounds serve several ecological functions, such as deterring herbivores, defending against pathogens, and facilitating interactions with pollinators. Importantly, these compounds have shown promise in the prevention and management of chronic conditions, including cardiovascular diseases, diabetes, and cancer (Liu et al., 2020). Secondary metabolites are generally classified as alkaloids, terpenoids, flavonoids, phenolic acids, saponins, tannins, and others (Nwachukwu & Babalola, 2022). While these compounds are primarily produced by plants for survival, research over the past few decades has shown that secondary metabolites can offer significant health benefits when consumed by humans. For example, flavonoids are among the most prominent groups of secondary metabolites found in staple crops such as legumes, cereals, and tubers (Table 14.1). They possess potent antioxidant properties that aid in mitigating oxidative stress, a major contributor to the onset of chronic diseases like cardiovascular diseases and some cancers (Kumar & Pandey, 2013). Similarly, phenolic acids, another group of secondary metabolites present in wheat and rice, have shown anti-inflammatory, anticancer, and antidiabetic

DOI: 10.1201/9781003518358-14

TABLE 14.1
Major Staple Crops and Their Associated Secondary Metabolites

Staple Crop	Secondary Metabolites	References
Quinoa (*Chenopodium quinoa*)	Phenolic acids (e.g., ferulic acid, gallic acid), flavonoids (e.g., quercetin), saponins, terpenoids	Lin et al. (2019); Wikipedia (Accessed on 2024)
Wheat (*Triticum aestivum*)	Phenolic compounds, alkaloids (e.g., avenin), terpenes	Fan and Shi (2024); Hussein et al. (2021)
Rice (*Oryza sativa*)	Phenolic acids (e.g., ferulic acid, flavonoids, γ-oryzanol, alkaloids	Fan and Shi (2024); Hussein et al. (2021)
Maize (*Zea mays*)	Phenolic acids, flavonoids, terpenoids (e.g., zeaxanthin), alkaloids (e.g., nicotine)	Fan and Shi (2024); Hussein et al. (2021)
Potato (*Solanum tuberosum*)	Glycoalkaloids (e.g., solanine), phenolic compounds, flavonoids	Fan and Shi (2024); Hussein et al. (2021)

effects in numerous studies (Martini et al., 2017). These metabolites act through mechanisms such as scavenging free radicals, modulating inflammatory pathways, and enhancing the body's antioxidant defenses. In the context of staple crops, the presence of these bioactive compounds elevates their status from simple energy sources to functional foods capable of preventing and managing various diseases. Recent research has increasingly focused on these secondary metabolites, not only in identifying their health-promoting properties but also in understanding how agricultural practices and breeding techniques can enhance their concentrations in staple crops (Nwachukwu & Babalola, 2022).

This chapter seeks to investigate quinoa's potential as a rich source of secondary metabolites that support human health. It emphasizes identifying the various secondary metabolites found in quinoa, their biological functions, and the health benefits they offer. Additionally, it highlights the role of bioactive compounds in delivering antioxidant, anti-inflammatory, and anticancer effects.

Beyond examining the scientific foundation for the health benefits of secondary metabolites, the chapter will also explore the potential for enhancing their concentrations through genetic modification and selective breeding. Advances in agricultural biotechnology have enabled the development of crop varieties with higher concentrations of specific bioactive compounds, thereby amplifying their therapeutic potential (Nwachukwu & Babalola, 2022). This is particularly relevant in addressing global health challenges, as biofortification of staple crops with secondary metabolites provides a sustainable and cost-effective solution to improve dietary quality, especially in regions with limited access to diverse nutrition sources. Recent research has highlighted the significant health benefits of secondary metabolites in staple crops. For example, a 2021 study in the *Journal of Agricultural and Food Chemistry* reported that phenolic compounds present in whole grains, such as wheat and rice, are vital in reducing the risk of cardiovascular diseases by enhancing vascular function and decreasing cholesterol levels (Martini et al., 2017). Similarly, a study published in *Nutrients* emphasized the antidiabetic potential of polyphenols in legumes, which help regulate blood glucose levels by inhibiting enzymes involved in carbohydrate digestion (Ganesan & Xu, 2017).

14.2 QUINOA: A TRENDING STAPLE CROP

Quinoa (*Chenopodium quinoa Willd.*), a pseudocereal originating from the Andean region of South America, including countries like Bolivia, Peru, and Ecuador (Jacobsen, 2003), was a staple food for pre-Columbian civilizations such as the Incas, who revered it as the "mother of all grains." Despite its ancient origins, quinoa has gained worldwide recognition in recent years due to its exceptional nutritional value and adaptability to diverse growing conditions. This crop thrives in harsh environments, including high altitudes, saline soils, and drought-prone areas, making it increasingly important for regions dealing with the impacts of climate change (Bazile et al., 2016).

In the early 2000s, quinoa attracted global attention after the United Nations Food and Agriculture Organization (FAO) declared 2013 the "International Year of Quinoa," which significantly boosted its cultivation beyond its native regions. Today, countries like the United States, Canada, France, India, and China have embraced quinoa cultivation (Oustani et al., 2023). Thanks to its versatility and resilience, quinoa is now grown in over 70 countries, as it can flourish in various climates, from sea level to high-altitude areas (Agregán et al., 2023).

14.2.1 Nutritional Profile and Health Benefits

Quinoa is widely recognized for its outstanding nutritional composition, distinguishing it from other staple crops (Alvarez-Jubete et al., 2009). It is one of the few plant-based foods that provide a complete protein, containing all nine essential amino acids, making it especially appealing to vegetarians, vegans, and those seeking alternative protein sources (Xi et al., 2024). Moreover, quinoa is an excellent source of dietary fiber, essential vitamins (B and E), minerals (including magnesium, phosphorus, iron, and zinc), and bioactive compounds such as flavonoids and saponins, which offer antioxidant, anti-inflammatory, and anticancer properties (Ehsan Safdar et al., 2011). Quinoa's fiber content surpasses that of most grains, promoting digestive health and aiding in the prevention of chronic diseases such as type 2 diabetes, cardiovascular disease, and obesity (Tang et al., 2015). Additionally, quinoa is naturally gluten-free, making it an ideal choice for individuals with celiac disease or gluten intolerance. Its low glycemic index aids in controlling blood sugar levels, offering significant benefits for those managing diabetes (Paśko et al., 2010).

The global rise in quinoa's popularity is driven by a shift towards healthier diets and the increasing demand for plant-based, gluten-free, and nutrient-rich foods. Its media portrayal as a "superfood," along with endorsements from health professionals and organizations like the FAO, has further propelled its global consumption. From 2010 to 2020, global quinoa production surged by over 300%, largely fueled by exports from Bolivia and Peru (Alandia et al., 2020). In regions such as North America, Europe, and parts of Asia, quinoa has found its way into a wide range of culinary uses, from salads and grain bowls to snacks and breakfast cereals. Its growing popularity is also linked to its sustainability, offering high yields with minimal environmental impact, particularly in water-scarce areas (Ruiz et al., 2014).

14.3 SECONDARY METABOLITES IN QUINOA: TYPES, ROLES, AND METHODS OF EXTRACTION

Quinoa (*Chenopodium quinoa Willd.*), frequently celebrated as a "superfood," has gained worldwide acclaim for its remarkable nutritional value and its abundance of bioactive compounds, especially secondary metabolites, as illustrated in Figure 14.1. Although these metabolites are not required for the plant's primary metabolic activities, they are vital for its defense systems and provide a range of health benefits for humans (Lin et al., 2019). Quinoa contains a diverse range of secondary metabolites, including flavonoids, phenolic acids, saponins, phytosterols, and alkaloids, which exhibit antioxidant, anti-inflammatory, and antimicrobial properties (Lu et al., 2015). This review explores the types of secondary metabolites found in quinoa, their roles in plant defense and human health, and the techniques used for their extraction and analysis.

14.3.1 Types of Secondary Metabolites in Quinoa

Polyphenols are a diverse group of bioactive secondary metabolites synthesized by plants, distinguished by the presence of one or more aromatic rings with hydroxyl groups. These compounds are broadly divided into two main categories: flavonoids and phenolic acids (Abbas et al., 2017).

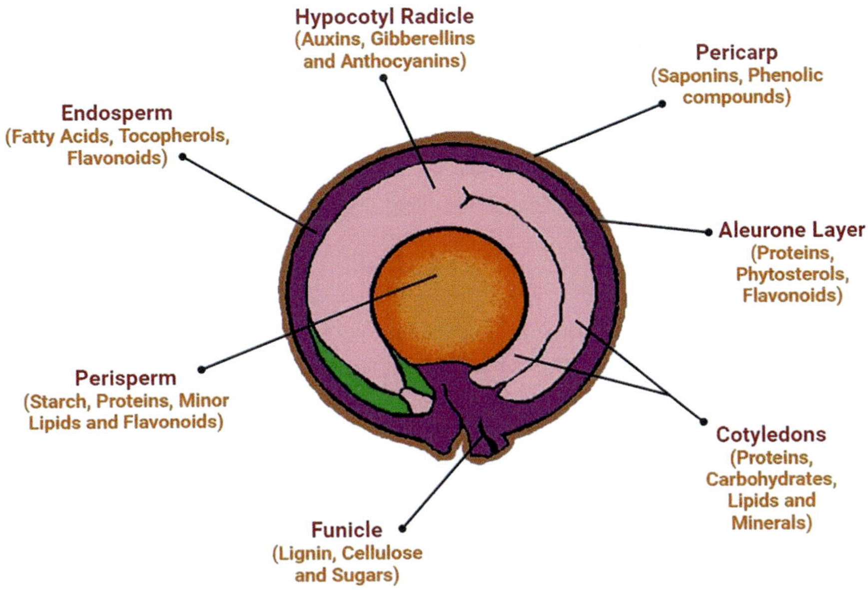

FIGURE 14.1 Anatomy of quinoa seed where secondary metabolites are present.

Flavonoids are further classified into subgroups such as flavones, flavanones, flavonols, flavanols, isoflavones, and anthocyanins, while phenolic acids are generally categorized as hydroxybenzoic or hydroxycinnamic acids. The type and concentration of polyphenols differ across quinoa varieties and their respective colors (Repo-Carrasco-Valencia et al., 2010). For example, in red, white, and black quinoa, the total phenolic content (TPC) and total flavonoid content (TFC) range from 514.03 to 1409.54 mg gallic acid equivalent (GAE)/100 g and 177.49 to 407.75 mg rutin equivalent (RE)/100 g, respectively. Red quinoa has the highest TPC and TFC levels, alongside superior ABTS and DPPH radical scavenging activity. Most of the TPC in red and black quinoa is in a bound form, with protocatechuic acid being the most prevalent phenolic acid, whereas in white quinoa, these acids are primarily found in a free form (Liu et al., 2020). Quinoa contains at least 23 identified polyphenols, predominantly phenolic acids such as vanillic acid, and ferulic acid, and their derivatives, as well as major flavonoids like quercetin, kaempferol, and glycosides. Notably, darker quinoa varieties often exhibit higher levels of phenolics and enhanced antioxidant activity (Albuquerque-Silva et al., 2024).

14.3.1.1 Flavonoids

Quinoa contains a variety of flavonoids, with quercetin, kaempferol, and isorhamnetin being the most extensively studied in this crop. These flavonoids are well-known for their potent antioxidant abilities, helping to neutralize free radicals and alleviate oxidative stress within the body (Tang & Tsao, 2017). Quinoa contains significantly higher levels of quercetin and kaempferol compared to many other grains, and these compounds exhibit diverse biological activities, including anti-inflammatory, antiviral, and anticancer effects, highlighting quinoa's value as a beneficial dietary component (Das et al., 2021).

14.3.1.2 Phenolic Acids

Phenolic acids, including ferulic acid, p-coumaric acid, and gallic acid, represent another significant category of secondary metabolites present in quinoa. These compounds are essential for the plant's defense, helping it withstand environmental stress, while also providing substantial health advantages for humans. Phenolic acids play a vital role in regulating inflammation, reducing oxidative stress, and preventing conditions such as cardiovascular diseases and cancer (Morya et al., 2022). Among the most prevalent phenolic acids in quinoa-ferulic acid, p-coumaric acid, and gallic acid—these are well-known for their antioxidant capabilities, which protect cells from oxidative damage, a key contributor to chronic diseases like cancer and cardiovascular disorders (Lu et al., 2015). Furthermore, these acids exhibit anti-inflammatory effects by inhibiting enzymes such as lipoxygenase (LOX) and cyclooxygenase (COX), which are central to the body's inflammatory processes. As a result, phenolic acids offer significant benefits in managing and preventing inflammatory conditions like asthma and arthritis (Agregán et al., 2023).

14.3.1.3 Saponins

Saponins are glycosides primarily concentrated in the outer seed coat of quinoa, where they contribute to its bitter taste and act as a robust defense mechanism against herbivores and pathogens (El Hazzam et al., 2020; Gómez-Caravaca et al., 2012; Yao et al., 2014). Although traditionally viewed as undesirable due to their bitterness, recent research has revealed their potential health benefits, including cholesterol-lowering effects, immunomodulatory functions, and anticancer properties (Repo-Carrasco-Valencia & Serna, 2011). Quinoa saponins are triterpene glycosides, composed of hydrophilic oligosaccharides linked to hydrophobic aglycones at the C-3 and C-28 positions. These oligosaccharides commonly include glucose, galactose, arabinose, glucuronic acid, and xylose (Sparg et al., 2004).

Saponins are among the most abundant secondary metabolites in quinoa, primarily located in the seed coat, with concentrations ranging from 3.81 to 27.1 mg/g (Liu et al., 2016). According to Liu et al., quinoa seeds contain significantly higher levels of sapogenin compared to the stems, leaves, and roots, with the roots exhibiting the highest total saponin content (13.39 g/100 g), followed by quinoa bran. Studies have shown that quinoa saponins can lower blood cholesterol levels and display strong anticancer potential by promoting apoptosis in cancer cells (Medina-Meza et al., 2016).

14.3.1.4 Other compounds

Quinoa contains small amounts of compounds such as phytic acid, oxalates, betanin, phytosterols, and phytoecdysteroids. Phytic acid, a common antinutrient in plant-based foods, is well-known for its strong chelating properties, enabling it to bind with minerals and other cationic molecules. However, excessive consumption of phytic acid can impede the absorption of key minerals like iron, calcium, magnesium, and zinc. In quinoa, phytic acid levels range between 1.03 and 1.22 g/100 g across different color varieties, but these levels decrease during germination due to increased phytase activity (Maldonado-Alvarado et al., 2023). Phytosterols, such as β-sitosterol, stigmasterol, and campesterol, are another significant group of secondary metabolites found in quinoa. Structurally similar to cholesterol, phytosterols reduce cholesterol levels in humans by inhibiting their intestinal absorption (Acosta-Estrada et al., 2019). Moreover, phytosterols exhibit anti-inflammatory and antioxidant properties, offering additional cardiovascular health benefits.

Though less studied, alkaloids in quinoa are another noteworthy group of secondary metabolites. These compounds play a vital role in plant defense due to their toxicity to herbivores and pathogens. In humans, alkaloids can have pharmacological benefits, including neuroprotective, anti-inflammatory, and analgesic effects (Lemya & Ahlem, 2024). Overall, the secondary metabolites in quinoa make it a functional food with significant therapeutic potential, as shown in Figure 14.2.

14.4 FLAVONOIDS IN QUINOA: TYPES, FUNCTIONS, AND HEALTH BENEFITS

14.4.1 Types of Flavonoids in Quinoa

Flavonoids, a group of polyphenolic compounds, are abundantly found in plants, including quinoa. These secondary metabolites are vital for plant defense mechanisms and are highly regarded for their health-promoting properties. Quinoa is particularly rich in quercetin, kaempferol, and

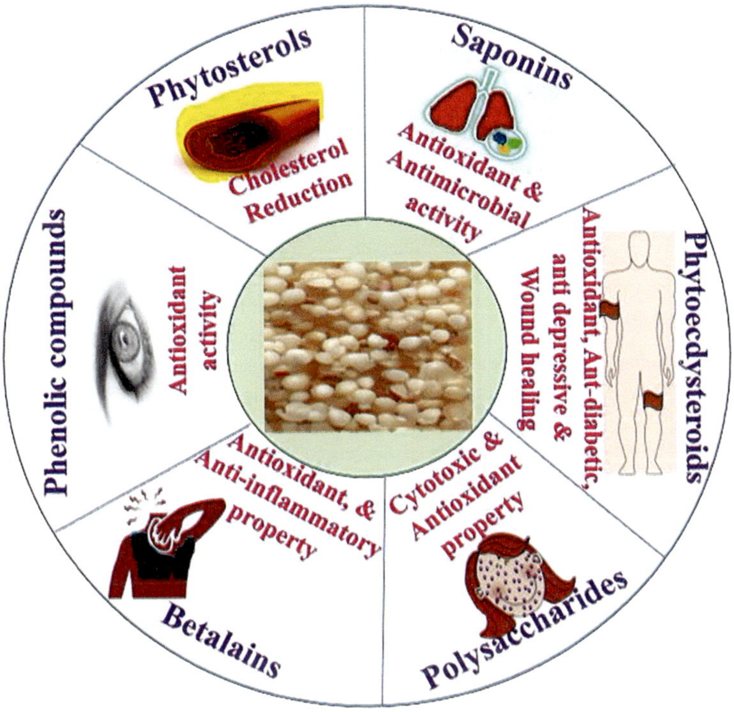

FIGURE 14.2 Bioactive compounds in foods and their health benefits.

isorhamnetin, which are among its most notable flavonoids. Quercetin, widely recognized for its powerful antioxidant properties, is present in significant amounts in quinoa, making it a superior source compared to many fruits and vegetables (Tang & Tsao, 2017). Kaempferol, another major flavonoid in quinoa, is valued for its potent antioxidant and anti-inflammatory activities, offering health benefits such as cancer prevention and cardiovascular support (Akram and Khan, 2024). . Isorhamnetin, a methylated form of quercetin, adds to quinoa's bioactive profile and has demonstrated potential in studies on the prevention of cancer and cardiovascular diseases (Tang et al., 2015). The bioactive compounds in quinoa and their roles in plant defense and associated health benefits are summarized in Table 14.2.

14.4.2 Functions and Health Benefits of Flavonoids

Flavonoids perform multiple functions in plants, including protecting from ultraviolet (UV) radiation, pathogens, and herbivores. Their antioxidant activity helps combat oxidative damage by scavenging reactive oxygen species (ROS), thereby enhancing the plant's defense against environmental stressors.

Flavonoids have attracted considerable interest for their antioxidant, anti-inflammatory, and anti-carcinogenic effects on human health. Regular intake of flavonoid-rich foods like quinoa has been associated with a reduction in oxidative stress, a critical factor contributing to the onset of chronic diseases such as cancer, cardiovascular ailments, and neurodegenerative conditions (Kumar & Pandey, 2013). Studies have indicated that quercetin and kaempferol can suppress the proliferation of cancer cells and may reduce the risk of cancers, including lung, prostate, and colon cancer (Lu et al., 2015). Additionally, the flavonoids in quinoa improve endothelial function and help lower blood pressure, promoting cardiovascular health.

TABLE 14.2
Bioactive Compounds in Quinoa: Roles in Plant Defense and Associated Human Health Benefits

Bioactive Compound	Range in Quinoa	Role in Plant Defense	Human Health Benefits	References
Flavonoids	High	Antioxidant defense against UV radiation, drought, and pathogens	Reduces risk of cardiovascular diseases, cancer, diabetes; anti-inflammatory properties	Tang and Tsao (2017); (Akram and Khan, 2024)
Phenolic acids	Moderate	Antioxidants aids in environmental stress resistance	Reduces oxidative stress, supports gut health	Tang and Tsao (2017)
Saponins	Variable	Deters herbivores/insects with detergent-like properties	Lowers cholesterol, anticancer, immunomodulatory	Mäkinen et al. (2021); Melini and Melini (2021); Mohyuddin (2019); Yang et al. (2024)
Alkaloids	Low	Deters herbivores and pathogens	Analgesic, anti-inflammatory, neuroprotective	Lemya and Ahlem (2024); Gupta and Morya (2022)
Phytosterols	Moderate	Stabilizes cell membranes, enhances resistance to stress	Lowers LDL, anti-inflammatory	Acosta-Estrada et al. (2019); Pourshahidi et al. (2020)
Phytoecdysteroids	Moderate	Protects against insect herbivores, resilient to drought, salinity	Anabolic effects improve glucose metabolism, neuroprotective	Dini et al. (2020); Isenmann et al. (2019)
Quercetin (a flavonoid)	Moderate	Shields from oxidative damage deter pathogens	Antioxidant, cancer prevention	Tang and Tsao (2017); Kumar and Pandey (2013)
Kaempferol (a flavonoid)	Moderate	Protects against UV and pathogens	Cardiovascular protection, anti-inflammatory, reduces cancer risk	Tang and Tsao (2017); Lu et al. (2015)
Isorhamnetin (a flavonoid)	Low	Protects from oxidative damage	Potential in cancer and cardiovascular disease prevention	Tang et al. (2015)
20-Hydroxyecdysone (phytoecdysteroid)	Moderate	Enhances plant resilience to environmental stress	Supports muscle growth, improves metabolic health	Dini et al. (2020)

Flavonoids in quinoa exhibit potent anti-inflammatory effects by influencing crucial signaling pathways such as nuclear factor-kappa B (NF-κB) and mitogen-activated protein kinases (MAPKs), which are central to the body's inflammatory processes (Tang & Tsao, 2017). Consequently, these bioactive compounds may aid in preventing or managing inflammatory conditions like rheumatoid arthritis and lowering the risk of chronic inflammatory diseases, including diabetes.

14.5 SAPONINS: ANTIOXIDANT PROPERTIES AND POTENTIAL THERAPEUTIC USES

Saponins are glycosides known for their surfactant properties, which contribute to quinoa's characteristic bitter taste. Primarily found in the outer seed coat, saponins serve as a natural protective strategy by discouraging herbivores and preventing pathogen attacks. The types and concentrations of saponins in quinoa vary depending on the specific variety or ecotype (Melini and Melini, 2021). From a pharmacological and nutritional perspective, saponins hold potential, particularly for their enhanced activity in the intestines (Mohyuddin, 2019). Much like flavonoids, quinoa saponins demonstrate strong antioxidant activity by neutralizing free radicals and reducing oxidative stress, thereby safeguarding cells from damage induced by ROS. They also enhance the body's antioxidant defense mechanisms by increasing the activity of endogenous enzymes such as superoxide dismutase (SOD) and catalase (CAT) (Stuardo et al., 2008). Saponins are widely acknowledged for their diverse therapeutic benefits, including their cholesterol-lowering (hypocholesterolemic), anti-inflammatory, and anticancer properties (Mustafa et al., 2022; Le, 2009; Ruales & Nair, 1992). Among their most notable health advantages is their ability to reduce cholesterol levels by blocking its absorption in the intestines, making quinoa an excellent dietary choice for preventing or managing cardiovascular diseases (Yang et al., 2024).

Additionally, saponins have demonstrated anticancer potential. Research has shown that quinoa saponins can trigger programmed cell death in cancer cells by activating key pathways involved in programmed cell death, suggesting their role as potential chemopreventive agents for various cancers (Medina-Meza et al., 2016). Beyond their cholesterol-lowering and anticancer effects, saponins also possess immunomodulatory properties. By stimulating immune cells and promoting the production of cytokines, quinoa saponins can enhance immune responses, making them beneficial for the prevention and treatment of infectious diseases (Yendo et al., 2010).

14.6 PHYTOECDYSTEROIDS: IMPACT ON HEALTH AND DISEASE PREVENTION

Phytoecdysteroids are plant-derived compounds that resemble insect molting hormones (ecdysteroids) in structure. Quinoa is among the rare plants capable of producing significant quantities of these compounds, with 20-hydroxyecdysone being the most prevalent (Dini et al., 2020). In quinoa, phytoecdysteroids act as a protective mechanism by repelling insect herbivores and enhancing the plant's ability to withstand environmental challenges such as drought and high salinity. In humans, these compounds show potential in various therapeutic areas. Phytoecdysteroids are known for their anabolic effects, which promote muscle growth and enhance physical performance, positioning quinoa as an essential dietary choice for athletes and those seeking to increase muscle mass and strength (Isenmann et al., 2023). Furthermore, phytoecdysteroids have been investigated for their potential in addressing metabolic disorders. Studies suggest that these compounds can enhance glucose metabolism and improve insulin sensitivity, providing advantages for individuals with diabetes or metabolic syndrome (Dini et al., 2020). Additionally, they exhibit neuroprotective effects, which may help in preventing neurodegenerative diseases such as Alzheimer's and Parkinson's (Daliri et al., 2017).

14.7 ANTIOXIDANT AND CANCER PREVENTION

a. Antioxidant Activity: One of the key mechanisms through which quinoa's secondary metabolites exhibit anti-cancer effects is their antioxidant activity. Flavonoids and phenolic acids in quinoa counteract free radicals and alleviate oxidative stress, a crucial factor in cancer development. By scavenging ROS, these compounds help prevent DNA damage that could lead to mutations and tumor formation (Ren et al., 2017). Quercetin and kaempferol, for example, boost the body's antioxidant defense system by increasing the activity of endogenous enzymes like SOD and CAT (Jeong et al., 2007).

b. Induction of Apoptosis: Quinoa's secondary metabolites can trigger apoptosis, also known as programmed cell death, in cancer cells. Saponins, for example, have shown the ability to trigger apoptotic pathways in various cell lines by activating caspases—proteins essential for apoptosis—leading to the selective death of malignant cells while sparing normal ones (Nadan et al., 2024). This selective induction of apoptosis is a valuable feature of potential anti-cancer agents.
c. Inhibition of Tumor Growth: Quinoa's secondary metabolites also possess the ability to inhibit tumor growth. Phenolic compounds, such as ferulic acid and gallic acid, have been found to suppress cancer cell proliferation by modulating signaling pathways related to cell cycle regulation. These compounds inhibit key proteins that drive cell division, thereby slowing tumor growth (Martini et al., 2017).
d. Anti-Inflammatory Benefits: Chronic inflammation is widely recognized as a significant risk factor for cancer. The secondary metabolites in quinoa possess anti-inflammatory properties that aid in mitigating this risk. For example, flavonoids suppress pro-inflammatory cytokines and inhibit enzymes such as COX and LOX, which are key players in the body's inflammatory processes (Yao et al., 2014). By reducing inflammation, these compounds may lower the risk of cancer initiation and progression.
e. Modulation of Signaling Pathways: Quinoa's bioactive compounds can modulate various signaling pathways associated with cancer development. For instance, research indicates that flavonoids can suppress the PI3K/Akt/mTOR signaling pathway, a mechanism often activated in cancer cells to support their growth and survival (Akram and Khan, 2024). By targeting these pathways, quinoa's secondary metabolites may help to disrupt tumorigenesis.

14.7.1 Clinical Studies

a. Clinical Trials on Flavonoids: A clinical study investigating the impact of quercetin supplementation on prostate cancer patients revealed encouraging results, including a decrease in tumor markers and enhanced quality of life (Fontana et al., 2020; Gonzalez et al., 2020). Individuals treated with quercetin experienced a notable reduction in serum prostate-specific antigen (PSA) levels compared to those in the control group.
b. Saponins and Cancer Cell Lines: Laboratory research has indicated that saponins derived from quinoa can trigger apoptosis in several cancer cell lines, such as breast cancer (MCF-7) and colon cancer (HT-29). One study demonstrated that quinoa saponin treatment led to increased caspase activity and DNA fragmentation, both markers of apoptosis (Mhada et al., 2020).
c. Phenolic Compounds: Research indicates that phenolic compounds derived from quinoa can inhibit the proliferation of human breast cancer cells by inducing cell cycle arrest at the G2/M phase. This phenomenon was linked to the upregulation of cyclin-dependent kinase inhibitors (CDKIs), which play a crucial role in controlling the cell cycle (Martini et al., 2017).
d. Dietary Interventions: A study focusing on dietary interventions, which included quinoa consumption among individuals at high risk for colorectal cancer, found that regular intake could lead to beneficial shifts in gut microbiota composition and a decrease in inflammatory markers linked to cancer risk (Navruz-Varli & Sanlier, 2016).
e. Gut Microbiota Modulation: Research involving healthy adults who consumed whole grain products containing quinoa observed significant alterations in fecal microbiota composition after four weeks. The study found an increase in populations of **Bifidobacterium** and **Lactobacillus**, suggesting that quinoa acts as a prebiotic food (Paśko et al., 2010).
f. Inflammatory Bowel Disease (IBD): Studies indicate that diets abundant in whole grains, such as quinoa, may assist in alleviating symptoms associated with IBD by promoting beneficial microbiota and reducing inflammatory markers (Nwachukwu & Babalola, 2022).

g. Metabolic Health: In individuals with metabolic syndrome, frequent quinoa consumption has been associated with improved gut health indicators, including elevated short-chain fatty acid (SCFA) levels and reduced systemic inflammation (Ganesan & Xu, 2017).

14.8 PREBIOTIC EFFECTS OF QUINOA

Prebiotics are non-digestible food components that selectively enhance the growth and activity of beneficial gut microbes. The dietary fiber in quinoa acts as a prebiotic by serving as a nutrient source for beneficial gut bacteria, such as *Bifidobacteria* and *Lactobacilli*. Research indicates that quinoa and amaranth exhibit prebiotic properties, stimulating the production of SCFAs, lactate, and mate during fermentation (Gullón et al., 2016). These fermentation by-products are crucial substrates for gut flora metabolism. Elevated levels of acetate, derived from these substrates, have been associated with an increase in beneficial bacteria populations, including *Bifidobacterium*, *Proteus mirabilis*, and *Bacteroides*, highlighting quinoa's potential to positively influence gut microbiota.

A study by Bianchi et al. (2014) demonstrated that a symbiotic beverage made from a blend of quinoa and soybean extracts (30% quinoa and 70% soybean) had beneficial effects on colon health. The drink promoted the growth of beneficial bacteria like *Lactobacillus* and *Bifidobacterium* while reducing harmful genera such as *Bacteroides* and *Clostridium*. Additionally, it significantly lowered ammonia ion production and increased the diversity and abundance of Lactobacillus species without affecting their activity. These findings indicate that quinoa, when combined with soy, can effectively deliver probiotics to the gut microbiota, enhancing intestinal composition and overall health.

In summary, studies have shown that whole quinoa extracts can improve gut microbiota by generating fermentation products that support microbial metabolism, effectively acting as prebiotics (Huang et al., 2024). Animal studies further support these findings, showing an increase in beneficial bacteria and a reduction in harmful ones. Future research may delve into the molecular mechanisms behind quinoa's effects on gut flora, potentially expanding its application. Quinoa's secondary metabolites influence gut microbiota through various mechanisms, enhancing its prebiotic potential:

a. Antioxidant Activity: Quinoa's flavonoids and phenolic acids possess powerful antioxidant properties that help reduce oxidative stress in the gut. By lowering oxidative stress, they promote a healthier environment that fosters the growth of beneficial bacteria like *Bifidobacteria* and *Lactobacilli*, while suppressing the proliferation of harmful bacteria. Oxidative stress, which can damage cells and tissues, often promotes the growth of pathogenic microbes. The antioxidant compounds in quinoa help mitigate this by neutralizing free radicals, thus contributing to gut microbial balance (Ballester-Sánchez et al., 2019).
b. Anti-inflammatory Effects: Quinoa's secondary metabolites, particularly phenolic acids, have been shown to reduce inflammation within the gastrointestinal tract. These compounds can reduce the production of pro-inflammatory cytokines and block enzymes involved in inflammatory pathways. By alleviating inflammation, quinoa creates a gut environment conducive to the growth of beneficial bacteria. Chronic gut inflammation is linked to numerous gastrointestinal disorders, and managing this inflammation with dietary compounds, such as those present in quinoa, may contribute to improved gut health (Martini et al., 2017).
c. Improvement of Gut Barrier Function: The gut barrier acts as a defense mechanism, preventing harmful bacteria and toxins from passing into the bloodstream. Certain bioactive compounds in quinoa, such as flavonoids, may enhance the strength and integrity of this barrier, thereby protecting against harmful bacteria and maintaining overall gut health. When the gut barrier is compromised, conditions such as leaky gut syndrome can occur, leading to systemic inflammation and other health issues. By promoting gut barrier integrity, quinoa supports the maintenance of homeostasis in the gut environment (Zeyneb et al., 2021).

d. Promotion of Short-Chain Fatty Acid (SCFA) Production: The dietary fibers in quinoa are metabolized by gut bacteria, resulting in the production of SCFAs like acetate, propionate, and butyrate. These SCFAs are essential for maintaining gut health as they lower intestinal pH, which suppresses the growth of harmful microorganisms. Moreover, SCFAs aid in mucosal repair and serve as an energy source for colon cells, further enhancing gut health. They also help regulate inflammation and encourage the proliferation of beneficial bacteria, fostering a healthier gut microbiota (Ganesan & Xu, 2017).

14.9 APPLICATIONS OF QUINOA SECONDARY METABOLITES IN FOOD AND PHARMACEUTICALS

In recent years, quinoa (*Chenopodium quinoa*) has gained significant recognition for its remarkable nutritional value and the abundance of various secondary metabolites. These bioactive compounds not only boost quinoa's nutritional qualities but also offer numerous health benefits, establishing it as a key component in the food and pharmaceutical sectors. This section explores quinoa-based functional foods, nutraceuticals, the creation of supplements and herbal products, and the role of quinoa's secondary metabolites in pharmaceutical applications.

14.9.1 Functional Foods and Nutraceuticals

Quinoa is often labeled a "superfood" due to its rich content of essential nutrients, including amino acids, vitamins, minerals, and various bioactive compounds, as detailed in Table 14.3. Among the secondary metabolites found in quinoa are phenolic acids, flavonoids, saponins, and terpenoids. These bioactive components are renowned for their strong antioxidant, anti-inflammatory, and antimicrobial properties, which reinforce quinoa's status as a functional food with numerous health-enhancing benefits (Mu et al., 2023).

14.9.2 Nutritional Profile

Quinoa seeds are notably rich in protein, comprising 10%–18% of their composition, and contain a significant amount of dietary fiber, ranging from 1% to 14%. The protein in quinoa is of exceptional quality, providing all essential amino acids, making it a highly valued protein source, particularly for those following vegetarian or vegan diets. Beyond its protein content, quinoa is a source of beneficial essential fatty acids, further enhancing its nutritional value. A prominent bioactive compound in quinoa, saponins, contributes to the seeds' distinct bitter flavor but also delivers health benefits, such as cholesterol-lowering properties and potential anti-cancer effects (Sobota et al., 2020).

14.9.3 Bioactive Compound

Quinoa is abundant in polyphenols and other bioactive compounds renowned for their powerful antioxidant effects. These compounds aid in neutralizing free radicals in the body, helping to reduce oxidative stress and prevent inflammation. Among them, quinoa is particularly rich in phenolic acids like ferulic acid and quercetin, which are well-known for their robust antioxidant properties. These compounds are essential in safeguarding cells from oxidative damage, promoting overall health, and preventing the development of chronic diseases (Ranjan et al., 2023).

14.9.4 Health Benefits

Consuming quinoa regularly is associated with a range of health benefits. It has been shown to enhance metabolic health, potentially reducing the risk of chronic conditions such as type 2 diabetes and heart disease. This is primarily attributed to its high fiber content, which promotes healthy

TABLE 14.3
Qunioa-Based Supplements with Their Health Benefits, Description, and Products

Supplement	Description	Health Benefits	Products	References
Quinoa protein hydrolysates	Derived from enzymatic hydrolysis of quinoa proteins, resulting in smaller peptides.	Anti-diabetic properties by inhibiting DPP-IV, aiding in blood sugar regulation.	Quinoa protein powders, protein bars	Xi et al. (2024) López-Moreno et al. (2023)
Quinoa extracts rich in phenolic compounds	Concentrated extracts from quinoa seeds and sprouts rich in phenolic compounds like ferulic acid and quercetin.	Antioxidant effects, reducing oxidative stress and lowering Malondialdehyde (MDA) levels.	Antioxidant supplements, health drinks	Xi et al. (2024)
Saponin-rich supplements	Supplements formulated with saponins extracted from quinoa for cholesterol management.	Lower cholesterol absorption, promoting cardiovascular health and aiding weight management.	Cholesterol-lowering supplements	Yao et al. (2014)
Quinoa-based probiotic products	Fermented quinoa products that combine quinoa's fiber with probiotics for gut health.	Supports digestive health by enhancing beneficial gut bacteria growth.	Quinoa yogurt, probiotic beverages	López-Moreno et al. (2023); Canaviri-Paz et al. (2021)
Anti-inflammatory supplements	Supplements utilizing quinoa's anti-inflammatory secondary metabolites to manage chronic inflammation.	Reduces pro-inflammatory cytokines, beneficial for conditions like arthritis and metabolic syndrome.	Herbal anti-inflammatory capsules	Yao et al. (2014)
Anti-cancer supplements	Extracts from black quinoa have been found to demonstrate cytotoxic effects against several cancer cell lines.	Potential anticancer properties through high antioxidant content.	Cancer prevention supplements	Xi et al. (2024)

digestion, helps regulate blood sugar levels, and lowers cholesterol. Furthermore, quinoa's secondary metabolites, including phenolic acids and flavonoids, possess significant anti-inflammatory properties, which may boost immune function and aid in disease prevention. For instance, research suggests that quinoa's anti-inflammatory effects may help protect against chronic inflammation, a major contributor to the development of various diseases (Scalbert et al., 2005).

14.9.5 Functional Food and Nutraceutical Development

Given quinoa's rich nutrient content and bioactive compounds, it has been increasingly used in the development of functional foods and nutraceutical products (Ashaolu & Adeyeye, 2024). These are foods that provide additional health benefits beyond basic nutrition, such as reducing disease risk or improving specific bodily functions. Quinoa-based products, including energy bars, breakfast cereals, and protein supplements, are now widely available in the market. The development of herbal remedies and supplements derived from quinoa, particularly those that focus on its secondary metabolites, holds the potential for addressing a variety of health concerns, including inflammatory conditions, oxidative stress, and gut health.

14.9.6 Role in Pharmaceutical Formulations

Quinoa's secondary metabolites also show great promise in pharmaceutical applications. Compounds like flavonoids, saponins, and phenolic acids are being explored for their potential use in pharmaceutical formulations aimed at managing diseases linked to inflammation, oxidative stress, and metabolic disorders. These compounds can be incorporated into drug development strategies to create natural remedies or supplements that support overall health and wellness. As research into quinoa continues, its secondary metabolites may serve as valuable components in future therapeutic approaches, potentially leading to new treatments for a wide range of health conditions.

14.9.7 Development of Supplements and Herbal Remedies

The growing demand for natural health products has led to the creation of various supplements derived from quinoa, tapping into its rich nutritional and medicinal properties.

14.9.7.1 Nutraceutical Products

Quinoa extracts are now being incorporated into nutraceutical products, often in the form of capsules or powders. These supplements are designed to concentrate quinoa's beneficial bioactive compounds, such as phenolic acids, flavonoids, and saponins, allowing individuals to obtain the health advantages of quinoa's secondary metabolites without needing to consume large quantities of the whole grain (Graf et al., 2015; Ranjan et al., 2023). Historically, quinoa has played a significant role in Andean culture, where products made from the plant have long been used to boost energy and endurance. For instance, Andean farmers used a traditional preparation called "llipta," which was made by mixing the pungent ashes from quinoa stems with coca leaves (*Erythroxylum coca*) and chewed to help maintain energy levels during long work hours (Martindale, 1894). Furthermore, "war balls," a combination of quinoa and fat, were used by Incan soldiers during their marches across the Andes, offering them prolonged energy and sustenance (Small, 2013).

14.9.7.2 Herbal Remedies

Quinoa has a rich history of use in traditional herbal medicine, particularly among Andean communities. It has been utilized to treat various health issues, ranging from digestive problems to wound healing, and has long been considered a natural remedy to boost vitality. In modern times, quinoa extracts are increasingly featured in herbal treatments due to their anti-inflammatory and antioxidant benefits (Graf et al., 2015; Ranjan et al., 2023). Historical accounts also highlight quinoa's wide-ranging medicinal applications. It was used to promote wound healing, aid in fracture recovery, and support digestive health (Mujica, 1994; Bhargava et al., 2006; FAO, 2011). Additionally, quinoa has traditionally been valued as a nourishing food that enhances overall well-being and endurance, playing a crucial role in diets designed to boost energy and maintain good health (Popenoe et al., 1989; Kokoska & Janovska, 2009; FAO, 2011).

14.9.7.3 Functional Beverages

Quinoa is also gaining popularity as an ingredient in functional beverages, such as smoothies, energy drinks, and fermented products. These beverages capitalize on the nutritional benefits of quinoa, including its high protein and fiber content, while appealing to consumers who are looking for health-conscious options (Graf et al., 2015). A traditional fermented drink made from quinoa, known as "chicha," has been consumed in Andean cultures for centuries. Chicha is a ceremonial alcoholic beverage typically prepared by fermenting quinoa or maize, and it continues to hold cultural significance in South America (Healy, 2001; FAO, 2011). Beyond traditional uses, quinoa seeds can also be fermented to produce beer, offering an alternative to conventional grain-based alcoholic beverages.

14.10 ROLE OF QUINOA SECONDARY METABOLITES IN PHARMACEUTICAL FORMULATION

Quinoa's secondary metabolites have attracted interest for their potential in pharmaceutical applications due to their therapeutic properties. These compounds offer unique benefits in drug delivery, antimicrobial solutions, and cancer treatment, making quinoa a valuable resource in the development of modern medicines.

14.10.1 Drug Delivery Systems

Recent research has focused on the use of hydrolyzed quinoa proteins as a novel carrier for delivering anti-inflammatory drugs. This innovative approach enhances the stability of medications as they pass through the gastrointestinal tract, ensuring that they remain effective until they reach the target area of inflammation. The micellar structures formed by hydrolyzed quinoa proteins create a protective environment for the drugs, improving their absorption and overall efficacy (Xi et al., 2024). Additionally, these structures facilitate the controlled release of medications, allowing for a more targeted delivery to inflamed tissues, which can reduce side effects and increase therapeutic outcomes.

14.10.2 Antimicrobial Agents

Quinoa's secondary metabolites exhibit potent antimicrobial properties, positioning them as promising candidates for developing natural preservatives and antimicrobial agents in the pharmaceutical industry. Saponins, a major bioactive compound in quinoa, have demonstrated considerable effectiveness against a range of pathogens, including bacteria, fungi, and viruses. This has sparked growing interest in their potential as natural antibiotics or preservatives in medical formulations (Lin et al., 2019). By harnessing these natural compounds, pharmaceutical companies can reduce their dependence on synthetic chemicals and create more sustainable, biocompatible products for infection prevention and drug stability.

14.10.3 Anti-Cancer Properties

Several components derived from quinoa have demonstrated promising anti-cancer properties, with triterpenoid saponins being particularly noteworthy. Research has indicated that these saponins can suppress the growth of various cancer cell lines by triggering apoptosis, a form of programmed cell death that plays a crucial role in removing cancer cells. The mechanism involves the activation of pro-apoptotic proteins, which trigger cancer cell destruction through regulated cellular pathways (Seifati et al., 2024).

This cytotoxic effect on cancer cells has sparked interest in further exploring quinoa as a potential source of natural compounds for cancer therapy. With ongoing research, these bioactive compounds could be integrated into new therapeutic strategies aimed at treating various types of cancer, offering a more natural alternative to traditional chemotherapy.

14.11 POTENTIAL FOR FUTURE RESEARCH AND DEVELOPMENT

The growing body of research on quinoa's secondary metabolites highlights their versatility in pharmaceutical applications. Quinoa's potential extends beyond its use in conventional medicine, offering opportunities for developing innovative drug delivery systems, natural antimicrobial agents, and cancer treatments. As scientific understanding of these compounds deepens, quinoa may become a key ingredient in future pharmaceutical advancements, paving the way for more effective, natural, and sustainable treatment options.

14.12 CHALLENGES AND FUTURE PERSPECTIVE

The use of quinoa (*Chenopodium quinoa*) as a source of secondary metabolites presents various challenges, particularly due to the variability in metabolite content and issues with standardization. One of the main challenges is the considerable fluctuation in the levels of bioactive compounds, which are affected by factors such as environmental conditions, farming practices, and genetic diversity. Quinoa is grown in diverse climates, which leads to differences in metabolite profiles. For example, studies show that phenolic acid levels can significantly vary based on soil type, altitude, and temperature (Bautista et al., 2016). These variations can influence the effectiveness of quinoa-based supplements and functional foods. Moreover, different quinoa varieties have distinct secondary metabolite compositions. Research suggests that red quinoa generally shows greater. Due to its higher content, it exhibits stronger antioxidant activity in comparison to white quinoa concentrations of flavonoids and phenolic acids (Złotek et al., 2019). This genetic diversity requires careful selection of quinoa varieties to target specific health benefits. Additionally, the absence of standardized extraction methods and quality control processes complicates the use of quinoa-derived metabolites. Without consistent extraction and analysis procedures, it is difficult to guarantee the potency and effectiveness of quinoa-based supplements. Thus, establishing standardized methods for quantifying key metabolites is essential for product development and ensuring consumer safety.

Safety and toxicity are also critical considerations when developing quinoa-derived supplements. Quinoa contains saponins, which can be harmful in large quantities. While saponins provide health benefits such as lowering cholesterol and offering antimicrobial properties, they can also irritate the gastrointestinal tract (Afzal et al., 2023). Therefore, determining safe consumption levels is crucial for supplement development. Additionally, the regulation of dietary supplements varies across regions, and in many countries, supplements do not undergo the same rigorous testing as pharmaceuticals. This lack of oversight can lead to inconsistencies in product quality and effectiveness. Regulatory authorities must establish clear guidelines for approving and labeling quinoa-based products to ensure consumer safety. There is also a growing demand for transparency in health supplement ingredients. Manufacturers need to provide detailed information about the sources and concentrations of bioactive compounds in their products and educate consumers about potential risks, such as excessive saponin intake or other metabolites.

To address these challenges and improve the use of quinoa as a source of secondary metabolites, several research areas should be explored. Advanced metabolomic techniques could be used to profile secondary metabolites across different quinoa cultivars and growing conditions (Ranjan et al., 2023). This research could identify specific varieties with enhanced health benefits, aiding production through selective breeding or biotechnology by uncovering the metabolic pathways responsible for synthesizing these compounds. Studying the bioavailability of quinoa's secondary metabolites is also crucial for understanding their effectiveness in human health. Research should focus on how processing methods, such as cooking or fermentation, affect the absorption and activity of these compounds (Melini and Melini., 2021). Such studies will help optimize food formulations to maximize health benefits. Additionally, well-designed clinical trials are necessary to provide solid evidence of the health benefits of quinoa-based supplements, assessing both effectiveness and safety over extended periods. Collaboration between researchers, healthcare professionals, and supplement manufacturers will help facilitate this process.

The rising interest in functional foods offers many opportunities for innovation with quinoa's secondary metabolites. Quinoa can be used in a variety of functional food products, such as snacks, beverages, and fortified foods. Innovations could include the creation of gluten-free products enriched with quinoa protein or ready-to-eat meals that highlight its nutritional profile (Ranjan et al., 2023). Furthermore, combining quinoa extracts with other bioactive ingredients, like probiotics or omega-3 fatty acids, could enhance health benefits while meeting specific consumer needs, such as promoting gut health or supporting cardiovascular health (Abdelshafy et al., 2022). This synergistic approach could result in more effective nutraceutical products. As demand for plant-based supplements grows, sustainable agricultural practices must also be prioritized. Research into organic farming methods or regenerative agriculture could improve quinoa production while maintaining its nutritional quality (Afzal et al., 2023). Exploring innovative cultivation techniques, such as hydroponics, may further increase yields while reducing environmental impact. By tackling these challenges and pursuing new avenues for innovation, quinoa can be positioned not only as a staple food but also as a valuable source of health-promoting compounds.

14.13 CONCLUSION

Quinoa, a superfood, has immense potential as a source of secondary metabolites with significant health benefits. These bioactive compounds, including flavonoids, saponins, and phenolic acids, contribute to antioxidant, anti-inflammatory, and antimicrobial effects, making quinoa a valuable crop for improving human health. The chapter elucidates the role of bioactive compounds in the functional properties of quinoa, particularly its application in food, pharmaceuticals, and nutraceuticals. Advanced extraction and analytical techniques could increase the ability to harness these compounds effectively. Moving forward, addressing the challenges of standardization, variability in metabolite content, and safety concerns will be vital. By leveraging modern agricultural practices and biotechnological advancements, quinoa can have a significant role in addressing global health and nutrition challenges, particularly in promoting functional foods and sustainable agriculture.

REFERENCES

Abbas, M., Saeed, F., Anjum, F. M., Afzaal, M., Tufail, T., Bashir, M. S., … & Suleria, H. A. R. (2017). Natural polyphenols: An overview. *International Journal of Food Properties*, *20*(8), 1689–1699.

Abdelshafy, A. M., El-Naggar, E. A., & Kenawi, M. N. (2022). Enhancing of nutritional properties of quinoa fermented by probiotics. *Discover Food*, *2*(1), 21.

Acosta-Estrada, B. A., Gutiérrez-Uribe, J. A., & Serna-Saldivar, S. O. (2019). Minor constituents and phytochemicals of the kernel. In S. O. Serna-Saldivar (Ed.), *Corn* (pp. 369–403). St. Paul, MN: AACC International Press.

Acosta-Estrada, B. A., Gutiérrez-Uribe, J. A., & Serna-Saldivar, S. O. (2019). Minor constituents and phytochemicals of the kernel. In S. O. Serna-Saldivar (Ed.), *Corn* (pp. 369–403). St. Paul, MN: AACC International Press.

Afzal, I., Haq, M. Z. U., Ahmed, S., Hirich, A., & Bazile, D. (2023). Challenges and perspectives for integrating quinoa into the agri-food system. *Plants*, *12*(19), 3361.

Agregán, R., Guzel, N., Guzel, M., Bangar, S. P., Zengin, G., Kumar, M., & Lorenzo, J. M. (2023). The Effects of processing technologies on nutritional and anti-nutritional properties of pseudocereals and minor Cereals. *Food and Bioprocess Technology*, *16*(5), 961–986.

Alandia, G., Rodriguez, J. P., Jacobsen, S. E., Bazile, D., & Condori, B. (2020). Global expansion of quinoa and challenges for the Andean region. *Global Food Security*, *26*, 100429.

Albuquerque-Silva, M. M., de Oliveira, M. T., de Oliveira, A. F. M., & Albuquerque, U. P. (2024). *Is there a relationship between the concentration of phenolic compounds and the versatility of medicinal plants in the Caatinga biome?* Preprints. https://www.preprints.org/manuscript/202405.0134/v1

Alvarez-Jubete, L., Arendt, E. K., & Gallagher, E. (2009). Nutritive value and chemical composition of pseudocereals as gluten-free ingredients. *International Journal of Food Sciences and Nutrition*, *60*(sup4), 240–257.

Ashaolu, T. J., & Adeyeye, S. A. (2024). African functional foods and beverages: A review. *Journal of Culinary Science & Technology*, *22*(1), 142–177.

Ballester-Sánchez, J., Gil, J. V., Haros, C. M., & Fernández-Espinar, M. T. (2019). Effect of incorporating white, red, or black quinoa flours on the free and bound polyphenol content, antioxidant activity, and color of bread. *Plant Foods for Human Nutrition*, *74*, 185–191.

Bautista, I., Boscaiu, M., Lidón, A., Llinares, J. V., Lull, C., Donat, M. P., ... & Vicente, O. (2016). Environmentally induced changes in antioxidant phenolic compound levels in wild plants. *Acta Physiologiae Plantarum*, *38*, 1–15.

Bazile, D., Jacobsen, S. E., & Verniau, A. (2016). The global expansion of quinoa: Trends and limits. *Frontiers in Plant Science*, *7*, 622.

Bhargava, A., Shukla, S., & Ohri, D. (2006). *Chenopodium quinoa*—An Indian perspective. *Industrial Crops and Products*, *23*(1), 73–87.

Bianchi, F., Rossi, E. A., Sakamoto, I. K., Adorno, M. A. T., Van de Wiele, T., & Sivieri, K. (2014). Beneficial effects of fermented vegetal beverages on the human gastrointestinal microbial ecosystem in a simulator. *Food Research International*, *64*, 43–52.

Canaviri-Paz, P., Oscarsson, E., Kjellström, A., Olsson, H., Jois, C., & Håkansson, Å. (2021). Effects on microbiota composition after consumption of quinoa beverage fermented by a novel xylose-metabolizing *L. plantarum* strain. *Nutrients*, *13*(10), 3318.

Daliri, E. B. M., Oh, D. H., & Lee, B. H. (2017). Bioactive peptides. *Foods*, *6*(5), 32.

Das, S., Morya, S., Neumann, A., & Chattu, V. K. (2021). A review of the pharmacological and nutraceutical properties of *Cynodon dactylon. Pharmacognosy Research, 13*(3), **104–111**. https://www.phcogres.com/article/2021/13/3/105530pres1331

Ehsan Safdar, M., Asif, M., Ali, A., Aziz, A., Yasin, M., Aziz, M., ... & Ali, A. (2011). Eficacia Comparativa de Distintas estrategias en el Manejo de Malezas en Trigo. *Chilean Journal of Agricultural Research*, *71*(2), 195–204.

El Hazzam, K., Hafsa, J., Sobeh, M., Mhada, M., Taourirte, M., El Kacimi, K., & Yasri, A. (2020). An insight into saponins from Quinoa (*Chenopodium quinoa* Willd): A review. *Molecules*, *25*(5), 1059.

Fan, Y., & Shi, B. (2024). Endophytic fungi from the four staple crops and their secondary metabolites. *International Journal of Molecular Sciences*, *25*(11), 6057.

Fanzo, J., Covic, N., Dobermann, A., Henson, S., Herrero, M., Pingali, P., & Staal, S. (2020). A research vision for food systems in the 2020s: Defying the status quo. *Global Food Security*, *26*, 100397.

FAO. (2011). Quinoa: An Ancient Crop to Contribute to World Food Security. https://www.fao.org/docrep/017/aq287e/aq287e.pdf. (Accessed August 8, 2024).

FAO. (2013) International Year of Quinoa 2013. https://www.fao.org/quinoa-2013/what-is-quinoa/nutritional-value/en/?no_mobile=1 (Accessed September 28th, 2024).

Ganesan, K., & Xu, B. (2017). Polyphenol-rich lentils and their health-promoting effects. *International Journal of Molecular Sciences*, *18*(11), 2390.

Gómez-Caravaca, A. M., Iafelice, G., Lavini, A., Pulvento, C., Caboni, M. F., & Marconi, E. (2012). Phenolic compounds and saponins in quinoa samples (*Chenopodium quinoa Willd.*) grown under different saline and nonsaline irrigation regimens. *Journal of Agricultural and Food Chemistry*, *60*(18), 4620–4627.

Graf, B. L., Rojas-Silva, P., Rojo, L. E., Delatorre-Herrera, J., Baldeón, M. E., & Raskin, I. (2015). Innovations in health value and functional food development of quinoa (*Chenopodium quinoa* Willd.). *Comprehensive Reviews in Food Science and Food Safety*, *14*(4), 431–445.

Gullón, B., Gullón, P., Tavaria, F. K., & Yáñez, R. (2016). Assessment of the prebiotic effect of quinoa and amaranth in the human intestinal ecosystem. *Food & Function*, *7*(9), 3782–3788.

Gupta, N., & Morya, S. (2022). Bioactive and pharmacological characterization of *Chenopodium quinoa*, *Sorghum bicolor* and *Linum usitassimum*: A review. *Journal of Applied and Natural Science*, *14*(3), 1067–1084.

Healy, K. (2001). The quinoa trail: From South American salt flats to Western health food stores. In N. P. Sellers (Ed.), *Llamas, weavings, and organic chocolate: Multicultural grassroots development in the Andes and Amazon of Bolivia* (pp. 155–188). Notre Dame, IN: University of Notre Dame Press.

Huang, H., Jia, C., Chen, X., Zhang, L., Jiang, Y., Meng, X., & Liu, X. (2024). Progress in research on the effects of quinoa (*Chenopodium quinoa*) bioactive compounds and products on intestinal flora. *Frontiers in Nutrition*, *11*, 1308384.

Hussain, M. I., Farooq, M., Syed, Q. A., Ishaq, A., Al-Ghamdi, A. A., & Hatamleh, A. A. (2021). Botany, nutritional value, phytochemical composition and biological activities of quinoa. *Plants*, *10*(11), 2258.

Jacobsen, S. E. (2003). The worldwide potential for quinoa (*Chenopodium quinoa* Willd.). *Food Reviews International*, *19*(1–2), 167–177.

Jeong, H. J., Jeong, J. B., Kim, D. S., Park, J. H., Lee, J. B., Kweon, D. H., … & de Lumen, B. O. (2007). The cancer-preventive peptide lunasin from wheat inhibits core histone acetylation. *Cancer Letters*, *255*(1), 42–48.

Kokoska, L., & Janovska, D. (2009). Chemistry and pharmacology of *Rhaponticum carthamoides*: A review. *Phytochemistry*, *70*(7), 842–855.

Kumar, S., & Pandey, A. K. (2013). Chemistry and biological activities of flavonoids: An overview. *The Scientific World Journal*, 2013(1), 162750.

Le, A. J. (2009). Quinoa (*Chenopodium quinoa* Willd.): Composition, chemistry, nutritional, and functional properties. *Advances in Food and Nutrition Research*, *58*, 1–31.

Lemya, D. E. R. O. U. I. C. H. E., & Ahlem, L. A. R. A. B. I. (2024). Phytoécologie de l'espèce Ephedra altissima Host, 1831 plantes utilisées pour ses vertus médicinales et cosmétologiques (Doctoral dissertation, University of Tlemcen).

Lin, M., Han, P., Li, Y., Wang, W., Lai, D., & Zhou, L. (2019). Quinoa secondary metabolites and their biological activities or functions. *Molecules*, *24*(13), 2512.

Liu, M., Zhu, K., Yao, Y., Chen, Y., Guo, H., Ren, G., … & Li, J. (2020). Antioxidant, anti-inflammatory, and antitumor activities of phenolic compounds from white, red, and black *Chenopodium quinoa* seed. *Cereal Chemistry*, *97*(3), 703–713.

Liu, R., Zhao, S., Regenstein, J. M., Liu, Q., Yang, H., & Xiong, S. (2016). Gelling properties of fish/pork mince mixtures. *Journal of Food Science*, *81*(2), C301–C307.

López-Moreno, M., Jiménez-Moreno, E., Gallego, A. M., Pasamontes, G. V., Uranga Ocio, J. A., Garcés-Rimón, M., & Miguel-Castro, M. (2023). Red quinoa hydrolysates with antioxidant properties improve cardiovascular health in spontaneously hypertensive rats. *Antioxidants*, *12*(6), 1291.

Lu, Y. H., Huang, J. H., Li, Y. C., Ma, T. T., Sang, P., Wang, W. J., & Gao, C. Y. (2015). Variation in nutritional compositions, antioxidant activity, and microstructure of *Lycopus lucidus* Turcz. Root at different harvest times. *Food Chemistry*, *183*, 91–100.

Maldonado-Alvarado, P., Pavón-Vargas, D. J., Abarca Robles, J., Valencia-Chamorro, S., & Haros, C. M. (2023). Effect of germination on the nutritional properties, phytic acid content, and phytase activity of quinoa (*Chenopodium quinoa* Willd). *Foods*, *12*(2), 389.

Martindale, W. (1894). *Coca and cocaine: Their history, medical and economic uses, and medicinal preparations*. London: H. K. Lewis.

Martini, S., Conte, A., & Tagliazucchi, D. (2017). Phenolic compounds profile and antioxidant properties of six sweet cherry (*Prunus avium*) cultivars. *Food Research International*, *97*, 15–26.

Medina-Meza, I. G., Aluwi, N. A., Saunders, S. R., & Ganjyal, G. M. (2016). GC–MS profiling of triterpenoid saponins from 28 quinoa varieties (*Chenopodium quinoa Willd.*) grown in Washington State. *Journal of Agricultural and Food Chemistry*, *64*(45), 8583–8591.

Mohyuddin, S. G. (2019). Quinoa is beneficial to the comprehensive nutritional value of potential health. *Pakistan Journal of Science, 71*(2), **90–94**. https://nja.pastic.gov.pk/PJS/index.php/PJS/article/view/1479

Morya, S., Menaa, F., Jiménez-López, C., Lourenço-Lopes, C., Bin Mowyna, M. N., & Alqahtani, A. (2022). Nutraceutical and pharmaceutical behavior of bioactive compounds of miracle oilseeds: An overview. *Foods*, *11*(13), 1824.

Mu, H., Xue, S., Sun, Q., Shi, J., Zhang, D., Wang, D., & Wei, J. (2023). Research progress of quinoa seeds (*Chenopodium quinoa* Wild.): Nutritional components, technological treatment, and application. *Foods*, *12*(10), 2087.

Mujica A. (1994). Andean grains and legumes. In: Hernando Bermujo JE, Leon J, editors. *Neglected Crops: 1492 from a Different Perspective* (Vol. 26, pp. 131–48). Rome, Italy: FAO. https://www.hort.purdue.edu/newcrop/1492/grains.html (Accessed October 1, 2024).

Nwachukwu, B. C., & Babalola, O. O. (2022). Metagenomics: A tool for exploring key microbiome with the potentials for improving sustainable agriculture. *Frontiers in Sustainable Food Systems*, *6*, 886987.

Oustani, M., Mehda, S., Halilat, M. T., & Chenchouni, H. (2023). Yield, growth development, and grain characteristics of seven Quinoa (*Chenopodium quinoa* Willd.) genotypes grown in open-field production systems under hot-arid climatic conditions. *Scientific Reports*, 13(1), 1991.

Paśko, P., Zagrodzki, P., Bartoń, H., Chłopicka, J., & Gorinstein, S. (2010). Effect of quinoa seeds (*Chenopodium quinoa*) in the diet on some biochemical parameters and essential elements in blood of high fructose-fed rats. *Plant Foods for Human Nutrition*, *65*, 333–338.

Popenoe, H., King, S. R., León, J., & Kalinowski, L. S. (1989). *Lost Crops of the Incas. Little-Known Plants of the Andes with Promise for Worldwide Cultivation* (pp. 139–161). Washington, DC: National Academy Press.

Pourshahidi, L. K., Caballero, E., Osses, A., Hyland, B. W., Ternan, N. G., & Gill, C. I. (2020). Modest improvement in CVD risk markers in older adults following quinoa (*Chenopodium quinoa Willd.*) consumption: A randomized-controlled crossover study with a novel food product. *European Journal of Nutrition*, *59*, 3313–3323.

Ranjan, S., Sow, S., Ghosh, M., Padhan, S. R., Kumar, S., Gitari, H., ... & Nath, D. (2023). Nutraceutical properties and secondary metabolites of quinoa (*Chenopodium quinoa* Willd.): A review. *International Journal of Food Properties*, *26*(2), 3477–3491.

Ren, G., Zhu, Y., Shi, Z., & Li, J. (2017). Detection of lunasin in quinoa (*Chenopodium quinoa* Willd.) and the in vitro evaluation of its antioxidant and anti-inflammatory activities. *Journal of the Science of Food and Agriculture*, *97*(12), 4110–4116.

Repo-Carrasco-Valencia, R., Hellström, J. K., Pihlava, J. M., & Mattila, P. H. (2010). Flavonoids and other phenolic compounds in Andean Indigenous grains: Quinoa (*Chenopodium quinoa*), kañiwa (*Chenopodium pallidicaule*), and kiwicha (*Amaranthus caudatus*). *Food Chemistry*, *120*(1), 128–133.

Repo-Carrasco-Valencia, R. A. M., & Serna, L. A. (2011). Quinoa (*Chenopodium quinoa*, Willd.) is a source of dietary fiber and other functional components. *Food Science and Technology*, *31*, 225–230.

Ruales, J., & Nair, B. M. (1992). Nutritional protein quality in quinoa (*Chenopodium quinoa*, Willd) seeds. *Plant Foods for Human Nutrition*, *42*, 1–11.

Ruiz, K. B., Biondi, S., Oses, R., Acuña-Rodríguez, I. S., Antognoni, F., Martinez-Mosqueira, E. A., ... & Molina-Montenegro, M. A. (2014). Quinoa biodiversity and sustainability for food security under climate change. A review. *Agronomy for Sustainable Development*, *34*, 349–359.

Scalbert, A., Manach, C., Morand, C., Rémésy, C., & Jiménez, L. (2005). Dietary polyphenols and the prevention of diseases. *Critical Reviews in Food Science and Nutrition*, *45*(4), 287–306.

Secondary metabolite. (2024, September 22). Wikipedia. https://en.wikipedia.org/wiki/Secondary_metabolite

Seifati, S. E., Ranaiezadeh, M. R., Daneshmand, F., Karimi Bekr, Z., & Rezaee, M. B. (2024). The phytochemical content and cytotoxic effects of quinoa commercial cultivars (*Chenopodium quinoa Willd.*) on MCF7 cell line of breast cancer. *Journal of Medicinal Plants and By-Products*, *14*(2), 112--120.

Small, E. (2013). 42. Quinoa–is the United Nations featured crop of 2013 bad for biodiversity? *Biodiversity*, *14*(3), 169–179.

Sobota, A., Świeca, M., Gęsiński, K., Wirkijowska, A., & Bochnak, J. (2020). Yellow-coated quinoa (*Chenopodium quinoa* Willd)–physicochemical, nutritional, and antioxidant properties. *Journal of the Science of Food and Agriculture*, *100*(5), 2035–2042.

Sparg, S., Light, M. E., & Van Staden, J. (2004). Biological activities and distribution of plant saponins. *Journal of Ethnopharmacology*, *94*(2–3), 219–243.

Stuardo, M., & San Martín, R. (2008). Antifungal properties of quinoa (*Chenopodium quinoa* Willd) alkali-treated saponins against *Botrytis cinerea*. *Industrial Crops and Products*, *27*(3), 296–302.

Tang, Y., Li, X., Chen, P. X., Zhang, B., Hernandez, M., Zhang, H., ... & Tsao, R. (2015). Characterization of fatty acid, carotenoid, tocopherol/tocotrienol compositions and antioxidant activities in seeds of three *Chenopodium quinoa* Willd. Genotypes. *Food Chemistry*, *174*, 502–508.

Tang, Y., Li, X., Zhang, B., Chen, P. X., Liu, R., & Tsao, R. (2015). Characterization of phenolics, betanins, and antioxidant activities in seeds of three *Chenopodium quinoa* Willd. Genotypes. *Food Chemistry*, *166*, 380–388.

Tang, Y., & Tsao, R. (2017). Phytochemicals in quinoa and amaranth grains and their antioxidant, anti-inflammatory, and potential health beneficial effects: A review. *Molecular Nutrition & Food Research*, *61*(7), 1600767.

Upadhyay, S. K., & Singh, S. P. (2021). *Bioprospecting of Plant Biodiversity for Industrial Molecules*. Wiley. DOI: 10.1002/9781119718017

Upadhyay, S. K., & Singh, S. P. (2023). *Plants as Bioreactors for Industrial Molecules*. Wiley. DOI: 10.1002/9781119875116

Upadhyay, S. K., & Singh, S. P. (2024). *Herbal Nutraceuticals: Products and Processes*. Wiley. DOI: 10.1002/9781394241576

Xi, X., Fan, G., Xue, H., Peng, S., Huang, W., & Zhan, J. (2024). Harnessing the potential of quinoa: Nutritional profiling, bioactive components, and implications for health promotion. *Antioxidants, 13*(7), 829.

Yang, F., Guo, T., Zhou, Y., Han, S., Sun, S., & Luo, F. (2024). Biological functions of active ingredients in quinoa bran: Advance and prospective. *Critical Reviews in Food Science and Nutrition, 64*(13), 4101–4115.

Yao, Y., Yang, X., Shi, Z., & Ren, G. (2014). Anti-inflammatory activity of saponins from quinoa (*Chenopodium quinoa* Willd.) seeds in lipopolysaccharide-stimulated RAW 264.7 macrophage cells. *Journal of Food Science, 79*(5), H1018–H1023.

Yendo, A. C., de Costa, F., Gosmann, G., & Fett-Neto, A. G. (2010). Production of plant bioactive triterpenoid saponins: Elicitation strategies and target genes to improve yields. *Molecular Biotechnology, 46*, 94–104.

Zeyneb, H., Pei, H., Cao, X., Wang, Y., Win, Y., & Gong, L. (2021). In vitro study of the effect of quinoa and quinoa polysaccharides on human gut microbiota. *Food Science & Nutrition, 9*(10), 5735–5745.

Złotek, U., Gawlik-Dziki, U., Dziki, D., Świeca, M., Nowak, R., & Martinez, E. (2019). Influence of drying temperature on phenolic acids composition and antioxidant activity of sprouts and leaves of white and red quinoa. *Journal of Chemistry, 2019*(1), 7125169.

Muhammad Akram, Fahad Said Khan, (2024), Pharmacology and Medicinal uses of Quinoa, Clinical Research and Clinical Reports, 3(5); DOI:10.31579/2835-8325/090

Melini, V., & Melini, F. (2021). Functional components and anti-nutritional factors in gluten-free grains: A focus on quinoa seeds. *Foods, 10*(2), 351.

Mustafa, A. M., Abouelenein, D., Acquaticci, L., Alessandroni, L., Angeloni, S., Borsetta, G., ... & Vittori, S. (2022). Polyphenols, saponins and phytosterols in lentils and their health benefits: an overview. *Pharmaceuticals, 15*(10), 1225.

Dini, I. (2020). Kancolla Seeds: High Nutritional Foods With Nutraceutical Properties. In *Nuts and Seeds in Health and Disease Prevention* (pp. 211-227). Academic Press.

Isenmann, E., Yuliandra, T., Touvleliou, K., Broekmann, M., Torre, X. D. L., Botrè, F., ... & Parr, M. K. (2023). Quinoa as Functional Food? Urinary elimination of ecdysterone after consumption of quinoa alone and in combination with spinach. *bioRxiv*, 2023-10.

Nandan, A., Koirala, P., Tripathi, A. D., Vikranta, U., Shah, K., Gupta, A. J., ... & Nirmal, N. (2024). Nutritional and functional perspectives of pseudocereals. *Food chemistry*, 139072.

Fontana, F., Raimondi, M., Marzagalli, M., Di Domizio, A., & Limonta, P. (2020). Natural compounds in prostate cancer prevention and treatment: mechanisms of action and molecular targets. *Cells, 9*(2), 460.

Mhada, M., Hafsa, J., Jdey, A., Debouba, M., & Ghaffari, T. (2020). *Variations of Saponins, Minerals and Total Phenolic Compounds Due to Processing and Cooking of Quinoa Seeds (Chenopodium quinoa Willd.).* **Foods, 9**(5), 660. https://doi.org/10.3390/foods9050660

Navruz-Varli, S., & Sanlier, N. (2016). Nutritional and health benefits of quinoa (Chenopodium quinoa Willd.). *Journal of cereal science, 69*, 371-376.

15 Algae
Sources of Secondary Metabolites and Their Applications

Sachin Kumar Mandotra[1] *and Shalom Akhai*[2]
[1]Department of Bio-Sciences and Technology, MMEC, Maharishi Markandeshwar (Deemed to be University), Mullana, Ambala, India
[2]Department of Mechanical Engineering, MMEC, Maharishi Markandeshwar (Deemed to be University), Mullana, Ambala, India

15.1 INTRODUCTION

Natural products are compounds with a remarkable variety of biological activities and have been instrumental in drug discovery. They often possess inherent pharmacological properties, including their ability to penetrate cells or bind to specific biological targets. Unlike primary metabolism, which is essential for basic cellular processes like growth and development, secondary metabolism is not vital for fundamental cellular functions. Secondary metabolites are highly diverse and often unique to particular species or even individual isolates (Sasso et al., 2012). These metabolites frequently facilitate interactions among organisms and their living or non-living surroundings, such as acting as signaling molecules, and hold significant potential for pharmaceutical applications.

Beyond plants, numerous well-documented metabolites have also been identified from microorganisms like algae. Microalgae produce a wide array of metabolites, ranging from simple primary compounds to intricate secondary metabolites. These microorganisms generate various bioactive substances, including vitamins, enzymes, carotenoids, fatty acids, antioxidants, toxins, polymers, sterols, and peptides, many of which have potential commercial applications (Bajhaiya et al., 2017). To date, over 15,000 distinct algal compounds have been identified and chemically characterized. Consequently, several microalgae species have been utilized for the natural synthesis of unprocessed materials (Cardozo et al., 2007). The diverse feeding behaviors of algae are mirrored by their remarkable biochemical variety, leading to the generation of numerous secondary metabolites. A significant number of such compounds exhibit notable biological activities, including effects on viruses, cancerous cell lines, larvae, fungi, and bacteria (Orefice et al., 2023). Metabolites produced by algae, such as polysaccharides, phenolics, phytosterols, alkaloids, carotenoids, and terpenoids, have gained significant interest in medicinal chemistry for their unique structures and wide-ranging functional properties (Hannan et al., 2020).

Algae, which may exist as single-celled or multicellular forms, are some of the earliest living creatures on Earth (Upadhyay et al., 2023). Microalgae flourished in the planet's oceans about 3 billion years back, as Earth's environment began to develop. The diversity of microalgae is immense, but much of it remains unexplored. There are over 50,000 species of microalgae found in oceans and freshwater sources, though merely 30,000 species have been documented so far (Bhardwaj et al., 2021; Richmond, 2004). Since ancient times, humans have utilized microalgae as a food source. Notably, algae can efficiently transform light energy from the sun into other chemical forms by fixing carbon dioxide, a process that is ten times more effective than higher plants. Algae are classified into two main groups, macro and microalgae, depending on their size. They are eukaryotic entities characterized by structures resembling higher plants, and they commonly exhibit green, brown,

 DOI: 10.1201/9781003518358-15

or red coloration. Whereas microalgae are microscopic and lack differentiation into stems, roots, or leaves, encompassing various types, including diatoms, green algae, and red algae. Microalgae exhibit greater photosynthetic efficiency compared to terrestrial plants, experience rapid cell division cycles ranging from 6 to 12 h under optimal conditions, and can thrive or persist even in challenging environments like high temperatures and elevated salinity levels (Park et al., 2022). Therefore, this review highlights the potential benefits and prospects of algae-derived metabolites, focusing on their diverse biological properties, chemical structures, applications, and possible mechanisms of action.

15.2 METABOLITES FROM ALGAL BIOMASS

15.2.1 Primary Metabolites

Several primary metabolites, including proteins, amino acids, lipids, and fatty acids, are produced by microalgae.

15.2.1.1 Fatty Acids and Lipids Derived from Algal Biomass

All plant and microalgal cells are made up of lipids and fatty acids, which are also utilized as energy sources, primary metabolites, storage products, and components of membranes (Mandotra et al., 2020). Certain microalgal species are special in that they have the capacity to synthesize unique long-chain unsaturated fatty acids (Table 15.1). The key distinction with higher plants is that microalgae respond to growing circumstances and different kinds of stress by drastically altering the fatty acid content of their oil (Sathasivam et al., 2019). Lipid biosynthesis takes place when photosynthetic activity declines in response to different stressors. When different types of stress stimulate microalgae, they typically produce acylglycerol, a molecule rich in lipid components, at their maximal rate in two stages. To maximize the lipid content, nitrogen deprivation or

TABLE 15.1
Algal Fatty Acids and Vitamins with Applications

S.No.	Fatty Acid	Algae	Application	References
1.	Hexadecatetraenoic acid	*Ulva pertusa*, *Chlamydomonas* spp., *D. tertiolecta*, *T. cylindrical*, *C. moewusii*, *T. suecica*, *D. bardawil*, *Ankitrodesmus* spp.	Nutritional supplements	Lordan et al. (2011); Arisz et al. (2000); D'Souza and Loneragan (1999); Becker (2004); Ben-Amotz (1993); Fried et al. (1982); Gouveia and Oliveira (2009)
2.	γ-Linolenic acid	*Spirulina* sp., *D. primolecta*, *C. homosphaera*, *Chlorococcum* sp.	Nutritional supplements	Yap and Chen (2001); Ohta et al. (1995); Chu (2012); Chiang et al. (2004)
3.	Arachidonic acid	*P. boryanum*, *Porphyridium* sp., *N. atomus*	Nutritional supplements	Zhang et al. (2002); Reitan et al. (1994)
4.	Eicosapentaenoic acid	*T. suecica*, *Pavlova* sp., *Tetraselmis* sp., *Nannochloropsis* sp., *N. atomus*, *Monodus* sp., *Nannochloris* sp., *Phaeodactylum* sp., *C. vulgaris*, *C. minutissima*	Aquaculture feed, Nutritional supplements	D'Souza and Loneragan (1999); Chu (2012); Tzovenis et al. (2009); Seto et al. (1984); Patil et al. (2007); Ben-Amotz (1993); Reitan et al. (1994)

(Continued)

TABLE 15.1 (*Continued*)
Algal Fatty Acids and Vitamins with Applications

S.No.	Fatty Acid	Algae	Application	References
5.	Docosahexaenoic acid	*Spirulina* sp., *Pyramimonas* sp., *Crypthecodinium* sp., *Schizochytrium* sp.	Aquaculture feed, Infant formula, Nutritional supplements	Sijtsma and de Swaaf (2004); Tzovenis et al. (2009); Chu (2012)
6.	Stearidonic acid	*T. suecica, M. pusilla*	Increase EPA content in the cell	D'Souza and Loneragan (1999)
7.	α-linolenic acid	*M. pusilla, Ankitrodesmus* sp., *D. tertiolecta, Botryococcus* sp., *D. salina, Chlamydomonas* sp., *D. bardawil, C. moewusii, C. vulgaris*	Works against cardiovascular diseases	Reitan et al. (1994); Ben-Amotz (1993); Martínez-Fernández et al. (2006); Chiang et al. (2004); Fried et al. (1982); Poerschmann et al. (2004); Piorreck et al. (1984); Arisz et al. (2000); Ahlgren et al. (1992)
8.	Linolenic acid	*D. tertiolecta, B. braunii, D. primolecta, C. moewusii, D. bardawil, C. protothecoides, Chlorococcum* sp., *C. vulgaris*	Retain moisture in the skin, Anti-inflammatory and prevent acne	Reitan et al. (1994); Chiang et al. (2004); Gouveia and Oliveira (2009); Arisz et al. (2000); Fried et al. (1982); Day et al. (2009); Piorreck et al. (1984)
9.	Vitamin C	*D. tertiolecta, C. reinhardtii, C. vulgaris, Chlorella* sp., *C. protothecoides*	Protect the skin from wrinkles; improve cardiovascular health; Help prevent eye-related conditions; support prenatal health	Barbosa et al. (2005); Borowitzka (1988); Becker (2004); Running et al. (2002)
10.	Vitamin B	*S. quadricauda, Chlamydomonas* sp., *S. obliquus, C. eugametos, S. acutus*	Anticancerous in nature; reduce fatigue; skin protective; Antidepressant; reduce the risk of cardiovascular disease	Becker (2004); Vilchez et al. (1997); Borowitzka (1988); Uhlik and Gowans (1974)
11.	Vitamin E	*D. tertiolecta, C. reinhardtii, C. vulgaris, C. pyrenoidosa, C. sorokiniana*	Antidiabetic in nature; Defend against harmful pollutants; Anti Alzheimer's potential; Support eye health	Carballo-Cárdenas et al. (2003); Borowitzka (1988); Matsukawa et al. (2000); Becker (2004)

similar stress conditions are often recommended. This is followed by maximizing the biomass content through fast growth under ideal culture conditions (Sharma et al., 2021). Lipids are categorized based on their polarity, which is determined by polar moieties like alcohols, carboxyl groups, sugars, and nonpolar carbon chains (Kumar et al., 2021). Typically, lipophilic organic solvents like chloroform and petroleum ether extract lipids from microalgal biomass. In microalgae, free fatty acids and triglycerides make up the majority of nonpolar lipids, whereas glycolipids and phospholipids which have one or more fatty acids replaced by polar groups make up the majority of polar lipids. Algal lipids have properties similar to the lipids extracted from other sources like fish and vegetable oils (Bag et al., 2019). As a result, microalgal lipids can be used for the manufacturing of cosmetics and petroleum products in place of vegetable oil. Furthermore, there are therapeutic uses and applications for certain important unsaturated fatty acids present in algal lipids.

15.2.1.2 Amino Acids and Proteins Derived from Microalgae

Some microalgal species may be used as protein sources due to their high protein concentrations. The availability, ratios, and amounts of amino acids are important in accurately assessing the dietary benefits of these proteins (Matos et al., 2024). Only a few, mostly non-essential amino acids can be synthesized by humans and animals; therefore, the essential amino acids must be obtained from dietary sources (Echave et al., 2022). However, plants and microalgae are able to synthesize all of the necessary amino acids. Since microalgae are also thought to be a biological source of next-generation proteins, the amino acid content of various microalgae is very significant in this respect (Vieira et al., 2018).

15.2.2 Secondary Metabolites from Algae

Algae are regarded as special phototrophic cell factories capable of producing primary and secondary metabolites (Figure 15.1) from carbon dioxide and organic substances containing carbon. Secondary metabolites like sterols, isoprenoids, pigments, mycosporine-like amino acids (MAAs), and alkaloids are synthesized in the algal growth cycle's stationary phase. These metabolites are necessary for ancillary functions for instance signaling, stress responses, plant defense mechanisms, and so forth (Jeong et al., 2020). The several groups of secondary metabolites that algae produce are mentioned in Table 15.2.

15.2.2.1 Pigments

Three major groups of bio-pigment, *viz.* carotenoids, phycobilins, and chlorophylls are present in microalgae. These light-absorbing materials, known as bio-pigments, capture sunlight at various visible spectrum wavelengths. In microalgae, chlorophyll is the primary pigment that absorbs light, and its structure is made up of 4 pyrrole rings arranged into a tetrapyrrole ring, also known as porphyrin (Chakdar and Pabbi, 2017).

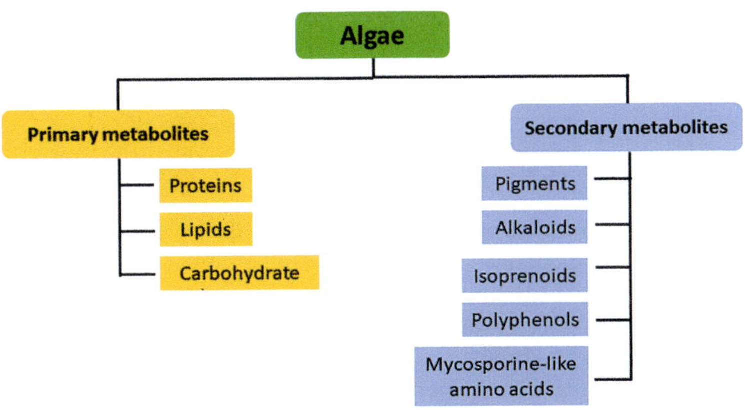

FIGURE 15.1 Primary and secondary metabolites from algae.

TABLE 15.2
Algal Secondary Metabolites with Applications

S.No.	Secondary Metabolite	Algae	Application	References
1.	Polyphenols, Flavonoid	*Chlorella Vulgaris*	Inhibits metastasis of lung cancer cell	Wang et al. (2010)
2.	Beta-carotene	*Dunaliella salina*	Decrease cell proliferation, arrest cell cycle at G_0 and G_1 phase, and initiate apoptosis in lung cancer cell	Sheu et al. (2008)
3.	Lycopene	*Chlorella zofingiensis*	decrease mRNA and protein levels of cyclin E in lung cancer cell	Mein et al. (2008)
4.	Phycoerythrin	*Lyngbya* sp.	Cell cycle arrest at G_0 and G_1 phase, decreases cell viability in lung cancer cell	Stengel et al. (2011)
5.	Fucoxanthin	*Isochrysis sp., Phaeodactylum tricornutum*	Anti-oxidant, Anti-inflammatory, anticancer, anti-obesity	Bule et al. (2018)
6.	Astaxanthin	*Chlorococcum* sp., *Haematococcus pluvialis, Chlorella vulgaris*	anti-diabetic, Anti-inflammatory, antiageing, Anti-oxidant	Bule et al. (2018)
7.	Lutein	*Chlamydomonas acidophila, Scenedesmus almeriensis, C. pyrenoidosa*	Prevents cataract, Anticancer, Antioxidant property, Prevents cardiovascular diseases	Garbayo et al. (2008); Leya et al. (2009); Sánchez et al. (2008)
8.	Cantaxanthin	*C. nivalis, Neospongiococcum* sp., *Chlorococcum* sp., *C. striolata, Chlamydocapsa* sp.	Antioxidant property	Chattopadhyay et al. (2008); Leya et al. (2009)
9.	Alpha-carotene	*C. sorokiniana*	Lower risk of premature death	Matsukawa et al. (2000)
10.	Fucosterol	*Eisenia bicyclis*	Prevent oxidative stress	Jung et al. (2013)
11.	Dieckol (phlorotannin)	*E. cava*	Anti-inflammation and antioxidation	Jung et al. (2009)
12.	Laminarin	*Laminaria digitata*	Antitumor and antioxidant properties, antibacterial, immunostimulatory	Lomartire and Gonçalves (2022); Peñalver et al. (2020)

Microalgae have colored proteins called phycobilins that absorb solar light in the visible spectrum, whereas chlorophyll has poor absorption. These protein-bound pigments exhibit a significant drop in response to heavy metals and nutritional deficiency stress, in contrast to other secondary metabolites. However, their production rates can be increased to a considerable degree by increasing the salt concentration (Saini et al., 2020).

Carotenoids are pigments that absorb sunlight and are mainly present alongside chlorophylls. Since chlorophyll cannot absorb electromagnetic radiation, carotenoids mostly absorb it in the visual range. The 40-carbon isoprene units that make up carotenoids' structure are separated into various categories according on whether or not oxygen is present at the end. Carotene and xanthophylls are the nonoxygenated derivatives and their oxygenated derivatives, respectively. Carotenoids are widely recognized for their anti-cancer and antioxidant qualities. A variety of carotenoids, including fucoxanthin, β-carotene, astaxanthin, lycopene, lutein, etc., are generated by microalgae (Vuppaladadiyam et al., 2018). Under severe stress circumstances, primary carotenoids like lutein can degrade, while secondary carotenoids like β-carotene accumulate in large quantities. The production of astaxanthin depends on a number of factors, including the intensity of the light, availability of nutrients, nitrogen supply, presence of heavy metals, and concentration of salt (Sathasivam et al., 2019).

15.2.2.2 Alkaloids

Alkaloids are yet another kind of nitrogen-containing secondary metabolites produced by variety of species, including algae and fungi. In nature, it is challenging to distinguish these substances from other nitrogen compounds. While most of the metabolites have commercial value, some may have toxic qualities that may kill some fish and animals (López-Hernández et al., 2020). These metabolites target calcium channels and acetylcholine receptors and have fatal effects. Various microalgal species have been reported for the accumulation and production of alkaloids. Saxitoxin is one of the potent alkaloid neurotoxins that affect voltage-gated sodium channels (Bule et al., 2018). Both freshwater cyanobacteria and marine dinoflagellates have been reported to synthesize natural compounds of the saxitoxin category. Zooxanthellamine, an alkaloid found in dinoflagellates shows structural resemblance with the nitrogen compounds found in soft corals, suggesting a symbiotic interaction between these species, with the soft corals perhaps acquiring these chemicals through the food chain (Patil and Kaliwal, 2019).

15.2.2.3 Mycosporine-Like Amino Acids (MAAs)

A family of secondary metabolites known as MAAs, also called microbial sunscreen, are synthesized by a variety of fungi and algae when exposed to high light intensities. Under intense UV light, the chemical was initially described in cyanobacteria *Gloeocapsa* sp. strain. MAAs range in molecular weight from 188 to 1,050 Da (Hartmann et al., 2015). The structure has a glycine moiety at position C3 and an amino alcohol at position C1 of the chromophore ring. They are extremely stable and exhibit zwitterion production and resonance structures (Sathasivam et al., 2019).

Because of their ability to accomodate free radicals inside their core ring structure, MAAs are widely recognized for their ability to absorb UV (ultraviolet) radiation. Therefore, MAAs are regarded as the scavengers of free radicals, capable of absorbing reactive oxygen species (ROS) to lessen the harm that oxidative stress causes to cells. MAAs are produced by a variety of marine bacteria, fungi, and algae and are involved in several of their stress defense and survival mechanisms (Karsten et al., 2007). These MAAs shield certain microalgae from UVB light exposure, which inhibits their motility. By controlling osmotic equilibrium, MAAs help shield microalgae from salt stress in addition to their well-known photoprotective properties (Nowruzi et al., 2020).

15.2.2.4 Isoprenoids

Isoprenoids, also known as terpenoids, are one of the biggest classes of natural compounds. They include both primary and secondary metabolites and are present in all plants (Kirby and Keasling, 2009). Most isoprenoids in higher plants serve the purpose of competing with other plants and defending against diseases or herbivores. Isoprenoids make up the majority of the natural compounds found in macroalgae (Maschek and Baker, 2008). Sesquiterpenes (C_{15}), diterpenes (C_{20}), and triterpenes (C_{30}) are a few of the significant categories in this class. Synthesis of isoprenoids is carried out by the biosynthesis of C_5 precursors and the production of polyprenyl pyrophosphates, followed by their subsequent processing to the final product (Sasso et al, 2012).

15.2.2.5 Vitamins from Algae

Algae, particularly seaweeds, are an excellent source of the vitamins C, E, and B complex (Table 15.1). A variety of factors, including species, location, time of year, and environmental conditions, affect the vitamin content of algae (Bekah et al., 2023). *Spirulina* species synthesize a high concentration of vitamin B_{12}, highlighting their evolutionary link to bacteria through similar biosynthetic pathways (Park et al., 2022). A challenge in utilizing microalgal vitamins lies in the reduction of certain vitamin levels during drying and subsequent processing. Nicotinic acid and vitamins B1, B2, and C are generally heat-sensitive and experience significant drops in concentrations throughout the drying process (Sultana et al., 2023).

15.3 PHARMACEUTICALS APPLICATIONS OF ALGAE

Microalgae are a diverse and fascinating group of microscopic entities known for their high protein and lipid content, along with moderate levels of carotenoids and vitamins. They possess distinct physiological and biochemical properties that set them apart from other plants and organisms. Additionally, microalgae are a rich source of biologically active novel primary and secondary metabolites (Rosa et al., 2019). Extracts derived from microalgae have demonstrated various biological properties, including anticancer, antiviral, anti-Alzheimer, antioxidant, antiprotozoal, and cytotoxic effects against human cancer cell lines (Figure 15.2). Carotenoids are powerful pigments with notable antioxidant properties that contribute to cancer prevention and overall human health (Shah et al., 2020). They also play a crucial role in protecting skin against UV radiation. Studies have shown that the use of carotenoids in cosmetics significantly enhances skin elasticity and promotes pronounced skin hydration. Fucoxanthin, a specific type of carotenoid, is particularly effective as a UV protector in sunscreens and cosmetics, helping to delay the effects of photoaging. Microalgae contain a significant amount of tocopherol, one of the essential compounds. It is widely utilized as a sunscreen in cosmetics and as a food preservative in several consumer goods (Salehi et al., 2019). Green algae, in particular, may be utilized in a wide range of applications due to its unique qualities and vast chemical variety, such as antioxidants, antibacterial, antiviral, anticancer, and cosmetics (Mobin et al., 2019).

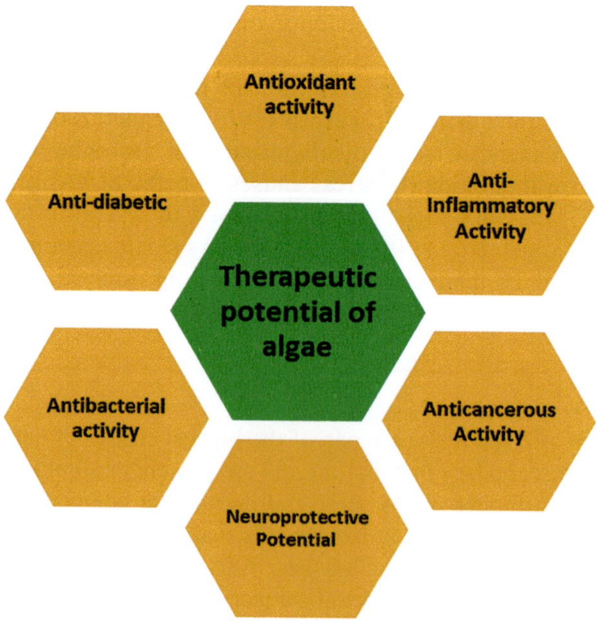

FIGURE 15.2 Therapeutic applications of algae.

15.3.1 Antioxidant Activity

The body's biochemical processes can generate intermediate compounds known as ROS. Some of these ROS, particularly free radicals, are highly reactive due to the presence of one or more unpaired electrons. When ROS levels become excessive, it can disrupt the body's metabolic processes and its ability to neutralize or detoxify these harmful compounds, leading to oxidative stress (Jesumani et al., 2020). This condition arises from an imbalance between the production of ROS and the effectiveness of the body's antioxidant defenses (Pirian et al., 2017). Essentially, oxidative stress occurs when the generation of reactive oxygen and nitrogen species outweighs the mechanisms that neutralize them. Seaweeds, often exposed to harsh environmental conditions, naturally produce polyphenols, which help them neutralize ROS, manage stress, prevent oxidative damage, and protect DNA from harm.

Microalgal biomass functions as a complex antioxidant system, often demonstrating enhanced effectiveness due to the synergistic interactions among its various antioxidants. Algae are a rich source of vitamins and can accumulate fat-soluble phenols and vitamin E, both known for their antioxidant properties. Vitamin E is widely utilized in the treatment of several medical conditions, including cardiovascular diseases, cancer, eye disorders, Parkinson's disease, and Alzheimer's disease. Certain microalgae species, such as *Tetraselmis suecica*, have been shown to increase vitamin E production under nitrogen-deprivation cultured conditions (Pham-Huy et al., 2008). Vitamin B_{12} (cobalamin), a water-soluble nutrient that certain microalgae are capable of producing. For instance, in comparison to typical plant or animal sources, *Spirulina* (blue-green algae) species generate higher levels of vitamin B_{12}. Additionally, some microalgae are known to accumulate Vitamin C (ascorbic acid), another water-soluble vitamin with antioxidant properties. Vitamin C plays a critical role in the synthesis of carnitine, collagen, and neurotransmitters. It is commonly used in the food industry and is associated with numerous health benefits, including its potential to prevent cancer, as well as its role in modulating the immune system (Barbosa et al., 2005). A carotenoid found in *Sargassum siliquastrum* called fucoxanthin reduced DNA damage brought on by oxidative stress, it also protected against hydrogen peroxide (H_2O_2)-induced DNA damage, a protective effect linked to enhanced glutathione (GSH) production and increased expression of superoxide dismutase (SOD) (Heo et al., 2008). Fucosterol, a compound commonly found in marine algae, inhibited the production of ROS in macrophages. Additionally, it protected cells from oxidative stress by boosting GSH levels and enhanced the expression of SOD and catalase in lung epithelial cells (Fernando et al., 2019).

15.3.2 Anti-inflammatory Activity

Inflammation plays a crucial role in the body's defense mechanism and can be triggered by a range of factors such as exposure to ultraviolet light, physical injury, microbial infections, and immune system activity. Common signs of inflammation include heat, redness, swelling, and pain (Chen et al., 2017b). While inflammation is essential for healing, persistent or chronic inflammation can be detrimental, contributing to the development of various diseases like multiple sclerosis, chronic asthma, cancer, and psoriasis. Exposure to UV radiation triggers various inflammatory reactions in human skin. The acute stage of inflammation is characterized by fluid buildup, increased blood flow, enhanced leukocyte activity, and greater vascular permeability. In contrast, the chronic stage of inflammation is linked to the activation of the immune response (Xu et al., 2018).

Air pollution, which causes severe inflammatory reactions in human skin, is largely caused by fine dust particles. A study conducted on diphlorethohydroxycarmalol (DPHC), derived from the brown alga *Ishige okamurae*, exhibits promising anti-inflammatory properties in countering inflammation caused by fine dust exposure (Fernando et al., 2019). Atopic dermatitis (AD) is a complex inflammatory skin condition often associated with allergies, marked by increased eosinophil counts and elevated levels of serum immunoglobulin E (IgE). Research has shown that extracts rich in polyphenolic compounds derived from brown algae, particularly species like *Eisenia* sp., *Ecklonia* sp., and *Sargassum* sp., exhibit notable anti-inflammatory effects and potential benefits for managing AD (Mihindukulasooriya et al., 2022). Numerous compounds derived from algae have been identified for their anti-inflammatory properties. Fucoxanthin, a carotenoid commonly found in brown algae, has been shown to reduce inflammation and oxidative stress in glial cells. Fucoxanthin has also been shown to decrease the production of prostaglandin E2 (PGE2) and suppress the expression of the cyclooxygenase-2 (COX-2) protein. Similarly, the anti-inflammatory effects of fucosterol have been highlighted in recent studies, demonstrating its ability to mitigate LPS-induced inflammation in alveolar macrophages and RAW 264.7 macrophages (Brandhorst et al., 2015).

15.3.3 Anticancerous Activity

Cancer is characterized by the abnormal behavior of cells, where they no longer respond appropriately to chemical signals from their surroundings. Instead of halting their growth, these cells persist in dividing uncontrollably. Such abnormal cell growth can harm the body by invading and disrupting normal tissues. Dysregulation in cellular processes, such as a reduction in signals promoting apoptosis or an increase in signals preventing it, can contribute to the development of cancer and other pathological conditions. As a result, promoting apoptosis has become a key focus in anticancer treatments, emphasizing the need to identify molecules that can effectively induce this process (Hamed et al., 2015; Taylor, 2011). Understanding the mechanisms through which various molecules operate is essential for developing cancer therapies and improving the well-being of individuals at risk. Molecules can influence processes such as cell cycle arrest, tumor angiogenesis inhibition, apoptosis or necrosis induction, and immune system activation. The functional properties of bioactive compounds from algae are primarily influenced by their chemical nature and characteristics.

Algae-derived bioactive compounds induce apoptosis through various pathways. It has been reported that polysaccharides from red alga *Gracilariopsis lemaneiformis* promote apoptosis through the Fas/FasL signaling pathway (Kang et al., 2017). In contrast, polysaccharides from brown alga *Sargassum fusiforme* were found to reduce the expression of VEGF-A, CD31, and tumor microvessel density (Chen et al., 2017a). In the SKH-1 hairless mouse skin model, research showed that polyphenols derived from brown algae efficiently inhibited UV-B-induced skin carcinogenesis (Hwang et al., 2006). In another study, phycocyanin has been shown to inhibit tumor cell proliferation by inducing cell cycle arrest at the G_0/G_1 phase and disrupting DNA synthesis. It also promotes apoptosis by interacting with cellular proteins through intrinsic and extrinsic pathways (Li et al., 2015).

15.3.4 Neuroprotective Potentials of Algae

One of the main causes of dementia in the elderly is neurodegenerative illnesses, such as Parkinson's and Alzheimer's diseases. These conditions are of significant public health concern. The key pathological features of Alzheimer's disease involve the accumulation of amyloid plaques outside neurons and the formation of neurofibrillary tangles (NFTs) within neurons. In contrast, Parkinson's disease is marked by the loss of dopaminergic neurons in the substantia nigra, accompanied by the buildup of α-synuclein aggregates inside neurons (Hou et al., 2019). The precise mechanisms underlying these brain disorders are not fully understood. However, research has shown that neuroinflammation, oxidative stress, protein misfolding, and mitochondrial dysfunction play significant roles in their progression. While neuroinflammation and oxidative stress are distinct pathological processes, they are closely interconnected throughout the disease. Consequently, suppressing neuroinflammation could potentially alleviate oxidative stress (Uddin et al., 2020; Hannan et al., 2020).

Because of this, researchers have been searching for new neuroprotective compounds to stop the central nervous system (CNS) from degenerating, malfunctioning, and experiencing apoptosis. Despite their function as neuroprotectives, synthetic drugs include adverse effects such as drowsiness, anxiety, and anxiousness (Hamed et al., 2015).

Marine algae have emerged as promising candidates for neuroprotective agents. A study carried out on *Ulva conglobata* demonstrated its neuroprotective potential. Extracts of this alga were tested on murine hippocampal HT22 cells exposed to glutamate-induced neurotoxicity and found that the extracts effectively reduced the neurotoxic effects. In another study, the inhibitory properties of extracts from seaweeds *Ulva reticulata* and *Hypnea valentiae* were investigated in vitro by incubating their different concentrations with acetylcholinesterase (AChE) or butyrylcholinesterase. The results revealed that the inhibitory activities of these extracts were comparable to those of the standard drug Donepezil (Jin et al., 2006). Several metabolites derived from algae have demonstrated significant potential in promoting neurite outgrowth under cell culture conditions. For example, by activating phosphatidylinositol-3 kinase, sargachromenol from *Sargassum macrocarpum* boosted the survival of PC12D cells and encouraged nerve growth factor-dependent neuronal development. Another metabolite of *S. macrocarpum*, argaquinoic acid, stimulated neuritogenesis in PC12D cells (Tsang et al., 2005; Kamei and Tsang, 2003).

15.3.5 Antibacterial Activity of Algae

The rise of antimicrobial resistance among certain pathogens poses a significant public health challenge that requires urgent attention (Mancuso et al., 2021). Many infectious diseases caused by various pathogens such as viruses, bacteria, fungi, and parasites are becoming harder to prevent and treat due to these organisms developing mechanisms to counteract commonly used medications (Orefice et al., 2023). Bacterial resistance to drugs is one of the most critical microbe-driven adaptation strategies, posing significant threats to human health. Many bacterial species have developed resistance to antibiotics, leading to severe health consequences. Among the most concerning are multidrug-resistant strains of species such as *Klebsiella pneumoniae, Staphylococcus aureus, Pseudomonas aeruginosa, Escherichia coli, Streptococcus pneumoniae,* and *Enterococcus faecium* (Prestinaci et al., 2015).

Algae have long been utilized as food and for the production of hydrocolloids like alginate and carrageenan. Recent advancements in screening techniques, however, have uncovered antibacterial compounds within the secondary metabolites of various algal classes, including Chlorophyceae, Phaeophyceae, Bacillariophyceae, Rhodophyceae, and Chrysophyceae (Rizzo et al., 2016; Blunt et al., 2015). These compounds contain functional groups with antibacterial properties. Phlorotannins are found within algal cells in specialized structures known as fucosan granules. Their antibacterial properties are attributed to their ability to disrupt oxidative phosphorylation and bind to bacterial proteins, including enzymes and cell membrane components, ultimately leading to cell lysis (Sahoo and Seckbach, 2015; Wang et al., 2009). Several terpene compounds derived from algae, including diterpene-benzoate bromophycolides, have demonstrated antibacterial activity (Shannon and Abu-Ghannam, 2016; Lane et al., 2009). Bromophycolides extracted from the red alga *Callophycus serratus* have been shown to effectively inhibit the growth of methicillin-resistant *Staphylococcus aureus* (MRSA) and vancomycin-resistant *Enterococcus faecium* (VRE) (Rodrigues et al., 2015). This inhibitory effect is attributed to the hydrophobic nature and structural rigidity of the tetrahydropyran ring. Phlorotannins exhibit strong antibacterial properties, showing greater efficacy against Gram-positive bacteria compared to Gram-negative ones (Lopes et al., 2012). These compounds can interact with bacterial proteins, including enzymes and membrane proteins, ultimately leading to cell lysis (Moubayed et al., 2017). The aromatic rings and hydroxyl groups in phloroglucinol units form hydrogen bonds and hydrophobic interactions with the amine groups of bacterial proteins (Besednova et al., 2020).

15.4 CONCLUSIONS AND FUTURE PERSPECTIVES

Though algae have historically been consumed as food and offer notable nutritional and health benefits, significant challenges remain regarding their use as a food source. These challenges include production and supply limitations, high costs, environmental impacts, health and safety concerns, regulatory hurdles for approval and commercialization, and consumer misconceptions about aspects like the color, smell, taste, and texture of seaweed products (Mendes et al., 2022). Researchers have continuously explored natural products derived from algae to advance their application in medicine. Despite this, the compounds in green algae have often been overlooked due to several obstacles. These include the lack of efforts to isolate or analyze the structures of crude green algae extracts, which limits their potential pharmaceutical applications. While these challenges impede the clinical utilization of algae-derived biomolecules, adopting alternative approaches such as large-scale collection and extraction to identify more bioactive metabolites can help overcome these barriers (Sugumaran et al., 2022).

Most research on algae has predominantly focused on primary metabolites, mainly lipid metabolism, while secondary metabolites remain underexplored across various fields. Expanding research on algal secondary metabolites is essential, with advanced strategies needed to address current challenges. Traditionally, applied research has relied on random genome insertion followed by screening as the primary engineering approach. However, the advent of precise gene-editing technologies has shifted the focus toward target-specific modifications. Advanced tools such as Clustered Regularly Interspaced Short Palindromic Repeats (CRISPR), Transcription Activator-Like Effectors, and Zinc Finger Nucleases can precisely modify key enzymes involved in rate-limiting steps, offering new possibilities for metabolic engineering and biotechnological applications (Belcour et al., 2020). The availability of genome sequences serves as an essential tool for advancing algal research, highlighting the vast potential yet to be explored in secondary metabolism. These sequences enable the identification of specific genes or entire pathways linked to known compounds, significantly enhancing regulatory and functional studies. Furthermore, genome data suggest that microalgae may synthesize previously unknown compounds, opening avenues for new discoveries (Sasso et al., 2012).

REFERENCES

Ahlgren, G., Gustafsson, I. B., & Boberg, M. (1992). Fatty acid content and chemical composition of freshwater microalgae 1. *Journal of phycology*, *28*(1), 37–50.

Arisz, S. A., van Himbergen, J. A., Musgrave, A., van den Ende, H., & Munnik, T. (2000). Polar glycerolipids of *Chlamydomonas moewusii*. *Phytochemistry*, *53*(2), 265–270.

Bag, P., Ansolia, P., Mandotra, S. K., & Bajhaiya, A. K. (2019). Potential of blue-green algae in wastewater treatment. In *Application of Microalgae in Wastewater Treatment: Volume 1: Domestic and Industrial Wastewater Treatment*, Editors: Sanjay Kumar Gupta, Faizal Bux, (pp. 363–381). Springer Cham.

Bajhaiya, A. K., Mandotra, S. K., Ansolia, A., & Barsana, A. (2017). Recent advances in improving ecophysiology of microalgae for biofuels. In *Algal Biofuels: Recent Advances and Future Prospects*, Editors: Sanjay Kumar Gupta, Anushree Malik, Faizal Bux, (pp. 141–162). Springer Cham.

Barbosa, M. J., Zijffers, J. W., Nisworo, A., Vaes, W., Van Schoonhoven, J., & Wijffels, R. H. (2005). Optimization of biomass, vitamins, and carotenoid yield on light energy in a flat-panel reactor using the A-stat technique. *Biotechnology and Bioengineering*, *89*(2), 233–242.

Becker, W. (2004). Microalgae in human and animal nutrition. In *Handbook of Microalgal Culture: Biotechnology and Applied Phycology*. Editor: Amos Richmond, 312–351. Wiley Blackwell.

Bekah, D., Thakoor, A. D., Ramanjooloo, A., Phul, I. C., Botte, S., Roy, P., … & Bhaw-Luximon, A. (2023). Vitamins, minerals and heavy metals profiling of seaweeds from Mauritius and Rodrigues for food security. *Journal of Food Composition and Analysis*, *115*, 104909.

Belcour, A., Girard, J., Aite, M., Delage, L., Trottier, C., Marteau, C., … & Markov, G. V. (2020). Inferring biochemical reactions and metabolite structures to understand metabolic pathway drift. *Iscience*, *23*(2), 100849.

Ben-Amotz, A. (1993). Production of β-carotene and vitamins by the halotolerant alga *Dunaliella*. In *Pharmaceutical and Bioactive Natural Products*, Editor: David H. Attaway, Oskar R. Zaborsky, (pp. 411–417). Boston, MA: Springer US.

Besednova, N. N., Andryukov, B. G., Zaporozhets, T. S., Kryzhanovsky, S. P., Kuznetsova, T. A., Fedyanina, L. N., … & Zvyagintseva, T. N. (2020). Algae polyphenolic compounds and modern antibacterial strategies: Current achievements and immediate prospects. *Biomedicines*, *8*(9), 342.

Bhardwaj, N., Sharma, C., Mandotra, S. K., & Ahluwalia, A. S. (2021). Potential of golden brown algae in forensic analysis: A review. In *Algae: Multifarious Applications for a Sustainable World*, Editors: Sachin Kumar Mandotra, Atul Kumar Upadhyay, Amrik Singh Ahluwalia, (pp. 353–373). Springer Singapore.

Blunt, J. W., Copp, B. R., Keyzers, R. A., Munro, M. H., & Prinsep, M. R. (2015). Marine natural products. *Natural Product Reports*, *32*(2), 116–211.

Borowitzka, M. A. (1988). Vitamins and fine chemicals from micro-algae. In *Micro-Algal Biotechnology*, Editors: M. A. Borowitzka and L. J. Borowitzka, (pp. 153–196). Cambridge University Press, Cambridge.

Brandhorst, S., Choi, I. Y., Wei, M., Cheng, C. W., Sedrakyan, S., Navarrete, G., … & Longo, V. D. (2015). A periodic diet that mimics fasting promotes multi-system regeneration, enhanced cognitive performance, and healthspan. *Cell Metabolism*, *22*(1), 86–99.

Bule, M. H., Ahmed, I., Maqbool, F., Bilal, M., & Iqbal, H. M. (2018). Microalgae as a source of high-value bioactive compounds. *Frontiers in Bioscience*, *10*(2), 197–216.

Carballo-Cárdenas, E. C., Tuan, P. M., Janssen, M., & Wijffels, R. H. (2003). Vitamin E (α-tocopherol) production by the marine microalgae *Dunaliella tertiolecta* and *Tetraselmis suecica* in batch cultivation. *Biomolecular Engineering*, *20*(4–6), 139–147.

Cardozo, K. H., Guaratini, T., Barros, M. P., Falcão, V. R., Tonon, A. P., Lopes, N. P., … & Pinto, E. (2007). Metabolites from algae with economical impact. *Comparative Biochemistry and Physiology Part C: Toxicology & Pharmacology*, *146*(1–2), 60–78.

Chakdar, H., & Pabbi, S. (2017). Algal pigments for human health and cosmeceuticals. In *Algal Green Chemistry*, Editors: Rajesh Prasad Rastogi, Datta Madamwar, Ashok Pandey, (pp. 171–188). Elsevier, Amsterdam, Netherlands.

Chattopadhyay, P., Chatterjee, S., & Sen, S. K. (2008). Biotechnological potential of natural food grade biocolorants. *African Journal of Biotechnology*, *7*(17), 2972–2985.

Chen, H., Zhang, L., Long, X., Li, P., Chen, S., Kuang, W., & Guo, J. (2017a). *Sargassum fusiforme* polysaccharides inhibit VEGF-A-related angiogenesis and proliferation of lung cancer in vitro and in vivo. *Biomedicine & Pharmacotherapy*, *85*, 22–27.

Chen, L., Deng, H., Cui, H., Fang, J., Zuo, Z., Deng, J., … & Zhao, L. (2017b). Inflammatory responses and inflammation-associated diseases in organs. *Oncotarget*, *9*(6), 7204.

Chiang, I. Z., Huang, W. Y., & Wu, J. T. (2004). Allelochemicals of *Botryococcus braunii* (Chlorophyceae) 1. *Journal of Phycology*, *40*(3), 474–480.

Chu, W. L. (2012). Biotechnological applications of microalgae. *IeJSME*, *6*(1), S24–S37.

Day, A. G., Brinkmann, D., Franklin, S., Espina, K., Rudenko, G., Roberts, A., & Howse, K. S. (2009). Safety evaluation of a high-lipid algal biomass from *Chlorella protothecoides*. *Regulatory Toxicology and Pharmacology*, *55*(2), 166–180.

D'souza, F. M. L., & Loneragan, N. R. (1999). Effects of monospecific and mixed-algae diets on survival, development and fatty acid composition of penaeid prawn (Penaeus spp.) larvae. *Marine Biology*, *133*(4), 621–633.

Echave, J., Otero, P., & Garcia-Oliveira, P. (2022). Seaweed-derived proteins and peptides: Promising marine bioactives. *Antioxidants*, *11*(1), 176.

Fernando, I. S., Jayawardena, T. U., Kim, H. S., Lee, W. W., Vaas, A. P. J. P., De Silva, H. I. C., … & Jeon, Y. J. (2019). Beijing urban particulate matter-induced injury and inflammation in human lung epithelial cells and the protective effects of fucosterol from *Sargassum binderi* (Sonder ex J. Agardh). *Environmental Research*, *172*, 150–158.

Fried, A., Tietz, A., Ben-Amotz, A., & Eichenberger, W. (1982). Lipid composition of the halotolerant alga, *Dunaliella bardawil*. *Biochimica et Biophysica Acta (BBA)-Lipids and Lipid Metabolism*, *713*(2), 419–426.

Garbayo, I., Cuaresma, M., Vílchez, C., & Vega, J. M. (2008). Effect of abiotic stress on the production of lutein and β-carotene by *Chlamydomonas acidophila*. *Process Biochemistry*, *43*(10), 1158–1161.

Gouveia, L., & Oliveira, A. C. (2009). Microalgae as a raw material for biofuels production. *Journal of Industrial Microbiology and Biotechnology*, *36*(2), 269–274.

Hamed, I., Özogul, F., Özogul, Y., & Regenstein, J. M. (2015). Marine bioactive compounds and their health benefits: A review. *Comprehensive Reviews in Food Science and Food Safety*, *14*(4), 446–465.

Hannan, M. A., Dash, R., Haque, M. N., Mohibbullah, M., Sohag, A. A. M., Rahman, M. A., … & Moon, I. S. (2020). Neuroprotective potentials of marine algae and their bioactive metabolites: Pharmacological insights and therapeutic advances. *Marine Drugs*, *18*(7), 347.

Hartmann, A., Becker, K., Karsten, U., Remias, D., & Ganzera, M. (2015). Analysis of mycosporine-like amino acids in selected algae and cyanobacteria by hydrophilic interaction liquid chromatography and a novel MAA from the red alga *Catenella repens*. *Marine Drugs*, *13*(10), 6291–6305.

Heo, S. J., Ko, S. C., Kang, S. M., Kang, H. S., Kim, J. P., Kim, S. H., … & Jeon, Y. J. (2008). Cytoprotective effect of fucoxanthin isolated from brown algae *Sargassum siliquastrum* against H_2O_2-induced cell damage. *European Food Research and Technology*, *228*, 145–151.

Hou, Y., Dan, X., Babbar, M., Wei, Y., Hasselbalch, S. G., Croteau, D. L., & Bohr, V. A. (2019). Ageing as a risk factor for neurodegenerative disease. *Nature Reviews Neurology*, *15*(10), 565–581.

Hwang, H., Chen, T., Nines, R. G., Shin, H. C., & Stoner, G. D. (2006). Photochemoprevention of UVB-induced skin carcinogenesis in SKH-1 mice by brown algae polyphenols. *International Journal of Cancer*, *119*(12), 2742–2749.

Jeong, Y., Cho, S. H., Lee, H., Choi, H. K., Kim, D. M., Lee, C. G., … & Cho, B. K. (2020). Current status and future strategies to increase secondary metabolite production from cyanobacteria. *Microorganisms*, *8*(12), 1849.

Jesumani, V., Du, H., Pei, P., Aslam, M., & Huang, N. (2020). Comparative study on skin protection activity of polyphenol-rich extract and polysaccharide-rich extract from *Sargassum vachellianum*. *PLoS One*, *15*(1), e0227308.

Jin, D. Q., Lim, C. S., Sung, J. Y., Choi, H. G., Ha, I., & Han, J. S. (2006). Ulva conglobata, a marine algae, has neuroprotective and anti-inflammatory effects in murine hippocampal and microglial cells. *Neuroscience Letters*, *402*(1–2), 154–158.

Jung, H. A., Jin, S. E., Ahn, B. R., Lee, C. M., & Choi, J. S. (2013). Anti-inflammatory activity of edible brown alga *Eisenia bicyclis* and its constituents fucosterol and phlorotannins in LPS-stimulated RAW264. 7 macrophages. *Food and Chemical Toxicology*, *59*, 199–206.

Jung, W. K., Heo, S. J., Jeon, Y. J., Lee, C. M., Park, Y. M., Byun, H. G., … & Choi, I. W. (2009). Inhibitory effects and molecular mechanism of dieckol isolated from marine brown alga on COX-2 and iNOS in microglial cells. *Journal of Agricultural and Food Chemistry*, *57*(10), 4439–4446.

Kamei, Y., & Tsang, C. K. (2003). Sargaquinoic acid promotes neurite outgrowth via protein kinase A and MAP kinases-mediated signaling pathways in PC12D cells. *International Journal of Developmental Neuroscience*, *21*(5), 255–262.

Kang, Y., Wang, Z. J., Xie, D., Sun, X., Yang, W., Zhao, X., & Xu, N. (2017). Characterization and potential antitumor activity of polysaccharide from *Gracilariopsis lemaneiformis*. *Marine Drugs*, *15*(4), 100.

Karsten, U., Lembcke, S., & Schumann, R. (2007). The effects of ultraviolet radiation on photosynthetic performance, growth and sunscreen compounds in aeroterrestrial biofilm algae isolated from building facades. *Planta*, *225*, 991–1000.

Kirby, J., & Keasling, J. D. (2009). Biosynthesis of plant isoprenoids: Perspectives for microbial engineering. *Annual Review of Plant Biology*, *60*(1), 335–355.

Kumar, S., Bhardwaj, N., Mandotra, S. K., & Ahluwalia, A. S. (2021). Recent insights of algal-based bioremediation and energy production for environmental sustainability. In *Phycology-Based Approaches for Wastewater Treatment and Resource Recovery*, Editors: Pradeep Verma, Maulin P. Shah, (pp. 263–288). CRC Press, Boca Raton.

Lane, A. L., Stout, E. P., Lin, A. S., Prudhomme, J., Le Roch, K., Fairchild, C. R., … & Kubanek, J. (2009). Antimalarial bromophycolides J–Q from the Fijian red alga *Callophycus serratus*. *The Journal of Organic Chemistry*, *74*(7), 2736–2742.

Leya, T., Rahn, A., Lütz, C., & Remias, D. (2009). Response of arctic snow and permafrost algae to high light and nitrogen stress by changes in pigment composition and applied aspects for biotechnology. *FEMS Microbiology Ecology*, *67*(3), 432–443.

Li, B., Gao, M. H., Chu, X. M., Teng, L., Lv, C. Y., Yang, P., & Yin, Q. F. (2015). The synergistic antitumor effects of all-trans retinoic acid and C-phycocyanin on the lung cancer A549 cells in vitro and in vivo. *European Journal of Pharmacology*, *749*, 107–114.

Lomartire, S., & Gonçalves, A. M. (2022). An overview of potential seaweed-derived bioactive compounds for pharmaceutical applications. *Marine Drugs*, *20*(2), 141.

Lopes, G., Sousa, C., Silva, L. R., Pinto, E., Andrade, P. B., Bernardo, J., … & Valentão, P. (2012). Can phlorotannins purified extracts constitute a novel pharmacological alternative for microbial infections with associated inflammatory conditions?. *PLoS One*, *7*(2), e31145.

López-Hernández, J. F., García-Alamilla, P., Palma-Ramírez, D., Álvarez-González, C. A., Paredes-Rojas, J. C., & Márquez-Rocha, F. J. (2020). Continuous microalgal cultivation for antioxidants production. *Molecules*, *25*(18), 4171.

Lordan, S., Ross, R. P., & Stanton, C. (2011). Marine bioactives as functional food ingredients: Potential to reduce the incidence of chronic diseases. *Marine Drugs*, *9*(6), 1056–1100.

Mancuso, G., Midiri, A., Gerace, E., & Biondo, C. (2021). Bacterial antibiotic resistance: The most critical pathogens. *Pathogens*, *10*(10), 1310.

Mandotra, S. K., Upadhyay, A. K., & Ahluwalia, A. S. (Eds.). (2020). *Algae: Multifarious Applications for a Sustainable World*. Springer Singapore.

Martínez-Fernández, E., Acosta-Salmón, H., & Southgate, P. C. (2006). The nutritional value of seven species of tropical microalgae for black-lip pearl oyster (*Pinctada margaritifera*, L.) larvae. *Aquaculture*, *257*(1–4), 491–503.

Maschek, J. A., & Baker, B. J. (2008). The chemistry of algal secondary metabolism. In *Algal Chemical Ecology*, Editor: Charles D. Amsler (pp. 1–24). Berlin, Heidelberg: Springer.

Matos, J., Cardoso, C., Serralheiro, M. L., Bandarra, N. M., & Afonso, C. (2024). Seaweed bioactives potential as nutraceuticals and functional ingredients: A review. *Journal of Food Composition and Analysis*, 133, 106453.

Matsukawa, R., Hotta, M., Masuda, Y., Chihara, M., & Karube, I. (2000). Antioxidants from carbon dioxide fixing *Chlorella sorokiniana*. *Journal of Applied Phycology*, *12*, 263–267.

Mein, J. R., Lian, F., & Wang, X. D. (2008). Biological activity of lycopene metabolites: Implications for cancer prevention. *Nutrition Reviews*, *66*(12), 667–683.

Mendes, M. C., Navalho, S., Ferreira, A., Paulino, C., Figueiredo, D., Silva, D., … & Speranza, L. G. (2022). Algae as food in Europe: An overview of species diversity and their application. *Foods*, *11*(13), 1871.

Mihindukulasooriya, S. P., Kim, H. J., Cho, J., Herath, K. H. I. N. M., Yang, J., Dinh, D. T. T., … & Jee, Y. (2022). Polyphenol-rich *Sargassum horneri* alleviates atopic dermatitis-like skin lesions in NC/Nga mice by suppressing Th2-mediated cytokine IL-13. *Algae*, *37*(4), 331–347.

Mobin, S. M., Chowdhury, H., & Alam, F. (2019). Commercially important bioproducts from microalgae and their current applications–A review. *Energy Procedia*, *160*, 752–760.

Moubayed, N. M., Al Houri, H. J., Al Khulaifi, M. M., & Al Farraj, D. A. (2017). Antimicrobial, antioxidant properties and chemical composition of seaweeds collected from Saudi Arabia (Red Sea and Arabian Gulf). *Saudi Journal of Biological Sciences*, *24*(1), 162–169.

Nowruzi, B., Sarvari, G., & Blanco, S. (2020). The cosmetic application of cyanobacterial secondary metabolites. *Algal Research*, *49*, 101959.

Ohta, S., Shiomi, Y., Kawashima, A., Aozasa, O., Nakao, T., Nagate, T., … & Miyata, H. (1995). Antibiotic effect of linolenic acid from *Chlorococcum* strain HS-101 and *Dunaliella primolecta* on methicillin-resistant *Staphylococcus aureus*. *Journal of Applied Phycology*, *7*, 121–127.

Orefice, I., Balzano, S., Romano, G., & Sardo, A. (2023). *Amphidinium* spp. as a source of antimicrobial, antifungal, and anticancer compounds. *Life*, *13*(11), 2164.

Park, Y. H., Han, S. I., Oh, B., Kim, H. S., Jeon, M. S., Kim, S., & Choi, Y. E. (2022). Microalgal secondary metabolite productions as a component of biorefinery: A review. *Bioresource Technology*, *344*, 126206.

Patil, L., & Kaliwal, B. B. (2019). Microalga *Scenedesmus bajacalifornicus* BBKLP-07, a new source of bioactive compounds with in vitro pharmacological applications. *Bioprocess and Biosystems Engineering*, *42*(6), 979–994.

Patil, V., Källqvist, T., Olsen, E., Vogt, G., & Gislerød, H. R. (2007). Fatty acid composition of 12 microalgae for possible use in aquaculture feed. *Aquaculture International*, *15*, 1–9.

Peñalver, R., Lorenzo, J. M., Ros, G., Amarowicz, R., Pateiro, M., & Nieto, G. (2020). Seaweeds as a functional ingredient for a healthy diet. *Marine Drugs*, *18*(6), 301.

Pham-Huy, L. A., He, H., & Pham-Huy, C. (2008). Free radicals, antioxidants in disease and health. *International Journal of Biomedical Science: IJBS*, *4*(2), 89.

Piorreck, M., Baasch, K. H., & Pohl, P. (1984). Biomass production, total protein, chlorophylls, lipids and fatty acids of freshwater green and blue-green algae under different nitrogen regimes. *Phytochemistry*, *23*(2), 207–216.

Pirian, K., Moein, S., Sohrabipour, J., Rabiei, R., & Blomster, J. (2017). Antidiabetic and antioxidant activities of brown and red macroalgae from the Persian Gulf. *Journal of Applied Phycology*, *29*, 3151–3159.

Poerschmann, J., Spijkerman, E., & Langer, U. (2004). Fatty acid patterns in *Chlamydomonas* sp. as a marker for nutritional regimes and temperature under extremely acidic conditions. *Microbial Ecology*, *48*, 78–89.

Prestinaci, F., Pezzotti, P., & Pantosti, A. (2015). Antimicrobial resistance: A global multifaceted phenomenon. *Pathogens and Global Health*, *109*(7), 309–318.

Reitan, K. I., Rainuzzo, J. R., & Olsen, Y. (1994). Effect of nutrient limitation on fatty acid and lipid content of marine microalgae 1. *Journal of Phycology*, *30*(6), 972–979.

Richmond, A. (Ed.). (2004). *Handbook of Microalgal Culture: Biotechnology and Applied Phycology* (Vol. 577). Oxford: Blackwell science.

Rizzo, L., Fraschetti, S., Alifano, P., Tredici, M. S., & Stabili, L. (2016). Association of *Vibrio* community with the Atlantic Mediterranean invasive alga *Caulerpa cylindracea*. *Journal of Experimental Marine Biology and Ecology*, *475*, 129–136.

Rodrigues, D., Alves, C., Horta, A., Pinteus, S., Silva, J., Culioli, G., ... & Pedrosa, R. (2015). Antitumor and antimicrobial potential of bromoditerpenes isolated from the red alga, *Sphaerococcus coronopifolius*. *Marine Drugs*, *13*(2), 713–726.

Rosa, G. P., Tavares, W. R., Sousa, P. M., Pagès, A. K., Seca, A. M., & Pinto, D. C. (2019). Seaweed secondary metabolites with beneficial health effects: An overview of successes in in vivo studies and clinical trials. *Marine Drugs*, *18*(1), 8.

Running, J. A., Severson, D. K., & Schneider, K. J. (2002). Extracellular production of L-ascorbic acid by *Chlorella protothecoides*, *Prototheca* species, and mutants of *P. moriformis* during aerobic culturing at low pH. *Journal of Industrial Microbiology and Biotechnology*, *29*(2), 93–98.

Sahoo, D., & Seckbach, J. (Eds.). (2015). *The Algae World* (No. 11564). Dordrecht: Springer Netherlands.

Saini, D. K., Chakdar, H., Pabbi, S., & Shukla, P. (2020). Enhancing production of microalgal biopigments through metabolic and genetic engineering. *Critical Reviews in Food Science and Nutrition*, *60*(3), 391–405.

Salehi, B., Sharifi-Rad, J., Seca, A. M., Pinto, D. C., Michalak, I., Trincone, A., ... & Martins, N. (2019). Current trends on seaweeds: Looking at chemical composition, phytopharmacology, and cosmetic applications. *Molecules*, *24*(22), 4182.

Sánchez, J. F., Fernández-Sevilla, J. M., Acién, F. G., Cerón, M. C., Pérez-Parra, J., & Molina-Grima, E. (2008). Biomass and lutein productivity of *Scenedesmus almeriensis*: Influence of irradiance, dilution rate and temperature. *Applied Microbiology and Biotechnology*, *79*, 719–729.

Sasso, S., Pohnert, G., Lohr, M., Mittag, M., & Hertweck, C. (2012). Microalgae in the postgenomic era: A blooming reservoir for new natural products. *FEMS Microbiology Reviews*, *36*(4), 761–785.

Sathasivam, R., Radhakrishnan, R., Hashem, A., & Abd_Allah, E. F. (2019). Microalgae metabolites: A rich source for food and medicine. *Saudi Journal of Biological Sciences*, *26*(4), 709–722.

Seto, A., Wang, H. L., & Hesseltine, C. W. (1984). Culture conditions affect eicosapentaenoic acid content of *Chlorella minutissima*. *Journal of the American Oil Chemists' Society*, *61*(5), 892–894.

Shah, S. A. A., Hassan, S. S. U., Bungau, S., Si, Y., Xu, H., Rahman, M. H., ... & Nemeth, S. (2020). Chemically diverse and biologically active secondary metabolites from marine *Phylum chlorophyta*. *Marine Drugs*, *18*(10), 493.

Shannon, E., & Abu-Ghannam, N. (2016). Antibacterial derivatives of marine algae: An overview of pharmacological mechanisms and applications. *Marine Drugs*, *14*(4), 81.

Sharma, C., Kumar, S., Bhardwaj, N., Mandotra, S. K., & Ahluwalia, A. S. (2021). Mitigation of heavy metals utilizing algae and its subsequent utilization for sustainable fuels. In *Algae: Multifarious Applications for a Sustainable World* (pp. 41–62).

Sheu, M. J., Huang, G. J., Wu, C. H., Chen, J. S., Chang, H. Y., Chang, S. J., & Chung, J. G. (2008). Ethanol extract of *Dunaliella salina* induces cell cycle arrest and apoptosis in A549 human non-small cell lung cancer cells. *in vivo*, *22*(3), 369–378.

Sijtsma, L., & De Swaaf, M. E. (2004). Biotechnological production and applications of the ω-3 polyunsaturated fatty acid docosahexaenoic acid. *Applied Microbiology and Biotechnology*, *64*, 146–153.

Stengel, D. B., Connan, S., & Popper, Z. A. (2011). Algal chemodiversity and bioactivity: Sources of natural variability and implications for commercial application. *Biotechnology Advances*, *29*(5), 483–501.

Sugumaran, R., Padam, B. S., Yong, W. T. L., Saallah, S., Ahmed, K., & Yusof, N. A. (2022). A retrospective review of global commercial seaweed production—current challenges, biosecurity and mitigation measures and prospects. *International Journal of Environmental Research and Public Health*, *19*(12), 7087.

Sultana, F., Wahab, M. A., Nahiduzzaman, M., Mohiuddin, M., Iqbal, M. Z., Shakil, A., ... & Asaduzzaman, M. (2023). Seaweed farming for food and nutritional security, climate change mitigation and adaptation, and women empowerment: A review. *Aquaculture and Fisheries*, *8*(5), 463–480.

Taylor, S. (2011). *Marine Medicinal Foods: Implications and Applications, Macro and Microalgae*. Academic Press, *London, United Kingdom.*

Tsang, C. K., Ina, A., Goto, T., & Kamei, Y. (2005). Sargachromenol, a novel nerve growth factor-potentiating substance isolated from *Sargassum macrocarpum*, promotes neurite outgrowth and survival via distinct signaling pathways in PC12D cells. *Neuroscience*, *132*(3), 633–643.

Tzovenis, I., Fountoulaki, E., Dolapsakis, N., Kotzamanis, I., Nengas, I., Bitis, I., ... & Economou-Amilli, A. (2009). Screening for marine nanoplanktic microalgae from Greek coastal lagoons (Ionian Sea) for use in mariculture. *Journal of Applied Phycology*, *21*, 457–469.

Uddin, M. S., Kabir, M. T., Al Mamun, A., Barreto, G. E., Rashid, M., Perveen, A., & Ashraf, G. M. (2020). Pharmacological approaches to mitigate neuroinflammation in Alzheimer's disease. *International Immunopharmacology*, *84*, 106479.

Uhlik, D. J., & Gowans, C. S. (1974). Synthesis of nicotinic acid in *Chlamydomonas eugametos*. *International Journal of Biochemistry*, *5*(1), 79–84.

Upadhyay, A. K., Mallick, S., Singh, R., Singh, L., Singh, N., Mandotra, S. K., … & Saxena, G. (2023). Novel cost-effective design for bio-volatilization studies in photosynthetic microalgae exposed to arsenic with emphasis on growth and glutathione modulation. *Frontiers in Microbiology*, *14*, 1170740.

Vieira, E. F., Soares, C., Machado, S., Correia, M., Ramalhosa, M. J., Oliva-Teles, M. T., … & Delerue-Matos, C. (2018). Seaweeds from the Portuguese coast as a source of proteinaceous material: Total and free amino acid composition profile. *Food Chemistry*, *269*, 264–275.

Vilchez, C., Garbayo, I., Lobato, M. V., & Vega, J. (1997). Microalgae-mediated chemicals production and wastes removal. *Enzyme and Microbial Technology*, *20*(8), 562–572.

Vuppaladadiyam, A. K., Prinsen, P., Raheem, A., Luque, R., & Zhao, M. (2018). Microalgae cultivation and metabolites production: A comprehensive review. *Biofuels, Bioproducts and Biorefining*, *12*(2), 304–324.

Wang, H. M., Pan, J. L., Chen, C. Y., Chiu, C. C., Yang, M. H., Chang, H. W., & Chang, J. S. (2010). Identification of anti-lung cancer extract from *Chlorella vulgaris* CC by antioxidant property using supercritical carbon dioxide extraction. *Process Biochemistry*, *45*(12), 1865–1872.

Wang, Y., Xu, Z. B. S. J., Bach, S. J., & McAllister, T. A. (2009). Sensitivity of *Escherichia coli* to seaweed (*Ascophyllum nodosum*) phlorotannins and terrestrial tannins. *Asian-Australasian Journal of Animal Sciences*, *22*(2), 238–245.

Xu, H., Zheng, Y. W., Liu, Q., Liu, L. P., Luo, F. L., Zhou, H. C., … & Li, Y. M. (2018). Reactive oxygen species in skin repair, regeneration, aging, and inflammation. *Reactive Oxygen Species (ROS) in Living Cells*, *8*, 69–88.

Yap, C. Y., & Chen, F. (2001, November). Polyunsaturated fatty acids: Biological significance, biosynthesis, and production by microalgae and microalgae-like organisms. In *Algae and their Biotechnological Potential: Proceedings of the 4th Asia-Pacific Conference on Algal Biotechnology*, 3–6 July 2000 in Hong Kong (pp. 1–32). Dordrecht: Springer Netherlands.

Zhang ChengWu, Z. C., Cohen, Z., Khozin-Goldberg, I., & Richmond, A. (2002). Characterization of growth and arachidonic acid production of *Parietochloris incisa* comb. nov (Trebouxiophyceae, Chlorophyta). *Journal of Applied Phycology*, *14*, 453–460.

16 Flavonoids and Their Advancement in Pharmaceutical Sciences

Ysrafil Ysrafil[1], *Nisa Kartika Komara*[2], *Meiyanti Ratna Kumalasari*[3], *and Firzan Nainu*[4]

[1]Faculty of Medicine, Department of Pharmacotherapy, Universitas Palangka Raya, Palangka Raya, Central Kalimantan, Indonesia

[2]Faculty of Medicine, Department of Physiology, Universitas Palangka Raya, Palangka Raya, Central Kalimantan, Indonesia,

[3]Faculty of Mathematics and Natural Science, Department of Chemistry, Universitas Palangka Raya, Palangka Raya, Central Kalimantan, Indonesia

[4]Faculty of Pharmacy, Department of Pharmacy, Universitas Hasanuddin, Makassar, Indonesia

16.1 INTRODUCTION

Flavonoids are a class of natural plant-derived secondary metabolites that belong to the polyphenol family. They are distinguished by their small molecular weight and the presence of a benzopyrone ring in their structure. Flavonoids are predominantly found in various parts of plants, including fruits, barks, roots, herbs, stems, cereals, nuts, vegetables, flowers, and seeds. These compounds are classified into subgroups based on the degree of oxidation and saturation of the central C ring, as well as the position of the B ring attachment. The major subgroups include flavones, flavonols, flavanonols, flavanols (catechins), anthocyanins, and chalcones [1,2].

The biosynthesis of flavonoids occurs via the phenylpropanoid pathway, where phenylalanine is changed into 4-coumaroyl-CoA, which subsequently enters the flavonoid biosynthetic pathway. Several enzymes are reported to play crucial roles in flavonoid biosynthesis, starting with chalcone synthase (CHS), which produces chalcone scaffolds. Additional enzymes, such as isomerases, reductases, hydroxylases, and Fe^{2+}/2-oxoglutarate-dependent dioxygenases change the fundamental flavonoid skeleton, leading to diverse flavonoid subclasses across different plant species. Finally, transferases modify the flavonoid backbone by adding sugar, methyl, and/or acyl groups, which alter the solubility, reactivity, and interactions of the resulting flavonoids with cellular targets. These modifications ultimately influence the physiological activity of flavonoids [3].

Flavonoids have been reported to exert a wide range of biological effects that can influence pathological conditions and provide therapeutic benefits for non-communicable diseases. These effects include anti-cholinesterase, anti-inflammatory, antioxidant, anticancer, antidiabetic, anti-obesity, antihypertensive, immunomodulatory, cardioprotective, anti-Alzheimer's disease, and anti-Parkinson's disease properties [2]. The therapeutic effects of flavonoids are mediated through various mechanisms, such as inhibiting key proteins (e.g., NF-κB in inflammation and cancer), regulating gluconeogenesis (in type 2 diabetes mellitus), and inhibiting amyloid plaque formation by binding to Aβ and preventing its aggregation (in Alzheimer's disease). In addition to their role in noncommunicable diseases, flavonoids are also effective against infectious diseases, exhibiting antifungal, antibacterial, and antiviral properties through multiple mechanisms. For instance, apigenin

 DOI: 10.1201/9781003518358-16

has demonstrated significant antibacterial activity against fluoroquinolone-resistant *Staphylococcus aureus* strains with quinolone-resistant S84L mutations in DNA gyrase [4]. Furthermore, flavonoids, such as flavones, have shown activity against *Candida* species, with minimum inhibitory concentrations (MICs) ranging from 62.5 to 83 μg/mL. The antifungal action is achieved through mechanisms such as membrane disruption, inhibition of cell division and cell wall synthesis, and the suppression of protein, RNA, and DNA synthesis [5].

However, despite their therapeutic potential, the effectiveness of flavonoids in vivo and in clinical trials remains debated. One of the primary reasons for this is their unfavorable pharmacokinetic profile. Flavonoids exhibit low bioavailability and bioactivity caused by poor water solubility, extensive metabolism, and restricted systemic absorption. Additionally, flavonoids undergo significant metabolic processing in the small intestine and liver, further reducing their bioavailability and diminishing their therapeutic effects. To address these limitations, researchers have explored various strategies, including structural modification of flavonoids and the creation of effective delivery mechanisms. Several platforms were created to enhance flavonoid delivery, such as polymer-based nanoparticles (e.g., chitosan), micelles, liposomes, solid lipid nanoparticles (SLNs), nanostructured lipid carriers (NLCs), nanosuspensions, and nanoemulsions. Numerous *ex vivo*, *in vitro*, and *in vivo* studies have demonstrated that nano-formulations of flavonoids offer enhanced therapeutic potential, improving their targeted delivery and efficacy in antibacterial, antidiabetic, and anticancer applications [6,7]. One of the most recent innovations is the use of microneedles (MNs) for continuous transdermal drug delivery. For example, quercetin-loaded MNs have shown good biocompatibility and sufficient mechanical strength to pierce the skin, enabling long-term drug delivery [8]. The aim of this chapter is to provide a comprehensive review of flavonoids, their biosynthesis, and the latest research on their therapeutic applications and development. We also seek to highlight recent advancements in flavonoid formulation technologies, which hold promise for their acceptance as modern therapeutic candidates for combating various diseases. The advancement of knowledge and research on flavonoids for pharmaceutical applications paves the path to the progress of novel drug candidates with promising effectiveness and fewer adverse effects. By utilizing sophisticated delivery methods, including nanoparticles and microparticles, flavonoids can be modified to enhance their solubility, bioavailability, and targeted delivery. These innovations enable more effective interaction with their sites of action, potentially leading to more satisfactory therapeutic outcomes.

16.2 FLAVONOID STRUCTURE AND BIOSYNTHESIS

Flavonoids, a diverse group of naturally occurring polyphenolic substances, are essential to human health and plant physiology. These compounds are widely distributed across the plant kingdom, contributing to the vibrant colors of fruits, flowers, and leaves, while also serving as important mediators in plant-environment interactions [1]. The biosynthesis of flavonoids involves a complex network of enzymatic reactions, resulting in the formation of various flavonoid subclasses, each characterized by distinct chemical structures and biological functions [9]. Flavonoid biosynthesis is a sophisticated biochemical process in plants, producing polyphenolic compounds through a series of enzymatic steps. Flavonoids are made up of fifteen carbon atoms grouped in a C6–C3–C6 backbone, consists of a three-carbon ring (C) joining two benzene rings (A and B) [1]. They are typically categorized as aglycones, glycosides, or methylated derivatives. Glycosylation enhances the water solubility and stability of flavonoids, but it may also affect their biological activity. The flavonoid structure is defined by its 15-carbon skeleton, with two benzene rings (A and B) connected by a three-carbon chain. This C6–C3–C6 arrangement is commonly used to describe flavonoids. Flavonoids can be classified into major subgroups, including chalcones, flavanones, dihydroflavonols, flavanols, flavones, isoflavones, leucoanthocyanidins, anthocyanins, and proanthocyanidins [9,10].

The basic phenylchromene-4-one scaffold of flavonoids can be changing with a phenyl ring at both the C2 or C3 position, forming the structures of flavones and isoflavones, respectively. Further structural changes, including the addition of a ketone group at the C4 position, saturation or olefinic bonding at the C2–C3 positions, and the addition of hydroxy or methoxy groups on either the phenyl chromene-4-one core or the attached phenyl ring, result in a wide variety of flavonoid types. These include flavanones, flavanols, flavonols, flavones, anthocyanins, and isoflavones. When the C ring does not form, the compound is classified as a chalcone, which serves as a precursor for several other flavonoid types [11].

Flavonoid biosynthesis is a complex process involving multiple biochemical pathways that contribute to the formation of various flavonoid subclasses. Three key pathways are central to this process: the phenylpropanoid pathway (also known as the acetate pathway), responsible for ring A formation; the shikimate pathway, which synthesizes ring B and the connecting C6–C3 structure; and the flavonoid pathway [9,10]. Ring A is synthesized from three malonyl-CoA molecules, which are derived from glucose metabolism, while ring B is formed from 4-coumaroyl-CoA, which is derived from phenylalanine via the shikimate pathway [12,13]. The condensation of rings A and B yields chalcones, which subsequently undergo isomerization to form flavanones, serving as precursors for the synthesis of various flavonoids [12,13].

All flavonoids share a common biosynthetic pathway, resulting in a similar structural framework. In flavonoids with a C ring, the basic structural unit is 2-phenylchroman (flavan). Structural variations among flavonoids arise from the degree of unsaturation and oxidation on the C ring. For instance, these variations include 2-phenylchromenes (e.g., anthocyanins and anthocyanidins), 2-phenylchromones (e.g., flavones, flavonols, flavanones, and dihydroflavonols), 2-phenylchromans (e.g., flavan, flavan-3-ol, flavan-3,4-diol, or proanthocyanidin), as well as chalcones, aurones, and other hydrocarbons [14].

A comprehensive understanding of these biosynthetic pathways is crucial for elucidating how plants produce these bioactive compounds, which not only support plant survival and adaptation but also confer substantial health benefits to humans. Flavonoid biosynthesis is modulated by environmental factors such as light, temperature, and pathogen attacks, which can either induce or repress the expression of genes involved in this pathway. As a result, flavonoids are produced, serving essential roles in plant growth and development, while also exhibiting significant pharmacological properties in humans, which include anti-inflammatory, anti-cancer, and antioxidant properties [15].

16.2.1 Shikimate Pathway

The shikimate pathway begins with the condensation of phosphoenolpyruvate (PEP) and D-erythrose-4-phosphate (E4P), catalyzed by DAHP synthase, to form 3-deoxy-D-arabino-heptulosonic acid-7-phosphate (DAHP). DAHP undergoes several enzymatic transformations, first being converted into 3-dehydroquinic acid by the action of dehydroquinate synthase (DHQS). This intermediate is then dehydrated to form 3-dehydroshikimic acid through the action of DHQ dehydratase, which removes water molecules. 3-dehydroshikimic acid is further reduced by shikimate dehydrogenase to produce shikimic acid [16]. Shikimic acid is then phosphorylated by shikimate kinase, resulting in the formation of shikimate-3-phosphate. This molecule reacts with another PEP, catalyzed by 5-enolpyruvylshikimate-3-phosphate (EPSP) synthase, yielding 5-EPSP, or enolpyruvyl shikimate-3-phosphate. EPSP is then converted into chorismic acid by chorismate synthase [16,17]. Chorismic acid, a key branching point in the pathway, is directed into various biosynthetic routes. One of these routes involves its conversion into prephenic acid by chorismate mutase. Prephenic acid then undergoes oxidative decarboxylation to form phenylpyruvate, which is subsequently transaminated by phenylalanine aminotransferase to yield phenylalanine. Phenylalanine can then enter the phenylpropanoid pathway, where it is deaminated by phenylalanine ammonia-lyase (PAL) to form trans-cinnamic acid, a precursor in the biosynthesis of flavonoids within the phenylpropanoid pathway [16,18].

16.2.2 Phenylpropanoid Pathway

The phenylpropanoid pathway begins with the amino acid phenylalanine. The enzyme PAL catalyzes the non-oxidative deamination of phenylalanine, resulting in the formation of cinnamic acid [10]. Cinnamic acid is subsequently hydroxylated by C4H, or cinnamonate 4-hydroxylase, to generate p-coumaric acid. The activation of p-coumaric acid with coenzyme A is then catalyzed by 4-coumarate ligase (4CL), forming p-coumaroyl-CoA. This compound serves as a primary precursor for flavonoid biosynthesis. In a reaction catalyzed by CHS, p-coumaroyl-CoA combines with three molecules of malonyl-CoA to create chalcone, the first flavonoid formed in the pathway. Chalcone isomerizes into flavanone under the catalytic action of chalcone isomerase (CHI) [14,16,17,19]. Flavanone is a key intermediate in the flavonoid biosynthesis pathway, acting as a branching point from which various flavonoid subclasses—such as flavones, flavonols, flavanols, anthocyanins, and isoflavones—are derived [9,10].

16.3 CHEMICAL CLASSIFICATION OF FLAVONOIDS

Flavonoids are classified into seven primary subclasses based on their chemical structures: flavonols, flavones, flavanones, isoflavones, anthocyanins, flavanols, and chalcones. The classification is primarily determined by the core heterocyclic ring's oxidation state. Furthermore, structural diversity within these subclasses arises from modifications, such as glycosylation and acylation, which result from variations in the methyl and hydroxyl groups attached to the other two rings [9].

16.3.1 Chalcone

Chalcones, a subclass of flavonoids, are characterized by their open-chain structure, lacking the "C ring" that is typical of other flavonoid classes. Examples of chalcones include phloridzin, arbutin, phloretin, and chalconaringenin [2]. These compounds are commonly found in various plant families, including Fabaceae, Moraceae, Zingiberaceae, and Cannabaceae. Chalcones also serve as key intermediates in the production of isoflavonoids and other flavonoids. Their core structures can be synthesized from simple aromatic precursors [13,20]. Table 16.1 provides a detailed classification of these flavonoid-derived compounds.

TABLE 16.1
Sources and Classification of Flavonoids

Classification	Compound	Source
Flavonols	Kaempferol, icarin, quercetin, fisetin, rutin, myricetin, galangin	Broccoli, onions, asparagus, apples, green tea, red wine
Flavones	Apigenin, chrysin, tangeretin, luteolin	Celery, parsley, red peppers, chamomile, mint, ginkgo biloba, tea, red pepper, grapes, apples, tomatoes, oranges
Isoflavones	Genistein, daidzein, glycitein, biochanin A, formononetin,	Soybean, flaxseed, red clove, alfafa, chickpeas
Anthocyanins	Delphinidin, cyanidin, petunidin, peonidin, malvidin, seranine, hyacinthin, pelargonidin	blueberries, strawberries, blueberries, red cabbage, tomatoes, purple sweet potatoes, eggplant
Flavanones	naringin, hesperidin, hesperetin, eriodictyol, naringenin	Orange, pomelo, lime, lemon
Flavanols	catechin, epicatechin, epicatechin gallate, gallocatechin, epigallocatechin	Apples, cherries, plums, apricots, and berries, especially in their skins, such as apple and grape skins
Chalcones	Phloridzin, arbutin, phloretin, and chalconaringenin	Coffee, tea, red bean, wine, beer

Chalcone is an essential compound in the biosynthesis of flavonoids, serving as a fundamental structural framework for various flavonoid types [9]. The synthesis of chalcone initiates with the incorporation of p-coumaroyl-CoA into the flavonoid biosynthetic pathway. This process involves three molecules of malonyl-CoA and one molecule of p-coumaroyl-CoA, which are derived from acetyl-CoA through the enzymatic action of acetyl-CoA carboxylase (ACCase). The resulting product, naringenin chalcone (4,20,40,60-tetrahydroxychalcone [THC]), is synthesized via the catalytic activity of CHS. CHS is an enzyme that belongs to the type III polyketide family and catalyzes critical reactions in the flavonoid biosynthetic pathway [15]. Structurally, chalcone is characterized by two rings of aromatic compounds joined by a three-carbon unsaturated carbonyl system (α, β), classifying it as a derivative of 1,3-diphenyl-2-propen-1-one. Chalcone is among the most structurally diverse forms of flavonoids. Chalcone derivatives exhibit a broad range of beneficial medicinal activities, including anticancer, antioxidant, anti-inflammatory, antidiabetic, and antimalarial properties, alongside numerous other pharmacological effects. Furthermore, chalcone plays a significant role in the formation of yellow pigments in plants [9]. Several chalcone derivatives are essential in the fields of pharmacy and medicine. Notable examples of chalcone derivatives include isoliquiritigenin, butein, and isobavachalcone [21].

16.3.2 Flavanones

Flavanones are synthesized through the action of the CHI, which catalyzes the conversion of chalcones into flavanones. There are four types of CHI enzymes: type 1, type 2, type 3, and type 4. Among these, CHI type 1 plays the most significant role in flavonoid biosynthesis. The 2'-hydroxyl (2'OH) group on chalcone is essential for the intramolecular cyclization mediated by CHI enzymes. The mechanism of flavonoid formation begins with the generation of a 2'-oxyanion, which occurs due to the loss of a proton from the 2'OH group of chalcone [21]. Subsequently, the freshly generated 2'-oxyanion targets the substrate's β-unsaturated double bond, leading to the formation of an acidic 2'OH group. The deprotonation mechanism of such 2'-chalcone hydroxyl groups is pH-dependent. The presence of van der Waals forces and extensive hydrogen bonding within the active site facilitates the formation of flavanones [15]. Flavanones can subsequently give rise to several other types of flavonoids with the assistance of various enzymes. They are known to serve as free radical scavengers and antioxidants [22]. Additionally, flavanones play a role in reducing the risk of chronic diseases, preventing heart-vascular-related diseases, and cancers. Moreover, flavanones exhibit antiviral, anti-inflammatory, and anti-allergic properties [22]. Several derivatives of flavanones are significant in the fields of pharmacy and medicine. Notable examples of flavanone derivatives include naringenin, eriodictyol, and liquiritigenin [9].

16.3.3 Flavones

Flavones, another major group of flavonoids, are characterized by a basic 4H-chromene-4-one structure containing a phenyl group substitution at the 2-position [13]. They are commonly found in the glycoside form appear in the fruits, flowers, and foliage of various plants. Important food sources of flavones include red peppers, chamomile, mint, celery, parsley, *Ginkgo biloba*, tea, red grapes, apples, tomatoes, and oranges [23]. The most common flavonoids in the diet are apigenin and luteolin [13]. Apigenin features a tricyclic core connected by a carbon or hydroxyl group to a sugar moiety, while other notable flavones include tangeretin and chrysin [24].

Flavones, or 2-aryl-4H-chromen-4-ones, represent the largest group of flavonoid derivatives, formed in various plants by various changes, including acylation, hydroxylation, glycosylation, and methylation. They are synthesized from flavanones with the aid of flavone synthase (FNS), which acts as a catalyst [15,25]. FNS is responsible for catalyzing the establishment of a double bond between the flavanone carbon chain's C-2 and C-3 locations. FNS is classified into two categories: FNSI and FNSII. FNSI is a water-soluble 2-oxoglutarate-dependent dioxygenase, whereas FNSII is a cytochrome P-450-dependent enzyme. Both FNSI and FNSII facilitate the conversion of flavanones into flavones [25,26].

Flavones are acknowledged for their potent antioxidant, antimicrobial, and anti-inflammatory activity, rendering them valuable as health supplements and potential alternatives to synthetic nutraceuticals. As antibacterial agents, flavones disrupt bacterial growth by targeting cell walls and membranes and inhibiting essential enzymes necessary for bacterial replication. This mechanism makes them effective against a variety of pathogens, including those resistant to conventional antibiotics. Research indicates that flavones, such as luteolin and apigenin, can effectively combat bacteria commonly associated with infections, including *Staphylococcus aureus* and *Escherichia coli* [27,28].

The anti-inflammatory properties of flavones arise from their capacity to suppress pro-inflammatory cytokines and inhibit mediators of inflammation, including NF-κB and COX-2. Flavones, including luteolin and apigenin, modulate immune responses by decreasing tumor necrosis factor-alpha (TNF-α) and interleukins production, which are critical contributors to chronic inflammation. This capacity to regulate inflammation without inducing significant side effects is essential for treating diseases linked to excessive immune responses, like inflammatory bowel disease and arthritis [27].

Furthermore, flavones function as antioxidants by preventing oxidative damage to cells and tissues by neutralizing reactive oxygen species (ROS). Their ability to donate electrons helps prevent lipid peroxidation and protects cellular integrity, thereby reducing the likelihood of illnesses like cancer and cardiovascular-related disease. Apigenin and luteolin are particularly recognized for enhancing the body's antioxidant defense systems by upregulating enzymes like superoxide dismutase (SOD) and catalase [27,28]. Due to their combined antibacterial, anti-inflammatory, and antioxidant properties, flavones are acknowledged as promising natural supplements that can either replace or complement synthetic nutraceuticals. Incorporating flavones into diets can enhance health outcomes, particularly in the treatment and prevention of inflammation-related disorders, oxidative stress, and microbial infections. Examples of flavone derivatives include luteolin, apigenin, chrysin, acacetin, baicalein, and wogonin [10,27].

16.3.4 Dihydroflavonol

Dihydroflavonols are synthesized from the precursor flavanone. The formation of dihydroflavonols is facilitated by the enzyme flavanone 3-hydroxylase (F3H), which requires 2-oxoglutarate and Fe^{2+}. F3H belongs to the dioxygenase family and catalyzes the hydroxylation at the C-3 position of flavanone. This enzyme adds a hydroxyl group (–OH) to carbon atom number 3 of flavanone, resulting in the production of dihydroflavonol. Subsequently, these dihydroflavonols can undergo further transformations, such as reduction or glycosylation, leading to the formation of various other flavonoid compounds, including flavonols and anthocyanins [29,30]. Dihydroflavonols serve as crucial intermediates in the biosynthesis of several flavonoids, including flavonols, which play significant roles in plant defense against pathogens, growth regulation, and the development of color pigments in flowers and fruits. Examples of dihydroflavonol derivatives include taxifolin [30], macarecurvatin A, macarecurvatin B, and 6,8-diisoprenylaromadendrin [31].

16.3.5 Flavonols

Flavonols, a subclass of flavonoids, contain ketone groups and serve as critical components of proanthocyanidins [23]. They are characterized by a higher number of hydroxyl (–OH) groups compared to other flavonoid types. Primarily located in the epidermal tissues of plants, flavonols protect plant DNA from ultraviolet (UV) radiation. Notable flavonol compounds include kaempferol, quercetin, fisetin, myricetin, and galangin, with rutin, a glycoside of quercetin, being the most abundant flavonol glycoside. Dietary sources rich in flavonols include broccoli, onions, asparagus, and various fruits such as apples. Tea and red wine also provide significant flavonol content [13,23,32].

Flavonols are produced through the hydroxylation of dihydroflavonols at the C-3 position of the C ring. This position is particularly susceptible to glycosylation, which is why flavonols are frequently found in glycoside form within plant cells. The conversion of dihydroflavonols to flavonols is facilitated by the enzyme flavonol synthase (FLS). FLS, an FeII/2-oxoglutarate-dependent dioxygenase, serves as an important rate-limiting enzyme in the flavonol biosynthetic [9]. This enzyme catalyzes the desaturation of dihydroflavonols, resulting in the formation of C-2 and C-3 double bonds on the C ring. The beneficial effects of flavonols include their antioxidant, anti-inflammatory, cardioprotective, and anticancer properties. Examples of flavonol derivatives include quercetin, morin, kaempferol, and myricetin, with quercetin being one of the most widely distributed flavonol derivatives [10,27].

16.3.6 Anthocyanins

Anthocyanins are in charge of the red, purple, and blue pigmentation in the flowers and fruits of higher plants, with the coloration attributed to their aglycone groups [13]. Chemically, anthocyanins are glycosides derived from polyhydroxy and polymethoxy compounds, known as anthocyanidin O-glycosides [19]. The variation in the number and positioning of hydroxyl and methoxyl groups within their molecular structure leads to a diversity of anthocyanin types. The color spectrum produced by anthocyanins ranges from red to blue, based on the environment's pH—blue under alkaline conditions, purple under neutral pH, and red under acidic conditions—due to the ionic nature of the anthocyanin molecules. Like other flavonoids, their stability is significantly influenced by factors such as pH, light, and temperature [33]. To date, over 650 distinct anthocyanin compounds have been identified in plants. The most common anthocyanins include delphinidin, cyanidin, petunidin, peonidin, malvidin, seranin, hyacinthin, and pelargonidin. These compounds are prevalent in various fruits and vegetables, such as blueberries, strawberries, red cabbage, tomatoes, purple sweet potatoes, and eggplant [13,23].

In the flavonoid biosynthetic pathway, dihydroflavonol 4-reductase (DFR) is a crucial NADPH-dependent enzyme that plays a significant role in the formation of anthocyanidins and proanthocyanidins. DFR catalyzes the reduction of dihydroflavonols—such as dihydrokaempferol (DHK), dihydroquercetin (DHQ), and dihydromyricetin (DHM)—to leukoanthocyanidins, also known as flavan-3,4-diols. These leukoanthocyanidins serve as direct precursors for the synthesis of anthocyanidins and proanthocyanidins. In this process, the enzyme anthocyanidin synthase (ANS), also referred to as leucoanthocyanidin dioxygenase (LDOX), converts colorless leukoanthocyanidins into colored anthocyanidins. Once the anthocyanidins are formed, they are stabilized into anthocyanins—the glycosylated forms of anthocyanidins—by the enzyme UDP-glucose flavonoid 3-glucosyltransferase (UFGT). This process produces compounds such as pelargonidin-3-glucoside, cyanidin-3-glucoside, and delphinidin-3-glucoside. Furthermore, the enzyme O-methyltransferase (OMT) catalyzes the conversion of cyanidin-3-glucoside to peonidin glycoside and delphinidin-3-glucoside to either petunidin glycoside or malvidin glycoside. These six primary categories of anthocyanins can subsequently undergo various modifications, including acylation, glycosylation, and methylation, resulting in a diverse array of anthocyanins. The resulting anthocyanins play a crucial role as pigments, imparting red, purple, and blue colors to flowers, fruits, and leaves while also serving as antioxidants that protect plants from various forms of environmental stress [9]. Anthocyanidins possess potent antioxidant properties, helping to protect body cells from oxidative damage caused by free radicals. Research indicates that anthocyanidins may lower the risk of heart disease and cancer. Furthermore, these compounds can mitigate inflammation in the body, which is advantageous for preventing and managing chronic inflammatory conditions such as arthritis. Examples of anthocyanin derivatives include cyanidin, delphinidin, pelargonidin, petunidin, and malvidin [10].

16.3.7 Isoflavones

Isoflavones represent a unique group of flavonoids characterized by a chemical structure based on the skeleton of 3-phenyl chromene-4-one [13,23]. The most recognized isoflavones include genistein, daidzein, and biochanin A. These compounds are predominantly found in legume plants, with soybeans serving as the primary source. Isoflavones are crucial precursors for phytoalexin production during plant-microbe interactions. In fermented soy products, isoflavones typically exist in their aglycone form due to the fermentation process, which removes sugar molecules from isoflavone glycosides [32,33]. Additionally, isoflavones are found in smaller quantities in alfalfa, chickpeas, and certain fruits and vegetables. Some isoflavonoids have also been reported in microbial organisms [13,23].

Structurally, isoflavones are characterized by a unique arrangement in which the B ring is positioned at the C-3 location of the C ring, in contrast to other flavonoids, which have the B ring at the C-2 position [10]. It is commonly recognized that isoflavones have estrogenic action, which supports a number of human health aspects. The formation of isoflavones is catalyzed by the enzyme Isoflavone Synthase (IFS), a cytochrome P450 (CYP93C) enzyme that facilitates the hydroxylation and aryl mutation reactions of flavanones (such as naringenin or liquiritigenin) to produce isoflavones. This process involves relocating the B ring from the C-2 position to the C-3 position on the C ring [10]. The biological effects of isoflavones include anti-inflammatory, anti-cancer, antibacterial, and antioxidant properties. Isoflavones may help lower the risk of osteoporosis, cardiovascular disease (CVDs), and the rate of bone resorption, according to several research. Examples of isoflavone derivatives include genistein, daidzein, glycitein, and formononetin [10].

16.3.8 Flavanols

Flavanols, also known as flavan-3-ols, constitute an important class of flavonoids widely present in the human diet. These compounds are derivatives of flavanes and are specified by their 2-phenyl-3,4 -dihydro-2H-chromene-3-ol structure. Flavanols serve as the foundational components of oligomers and polymers of proanthocyanidins, forming a distinct subgroup of flavonoids that lack carbonyl groups. Their diversity arises from the number and arrangement of hydroxyl groups on the A, B, and C rings [13,19,34]. Flavanols frequently occur in polymeric forms, known as tannins, which are composed of catechin molecules. Prominent flavanol compounds include catechin, epicatechin, epicatechin gallate (EGCG), gallocatechin, and epigallocatechin [10]. Numerous fruits, including apples, cherries, plums, apricots, and berries, have been reported abandon of these substances, which are particularly prevalent in apple and grape skins [13,19,34].

16.4 THERAPEUTICAL POTENTIAL AND THEIR UNDERLIYING MECHNISMS

16.4.1 Non-Communicable Diseases

16.4.1.1 Flavonoids as Antidiabetic Agents

Flavonoids, a class of polyphenolic compounds found in fruits, vegetables, and certain medicinal plants, have shown substantial potential as therapeutic agents for diabetes mellitus, particularly Type 2 diabetes Mellitus (T2DM). Their antidiabetic effects involve various biochemical processes, including anti-inflammatory actions, antioxidant activity, and the regulation of glucose metabolism [35]. One of the key ways by which flavonoids show positive effects in T2DM is by enhancing insulin sensitivity. Hyperglycemia prompts insulin release from the pancreas, which activates insulin signaling pathways in the liver, muscle, and fat tissues, facilitating the transport of the GLUT4 protein to the cell membrane for glucose absorption. Flavonoids such as puerarin, anthocyanins, baicalin, naringin, scutellarin, apigenin, and luteolin contribute to reducing insulin resistance by activating these insulin signaling pathways [36].

Puerarin, as a key isoflavone in *Pueraria lobata*, has shown significant effects in improving skeletal muscle insulin sensitivity [37]. It activates the μ-opioid receptor in muscles, thereby increasing insulin signaling [38]. Additionally, puerarin regulates insulin resistance via the B-AMPK/ACC signaling pathway in T2DM mouse models [39]. Baicalin, another flavonoid, has been found to reduce insulin resistance by stimulating the GALR2-GLUT4 signaling pathway, although this mechanism requires further exploration. Baicalin is also effective in reducing lipid accumulation in muscle tissue by altering enzyme phosphorylation involved in lipid metabolism. It activates AMPK, phosphorylates ACC at its inhibitory site, and reduces malonyl-CoA levels, thereby promoting fatty acid oxidation (FAO) in mitochondria and decreasing lipid levels. Furthermore, baicalin increases the phosphorylation of insulin-related molecules such as PKB/Akt and GLUT4, improving insulin sensitivity and glucose uptake in muscles. In adipose tissue, baicalin enhances insulin action by boosting mitochondrial biogenesis and metabolic activity while reducing fat content and inflammatory markers. In the liver, baicalin promotes fatty acid degradation and suppresses synthesis, which reduces lipid accumulation. It also increases the level of GLUT1 and GLUT4 while decreasing glucose-forming enzymes, contributing to lower blood glucose levels [36].

Flavonoids act as powerful antioxidants, mitigating stress oxidative as a crucial contributor to the progression of diabetes and its complications. Hyperglycemia raises the levels of ROS, leading to cellular and tissue damage. By scavenging free radicals, flavonoids like quercetin and naringenin protect pancreatic β-cells and enhance their functionality [35]. This antioxidant activity is crucial in preventing vascular complications associated with diabetes, including nephropathy and retinopathy [36]. Additionally, flavonoids modulate inflammatory pathways associated with diabetes. Chronic inflammation is a hallmark of T2DM, since high amounts of inflammatory cytokines can lead to insulin resistance and β-cell malfunction [37].

16.4.1.2 Flavonoids against Obesity

Flavonoids and their derivatives have been widely investigated for their therapeutic potential in managing obesity, primarily by regulating lipid metabolism, inflammation, and energy homeostasis. One of the key mechanisms through which flavonoids influence energy balance is by increasing energy expenditure. This is achieved by enhancing brown adipose tissue (BAT) activity and inducing the browning of white adipose tissue (WAT), which increases thermogenesis. The browning process is driven by the upregulation of uncoupling protein 1 (UCP1) in adipocytes, as observed in studies involving anthocyanins and other flavonoid subclasses, such as flavanols, flavanones, and chalcones [40].

Anthocyanins have been particularly effective in combating diet-induced obesity by regulating lipolysis, FAO, and adipogenesis, while also improving overall adipose tissue function [41–44]. These compounds induce the browning of WAT, increase BAT activity, and reduce inflammation associated with obesity [45,46]. For example, anthocyanin-rich extracts from bilberries and *Aronia melanocarpa* have been demonstrated to increase insulin sensitivity and lower WAT mass [43,47]. These effects are mediated by the activation of AMP-activated protein kinase (AMPK), which leads to increased expression of glucose transporter type 4 (GLUT4) [47]. Additionally, *Aristotelia chilensis* consumption has been linked to improved body weight management and glucose metabolism through the modulation of genes such as carbohydrate-responsive element-binding protein β (Chrebpβ), fibroblast growth factor 21 (Fgf21), and adiponectin, which play key roles in lipogenesis and FAO [48]. Anthocyanins also exhibit anti-inflammatory properties in WAT, as demonstrated in both mice and humans, with treatments involving black soybean testa (BBT) and other anthocyanin sources leading to reduced fat accumulation and improved lipid profiles. These effects are partly due to the suppression of inflammatory markers like IL-6 and TNF-α in high-fat diet-fed mice [41,46,49]

Flavanols, which are abundant in tea, cocoa, and fruits like grapes, also reduce adiposity and induce WAT browning. This is achieved through the regulation of lipid metabolism genes, with flavanols downregulating lipogenesis genes (e.g., Acc1, Fasn, Pparγ) and upregulating genes involved in lipolysis and FAO (Hsl, Cpt1a) [50]. Their molecular mechanism involves the activation of AMPK, a crucial enzyme in energy homeostasis. Furthermore, flavanols and flavanones enhance mitochondrial biogenesis and thermogenesis by upregulating proteins like PGC-1α, UCP1, and SIRT1, thereby promoting energy expenditure and the brown-like phenotype in WAT [51,52]. Chalcones, a less abundant flavonoid subclass found in certain vegetables and hops, similarly induce browning via the AMPK-SIRT1 pathway. Compounds like butein and xanthohumol have been shown to upregulate thermogenic genes such as UCP1, cytochrome C oxidase 8b, Cidea, and Prdm16, as well as mitochondrial markers like Cpt1a [53,54]. These molecular pathways highlight flavonoids' potential as therapeutic agents in reducing adiposity and managing obesity.

Moreover, flavonoids influence appetite and satiety by modulating hormones such as leptin and ghrelin. Naringenin, a citrus-derived flavonoid, has been shown to increase leptin sensitivity, enhancing satiety signals and thereby reducing food intake. This mechanism helps control caloric intake and prevents excessive weight gain [55]. Additionally, *Caralluma fimbriata*, a succulent cactus native to Asia, has been recognized for its ability to suppress hunger and quench thirst. The plant contains various phytochemicals, including pregnane glycosides and flavone glycosides, which contribute to its appetite-suppressing effects. Pregnane glycosides, in particular, have been linked to the downmodulation of ghrelin production in the stomach as well as neuropeptide Y (NPY) in the hypothalamus, both of which regulate hunger [56].

16.4.1.3 Flavonoids against Cardiovascular Diseases

Flavonoids are a wide class of plant-based sencodary metabolites, have emerged as major actors in the prevention and management of CVDs due to their broad biological activities, including antioxidant, anti-inflammatory, and vasodilatory effects. CVDs such as coronary heart disease, hypertension, and atherosclerosis are major contributors to global mortality, and the therapeutic potential of flavonoids in mitigating these conditions is widely studied. One prominent flavonoid, quercetin, has shown efficacy in improving endothelial function and modulating the renin-angiotensin-aldosterone system (RAAS), which is crucial for regulating blood pressure [57]. The flavonoid components found in red wine, including anthocyanins, flavan-3-ols, catechins, and resveratrol, play a key role in inhibiting AngII, thus mitigating oxidative stress and maintaining vascular health. Similarly, other flavonoids, including proanthocyanidin, have been demonstrated to possess comparable antihypertensive effects, particularly in studies involving pregnant rats [58].

Atherosclerosis, a condition marked by the accumulation of fatty deposits in arterial walls, results in the narrowing of arteries, heightening the risk of heart attacks and strokes. Flavonoids have been shown to inhibit the progression of atherosclerosis by counteracting oxidative stress, inflammation, and lipid accumulation within arterial walls. Flavonols like quercetin and kaempferol play pivotal roles in reducing the oxidation of low-density lipoprotein (LDL), a key event in the formation of atherosclerotic plaques. By preventing lipid peroxidation, flavonoids inhibit cholesterol buildup in arteries and promote its removal via reverse cholesterol transport [59]. Another flavonoid, epigallocatechin gallate (EGCG), a catechin derived from green tea, has demonstrated its efficacy in reducing atherosclerotic plaque formation by inhibiting pro-inflammatory cytokines and oxidative stress [60]. Furthermore, resveratrol contributes to atherosclerosis prevention through the activation of sirtuin-1 (SIRT1), a protein that regulates cellular stress responses and lipid metabolism [61].

Apigenin, a flavonoid present in citrus fruits like oranges and grapefruit, exerts a range of biological effects, including neuroprotection, antioxidant properties, and anti-tumor activity [62]. Recent studies indicate that apigenin exerts anti-atherosclerosis effects by intervening at various stages of the disease. Hyperlipidemia, often caused by high-fat diet, is a major trigger for atherosclerosis. A study conducted on Sprague-Dawley rats, apigenin administration at a dose of 8 g/kg for two weeks significantly reduced the thickening of the aorta's inner layer and improved lipid profiles by decreasing total cholesterol (TC), triglycerides (TG), and LDL cholesterol (LDL-C), while raising HDL cholesterol (HDL-C) levels [63]. Additionally, apigenin mitigates endothelial dysfunction by reducing the expression of LOX-1, VCAM-1, and E-selectin, which are markers of endothelial cell dysfunction under hyperlipidemic conditions [64]. Further studies using human endothelial cell models have confirmed that apigenin protects against vascular damage through multiple signaling pathways [65–67]. This reinforces apigenin's role as a promising therapeutic agent for preventing atherosclerosis.

Quercetin plays a significant role in regulating lipid metabolism and reducing atherosclerosis symptoms through multiple molecular pathways. It modulates the expression of genes such as PPARγ, LXR-α, and ABCA1, which are involved in cholesterol homeostasis, while concurrently reducing the expression of PCSK9 and CD36. This regulation results in decreased levels of TC, LDL-C, oxidized LDL (oxLDL), and intracellular lipid droplets, leading to the amelioration of atherosclerotic symptoms [68]. Coronary heart disease, often myocardial infarction is primarily caused by myocardial ischemia, a type of necrosis caused by prolonged coronary ischemia and hypoxia. Flavonoids, including quercetin, have been associated with reduced risk of coronary heart disease by inhibiting platelet aggregation and thrombosis in patients with dyslipidemia [69]. When platelets are activated, protein disulfide isomerase (PDI) is released and binds to αIIbβ3 integrins on the platelet surface, promoting platelet adhesion and thrombosis. Flavonoids act as PDI inhibitors, preventing the rapid activation of platelets by regulating calcium ion levels, thereby reducing thrombotic events [70,71]. Additionally, flavonoids protect against myocardial ischemia-reperfusion injury by regulating the expression of NF-κB and the epigenetic regulatory gene ANRIL [72]. This highlights quercetin's therapeutic potential in managing atherosclerosis and CVDs.

Naringin, a flavonoid predominantly found in grapefruit, has been identified as a key regulator of atherosclerosis through its interaction with the gut microbiota. Its hydrophilic properties, combined with the absence of specific hydrolases required for digestion and absorption in the small intestine, allow it to remain intact as it reaches the colon. This unique characteristic enables naringin to significantly influence the composition of the gut microbiota upon oral intake. Studies have shown that naringin alters the gut microbial community by decreasing the relative abundance of beneficial gut bacteria such as *Bacteroides*, *Bifidobacterium*, and *Lactococcus*, alongside a reduction in bile salt hydrolase (BSH) content. These microbial changes have a direct impact on bile acid metabolism, promoting the degradation of free bile acids. This, in turn, regulates cholesterol metabolism in the body, leading to increased excretion of bile acids and neutral sterols by 1.6-fold and 4.3-fold, respectively. These shifts in gut microbiota composition contribute to significant reductions in serum and liver cholesterol levels, further supporting the hypothesis that naringin alleviates atherosclerosis through modulation of the gut-liver-cholesterol axis [73]. This highlights the potential of naringin as a therapeutic agent in managing cholesterol-related CVDs, potentially offering an alternative strategy to control atherosclerosis through gut microbiome regulation.

16.4.1.4 Flavonoids against Alzheimer's Disease

Flavonoids and their derivatives have been identified as powerful neuroprotective agents against Alzheimer's disease (AD) through multiple molecular mechanisms. Found abundantly in plant-based foods such as cocoa, berries, and tea, these compounds have been shown to improve cognitive function and slow the progression of AD by targeting several key processes involved in the disease, including amyloid-beta (Aβ) aggregation, neurofibrillary tangles (NFTs), oxidative stress, and neuroinflammation [74]. Aβ is a major contributor to AD pathology, and flavonoids have been shown to modulate Aβ production, aggregation, and toxicity. For instance, anthocyanins from bilberries and blackcurrants have been demonstrated to reduce Aβ precursor protein (APP) C-terminal fragments in the brains of transgenic mice models, alleviating behavioral symptoms of AD [75]. Similarly, grape-derived polyphenols, primarily catechins and epicatechins, have been shown to prevent Aβ oligomerization and mitigate cognitive impairments in AD mouse models like the Tg2576 and APP/PS1 [76,77]. Long-term use of *Ginkgo biloba* extracts, which contain flavonols such as quercetin, kaempferol, and isorhamnetin, has also been found to significantly reduce APP levels in Tg2576 cortex of mice [78]. EGCG, a well-studied flavonoid from tea, has been shown to regulate APP processing and reduce Aβ levels in the brain [79]. EGCG also interacts with unfolded polypeptides, preventing the formation of β-sheet-rich amyloid fibrils and degrading large Aβ fibrils into smaller, non-toxic aggregates [80]. Additionally, flavonoids like wogonin have been shown to reduce β-secretase levels, inhibiting Aβ production [81]. Overall, flavonoids exert their anti-amyloidogenic effects by modulating APP processing, preventing Aβ aggregation, and degrading toxic Aβ fibrils into smaller, non-toxic forms.

Flavonoids have been reported to play a crucial role in modulating tau proteins, which are central to AD pathology. Tau proteins, when hyperphosphorylated, aggregate into NFTs, leading to synaptic dysfunction and neurodegeneration by disrupting their normal function in stabilizing microtubules [82]. The aggregation of hyper-phosphorylated tau is a hallmark of AD, and flavonoids have been shown to counteract this process through several mechanisms. For instance, flavonoids like myricetin and epicatechin-5-gallate inhibit tau aggregation and hyperphosphorylation, reducing tau-related toxicity [83]. Flavonoids can reduce tau aggregation by modulating key kinases involved in tau phosphorylation, such as glycogen synthase kinase-3β (GSK-3β) and cyclin-dependent kinase-5 (CDK-5) [84]. These kinases are implicated in the abnormal phosphorylation of tau proteins, which leads to tau fibril formation. By inhibiting the activity of GSK-3β and CDK-5, flavonoids prevent tau from adopting its pathological form, thereby preserving neuronal integrity and reducing the formation of toxic tau aggregates [82,84].

Quercetin is another potent inhibitor of two key enzymes involved in AD pathology: acetylcholinesterase (AChE) and butyrylcholinesterase (BChE) [85,86]. These enzymes are responsible for the hydrolytic degradation of acetylcholine (ACh), a neurotransmitter critical for neuronal communication, synaptic plasticity, and the regulation of neuron excitability [87]. In AD, reduced ACh levels impair communication between neurons, contributing to cognitive decline. Quercetin, through its hydroxyl (OH) groups on the phenyl ring, forms hydrogen bonds with specific amino acids in the active site of AChE, effectively inhibiting its activity [85]. By inhibiting both AChE and BChE, quercetin enhances cholinergic activity and promotes improved neuronal communication, potentially alleviating memory loss symptoms [88,89]. Moreover, BChE is involved in the transformation of amyloid plaques into more pathogenic structures, which are linked to dementia and AD progression. Therefore, by inhibiting BChE, quercetin may play an additional protective role in reducing amyloid plaque formation, further contributing to neuroprotection in AD treatment [88,90].

16.4.1.5 Flavonoids against Immune-Related Diseases

Flavonoids, a diverse group of polyphenolic compounds found abundantly in fruits, vegetables, tea, and various plant-based foods, have attracted significant attention for their role in modulating immune function and addressing immune-related diseases. These natural compounds are recognized for their potent antioxidant, anti-inflammatory, and immunomodulatory properties, which underlie their therapeutic potential in immune-related disorders, including autoimmune diseases and chronic inflammatory conditions. At the molecular level, flavonoids exert their effects through several key pathways that regulate immune responses [91–93].

16.4.1.6 Flavonoids Enhance Innate Immune Responses by Activating NK Cells

Flavonoids play a crucial role in modulating the innate immune system, particularly by increasing the activity of natural killer (NK) cells. NK cells are a critical component of the body's first line of defense, responsible for identifying and destroying virally infected cells and tumor cells. Flavonoids, with their diverse biological activities, can activate NK cells, thereby boosting the innate immune response. One mechanism by which flavonoids enhance NK cell activity is by influencing cytokine production. For instance, flavonoids such as quercetin and EGCG have been shown to increase the secretion of key cytokines like interferon-gamma (IFN-γ) and interleukin-12 (IL-12) [94]. These cytokines are essential for NK cell activation and proliferation. IFN-γ, in particular, enhances the cytotoxic activity of NK cells, allowing them to more effectively target and kill infected or malignant cells [95].

Flavonoids can also directly modulate signaling pathways within NK cells. For example, some flavonoids enhance the expression of activating receptors on NK cells, such as NKG2D and NKp46 [96]. These receptors are crucial for recognizing stress-induced ligands on abnormal cells. Upregulation of these receptors makes NK cells more sensitive to identifying and eliminating dangerous cells, thereby improving the overall immune response. One study demonstrated that flavonoids from *Astragalus complanatus* increased NK cell cytotoxicity through upregulation of NKG2D and NKp44 [97]. Moreover, in vivo studies have shown that flavonoids from *Hippophae rhamnoides* L enhance the expression of NKp44 and NKp46 [98].

16.4.1.7 Flavonoids Reducing the Maturation and Differentiation of DCs

Flavonoids can inhibit the activation of M1 macrophages and promote their conversion to the M2 phenotype [99]. Macrophages, essential components of the immune system, exist in two primary functional states: M1 macrophages, which are involved in inflammatory responses and pathogen elimination, and M2 macrophages, which are associated with tissue repair and resolution of inflammation [100]. Flavonoids exert immunomodulatory effects by both inhibiting M1 macrophage activation and facilitating their polarization into the M2 phenotype. This transition helps resolve chronic inflammation and promotes tissue healing [101].

M1 macrophages are typically activated by pro-inflammatory signals such as interferon-gamma (IFN-γ) and lipopolysaccharides (LPS), leading to the production of pro-inflammatory cytokines like tumor necrosis factor-alpha (TNF-α), interleukin-6 (IL-6), and interleukin-1β (IL-1β). Sustained activation of M1 macrophages can contribute to chronic diseases, including autoimmune disorders, CVDs, and obesity-related inflammation [102]. Flavonoids such as quercetin, kaempferol, and apigenin have been shown to inhibit M1 macrophage activation by suppressing pro-inflammatory cytokine production and downregulating key signaling pathways, including NF-κB and MAPK, which are crucial for M1 activation [34].

Flavonoids also exert distinct effects on different macrophage phenotypes. Pentamethoxyflavanone (Pen), a polymethoxylated flavonoid found in Citrus species, inhibits M1 macrophage polarization in mice by reducing p65 expression [103]. Pen also diminishes M1 polarization by downregulating p-STAT1, while promoting M2-associated molecules by upregulating p-STAT6 [101]. Similarly, Baicalin, a flavone glycoside, enhances the polarization of M2 macrophages in mice by increasing the expression of the macrophage c-Mer tyrosine kinase (MERTK) receptor, which facilitates the phagocytosis of apoptotic cells [104]. Another flavonoid, hesperetin, inhibits M1 macrophage activation by reducing the production of inflammatory mediators such as nitric oxide (NO) and TNF-α in LPS-stimulated macrophages. At the same time, hesperetin promotes the expression of M2 markers, contributing to the resolution of inflammation in a model of liver injury [105].

16.4.1.8 Flavonoids Regulate Adaptive Immune Responses

Flavonoids control adaptive immune responses by promoting Tregs cell proliferation, enhancing the activity of cytotoxic T lymphocytes (CTLs), and modulating various transcription factors, inflammatory cytokines, and signal transduction pathways. Simultaneously, they inhibit the activation and proliferation of T helper (Th) cells [99]. Flavonoids promote Treg proliferation, which helps maintain immune balance and prevents autoimmune reactions. They achieve this by modulating transcription factors such as Foxp3, essential for Treg development and function. Additionally, flavonoids influence signaling pathways like TGF-β (Transforming Growth Factor-β) and IL-10, both of which are anti-inflammatory cytokines that support Treg activity [106]. For instance, baicalin, a compound abundant in traditional Chinese herbal medicine, not only promotes the release of TGF-β and IL-10 but also enhances Foxp3 expression in vivo, likely through its effect on mTOR activation [107,108]. Furthermore, quercetin increases Foxp3 expression by targeting T-cell immunoglobulin and mucin domain-containing protein 3 (Tim-3), which is known to induce Treg cell differentiation [109]. Several studies have shown that EGCG reduces airway inflammation by increasing Treg cells and decreasing pro-inflammatory cytokines through the downregulation of the TGF-β1 signaling pathway [110]. Moreover, the regulatory effects of flavonoids on Treg cells may be linked to the induction of aryl hydrocarbon receptor (AhR) expression, inhibition of DNA methyltransferase (DNMT), and suppression of the Notch1/Jagged1 signaling pathway [106,109,110].

Flavonoids, such as luteolin, enhance CTL activity by modulating signaling pathways like NF-κB and MAPK (Mitogen-Activated Protein Kinase), which are essential for CTL proliferation and function. They also facilitate the production of cytokines like IFN-γ (Interferon-gamma), which aids in CTL activation [111]. Additionally, flavonoids can inhibit the activation and proliferation of T helper cells, which play a key role in coordinating immune responses. They suppress signaling pathways like STAT (Signal Transducer and Activator of Transcription), particularly p-STAT3, which is involved in the differentiation of Th cells into inflammatory profiles such as Th1 or Th17. This reduction in Th activity helps to reduce inflammation and limit excessive autoimmune responses. For example, apigenin, a flavonoid, has been shown to reduce the activity of Th1 and Th17 cells, which are implicated in autoimmune conditions like lupus. Apigenin suppresses these cells by downregulating key cytokines such as IFN-γ and IL-17, and inhibiting STAT3 signaling, thereby dampening the inflammatory response [112].

16.4.1.9 Flavonoids and Autoimmune Diseases

Autoimmune diseases occur when the immune system loses the ability to distinguish between the body's own tissues and foreign invaders, resulting in self-destruction. These conditions are characterized by inflammation, loss of immune tolerance, and the production of autoantibodies [113]. T cells, particularly autoreactive CD4+ T cells, play a central role in the onset of autoimmune diseases. Normally, mechanisms such as negative selection in the thymus help eliminate these harmful T cells. However, when these processes fail, autoreactive T cells escape into the periphery, triggering systemic inflammation and causing damage to tissues and organs. Regulatory T cells (Tregs) are crucial for maintaining immune tolerance, and any dysfunction in their regulation can exacerbate autoimmune responses [114,115].

Flavonoids, a group of natural compounds found in plants, have been shown to modulate immune responses and offer potential therapeutic benefits for treating autoimmune diseases. In systemic lupus erythematosus (SLE), characterized by the activation of autoreactive CD4+ T cells and excessive autoantibody production, flavonoids like quercitrin have been found to inhibit CD4+ T-cell activation, reduce inflammatory cytokine expression, and mitigate disease progression [105]. Naringenin, another flavonoid, has demonstrated the ability to reduce autoimmune responses in lupus-prone mice by decreasing pro-inflammatory cytokines such as TNF-α, IFN-γ, and IL-6, while increasing the anti-inflammatory cytokine IL-10 [116]. Furthermore, baicalein has shown effectiveness in improving renal function in lupus nephritis by inhibiting the NLRP3 inflammasome and activating the Nrf2/HO-1 pathway [108]. Additionally, flavonoids like fisetin, found in strawberries, apples, and other plants, have shown promise in inhibiting the development of lupus nephritis through the regulation of Th17 differentiation and the CXCL signaling pathway [117].

In type 1 diabetes mellitus (T1DM), where autoimmune responses lead to the destruction of pancreatic β-cells and decreased insulin secretion, flavonoids like baicalin have been shown to protect β-cells by reducing apoptosis through the activation of the PI3K/Akt pathway and inhibition of the NF-κB pathway [118]. Similarly, icariin has been found to prevent β-cell death by inhibiting NF-κB signaling [119]. Another autoimmune disease is rheumatoid arthritis (RA), a complex condition characterized by uncontrolled inflammatory reactions to self-antigens, systemic inflammation, and chronic synovitis [120]. Quercetin was found to reduce inflammation by suppressing neutrophil activity and inducing apoptosis in activated neutrophils [121]. Moreover, genkwanin, a flavonoid of *Daphne genkwa,* decreased the survival of fibroblast-like synoviocytes, which are important pathogenic characteristics of RA, and significantly inhibited the production of IL-6 and NO [122].

16.4.1.10 Anticancer Activities of Flavonoids

Cancer is a disease defined by abnormal and uncontrolled cell growth, as well as dysregulation of the cell cycle, resulting in aberrant cell proliferation with the capacity to metastasis and infect other sections of the body. It arises from genetic and molecular changes, including alterations in gene expression, which drive the progression of cancer. Various treatment strategies have been developed, including the use of drugs either as monotherapy or in combination, depending on the type, stage, and location of the cancer. However, the growing evidence of side effects associated with these drugs, along with the emergence of multidrug resistance, has prompted researchers to develop new therapeutic agents, including those based on natural compounds [123–125]. Flavonoids, secondary metabolites found in plants, have shown promising effects against various degenerative diseases, including several types of cancer. Literature reports indicate that flavonoids can inhibit cancer cell growth through multiple mechanisms, such as suppressing proliferation, migration, invasion, and angiogenesis, inducing apoptotic and autophagic cell death, and modulating ROS-scavenging enzyme activities, which may help overcome chemoresistance. Flavonoids have also been shown to scavenge free radicals, reduce oxidative stress, and regulate cellular metabolism [123].

A study by Kaushik et al. [126] reported that genistein can synergistically enhance the effects of centchroman in reducing breast cancer cell growth, both in vitro and in vivo. The mechanism of action is believed to involve the increasing of Bax expression and downmodulation of anti-apoptotic proteins, like Bcl-2 and Bcl-xL. The combination also leads to decreased phosphorylation of PI3K, Akt, and NF-κB. Moreover, another flavonoid, EGCG, has been proven for its chemopreventive and apoptotic activity in chronic myeloid leukemia and glioblastoma cells. These effects are mediated through modulation of the Abl/Bcr-mediated p38-JAK2/STAT3/Akt and MAPK/JNK signaling pathways, respectively [127]. EGCG also inhibits cancer cell growth by interfering with the EGFR signaling pathway, binding to EGFR receptors, and inducing blockade. A study by Filippi et al. showed that pretreatment with 100 and 200 μM EGCG partially inhibited EGFR phosphorylation in an epidermoid carcinoma cell line [128].

At the onset of cancer, cancer cells create ways to avoid apoptosis by upregulation of anti-apoptotic proteins such as Bcl-2 and Bcl-xL. Several flavonoids counteract this resistance to programmed cell death. Hesperetin, for example, has been reported to reduce the Bax-to-Bcl-2 ratio, thereby promoting apoptosis in various cancer cell lines, including Eca109, HT-29, MCF-7, and MDA-MB-231 [125]. Other flavonoid compounds have also demonstrated anticancer effects through diverse mechanisms. Quercetin, for instance, inhibits matrix metallopeptidases MMP2 and MMP9 in human glioblastoma cells and downregulates the phosphorylation of Akt and PDK1. This suppression leads to reduced phosphorylation of downstream survival factors such as Bad, GSK-3β, mTOR, and NF-κB in Dalton's lymphoma-bearing mice [129].

16.4.2 Communicable Disease

Recent evidence suggests that flavonoids possess significant therapeutic potential against communicable diseases by inhibiting viral, bacterial (including resistant strains), and fungal pathogens associated with human diseases.

16.4.2.1 Antibacterial Activity of Flavonoids

Many plants that produce flavonoids exhibit strong antibacterial activity, often as a defense mechanism against plant-associated pathogens, and this activity can extend to human pathogens as well [130]. Plant-derived flavonoids display antibacterial effects through diverse mechanisms, which are distinct from those of conventional drugs. Notably, these mechanisms reduce the likelihood of bacteria developing resistance, as natural compounds are typically not encoded by resistance genes, an important factor in improving antibacterial therapies.

16.4.2.1.1 Inhibition of DNA Gyrase

DNA gyrase inhibition is a well-established mechanism for antibacterial agents. Conventional drugs such as quinolones (e.g., levofloxacin, ciprofloxacin, moxifloxacin, norfloxacin, nalidixic acid, and ofloxacin) target this enzyme by stabilizing the DNA gyrase cleavage complex, which inhibits the replication fork and slows bacterial growth. Several flavonoids, such as quercetin and genistein, have been demonstrated to influence DNA gyrase activity. For instance, genistein stabilizes the topoisomerase II-DNA cleavage complex, leading to inhibition of *Vibrio harveyi* growth. Additionally, ellagic acid has been identified as a promising inhibitor of *Escherichia coli* DNA gyrase, and quercetin, apigenin, and 3,6,7,3',4'-pentahydroxyflavone have been shown to effectively inhibit the enzyme in *E. coli* with IC50 values of 47, 67.6, and 55 μg/mL, respectively [130,131]. Quercetin binds to the ATP-binding pocket of GyrB with a dissociation constant (KD) value of 15 μM, inhibiting the ATPase activity of GyrB [132]. Another study reported that apigenin, a flavonoid, effectively inhibits DNA gyrase harboring the S84L mutation in gyrA [4].

The inhibitory activity of flavonoids on DNA gyrase appears to be influenced by their structural characteristics. Structure-activity relationships (SAR) are crucial for understanding how flavonoids exert their antibacterial effects. Wu et al. [133] demonstrated that the presence of hydroxyl groups at C3' in quercetin and kaempferol, as well as at C5' and C6' in myricetin, quercetin, and baicalein, decreases their inhibitory activity on DNA gyrase. In contrast, lipophilic groups (e.g., OCH3) at positions C6 or C8 enhance antibacterial activity. Furthermore, hydroxylation at C5 increases activity, as observed with baicalein, whereas a methoxy group at the same position decreases activity, as seen in 5,6,7,4-tetramethoxyflavone. The authors concluded that the presence of 3-hydroxyl, 5-hydroxyl, 7-hydroxyl, and 4-carbonyl groups plays a crucial role in the interaction of flavonoids with key residues of GyrB, significantly enhancing their antibacterial efficacy against *E. coli*.

16.4.2.1.2 Inhibition of Bacterial Motility

Several flavonoids have been shown to inhibit bacterial swarming and twitching motility, thus preventing bacterial adhesion and colonization [130]. Swarming and twitching motility refer to bacterial translocation, both dependent and independent of flagella, into tissues [130]. Faleye et al. [134] demonstrated that flavonoid-derived compounds, 3,2-dihydroxyflavone (3,2-DHF) and 2,2-dihydroxy-4-methoxybenzophenone (DHMB), exhibit strong activity against *Vibrio* planktonic cells, with MICs of 20 and 50 μg/mL, respectively. Both compounds disrupt pre-formed biofilms and suppress virulence factors, including motility, cell hydrophobicity, and aggregation. Additionally, quercetin has been reported to decrease biofilm formation by 95% and inhibit twitching motility in *Pseudomonas aeruginosa*, with an IC50 value of 0.085 μg/mL [135]. A recent study by Tao et al. [136] demonstrated that an extract from *Potentilla kleiniana* Wight et Arn, rich in flavonoids, significantly reduced the swimming ability of *P. aeruginosa* by 77.21% ± 5.39% and 52.84% ± 4.86% at concentrations of 1600 and 800 μg/mL, respectively. The extract also inhibited autoaggregation of *P. aeruginosa* by 56.23% ± 4.82% and 34.69% ± 2.47% at the same concentrations.

16.4.2.1.3 Inhibition of MRSA

Staphylococcus aureus is highly virulent pathogen responsible for various infections, including food poisoning, skin and soft tissue infections, and sepsis. The emergence and rapid spread of multidrug-resistant *S. aureus*, particularly methicillin-resistant *S. aureus* (MRSA), have significantly worsened the control of *S. aureus*-associated infections, leading to an increase in mortality rates from infection-related diseases. Growing evidence suggests that flavonoids hold promise in combating MRSA-related infections [137]. Tao et al. [138] reported that *Potentilla kleiniana* Wight, rich in flavonoids, exhibited a satisfactory inhibitory effect on MRSA, with a MIC of 20 μg/mL. This inhibitory effect was attributed to multiple mechanisms, including the inhibition of autoaggregation, reduction of α-hemolysin-induced hemolysis, and decreased α-toxin protein levels. The plant extract also reduced surface hydrophobicity, extracellular polymeric substance (EPS) production, and adhesion-related gene expression, leading to significant inhibition of aggregation and biofilm maturation in MRSA. Other studies have shown that glabrol compounds, at concentrations of 4–16 μg/mL, effectively kill MRSA and disinfect cooking utensils within 1–3 h of exposure, both in vitro and in vivo [137].

16.4.2.1.4 Membrane Disruption

The plasma membrane is essential for bacterial growth, playing a crucial role in osmoregulation, respiration, transport, peptidoglycan biosynthesis, and lipid biosynthesis. Disruption of the membrane can cause metabolic malfunction and lead to bacterial death. Flavonoids have been suggested to destabilize and disrupt the bacterial cell membrane, a mechanism supported by several flavonoid-derived compounds. For instance, catechins are hypothesized to cause membrane disruption by inserting between lipid bilayers, increasing lateral expansion, and decreasing the barrier capacity of the bacterial cell membrane [139]. Other studies indicate that flavonoids can break down bacterial membranes by binding to the lipid bilayer and inhibiting intracellular and extracellular enzyme synthesis [140].

16.4.2.1.5 Inhibition of Electron Transport Chain and ATP Synthesis

Microorganisms require membrane potential for survival, which is essential for energy production in the form of ATP. This process involves the electron transport chain (ETC), supported by a series of enzyme complexes (Complex I to Complex IV). These complexes transfer electrons from electron donors to electron acceptors through redox reactions, resulting in proton transfer across the membrane to create an electrochemical proton gradient, which drives ATP synthesis. Complex I of the ETC is a major source of superoxide production in mitochondria. Interfering with this pathway disrupts bacterial processes, leading to cell death. Therefore, targeting the ETC and ATP synthesis has been identified as an effective approach to combat bacterial infections. For instance, Bedaquiline is an approved ATP synthase inhibitor used against *Mycobacterium tuberculosis*, a microorganism that relies on ATP synthase for growth, and oligomycin is a specific ATP synthase inhibitor [141].

Recent evidence has highlighted the potential of natural products in targeting this mechanism. Several studies have reported that flavonoids such as quercetin, epicatechin, kaempferol, and apigenin can affect the ETC by interacting with Complex I and reducing cytochrome c, ultimately leading to bacterial death [130,142,143]. Additionally, quercetin and cirsiliol have been identified as antioxidant agents due to their ability to inhibit the F1Fo ATP synthase, which is responsible for ATP production in the bacterial lifecycle. Both compounds can bind to a common site on the inner surface of the N-terminal domains of the α and β subunits of the F1 moiety, blocking the rotation of the gamma subunit of ATP synthase [144]. A study by Ravera et al. analyzed the effects of these compounds on ATP synthase activity in clinical isolates of MRSA and MRSE. The results demonstrated that cirsiliol and quercetin (100 μM) significantly inhibited ATP synthesis in both MRSA and MRSE, even more effectively than oligomycin at 50 μM. The inhibition was dose-dependent, with maximum effect observed as early as 0.5 h after incubation, leading to a reduction in ATP production [145].

16.4.2.2 Flavonoids as Antiviral Agents

Recent studies have identified flavonoids as promising antiviral candidates due to their ability to affect various stages of the viral lifecycle. These compounds can target a broad range of viruses, including influenza, dengue virus type-2, SARS-CoV-2, Chikungunya virus (CHIKV), hepatitis C virus (HCV), HIV-1, and enterovirus A71. Their antiviral effects occur through several mechanisms, such as inhibiting viral attachment and entry into host cells, interfering with viral DNA replication, and disrupting protein translation and polyprotein processing. Additionally, flavonoids have immunomodulatory properties that can help reduce viral load [146].

One such flavonoid, epigallocatechin (–)-EGC, found in green tea, has been extensively researched and proven to inhibit influenza viruses. This compound can block viral replication during the early stages of infection for both influenza A and B. Moreover, EGC exhibits broad-spectrum antiviral activity, inhibiting herpes simplex virus (HSV), adenovirus (Adenoviridae), human papillomavirus (HPV), and hepatitis B virus (HBV). Beyond DNA viruses, EGC also targets (+)-RNA viruses such as hepatitis C virus (HCV), Zika virus (ZIKV), dengue virus (DENV), West Nile virus (WNV), Chikungunya virus (CHIKV), and Porcine Reproductive and Respiratory Syndrome virus (PRRS; Atteriviridae), as well as (-)-RNA viruses like human immunodeficiency virus (HIV) and Ebola virus (EBOV) [147].

16.4.2.3 Flavonoids as Antifungal Agents

Fungi are eukaryotic organisms with structures similar to mammalian cells, including a cell nucleus, DNA, plasma membrane, cytoplasm, mitochondria, endoplasmic reticulum, and Golgi apparatus. These characteristics make fungal cells potential drug targets [148]. Baicalein, isolated from *Scutellaria baicalensis*, has shown significant inhibitory effects against *Candida* species, particularly when combined with fluconazole, yielding an MIC of 64 μg/mL. The inhibitory effect of baicalein is attributed to its ability to alter mitochondrial membrane potential, decrease intracellular ROS, and modulate redox-related genes, ultimately inducing apoptosis in the fungal cells [148,149].

Other flavonoid compounds have been reported to inhibit cell wall formation in various fungal species. For instance, glabridin, an isoflavone isolated from *Glycyrrhiza glabra*, has been shown to disrupt the cell wall integrity of *Candida albicans*, *Candida tropicalis*, *Cryptococcus neoformans*, and *Candida glabrata*, leading to inhibited fungal growth. Moreover, these compounds have been found to increase the expression of apoptosis-related genes in *C. glabrata*. Flavonoids exert antifungal activity through several mechanisms, including inducing mitochondrial dysfunction, inhibiting fungal cell wall formation, and interfering with RNA/DNA and protein synthesis [148].

16.5 PHARMACOKINETIS AND INTERACTION OF FLAVONOIDS

16.5.1 Absorption

Plant secondary metabolites, including flavonoids, generally provide therapeutic effects only if they are adequately absorbed after oral administration. Enhanced absorption leads to higher bioavailability, which in turn increases the compound's therapeutic potential. Flavonoids are typically absorbed in the colon via proton-coupled active transport and passive transcellular diffusion. However, flavonoids belong to the Biopharmaceutical Classification System (BCS) Class 3, characterized by high solubility but low permeability. This low permeability is partly because flavonoids often exist as glycosides, which are highly water-soluble and thus unable to diffuse easily across cellular membranes. For effective absorption, they must first undergo deglycosylation to their more hydrophobic aglycone forms, which can be passively absorbed through diffusion in colon cells. The deglycosylation process is assisted by key hydrolase enzymes such as β-glucosidases, produced by colonic microflora. Other enzymes produced by the approximately 400 species of human intestinal microflora, such as α-arabinofuranosidase, α-fucosidase, β-fucosidase, β-glucosidase, β-glucuronidase, and α-rhamnosidase, also contribute to the hydrolysis of flavonoid glycosides [150,151].

Zhang and colleagues [152] compared the cellular uptake and transport efficiencies of two flavonoids, cyanidin/cyanidin 3-O-glucoside (cyanidin 3,5-diglucoside) and quercetin/quercetin 3-β-D-glucoside, using Caco-2 BBe1 cells. Their results showed that the aglycone forms had higher concentrations in human intestinal cells and basolateral compartments, indicating more efficient absorption compared to their respective glycosides. Nevertheless, some flavonoid glycosides can still be detected in intestinal cells and the basolateral side. This may be because certain flavonoid compounds, like anthocyanidins from red fruits, can be absorbed directly and appear in plasma within 30 min of ingestion [153]. This direct absorption may occur via active transport mechanisms in the intestinal epithelium [152]. For example, quercetin 4′-β-glucoside and quercetin-3-glucoside can be absorbed into intestinal epithelial cells via the sodium-glucose co-transporter 1 (SGLT-1). Additionally, polar substances are more easily absorbed at low pH through solute carrier (SLC) proteins, such as organic anion transport polypeptides (OATPs), expressed in the intestinal epithelium. Quercetin, in particular, is a high-affinity substrate for OATP-B, facilitating its uptake at lower pH values. At higher pH, quercetin can also be absorbed through passive diffusion [154].

The absorption of flavonoids is influenced by whether they are in their glycosidic or aglycone form. Some glycosides are absorbed via SGLT1 or are first hydrolyzed by lactase phlorizin hydrolase (LPH) before being absorbed as aglycones, which are more easily absorbed in the small intestine. Moreover, before entering the bloodstream, flavonoids undergo significant phase II metabolism, including glucuronidation by uridine diphosphate glucuronosyltransferases (UDPGT) and subsequent sulfation by sulfotransferase (SULT) in both enterocytes and hepatocytes [155].

Other factors can influence flavonoid absorption, including the presence of transporters such as P-glycoproteins and conditions related to multidrug resistance. The transcellular passage of flavonoids across the epithelium is facilitated by ATP-binding cassette (ABC) transporters, such as P-glycoproteins, multidrug resistance proteins, and breast cancer resistance proteins. These transporters can efflux bioactive substances back into the intestinal lumen, reducing bioavailability, or into the basolateral membrane for absorption into the portal circulation. Paracellular transport has also been observed, and its contribution to overall bioavailability may depend on the dosage and isoform [156].

Another important factor in absorption is lipid solubility. When flavonoids, such as quercetin, are ingested, their absorption improves when the sugar component is hydrolyzed, increasing their fat solubility. Consequently, dietary fat can enhance absorption. In a crossover study by Guo et al. [157], participants ingested muffins that were fat-free, low-fat, or high-fat, along with a quercetin aglycone supplement. The results showed that after the low-fat trial, total plasma quercetin levels increased by 12%, whereas after the high-fat trial, they increased by 45% ($p<0.05$). The enhanced absorption in the high-fat trial was attributed to increased incorporation into micelles, which are fat-soluble droplets that facilitate absorption in the small intestine.

16.5.2 Distribution

After passing through the intestinal barrier, flavonoids are transported into the bloodstream and distributed throughout the body, including to their target tissues. The biological activity of flavonoids can be influenced by their ability to bind to plasma proteins, which serve as carriers and reservoirs for these compounds. Among plasma proteins, human serum albumin (HSA) is the most abundant and plays a crucial role in binding flavonoids. This protein binding can significantly affect the pharmacokinetics of flavonoids, as the bound fraction may be less available to exert biological effects once in circulation. The extent of binding to plasma proteins is influenced by the structure of the flavonoids. For instance, the presence of sugar groups on flavonoids reduces their binding affinity to albumin. Studies have shown that the binding constant of flavonoids to bovine serum albumin (BSA) decreases by 1–3-fold depending on the site and class of sugar. For example, a study comparing the binding constants of quercetin-BSA with its derivatives rutin-BSA and isoquercitrin-BSA found that the binding constants for quercetin were 11.97 and 7.10 times lower, respectively. This decrease in binding affinity was attributed to the presence of a sugar group at the C-3 position of quercetin, which weakens its affinity for BSA. Additionally, it was found that the binding constant of quercetin to HSA was significantly higher than that of rutin, reaching 478.6 times greater [151,158,159].

Several studies have shown that flavonoids are widely distributed across various organs, including the gastrointestinal tract. Sun et al. [160] observed that pinostrobin was most abundantly present in the stomach area 2 h after ingestion. Other compartments where flavonoids are deposited include the small intestine, large intestine, liver, kidney, heart, lung, and spleen, in decreasing order of deposition. In contrast to other flavonoid compounds, baicalein was found to accumulate more in the lungs than in the kidneys or liver, while wogonin showed higher deposition in the liver than in the lungs and kidneys [161]. Kaempferol, for example, is metabolized in the kidneys, resulting in the production of ubiquinone within kidney cells. Furthermore, the concentration of unbound kaempferol in urine was found to be significantly lower than in the blood, suggesting that the kidneys metabolize part of the aglycone before it is excreted [20].

16.5.3 Metabolism

Flavonoid metabolism begins in the small intestine and continues in the liver, involving both phase I and phase II metabolic processes, including hydrolysis, oxidation, reduction, glucuronidation, methylation, and sulfation [162]. In the small intestine, flavonoids undergo extensive phase I and II metabolism, particularly hydrolysis, glucuronidation, sulfation, and methylation, which significantly affects their absorption and bioavailability. Approximately 90% of unabsorbed flavonoids undergo this metabolic process. Several studies have supported this finding. For instance, methylation of the hydroxyl groups on flavonols has been shown to significantly enhance their passage through intestinal cell models. Furthermore, the degree of glycosylation and the type of glycoside on flavonoids have been reported to markedly influence their bioavailability compared to their aglycone counterparts. In vivo studies have demonstrated that quercetin-3-O-glucoside results in higher plasma levels of quercetin metabolites compared to rutin (quercetin-3-O-glucorhamnoside), which remains almost undetectable in plasma up to 3 h post-administration. This delayed peak in plasma concentration is due to the deglycosylation of rutin by gut microbiota in the duodenum, as well as the absence of rhamnosidase in the intestinal epithelium [156,163]. Extensive phase II metabolism, particularly glucuronidation, is often associated with poor bioavailability of flavonoids. Approximately 80% of flavonoids undergo glucuronidation by several isoforms of UDP-glucuronosyltransferase (UGT), a process that further reduces their bioavailability [164,165]. Besides glucuronidation, sulfation by sulfotransferases (SULTs) can lead to the efflux of flavonoids back into the intestinal lumen, thus inhibiting absorption and decreasing bioavailability.

After metabolism in the intestines, the resulting metabolites are transported to the liver via the portal vein, where further phase I and II biotransformation occurs. In the liver, phase I metabolism involves oxidation through cytochrome P450 (CYP450) enzymes, contributing to a minor but significant clearance of flavonoids. Various CYP isoforms, such as CYP2C9, CYP1A3, CYP1A2, and CYP1A1, have been identified in the oxidation of flavonoids like galangin. Additionally, CYP1A1, CYP2B, and CYP2E are involved in the metabolism of apigenin, producing three monohydroxylated derivatives [162]. Another CYP isoform, CYP2A6, is reported to play a role in hydroxylation of flavone at the 4′- and 3′-positions, producing OHFM6 (4′OHF) and OHFM9 (3′OHF) as well as flavanone 2′-hydroxylation [166,167]. Phase 2 metabolism, such as sulfation and glucoronidation of flavonoid glycosides, are commonly performed to improve their molecular weight and solubility [151]. Both the intestine and liver microsomes are involved in glucuronidation, with the UGT1 and UGT2 families being the primary enzymes responsible. UGT1A and UGT2B isoforms are particularly efficient in glucuronidating a variety of endogenous and exogenous compounds, including flavonoids. For instance, metabolism of baicalein and baicalin in rat liver microsomes results in the production of baicalin, baicalein 7-O-glucuronide, and baicalein 6,7-di-O-glycopyranuronoside [159].

Interestingly, a study suggests that hepatic glucuronidation is more active than intestinal glucuronidation [168]. Kaempferol, for example, undergoes extensive metabolism in the liver, where it is conjugated to methyl, sulfate, or glucuronide forms before entering the bloodstream [20]. In addition, approximately 93% of quercetin aglycone or its derivatives are metabolized in the intestine. Through enzymes such as UGT, catechol O-methyltransferases (COMT), and SULTs, quercetin undergoes effective glucuronidation in enterocytes. The conjugated quercetin metabolites, including quercetin 3-O-glucuronide and quercetin 3′-O-sulfate, are the primary metabolites that enter the bloodstream. These metabolites are transported from the enterocyte to the liver via the portal vein. Notably, there have been no reports of enzymatic reactions of quercetin metabolites in plasma [169].

16.5.4 Excretion

Flavonoids are predominantly excreted through urine as their metabolic products, with some studies also reporting excretion through feces, particularly for anthocyanin metabolites. Approximately 60%–80% of anthocyanin metabolites are excreted via urine, while the remaining portion is eliminated through feces [153]. A study conducted by Chomphen et al. [155] investigated the excretion of flavonoid metabolites in urine over different time periods following the ingestion of flavonoid-rich diets. The results demonstrated that flavonoids are excreted in various conjugated forms. For instance, compounds such as kaempferol and quercetin were excreted as both glucuronide and sulfate conjugates. Quercetin metabolites found in urine include quercetin-diglucuronide, quercetin-3′-glucuronide, isorhamnetin-glucuronide, quercetin-glucuronide sulfate, and quercetin-methyl diglucuronide [170].

In contrast, kaempferol is found to excreted in urine as kaempferol-3-glucuronide, which is the primary metabolite found, alongside kaempferol monosulfate and disulfates [20]. For other flavonoids like luteolin and myricetin, sulfate conjugates were the sole form detected. Glucuronide metabolites were identified between 6 and 8 h after ingestion, whereas sulfate metabolites appeared earlier, within 4–6 h [155].

16.5.5 Interactions

Flavonoids can interact with various compounds, including drugs, leading to changes in the pharmacokinetics, pharmacodynamics, and overall pharmaceutical profile of both the flavonoid and the interacting drug. These interactions can significantly affect the bioavailability of pharmaceuticals. Flavonoids are known to both stimulate and inhibit cytochrome P450 (CYP450) enzymes, which play a crucial role in drug metabolism within the liver and gastrointestinal system. Additionally, flavonoids can alter the function of facilitative transporters and xenobiotic efflux transporters in the liver, GI tract, and other organs. Such changes in pharmacokinetics are particularly important in patients undergoing organ transplants or those on medications with narrow therapeutic indices, such as warfarin and digoxin. A study has demonstrated that flavonoids inhibit specific CYP450 enzymes. For example, CYP1A2 activity was inhibited by hesperetin, pinocembrin, chrysin, isorhamnetin, and morin; CYP2A6 activity was inhibited by apigenin, tangeretin, galangin, and isorhamnetin; and CYP2C8 activity was inhibited by chrysin, chrysin-dimethylether, and galangin. However, none of the flavonoids analyzed showed inhibition of CYP2D6. This study primarily found flavonoids to inhibit CYP1A2 and CYP2A6, with mixed-type (both reversible and irreversible) inhibition of CYP2C8 [171].

16.6 RECENT ADVANCES IN FLAVONOID-BASED MICRO AND NANOCARRIERS

Numerous studies have highlighted the therapeutic efficacy of flavonoids, including epigallocatechin-3-gallate, quercetin, genistein, apigenin, naringenin, silibinin, and kaempferol, in addressing various diseases. Despite their promising potential, the clinical application of flavonoids is hindered by several challenges, such as their low solubility, poor absorption, and rapid metabolism [172]. To overcome these limitations, technological advancements have focused on modifying the particle size of flavonoids. Microcarrier systems (ranging from 1 to 1,000 μm) and nanocarrier systems (ranging from 1 to 1,000 nm) have emerged as promising solutions. These carrier systems include polymer-based carriers, as well as metallic micro- and nanoparticles such as gold and solid lipid carriers. Such innovations aim to enhance the bioavailability and therapeutic efficacy of flavonoids in medicinal treatments [172,173].

16.6.1 Flavonoid-Based Microcarriers

The development of flavonoid-based nanocarriers has made significant strides, owing to their smaller size and enhanced penetration capabilities [174]. However, microcarrier systems continue to hold significant potential for improving the pharmacological properties of flavonoids. Studies have shown that flavonoid-based microcarriers can enhance solubility, stability, and bioavailability, making them a promising option for drug delivery.

Various techniques have been successfully applied to develop flavonoid-based microcarriers for oral delivery. For instance, Morelo et al. and Sansone et al. employed spray-drying techniques to create flavonoid microparticles, which improved both the stability and antioxidant activity of flavonoids [175,176]. A Self-Emulsifying Drug Delivery System (SEDDS) was used by Zhang et al. [173] to create tectorigenin-based microcarriers, which improved dissolution rates and oral bioavailability in mice. Similarly, Shuai et al. [177] developed a solid dispersion of tectorigenin with polyvinylpyrrolidone (PVP) and PEG4000, enhancing its oral absorption compared to pure tectorigenin.

Flavonoid-based microcarriers have also shown promise in topical and pulmonary delivery applications. Scalia et al. [178] formulated quercetin microparticles for lung administration using an oil/water emulsification technique combined with phase inversion, employing tristearin as the lipid component and phosphatidylcholine as the emulsifier. Hariyadi et al. fabricated quercetin solid lipid microparticles for lung delivery via inhalation, demonstrating their potential for direct lung administration [179]. Additionally, Vu et al. [180] developed a genistein-based microcarrier using a microemulsion technique, enhancing topical bioavailability and skin-whitening effects. Their study showed that genistein effectively inhibited UVB-induced skin carcinogenesis and reduced erythema caused by UVB exposure in human skin.

While the development of flavonoid-based microcarriers may not be advancing as rapidly as nanocarrier systems, micro-scale modifications still offer significant advantages, particularly in enhancing solubility and stability. Research has shown that flavonoid-based microcarriers can effectively improve the biological activity of flavonoids, enhancing their anti-inflammatory and analgesic properties and promoting better oral absorption. A recent innovation in flavonoid-based delivery systems is the use of MNs for transdermal drug delivery. MNs provide a promising platform for continuous transdermal drug delivery through their needle tips. For instance, quercetin-loaded MNs have demonstrated good biocompatibility and sufficient mechanical strength to pierce the skin, enabling long-term drug delivery [8].

16.6.2 Flavonoid-Based Nanocarrier

Flavonoid-based nanocarriers offer significant advantages in therapeutic applications by enhancing flavonoids' stability, bioavailability, and therapeutic efficacy. Various types of nanocarriers, including nanoparticles, liposomes, and micelles, have been demonstrated to improve the solubility, stability, and targeted delivery of these bioactive compounds [174]. These nanocarriers are specifically designed to protect flavonoids from degradation, enhance their pharmacokinetic profiles, and facilitate their accumulation at the intended site of action.

Recent research has explored the development of flavonoid-based nanocarriers, showcasing their potential in a range of biomedical applications. For instance, quercetin-loaded nanoparticles have exhibited superior anti-inflammatory and antioxidant activities compared to free quercetin, with enhanced cellular uptake and improved therapeutic efficacy [181,182]. Similarly, liposomes encapsulating curcumin have demonstrated improved solubility, stability, and targeted delivery to tumor sites, significantly enhancing anticancer effects [183,184].

Wu et al. [185] developed genistein-loaded nanoparticles that showed greater antitumor efficacy in liver cancer than both linear poly(D, L-lactide-co-glycolide)-d-α-tocopheryl polyethylene glycol 1000 succinate (PLGA-TPGS) nanoparticles and PLGA nanoparticles in both in vitro and in vivo tests. Additionally, Zarrabi et al. [184] developed biocompatible stealth nanoliposomes for delivering curcumin in cancer therapy. Their findings showed that stealth liposomes provided more controlled drug release compared to non-coated liposomes, enhancing the cytotoxicity of the drug-loaded carrier against cancer cells. These results highlight the potential of stealth nanoliposomes as a promising candidate for cancer treatment.

Flavonoid-based nanocarriers have also been applied in tissue engineering, particularly in the treatment of degenerative bone diseases. Zaghian et al. [7] reported that hesperetin, when loaded into gliadin nanoparticles, could enhance osteoconductivity. Huang et al. [186] utilized a metal-organic framework-based delivery system for baicalin, which reduced inflammation by regulating macrophage polarization from M1 to M2 in osteoarthritis therapy. This nanosystem demonstrated higher antioxidant capacity compared to free baicalin and increased baicalin accumulation at the target site, more effectively alleviating inflammatory arthritis in osteoarthritis mice.

Beyond healthcare, flavonoid-based nanocarriers have promising applications in functional foods. Rutin-loaded nanocarriers, for example, have been widely incorporated into functional food products such as yogurt, milk, cheese, tea, wine, orange juice, and apple juice [187]. Similarly, black rice anthocyanin nanoparticles have shown great potential in the development of nutraceuticals and functional foods. Research indicates that these nanoparticles significantly enhance the biological activity of anthocyanins. Additionally, hyaluronic acid/oxidized hyaluronic acid-coated black rice anthocyanin nanoparticles could serve as effective carrier systems in nutraceutical and functional food applications [188].

REFERENCES

1. Roy A, Khan A, Ahmad I, Alghamdi S, Rajab BS, Babalghith AO, et al. Flavonoids a bioactive compound from medicinal plants and its therapeutic applications. *BioMed Research International*. 2022;2022:5445291.
2. Panche AN, Diwan AD, Chandra SR. Flavonoids: An overview. *Journal of Nutritional Science*. 2016;5:e47.
3. Falcone Ferreyra ML, Rius SP, Casati P. Flavonoids: Biosynthesis, biological functions, and biotechnological applications. *Frontiers in Plant Science*. 2012;3:222.
4. Morimoto Y, Aiba Y, Miyanaga K, Hishinuma T, Cui L, Baba T, et al. CID12261165, a flavonoid compound as antibacterial agents against quinolone-resistant *Staphylococcus aureus*. *Scientific Reports*. 2023;13(1):1725.
5. Lee JH, Kim YG, Park I, Lee J. Antifungal and antibiofilm activities of flavonoids against *Candida albicans*: Focus on 3,2′-dihydroxyflavone as a potential therapeutic agent. *Biofilm*. 2024;8:100218.
6. Wu Y, Gao H, Liu J, Liang H. Chitosan nanoparticles efficiently enhance the dispersibility, stability and selective antibacterial activity of insoluble isoflavonoids. *International Journal of Biological Macromolecules*. 2023;232:123420.
7. Zaghian M, Varshosaz J, Rostami M, Mirian M. A forsterite-reinforced polypropylene fumarate/methoxy polyethylene glycol-hydrogel enriched with flavonoid nanoparticles enhances osteoconductivity. *Materials Advances*. 2024;5(10):4324–44.
8. Yi Y, Yang Z, Zhou C, Yang Y, Wu Y, Zhang QJNT. Quercetin-encapsulated GelMa hydrogel microneedle reduces oxidative stress and facilitates wound healing. *Nano TransMed*. 2024;3:100030.
9. Liu W, Feng Y, Yu S, Fan Z, Li X, Li J, et al. The flavonoid biosynthesis network in plants. *International Journal of Molecular Sciences*. 2021;22(23):12824.
10. Liga S, Paul C, Péter F. Flavonoids: Overview of biosynthesis, biological activity, and current extraction techniques. Plants. 2023;12(14):2732.
11. Safe S, Jayaraman A, Chapkin RS, Howard M, Mohankumar K, Shrestha R. Flavonoids: Structure-function and mechanisms of action and opportunities for drug development. *Toxicological Research*. 2021;37(2):147–62.
12. Alseekh S, Perez de Souza L, Benina M, Fernie AR. The style and substance of plant flavonoid decoration; towards defining both structure and function. *Phytochemistry*. 2020;174:112347.
13. Shen N, Wang T, Gan Q, Liu S, Wang L, Jin B. Plant flavonoids: Classification, distribution, biosynthesis, and antioxidant activity. *Food Chemistry*. 2022;383:132531.
14. Nabavi SM, Šamec D, Tomczyk M, Milella L, Russo D, Habtemariam S, et al. Flavonoid biosynthetic pathways in plants: Versatile targets for metabolic engineering. *Biotechnology Advances*. 2020;38:107316.
15. Kaur R, Aslam L, Hussain S, Kapoor N, Mahajan R. Flavonoid biosynthetic pathway: Genetics and biochemistry. *Biosciences Biotechnology Research Asia*. 2021;18(2):271.
16. Rehan M. Biosynthesis of diverse class flavonoids via shikimate and phenylpropanoid pathway. In Leila Queiroz Zepka, Tatiele Casagrande do Nascimento and Eduardo Jacob-Lopes (Eds.), *Bioactive Compounds-Biosynthesis, Characterization Applications*. IntechOpen London, IntechOpen Limited, (pp. 87–104).
17. Tariq H, Asif S, Andleeb A, Hano C, Abbasi BH. Flavonoid production: Current trends in plant metabolic engineering and de novo microbial production. *Metabolites*. 2023;13(1):124.
18. Qian Y, Lynch JH, Guo L, Rhodes D, Morgan JA, Dudareva N. Completion of the cytosolic post-chorismate phenylalanine biosynthetic pathway in plants. *Nature Communications*. 2019;10(1):15.
19. Dias MC, Pinto D, Silva AMS. Plant flavonoids: Chemical characteristics and biological activity. *Molecules*. 2021;26(17):5377.

20. Chen J, Zhong H, Huang Z, Chen X, You J, Zou T. A critical review of kaempferol in intestinal health and diseases. *Antioxidants*. 2023;12(8).
21. Ouyang Y, Li J, Chen X, Fu X, Sun S, Wu Q. Chalcone derivatives: Role in anticancer therapy. *Biomolecules*. 2021;11(6):894.
22. Speisky H, Shahidi F, Costa de Camargo A, Fuentes J. Revisiting the oxidation of flavonoids: Loss, conservation or enhancement of their antioxidant properties. *Antioxidants*. 2022;11(1):133.
23. Gahlawat SK, Salar RK, Siwach P, Duhan JS, Kumar S, Kaur P. *Plant Biotechnology: Recent Advancements and Developments*. Springer; 2017.
24. Mutha RE, Tatiya AU, Surana SJ. Flavonoids as natural phenolic compounds and their role in therapeutics: An overview. *Future Journal of Pharmaceutical Sciences*. 2021;7(1):25.
25. Wu J, Wang XC, Liu Y, Du H, Shu QY, Su S, et al. Flavone synthases from *Lonicera japonica* and *L. macranthoides* reveal differential flavone accumulation. *Scientific Reports*. 2016;6:19245.
26. Jiang N, Doseff AI, Grotewold E. Flavones: From biosynthesis to health benefits. Plants. 2016;5(2):27.
27. Chagas M, Behrens MD, Moragas-Tellis CJ, Penedo GXM, Silva AR, Gonçalves-de-Albuquerque CF. Flavonols and flavones as potential anti-inflammatory, antioxidant, and antibacterial compounds. *Oxidative Medicine and Cellular Longevity*. 2022;2022:9966750.
28. Hostetler GL, Ralston RA, Schwartz SJ. Flavones: Food sources, bioavailability, metabolism, and bioactivity. *Advances in Nutrition*. 2017;8(3):423–35.
29. Song X, Diao J, Ji J, Wang G, Guan C, Jin C, et al. Molecular cloning and identification of a flavanone 3-hydroxylase gene from *Lycium chinense*, and its overexpression enhances drought stress in tobacco. *Plant Physiology and Biochemistry: PPB*. 2016;98:89–100.
30. Wang Y, Mi X, Du Y, Li S, Yu L, Gao M, et al. Design, synthesis, and anti-inflammatory activities of 12-dehydropyxinol derivatives. *Molecules*. 2023;28(3):1307.
31. Tanjung M, Hakim EH, Elfahmi, Latip J, Syah YM. Dihydroflavonol and flavonol derivatives from *Macaranga recurvata*. *Natural Product Communications*. 2012;7(10):1309–10.
32. Arifin B, Ibrahim SJJZ. Struktur, bioaktivitas dan antioksidan flavonoid. *Jurnal zarah*. 2018;6(1):21–9.
33. Krysa M, Szymańska-Chargot M, Zdunek A. FT-IR and FT-Raman fingerprints of flavonoids – A review. *Food Chemistry*. 2022;393:133430.
34. Wahnou H, Limami Y, Oudghiri M. Flavonoids and flavonoid-based nanoparticles for osteoarthritis and rheumatoid arthritis management. *Biochemical and Biophysical Research Communications*. 2024;4(1):38–61.
35. Ayoka TO, Ezema BO, Eze CN, Nnadi CO. Antioxidants for the prevention and treatment of non-communicable diseases. *Journal of Exploratory Research in Pharmacology*. 2022;7(3):179–89.
36. Yi X, Dong M, Guo N, Tian J, Lei P, Wang S, et al. Flavonoids improve type 2 diabetes mellitus and its complications: A review. *Frontiers in Nutrition*. 2023;10:1192131.
37. Chen X, Yu J, Shi J. Management of diabetes mellitus with puerarin, a natural isoflavone from *Pueraria lobata*. *The American Journal of Chinese Medicine*. 2018;46(8):1771–89.
38. Chen X, Wang L, Fan S, Song S, Min H, Wu Y, et al. Puerarin acts on the skeletal muscle to improve insulin sensitivity in diabetic rats involving μ-opioid receptor. *European Journal of Pharmacology*. 2018;818:115–23.
39. Gao J, Liu M, Guo Z, Hu C, Feng Z, Yan J. [Puerarin alleviates insulin resistance in type 2 diabetic mice by modulating fetuin B-AMPK/ACC signaling pathway in the liver]. *Nan fang yi ke da xue xue bao = Journal of Southern Medical University*. 2021;41(6):839–46.
40. Sandoval V, Sanz-Lamora H, Arias G, Marrero PF, Haro D, Relat J. Metabolic impact of flavonoids consumption in obesity: From central to peripheral. *Nutrients*. 2020;12(8):2393.
41. Kim SY, Wi H-R, Choi S, Ha TJ, Lee BW, Lee M. Inhibitory effect of anthocyanin-rich black soybean testa (*Glycine max* (L.) Merr.) on the inflammation-induced adipogenesis in a DIO mouse model. *Journal of Functional Foods* 2015;14:623–33.
42. Lee M, Sorn SR, Park Y, Park HK. Anthocyanin rich-black soybean testa improved visceral fat and plasma lipid profiles in overweight/obese Korean adults: A randomized controlled trial. *Journal of Medicinal Food*. 2016;19(11):995–1003.
43. Lim SM, Lee HS, Jung JI, Kim SM, Kim NY, Seo TS, et al. Cyanidin-3-O-galactoside-enriched *Aronia melanocarpa* extract attenuates weight gain and adipogenic pathways in high-fat diet-induced obese C57BL/6 mice. *Nutrients*. 2019;11(5):1190.
44. Solverson PM, Rumpler WV, Leger JL, Redan BW, Ferruzzi MG, Baer DJ, et al. Blackberry feeding increases fat oxidation and improves insulin sensitivity in overweight and obese males. *Nutrients*. 2018;10(8):1048.

45. Gomes JVP, Rigolon TCB, Souza M, Alvarez-Leite JI, Lucia CMD, Martino HSD, et al. Antiobesity effects of anthocyanins on mitochondrial biogenesis, inflammation, and oxidative stress: A systematic review. *Nutrition*. 2019;66:192–202.
46. Wu T, Gao Y, Guo X, Zhang M, Gong L. Blackberry and blueberry anthocyanin supplementation counteract high-fat-diet-induced obesity by alleviating oxidative stress and inflammation and accelerating energy expenditure. *Oxidative Medicine and Cellular Longevity*. 2018;2018:4051232.
47. Takikawa M, Inoue S, Horio F, Tsuda T. Dietary anthocyanin-rich bilberry extract ameliorates hyperglycemia and insulin sensitivity via activation of AMP-activated protein kinase in diabetic mice. *The Journal of Nutrition*. 2010;140(3):527–33.
48. Termkwancharoen C, Malakul W, Phetrungnapha A, Tunsophon S. Naringin ameliorates skeletal muscle atrophy and improves insulin resistance in high-fat-diet-induced insulin resistance in obese rats. *Nutrients*. 2022;14(19):4120.
49. Fan D, Alamri Y, Liu K, MacAskill M, Harris P, Brimble M, et al. Supplementation of blackcurrant anthocyanins increased cyclic glycine-proline in the cerebrospinal fluid of Parkinson patients: Potential treatment to improve insulin-like growth factor-1 function. *Nutrients*. 2018;10(6):714.
50. Li F, Gao C, Yan P, Zhang M, Wang Y, Hu Y, et al. EGCG reduces obesity and white adipose tissue gain partly through AMPK activation in mice. *Frontiers in Pharmacology*. 2018;9:1366.
51. Kwon EY, Choi MS. Dietary eriodictyol alleviates adiposity, hepatic steatosis, insulin resistance, and inflammation in diet-induced obese mice. *International Journal of Molecular Sciences*. 2019;20(5):1227.
52. Varela CE, Rodriguez A, Romero-Valdovinos M, Mendoza-Lorenzo P, Mansour C, Ceballos G, et al. Browning effects of (-)-epicatechin on adipocytes and white adipose tissue. *European Journal of Pharmacology*. 2017;811:48–59.
53. Song NJ, Chang SH, Kim S, Panic V, Jang BH, Yun UJ, et al. PI3Ka-Akt1-mediated Prdm4 induction in adipose tissue increases energy expenditure, inhibits weight gain, and improves insulin resistance in diet-induced obese mice. *Cell Death & Disease*. 2018;9(9):876.
54. Song NJ, Choi S, Rajbhandari P, Chang SH, Kim S, Vergnes L, et al. Prdm4 induction by the small molecule butein promotes white adipose tissue browning. *Nature Chemical Biology*. 2016;12(7):479–81.
55. López-Almada G, Domínguez-Avila JA, Mejía-León ME, Robles-Sánchez M, González-Aguilar GA, Salazar-López NJ. Could naringenin participate as a regulator of obesity and satiety? *Molecules*. 2023;28(3):1450.
56. Shaik Mohamed Sayed UF, Moshawih S, Goh HP, Kifli N, Gupta G, Singh SK, et al. Natural products as novel anti-obesity agents: Insights into mechanisms of action and potential for therapeutic management. *Frontiers in Pharmacology*. 2023;14:1182937.
57. Larson AJ, Symons JD, Jalili T. Therapeutic potential of quercetin to decrease blood pressure: Review of efficacy and mechanisms. *Advances in Nutrition*. 2012;3(1):39–46.
58. Zhu FQ, Hu J, Lv FH, Cheng P, Gao S. Effects of oligomeric grape seed proanthocyanidins on L-NAME-induced hypertension in pregnant mice: Role of oxidative stress and endothelial dysfunction. *Phytotherapy Research: PTR*. 2018;32(9):1836–47.
59. Ciumărnean L, Milaciu MV, Runcan O, Vesa Ş C, Răchişan AL, Negrean V, et al. The effects of flavonoids in cardiovascular diseases. *Molecules*. 2020;25(18):4320.
60. Quan J, Jia Z, Lv T, Zhang L, Liu L, Pan B, et al. Green tea extract catechin improves cardiac function in pediatric cardiomyopathy patients with diastolic dysfunction. *Journal of Biomedical Science*. 2019;26(1):32.
61. Łanoszka K, Vlčková N. Natural Sirtuin1 activators and atherosclerosis: An overview. *Current Atherosclerosis Reports*. 2023;25(12):979–94.
62. Shukla S, Gupta S. Apigenin: A promising molecule for cancer prevention. *Pharmaceutical Research*. 2010;27(6):962–78.
63. Xu Q, Li YC, Du C, Wang LN, Xiao YH. Effects of apigenin on the expression of LOX-1, Bcl-2, and bax in hyperlipidemia rats. *Chemistry & Biodiversity*. 2021;18(8):e2100049.
64. Li RL, Wang LY, Liu S, Duan HX, Zhang Q, Zhang T, et al. Natural flavonoids derived from fruits are potential agents against atherosclerosis. *Frontiers in Nutrition*. 2022;9:862277.
65. Qin W, Ren B, Wang S, Liang S, He B, Shi X, et al. Apigenin and naringenin ameliorate PKCβII-associated endothelial dysfunction via regulating ROS/caspase-3 and NO pathway in endothelial cells exposed to high glucose. *Vascular Pharmacology*. 2016;85:39–49.
66. Yamagata K, Hashiguchi K, Yamamoto H, Tagami M. Dietary apigenin reduces induction of LOX-1 and NLRP3 expression, leukocyte adhesion, and acetylated low-density lipoprotein uptake in human endothelial cells exposed to trimethylamine-N-oxide. *Journal of Cardiovascular Pharmacology*. 2019;74(6):558–65.

67. Zhang K, Song W, Li D, Jin X. Apigenin in the regulation of cholesterol metabolism and protection of blood vessels. *Experimental and Therapeutic Medicine*. 2017;13(5):1719–24.
68. Jia Q, Cao H, Shen D, Li S, Yan L, Chen C, et al. Quercetin protects against atherosclerosis by regulating the expression of PCSK9, CD36, PPARγ, LXRα and ABCA1. *International Journal of Molecular Medicine*. 2019;44(3):893–902.
69. Jiang YH, Jiang LY, Wang YC, Ma DF, Li X. Quercetin attenuates atherosclerosis via modulating oxidized LDL-induced endothelial cellular senescence. *Frontiers in Pharmacology*. 2020;11:512.
70. Flaumenhaft R, Furie B, Zwicker JI. Therapeutic implications of protein disulfide isomerase inhibition in thrombotic disease. *Arteriosclerosis, Thrombosis, and Vascular Biology*. 2015;35(1):16–23.
71. Vaiyapuri S, Roweth H, Ali MS, Unsworth AJ, Stainer AR, Flora GD, et al. Pharmacological actions of nobiletin in the modulation of platelet function. *British Journal of Pharmacology*. 2015;172(16):4133–45.
72. Han Y, Wang H, Wang Y, Dong P, Jia J, Yang S. Puerarin protects cardiomyocytes from ischemia-reperfusion injury by upregulating LncRNA ANRIL and inhibiting autophagy. *Cell and Tissue Research*. 2021;385(3):739–51.
73. Wang F, Zhao C, Tian G, Wei X, Ma Z, Cui J, et al. Naringin alleviates atherosclerosis in ApoE(-/-) mice by regulating cholesterol metabolism involved in gut microbiota remodeling. *Journal of Agricultural and Food Chemistry*. 2020;68(45):12651–60.
74. Li J, Sun M, Cui X, Li C. Protective effects of flavonoids against Alzheimer's disease: Pathological hypothesis, potential targets, and structure-activity relationship. *International Journal of Molecular Sciences*. 2022;23(17):10020.
75. Vepsäläinen S, Koivisto H, Pekkarinen E, Mäkinen P, Dobson G, McDougall GJ, et al. Anthocyanin-enriched bilberry and blackcurrant extracts modulate amyloid precursor protein processing and alleviate behavioral abnormalities in the APP/PS1 mouse model of Alzheimer's disease. *The Journal of Nutritional Biochemistry*. 2013;24(1):360–70.
76. Wang J, Ho L, Zhao W, Ono K, Rosensweig C, Chen L, et al. Grape-derived polyphenolics prevent Abeta oligomerization and attenuate cognitive deterioration in a mouse model of Alzheimer's disease. *The Journal of Neuroscience: The Official Journal of the Society for Neuroscience*. 2008;28(25):6388–92.
77. Wang YJ, Thomas P, Zhong JH, Bi FF, Kosaraju S, Pollard A, et al. Consumption of grape seed extract prevents amyloid-beta deposition and attenuates inflammation in brain of an Alzheimer's disease mouse. *Neurotoxicity Research*. 2009;15(1):3–14.
78. Augustin S, Rimbach G, Augustin K, Schliebs R, Wolffram S, Cermak R. Effect of a short- and long-term treatment with *Ginkgo biloba* extract on amyloid precursor protein levels in a transgenic mouse model relevant to Alzheimer's disease. *Archives of Biochemistry and Biophysics*. 2009;481(2):177–82.
79. Rezai-Zadeh K, Shytle D, Sun N, Mori T, Hou H, Jeanniton D, et al. Green tea epigallocatechin-3-gallate (EGCG) modulates amyloid precursor protein cleavage and reduces cerebral amyloidosis in Alzheimer transgenic mice. *The Journal of Neuroscience: The Official Journal of the Society for Neuroscience*. 2005;25(38):8807–14.
80. Ehrnhoefer DE, Bieschke J, Boeddrich A, Herbst M, Masino L, Lurz R, et al. EGCG redirects amyloidogenic polypeptides into unstructured, off-pathway oligomers. *Nature structural & Molecular Biology*. 2008;15(6):558–66.
81. Huang DS, Yu YC, Wu CH, Lin JY. Protective effects of wogonin against Alzheimer's disease by inhibition of amyloidogenic pathway. *Evidence-Based Complementary and Alternative Medicine: eCAM*. 2017;2017:3545169.
82. Martin L, Latypova X, Wilson CM, Magnaudeix A, Perrin ML, Yardin C, et al. Tau protein kinases: Involvement in Alzheimer's disease. *Ageing Research Reviews*. 2013;12(1):289–309.
83. Calderaro A, Patanè GT, Tellone E, Barreca D, Ficarra S, Misiti F, et al. The neuroprotective potentiality of flavonoids on Alzheimer's disease. *International Journal of Molecular Sciences*. 2022;23(23):14835.
84. Baptista FI, Henriques AG, Silva AM, Wiltfang J, da Cruz e Silva OA. Flavonoids as therapeutic compounds targeting key proteins involved in Alzheimer's disease. *ACS Chemical Neuroscience*. 2014;5(2):83–92.
85. Ademosun AO, Oboh G, Bello F, Ayeni PO. Antioxidative properties and effect of quercetin and its glycosylated form (rutin) on acetylcholinesterase and butyrylcholinesterase activities. *Journal of Evidence-Based Complementary & Alternative Medicine*. 2016;21(4):NP11–7.
86. Khan MT, Orhan I, Senol FS, Kartal M, Sener B, Dvorská M, et al. Cholinesterase inhibitory activities of some flavonoid derivatives and chosen xanthone and their molecular docking studies. *Chemico-Biological Interactions*. 2009;181(3):383–9.

87. Nordberg A, Ballard C, Bullock R, Darreh-Shori T, Somogyi M. A review of butyrylcholinesterase as a therapeutic target in the treatment of Alzheimer's disease. *The Primary Care Companion for CNS Disorders.* 2013;15(2):26731.
88. Guillozet AL, Smiley JF, Mash DC, Mesulam MM. Butyrylcholinesterase in the life cycle of amyloid plaques. *Annals of Neurology.* 1997;42(6):909–18.
89. Mehta M, Adem A, Sabbagh M. New acetylcholinesterase inhibitors for Alzheimer's disease. *International Journal of Alzheimer's Disease.* 2012;2012(1):728983.
90. Darvesh S, Cash MK, Reid GA, Martin E, Mitnitski A, Geula C. Butyrylcholinesterase is associated with β-amyloid plaques in the transgenic APPSWE/PSEN1dE9 mouse model of Alzheimer disease. *Journal of Neuropathology and Experimental Neurology.* 2012;71(1):2–14.
91. Cárdenas C, Quesada AR, Medina MA. Anti-angiogenic and anti-inflammatory properties of kahweol, a coffee diterpene. *PLoS One.* 2011;6(8):e23407.
92. Middleton E, Jr., Kandaswami C, Theoharides TC. The effects of plant flavonoids on mammalian cells: Implications for inflammation, heart disease, and cancer. *Pharmacological Reviews.* 2000;52(4):673–751.
93. Ravishankar D, Rajora AK, Greco F, Osborn HM. Flavonoids as prospective compounds for anti-cancer therapy. *The International Journal of Biochemistry & Cell Biology.* 2013;45(12):2821–31.
94. Al-Khayri JM, Sahana GR, Nagella P, Joseph BV, Alessa FM, Al-Mssallem MQ. Flavonoids as potential anti-inflammatory molecules: A review. *Molecules.* 2022;27(9).
95. Tanaka S, Furuya K, Yamamoto K, Yamada K, Ichikawa M, Suda M, et al. Procyanidin B2 gallates inhibit IFN-γ and IL-17 production in T cells by suppressing T-bet and RORγt expression. *International Immunopharmacology.* 2017;44:87–96.
96. Burkard M, Leischner C, Lauer UM, Busch C, Venturelli S, Frank J. Dietary flavonoids and modulation of natural killer cells: Implications in malignant and viral diseases. *The Journal of Nutritional Biochemistry.* 2017;46:1–12.
97. Han R, Wu WQ, Wu XP, Liu CY. Effect of total flavonoids from the seeds of *Astragali complanati* on natural killer cell function. *Journal of Ethnopharmacology.* 2015;173:157–65.
98. Hou D, Wang D, Ma X, Chen W, Guo S, Guan H. Effects of total flavonoids of sea buckthorn (*Hippophae rhamnoides* L.) on cytotoxicity of NK92-MI cells. *International Journal of Immunopathology and Pharmacology.* 2017;30(4):353–61.
99. Han L, Fu Q, Deng C, Luo L, Xiang T, Zhao H. Immunomodulatory potential of flavonoids for the treatment of autoimmune diseases and tumour. *Scandinavian Journal of Immunology.* 2022;95(1):e13106.
100. Bardi GT, Smith MA, Hood JL. Melanoma exosomes promote mixed M1 and M2 macrophage polarization. *Cytokine.* 2018;105:63–72.
101. Feng L, Song P, Zhou H, Li A, Ma Y, Zhang X, et al. Pentamethoxyflavanone regulates macrophage polarization and ameliorates sepsis in mice. *Biochemical Pharmacology.* 2014;89(1):109–18.
102. Strizova Z, Benesova I, Bartolini R, Novysedlak R, Cecrdlova E, Foley LK, et al. M1/M2 macrophages and their overlaps – Myth or reality? *Clinical Science.* 2023;137(15):1067–93.
103. Feng LL, Xu LS, Guo MM, Huang W, Zhu JZ, Kong LD, et al. 5, 7, 2′, 4′, 5′-Pentamethoxyflavanone regulates M1/M2 macrophage phenotype and protects the septic mice. *Chinese Journal of Natural Medicines.* 2019;17(5):363–71.
104. Lai YS, Putra R, Aui SP, Chang KT. M2(C) polarization by baicalin enhances efferocytosis via upregulation of MERTK receptor. *The American Journal of Chinese Medicine.* 2018;46(8):1899–914.
105. Schmidt A, Éliás S, Joshi RN, Tegnér J. *In vitro* differentiation of human CD4+ FOXP3+ induced regulatory T cells (iTregs) from Naïve CD4+ T cells using a TGF-β-containing protocol. *Journal of Visualized Experiments: JoVE.* 2016;30(118):55015.
106. Zhu W, Chen X, Yu J, Xiao Y, Li Y, Wan S, et al. Baicalin modulates the Treg/Teff balance to alleviate uveitis by activating the aryl hydrocarbon receptor. *Biochemical Pharmacology.* 2018;154:18–27.
107. Shan L, Kang X, Liu F, Cai X, Han X, Shang Y. Epigallocatechin gallate improves airway inflammation through TGF-β1 signaling pathway in asthmatic mice. *Molecular Medicine Reports.* 2018;18(2):2088–96.
108. Yang J, Yang X, Yang J, Li M. Baicalin ameliorates lupus autoimmunity by inhibiting differentiation of Tfh cells and inducing expansion of Tfr cells. *Cell Death & Disease.* 2019;10(2):140.
109. Lv Q, Shi C, Qiao S, Cao N, Guan C, Dai Y, et al. Alpinetin exerts anti-colitis efficacy by activating AhR, regulating miR-302/DNMT-1/CREB signals, and therefore promoting Treg differentiation. *Cell Death & Disease.* 2018;9(9):890.
110. Xu S, Kong YG, Jiao WE, Yang R, Qiao YL, Xu Y, et al. Tangeretin promotes regulatory T cell differentiation by inhibiting Notch1/Jagged1 signaling in allergic rhinitis. *International Immunopharmacology.* 2019;72:402–12.
111. Huang L, Kim MY, Cho JY. Immunopharmacological activities of luteolin in chronic diseases. *International Journal of Molecular Sciences.* 2023;24(3):1953.

112. Hosseinzade A, Sadeghi O, Naghdipour Biregani A, Soukhtehzari S, Brandt GS, Esmaillzadeh A. Immunomodulatory effects of flavonoids: Possible induction of T CD4+ regulatory cells through suppression of mTOR pathway signaling activity. *Frontiers in Immunology*. 2019;10:51.
113. Kumar BV, Connors TJ, Farber DL. Human T cell development, localization, and function throughout life. *Immunity*. 2018;48(2):202–13.
114. Richards DM, Delacher M, Goldfarb Y, Kägebein D, Hofer A-C, Abramson J, et al. Treg cell differentiation: From thymus to peripheral tissue. *Progress In Molecular Biology and Translational Science*. 2015;136:175–205.
115. Salinas GF, Braza F, Brouard S, Tak P-P, Baeten D. The role of B lymphocytes in the progression from autoimmunity to autoimmune disease. *Clinical Immunology*. 2013;146(1):34–45.
116. Abrego-Peredo A, Romero-Ramírez H, Espinosa E, López-Herrera G, García-García F, Flores-Muñoz M, et al. Naringenin mitigates autoimmune features in lupus-prone mice by modulation of T-cell subsets and cytokines profile. *PLoS One*. 2020;15(5):e0233138.
117. Xu SP, Li YS. Fisetin inhibits pristine-induced systemic lupus erythematosus in a murine model through CXCLs regulation. *International Journal of Molecular Medicine*. 2018;42(6):3220–30.
118. You W, Wang K, Yu C, Song L. Baicalin prevents tumor necrosis factor-α-induced apoptosis and dysfunction of pancreatic β-cell line Min6 via upregulation of miR-205. *Journal of Cellular Biochemistry*. 2018;119(10):8547–54.
119. Zhong S, Ge J, Yu JY. Icariin prevents cytokine-induced β-cell death by inhibiting NF-κB signaling. *Experimental and Therapeutic Medicine*. 2018;16(3):2756–62.
120. Scott DL, Wolfe F, Huizinga TW. Rheumatoid arthritis. *Lancet*. 2010;376(9746):1094–108.
121. Yuan K, Zhu Q, Lu Q, Jiang H, Zhu M, Li X, et al. Quercetin alleviates rheumatoid arthritis by inhibiting neutrophil inflammatory activities. The *Journal of Nutritional Biochemistry*. 2020;84:108454.
122. Sun YW, Bao Y, Yu H, Chen QJ, Lu F, Zhai S, et al. Anti-rheumatoid arthritis effects of flavonoids from *Daphne genkwa*. *International Immunopharmacology*. 2020;83:106384.
123. Mir SA, Dar A, Hamid L, Nisar N, Malik JA, Ali T, et al. Flavonoids as promising molecules in the cancer therapy: An insight. *Current Research in Pharmacology and Drug Discovery*. 2024;6:100167.
124. Brown JS, Amend SR, Austin RH, Gatenby RA, Hammarlund EU, Pienta KJ. Updating the definition of cancer. *Molecular Cancer Research: MCR*. 2023;21(11):1142–7.
125. Kopustinskiene DM, Jakstas V, Savickas A, Bernatoniene J. Flavonoids as anticancer agents. *Nutrients*. 2020;12(2):457.
126. Kaushik S, Shyam H, Agarwal S, Sharma R, Nag TC, Dwivedi AK, et al. Genistein potentiates Centchroman induced antineoplasticity in breast cancer *via* PI3K/Akt deactivation and ROS dependent induction of apoptosis. *Life Sciences*. 2019;239:117073.
127. Grube S, Ewald C, Kögler C, Lawson McLean A, Kalff R, Walter J. Achievable central nervous system concentrations of the green tea catechin EGCG induce stress in glioblastoma cells in vitro. *Nutrition and Cancer*. 2018;70(7):1145–58.
128. Filippi A, Picot T, Aanei CM, Nagy P, Szöllősi J, Campos L, et al. Epigallocatechin-3-O-gallate alleviates the malignant phenotype in A-431 epidermoid and SK-BR-3 breast cancer cell lines. *International Journal of Food Sciences and Nutrition*. 2018;69(5):584–97.
129. Maurya AK, Vinayak M. Quercetin attenuates cell survival, inflammation, and angiogenesis via modulation of AKT signaling in murine T-cell lymphoma. *Nutrition and Cancer*. 2017;69(3):470–80.
130. Biharee A, Sharma A, Kumar A, Jaitak V. Antimicrobial flavonoids as a potential substitute for overcoming antimicrobial resistance. *Fitoterapia*. 2020;146:104720.
131. Ohemeng K, Schwender C, Fu K, Barrett JJB, Letters MC. DNA gyrase inhibitory and antibacterial activity of some flavones (1). *Bioorganic & Medicinal Chemistry Letters*. 1993;3(2):225–30.
132. Verma K, Mahalapbutr P, Suriya U, Somboon T, Aiebchun T, Shi L, et al. In silico screening of DNA gyrase B potent flavonoids for the treatment of clostridium difficile infection from phytoHub database. *Brazilian Archives of Biology and Technology*. 2021;64:e21200402.
133. Wu T, Zang X, He M, Pan S, Xu X. Structure-activity relationship of flavonoids on their anti-Escherichia coli activity and inhibition of DNA gyrase. *Journal of Agricultural and Food Chemistry*. 2013;61(34):8185–90.
134. Faleye OS, Lee J-H, Lee JJB. Selected flavonoids exhibit antibiofilm and antibacterial effects against Vibrio by disrupting membrane integrity, virulence and metabolic activities. *Biofilm*. 2023;6:100165.
135. Pejin B, Ciric A, Markovic JD, Glamoclija J, Nikolic M, Stanimirovic B, et al. Quercetin potently reduces biofilm formation of the strain *Pseudomonas aeruginosa* PAO1 in vitro. *Current Pharmaceutical Biotechnology*. 2015;16(8):733–7.

136. Tao J, Yan S, Wang H, Zhao L, Zhu H, Wen Z. Antimicrobial and antibiofilm effects of total flavonoids from *Potentilla kleiniana* Wight et Arn on *Pseudomonas aeruginosa* and its potential application to stainless steel surfaces. *LWT.* 2022;154:112631.
137. Wu SC, Yang ZQ, Liu F, Peng WJ, Qu SQ, Li Q, et al. Antibacterial effect and mode of action of flavonoids from licorice against methicillin-resistant *Staphylococcus aureus. Frontiers in Microbiology.* 2019;10:2489.
138. Tao J, Yan S, Zhou C, Liu Q, Zhu H, Wen Z. Total flavonoids from Potentilla kleiniana Wight et Arn inhibits biofilm formation and virulence factors production in methicillin-resistant *Staphylococcus aureus* (MRSA). *Journal of Ethnopharmacology.* 2021;279:114383.
139. Wu M, Brown AC. Applications of catechins in the treatment of bacterial infections. Pathogens. 2021;10(5):546.
140. Górniak I, Bartoszewski R, Króliczewski J. Comprehensive review of antimicrobial activities of plant flavonoids. *Phytochemistry Reviews.* 2019;18:241–72.
141. Vestergaard M, Bald D, Ingmer H. Targeting the ATP synthase in bacterial and fungal pathogens: Beyond *Mycobacterium tuberculosis. Journal of Global Antimicrobial Resistance.* 2022;29:29–41.
142. Lagoa R, Graziani I, Lopez-Sanchez C, Garcia-Martinez V, Gutierrez-Merino C. Complex I and cytochrome c are molecular targets of flavonoids that inhibit hydrogen peroxide production by mitochondria. *Biochimica et Biophysica Acta.* 2011;1807(12):1562–72.
143. Zymone K, Benetis R, Trumbeckas D, Baseviciene I, Trumbeckaite S. Different effects of quercetin glycosides and quercetin on kidney mitochondrial function-uncoupling, cytochrome C reducing and antioxidant activity. *Molecules.* 2022;27(19):6377.
144. Gledhill JR, Montgomery MG, Leslie AG, Walker JE. Mechanism of inhibition of bovine F1-ATPase by resveratrol and related polyphenols. *Proceedings of the National Academy of Sciences of the United States of America.* 2007;104(34):13632–7.
145. Ravera S, Tancreda G, Vezzulli L, Schito AM, Panfoli I. Cirsiliol and quercetin inhibit ATP synthesis and decrease the energy balance in methicillin-resistant Staphylococcus aureus (MRSA) and methicillin-resistant Staphylococcus epidermidis (MRSE) strains isolated from patients. *Molecules.* 2023;28(17):6183.
146. Badshah SL, Faisal S, Muhammad A, Poulson BG, Emwas AH, Jaremko M. Antiviral activities of flavonoids. *Biomedicine & Pharmacotherapy = Biomedecine & Pharmacotherapie.* 2021;140:111596.
147. Kaihatsu K, Yamabe M, Ebara Y. Antiviral mechanism of action of epigallocatechin-3-O-gallate and its fatty acid esters. *Molecules.* 2018;23(10):2475.
148. Aboody MSA, Mickymaray S. Anti-fungal efficacy and mechanisms of flavonoids. *Antibiotics.* 2020;9(2):45.
149. Dai BD, Cao YY, Huang S, Xu YG, Gao PH, Wang Y, et al. Baicalein induces programmed cell death in Candida albicans. *Journal of Microbiology and Biotechnology.* 2009;19(8):803–9.
150. Murota K, Nakamura Y, Uehara M. Flavonoid metabolism: The interaction of metabolites and gut microbiota. *Bioscience, Biotechnology, and Biochemistry.* 2018;82(4):600–10.
151. Xiao J. Dietary flavonoid aglycones and their glycosides: Which show better biological significance? *Critical Reviews in Food Science and Nutrition.* 2017;57(9):1874–905.
152. Zhang H, Hassan YI, Liu R, Mats L, Yang C, Liu C, et al. Molecular mechanisms underlying the absorption of aglycone and glycosidic flavonoids in a Caco-2 BBe1 cell model. *ACS Omega.* 2020;5(19):10782–93.
153. Gonçalves AC, Nunes AR, Falcão A, Alves G, Silva LR. Dietary effects of anthocyanins in human health: A comprehensive review. Pharmaceuticals. 2021;14(7):690.
154. Hai Y, Zhang Y, Liang Y, Ma X, Qi X, Xiao J, et al. Advance on the absorption, metabolism, and efficacy exertion of quercetin and its important derivatives: Absorption, metabolism and function of quercetin. *Food Frontiers.* 2020;1(4):420–34.
155. Chomphen L, Yamanont P, Morales NP. Flavonoid metabolites in serum and urine after the ingestion of selected tropical fruits. *Nutrients.* 2024;16(1):161.
156. Cassidy A, Minihane AM. The role of metabolism (and the microbiome) in defining the clinical efficacy of dietary flavonoids. *The American Journal of Clinical Nutrition.* 2017;105(1):10–22.
157. Guo Y, Mah E, Davis CG, Jalili T, Ferruzzi MG, Chun OK, et al. Dietary fat increases quercetin bioavailability in overweight adults. *Molecular Nutrition & Food Research.* 2013;57(5):896–905.
158. Dangles O, Dufour C, Bret S, Perkin Transactions 2. Flavonol–serum albumin complexation. Two-electron oxidation of flavonols and their complexes with serum albumin. *Journal of the Chemical Society, Perkin Transactions.* 1999(4):737–44.

159. Xiao J, Mao F, Yang F, Zhao Y, Zhang C, Yamamoto K. Interaction of dietary polyphenols with bovine milk proteins: Molecular structure-affinity relationship and influencing bioactivity aspects. *Molecular Nutrition & Food Research.* 2011;55(11):1637–45.
160. Sun X, Liu X, Chen S. The pharmacokinetics, tissue distribution, metabolism, and excretion of pinostrobin in rats: Ultra-high-performance liquid chromatography coupled with linear trap quadrupole orbitrap mass spectrometry studies. *Frontiers in Pharmacology.* 2020;11:574638.
161. Hou YC, Lin SP, Tsai SY, Ko MH, Chang YC, Chao PD. Flavonoid pharmacokinetics and tissue distribution after repeated dosing of the roots of Scutellaria baicalensis in rats. *Planta Medica.* 2011;77(5):455–60.
162. Ysrafil Y, Sapiun Z, Slamet NS, Mohamad F, Hartati H, Damiti SA, et al. Anti-inflammatory activities of flavonoid derivates. *ADMET & DMPK.* 2023;11(3):331–59.
163. Gonzales GB, Smagghe G, Grootaert C, Zotti M, Raes K, Van Camp J. Flavonoid interactions during digestion, absorption, distribution and metabolism: A sequential structure-activity/property relationship-based approach in the study of bioavailability and bioactivity. *Drug Metabolism Reviews.* 2015;47(2):175–90.
164. Wu B, Kulkarni K, Basu S, Zhang S, Hu M. First-pass metabolism via UDP-glucuronosyltransferase: A barrier to oral bioavailability of phenolics. *Journal of Pharmaceutical Sciences.* 2011;100(9):3655–81.
165. Wu B, Morrow JK, Singh R, Zhang S, Hu M. Three-dimensional quantitative structure-activity relationship studies on UGT1A9-mediated 3-O-glucuronidation of natural flavonols using a pharmacophore-based comparative molecular field analysis model. *The Journal of Pharmacology and Experimental Therapeutics.* 2011;336(2):403–13.
166. Nagayoshi H, Murayama N, Takenaka S, Kim V, Kim D, Komori M, et al. Roles of cytochrome P450 2A6 in the oxidation of flavone, 4'-hydroxyflavone, and 4'-, 3'-, and 2'-methoxyflavones by human liver microsomes. *Xenobiotica; the Fate of Foreign Compounds in Biological Systems.* 2021;51(9):995–1009.
167. Nagayoshi H, Murayama N, Kakimoto K, Takenaka S, Katahira J, Lim YR, et al. Site-specific oxidation of flavanone and flavone by cytochrome P450 2A6 in human liver microsomes. *Xenobiotica; the Fate of Foreign Compounds in Biological Systems.* 2019;49(7):791–802.
168. Zeng M, Sun R, Basu S, Ma Y, Ge S, Yin T, et al. Disposition of flavonoids via recycling: Direct biliary excretion of enterically or extrahepatically derived flavonoid glucuronides. *Molecular Nutrition & Food Research.* 2016;60(5):1006–19.
169. Muñoz-Reyes D, Morales AI, Prieto M. Transit and metabolic pathways of quercetin in tubular cells: Involvement of its antioxidant properties in the kidney. *Antioxidants.* 2021;10(6):909.
170. Dabeek WM, Marra MV. Dietary quercetin and kaempferol: Bioavailability and potential cardiovascular-related bioactivity in humans. *Nutrients.* 2019;11(10):2288.
171. Bojić M, Kondža M, Rimac H, Benković G, Maleš Ž. The effect of flavonoid aglycones on the CYP1A2, CYP2A6, CYP2C8 and CYP2D6 enzymes activity. *Molecules.* 2019;24(17):3174.
172. Espíndola C. Some nanocarrier's properties and chemical interaction mechanisms with flavones. *Molecules.* 2023;28(6):2864.
173. Zhang Y, He L, Yue S, Huang Q, Zhang Y, Yang J. Characterization and evaluation of a self-microemulsifying drug delivery system containing tectorigenin, an isoflavone with low aqueous solubility and poor permeability. *Drug Delivery.* 2017;24(1):632–40.
174. Rahman HS, Othman HH, Hammadi NI, Yeap SK, Amin KM, Abdul Samad N, et al. Novel drug delivery systems for loading of natural plant extracts and their biomedical applications. *International Journal of Nanomedicine.* 2020;15:2439–83.
175. Morelo G, Márquez-Ruiz G, Holgado F, Giménez B, Robert P. Design of flavonoid microparticles with channel forming properties to improve oxidative stability of sunflower oil. *European Journal of Lipid Science & Technology.* 2017;119(10):1700041.
176. Sansone F, Picerno P, Mencherini T, Villecco F, D'Ursi AM, Aquino RP, et al. Flavonoid microparticles by spray-drying: Influence of enhancers of the dissolution rate on properties and stability. *Journal of Food Engineering.* 2011;103(2):188–96.
177. Shuai S, Yue S, Huang Q, Wang W, Yang J, Lan K, et al. Preparation, characterization and in vitro/vivo evaluation of tectorigenin solid dispersion with improved dissolution and bioavailability. *European Journal of Drug Metabolism and Pharmacokinetics.* 2016;41(4):413–22.
178. Scalia S, Haghi M, Losi V, Trotta V, Young PM, Traini D. Quercetin solid lipid microparticles: A flavonoid for inhalation lung delivery. *European Journal of Pharmaceutical Sciences: Official Journal of the European Federation for Pharmaceutical Sciences.* 2013;49(2):278–85.

179. Hariyadi DM, Sairiyah SN, Rahman F, Rijal MAS, Rosita N, Rahmadi M. Quercetin solid lipid microparticle stability and deposition in rat lungs: A study of surfactant effect. *Science Technology Indonesia.* 2024;9(3):682–96.
180. Vu QL, Fang CW, Suhail M, Wu PC. Enhancement of the topical bioavailability and skin whitening effect of genistein by using microemulsions as drug delivery carriers. *Pharmaceuticals.* 2021;14(12):1233.
181. Parhi B, Bharatiya D, Swain SK. Application of quercetin flavonoid based hybrid nanocomposites: A review. *Saudi Pharmaceutical Journal: SPJ: The Official Publication of the Saudi Pharmaceutical Society.* 2020;28(12):1719–32.
182. Tripathi N, Verma S, Vyas M, Yadav NS, Gain S, Khatik GL. Nanoformulations of quercetin: A potential phytochemical for the treatment of uv radiation induced skin damages. *Brazilian Journal of Pharmaceutical Sciences.* 2022;58:e18744.
183. Feng T, Wei Y, Lee RJ, Zhao L. Liposomal curcumin and its application in cancer. *International Journal of Nanomedicine.* 2017;12:6027–44.
184. Zarrabi A, Zarepour A, Khosravi A, Alimohammadi Z, Thakur VK. Synthesis of curcumin loaded smart pH-responsive stealth liposome as a novel nanocarrier for cancer treatment. *Fibers.* 2021;9(3):19.
185. Wu B, Liang Y, Tan Y, Xie C, Shen J, Zhang M, et al. Genistein-loaded nanoparticles of star-shaped diblock copolymer mannitol-core PLGA-TPGS for the treatment of liver cancer. *Materials Science & Engineering C, Materials for Biological Applications.* 2016;59:792–800.
186. Huang L, Yao Y, Ruan Z, Zhang S, Feng X, Lu C, et al. Baicalin nanodelivery system based on functionalized metal-organic framework for targeted therapy of osteoarthritis by modulating macrophage polarization. *Journal of Nanobiotechnology.* 2024;22(1):221.
187. Rashidinejad A, Dima C, Karaca AC, Jafari SM. A review on rutin-loaded nanocarriers: Fundamentals, bioavailability, application in functional foods, and challenges. *European Polymer Journal.* 2024;219(August):113385.
188. Zhang M, Zhou N, Zhao L, Zhao L. Black rice anthocyanins nanoparticles based on bovine serum albumin and hyaluronic acid: Preparation, characterization, absorption and intestinal barrier function protection in Caco-2 monolayers. *International Journal of Biological Macromolecules.* 2024;267(Pt 2):131325.

17 Insight into the Regulation of Plant Secondary Metabolism by microRNAs

Rumdeep Kaur Grewal[1] *and Amit Roy*[2]
[1]Department of Botany, Bhairab Ganguly College, Kolkata, India
[2]Faculty of Forestry and Wood Sciences, Czech University of Life Sciences Prague, Prague,Czechia

17.1 INTRODUCTION

Plants churn out a wide range of organic compounds that appear not to contribute directly to growth and development, these substances are known as secondary metabolites (SMs). The term "secondary metabolites" was first coined by Albrecht Kossel in 1910, and later it was used by Polish botanist Friedrich Czapek to describe end products of nitrogen metabolism (Bourgaud et al., 2001). These compounds are produced in low quantities and are usually stored in specialized cells or organs. The site of production and accumulation may be different. SMs may undergo short- or long-distance transportation from production to storage sites (Selmar, 1989). Lupine alkaloids are produced in mesophyll cell chloroplasts and are transported to epidermal cells through phloem for storage. Lipophilic SMs accumulate in vesicles, dead cells, or extracellular sites whereas hydrophilic SMs are stored in aqueous medium, primarily in vacuoles of living cells. Four Secondary metabolite biosynthesis pathways are functional in plants: Methylerythritol-phosphate pathway, Malonate pathway, Mevalonate pathway, and Shikimate pathway. Secondary metabolic pathways are often initiated from primary metabolic pathways. Products or intermediates of primary metabolism may be used as substrate/precursor or may be modified to form SMs. SMs produced due to such modifications are found to be more widely distributed than those that require multiple steps in their production. Harmine (Beta-carboline alkaloid) is reported from more than 20 diverse plant families whereas Quinine is limited to only a few genera or Rubiaceae (Seigler, 2012). Secondary metabolic pathways involve the coordinated expression of numerous genes, i.e enzymes and associated proteins. Moreover, the expression of SM-associated genes is also tissue- and condition-specific. Therefore, the expression of these genes is carefully controlled both at the transcriptional and post-transcriptional stages through a variety of molecular processes that include transcription factors and miRNAs.

MicroRNAs (miRNAs) are non-coding RNA molecules that range from 19 to 24 nucleotides in length and play a role in regulating gene expression after transcription. The initial form of miRNAs (pri-miRNAs) adopts a hairpin shape, which is identified and processed by Dicer-like RNase III endonucleases (DCLs) to generate precursor miRNAs (pre-miRNAs) (X. Chen, 2005). The pre-miRNA is transformed into a duplex structure that features a 3' overhang and is methylated at the 3' end. It is then transported to the cytoplasm while still in this duplex form (Park et al., 2005). Once in the cytoplasm, one strand of the duplex undergoes degradation by the exosome, while the other strand (the mature miRNA) associates with the RNA-induced silencing complex (RISC) and is incorporated into an ARGONAUTE (AGO) complex. The AGO complex identifies its target transcript through sequence complementarity with the miRNA, leading to the cleavage of the target (Figure 17.1).

DOI: 10.1201/9781003518358-17

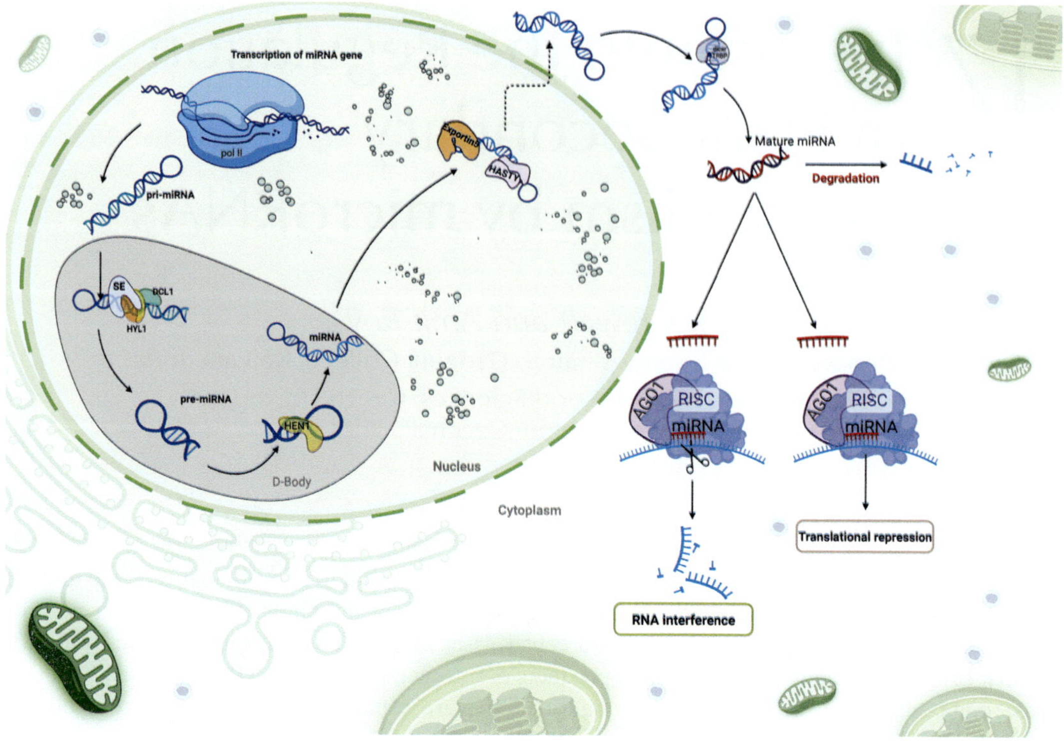

FIGURE 17.1 miRNA synthesis and mode of action *in vivo.* miRNA genes (MIR genes) are transcribed by RNA polymerase II in the nucleus, forming a primary miRNA (pri-miRNA) transcript. The pri-miRNA is typically capped and polyadenylated. The pri-miRNA is processed by a protein complex containing DICER-LIKE 1 (DCL1), an RNase III-like enzyme, along with accessory proteins like HYPONASTIC LEAVES1 (HYL1) and SERRATE (SE). DCL1 processes the pri-miRNA into a precursor miRNA (pre-miRNA) by cutting it into a smaller, stem-loop structure. DCL1 then further cleaves the pre-miRNA to release the mature miRNA/miRNA* duplex. The miRNA/miRNA* duplex is methylated at the 3' ends by HUA ENHANCER1 (HEN1) to protect it from degradation. The duplex is then exported from the nucleus to the cytoplasm by a nuclear export receptor called HASTY in plants or Exportin-5 in animals. Once the miRNAs are mature and have reached the cytoplasm, they regulate gene expression primarily through two mechanisms: mRNA cleavage and translation inhibition. Precisely, in the cytoplasm, the duplex is unwound by the AGO1, allowing the mature miRNA to join the RNA-induced silencing complex (RISC). The complementary strand is mostly degraded. The template strand of the duplex guides the RISC to bind to complementary messenger RNA (mRNA) targets, leading to mRNA silencing through cleavage or translational repression.

Plant miRNAs may regulate secondary metabolic pathways by targeting transcription factors, structural genes, or/and signaling pathways (hormone or environmental) that govern the expression of genes related to these pathways (Jeena et al., 2022; Owusu Adjei et al., 2021). The role of miRNAs in various important secondary metabolite-producing plants is well documented (Robert-Seilaniantz et al., 2011; N. Singh, Srivastava, Shasany, et al., 2016; Vashisht et al., 2015; Zhao et al., 2020). This chapter aims to sum up the latest progress in research on the regulation of plant secondary metabolism by miRNAs.

17.2 ROLE OF SECONDARY METABOLITES IN PLANTS

SMs are known to have a role in signaling, in attracting pollinators, and acting as allelopathic agents. They are known to protect against toxicity, UV damage, pathogens, herbivores, and several environmental stresses. Plant defense is multilayered, involving physical, and chemical barriers, constitutive compounds and induced specific response compounds. Accumulation of constitutive compounds, i.e. tannins and simple phenols, provides a chemical barrier to infection. The induction of the phenylpropanol pathway in response to infection is important for defense against pathogens (Lamb et al., 1992). Herbivory and pathogen infection also increase the production of several plant SMs, including phytoalexins, terpenoids, and alkaloids. For instance, *Senecio jacobaea* produces senecionine, a pyrrolizidine alkaloid known to deter mammalian herbivores. Similarly, some varieties of *Trifolium repens* produce hydrocyanic acid to deter slugs and snail herbivores (Angseesing, 1974). Terpene derivatives like pyrethrin-1 (Tattersfield et al., 1929) and azadirachtin (Jennifer Mordue (Luntz) et al., 1998) are specifically toxic to insects. Sesquiterpene lactones, saponins, iridoids, and non-protein amino acids also act as insect-deterrents. Triterpene derivatives such as cardenolide (B. Singh & Rastogi, 1970), is a mammal-specific cardiac toxins. Research indicates that under nutrient-deficient conditions production of non-nitrogenous SMs associated with the shikimic acid pathway increases whereas under high light intensity, accumulation of phenolic compounds and terpenoids is enhanced. Similarly, water stress enhances the production of terpenoids, alkaloids, glycosides, glucosinolates, and condensed tannins. SMs shield the plants against the harmful effects of incident UV radiation. The plant surface is covered by resins, waxes, and lipophilic compounds, i.e. flavonoids that are capable of screening out harmful radiation. Other SMs, i.e phenyl amides act as anti-oxidants to remove free radicals arising through environmental stress (Edreva et al., 2007).

Activation of the phenylpropanol pathway leads to the accumulation of several SMs, i.e., chlorogenic acid, coumarins, cinnamic acid esters, lignin, suberin, and phytoalexins. Phytoalexins are compounds formed in response to attack by pathogens (Ebel, 1986). Chemically different compounds are represented as phytoalexins in different plant groups, i.e isoflavonoids in Fabaceae and sesquiterpenes in Solanaceae are phytoalexins (Rhodes, 1994). Many SMs, like benzoic acid and cinnamic acid derivatives have been shown to have allelopathic activity for example *Juglans nigra* produces the germination inhibitor and phytotoxin, naphthoquinone juglone (Harborne, 2014). Plant SMs are also known to inhibit the growth of microorganisms, including root symbionts and detritivores (Horner et al., 1988; Kuiters, 1990), resulting in a change in nutrient turnover in soil. Polyphenolic compounds are known to cause acidification of soils, which diminishes the availability of nutrients, specifically nitrogen (Kuiters, 1990).

Flavonoids and carotenes are commonly responsible for flower and fruit color. In some cases, anthocyanins, betacyanins, and betaxanthins may be responsible for flower coloration. Differences in colour hues are achieved by differential compartmentalization of accessory pigments, use of co-pigments – metal ion chelators (Mg, Al) or modification of SM structure through methylation, oxidation, reduction and /or glycosylation. Differential coloration of parts of organs gives rise to various colored patterns some of these patterns may not be visible to the human eye but become apparent under UV light (Harborne & Nash, 1984). Many insects can detect radiation in the UV spectrum and are attracted to particular color patterns (Faegri & Pijl, 2013). Changes associated with fruit ripening are also associated with seed dispersal. Attractive aromas associated with different flowers result from complex mixing of low molecular mass SMs, i.e., monoterpenes, sesquiterpenes, cinnamic acid and benzoic acid derivatives, long-chain alkanes, alkanals, and alkanols, e.g., citronella, eugenol from *Osmium gratusimum*. The structures of some aromatic SMs mimic that of mating pheromones of potential pollinators (Borghi et al., 2017).

17.3 IMPORTANCE OF SECONDARY METABOLITES IN MEDICINE

Since prehistoric times, plants have been used as a source of medicinally active compounds. It was later discovered that specific SMs are responsible for the therapeutic properties of plants, and hence, plant SMs are of great value in the pharmaceutical industry. SMs in combination with other metabolites or alone are known to have antioxidant, antitumor, and antimicrobial properties (Wink, 2015). Phytochemicals have been shown to increase treatment efficacy in the case of various human diseases. Plant SMs with known anti-tumor activity are lupeol (Nyaboke et al., 2018), asiatic acid

(Hsu et al., 2005), ursolic acid, celastrol, aurapten, indole-3-carbinol, saidmanetin, and hypericin. These SMs affect cell proliferation by modulating cell signaling related to cell division, metastasis, apoptosis, and immune response (Gali-Muhtasib et al., 2015; Idris et al., 2009). Paclitaxel, an oxygenated tetracyclic diterpenoid, obtained from *Taxus brevifolia*, inhibits cell mitosis and is used for the treatment of various cancers (Atanasov et al., 2015).

Many SMs show anti-microbial activity against both fungi and bacteria and are used to cure microbial infections. Plant polyphenols have been shown to inhibit bacterial growth by oxidation of specific bacterial enzymes. Polyphenols may also bind to cell membranes and disrupt their structure. Monoterpenes are also known to disrupt membrane structure by interacting with phospholipid molecules (Concha et al., 2018; Vasconcelos et al., 2018).

Polyphenols show antioxidant properties (Martelli & Giacomini, 2018). Quercetin – a flavonol prevents the formation of reactive oxygen species (ROS) by binding to metal ions (Nile et al., 2018). Galantamine is an alkaloid obtained from *Galanthus woronowii* (Dewick, 2002) and is known to inhibit acetylcholinesterase (AChE) and regulate nicotinic acetylcholine receptor (nAChR). It has been authorized for use in treating Alzheimer's disease. Artemisinin, an antimalarial medication, is sourced from *Artemisia annua* and contains an active ingredient, sesquiterpene lactone, utilized for treating *Plasmodium falciparum* infections, which can cause severe cerebral malaria (Phillipson, 2001). Calanolide A is a dipyranocoumarin obtained from *Calophyllum lanigerum*is, an inhibitor of type-1 HIV non-nucleoside reverse transcriptase (NNRTi) and of AZT-resistant strains of HIV (Sorbera et al., 1999).

SMs present in food are now known to have health benefits as well. Flavonoids in food items, i.e., wine, have been shown to decrease the risk of atherosclerosis, as flavonoids prevent the oxidation of low-density lipoprotein (McCullough et al., 2012). Sweet potato (*Ipomoea batata* L.) is used to overcome Vitamin A deficiency as it contains vitamin A precursor beta-carotene (Ghasemzadeh et al., 2016). Liquorice (*Glycyrrhiza glabra*) root contains phytochemicals that help to regulate blood glucose level ((Pastorino et al., 2018)). *Ocimum sanctum* L. or Holy basil helps to manage inflammation and cancer (Siva et al., 2016). *Aronia melanocarpa* (Michx.) Elliott fruit juice containing four cyanidin 3-O-glycosides has been shown to have a neuroprotective effect against amyloid-β toxicity (Wen et al., 2020).

17.4 REGULATION OF PLANT SECONDARY METABOLISM BY MICRORNAS

Plant miRNAs have been linked to the regulation of secondary metabolism. Genes targeted by miRNAs may include transcription factors or key enzymes that play crucial roles in secondary metabolism (Jeena et al., 2022; Owusu Adjei et al., 2021). The impact of miRNAs on the regulation of genes related to various significant plant SMs is well documented (Robert-Seilaniantz et al., 2011; N. Singh, Srivastava, Shasany, et al., 2016; Vashisht et al., 2015; Zhao et al., 2020). Plant SMs are conventionally categorized into three groups: Nitrogen-containing compounds (Alkaloids), Terpenoids, and Phenolic (i.e. flavonoid) compounds. Alkaloids have nitrogen-containing heterocyclic ring in their structure and are usually water soluble. Terpenoids are isoprene unit-containing compounds generally insoluble in water. Phenolics include wide range of biomolecules containing aromatic ring(s) in their structure, and most are soluble in water or organic solvents some form insoluble polymers, i.e, lignins (Bourgaud et al., 2001).

17.4.1 Regulation of Terpenoid Metabolism by microRNAs

Terpenoids represent a diverse and abundant class of SMs. Terpenoids are crucial not only for plant functioning but also for ecological interactions between plants and animals. Plant terpenoids hold significant medicinal and industrial value. Numerous studies have highlighted their wide-ranging applications, particularly in medicine. Understanding the importance of miRNAs in terpenoid biosynthesis offers essential insights for boosting the production of these valuable compounds in plants.

Various miRNAs play critical role in terpenoid biosynthesis (Figure 17.2, Table 17.1). Fan et al. identified several miRNAs in *Xanthium strumarium* L., including miR5183 and miR5255, which target enzymes like gibberellin 3-oxidase and β-amyrin synthase, thereby influencing terpenoid production (Fan et al., 2015). Biswas et al. detected miR5538 in *Podophyllum hexandrum* (Biswas et al., 2016). It targets S-isoprenylcysteine O-methyltransferase required for terpenoid skeleton biosynthesis.

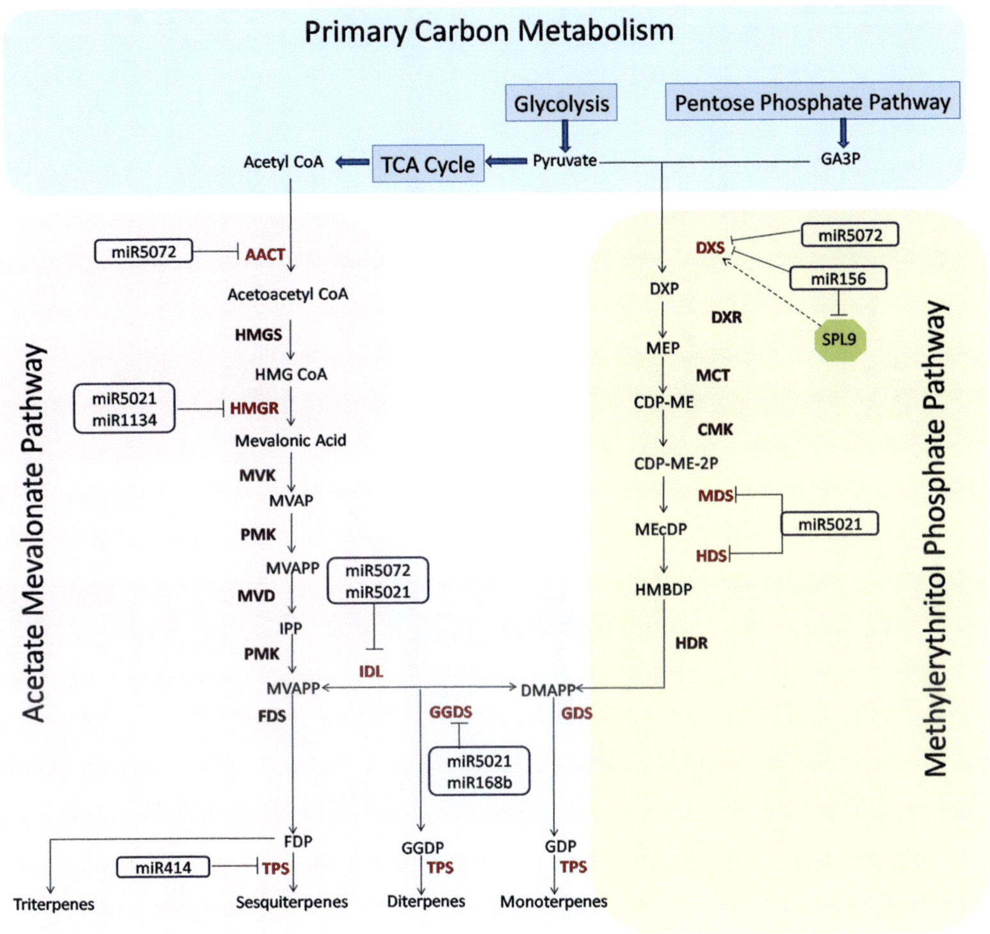

FIGURE 17.2 Regulation of Terpenoid Biosynthesis in plants. The compounds are: (S)-3-hydroxy -3-methylglutaryl-CoA (HGM CoA); mevalonate 5-phosphate(MVAP); mevalonate 5-diphosphate (MVAPP); sopentenyl diphosphate (IPP); farnesl diphosphate (FDP); glyceraldehyde-3-phosphate (GA3P); 1-deoxy-D-xylulose 5-phosphate (DXS); 2-C-methyl-erythritol 4-phosphate (MEP); 4-(cytidine 5'-diphospho)-2-methyl-D-erthritol phosphate (CDP-ME); 4-diphosphocytidyl-2C-methyl-D-erythritol 2-phosphate (CDP-ME-2P); 2-C-methyl-D-erythritol 2,4-cyclodiphosphate (MEcDP); (E)-4-hydroxy-3-methylbut-2-enyl diphosphate (HMBDP); dimethylallyl diphosphate (DMAPP); geranyl diphosphate (GDP); geranyl geranyl diphosphate (GGDP). Enzymes are: acetoacetyl-CoA thiolase (AACT); HMGS-CoA synthase (HMGS); HNG-CoA reductase (HMGR); mevalonate kinase (MVK); Phosphomevalonate kinase (PMK); mevalonate diphosphate decarboxylase (MVD); isopentenyl diphosphate isomerase (IDL); farnesyl diphosphate synthase (FDS); geranyl diphosphate synthase (GDS); Terpene synthase (TPS); DXOP synthase (DXS); DOXP reductoisomerase (DXR); 2-C-methyl-D-erythritol 4-phosphate synthase (MCT); 2-C-methyl-D-erythritol 2,4-cyclodiphosphate synthase (MDS); CDP-ME-kinase (CMK); (E)-4-hydroxy-3-methylbut-2-enyl diphosphatesynthase (HDS); (E)-4-hydroxy-3-methylbut-2-enyl diphosphate reductase (HDR); geranyl geranyl diphosphate synthase (GGDS); terpene synthase (TPS); geranyldiphosphate synthase (GDS). The transcription factors in green are: SQUAMOSA promoter-binding-like (SPL).

TABLE 17.1
List of miRNAs that Regulate Flavonoid Metabolism Pathways

miRNA	Target	Pathway	Plant	References
miR2911	Gamma-tocopherol methyltransferase	Tocopherol biosynthesis	*Helianthus annuus* L.	Barozai et al. (2012)
miR828a, miR948a	MYB12 lipoxygenase	Flavonoid biosynthesis	*Salvia sclarea* L.	Legrand et al. (2010)
miR2919	Flavanone synthase	Curcumin biosynthesis	*Curcuma longa* L.	Cho et al. (2013); N. Singh and Sharma (2017)
miR1168.2, miR156b				
miR1858				
ib-miR156	ibSPL	Anthocyanin biosynthesis	*Ipomoea batatas* (L.) Lam.	He et al. (2019)
miR858	AaF3H, AaLDOX, AaF3GT		*Actinidia arguta* (Siebold & Zucc.) Planch. ex Miq.	Y. Li et al. (2019)
miR395p-3p	bHLH, MYB	Proanthocyanidin biosynthesis	*Diospyros kaki* L. f.	Luo et al. (2015)
			miR858b	
miRNAs (U436803, U977315, U805963, U3938865, 4351355)	R2R3-MYB transcription factors	Flavonoid biosynthesis	*Lonicera japonica* Thunb.	J. Liu et al. (2017)
miR6194, miR858b	Flavanone 3-hydroxylase	Flavonol, anthocyanidin, and proanthocyanidin syntheses	*Halostachys caspica* C.A.Mey.	Yang et al. (2015)
miR156*	SPL9	Anthocyanin biosynthesis	*Arabidopsis thaliana* (L.) Heynh.	Gou et al. (2011)
miR858a*	R2R3-MYB transcription factors	Flavonoid biosynthesis	*Arabidopsis thaliana* (L.) Heynh.	D. Sharma et al. (2016)
miR156	SPL transcription factor, transcription factors R2R3-MYB	Flavonoid biosynthesis	*Arabidopsis thaliana* (L.) Heynh.	Gou et al. (2011)
miR396b	Kaempferol 3-O-beta-dgalactosyltransferase	Flavonol glycoside	*Rauvolfia serpentina* (L.) Benth. ex Kurz	Prakash et al. (2016)
miR828a	Anthocyanin regulatory C1 protein	Anthocyanin biosynthesis	*Rauvolfia serpentina* (L.) Benth. ex Kurz	Prakash et al. (2016)
miR1438	Caffeoyl-CoA O-methyl transferase	Lignin biosynthesis	*Podophyllum hexandrum* Royle	Biswas et al. (2016)
miR1873	Dihydroflavonol 4-reductase C	Flavonoid biosynthesis	*Podophyllum hexandrum* Royle	Biswas et al. (2016)
miR1873	Phenylalanine ammonia lyase (PAL)	Gingerol (phenolic) biosynthesis	*Zingiber officinale*	N. Singh, Srivastava, and Sharma (2016)
miR5532	2-Hydroxyisoflavanone dehydratase	Isoflavonoid biosynthesis		
miR1873/miR5532	Dihydroflavonol 4-reductase C/ -hydroxyisoflavanone dehydratase	Flavonoid/isoflavonoid biosynthesis		
miR172i	4-Coumarate–CoA ligase	Flavonoid biosynthesis		
miR829.1	Chalcone synthase			

(Continued)

TABLE 17.1 (*Continued*)
List of miRNAs that Regulate Flavonoid Metabolism Pathways

miRNA	Target	Pathway	Plant	References
CHS-siRNA	Chalcone synthase	Flavonoid biosynthesis	*Glycine max* (L.) Merr.	Tuteja et al. (2009)
miR1061-3p	Naringenin 3-dioxygenase	Flavonoid biosynthesis	*Pyrus bretschneideri* Rehder	J. Wu et al. (2014)
miR2911	?-Tocopherol methyl transferase	a-Tocopherol biosynthesis	*Helianthus annuus* L.	Barozai et al. (2012)
miR156, miR828, miR858	MYBs, bHLH, WD40. SPLs, ARF, EIN3, WRKY, MADS-box, sucrose synthase, sugar/inositol transporter ABC transporter	Anthocyanin biosynthesis	*Raphanus sativus* L.	Y. Sun et al. (2017)
miR828	MYB114, flavonoid	Flavonol biosynthesis	*Vitis vinifera* L.	Tirumalai et al. (2019)
miR858	3′-Monooxygenase (F3MO)			
csn-miR160a	Auxin response factor 18	Catechin biosynthesis	*Camellia sinensis* (L.) Kuntze	P. Sun et al. (2017)
csn-miR396a	Growth-regulating factor 7, calcium-transporting ATPase 13, DNA-directed RNA polymerase V subunit			
csn-miR167a	Auxin response factor 6 ARF6, CHI enzyme			
csn-miR4380a	DFR			
csn-miR3444b				
csn-miR5251	C4H			
csn-miR7777-5P.1				
miR2593e	ANR			
miR156	Basic helix-loop-helix (bHLH)	Flavone/flavonol biosynthesis	*Mentha* spp.	N. Singh, Srivastava, Shasany, et al. (2016)
miR6194	Flavanone 3b-hydroxylase (F3H)	Flavonols, anthocyanidins biosynthesis	*Halostachys caspica* C.A.Mey.	Yang et al. (2015)

In *Salvia miltiorrhiza,* biosynthesis of terpenoids and tanshinones is associated with miR5072 as it targets acetyl-CoA C-acetyltransferase (*ACAT*) (X. Xu et al., 2014). MiR838, miR5021, miR5251, miR2919, and miR5658 in *Ferula assa-foetida* are known to regulate sesquiterpene biosynthesis. In *Dryopteris fragrans,* Song et al. reported that miR160 and miR156 modulate terpenoid biosynthesis by targeting the DfARF and DISPL genes (Song et al., 2022). In *Arabidopsis* and *Pogostemon cablin,* overexpression of miR156 leads to a decrease of sesquiterpene levels. In both these plants, miR156 targets sesquiterpene synthesis, activating transcription factors SPL9 and SPL10 (Yu et al., 2015). Similarly, in *Cuminum cyminum,* Reyes-Calderon et al. discovered miR159, miR397, and miR156q target enzymes crucial for the terpenoid skeleton biosynthesis (Reyes-Calderón et al., 2023). Another study in *Paeonia lactiflora* (P. Xu et al., 2023). reported nine miRNAs from seven families that were linked to terpenoid biosynthesis through degradome sequencing (Table 17.1).

Research has also delved into miRNAs involved in specific terpenoid biosynthesis pathways, – artemisinin, ginsenosides, and paclitaxel. Researchers found miR414 in *Artemisia annua* regulates artemisinin biosynthesis by targeting enzymes like farnesyl pyrophosphate synthase (FPS) and Amorpha-4,11-diene synthase (ADS) (Pani et al., 2011). In another study, miR390 was linked to trichome development, a site for artemisinin production (Pérez-Quintero et al., 2012). In *Panax ginseng,* miR854e, miR1439b, and miR1439h target enzymes involved in ginsenoside biosynthesis. The first targets the enzyme FPS, while the latter two target beta-amyrin synthase (b-AS), respectively (Y. Xu et al., 2023). Similarly, through degradome sequencing, Hao et al. predicted that miR171 and miR164 regulate taxol biosynthesis in *Taxus chinensis* by targeting taxane 2α-O-benzoyltransferase(Hao et al., 2012). A study in *Salvia miltiorrhiza,* indicates miR168b regulates geranylgeranyl diphosphate synthase (*SmGGPPS1*), an enzyme of the tanshinone biosynthesis pathway. It produces geranylgeranyl diphosphate (GGPP) from isopentenyl diphosphate (IPP) and farnesyl diphosphate (FPP) (X. Xu et al., 2014). In *Catharanthus roseus,* miR5021 and miR5649 contribute to the synthesis of terpene skeleton. MiR5021 regulates GCPE protein and UDP-glucose cyclohexenyltransferase, cytochrome c oxidase subunit 1, an enzymes involved in oxidative phosphorylation, protein processing, and terpene biosynthesis in the endoplasmic reticulum (Pani & Mahapatra, 2013).

Research into the regulation of terpenoid pathways has indicated that miRNAs may influence terpenoid metabolism indirectly by targeting transcription factors required for terpenoid biosynthesis pathway gene expression regulation. In *Arabidopsis thaliana,* transcription factors such as ATHB13, SPL7, and SPL11 are modulated by miR5658 and miR5021 (Sobhani Najafabadi & Naghavi, 2018) (Figure 17.2). These miRNAs and their target enzymes and transcription factors form a regulatory network that modulates the biosynthesis of plant terpenoids.

17.4.2 Regulation of Flavonoid Metabolism by microRNAs

Flavonoids are essential SMs in plants, widely recognized for their diverse biological activities and significant roles in plant biology and nutrition (Y. Wu, Huang, et al., 2022). Flavonoids are vital in plant disease resistance, stress responses, and pigmentation, and their pharmacological properties make them indispensable in the pharmaceutical, cosmetic, food, and biotechnology industries (Chiorcea-Paquim, 2023). These compounds have garnered attention for their potential to improve various human health conditions, such as diabetes, cardiovascular diseases, and cancer. Understanding the influence of miRNA on flavonoid biosynthesis is crucial in unveiling the molecular mechanisms behind plant secondary metabolism. This research offers theoretical insights and has practical implications, such as enhancing plant stress resistance and improving the quality of medicinal plants.

Flavonoid biosynthesis is controlled by multiple transcription factors and miRNAs (Figure 17.3, Table 17.2). In *Camellia sinensis,* csn-miR828 and csn-miR156 were identified as crucial regulators of catechin biosynthesis (Fan et al., 2015). In *Podophyllum hexandrum,* miR829.1, miR172i, and miR1873 target chalcone synthase, 4-coumaroyl-CoA ligase, and dihydroflavonol 4-reductase, respectively, all involved in flavonoid biosynthesis (Biswas et al., 2016). Overexpression of

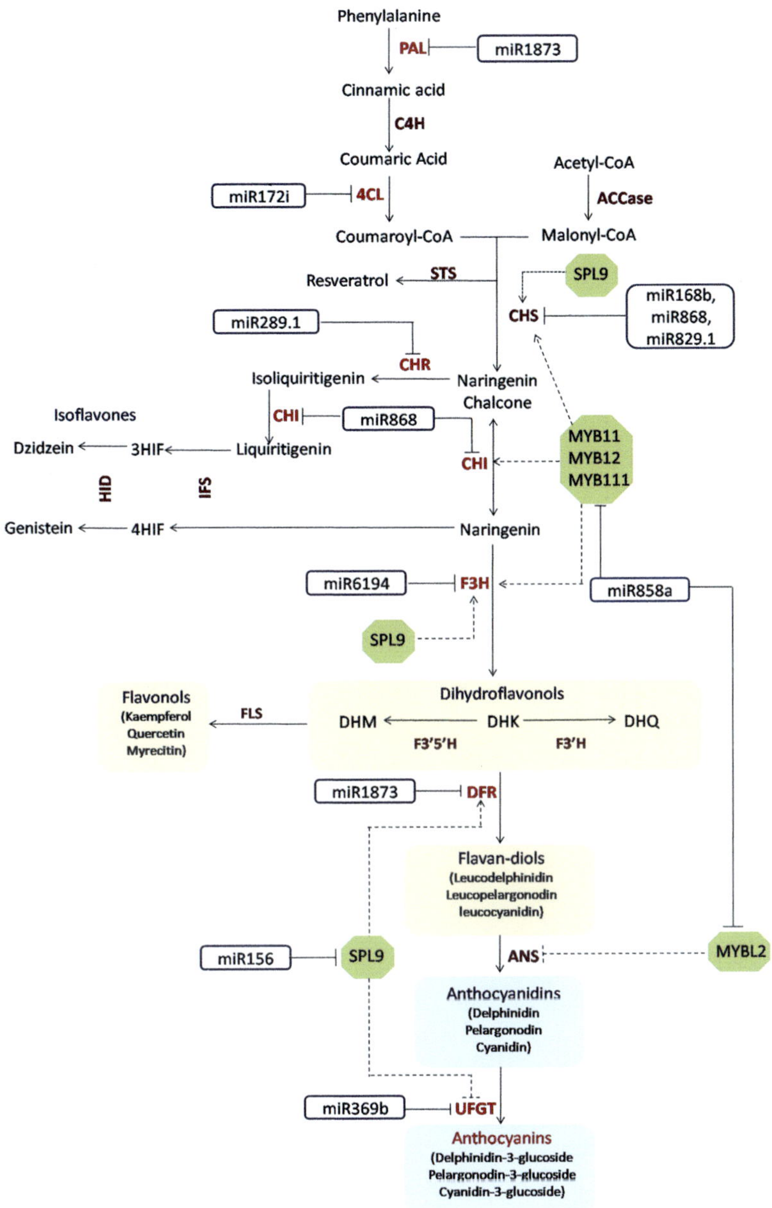

FIGURE 17.3 Regulation of flavonoid biosynthesis pathway. The compounds are: dihydrokaempferol (DHK); dihydroquercetin (DHQ); dihydromyricetin (DHM); trihydroxyisoflavanone (3HIF); tetrahydroxyisoflavanone(4HIF). The enzymes involved are: phenylalanine ammonia lyase (PAL); cinnamic acid 4-hydroxylase (C4H); 4-coumarate: CoA ligase (4CL); acetyl-CoA carboxylase (ACCase); stilbene synthase (STS); chalcone synthase (CHS); chalcone reductase (CHR); chalcone 2′-glucosyltransferase (CH2′GT); chalcone 4′-O-glucosyltransferase (CH4′GT); aureusidin synthase (AS); chalcone isomerase (CHI); flavone synthase (FNS); cinnamate–CoA ligase (CLL-7); flavonoid 6-hydroxylase (F6H); flavonoid 8-hydroxylase (F8H); isoflavone synthase(IFS); 2-hydroxyisoflavanone dehydratase (HID); flavanone 4-reductase (FNR); F3H, flavanone 3-hydroxylase (F3H); flavanone 3',5'-hydroxylase (F3'5'H) flavonol synthase (FLS); dihydroflavonol 4-reductase(DFR); anthocyanidin synthase (ANS); UDP-glucose flavonoid 3-Oglucosyltransferase (UFGT); O-methyl transferases (OMT); leucoanthocyanidin reductase (LAR); anthocyanidin reductase (ANR). The transcription factors in green are: SQUAMOSA promoter-binding-like (SPL); v-mybavian myeloblastosis viral oncogene homolog (MYB).

TABLE 17.2
List of miRNAs that Regulate Terpenoid Metabolism Pathways

miRNA	Target	Pathway	Plant	References
iRNA-4995	3-Deoxy-7-phosphoheptulonate synthase (DAHP synthase)	Terpenoid biosynthesis	*Picrorhiza kurroa* Royle ex Benth.	Vashisht et al. (2015)
mir-5021	MYB transcription factor, geranyl diphosphate synthase, GCPE protein, UDP-glucose iridoid glucosyltransferase	Isoprenoid/terpenoid biosynthesis	*Catharanthus roseus* (L.) G.Don	Pani and Mahapatra (2013)
miR2919, miR5251, miR838, miR5021, miR5658	SPL7, SPL11, ATHB13 TFs	Terpene biosynthesis	*Ferula gummosa* Boiss.	Sobhani Najafabadi and Naghavi (2018)
miR156*	SPL9	Sesquiterpenoid, triterpenoid biosynthesis	*Pogostemon cablin* (Blanco) Benth.	Yu et al. (2015)
miR156, miR414, and miR5021	Basic helix-loop-helix (bHLH) geranyl di-phosphate synthase subunit alpha-like protein (NACA)	Terpenoid backbone biosynthesis	*Mentha* spp.	N. Singh, Srivastava, Shasany, et al. (2016)
miR156*	1-Deoxy-D-xylulose 5-Phosphate synthase (DXS)	Terpenoid biosynthesis		
miR414	Terpene synthase 21 (TPS21)	Sesquiterpenoid, triterpenoid biosynthesis		
miR396b	Secologanin synthase	Secologanin	*Rauvolfia serpentina* (L.) Benth. ex Kurz	Prakash et al. (2016)
miR5072	Acetyl-CoA C-acetyl transferase	Tanshinones (abietane-type norditerpenoid quinones)	*Salvia miltiorrhiza*	X. Xu et al. (2014)
miR838	CYP71	Terpenoid metabolism	*Zingiber officinale* Roscoe	N. Singh, Srivastava, and Sharma (2016)
miR4995	3-Deoxy-7-phosphoheptulonate synthase (DAHP synthase)	Picroside biosynthesis	*Picrorhiza kurroa* Royle	Vashisht et al. (2015)
miR5538	Protein-S-isoprenylcysteine O-methyltransferase	Terpenoid backbone biosynthesis	*Podophyllum hexandrum* Royle	Biswas et al. (2016)

(Continued)

TABLE 17.2 (*Continued*)
List of miRNAs that Regulate Terpenoid Metabolism Pathways

miRNA	Target	Pathway	Plant	References
miR7539, miR5021, miR1134	Upstream genes of terpencid pathway	Terpenoid	*Xanthium strumarium* L.	Fan et al. (2015)
miR5491, miR5255	Beta-amyrin synthase	Xanthanolide (sesquiterpenoids)		
miR6449, miR5183	Squalene epoxidase			
miR7540	Ent-kaurene synthase, Gibberellin 3-oxidase, R-Linalool synthase			
miR7540, mmiR5183, miR6449, miR5255, miR5491, miR6435	R-Linalool synthase, gibberellin 3-oxidase, ent-kaurene synthase, squalene epoxidase, beta-amyrin synthase, germacrene A oxidase	Mono-, sesqui-, di-, and triterpenoid biosynthesis		
miR1134	3-Hydroxy-3-methylglutaryl coenzyme A reductase (HMGR)	Terpenoid backbone biosynthesis		
miR5183	Gibberellin 3-oxidase	Diterpenoid		
miR5255	Squalene epoxidase	Triterpenoid		
miR5491	Beta-amyrin synthase	Triterpenoid		
miR6435	Germacrene A oxidase	Sesquiterpenoid		
miR6449	Ent-kaurene synthase	Diterpenoid		
miR7539	1-Deoxy-D-xylulose 5-Phosphate synthase (DXS)	Terpenoid backbone		
miR7540	R-Linalool synthase	Monoterpenoid		
miR854e miR854b miR854c	Farnesyl diphosphate synthase, squalene epoxidase (SE)	Ginsenosides (triterpene)	*Panax ginseng* C.A.Mey.	Mathiyalagan et al. (2013)
miR1439b, miR1439h	Beta amyrin synthase			

(*Continued*)

TABLE 17.2 (*Continued*)
List of miRNAs that Regulate Terpenoid Metabolism Pathways

miRNA	Target	Pathway	Plant	References
miR390	A gene involved in trichome development	Artemisinin (sesquiterpene)	*Artemisia annua* Pall.	Pérez-Quintero et al. (2012)
miR5649a	Geranylgeranyl-diphosphate specific	Terpenoid backbone biosynthesis	*Curcuma longa* L.	N. Singh and Sharma (2017)
miR5021	Geranyl-diphosphate synthase, 1-Deoxy-D-xylulose-5-phosphate synthase (DXS), Geranylgeranyl-diphosphate synthase			
miR3630	MYC2	Terpenoid biosynthesis	*Camellia sinensis* (L.) Kuntze	Zhao et al. (2018)
miR33418	MYB.			
miR156f	SPL9			
miR156	AaERF1/2			
miR535				
miR858				
pmi-miR530 pmimiR6173, pmi-miR6300	MVD, probable sulphate transporter; Sesquiterpene synthase, HMGR, farnesyl diphosphate synthase 1	Terpenoid biosynthesis	*Persicaria minor* (Huds.) Opiz	Samad et al. (2019)
miR_2924, miR_218, miR_2642	HMGS, MK, MPDC, LPS	Terpene trilactones (TTLs) biosynthesis	*Ginkgo biloba* L.	Ye et al. (2020)
miR159a, miR159, miR_2432	LPS	Terpene trilactones (TTLs) biosynthesis	*Ginkgo biloba* L.	
miR159, miR172, miR166	Cytochrome P450 reductase; CYP71AV1 catalyzes	Artemisinin (ART) biosynthesis (sesquiterpene)	*Artemisia annua* Pall.	Khan et al. (2020)
miR1533, miR5021, miR5658	SPL7, SPL11, ATHB13	Terpene biosynthesis	*Ferula gummosa* Boiss.	Sobhani Najafabadi and Naghavi (2018)
miRstv_7*	UDP-glycosyl transferase 76G1 (UGT76G1)	Steviol glycoside biosynthesis	*Stevia rebaudiana* (Bertoni) Bertoni	Saifi et al. (2015)
miRstv_7*	Kaurenoic acid hydroxylase (KAH)	Steviol glycoside biosynthesis		
miRstv_7*	Kaurene oxidase (KO)	Steviol glycoside biosynthesis		

miR5298b and miR8154 in primary cultured cells of *Taxus chinensis* demonstrated their regulatory roles in phenylalanine, paclitaxel, and flavonoid biosynthesis (Gupta et al., 2017). In *Murraya koenigii*, miR168b targets chalcone synthase and miR858 was shown to target both chalcone isomerase and chalcone synthase, thereby regulating flavonoid production (Gutiérrez-García et al., 2022). Additionally, in *Podophyllum hexandrum*, miR1873, miR172i, and miR829.1 are reported to be involved in flavonoid biosynthesis. In *Halostachys caspica*, miR6194 has been implicated in the flavonoid 3-hydroxylase expression regulation (Yang et al., 2015). Flavonoid pathways are also indirectly regulated by miRNAs that target transcription factors of genes associated with flavonoid metabolism. In Arabidopsis, miR858a, a light-regulated miRNA, was found to target flavonoid biosynthesis regulating transcription factors R2R3-MYB. This suggests miR858a might function as a regulatory hub, coordinating flavonoid biosynthesis (D. Sharma et al., 2016). MicroRNA858a overexpression resulted in the down-regulation of several MYB transcription factors associated with flavonoid biosynthesis. Up-regulation of MYBs in MIM858 (artificial miRNA target mimic) lines redirected the metabolic flux towards the biosynthesis of flavonoids, resulting in decreased lignin synthesis (D. Sharma et al., 2016).

The flavonoid biosynthetic pathway subsequently leads to anthocyanin biosynthesis. Anthocyanin biosynthesis requires phenylalanine as a substrate and involves key enzymes like phenylalanine ammonia-lyase (PAL) in the phenylpropanoid and flavonoid pathways. Through several enzymatic steps, including those catalyzed by chalcone synthase (CHS), chalcone isomerase (CHI), and dihydroflavonol 4-reductase (DFR), flavanone is produced, which is further processed into anthocyanin pigments and is stabilized by proteins such as anthocyanin synthase (ANS) and glycosyltransferase (AGT) before being transported to vacuoles and various sites in the cytoplasm (Jiang et al., 2023; S. Liu et al., 2023). They are vital SMs with notable antioxidant, photoprotective, and anticancer activities (Y. Wu, Zhang, et al., 2022). In *Arabidopsis thaliana,* miR156 targets Squamosa Promoter Binding Protein-Like (SPL), particularly SPL9, which is a negative regulator of anthocyanin biosynthesis (S. Li et al., 2020). SPL9 suppresses the expression of enzymes involved in anthocyanin glycoside production by destabilizing the MBW transcriptional complex, essential for anthocyanin synthesis, and miR156 inhibits SPL9. Down-regulation of SPL9 also results in down-regulation of CHS and F3H genes. Thus miR156 not only controls the biosynthesis of anthocyanins but also influences flavonol production and is known to regulate ROS accumulation and defense responses in Arabidopsis (Wang et al., 2020; Yin et al., 2019). Furthermore, miR858a is another miRNA critical for anthocyanin glycoside synthesis and Arabidopsis seedling development (Badola et al., 2022). Together with the transcription factor HY5, miR858a modulates light-induced production of anthocyanin by regulating CHS and CHI expression. HY5 enhances anthocyanin synthesis by directly down-regulating MYBL2, a crucial suppressor of anthocyanin biosynthetic genes. MicroRNA858a further enhances anthocyanin synthesis by inhibiting MYBL2 translation (A. Sharma et al., 2022; Xiao et al., 2022) (Figure 17.3).

Additionally, pri-miRNA encodes peptides like miPEP858a that enhance miRNA expression, such as pri-miR858a in Arabidopsis. MYB transcription factors (MYB11, MYB12, and MYB111) bind to miPEP858a and regulate key enzymes (F3H, CHS, and CHI) involved in flavonoid anthocyanin production (A. Sharma et al., 2020). This intricate regulatory network underscores the importance of miRNAs in controlling anthocyanin biosynthesis and broader flavonoid metabolism.

17.4.3 Regulation of Alkaloid Metabolism by microRNAs

Alkaloids are SMs with significant pharmacological value and are particularly noted for their anticancer properties (Olofinsan et al., 2023). Alkaloids play a critical role in the development, growth, and defense mechanisms of plants (C. Guo et al., 2022). Many of them are toxic to pests and herbivores (Rui et al., 2022). Many complex alkaloids are prized for their pharmacological properties and are widely used as medicinal compounds (Vchislo & Verochkina, 2020).

Research has demonstrated that various miRNAs regulate the biosynthesis of plant alkaloids (Figure 17.4, Table 17.3). For example, Boke et al. discovered that pso-miR13, pso-miR408, and pso-miR2161 in *Papaver somniferum* target genes associated with benzylisoquinoline alkaloid (BIA) metabolism, thus contributing to their biosynthesis (Boke et al., 2015). Pso-miR13 targets 7-O-methyltransferase mRNA, the enzyme required to transform S-reticulin into morphine alkaloids. Pso-miR408 mRNA reticuline oxidase-like protein converts S-reticuline to (S)-scoulerine in the BIAs pathway (Gupta et al., 2017). Pso-miR2161 targets 3′-hydroxy-N-methylcoclaurine 4′-O-methyltransferase 2, a benzylisoquinoline biosynthesis pathway enzyme, responsible for transforming S-norcoclaurine into S-reticuline. In *Lycoris aurea*, miR396 targets tyrosine decarboxylase, and CPYP450 enzymes of alkaloid biosynthesis (S. Xu et al., 2016). Similarly, miR166, miR396, and miR397 target, key enzymes of the Piperine biosynthesis pathway, Cinnamoyl-CoA reductase, 4-coumarate-CoA ligase, and peroxidase superfamily protein, respectively in *Piper nigrum* L. (Ding et al., 2021) (Figure 17.4a).

In *Curcuma longa*, miR5021 regulates aromatic amino acid aminotransferase, an isoquinoline alkaloid biosynthesis pathway gene (N. Singh & Sharma, 2017). Curcumin is an essential diketone compound in turmeric rhizomes. Several miRNAs are predicted to regulate curcumin biosynthesis (Figure 17.4c). Notably, miR156b, miR1168.2, and miR1858 were predicted to modulate curcumin biosynthesis (N. Singh & Sharma, 2017). MiR1168.2 regulates the expression of CYP73A, while miR1858 and miR156b modulate the activity of 4-coumaroyl-CoA synthase (4CL), influencing the curcumin biosynthetic pathway. Curcumin itself can induce miRNA expression, as seen in its activation of the ROS/KEAP/NRF/miR-34a/b/c cascade that inhibits metastasis in the case of colorectal cancer (Y. Liu et al., 2023). Additionally, miRNAs regulate the biosynthesis of other phenolic compounds. For instance, Sm-miR858 down-regulates phenolic acid metabolism enzyme-encoding genes, while Sm-miR408 down-regulates laccase 3 to modulate salvianolic acid accumulation, thereby impacting secondary metabolite production in *Salvia miltiorrhiza* (Zhang et al., 2022).

Tobacco (*Nicotiana tabacum*) nicotine metabolism is regulated by several miRNAs – nta-miRX17 and nta-miRX27 target Quinolinate Phosphoribosyltransferase1 and Quinolinate Phosphoribosyltransferase2 enzymes, respectively (Fu et al., 2013); nta-miRX20 targets *CYP82E4*, an enzyme required for conversion of nicotine to nornicotine by demethylation (Figure 17.4b); nta-miRX13 targets PMT2 that converts Putrescine to N-methylputrescine. Transcription factors associated with nicotine biosynthesis were reported to be regulated by miRNAs as well. Expressions of NtNAC-R1 and NtNAC_148 were regulated by miR164 and nta-miRX19, respectively (F. Li et al., 2015).In *Taxus mairei* leaves, miR164 and miR171 target taxane 13α hydroxylase and taxane 2α-O-benzoyltransferase, respectively, both are paclitaxel biosynthesis genes.

17.5 FUTURE PROSPECTS

SMs empower plants to adapt to environmental stresses, thereby improving their survival in natural habitats. Their functions vary across different tissues and cells due to the diversity of their chemical structures. Recent studies in plants have highlighted the diverse origins of SMs, broadening research avenues and aiding the extraction of pharmacologically active components. The regulation of biosynthesis of these metabolites is highly complex; especially the role of miRNAs in controlling synthesis is not yet adequately explored. Understanding how miRNAs regulate secondary metabolite biosynthesis is essential for further progress in this field. With the advancement of technology, new tools are available to study the complex regulatory pathways behind secondary metabolite pathways.

The development of sequencing technologies and the whole-genome sequencing of plants has revolutionized the identification of SMs. These advancements have greatly enhanced our ability to obtain high-quality chromosome-level assembled plant genomes, which are crucial for exhaustive research in molecular genetics and for identifying genes involved in the biosynthesis of plant SMs

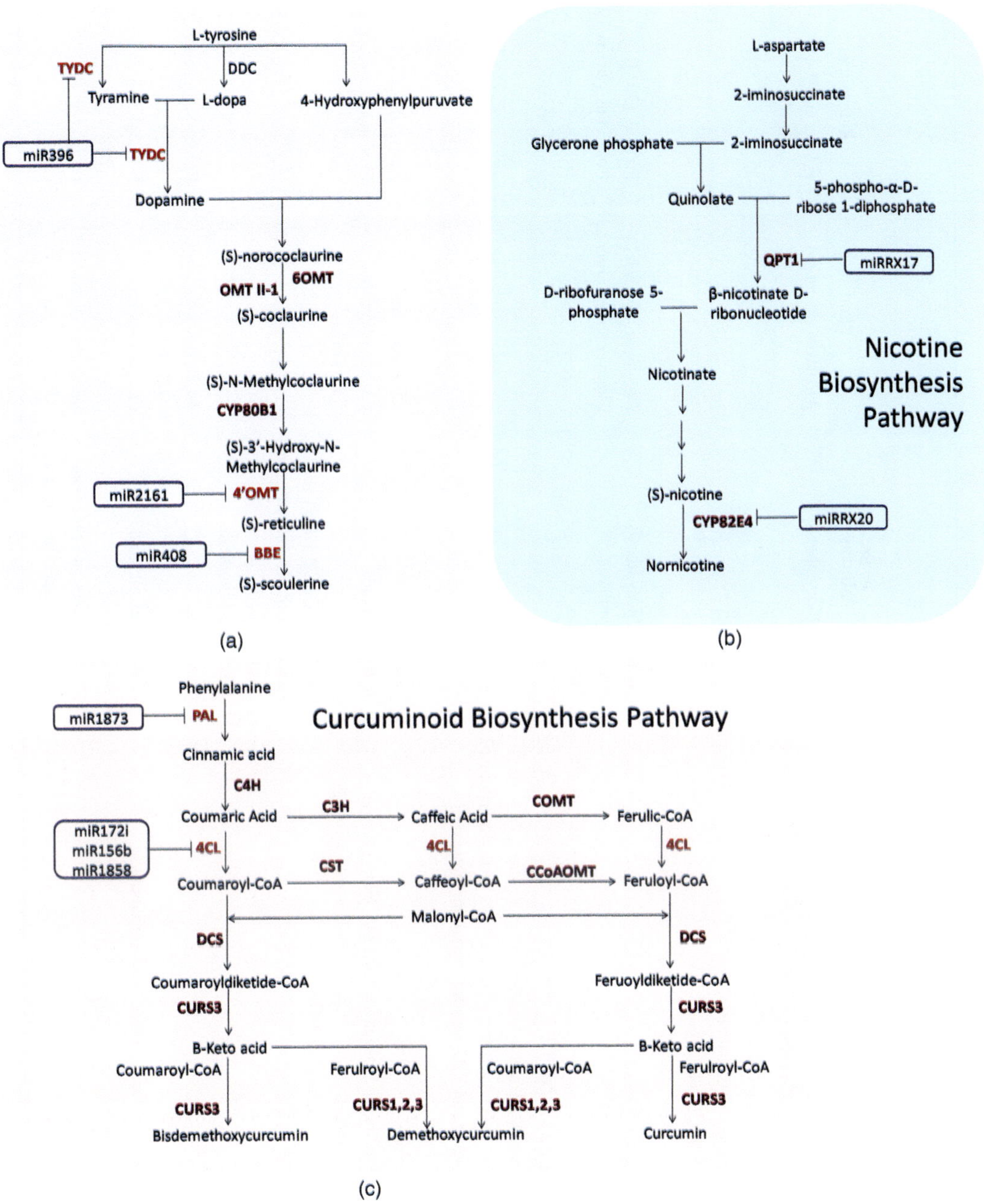

FIGURE 17.4 (a). Regulation of alkaloid biosynthesis pathway. The enzymes involved are: tyrosine/dopa decarboxylase (TYDC); norcoclaurine 6-O-methyltransferase (6OMT); O-methyltransferase II-1 (OMT II-1); (S)-N-methylcoclaurine 3'-hydroxylase (CYP80B1); 3'-hydroxy-N-methylcoclaurine 4'-O-methyltransferase (4'OMT); berberine bridge enzyme (BBE). (b). Regulation of nicotine biosynthesis in tobacco. The enzymes are: Quinolinate Phosphorobosyltransferasel (QPT1); nicotine N-demethylase (CYP82E4). (c). Regulation of curcuminoid biosynthesis pathway in turmeric. The enzymes involved are: phenylalanine ammonia lyase (PAL);cinnamate-4-hydroxylase (C4H); 4-coumarate-CoA ligase (4CL); p-coumaroyl shikimate transferase (CST); caffeoyl-CoA-O-methyltransferase (CCoAOMT); diketide-CoA synthase (DCS); curcumin synthases (CURS1, CURS2, and CURS3).

TABLE 17.3
List of miRNAs that Regulate Alkaloid Metabolism Pathways

miRNA	Target	Pathway	Plant	References
miRX13*	Putrescine methyltransferase 2 (PMT2)	Nicotine biosynthesis	*Nicotiana tabacum* L.	F. Li et al. (2015)
miRX17*	Quinolinate phosphoribosyl transferase 1 (QPT1)			
miRX20*	Cytochrome P450 monooxygenase (CYP82E4)			
miRX27*	Quinolinate phosphoribosyl-transferase 2 (QPT2)			
miRX17, miRX27, miRX20, miRX19	QPT1, QPT2, CYP82E4, PMT2			
miR164	NtNAC-R1	Nicotine biosynthesis	*Nicotiana tabacum* L.	Fu et al. (2013)
miR13	7-O-Methyltransferase (7-OMT)	BIA biosynthesis	*Papaver somniferum* L.	Boke et al. (2015)
miR408	Reticuline oxidase	Morphinan		
miR2161	4′-O-ethyltransferase 2 (4-OMT)			
pso-miR13, pso-miR2161, pso-miR408	7-O-methyltransferase, S-adenosyl-l-methionine:3′-hydroxy-N-methylcoclaurine 4′-O-methyltransferase 2/4′-O-methyltransferase2 (4-OMT)/FAD-binding and BBE domain-containing protein			
pso-miR13	7-OMT			
miR164	Taxane 13α hydroxylase	Taxol	*Taxus baccata* L.	Hao et al. (2012)
miR-5021	Two enzymes involved in the biosynthesis of TIAs, UDP-glucose iridoid glucosyltransferase	Indole alkaloids	*Catharanthus roseus* (L.) G.Don	Pani and Mahapatra (2013)
	cro-miR160, cro-miR164a, cro-miR164b	CrARF16, ORCA3, ORCA4, and BIS1 in TIA biosynthetic pathway		
	cro-miR393d	Alkaloid biosynthesis		
miR393	Auxin receptor	Camalexin	*Arabidopsis thaliana* (L.) Heynh.	Robert-Seilaniantz et al. (2011)
miR5021	Aspartate transaminase, aromatic-amino-acid transaminase	Isoquinoline alkaloid biosynthesis	*Curcuma longa* L.	N. Singh and Sharma (2017)
U3938865 and U4351355	MYB34	Glucosinolates and camalexin (indole alkaloid) biosynthesis	*Helianthus annuus* L.	J. Liu et al. (2017)

and their regulation. Phylogenetic studies and comparative genomics have provided new insights into identifying genes of secondary metabolism. For example, research in *Coptis chinensis* revealed gene families involved in BIA metabolism (Y. Liu et al., 2021), while studies on the *Camellia sinensis* (tea) genome uncovered genome duplication events that contributed to the production of caffeine and catechins (Xia et al., 2020). Moreover, tandem gene duplications have been linked to increased SM production in plants such as poppy, where these events enhance the biosynthesis of morphine (L. Guo et al., 2018). Functional genomics, including metabolomics and transcriptomics, is crucial in understanding plant SMs. Metabolomics helps to clarify the physiological roles of specific compounds and aids in identifying metabolic pathways, while transcriptomics reveals regulatory networks and gene expression patterns under various conditions. For instance, transcriptomic studies in *Glycyrrhiza uralensis* linked nitrogen utilization to SMs accumulation (Y. Chen et al., 2023), and similar work in *Ganoderma lucidum* showed that ethylene promotes SMs production (Meng et al., 2022).

Artificial microRNAs (amiRNAs) are widely used in plant research to silence genes. Unlike traditional methods like RNA interference (RNAi) or Virus-Induced Gene Silencing (VIGS) amiRNA can target a large number of closely linked genes, decreasing the chance of target miss and increasing efficiency. Its advantages include longer action duration, flexibility, and structural stability. By binding to target mRNA and by inducing RNA interference, amiRNA can degrade or inhibit translation, leading to gene silencing of crucial enzymes of secondary metabolic pathways. Furthermore, amiRNA technology enables the simultaneous production of multiple amiRNAs through the cis-antielectronic tRNA pre-amiRNA strategy, silencing numerous genes at once. Application of this technology showed that by silencing chalcone synthase using amiRNA, anthocyanin levels can be significantly reduced in transgenic plants compared to wild types (Niemeier et al., 2010). Similarly, phenylpropanoid biosynthesis may be regulated by targeting the PLA gene in *Populus trichocarpa* (Shi et al., 2010).

A powerful genome-editing tool, CRISPR/Cas9, is used to modify plant genes (Alok et al., 2020). It can disrupt miRNA biogenesis by deleting or inserting indels in the pre-miRNAs, affecting miRNA activity. CRISPR/Cas9 can be used to modify the expression of MIR genes by introducing indels into their transcription start sites or cis-regulatory elements. Furthermore, insertion of indels into target genes can block miRNA binding to its target, resulting in miRNA-induced gene cleavage failure (Ma et al., 2016). A notable application of CRISPR/Cas9 involved editing the miPEP858a MIR gene in *Arabidopsis*, which resulted in reduced lignin content and increased flavonoid and anthocyanin accumulation (A. Sharma et al., 2020). CRISPR/Cas9 can also be combined with multiple sgRNAs to enable multiplex genome editing, making it a versatile tool for promoting secondary metabolism in plants by targeting specific miRNAs that inhibit these processes.

Additionally, the Short Tandem Target Mimic (STTM) technique is used to downregulate entire miRNA families by causing miRNA degradation and functional loss. Both amiRNA and STTM can be employed for miRNA silencing experiments (C. Guo et al., 2022). For example, silencing *Arabidopsis* miR858 via STTM led to prolonged flowering and reduced plant growth (N. Singh, Srivastava, Shasany, et al., 2016). These advanced technologies offer novel ways to manipulate miRNA to boost secondary metabolite production.

17.6 CONCLUSION

In summary, the application of modern sequencing technologies, coupled with comparative and functional genomics, has opened new avenues for studying secondary metabolism in medicinal plants. Further, this has allowed the application of advanced molecular tools, i.e. amiRNA gene-silencing, STTM and CRISPR/Cas9-mediated genome-editing tools, to dissect secondary metabolism pathways. However, while promising, these advancements still require careful refinement and validation to overcome existing limitations. Continued research, leveraging high-tech tools in cultivation and pharmaceutical production, holds the potential to meet the growing demand for SMs while improving their quality and efficacy.

ACKNOWLEDGMENTS

RKG is thankful to Bhairab Ganguly College, Kolkata, India, for providing the infrastructural and financial assistance. AR acknowledges the "Excellent Team Grant" 2025–2026 from the Faculty of Forestry and Wood Sciences, Czech University of Life Sciences Prague, for financial support. Figure 17.1 was prepared using biorender.com. We acknowledge Mr. Saravanasakthi Sarath for his assistance in preparing the figure.

REFERENCES

Alok, A., Sandhya, D., Jogam, P., Rodrigues, V., Bhati, K. K., Sharma, H., & Kumar, J. (2020). The rise of the CRISPR/Cpf1 system for efficient genome editing in plants. *Frontiers in Plant Science, 11*. https://doi.org/10.3389/fpls.2020.00264

Angseesing, J. (1974). Selective eating of the acyanogenic form of *Trifolium repens*. *Heredity, 32*(1), 73–83.

Atanasov, A. G., Waltenberger, B., Pferschy-Wenzig, E.-M., Linder, T., Wawrosch, C., Uhrin, P., Temml, V., Wang, L., Schwaiger, S., Heiss, E. H., & others. (2015). Discovery and resupply of pharmacologically active plant-derived natural products: A review. *Biotechnology Advances, 33*(8), 1582–1614.

Badola, P. K., Sharma, A., Gautam, H., & Trivedi, P. K. (2022). MicroRNA858a, its encoded peptide, and phytosulfokine regulate Arabidopsis growth and development. *Plant Physiology, 189*(3), 1397–1415. https://doi.org/10.1093/plphys/kiac138

Barozai, M. Y. K., Baloch, I. A., & Din, M. (2012). Identification of microRNAs and their targets in *Helianthus*. *Molecular Biology Reports, 39*(3), 2523–2532. https://doi.org/10.1007/s11033-011-1004-y

Biswas, S., Hazra, S., & Chattopadhyay, S. (2016). Identification of conserved miRNAs and their putative target genes in *Podophyllum hexandrum* (Himalayan Mayapple). *Plant Gene, 6*, 82–89. https://doi.org/10.1016/j.plgene.2016.04.002

Boke, H., Ozhuner, E., Turktas, M., Parmaksiz, I., Ozcan, S., & Unver, T. (2015). Regulation of the alkaloid biosynthesis by miRNA in opium poppy. *Plant Biotechnology Journal, 13*(3), 409–420. https://doi.org/10.1111/pbi.12346

Borghi, M., Fernie, A. R., Schiestl, F. P., & Bouwmeester, H. J. (2017). The sexual advantage of looking, smelling, and tasting good: The metabolic network that produces signals for pollinators. *Trends in Plant Science, 22*(4), 338–350.

Bourgaud, F., Gravot, A., Milesi, S., & Gontier, E. (2001). Production of plant secondary metabolites: A historical perspective. *Plant Science, 161*(5), 839–851. https://doi.org/10.1016/S0168-9452(01)00490-3

Chen, X. (2005). microRNA biogenesis and function in plants. *FEBS Letters, 579*(26), 5923–5931. https://doi.org/10.1016/j.febslet.2005.07.071

Chen, Y., Bai, Y., Zhang, Z., Zhang, Y., Jiang, Y., Wang, S., Wang, Y., & Sun, Z. (2023). Transcriptomics and metabolomics reveal the primary and secondary metabolism changes in *Glycyrrhiza uralensis* with different forms of nitrogen utilization. *Frontiers in Plant Science, 14*. https://doi.org/10.3389/fpls.2023.1229253

Chiorcea-Paquim, A.-M. (2023). Electrochemistry of flavonoids: A comprehensive review. *International Journal of Molecular Sciences, 24*(21), 21. https://doi.org/10.3390/ijms242115667

Cho, Y. B., Jones, S. I., & Vodkin, L. (2013). The transition from primary siRNAs to amplified secondary siRNAs that regulate chalcone synthase during development of *Glycine max* seed coats. *PLoS One, 8*(10), e76954. https://doi.org/10.1371/journal.pone.0076954

Concha, E., Heipieper, H. J., Wick, L. Y., Ciudad, G. A., & Navia, R. (2018). Effects of limonene, n-decane and n-decanol on growth and membrane fatty acid composition of the microalga *Botryococcus braunii*. *AMB Express, 8*, 1–8.

Dewick, P. M. (2002). *Medicinal Natural Products: A Biosynthetic Approach*. John Wiley & Sons.

Ding, Y., Mao, Y., Cen, Y., Hu, L., Su, Y., Ma, X., Long, L., Hu, H., Hao, C., & Luo, J. (2021). Small RNA sequencing reveals various microRNAs involved in piperine biosynthesis in black pepper (*Piper nigrum* L.). *BMC Genomics, 22*, 838. https://doi.org/10.1186/s12864-021-08154-4

Ebel, J. (1986). Phytoalexin synthesis: The biochemical analysis of the induction process. *Annual Review of Phytopathology, 24*(1), 235–264.

Edreva, A. M., Velikova, V. B., & Tsonev, T. D. (2007). Phenylamides in plants. *Russian Journal of Plant Physiology, 54*(3), 287–301. https://doi.org/10.1134/S1021443707030016

Faegri, K., & Pijl, L. van der. (1979). *The Principles of Pollination Ecology.* (3rd ed.), Oxford: Pergamon Press.

Fan, R., Li, Y., Li, C., & Zhang, Y. (2015). Differential microRNA analysis of glandular trichomes and young leaves in *Xanthium strumarium* L. reveals their putative roles in regulating terpenoid biosynthesis. *PLoS One, 10*(9), e0139002. https://doi.org/10.1371/journal.pone.0139002

Fu, Y., Guo, H., Cheng, Z., Wang, R., Li, G., Huo, G., & Liu, W. (2013). NtNAC-R1, a novel NAC transcription factor gene in tobacco roots, responds to mechanical damage of shoot meristem. *Plant Physiology and Biochemistry*, *69*, 74–81. https://doi.org/10.1016/j.plaphy.2013.05.004

Gali-Muhtasib, H., Hmadi, R., Kareh, M., Tohme, R., & Darwiche, N. (2015). Cell death mechanisms of plant-derived anticancer drugs: Beyond apoptosis. *Apoptosis*, *20*(12), 1531–1562. https://doi.org/10.1007/s10495-015-1169-2

Ghasemzadeh, A., Talei, D., Jaafar, H. Z. E., Juraimi, A. S., Mohamed, M. T. M., Puteh, A., & Halim, M. R. A. (2016). Plant-growth regulators alter phytochemical constituents and pharmaceutical quality in Sweet potato (*Ipomoea batatas* L.). *BMC Complementary and Alternative Medicine*, *16*(1), 152. https://doi.org/10.1186/s12906-016-1113-1

Gou, J.-Y., Felippes, F. F., Liu, C.-J., Weigel, D., & Wang, J.-W. (2011). Negative regulation of anthocyanin biosynthesis in arabidopsis by a miR156-targeted SPL transcription factor. *The Plant Cell*, *23*(4), 1512–1522. https://doi.org/10.1105/tpc.111.084525

Guo, C., Chen, Y., Wu, D., Du, Y., Wang, M., Liu, C., Chu, J., & Yao, X. (2022). Transcriptome analysis reveals an essential role of exogenous brassinolide on the alkaloid biosynthesis pathway in *Pinellia ternata*. *International Journal of Molecular Sciences*, *23*(18), 18. https://doi.org/10.3390/ijms231810898

Guo, L., Winzer, T., Yang, X., Li, Y., Ning, Z., He, Z., Teodor, R., Lu, Y., Bowser, T. A., Graham, I. A., & Ye, K. (2018). The opium poppy genome and morphinan production. *Science*, *362*(6412), 343–347. https://doi.org/10.1126/science.aat4096

Gupta, O. P., Karkute, S. G., Banerjee, S., Meena, N. L., & Dahuja, A. (2017). Contemporary understanding of miRNA-based regulation of secondary metabolites biosynthesis in plants. *Frontiers in Plant Science*, *8*. https://doi.org/10.3389/fpls.2017.00374

Gutiérrez-García, C., Ahmed, S. S. S. J., Ramalingam, S., Selvaraj, D., Srivastava, A., Paul, S., & Sharma, A. (2022). Identification of microRNAs from medicinal plant Murraya koenigii by high-throughput sequencing and their functional implications in secondary metabolite biosynthesis. *Plants*, *11*(1), 1. https://doi.org/10.3390/plants11010046

Hao, D.-C., Yang, L., Xiao, P.-G., & Liu, M. (2012). Identification of Taxus microRNAs and their targets with high-throughput sequencing and degradome analysis. *Physiologia Plantarum*, *146*(4), 388–403. https://doi.org/10.1111/j.1399-3054.2012.01668.x

Harborne, J. B. (2014). *Introduction to Ecological Biochemistry*. **London and San Diego:** Academic Press.

Harborne, J. B., & Nash, R. J. (1984). Flavonoid pigments responsible for ultraviolet patterning in petals of the genus Potentilla. *Biochemical Systematics and Ecology*, *12*(3), 315–318.

He, L., Tang, R., Shi, X., Wang, W., Cao, Q., Liu, X., Wang, T., Sun, Y., Zhang, H., Li, R., & Jia, X. (2019). Uncovering anthocyanin biosynthesis related microRNAs and their target genes by small RNA and degradome sequencing in tuberous roots of sweetpotato. *BMC Plant Biology*, *19*(1), 232. https://doi.org/10.1186/s12870-019-1790-2

Horner, J. D., Gosz, J. R., & Cates, R. G. (1988). The role of carbon-based plant secondary metabolites in decomposition in terrestrial ecosystems. *The American Naturalist*, *132*(6), 869–883.

Hsu, Y.-L., Kuo, P.-L., Lin, L.-T., & Lin, C.-C. (2005). Asiatic acid, a triterpene, induces apoptosis and cell cycle arrest through activation of extracellular signal-regulated kinase and p38 mitogen-activated protein kinase pathways in human breast cancer cells. *Journal of Pharmacology and Experimental Therapeutics*, *313*(1), 333–344. https://doi.org/10.1124/jpet.104.078808

Idris, A. I., Libouban, H., Nyangoga, H., Landao-Bassonga, E., Chappard, D., & Ralston, S. H. (2009). Pharmacologic inhibitors of IκB kinase suppress growth and migration of mammary carcinosarcoma cells in vitro and prevent osteolytic bone metastasis in vivo. *Molecular Cancer Therapeutics*, *8*(8), 2339–2347. https://doi.org/10.1158/1535-7163.MCT-09-0133

Jeena, G. S., Singh, N., Shikha, & Shukla, R. K. (2022). An insight into microRNA biogenesis and its regulatory role in plant secondary metabolism. *Plant Cell Reports*, *41*(8), 1651–1671. https://doi.org/10.1007/s00299-022-02877-8

Jennifer Mordue (Luntz), A., Simmonds, M. S. J., Ley, S. V., Blaney, W. M., Mordue, W., Nasiruddin, M., & Nisbet, A. J. (1998). Actions of azadirachtin, a plant allelochemical, against insects. *Pesticide Science*, *54*(3), 277–284. https://doi.org/10.1002/(SICI)1096-9063(1998110)54:3<277::AID-PS801>3.0.CO;2-I

Jiang, J., Huang, H., Gao, Q., Li, Y., Xiang, H., Zeng, W., Xu, L., Liu, X., Li, J., Mi, Q., Deng, L., Yang, W., Zhang, J., Yang, G., & Li, X. (2023). Effects of editing DFR genes on flowers, leaves, and roots of tobacco. *BMC Plant Biology*, *23*(1), 349. https://doi.org/10.1186/s12870-023-04307-7

Khan, S., Ali, A., Saifi, M., Saxena, P., Ahlawat, S., & Abdin, M. Z. (2020). Identification and the potential involvement of miRNAs in the regulation of artemisinin biosynthesis in *A. annua*. *Scientific Reports*, *10*(1), 13614. https://doi.org/10.1038/s41598-020-69707-3

Kuiters, A. (1990). Role of phenolic substances from decomposing forest litter in plant-soil interactions. *Acta Botanica Neerlandica*, *39*(4), 329–348.

Lamb, C. J., Ryals, J. A., Ward, E. R., & Dixon, R. A. (1992). Emerging strategies for enhancing crop resistance to microbial pathogens. *Bio/Technology*, *10*(11), 1436–1445. https://doi.org/10.1038/nbt1192-1436

Legrand, S., Valot, N., Nicolé, F., Moja, S., Baudino, S., Jullien, F., Magnard, J.-L., Caissard, J.-C., & Legendre, L. (2010). One-step identification of conserved miRNAs, their targets, potential transcription factors and effector genes of complete secondary metabolism pathways after 454 pyrosequencing of calyx cDNAs from the Labiate *Salvia sclarea* L. *Gene*, *450*(1), 55–62. https://doi.org/10.1016/j.gene.2009.10.004

Li, F., Wang, W., Zhao, N., Xiao, B., Cao, P., Wu, X., Ye, C., Shen, E., Qiu, J., Zhu, Q.-H., Xie, J., Zhou, X., & Fan, L. (2015). Regulation of nicotine biosynthesis by an endogenous target mimicry of microRNA in tobacco. *Plant Physiology*, *169*(2), 1062–1071. https://doi.org/10.1104/pp.15.00649

Li, S., Li, L., Jiang, Y., Wu, J., Sun, H., Zhao, M., Jiang, Y., Zhu, L., Wang, Y., Su, Y., Wang, K., Wang, Y., & Zhang, M. (2020). *SQUAMOSA* promoter binding protein-like (*SPL*) gene family: TRANSCRIPTOME-wide identification, phylogenetic relationship, expression patterns and network interaction analysis in *Panax ginseng* C. A. Meyer. *Plants*, *9*(3), 3. https://doi.org/10.3390/plants9030354

Li, Y., Cui, W., Wang, R., Lin, M., Zhong, Y., Sun, L., Qi, X., & Fang, J. (2019). MicroRNA858-mediated regulation of anthocyanin biosynthesis in kiwifruit (*Actinidia arguta*) based on small RNA sequencing. *PLoS One*, *14*(5), e0217480. https://doi.org/10.1371/journal.pone.0217480

Liu, J., Yuan, Y., Wang, Y., Jiang, C., Chen, T., Zhu, F., Zhao, Y., Zhou, J., & Huang, L. (2017). Regulation of fatty acid and flavonoid biosynthesis by miRNAs in *Lonicera japonica*. *RSC Advances*, *7*(56), 35426–35437. https://doi.org/10.1039/C7RA05800D

Liu, S., Zhang, Q., Kollie, L., Dong, J., & Liang, Z. (2023). Molecular networks of secondary metabolism accumulation in plants: Current understanding and future challenges. *Industrial Crops and Products*, *201*, 116901. https://doi.org/10.1016/j.indcrop.2023.116901

Liu, Y., Singh, S. K., Pattanaik, S., Wang, H., & Yuan, L. (2023). Light regulation of the biosynthesis of phenolics, terpenoids, and alkaloids in plants. *Communications Biology*, *6*(1), 1–14. https://doi.org/10.1038/s42003-023-05435-4

Liu, Y., Wang, B., Shu, S., Li, Z., Song, C., Liu, D., Niu, Y., Liu, J., Zhang, J., Liu, H., Hu, Z., Huang, B., Liu, X., Liu, W., Jiang, L., Alami, M. M., Zhou, Y., Ma, Y., He, X., … Nie, J. (2021). Analysis of the *Coptis chinensis* genome reveals the diversification of protoberberine-type alkaloids. *Nature Communications*, *12*(1), 3276. https://doi.org/10.1038/s41467-021-23611-0

Luo, Y., Zhang, X., Luo, Z., Zhang, Q., & Liu, J. (2015). Identification and characterization of microRNAs from Chinese pollination constant non-astringent persimmon using high-throughput sequencing. *BMC Plant Biology*, *15*(1), 11. https://doi.org/10.1186/s12870-014-0400-6

Ma, X., Zhu, Q., Chen, Y., & Liu, Y.-G. (2016). CRISPR/Cas9 platforms for genome editing in plants: Developments and applications. *Molecular Plant*, *9*(7), 961–974. https://doi.org/10.1016/j.molp.2016.04.009

Martelli, G., & Giacomini, D. (2018). Antibacterial and antioxidant activities for natural and synthetic dual-active compounds. *European Journal of Medicinal Chemistry*, *158*, 91–105.

Mathiyalagan, R., Subramaniyam, S., Natarajan, S., Kim, Y. J., Sun, M. S., Kim, S. Y., Kim, Y.-J., & Yang, D. C. (2013). Insilico profiling of microRNAs in Korean ginseng (*Panax ginseng* Meyer). *Journal of Ginseng Research*, *37*(2), 227–247. https://doi.org/10.5142/jgr.2013.37.227

McCullough, M. L., Peterson, J. J., Patel, R., Jacques, P. F., Shah, R., & Dwyer, J. T. (2012). Flavonoid intake and cardiovascular disease mortality in a prospective cohort of US adults. *The American Journal of Clinical Nutrition*, *95*(2), 454–464. https://doi.org/10.3945/ajcn.111.016634

Meng, L., Zhou, R., Lin, J., Wang, Q., Wang, P., Wang, W., Wang, L., & Li, Z. (2022). Integrated transcriptomics and nontargeted metabolomics analysis reveal key metabolic pathways in *Ganoderma lucidum* in response to ethylene. *Journal of Fungi*, *8*(5), 5. https://doi.org/10.3390/jof8050456

Niemeier, S., Alves, L., & Merkle, T. (2010). Improvement of the design and generation of highly specific plant knockdown lines using primary synthetic microRNAs (pri-smiRNAs). *BMC Research Notes*, *3*(1), 59. https://doi.org/10.1186/1756-0500-3-59

Nile, S. H., Keum, Y. S., Nile, A. S., Jalde, S. S., & Patel, R. V. (2018). Antioxidant, anti-inflammatory, and enzyme inhibitory activity of natural plant flavonoids and their synthesized derivatives. *Journal of Biochemical and Molecular Toxicology*, *32*(1), e22002.

Nyaboke, H. O., Moraa, M., Omosa, L. K., Mbaveng, A. T., Vaderament-Alexe, N.-N., Masila, V., Okemwa, E., Heydenreich, M., Efferth, T., & Kuete, V. (2018). Cytotoxicity of lupeol from the stem bark of *Zanthoxylum gilletii* against multi-factorial drug resistant cancer cell lines. *Investigational Medicinal Chemistry and Pharmacology*, *1*(1), 10.

Olofinsan, K., Abrahamse, H., & George, B. P. (2023). Therapeutic role of alkaloids and alkaloid derivatives in cancer management. *Molecules*, *28*(14), 4. https://doi.org/10.3390/molecules28145578

Owusu Adjei, M., Zhou, X., Mao, M., Rafique, F., & Ma, J. (2021). MicroRNAs roles in plants secondary metabolism. *Plant Signaling & Behavior*, *16*(7), 1915590. https://doi.org/10.1080/15592324.2021.1915590

Pani, A., & Mahapatra, R. K. (2013). Computational identification of microRNAs and their targets in *Catharanthus roseus* expressed sequence tags. *Genomics Data*, *1*, 2–6. https://doi.org/10.1016/j.gdata.2013.06.001

Pani, A., Mahapatra, R. K., Behera, N., & Naik, P. K. (2011). Computational identification of sweet wormwood (*Artemisia Annua*) microRNA and their mRNA targets. *Genomics, Proteomics & Bioinformatics*, *9*(6), 200–210. https://doi.org/10.1016/S1672-0229(11)60023-5

Park, M. Y., Wu, G., Gonzalez-Sulser, A., Vaucheret, H., & Poethig, R. S. (2005). Nuclear processing and export of microRNAs in *Arabidopsis*. *Proceedings of the National Academy of Sciences of the United States of America*, *102*(10), 3691–3696. https://doi.org/10.1073/pnas.0405570102

Pastorino, G., Cornara, L., Soares, S., Rodrigues, F., & Oliveira, M. B. P. P. (2018). Liquorice (Glycyrrhiza glabra): A phytochemical and pharmacological review. *Phytotherapy Research*, *32*(12), 2323–2339. https://doi.org/10.1002/ptr.6178

Pérez-Quintero, Á. L., Sablok, G., Tatarinova, T. V., Conesa, A., Kuo, J., & López, C. (2012). Mining of miRNAs and potential targets from gene oriented clusters of transcripts sequences of the anti-malarial plant, *Artemisia annua*. *Biotechnology Letters*, *34*(4), 737–745. https://doi.org/10.1007/s10529-011-0808-0

Phillipson, J. D. (2001). Phytochemistry and medicinal plants. *Phytochemistry*, *56*(3), 237–243. https://doi.org/10.1016/S0031-9422(00)00456-8

Prakash, P., Rajakani, R., & Gupta, V. (2016). Transcriptome-wide identification of *Rauvolfia serpentina* microRNAs and prediction of their potential targets. *Computational Biology and Chemistry*, *61*, 62–74. https://doi.org/10.1016/j.compbiolchem.2015.12.002

Reyes-Calderón, A., Gutiérrez-García, C., Urióstegui-Pena, A. G., Srivastava, A., Aguilar-Marcelino, L., Iqbal, H. M. N., Ahmed, S. S. S. J., Paul, S., & Sharma, A. (2023). Identification of cumin (*Cuminum cyminum*) microRNAs through deep sequencing and their impact on plant secondary metabolism. *Plants*, *12*(9), 9. https://doi.org/10.3390/plants12091756

Rhodes, M. (1994). Physiological roles for secondary metabolites in plants: Some progress, many outstanding problems. *Plant Molecular Biology*, *24*, 1–20.

Robert-Seilaniantz, A., MacLean, D., Jikumaru, Y., Hill, L., Yamaguchi, S., Kamiya, Y., & Jones, J. D. G. (2011). The microRNA miR393 re-directs secondary metabolite biosynthesis away from camalexin and towards glucosinolates. *The Plant Journal*, *67*(2), 218–231. https://doi.org/10.1111/j.1365-313X.2011.04591.x

Rui, Y., Li, S., Luan, F., Li, D., Liu, R., & Zeng, N. (2022). Several alkaloids in Chinese herbal medicine exert protection in acute kidney injury: Focus on mechanism and target analysis. *Oxidative Medicine and Cellular Longevity*, *2022*(1), 2427802. https://doi.org/10.1155/2022/2427802

Saifi, M., Nasrullah, N., Ahmad, M. M., Ali, A., Khan, J. A., & Abdin, M. Z. (2015). *In silico* analysis and expression profiling of miRNAs targeting genes of steviol glycosides biosynthetic pathway and their relationship with steviol glycosides content in different tissues of *Stevia rebaudiana*. *Plant Physiology and Biochemistry*, *94*, 57–64. https://doi.org/10.1016/j.plaphy.2015.05.009

Samad, A. F. A., Rahnamaie-Tajadod, R., Sajad, M., Jani, J., Murad, A. M. A., Noor, N. M., & Ismail, I. (2019). Regulation of terpenoid biosynthesis by miRNA in *Persicaria minor* induced by *Fusarium oxysporum*. *BMC Genomics*, *20*(1), 586. https://doi.org/10.1186/s12864-019-5954-0

Seigler, D. S. (2012). *Plant Secondary Metabolism*. Springer Science & Business Media.

Selmar, D. (1989). Cyanogenesis and metabolization of cyanogenic compounds: An example for multifunctionality of secondary plant products. *Planta Medica*, *55*(07), 592–592. https://doi.org/10.1055/s-2006-962126

Sharma, A., Badola, P. K., Bhatia, C., Sharma, D., & Trivedi, P. K. (2020). Primary transcript of miR858 encodes regulatory peptide and controls flavonoid biosynthesis and development in *Arabidopsis*. *Nature Plants*, *6*(10), 1262–1274. https://doi.org/10.1038/s41477-020-00769-x

Sharma, A., Badola, P. K., Gautam, H., Gaddam, S. R., & Trivedi, P. K. (2022). HY5 regulates light-dependent expression and accumulation of miR858a-encoded peptide, miPEP858a. *Biochemical and Biophysical Research Communications*, *589*, 204–208. https://doi.org/10.1016/j.bbrc.2021.12.020

Sharma, D., Tiwari, M., Pandey, A., Bhatia, C., Sharma, A., & Trivedi, P. K. (2016). MicroRNA858 is a potential regulator of phenylpropanoid pathway and plant development. *Plant Physiology*, *171*(2), 944–959. https://doi.org/10.1104/pp.15.01831

Shi, R., Yang, C., Lu, S., Sederoff, R., & Chiang, V. L. (2010). Specific down-regulation of *PAL* genes by artificial microRNAs in *Populus trichocarpa*. *Planta*, *232*(6), 1281–1288. https://doi.org/10.1007/s00425-010-1253-3

Singh, B., & Rastogi, R. P. (1970). Cardenolides—Glycosides and genins. *Phytochemistry*, *9*(2), 315–331. https://doi.org/10.1016/S0031-9422(00)85141-9

Singh, N., & Sharma, A. (2017). Turmeric (*Curcuma longa*): miRNAs and their regulating targets are involved in development and secondary metabolite pathways. *Comptes Rendus Biologies*, *340*(11–12), 481–491. https://doi.org/10.1016/j.crvi.2017.09.009

Singh, N., Srivastava, S., & Sharma, A. (2016). Identification and analysis of miRNAs and their targets in ginger using bioinformatics approach. *Gene*, *575*(2, Part 2), 570–576. https://doi.org/10.1016/j.gene.2015.09.036

Singh, N., Srivastava, S., Shasany, A. K., & Sharma, A. (2016). Identification of miRNAs and their targets involved in the secondary metabolic pathways of *Mentha* spp. *Computational Biology and Chemistry*, *64*, 154–162. https://doi.org/10.1016/j.compbiolchem.2016.06.004

Siva, M., Shanmugam, K., Shanmugam, B., Venkata, S. G., Ravi, S., Sathyavelu, R., & Mallikarjuna, K. (2016). *Ocimum sanctum*: A review on the pharmacological properties. *International Journal of Basic & Clinical Pharmacology*, *5*(3), 558–565.

Sobhani Najafabadi, A., & Naghavi, M. R. (2018). Mining *Ferula gummosa* transcriptome to identify miRNAs involved in the regulation and biosynthesis of terpenes. *Gene*, *645*, 41–47. https://doi.org/10.1016/j.gene.2017.12.035

Song, C., Guan, Y., Zhang, D., Tang, X., & Chang, Y. (2022). Integrated mRNA and miRNA transcriptome analysis suggests a regulatory network for UV–B-controlled terpenoid synthesis in fragrant woodfern (*Dryopteris fragrans*). *International Journal of Molecular Sciences*, *23*(10), 10. https://doi.org/10.3390/ijms23105708

Sorbera, L., Leeson, P., & Castaner, J. (1999). Calanolide A. *Drugs of the Future*, *24*(3), 235–245.

Sun, P., Cheng, C., Lin, Y., Zhu, Q., Lin, J., & Lai, Z. (2017). Combined small RNA and degradome sequencing reveals complex microRNA regulation of catechin biosynthesis in tea (*Camellia sinensis*). *PLoS One*, *12*(2), e0171173. https://doi.org/10.1371/journal.pone.0171173

Sun, Y., Qiu, Y., Duan, M., Wang, J., Zhang, X., Wang, H., Song, J., & Li, X. (2017). Identification of anthocyanin biosynthesis related microRNAs in a distinctive Chinese radish (*Raphanus sativus* L.) by high-throughput sequencing. *Molecular Genetics and Genomics*, *292*(1), 215–229. https://doi.org/10.1007/s00438-016-1268-y

Tattersfield, F., Hobson, R. P., & Gimingham, C. T. (1929). Pyrethrin I and II: Their insecticidal value and estimation in Pyrethrum (*Chrysanthemum cinebariaefolium*). I. *The Journal of Agricultural Science*, *19*(2), 266–296. https://doi.org/10.1017/S002185960001128X

Tirumalai, V., Swetha, C., Nair, A., Pandit, A., & Shivaprasad, P. V. (2019). miR828 and miR858 regulate VvMYB114 to promote anthocyanin and flavonol accumulation in grapes. *Journal of Experimental Botany*, *70*(18), 4775–4792. https://doi.org/10.1093/jxb/erz264

Tuteja, J. H., Zabala, G., Varala, K., Hudson, M., & Vodkin, L. O. (2009). Endogenous, tissue-specific short interfering RNAs silence the chalcone synthase gene family in Glycine max seed coats. *The Plant Cell*, *21*(10), 3063–3077. https://doi.org/10.1105/tpc.109.069856

Vasconcelos, N., Croda, J., & Simionatto, S. (2018). Antibacterial mechanisms of cinnamon and its constituents: A review. *Microbial Pathogenesis*, *120*, 198–203.

Vashisht, I., Mishra, P., Pal, T., Chanumolu, S., Singh, T. R., & Chauhan, R. S. (2015). Mining NGS transcriptomes for miRNAs and dissecting their role in regulating growth, development, and secondary metabolites production in different organs of a medicinal herb, *Picrorhiza kurroa*. *Planta*, *241*(5), 1255–1268. https://doi.org/10.1007/s00425-015-2255-y

Vchislo, N. V., & Verochkina, E. A. (2020). Recent advances in total synthesis of alkaloids from α,β-unsaturated aldehydes. *ChemistrySelect*, *5*(31), 9579–9589. https://doi.org/10.1002/slct.202002872

Wang, Y., Liu, W., Wang, X., Yang, R., Wu, Z., Wang, H., Wang, L., Hu, Z., Guo, S., Zhang, H., Lin, J., & Fu, C. (2020). MiR156 regulates anthocyanin biosynthesis through SPL targets and other microRNAs in poplar. *Horticulture Research*, *7*, 118. https://doi.org/10.1038/s41438-020-00341-w

Wen, H., Cui, H., Tian, H., Zhang, X., Ma, L., Ramassamy, C., & Li, J. (2020). Isolation of neuroprotective anthocyanins from black chokeberry (*Aronia melanocarpa*) against amyloid-β-induced cognitive impairment. *Foods*, *10*(1), 63.

Wink, M. (2015). Modes of action of herbal medicines and plant secondary metabolites. *Medicines*, *2*(3), 251–286.

Wu, J., Wang, D., Liu, Y., Wang, L., Qiao, X., & Zhang, S. (2014). Identification of miRNAs involved in pear fruit development and quality. *BMC Genomics*, *15*(1), 953. https://doi.org/10.1186/1471-2164-15-953

Wu, Y., Huang, X., Zhang, S., Zhang, C., Yang, H., Lyu, L., Li, W., & Wu, W. (2022). Small RNA and degradome sequencing reveal the role of blackberry miRNAs in flavonoid and anthocyanin synthesis during fruit ripening. *International Journal of Biological Macromolecules*, *213*, 892–901. https://doi.org/10.1016/j.ijbiomac.2022.06.035

Wu, Y., Zhang, C., Huang, Z., Lyu, L., Li, W., & Wu, W. (2022). Integrative analysis of the metabolome and transcriptome provides insights into the mechanisms of flavonoid biosynthesis in blackberry. *Food Research International*, *153*, 110948. https://doi.org/10.1016/j.foodres.2022.110948

Xia, E., Tong, W., Hou, Y., An, Y., Chen, L., Wu, Q., Liu, Y., Yu, J., Li, F., Li, R., Li, P., Zhao, H., Ge, R., Huang, J., Mallano, A. I., Zhang, Y., Liu, S., Deng, W., Song, C., … Wan, X. (2020). The reference genome of tea plant and resequencing of 81 diverse accessions provide insights into its genome evolution and adaptation. *Molecular Plant*, *13*(7), 1013–1026. https://doi.org/10.1016/j.molp.2020.04.010

Xiao, Y., Chu, L., Zhang, Y., Bian, Y., Xiao, J., & Xu, D. (2022). HY5: A pivotal regulator of light-dependent development in higher plants. *Frontiers in Plant Science*, *12*. https://doi.org/10.3389/fpls.2021.800989

Xu, P., Li, Q., Liang, W., Hu, Y., Chen, R., Lou, K., Zhan, L., Wu, X., & Pu, J. (2023). A tissue-specific profile of miRNAs and their targets related to paeoniaflorin and monoterpenoids biosynthesis in *Paeonia lactiflora* Pall. By transcriptome, small RNAs and degradome sequencing. *PLoS One*, *18*(1), e0279992. https://doi.org/10.1371/journal.pone.0279992

Xu, S., Jiang, Y., Wang, N., Xia, B., Jiang, Y., Li, X., Zhang, Z., Li, Y., & Wang, R. (2016). Identification and differential regulation of microRNAs in response to methyl jasmonate treatment in Lycoris aurea by deep sequencing. *BMC Genomics*, *17*(1), 789. https://doi.org/10.1186/s12864-016-2645-y

Xu, X., Jiang, Q., Ma, X., Ying, Q., Shen, B., Qian, Y., Song, H., & Wang, H. (2014). Deep sequencing identifies tissue-specific microRNAs and their target genes involving in the biosynthesis of tanshinones in *Salvia miltiorrhiza*. *PLoS One*, *9*(11), e111679. https://doi.org/10.1371/journal.pone.0111679

Xu, Y., Liu, J., Ji, X., Zhao, G., Zhao, T., Wang, X., Wang, L., Gao, S., Hao, Y., Gao, Y., Gao, Y., Weng, X., Jia, L., & Chen, Z. (2023). Integrative analysis of microRNAs and mRNAs reveals the regulatory networks of triterpenoid saponin metabolism in Soapberry (*Sapindus mukorossi* Gaertn.). *Frontiers in Plant Science*, *13*. https://doi.org/10.3389/fpls.2022.1037784

Yang, R., Zeng, Y., Yi, X., Zhao, L., & Zhang, Y. (2015). Small RNA deep sequencing reveals the important role of microRNAs in the halophyte *Halostachys caspica*. *Plant Biotechnology Journal*, *13*(3), 395–408. https://doi.org/10.1111/pbi.12337

Ye, J., Zhang, X., Tan, J., Xu, F., Cheng, S., Chen, Z., Zhang, W., & Liao, Y. (2020). Global identification of *Ginkgo biloba* microRNAs and insight into their role in metabolism regulatory network of terpene trilactones by high-throughput sequencing and degradome analysis. *Industrial Crops and Products*, *148*, 112289. https://doi.org/10.1016/j.indcrop.2020.112289

Yin, H., Hong, G., Li, L., Zhang, X., Kong, Y., Sun, Z., Li, J., Chen, J., & He, Y. (2019). miR156/SPL9 regulates reactive oxygen species accumulation and immune response in *Arabidopsis thaliana*. *Phytopathology®*, *109*(4), 632–642. https://doi.org/10.1094/PHYTO-08-18-0306-R

Yu, Z.-X., Wang, L.-J., Zhao, B., Shan, C.-M., Zhang, Y.-H., Chen, D.-F., & Chen, X.-Y. (2015). Progressive regulation of sesquiterpene biosynthesis in *Arabidopsis* and Patchouli (*Pogostemon cablin*) by the miR156-targeted SPL transcription factors. *Molecular Plant*, *8*(1), 98–110. https://doi.org/10.1016/j.molp.2014.11.002

Zhang, S., Wu, Y., Huang, X., Wu, W., Lyu, L., & Li, W. (2022). Research progress about microRNAs involved in plant secondary metabolism. *International Journal of Biological Macromolecules*, *216*, 820–829. https://doi.org/10.1016/j.ijbiomac.2022.07.224

Zhao, S., Mi, X., Guo, R., Xia, X., Liu, L., An, Y., Yan, X., Wang, S., Guo, L., & Wei, C. (2020). The biosynthesis of main taste compounds is coordinately regulated by miRNAs and phytohormones in tea plant (*Camellia sinensis*). *Journal of Agricultural and Food Chemistry*, *68*(22), 6221–6236. https://doi.org/10.1021/acs.jafc.0c01833

Zhao, S., Wang, X., Yan, X., Guo, L., Mi, X., Xu, Q., Zhu, J., Wu, A., Liu, L., & Wei, C. (2018). Revealing of microRNA involved regulatory gene networks on terpenoid biosynthesis in *Camellia sinensis* in different growing time points. *Journal of Agricultural and Food Chemistry*, *66*(47), 12604–12616. https://doi.org/10.1021/acs.jafc.8b05345

18 Unraveling the Role of Secondary Metabolites in Influencing the Plant-Pollinator Interactions

Pratap A. Divekar[1], *Abhishek Chaudhary*[2], *Ram Kumar*[2], *Kajol Gupta*[2], *and Priyanka Shahi*[2]

[1]Division of Crop Protection, ICAR-Indian Institute of Vegetble Research, Varanasi, India

[2]Institute of Agricultural Sciences, Banaras Hindu University, Varanasi, India

18.1 INTRODUCTION

18.1.1 What Are Secondary Metabolites?

Eco-friendly insect pest management is crucial for achieving sustainable food production. Several environmentally conscious and sustainable approaches to pest control should be prioritized to protect crops while minimizing negative impacts on pollinator bees and beneficial organisms. The eco-friendly pest management approaches includes the application of plant-derived substances (Divekar et al., 2022; Divekar et al., 2024a), host plant resistance (HPR) (Divekar et al., 2019), compounds produced by plants as secondary metabolites (Divekar et al., 2022), biological control agents (Divekar et al., 2023a; Shinde et al., 2021), proteins involved in defense mechanisms (Divekar et al., 2023b), and more environmentally friendly chemical control alternatives (Kodandaram et al., 2024). As a natural defense mechanism against herbivorous insects, plants have developed various types of secondary metabolites (Khare et al., 2020).

Secondary metabolites are synthesized in plants that are non-essential for normal growth and development but serve other important functions in their life cycle. Plants, in particular, generate a wide array of these compounds, referred to as Plant Secondary Metabolites (PSMs). These substances play various roles in plant life, including deterring microorganisms and herbivores. While the primary purpose of PSMs is defense and protection against herbivores, some also function to attract animals for seed dispersal and pollination. For instance, certain PSMs use colors or fragrant terpenoids to draw in pollinating insects, which are then rewarded with nectar to prevent them from consuming the flowers (Wink, 2018). Scientists have identified and characterized approximately 200,000 PSMs, a relatively small number compared to the 391,000 known plant species (Divekar et al., 2022, Divekar et al., 2023d). These secondary metabolites are synthesized from simple chemical precursors derived from primary metabolism, allowing for a vast range of potential outcomes (Kessler and Kalske, 2018). Different roles played by the PSM are illustrated in Figure 18.1.

DOI: 10.1201/9781003518358-18

FIGURE 18.1 Different roles performed by plant secondary metabolites in plants.

18.1.2 Types of Secondary Metabolites

The secondary metabolites are a useful weapon of plants in both their defense from herbivorous insects and disease-causing microorganisms as well as in mediation of their pollination. Hence it becomes important for us to classify the PSMs. They are divided into the following categories:

1. Terpenes.
2. Phenolic compounds.
3. Sulphur containing secondary metabolites.
4. Nitrogen containing secondary metabolites (Divekar et al., 2022)

Intriguingly, metabolites typically associated with plant defense against herbivores are commonly found in floral nectar. The presence of toxic compounds or secondary metabolites in nectar has been documented across at least 21 distinct plant families (Adler 2000). A comprehensive analysis by Baker (1977) revealed that non-protein amino acids were detected in 55% (36 out of 66) of examined species, phenolics in 36% (191 of 528) of species, and alkaloids in 9% (50 of 567) of species. This widespread occurrence of secondary compounds in nectar is perplexing, considering that nectar's primary recognized role is to attract pollinators. Several theories have been proposed to explain the potential benefits of nectar defense (reviewed in Adler 2000). These include discouraging inefficient pollinators or nectar thieves, modifying pollinator behavior to enhance pollen transfer efficiency, and reducing microbial degradation of nectar.

Specialized organs known as nectaries are responsible for producing nectar. Floral nectar, specifically, is linked to the reproductive structures of plants (Nepi, 2017) and functions as a reward for animals that may assist in pollination during their visits to flowers (Nepi et al., 2018). It is difficult to ascertain whether the primary factor leading to the development of nectar was pollinator attraction (e.g. Bonnier, 1878). The word "nectar" describes the substance that honeybees gather from fields and store in their combs as honey. Nectar is viewed as an easily digestible and economical food reward offered by plants in exchange for animal-assisted pollination services due to its carbohydrate and amino acid content (Nicolson, 2007). Sugars make up as much as 90% of its dry weight (Lüttge, 1977), with the remaining 10% consisting of various elements such as amino acids, inorganic ions, proteins, lipids, and organic acids (Nicolson and Thornburg, 2007). Following sugars, amino acids and inorganic ions are the most prevalent components (e.g. Göttlinger et al., 2019). Despite being less abundant than in pollen, amino acids play a crucial role in enhancing nectar's nutritional value (e.g. Jervis et al., 2005), contributing to its taste and appeal (e.g. Carter et al., 2006), and influencing the foraging decisions of pollinators (e.g. Seo et al., 2019).

18.2 POLLEN CHEMISTRY AND POLLINATION

The defense of pollen by herbivores is consistent with optimal defense theory, given its role as the male gamete and vital plant tissue (Cook et al., 2013). It is not unexpected that secondary metabolite concentrations in pollen typically surpass those in nectar by factors of 10–100 (Palmer-Young et al., 2019). The uniform distribution of secondary metabolites across plant tissues indicates that compounds found in pollen are generally analogous to those used for defense in other parts of the plant. The Lupinus genus, which includes several food crops such as *Lupinus mutabilis* (Leguminosae) cultivated in South America, demonstrates this pattern. Various quinolizidine alkaloids, notably sparteine and lupanine, are produced by Lupinus species. While these compounds are present throughout the plant, they are most concentrated in the seeds, comprising 3% of their weight (Hatzold et al., 1983). This distribution is characteristic of the genus. Evidence suggests that these compounds primarily act as a defense against insect herbivory, with D-lupanine exhibiting both toxic and repellent properties towards insects, including beneficial arthropods that feed on herbivores (Kordan et al., 2012). Recent research has identified D-lupanine and its derivatives (13-O-tigloyloxy, 13-O-angeloyloxy, 13-hydroxy, and 4-O-hydroxy-13-O-angeloyloxy) in pollen at levels above 2 mg/g. Although these concentrations did not increase mortality in *Bombus terrestris*, a lupin pollinator, they resulted in experimental micro-colonies producing fewer and smaller males, which could affect long-term colony and population viability (Arnold et al., 2014). While the protective function of toxins in pollen is logical, causing significant impacts on key mutualists seems counterintuitive. In these studies, insects were only provided with treated pollen without alternatives, indicating that offering choices might provide a more effective experimental framework for understanding the role of pollen defense chemistry.

18.3 SECONDARY PLANT COMPOUNDS AND THEIR INFLUENCE ON PLANT-POLLINATOR RELATIONSHIPS

The significance of both pollinators and pollination activities in achieving sustainable agricultural production can be seen in terms of enhanced seed production. Most of the cross-pollinated vegetables are considered entomophilous, in which the pollination is, to a large extent, determined by the insects. The role of honeybees in enhancing the seed production of vegetables is well documented, and the use of bee attractants can augment bee pollination. Pollination with the help of honeybees is one of the economic and ecologically sustainable methods for increasing the yield of cross-pollinated vegetables (Divekar et al., 2023a,c). Bees are the most explored pollinators for cucurbit and cole crops worldwide (Divekar et al., 2024c). Cole crops are typically pollinated by honeybees, however the quality as well as the quantity of seeds will increase if wild pollinators are permitted to visit these crops.

The concept that secondary metabolites deter inefficient floral visitors from exploiting nectar was extended to suggest that these compounds might also prevent microbial waste of nectar (e.g. Verpoorte and Schripsema, 1994). Researchers have long recognized nectar as a potentially rich environment for microbial growth due to its sugar and amino acid content (e.g. Schoellhorn, 1919). Interestingly, nectar proteins, now known to safeguard nectar from microbial proliferation and protect plant tissues from pathogen infection (Nepi, 2017), were first discovered over 90 years ago (Buxbaum, 1927). Initially, these proteins were thought to serve a nutritional purpose, providing organic nitrogen to nectar consumers (Heil, 2011). The "antimicrobial" hypothesis (Adler, 2000) gained traction in subsequent years as more experimental evidence emerged to support it. For instance, Montenegro et al. (2013) found that honey derived from *Quillaja saponaria* exhibited biological activity against pathogens, attributed to phenols present in the plant's floral nectar, indicating an antimicrobial effect of these compounds.

Nectar secondary metabolites were initially viewed unfavorably (Baker and Baker, 1975) and later often considered toxic. Baker and Baker were pioneers in suggesting that certain nectar secondary metabolites, specifically non-protein amino acids, could be harmful to some flower visitors (Baker and Baker 1975). Concurrent research during this time unveiled the potential detrimental effects of several secondary metabolites on animal consumers in plant-herbivore interactions. These studies indicated that such compounds could negatively influence growth (e.g. Isman and Duffey, 1982), impair organ function (e.g. Berenbaum, 1988), and hinder nutrient absorption (e.g. Slansky, 1992). These discoveries prompted a reevaluation of the earlier assumption that the benefits of nectar secondary metabolites must outweigh their disadvantages. The notion of direct selection for nectar toxins was momentarily eclipsed by the hypothesis that their presence stemmed from prior evolutionary pressures or pleiotropic constraints, lacking adaptive functions related to pollination, or serving only as a secondary defense mechanism (Stevenson et al., 2017). This pleiotropic perspective suggests that nectar chemistry is primarily molded by co-evolutionary interactions with herbivores, and the occurrence of "undesirable" compounds in floral nectar is incidental and, overall, a minor cost for plant defense against the substantial threat of herbivory.

The presence of certain nectar secondary metabolites, which bear structural resemblance to key neurotransmitters, may offer advantages that surpass any potential negative effects of their consumption (Verpoorte, 2005). This concept was initially explored in the 1970s, with researchers suggesting that compounds such as alkaloids, glycosides, and phenols could exert significant influence on the central nervous system of flower-visiting organisms (Baker and Baker, 1975). A specific chemical's capacity to modulate neuronal signal transduction could potentially lead to changes in neurotransmitter levels and alter receptor activity or expression, possibly resulting in profound behavioral modifications in animals (Wink, 2018).

Beyond secondary metabolites that demonstrate strong biological effects due to their structural analogies with animal neurotransmitters (Verpoorte, 2005), an additional phenomenon involves nectar components functioning as direct environmental reservoirs of invertebrate neurotransmitters. This is exemplified by biogenic amines (Roeder, 1999, Farooqui, 2012), a class of compounds only recently identified in floral nectar (Barberis et al., 2023a, Barberis et al., 2023b).Tyramine and octopamine are the two biogenic amines identified in floral nectar thus far, functioning as invertebrate equivalents of vertebrate adrenergic transmitters and regulating the "fight or flight" response, which enables swift adaptation to energy-intensive situations (Roeder, 2005). The amino acid tyrosine undergoes decarboxylation to produce these compounds. Although tyramine serves as the biological precursor to octopamine, both are recognized as distinct neurotransmitters (Roeder,

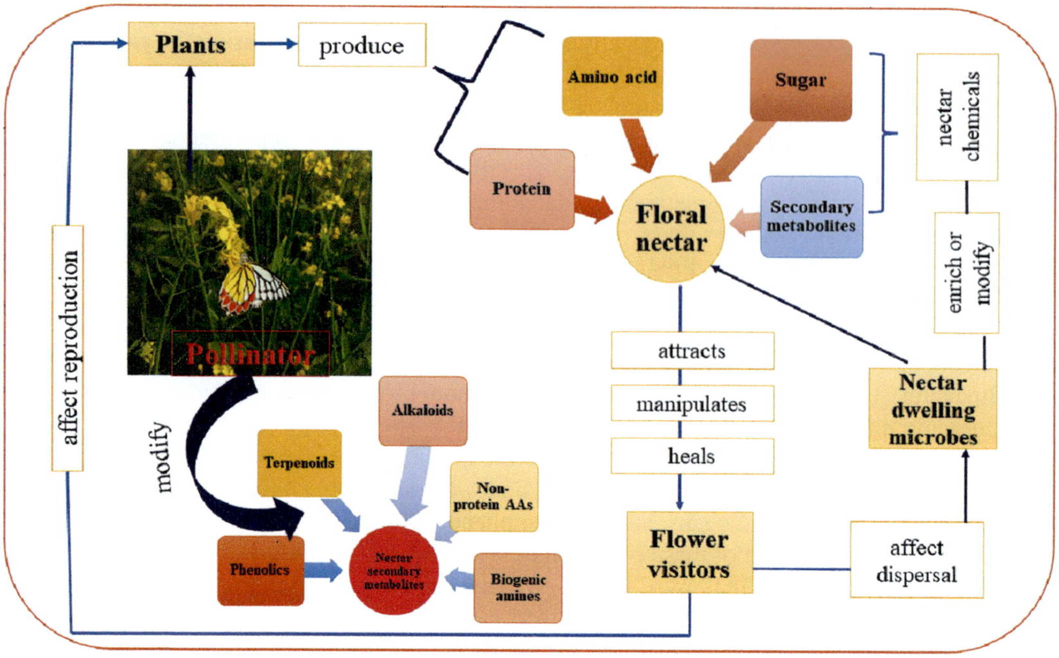

FIGURE 18.2 The complex relationship of plant-pollinators due to plant secondary metabolites present in nectar. (Modified after Barberis et al., 2023a.)

2005). Consuming these substances can affect various behavioral aspects, such as motivation (e.g. Farooqui, 2012), reward-seeking conduct (e.g. Peng et al., 2020), locomotion (e.g. Hardie et al., 2007), cognitive processes (e.g. Hammer and Menzel, 1998), and interpersonal dynamics (e.g. Linn et al., 2020).

Studies have shown that octopamine influences food-source communication and exploitation in bees. For instance, it increases the probability of honeybees performing dances (Barron et al., 2007), while reducing their likelihood of using social information from other foragers (Linn et al., 2020). This suggests that bees are more inclined to rely on their own experience, even when faced with a poor food source, rather than accepting information about a potentially better one. This finding supports the idea that nectar octopamine enhances bee loyalty to specific plant species, potentially benefiting the plants' reproductive success. Studies have revealed that octopamine influences the foraging activities of honeybee colonies, prompting foragers to prioritize less beneficial resources (Giray et al., 2007). This shift in behavior may be linked to the consumption of octopamine and tyramine, which decreases the sucrose concentration needed to initiate the proboscis extension reflex. Consequently, this alteration enhances the bees' evaluation of food source quality (e.g. Cabe et al., 2017). In terms of locomotion, a preliminary study examining the impact of tyramine in nectar-like concentrations on bumblebee flower visits revealed that bees given tyramine-enriched solutions spent less time foraging on individual flowers compared to those given a control solution. This observation implies that tyramine may induce more dynamic foraging behavior overall (Barberis et al., 2023b) (Figure 18.2).

18.4 PLANT SECONDARY METABOLITES AS ATTRACTANTS TO POLLINATORS

Plants and pollinators engage in a mutually beneficial relationship, providing advantages to one another. Their interactions have led to cross-pollination, which not only enhances food production but also increases genetic diversity among plants. The attractants and rewards offered by plants are closely interrelated. To draw pollinators, plants employ various strategies, including the flower's shape, size, and color, as well as scents, deception, and strategic flower placement. In exchange, plants offer rewards to visiting pollinators, such as pollen and nectar, which are nutritionally rich and energy-dense, compensating for the effort expended by pollinators in reaching the plants. Additional rewards include protection from predators, as well as breeding, egg-laying, and mating sites within the flowers. The ecological bond between plants and pollinators results in effective crop pollination when the attractants and incentives are significantly interdependent.

18.5 OLFACTORY CUES

Floral scents typically comprise numerous volatile compounds from various chemical groups, including terpenoids and aromatics (Knudsen and Gershenzon, 2006). Most, if not all, of these compounds can attract pollinators. Some plant species emit scents in specific patterns. Pollinators utilize different cues during the day and night. For day-flying pollinators, scents often function as landing cues, while for night-flying pollinators, they serve as initial attractants (Ômura et al., 1999; Raguso and Willis, 2002; Raguso and Willis, 2005).

18.6 EFFECT OF NECTAR COMPOSITION ON POLLINATOR BEHAVIOUR AND PREFERENCES

The nectar's composition plays a crucial role in influencing pollinator behavior and preferences. Variations within a species regarding the energy available to pollinating insects from nectar consumption can explain why pollinators favor certain plants within a crop. These variations may stem from differences in nectar sugar concentration, flowering duration, timing of anthesis, and the volume of nectar produced per flower. Research on wild *Brassica rapa* revealed a correlation between the quantity of floral volatile phenylacetaldehyde and floral rewards (sugar and pollen per flower). Additionally, bumblebees developed a positive response to this volatile compound after foraging on these plants (Knauer and Schiestl, 2015). Secondary metabolites having synergistic/mutualistic effects in plant-pollinator interactions are presented in Table 18.1.

TABLE 18.1
Synergist/Mutualist Effect of Plant Secondary Metabolite on Plant-Pollinator Interaction

Sl. No.	Phytochemical Compound	Synergistic Effect on Pollinators	References
	Caffeine	Improves retention of signals linked to favorable food rewards Augments pollen transfer by *Bombus* species Enhances bees' recruitment behaviors and foraging precision Lowers parasitic infestation in bees	Detzel and Wink (1993); Wright et al. (2013); Richardson et al. (2015); Couvillon et al. (2015); Thomson et al. (2015); Bernklau et al. (2019)
	Aconitine	Safeguarding against nectar theft by Bombus terrestris, complemented by the adaptive responses of the pollinating species Bombus hortorum.	Barlow et al. (2017)
	Linalool	Entices organisms that prey on plant consumers Beckons Thrips major to the flowers of Sambucus nigra for pollination assistance Lures Colobopsis leonardii to Hoya pottsii for pollination purposes.	Scott-Brown et al. (2019); Landrein et al. (2021)

(Continued)

TABLE 18.1 (*Continued*)
Synergist/Mutualist Effect of Plant Secondary Metabolite on Plant-Pollinator Interaction

Sl. No.	Phytochemical Compound	Synergistic Effect on Pollinators	References
	Nicotine	Decreased *Crithidia bombi* pathogen burdens in Bombus species.	Kessler and Baldwin (2007); Richardson et al. (2015)
	Grayanotoxin 1	Targeted toxicity to floral visitors allows preferential nectar access for *Bombus* species.	Tiedeken et al. (2016)
	Thymol	Lowers *Crithidia bombi* parasite burden in *Bombus impatiens* Effective against the parasite at environmentally relevant doses	Richardson et al. (2015); Palmer-Young et al. (2016); Biller et al. (2015)

18.7 MUTUALIST AND ANTAGONIST INTERACTIONS

18.7.1 Mutualist Interactions

Mutualistic relationships involve the exchange of benefits between two or more organisms. In these interactions, the partners typically receive unequal advantages. Although all participants gain from the relationship, their actions are not altruistic. Instead, each species involved in the mutualism acts in its own self-interest, and any advantages gained are unintentional byproducts of their behavior.

18.7.2 Pollination Mutualism

Plant-Pollinator interaction is one of the best example of Mutualism. Mutualism can be of two types:

1. Obligatory or Specialized Mutualism
2. Generalized Mutualism

18.7.2.1 Obligatory Mutualism

This form of mutualism indicates that flowers have evolved to accommodate their most effective pollinators. Several common examples of such mutualisms include bee pollination (melittophily), fly pollination (myophily), beetle pollination (cantharophily), bird pollination (ornithophily), and hawk-moth pollination (sphingophily) (Balamurali et al, 2015). For instance, flowers adapted for hawk-moth pollination typically exhibit white coloration, strong fragrances, elongated corolla tubes, and nighttime blooming. Subsequent research revealed that only 30% of plant pollinations adhere to this mutualistic pattern (Ollerton et al, 2009.). Nevertheless, a more recent investigation conducted by Rosas-Guerrero et al. (2014) examining 417 plant species suggested that pollination syndromes are widespread in the plant kingdom, with plant-pollinator co-evolution resulting in adaptations that enhance pollination efficiency.

18.7.2.2 Generalized Mutualism

The concept of mutualism described here challenges previous co-evolutionary theories by highlighting the interactions between plants and multiple pollinator species. Generalization benefits pollinators when reward consistency exists across plant species and their lifespan surpasses that of flowers. For plants, employing various insect pollinators serves as a safeguard against ineffective pollination and reproductive issues. Experimental work by Fontaine et al. (2006) revealed that pollinator diversity positively impacts plant seed production. This mutual relationship underscores the advantages of generalization for both plants and their pollinators in maintaining successful reproduction and ecological balance.

18.7.3 Effect of Different Plant Secondary Metabolites on Plant-Pollinator Interactions

18.7.3.1 Effect of Caffeine on Honeybee

Caffeine, a purine alkaloid, is considered a natural biopesticide due to its toxicity to invertebrates (Hollingsworth et al., 2002). Its insecticidal properties are manifested through the inhibition of phosphodiesterase activity and elevation of intracellular cyclic AMP levels, resulting in insect paralysis and intoxication. This compound is widespread in the plant kingdom, occurring in at least 5 families of flowering plants, including Rubiaceae, Rutaceae, Malvaceae, and Theaceae. High doses of caffeine can be detrimental to phytophagous insects, reducing their lifespan, deterring pollinators, and impairing their memory (Ashihara et al., 2008). Interestingly, certain plant species contain lower concentrations of caffeine in their nectar, which is utilized by pollinators rather than foragers. Koch and Stevenson (2017) observed that honey bees were attracted to *Tilia* spp. based on an odor-based memory system for food rewards, leading to the discovery of dead honey bees under these trees. This occurred because the bees continued foraging around the trees even after nectar production had ceased. Notably, the nectar of *Tilia* spp. contains caffeine (Naef et al., 2005; Mathona et al., 2014). Additionally, caffeine has been reported to possess antimicrobial properties that may reduce infections caused by *Nosema bombi* (Dissociodihaplophasida: Nosematidae) (Bernklau et al., 2019), suggesting an added benefit associated with its presence in plants.

18.7.4 Antagonist Effect of Plant Secondary Metabolites on Plant Pollinator Interaction

The consumption of nectar containing secondary metabolites, which can be detrimental to nectar-feeding animals, was initially deemed unintentional. However, the frequent observation of animals feeding on intoxicating sources led researchers to question the deliberateness of this behavior. An illustrative example is the hawkmoth *Manduca quinquemaculata*, which feeds on *Datura meteloides* nectar, a Solanaceae plant likely containing hallucinogenic substances present in its other tissues (Grant, 1983). These compounds have an intoxicating effect on insects, leading to sluggishness and disorientation (Grant, 1983). Grounded moths face increased predation risk (Grant, 1983). Bees encountering psychoactive or narcotic compounds in orchids such as *Epipactis helleborine* represent an analogous case (Jakubska et al., 2005). Ecologically, the presence of intoxicating substances in floral nectar could potentially boost pollination effectiveness by increasing the duration of animal visitations to the flowers (Jakubska et al., 2005).

Intoxication is not the only coercive method known to influence pollinator behavior. For example, some plants produce nectar containing nicotine, which causes bees to repeatedly visit the same food source, even when better alternatives are available (e.g. Baracchi et al., 2017a,b). This behavior aligns with Rhoades and Bergdahl's (1981) hypothesis that prolonging a pollinator's visit may enhance the transfer of pollen from the same species, thereby boosting the plant's reproductive success.

The illustrations provided above point to potentially detrimental effects on floral visitors resulting from nectar consumption. These instances have contributed to the development of the pollinator manipulation concept, a term employed by researchers since the early 2000s (e.g. Bailey et al., 2007). This idea gained widespread recognition following Pyke's (2016) formal introduction of the "manipulation" hypothesis, which was subsequently elaborated upon by Nepi et al. (2018). Notably, as early as 1981, Rhoades and Bergdahl addressed the presence of toxic compounds in nectar, asserting: "though at first sight the presence of these toxic substances seems incompatible with the reward function of nectar, they probably represent a mechanism to manipulate pollinator behaviour to the advantage of the plant and to exclude nectar thieves". Their hypothesis suggested that a combination of rewarding and defensive chemicals could shape insect visitation patterns in a manner that benefits plant fitness, extending beyond the pollinator fidelity hypothesis. Some of the secondary metabolites showing antagonistic effects in plant-pollinator interactions are presented in Table 18.2.

TABLE 18.2
Antagonist Effect of Plant Secondary Metabolites on Plant Pollinator Interaction

Species	Family	Effects	Reference
Aesculus californica	Hippocastanaceae	Toxic to bees	Eckert (1946); Mussen (1979)
Astragalus spp.	Fabaceae	Toxic to bees	Eckert (1946); Mussen (1979)
Cuscuta spp.	Convulvulaceae	Toxic to bees	Eckert (1946); Mussen (1979)
Cyrilla racemiflora	Cyrillaceae	Toxic to bees	Eckert (1946); Mussen (1979)
Gelsemium sempervirens	Loganiaceae	Toxic to bees	Eckert (1946); Mussen (1979)
Kalmia latifolia	Ericaceae	Toxic to bees	Eckert (1946); Mussen (1979)
Solanum nigrum	Solanaceae	Toxic to bees	Mamoudou et al. (2021)
Veratrum californicum	Liliaceae	Toxic to bees	Eckert (1946); Mussen (1979)
Zygadenus venesosus	Liliaceae	Toxic to bees	Irwin et al. (2014)
Cornycocarpus laevigata	Corynocarpaceae	Toxic to honeybees	Palmer-Jones and Line (1962)
Angelica triqueta	Apiaceae	Toxic to bees	Bell (1971)
Astragalus lentiginosus	Fabaceae	Toxic to bees	Soltani et al. (2021)
Camellia thea	Theaceae	Lethal to honeybee larvae	Zhang et al. (2024)
Erythrina fusca	Fabaceae	Deterred ants	Leroy et al. (2021)
Hippobroma longiflora	Campanulaceae	Deterred ants	Feinsinger and Swarm (1978)
Ochroma lagopus	Bombacaceae	Toxic to bees and other insects	Adler (2000)
Sophora microphylla	Fabaceae	Toxic to honeybees	Nguyen et al. (2021)
Tilia spp.	Tiliaceae	Toxic to bees and other insects	Jacquemart et al. (2018)
Veratrum californicum	Liliaceae	Toxic to bees	Cane et al. (2020)

18.8 PLANT SECONDARY METABOLITES AND MEDICATION IN POLLINATORS

Specific secondary metabolites found in nectar have been categorized as drugs when they induce pharmacological effects on the brains of organisms that consume nectar. As an illustration, Wright et al. (2013) explored the nectar alkaloid caffeine from this perspective, offering initial proof of its capacity to pharmacologically modify a pollinator's memory of rewards. Their investigation showed that honeybees given caffeine were three times more likely to recollect a previously acquired floral scent in comparison to those provided with only sucrose.

As a phenolic component of essential oils, thymol is prevalent in various plants, notably *Thymus vulgaris* (Lamiaceae). This compound exhibits biological effects on diverse insect species and holds promise as a potential commercial biopesticide, offering protection against invertebrate adversaries. Although Richardson et al. (2015) reported minimal activity of thymol against *Crithidia bombi* in *B. impatiens* at honey-level concentrations, Palmer-Young et al. (2016) demonstrated that thymol is more abundant in nectar and shows enhanced efficacy against *C. bombi* at these ecologically pertinent levels. As a result, thymol may confer benefits to bees in their natural environments by providing nectar sources and potentially reducing disease prevalence. Among various plant species, including the commonly cultivated *T. vulgaris*, thymol is found to be present. Richardson et al. (2015) demonstrated that this compound is less detrimental to bees compared to other tested substances, which can be directly toxic or lead to malaise behaviors and sublethal effects (Hurst et al., 2014; Oliver et al., 2005). Despite this, thymol was unable to prevent parasite acquisition, and its impact on *C. bombi* when isolated may be a result of indirect host-mediated effects from prolonged ingestion (Rothchild et al., 2018). As a result, the development of more consistently effective compounds is necessary before medicinal plants can be considered for use in field margins and flower strips to benefit bees.

18.9 ROLE OF MICROBES IN PLANT-POLLINATOR INTERACTIONS

Generally, microorganisms are viewed as detrimental to the quality of floral nectar (e.g. Herrera et al., 2009, Vannette et al., 2013), potentially jeopardizing the mutually beneficial relationship between plants and their pollinators. Studies have demonstrated that yeasts can reduce the nutritional content of floral nectar by lowering sugar (de Vega and Herrera, 2013) and amino acid concentrations (Pozo et al., 2015). Floral microbes are typically not considered beneficial for plants (Vannette, 2020). Both flower pathogens and certain nectar-dwelling bacteria can negatively impact plant fitness, either directly or by discouraging pollinator visits (e.g. Vannette et al., 2013). However, some research suggests that in particular cases, microorganisms may enhance pollination by producing volatile compounds that draw pollinators, indirectly benefiting plant fitness (e.g. Cullen et al., 2021). Nonetheless, even when nectar yeasts increase pollinator activity, this doesn't always translate to improved plant fitness. For example, Herrera et al. (2013a) found reduced seed production in *Helleborus foetidus* plants inhabited by yeasts, despite increased pollinator attraction. In addition to influencing pollinator attraction and visits through odor emissions or chemical alterations (e.g. Rering et al., 2018, Rering et al., 2020), the mere presence of microorganisms in nectar seems to create a preference in pollinators such as bumblebees for flowers containing yeast, as these insects have been shown to detect microbial presence in nectar (Herrera et al., 2013b, Schaeffer et al., 2014, Schaeffer et al., 2017). It is suggested that yeast cells in nectar supply insects with crucial nutritional elements, including vitamins, sterols, and minerals (Stefanini, 2018). The initial research in this field was conducted by Dharampal et al. (2019), who demonstrated the benefits of diverse communities of symbiotic microorganisms on pollen surfaces for honeybee larvae.

18.10 CONCLUSION

The chemical makeup of floral tissues significantly influences plant interactions with pollinators, pathogens, and herbivores, profoundly affecting the fitness of both plants and their visitors. These interactions are often dose-dependent and context-specific. Secondary metabolites, while not as critical for plant growth and development as primary metabolites, play a significant role in plants routine activities. These compounds are essential for plant survival and proper functioning, despite not being crucial for growth itself. In plant-pollinator interactions, secondary metabolites are of paramount importance. Produced by plants for various purposes, these organic compounds can function as both attractants and deterrents for pollinators. The specific type, concentration, and presentation of secondary metabolites found in nectar and floral aromas can shape pollinator behavior, fostering specialized or generalized mutualistic relationships. Certain secondary metabolites may repel pollinators or decrease nectar quality, while others enhance pollinator attraction and memory, ultimately benefiting both plants and their pollinators. Comprehending the intricate dynamics between secondary metabolites and pollinators is crucial for biodiversity conservation and ensuring the continued delivery of vital ecosystem services, such as pollination.

18.11 FUTURE PROSPECTS

The majority of nectar research has focused on its role as a food reward for pollinators in exchange for their services. While extensive studies have been conducted on nectar composition, further investigation is needed to understand how these compounds have changed through evolutionary processes. Another potential area of study is the concentration of plant secondary compounds in flower pollen, which serves as a primary food source for many pollinators. Research has shown that secondary metabolites are more concentrated in pollen compared to nectar. PSMs, particularly phenolics, exhibit anti-fungal and trypanocidal properties, suggesting a possible role in reducing *Nosema* protozoan infections in pollinators – a topic worthy of exploration. Current research on nectar secondary metabolites has primarily focused on their effects on bees and bumblebees. Future studies should expand to include other significant pollinating groups such as flies, butterflies, and various solitary bee species. Future research could explore how secondary compounds in nectar influence pollinator behavior through other floral characteristics, such as color. Moreover, investigations into the impact of floral chemistry on interactions between multiple species and its role in the co-evolution of plants and their pollinators would be valuable to the scientific community. The study of chemical ecology in plant-pollinator relationships is a growing field, and further exploration will undoubtedly provide greater insights into the ecology and evolution of species interactions.

REFERENCES

Adler LS. The ecological significance of toxic nectar. *Oikos*. **2000**, 91(3):409–20. https://doi.org/10.1034/j.1600-0706.2000.910301.x

Arnold SE, Idrovo ME, Arias LJ, Belmain SR, Stevenson PC. Herbivore defence compounds occur in pollen and reduce bumblebee colony fitness. *Journal of Chemical Ecology*. **2014**, 40:878–81. https://doi.org/10.1007/s10886-014-0467-4

Ashihara H, Sano H, Crozier A. Caffeine and related purine alkaloids: Biosynthesis, catabolism, function and genetic engineering. *Phytochemistry*. **2008**, 69(4):841–56.

Bailey SF, Hargreaves AL, Hechtenthal SD, Laird RA, Latty TM, Reid TG, Teucher AC, Tindall JR. Empty flowers as a pollination-enhancement strategy. *Evolutionary Ecology Research*. **2007**, 9(8):1245–62.

Baker HG. Non-sugar chemical constituents of nectar. *Apidologie*. **1977**, 8(4):349–56.

Baker HG, Baker I. Studies of nectar-constitution and pollinator-plant coevolution; Coevolution of animals and plants (Eds.), LE Gilbert and PH Raven, Austin: University of Texas Press. 1975. 100–140.

Baker HG, Baker I. Intraspecific constancy of floral nectar amino acid complements. *Botanical Gazette*. **1977**, 138(2):183–91. https://doi.org/10.1086/336914

Balamurali GS, Krishna S, Somanathan H. Senses and signals: Evolution of floral signals, pollinator sensory systems and the structure of plant–pollinator interactions. *Current Science*. **2015**, 108(10):1852–61.

Baracchi D, Lihoreau M, Giurfa M. Do insects have emotions? Some insights from bumble bees. *Frontiers in Behavioral Neuroscience*. **2017a**, 11:157. https://doi.org/10.3389/fnbeh.2017.00157

Baracchi D, Marples A, Jenkins AJ, Leitch AR, Chittka L. Nicotine in floral nectar pharmacologically influences bumblebee learning of floral features. *Scientific Reports*. **2017b**, 7(1):**1951**.

Barberis M, Bogo G, Bortolotti L, Flaminio S, Giordano E, Nepi M, Galloni M. Floral nectar and insect flower handling time change over the flowering season: Results from an exploratory study. *Acta Oecologica*. **2023a**, 120:103937. https://doi.org/10.1016/j.actao.2023.103937

Barberis M,-ogo G, Bortolotti L, Guarnieri M, Nepi M, Felicioli A, Galloni M. Nectar tyramine decreases the duration of bumblebee visits on flowers. *Arthropod-Plant Interactions*. **2023b**, 17(5):563–9. https://doi.org/10.1007/s11829-023-09976-7

Barlow SE, Wright GA, Ma C, Barberis M, Farrell IW, Marr EC, Brankin A, Pavlik BM, Stevenson PC. Distasteful nectar deters floral robbery. *Current Biology*. **2017**, 27(16):2552–8.

Barron AB, Maleszka R, Vander Meer RK, Robinson GE. Octopamine modulates honey bee dance behavior. *Proceedings of the National Academy of Sciences of the United States of America*. **2007**, 104(5):1703–70. https://doi.org/10.1073/pnas.0610506104

Bell CR. Breeding systems and floral biology of the Umbelliferae; or, evidence for specialization in unspecialized flowers. VH. Heywood (Eds.), The Biology and Chemistry of the Umbelliferae. **1971**:93–107.

. Berenbaum MR. Allelochemicals in insect-microbe-plant interactions. Agents provocateurs in the coevolutionary arms race, in P. Barbosa and D.K. Letourneau (Eds.), Novel Aspects of Insect-Plant Interactions. Wiley & Sons, New York 1988: 97–123.

Bernklau E, Bjostad L, Hogeboom A, Carlisle A, HS A. Dietary phytochemicals, honey bee longevity and pathogen tolerance. *Insects*. **2019**, 10(1):14.

Biller OM, Adler LS, Irwin RE, McAllister C, Palmer-Young EC. Possible synergistic effects of thymol and nicotine against *Crithidia bombi* parasitism in bumble bees. *PLoS One*. **2015**, 10(12):e0144668.

Bonnier MG. Étude sur l'anatomie et la physiologie des nectaires. *Bulletin de la Société Botanique de France*. **1878**, 25(8):262–71. https://doi.org/10.1080/00378941.1878.10827912

Buxbaum F. Zur Frage des Eiweissgehaltes des Nektars.(Vorläufige Mitteilung). *Planta*. **1927**, 4:818–21. https://doi.org/10.1007/BF01911267

Cabe SI, Ferro MW, Farina WM, Hrncir M. Dose-and time-dependent effects of oral octopamine treatments on the sucrose responsiveness in stingless bees (*Melipona scutellaris*). *Apidologie*. **2017**, 48:1–7. https://doi.org/10.1007/s13592-016-0442-x

Cane JH, Gardner DR, Weber M. Neurotoxic alkaloid in pollen and nectar excludes generalist bees from foraging at death-camas, *Toxicoscordion paniculatum* (Melanthiaceae). *Biological Journal of the Linnean Society*. **2020**, 131(4):927–35.

Carter C, Shafir S, Yehonatan L, Palmer RG, Thornburg R. A novel role for proline in plant floral nectars. *Naturwissenschaften*. **2006**, 93:72–9. https://doi.org/10.1007/s00114-005-0062-1

Cook D, Manson JS, Gardner DR, Welch KD, Irwin RE. Norditerpene alkaloid concentrations in tissues and floral rewards of larkspurs and impacts on pollinators. *Biochemical Systematics and Ecology*. **2013**, 48:123–31.

Couvillon MJ, Riddell Pearce FC, Accleton C, Fensome KA, Quah SK, Taylor EL, & Ratnieks, FL. Honey bee foraging distance depends on month and forage type. *Apidologie*. **2015**. 46: 61–70.

Cullen NP, Fetters AM, Ashman TL. Integrating microbes into pollination. *Current Opinion in Insect Science*. **2021**, 44:48–54. doi: 10.1016/j.cois.2020.11.002

de Vega C, Herrera CM. Microorganisms transported by ants induce changes in floral nectar composition of an ant-pollinated plant. *American Journal of Botany*. **2013**, 100(4):792–800.

Detzel A, Wink M. Attraction, deterrence or intoxication of bees (*Apis mellifera*) by plant allelochemicals. *Chemoecology*. **1993**, 4:8–18.

Dharampal PS, Carlson C, Currie CR, Steffan SA. Pollen-borne microbes shape bee fitness. *Proceedings of the Royal Society B*. 2019, 286(1904):20182894**.** https://doi.org/10.1098/rspb.2018.2894

Divekar P, Kumar P, Suby SB, **2019**. Screening of maize germplasm through antiobiosis mechanism of resistance against *Chilo partellus* (Swinhoe). *Journal of Entomology and Zoology Studies*. **2019**, 7:1111–4.

Divekar PA, Majumder S, Halder J, Kedar SC, Singh V. Sustainable pest management in cabbage using botanicals: Characterization, effectiveness and economic appraisal. *Journal of Plant Diseases and Protection*. **2024a**, 131(1):113–30.https://doi.org/10.1007/s41348-023-00848-z

Divekar PA, Narayana S, Divekar BA, Kumar R, Gadratagi BG, Ray A, Singh AK, Rani V, Singh V, Singh AK, Kumar A. Plant secondary metabolites as defense tools against herbivores for sustainable crop protection. *International journal of Molecular Sciences*. **2022**, 23(5):2690.

Divekar PA, Nebapure S, Majumder S, Verma CK. Honeybee and Crop Pollination: concepts, challenges and applications. **2021**. https://vikaspedia.in/agriculture/best-practices/sustainableagriculture/biodiversity/honeybee-and-crop-pollination. Accessed on 24th April, 2024.

Divekar PA, Patel SK, Manimurugan C, Singh V. Insect pollinator diversity and their foraging behaviour in Radish, *Raphanus sativus* L. *Emergent Life Sciences Research*. **2023a**, 9:68–76.

Divekar PA, Patel SK, Singh V, Singh J. Spinetoram, a selective novel insecticide able to check key lepidopteran pests in cabbage ecosystem. *Pakistan Journal of Zoology*. **2024b**, 56(2). https://doi.org/10.17582/journal.pjz/20220724130756

Divekar PA, Rani V, Majumder S, Karkute SG, Molla KA, Pandey KK, Behera TK, Govindharaj GP. Protease inhibitors: An induced plant defense mechanism against herbivores. *Journal of Plant Growth Regulation*. **2023b**, 42(10):6057–73. https://doi.org/10.1007/s00344-022-10767-2

Divekar PA, Singh K, Verma CK, Rai AB, Singh B, Yadav S, Karkute SG. Assessment of bee flora and development of a floral calendar in relation to pharmaceutical potential of honey and bee pollen in Eastern Uttar Pradesh, India. *Annals of Phytomedicine*. **2023c**, 12(1):844–55. https://doi.org/10.54085/ap.2023.12.1.55

Divekar PA, Mishra A, Singh AK. Plant Secondary Metabolites for Defense Against Herbivores. In: Mérillon, JM., Ramawat, K.G. (Eds.), Plant Specialized Metabolites. Reference Series in Phytochemistry. Springer, Cham. **2023d**. https://doi.org/10.1007/978-3-031-30037-0_24-1

Divekar PA, Singh K, Yadav S, Manimurugan C, Patel SK, et al. Diversity and foraging behaviour of insect pollinators in Cauliflower. *Ecology Environment & Conservation*. **2023d,** 29(2):971–78.

Divekar PA, Mishra A, Kumar R. Bee pollination in vegetables: current status, challenges and prospects. *Circular Agricultural Systems*. **2024c**. 4(1).

Eckert JE. Injury to bees by poisoning. In *The Hive and the Honey Bee* in R. A. Grout, editor 1946. Dadant and Sons, Hamilton, Illinois, USA, pp. 570–82.

Fontaine C, Dajoz I, Meriguet J, Loreau M. Functional Diversity of Plant–Pollinator Interaction Webs Enhances the Persistence of Plant Communities. *PLoS Biology*. **2006**. 4(1): e1. https://doi.org/10.1371/journal.pbio.0040001.

Farooqui T. Review of octopamine in insect nervous systems. *Open Access Insect Physiology*. **2012**, 5:1–7. https://doi.org/10.2147/OAIP.S20911

Feinsinger P, Swarm LE. How common are ant-repellent nectars? **1978**. *Biotropica*.10: 238–239.

Giray T, Galindo-Cardona A, Oskay D. Octopamine influences honey bee foraging preference. *Journal of Insect Physiology*. **2007**, 53(7):691–8. DOI: 10.1016/j.jinsphys.2007.03.016

Göttlinger T, Schwerdtfeger M, Tiedge K, Lohaus G. What do nectarivorous bats like? Nectar composition in Bromeliaceae with special emphasis on bat-pollinated species. *Frontiers in Plant Science*. **2019**, 10:205. https://doi.org/10.3389/fpls.2019.00205

Grant V. Behavior of hawkmoths on flowers of *Datura meteloides*. *Botanical Gazette*. **1983**, 144(2):280–4. https://doi.org/10.1086/337374

Hammer M, Menzel R. Multiple sites of associative odor learning as revealed by local brain microinjections of octopamine in honeybees. *Learning & Memory*. **1998**, 5(1):146–56.

Hardie SL, Zhang JX, Hirsh J. Trace amines differentially regulate adult locomotor activity, cocaine sensitivity, and female fertility in *Drosophila melanogaster*. *Developmental Neurobiology*. **2007**, 67(10):1396–405. https://doi.org/10.1002/dneu.20459

Hatzold T, Elmadfa I, Gross R, Wink M, Hartmann T, Witte L. Quinolizidine alkaloids in seeds of Lupinus mutabilis. *Journal of Agricultural and Food Chemistry*. **1983**, 31(5):934–8. https://doi.org/10.1021/jf00119a003

Heil M. Nectar: Generation, regulation and ecological functions. *Trends in Plant Science*. 2011, 16(4):191–200. https://doi.org/10.1016/j.tplants.2011.01.003

Herrera CM, De Vega C, Canto A, Pozo MI. Yeasts in floral nectar: A quantitative survey. *Annals of Botany*. **2009**, 103(9):1415–23. https://doi.org/10.1093/aob/mcp026

Herrera CM, Pozo MI, Medrano M. Bumble bees prefer nectar containing yeasts. *The Bulletin of the Ecological Society of America*. **2013a**, 94(2):186–9.

Herrera CM, Pozo MI, Medrano M. Yeasts in nectar of an early-blooming herb: Sought by bumble bees, detrimental to plant fecundity. *Ecology*. **2013b**, 94(2):273–9. https://doi.org/10.1890/12-0595.1

Hollingsworth RG, Armstrong JW, Campbell E. Caffeine as a repellent for slugs and snails. *Nature*. **2002**, 417(6892):915–6.

Hurst V, Stevenson PC, Wright GA. Toxins induce 'malaise' behaviour in the honeybee (*Apis mellifera*). *Journal of Comparative Physiology A*. **2014**, 200:881–90. https://doi.org/10.1007/s00359-014-0932-0

Irwin RE, Cook D, Richardson LL, Manson JS, Gardner DR, 2014. Secondary compounds in floral rewards of toxic rangeland plants: Impacts on pollinators. *Journal of Agricultural and Food Chemistry*. 62(30):7335–44.

Isman MB, Duffey SS. Toxicity of tomato phenolic compounds to the fruitworm, *Heliothis zea*. *Entomologia Experimentalis et Applicata*. **1982**, 31(4):370–6. https://doi.org/10.1111/j.1570-7458.1982.tb03162.x

Jacquemart AL, Moquet L, Ouvrard P, Quetin-Leclercq J, Hérent MF, Quinet M. Tilia trees: Toxic or valuable resources for pollinators? *Apidologie*. **2018**, 49:538–50.

Jakubska A, Przado D, Steininger M, Aniol-Kwiatkowska J, Kadej M. Why do pollinators become" sluggish"? nectar chemical constituents from *Epipactis helleborine* (L.) Crantz(orchidaceae). *Applied Ecology and Environmental Research*. **2005**, 3(2):29–38. https://doi.org/10.15666/aeer/0302_029038

Jervis MA, Boggs CL, Ferns PN. Egg maturation strategy and its associated tradeoffs: A synthesis focusing on Lepidoptera. *Ecological Entomology*. **2005,** 30:359–75

Kessler A, Kalske A. Plant secondary metabolite diversity and species interactions. *Annual Review of Ecology, Evolution, and Systematics*. **2018**, 49(1):115–38.

Kessler D, Baldwin IT. Making sense of nectar scents: The effects of nectar secondary metabolites on floral visitors of *Nicotiana attenuata*. *The Plant Journal*. **2007**, 49(5):840–54.

Khare S, Singh NB, Singh A, Hussain I, Niharika KM, Yadav V, Bano C, Yadav RK, Amist N. Plant secondary metabolites synthesis and their regulations under biotic and abiotic constraints. *Journal of Plant Biology*. **2020**, 63:203–16. https://doi.org/10.1007/s12374-020-09245-7

Knauer AC, Schiestl FP. Bees use honest floral signals as indicators of reward when visiting flowers. *Ecology Letters*. **2015**, 18(2):135–43.

Knudsen JT, Eriksson R, Gershenzon J. et al. Diversity and distribution of floral scent. The Botanical Review. **2006**. 72–1. https://doi.org/10.1663/00068101(2006)72[1:DADOFS]2.0.CO;2.

Koch H, Stevenson PC. Do linden trees kill bees? Reviewing the causes of bee deaths on silver linden (*Tilia tomentosa*). *Biology Letters*. 2017, 13(9):20170484.

Kodandaram MH, Divekar PA, Wangi N, Mohite NR, Rai AB. Optimizing sucking pest control in okra: An analysis of flupyradifurone 200 SL effectiveness, phytotoxicity, safety to natural enemies, pollinators and cost-efficiency. *Journal of Plant Diseases and Protection*. **2024**, 131(3):757–71.

Kordan B, Dancewicz K, Wroblewska A, Gabrys B. Intraspecific variation in alkaloid profile of four lupine species with implications for the pea aphid probing behaviour. *Phytochemistry Letters*. **2012**. 5: 71–77.

Landrein S, Zhou ZY, Song SJ. Pollinators of Hoya pottsii: Are the strongest the most effective?. *Flora*, **2021**. 274: 151734.

Leroy L, Mille C, Fogliani B. The common fruit-piercing moth in the Pacific Region: A Survey of the current state of a significant Worldwide Economic Pest, *Eudocima phalonia* (Lepidoptera: Erebidae), with a focus on New Caledonia. *Insects*. **2021**, 12(2):117.

Linn M, Glaser SM, Peng T, Grüter C. Octopamine and dopamine mediate waggle dance following and information use in honeybees. *Proceedings of the Royal Society B*. **2020**, 287(1936):20201950.

Lüttge U. Nectar composition and membrane transport of sugars and amino acids: A review on the present state of nectar research. *Apidologie*. **1977**, 8(4):305–19.

Mamoudou J, Topé SF, Basga E, Fohouo FN. Pollination efficiency of *Apis mellifera* (Hymenoptera: Apidae) on *Solanum nigrum* (Solanaceae) at Meskine (Maroua, Cameroon). *International Journal of Biological and Chemical Sciences*. **2021**, 15(3):1073–89.

Mathona C, Edder P, Christen P, Bieri S. Unexpected occurrence of caffeine in sleep-inducing herbal teas. *Chimia*. **2014**, 68(10):705.

Mc Cabe SIM, Ferro MWB, Farina WM, Hrncir M (**2017**). Dose- and time-dependent effects of oral octopamine treatments on the sucrose responsiveness in stingless bees (*Melipona scutellaris*). *Apidologie*. 48, 204–10. https://doi.org/10.1007/s13592-016-0442-x

Montenegro G, Díaz-Forestier J, Fredes C, Rodríguez S. Phenolic profiles of nectar and honey of *Quillaja saponaria* Mol.(Quillajaceae) as potential chemical markers. *Biological Research*. **2013**, 46(2):177–82. https://doi.org/10.4067/S0716-97602013000200009

Mussen EC. Buckeye poisoning. U.C. Apiaries: Univ. of California Cooperative Extension: 1979, pp. 1–4.

Naef R, Jaquier A, Velluz A, Bachofen B., **2005**. From the linden flower to linden honey–volatile constituents of linden nectar, the extract of bee-stomach and ripe honey. *Perspectives in Flavor and Fragrance Research*. 23:31–40.

Nepi M. New perspectives in nectar evolution and ecology: Simple alimentary reward or a complex multiorganism interaction? *Acta Agrobotanica*. **2017**, 70(1):1–2. https://doi.org/10.5586/aa.1704

Nepi M, Grasso DA, Mancuso S. Nectar in plant–insect mutualistic relationships: From food reward to partner manipulation. *Frontiers in Plant Science*. **2018**, 9:1063. https://doi.org/10.3389/fpls.2018.01063

Nguyen L, Bayne K, Altaner C. A review of kowhai (*Sophora* spp.) and its potential for commercial forestry. *New Zealand Journal of Forestry Science*. **2021**, 51. https://doi.org/10.33494/nzjfs512021x157x

Nicolson SW. Nectar consumers. In: Nicolson SW, Nepi M, Pacini E (Eds.) *Nectaries and Nectar*. **2007**. Springer, Dordrecht, pp. 289–342.

Nicolson SW, Thornburg RW. Nectar chemistry. In: *Nectaries and Nectar*. Nicolson, S. W., Nepi, M., & Pacini, E. (Eds.). **2007**. Springer Science & Business Media., pp. 215–64.

Oliver CJ, Softley S, Williamson SM, Stevenson PC, Wright GA, **2005**. Pyrethroids and nectar toxins have subtle effects on the motor function, grooming and wing fanning behaviour of honeybees (*Apis mellifera*). *PLoS One*. **2015**, 10(8):e0133733. https://doi.org/10.1371/journal.pone.0133733

Ollerton J, Alarcón R, Waser NM, Price MV, Watts S, Cranmer L, … & Rotenberry J. A global test of the pollination syndrome hypothesis. *Annals of Botany*. **2009**, 103(9):1471–80.

Ômura H, Honda K, Hayashi N. Chemical and chromatic bases for preferential visiting by the cabbage butterfly, *Pieris rapae*, to rape flowers. *Journal of Chemical Ecology*. **1999**, 25:1895–906.

Palmer-Jones T, Line LJS. Poisoning of honey bees by nectar from the karaka tree (*Corynocarpus laevigata* JR et G. Forst.). *New Zealand Journal of Agricultural Research*. **1962**, 5(5–6):433–6.

Palmer-Young EC, Farrell IW, Adler LS, Milano NJ, Egan PA, Junker RR, Irwin RE, Stevenson PC. Chemistry of floral rewards: Intra-and interspecific variability of nectar and pollen secondary metabolites across taxa. *Ecological Monographs*. **2019**, 89(1):e01335.https://doi.org/10.1002/ecm.1335

Palmer-Young EC, Sadd BM, Stevenson PC, Irwin RE, Adler LS. Bumble bee parasite strains vary in resistance to phytochemicals. *Scientific Reports*. **2016**, 6(1):37087.

Peng T, Schroeder M, Grüter C. Octopamine increases individual and collective foraging in a neotropical stingless bee. *Biology Letters*. **2020**, 16(6):20200238. https://doi.org/10.1098/rsbl.2020.0238

Pozo MI, Lievens B, Jacquemyn H. Impact of microorganisms on nectar chemistry, pollinator attraction and plant fitness. R.L.Nova Peck (Ed.), Nectar: Production, Chemical Composition and Benefits to Animals and Plants, 1, Science Publishers Inc. 2015, 41 **2015**, 41.

Pyke GH. Floral nectar: Pollinator attraction or manipulation? *Trends in Ecology & Evolution*. **2016**, 31(5):339–41. doi: 10.1016/j.tree.2016.02.013

Richardson LL, Adler LS, Leonard AS, Andicoechea J, Regan KH, Anthony WE, ... & Irwin, RE. Secondary metabolites in floral nectar reduce parasite infections in bumblebees. *Proceedings of the Royal Society B: Biological Sciences*. **2015.** 282(1803): 20142471.

Raguso RA, Willis MA. Synergy between visual and olfactory cues in nectar feeding by naıve hawkmoths, *Manduca sexta. Animal Behaviour.* **2002,** 64(5):685–95.

Raguso RA, Willis MA. Synergy between visual and olfactory cues in nectar feeding by wild hawkmoths, Manduca sexta. *Animal Behaviour.* **2005**, 69(2):407–18.

Rering CC, Beck JJ, Hall GW, McCartney MM, Vannette RL. Nectar-inhabiting microorganisms influence nectar volatile composition and attractiveness to a generalist pollinator. *New Phytologist.* **2018**, 220(3):750–9. https://doi.org/10.1111/nph.14809

Rering CC, Vannette RL, Schaeffer RN, Beck JJ. Microbial co-occurrence in floral nectar affects metabolites and attractiveness to a generalist pollinator. *Journal of Chemical Ecology.* **2020**, 46:659–67.

Rhoades DF, Bergdahl JC. Adaptive significance of toxic nectar. *The American Naturalist.* **1981**, 117(5):798–803.

Richardson LL, Adler LS, Leonard AS, Andicoechea J, Regan KH, Anthony WE, Manson JS, Irwin RE. Secondary metabolites in floral nectar reduce parasite infections in bumblebees. *Proceedings of the Royal Society B: Biological Sciences.* **2015**, 282(1803):20142471.

Roeder T. Octopamine in invertebrates. *Progress in Neurobiology.* **1999**, 59:533–61. https://doi.org/10.1016/S0301-0082(99)00016-7

Roeder T. Tyramine and octopamine: Ruling behavior and metabolism. *Annual Review of Entomology.* **2005**, 50(1):447–77. https://doi.org/10.1146/annurev.ento.50.071803.130404

Rosas-Guerrero V, Aguilar R, Martén-Rodríguez S, Ashworth L, Lopezaraiza-Mikel M, Bastida JM, Quesada M. A quantitative review of pollination syndromes: Do floral traits predict effective pollinators? *Ecology Letters.* **2014**, 17(3):388–400. https://doi.org/10.1111/ele.12224

Rothchild KW, Adler LS, Irwin RE, Sadd BM, Stevenson PC, Palmer-Young EC. Effects of short-term exposure to naturally occurring thymol concentrations on transmission of a bumble bee parasite. *Ecological Entomology.* **2018**, 43(5):567–77. https://doi.org/10.1111/een.12631

Schaeffer RN, Mei YZ, Andicoechea J, Manson JS, Irwin RE. Consequences of a nectar yeast for pollinator preference and performance. *Functional Ecology.* **2017**, 31(3):613–21. https://doi.org/10.1111/1365-2435.12762

Schaeffer RN, Phillips CR, Duryea MC, Andicoechea J, Irwin RE. Nectar yeasts in the tall larkspur *Delphinium barbeyi* (Ranunculaceae) and effects on components of pollinator foraging behavior. *PLoS One.* **2014**, 9(10):e108214. https://doi.org/10.1371/journal.pone.0108214

Schoellhorn, K. *Sur la fermentation de quelques levures des nectars des plantes d'hiver*, vol. 11. **1919**. Universite de Geneve., pp. 154–90.

Scott-Brown AS, Arnold SE, Kite GC, Farrell IW, Farman DI, Collins DW, Stevenson PC. Mechanisms in mutualisms: A chemically mediated thrips pollination strategy in common elder. *Planta.* **2019**, 250:367–79.

Seo HJ, Song J, Yoon HJ, Lee KY. Effects of nectar contents on the foraging activity of honeybee (*Apis mellifera*) on Asian pear (*Pyrus pyrifolia* Nakai). *Scientia Horticulturae.* **2019,** 245:185–92.

Shinde PG, Divekar PA, Singh DK, Pal DS, Nadaf A. Bio-pesticide management strategy for mustard aphid *Lipaphis erysimi* (Kaltenbach)(Homoptera: Aphididae). *The Pharma Innovation Journal.* **2021**, 10(7):397–400.

Slansky F. Allelochemical–nutrient interactions in herbivore nutritional. In G. A. Rosenthal and M. R. Berenbaum, editors. Herbivores: their interactions with secondary plant metabolites, Volume II, ecological and evolutionary processes. Academic Press, San Diego, California, USA **1992**.135-174.

Soltani E, Benakashani F, Baskin JM, Baskin CC. Reproductive biology, ecological life history/demography and genetic diversity of the megagenus *Astragalus* (Fabaceae, Papilionoideae). *The Botanical Review.* **2021**, 87:55–106.

Stefanini I. Yeast-insect associations: It takes guts. *Yeast.* **2018**, 35(4):315–30. https://doi.org/10.1002/yea.3309

Stevenson PC, Nicolson SW, Wright GA. Plant secondary metabolites in nectar: Impacts on pollinators and ecological functions. *Functional Ecology.* **2017**, 31(1):65–75.

Thomson JD, Draguleasa MA, Tan MG. Flowers with caffeinated nectar receive more pollination. *Arthropod-Plant Interactions.* **2015**, 9:1–7.

Tiedeken EJ, Egan PA, Stevenson PC, Wright GA, Brown MJ, Power EF, … & Stout JC. Nectar chemistry modulates the impact of an invasive plant on native pollinators. *Functional Ecology.* **2016**, 30(6):885–93.

Vannette RL. The floral microbiome: Plant, pollinator, and microbial perspectives. *Annual Review of Ecology, Evolution, and Systematics.* **2020**, 51(1):363–86. https://doi.org/10.1146/annurev-ecolsys-011720-013401

Vannette RL, Gauthier MP, Fukami T. Nectar bacteria, but not yeast, weaken a plant–pollinator mutualism. *Proceedings of the Royal Society B: Biological Sciences.* **2013**, 280(1752):20122601.

Verpoorte R. Encyclopedia of Analytical Science. Elsevier; Amsterdam, The Netherlands: 2005. ALKALOIDS: 56–61.

Verpoorte R, Schripsema J. Isolation, Identification, and Structure Elucidation of Alkaloids A General Overview. In: Linskens, H.F., Jackson, J.F. (eds) Alkaloids. Modern Methods of Plant Analysis. Springer, Berlin, Heidelberg. **1994**, (15):1-24. https://doi.org/10.1007/978-3-642-84226-9_1

Wink M. Plant secondary metabolites modulate insect behavior-steps toward addiction? *Frontiers in Physiology*. **2018**, 9:364. https://doi.org/10.3389/fphys.2018.00364

Wright GA, Baker DD, Palmer MJ, Stabler D, Mustard JA, Power EF, Borland AM, Stevenson PC. Caffeine in floral nectar enhances a pollinator's memory of reward. *Science*. **2013**, 339(6124):1202–4. https://doi.org/10.1126/science.1228806

Zhang C, Feng HH, Liu YL, Huang SQ. Lethal effects of tea-oil Camellia on honeybee larvae due to pollen toxicity. *Journal of Integrative Plant Biology*. **2024**, 66(11):2313–6.

Index

Note: **Bold** page numbers refer to tables and *italic* page numbers refer to figures.